TECHNIQUES OF CHEMISTRY

ARNOLD WEISSBERGER, *Editor*

VOLUME V, PART II

TECHNIQUE OF ELECTROORGANIC SYNTHESIS

TECHNIQUES OF CHEMISTRY

ARNOLD WEISSBERGER, *Editor*

TECHNIQUES OF CHEMISTRY

VOLUME V, PART II

TECHNIQUE OF ELECTROORGANIC SYNTHESIS

Edited by

NORMAN L. WEINBERG

Supervisor, Research Center
Hooker Chemicals & Plastics Corp.
Grand Island, New York

A WILEY-INTERSCIENCE PUBLICATION

JOHN WILEY & SONS

New York · London · Sydney · Toronto

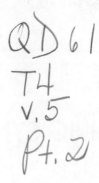

Library of Congress Cataloging in Publication Data:

Weinberg, Norman L.
  Technique of electroorganic synthesis.

  (Techniques of chemistry, v. 5)
  "A Wiley-Interscience publication."
  Includes bibliographical references.
  1. Electrochemistry. 2. Chemistry, Organic—
Synthesis. I. Title.

QD61.T4 vol. 5 [QD273]   540'.28s [547'.1'37]   73-18447

ISBN 0-471-93272-8

Printed in the United States of America

10 9 8 7 6 5 4 3 2 1

AUTHORS OF PART II

B. L. FUNT

Department of Chemistry, Simon Fraser University, Vancouver, British Columbia, Canada

L. D. MCKEEVER

The Dow Chemical Company, Physical Research Laboratory, Midland, Michigan

R. F. NELSON

Department of Chemistry, University of Georgia, Athens, Georgia

M. R. RIFI

Union Carbide Corporation, Chemicals and Plastics, River Road, Bound Brook, New Jersey

W. J. SETTINERI

The Dow Chemical Company, Physical Research Laboratory, Midland, Michigan

H. SIEGERMAN

Princeton Applied Research Corporation, Princeton, New Jersey

J. TANNER

Imperial Chemical Industries, Ltd., Petrochemical and Polymer Laboratory, Runcorn, Cheshire, England

N. L. WEINBERG

Research Center, Hooker Chemicals & Plastics Corporation, Grand Island, New York

INTRODUCTION TO THE SERIES

Techniques of Chemistry is the successor to the Technique of Organic Chemistry Series and its companion—Technique of Inorganic Chemistry. Because many of the methods are employed in all branches of chemical science, the division into techniques for organic and inorganic chemistry has become increasingly artificial. Accordingly, the new series reflects the wider application of techniques, and the component volumes for the most part provide complete treatments of the methods covered. Volumes in which limited areas of application are discussed can easily be recognized by their titles.

Like its predecessors, the series is devoted to a comprehensive presentation of the respective techniques. The authors give the theoretical background for an understanding of the various methods and operations and describe the techniques and tools, their modifications, their merits and limitations, and their handling. It is hoped that the series will contribute to a better understanding and a more rational and effective application of the respective techniques.

Authors and editors hope that readers will find the volumes in this series useful and will communicate to them any criticisms and suggestions for improvements.

Research Laboratories   ARNOLD WEISSBERGER
Eastman Kodak Company
Rochester, New York

PREFACE

For too long the field of electroorganic synthesis has been considered "more of an art than a science." This attitude was appropriate until quite recently. However, the progress made in electroanalytical methods, especially in polarography and cyclic voltammetry of organic compounds, coupled with product isolation studies has given us new insight into the electrochemical reaction variables (potential, current density, adsorption, electrode material, etc.) and their interdependency. The result is that the organic chemist has at his disposal a highly useful synthetic tool, with great potentialities inherent in the technique for developing *electrode specific syntheses* including products that are not easily attainable by other methods.

For many years, the excellent chapter of Professor Sherlock Swann, Jr., in *Technique of Organic Chemistry* (Vol. II, 1956) has served as a primary source and incentive for those entering the field; but the technique of electroorganic synthesis has undergone much change since the appearance of that work. It has now become customary to carry out electrosyntheses under controlled potential conditions using three electrodes, the third one being a reference electrode. This contrasts with earlier synthetic studies in which constant current conditions were usually employed. This does not negate the value of the earlier literature. It is still important to know the products and which electrode material, added catalytic substances, pH, temperature, and so on were found necessary in a particular synthesis. Much can be learned from this older work which is directly applicable today. The new techniques using potentiostatic methods, however, are found superior mainly because they lead to fewer side products.

An approach for a particular synthesis that the organic chemist with a minimum of electrochemical background can easily follow involves, as a start, examination of the literature for suitable starting materials and conditions that fulfill the requirements of both the chemical and electrochemical reaction variables. This is followed by

1. setting up an electrochemical cell, usually with three compartments separating anode, cathode, and reference electrodes.

2. selection of a suitable controlled potential by determining current-potential curves on the background solution (solvent plus supporting

electrolyte) and with added substrate. The curves should be obtained on solutions with the concentrations of materials intended for use. Such curves are readily derived using a constant current power supply and a vacuum tube voltmeter, although a potentiostat is a great convenience (Chapter II). Knowledge of oxidation and reduction potentials (Chapter XII) can greatly aid in making a judicious choice of the operating electrode potential.

The electrolysis is carried out, products are isolated, and, if need be, the reaction variables are revised (these are often the electrode material and the solvent/supporting electrolyte) if necessary with repetition of steps 1 and 2.

The above is only one possible approach. Another approach involves the complementary use of electroanalytical methods to help sort out the possible reaction pathways and establish appropriate mechanisms (Chapters III and V).

The two parts of Volume V include the following:

1. An introduction to the field, valuable to the organic chemist in the laboratory as well as to the engineer considering scale-up (Chapter II).
2. A brief review of electrochemistry and the electroanalytical approach presenting some fundamental principles as well as an understanding of current problems (Chapters II, III, and V).
3. A critical review of the various electrochemical reactions of organic compounds, containing extensive tables of data (Chapters IV–XI).
4. Discussions of important electrosyntheses, including the scope, limitations, technical problems, comparisons with chemical methods, and mechanistic considerations.
5. Actual examples in recipe style.
6. Oxidation and reduction potential data in a form useful to the organic chemist (Chapter XII).

I sincerely thank the contributing authors for their painstaking efforts, conscientiousness of purpose, and patience. These works were envisioned by Dr. A. Weissberger to whom especial thanks are due. I am greatly indebted to Dr. Gerhard Popp of the Eastman Kodak Research Laboratories for his considerable help in reviewing the manuscripts. Thanks are due also to Professor Jack Stocker of Louisiana State University in New Orleans, Professor Albert J. Fry of Wesleyan University, and other reviewers acknowledged by the authors of this work. Their primary reward as well as mine will be to see the technique of electroorganic synthesis adopted into the arsenal of methods of organic chemists everywhere.

NORMAN L. WEINBERG

CONTENTS

PART I

TECHNIQUES OF CHEMISTRY

ARNOLD WEISSBERGER, *Editor*

VOLUME V, PART II

TECHNIQUE OF ELECTROORGANIC SYNTHESIS

Chapter **VII**

ELECTROCHEMICAL HALOGENATION
OF ORGANIC COMPOUNDS

N. L. Weinberg

List of Symbols and Abbreviations:

The following symbols and abbreviations should be noted:

V	volt
A	ampere; mA, milliampere
i	current
CPE	controlled potential electrolysis
CD	current density
g	gram
Xg/Ygsm	X grams product per Y grams starting material
F	Faraday (96,500 coulombs)
CE	current efficiency (%)
SCE	saturated calomel electrode

1   ELECTROCHEMICAL HALOGENATION

The oxidation potentials of halogen species in several media are listed in Table 7.1. The relative ease of oxidation is in the order $I^- > Br^- > Cl^- > F^-$, spanning almost the total available anodic range in which oxidation reactions of organic substrates have been studied. Thus in the discharge of halide ions according to the equations

$$2X^- -2e \longrightarrow 2X\cdot \longrightarrow X_2$$

or

$$3X^- \longrightarrow X_3^- + 2e$$

and

$$X_3^- \longrightarrow \frac{3}{2} X_2 + e,$$

it would be expected that anodic halogenation of a substrate S should proceed according to one or more of the following possibilities:

Route A: Discharge of S to a cationic intermediate followed by reaction with X^-.

Route B: Reaction of electrogenerated X_2 with S.

Route C: Reaction of electrogenerated $X\cdot$ with S.

Route D: Reaction of electrogenerated X_2 with solvent and S.

At present there are few clearly established examples of *Route A*, and it appears that starting substrate, solvent, or water in trace amounts competes effectively with X^- for the cationic species derived from the substrate. *Route B* involves the chemical reaction of electrogenerated halogen with substrate. This is generally a slow reaction for X = I or Br compared to X = Cl or F. Recent work on the bromination of aromatics [4] demonstrates that halogenation of anthracene occurs very poorly at the discharge potential of the halide. In contrast, halogenation does occur when the electrolysis is carried out at potentials more positive than the oxidation potential of the hydrocarbon, suggesting among other mechanisms, a reaction of cationic substrate with halide ion, or the discharge of a substrate − halogen charge transfer complex at the anode. *Route C* is possible with all of the halogens, but in the absence of high temperature and irradiation with light, is more important with X = Cl and F. *Route D* is applicable to olefinic compounds in which the olefin reacts with electrogenerated halogen and a suitable solvent (see Part I, p. 368 for halofunctionalization of olefins). Thus in water, alcohols, or carboxylic acids, ROX [formed on discharge of X^- (R is H, alkyl or acyl)] may add to olefins. Alternatively the cyclic halonium ion species formed from olefin and X_2 may be attacked by solvent giving halofunctionalized products. Likewise amines and amides react with X_2 to give *N*-halo derivatives, which are known to add to certain olefins under various conditions. Acetonitrile is apparently iodinated during iodine oxidation to give *N*-iodo-acetonitriliumion ($CH_3 - \overset{\cup}{C} = NI$), which serves as an iodinatinge agent for aromatics [5].

Experimental Parameters (General)

It is by now well recognized that such effects as adsorption and high rates of mass transfer of substrate to the electrode can lead to unexpected reaction paths that are not simply described by *Routes A → D*. Moreover, with judicious choice of parameters, what are often minor reaction paths can be maximized. Some general comments on important variables in halogenation reactions follow.

For halogenations in Cl^-, Br^-, or I^- electrolytes, there are no special requirements for anode materials other than that they be inert to electrochemical oxidation and the presence of electrogenerated halogen. Usually Pt and carbon in their various forms are suitable.

In contrast, fluorinations in liquid HF require nickel or carbon anodes for perfluorination to occur. As will be evident in the discussion on electrofluorination, the reaction on nickel appears to proceed on a high valent nickel fluoride surface, whereas the mode of reaction on carbon may be with adsorbed fluorine or fluorine atoms. Fluorinations on Pt have been carried out in CH_3CN, but the nature of the products are characteristic of reaction paths involving direct oxidation of the organic. Carbon anodes are consumed in molten inorganic fluoride media with production of short-chain perfluorocarbons.

Improved yields and fewer side products have been demonstrated with the use of nonaqueous media in halogenations using Cl^-, Br^-, and I^- solutions. Electrochemical perfluorinations require almost anhydrous HF, since the OF_2 produced from traces of water is a powerful oxidizing agent and causes explosions when admixed with organic materials.

Table 7.1 Oxidation Potentials of Halogen Species

Halogen Couple	$E_{1/4}$ At Pt (V vs SCE)[a]			E_0(V vs NHE)[c]
	CH_3CN	DMF	DMSO	
I^-/I_2 (or I_3^-)	0.20	0.50	0.45	0.54
I_3^-/I_2	0.55	0.62	0.60	–
$I_2/I_2^{+\cdot}$(?)	2.2^b	–	–	–
Br^-/Br_2	0.70	0.65	0.60	1.09
Br_3^-/Br_2	0.95	0.85	0.85	–
Cl^-/Cl_2	1.15	1.02	1.00	1.36
F^-/F_2	–	–	–	2.87

[a]See Ref. 1.
[b]See Ref. 2.
[c]See Ref. 3.

Generally, radical halogenation reactions of lower specificity predominate with higher temperatures; but in some cases higher temperatures may be required to effect any reaction. For example, *n*-dodecane, which does not react with Cl_2 to any appreciable extent at room temperature, gives a mixture of primary monochlorides, secondary monochlorides, and dichlorides in 44% CE at 93°C [6].

Use of high current densities will require the use of higher potentials, resulting in electrooxidation of the halogenated product. This is especially true for iodinated and brominated products, which are readily electrooxidized [2]. Chlorinated and fluorinated products are relatively inert to further oxidation.

One important effect of high current density is to increase the concentration of radicals $(X\cdot)$ at the anode surface. Thus radical abstraction of hydrogen by halogen may be expected to become important. The effects of current density and potential are discussed in greater detail under the appropriate halogenation reaction.

Partial fluorination generally occurs most successfully with increasing concentration of substrate. In contrast, the formation of bromoform and iodoform from acetone or ethyl alcohol appears to proceed best at high concentration of substrate.

Porous electrodes have been used for introducing a high concentration of a gaseous substrate through the anode. Nagase [7] has reviewed the use of this method for fluorination of a wide range of gaseous or low boiling substances.

Electrochemical Fluorination

Credit for the first electrochemical fluorination has always been given by reviewers to Lebeau and Damiens [8], who isolated CF_4 during electrolysis of molten beryllium fluoride at a carbon anode. Actually Lyons and Broadwell [9] produced CF_4 more than two decades earlier by electrolyzing a melt (1000°C) of sodium and potassium fluorides at a graphite anode surrounded by carbonaceous material.

In contrast, the Simons method, discovered in 1941, may be described as the low-temperature electrolysis at a Ni anode of an organic compound, dissolved or suspended in liquid HF, to give a perfluorinated product. A large number of compounds have been subjected to the Simons procedure [7, 10-17], but results obtained for the same reaction by different laboratories show that reproducibility is apparently difficult to achieve.

Physical and Chemical Properties of Liquid HF

Hydrogen fluoride is an ideal solvent for electrochemical studies with a convenient boiling point (19.5°C), high dielectric constant, low viscosity, and powerful solubilizing ability for a wide range of organic as well as inorganic compounds. It has the disadvantage of dissolving glass. Fortunately, however, there are many materials available now that are useful, such as inert plastics (Teflon[R], polyethylene, polypropylene, Vitron[R], Kel-F[R], and FEP[R]), [18]; there is also Pt, which is not attacked, and Cu, Ni, Mg, and Al, which form protective fluoride coatings.

Many organic compounds dissolve in liquid HF to give conducting solutions. Weak bases such as ethanol, acetone, and acetic acid behave as strong electrolytes with conductances similar to that of fluorides [19]. Spectroscopic and other studies have established that these compounds are protonated in HF [20]:

$$ROH + 2HF \rightleftharpoons ROH_2{}^+ + HF_2{}^-$$

$$RCO_2H + 2HF \rightleftharpoons R-C\underset{OH}{\overset{OH}{\lessgtr}}{}^+ + HF_2{}^-$$

Aromatic hydrocarbons behave similarly [20a]:

$$Ar + 2HF \rightleftharpoons ArH^+ + HF_2{}^-$$

The least soluble materials in HF are saturated hydrocarbons and especially, perfluorinated hydrocarbons.

Liquid HF can cause partial fluorination to some extent with several classes of organic compounds. Addition occurs to olefins and acetylenes, while carboxylic acids, anhydrides, and chlorides are converted to the corresponding carboxylic acid fluorides.

Experimental Technique

The electrochemical variables for electrofluorination of organic compounds have traditionally been expressed as cell voltage (5-7 V) and current density (0.0008 - 0.02 A/cm^2). Only a few electrochemical studies, confined to the investigation of inorganic fluorination reactions, have been reported using reference electrodes. This unfortunate situation exists in spite of the fact that reference electrodes for liquid HF have been developed. These include the Hg/Hg_2F_2 electrode [21, 22] and the Cu/CuF_2 couple [23]. Not only could a clearer understanding of the reaction mechanism be possible, but control of the electrode potential could possibly result in higher yields of desired product and repoducibility of results.

The most practical anode material [24, 25] has been Ni or a Ni alloy such as Monel (66% Ni), although some studies have employed carbon or platinum under special conditions. Other anodes and anodes of metals of which the higher fluorides are used as chemical fluorinating agents were found unsiutable because of corrosion, passivation, or little or no fluorination product formed. Recently, the use of a porous carbon anode in a solution of composition KF.2HF at about 100°C has been shown by Fox et al. [25a] to give excellent results in partial fluorination reactions. An essential feature of this novel process is that the insoluble feed is kept confined to the pores of the porous carbon anode and is not permitted to break out into the bulk of the solution.

Further characteristics of electrofluorination include the following: Use of a rotating anode or pumping the HF solution through a porous anode (especially for gaseous or insoluble substrates) raises the yield of product substantially; since most perfluorinated compounds are inert to electrochemical reductions, diaphragms or separators are not required; cathode materials of Ni, Fe, Cu, or Pt are suitable, but those of Al, Zn, and Pb have a deleterious effect on the reaction; electrolysis temperatures near $0°C$ are most often used, but higher temperatures may be practical; conductivity additives such as NaF or KF are needed with noncunducting substrate-HF solutions; with conducting substrate-HF solutions the end of the electrolysis is signalled when the current has dropped to a low value, since perfluorinated compounds are usually nonconducting; often a preelectrolysis of the HF solution without the substrate is carried out to remove small amounts of water present in the solvent (Note. The oxygen difluoride that is formed is a powerful oxidizing agent and can be explosive when concentrated with organics).

A typical industrial fluorination cell is described in Chapter II, p. 135.

Experimental Examples

Dr. Shunji Nagase (Government Industrial Research Institute, Nagoya, Japan) has kindly provided the following examples which typify many aspects of the technique of electrofluorination.

PARTIAL FLUORINATIONS OF METHANE (PREPARATION OF FLUORINATED METHANES) [26]

Apparatus. The apparatus consists essentially of an electrolytic fluorination cell, hydrogen fluoride absorber, oxygen difluoride absorber, and a series of cold traps as shown in Fig. 7.1. The electrolytic cell is a cylindrical Monel vessel, 10 cm in diameter and 21 cm long, provided with a copper condenser at the top to condense vaporizing hydrogen fluoride. The electrodes consist of eight and nine pieces of nickel plates (0.5 mm in thickness) as anodes and cathodes, fastneed together alternately, suspended from the top, and insulated from each other by poly(tetrafluoroethylene) pieces. The distance between the plates is 1.7 mm and the effective surface area of the anodes and cathodes is 9.2 dm^2.

A copper spiral tubing is provided inside the cell to maintain the optimum temperature by circulation of ice water during the reaction. The bubbler that is placed at the bottom of the cell consists of a 6 X 15-mm polyethylene tube, with an internal diameter of 4 mm whose tip is enlarged to 8 mm in diameter and capped with a woven fabric made of poly(tetrafluoroethylene).

The hydrogen fluoride absorber is an iron tube containing a bed of sodium fluoride pellets that absorb uncondensed hydrogen fluoride escaping from the reflux condenser. As an oxygen difluoride absorber, gas-washing bottles containing an aqueous solution of sodium sulfite with a small amount of potassium iodide are used. (Anhydrous hydrogen fluoride has a strong affinity for water, and commercial hydrogen fluoride contains a trace of dissolved water

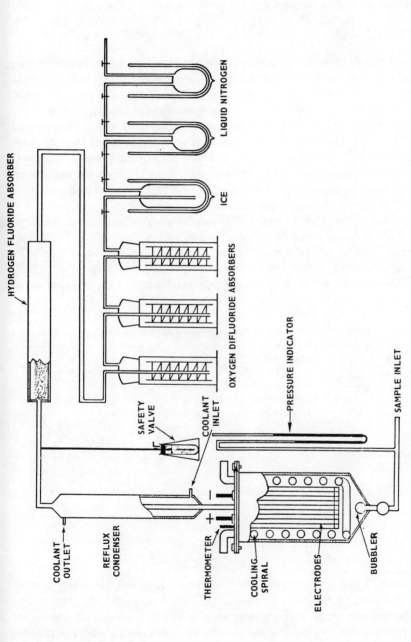

Fig. 7.1 Electrofluorination apparatus.

7

that can be removed by passing a current through it. Therefore, electrochemical fluorination is often accompanied by the formation of oxygen difluoride [27], especially in preliminary electrolysis.) Mercury in the sefety valve and in the pressure indicator is covered by fluorocarbon oil.

Procedure. Anhydrous hydrogen fluoride (1 liter) is placed in the cell and, in order to remove traces of impurities (mainly water), electrolysis is carried out in a nitrogen atmosphere, prior to the introduction of methane, with an anodic current density of 2.7 A/dm^2 and at 5-6°C, until the cell voltage rises to 6.0 V. Then 10 g of sodium fluoride (conductivity additive) is added, and metered methane is introduced from the bottom of the cell into hydrogen fluoride maintained at 5-6°C, through the bubbler at a flow rate of 65 ml/min. Electrolysis is conducted at an anodic current density of 2.2 A/dm^2. When the cell voltage becomes almost steady at 6-6.5 V, the traps are placed in ice and liquid nitrogen, respectively, and electrolysis is carried out for 71 A hr.

Fluorinated gas evolved from the cell passes through a reflux condenser kept at −20°C and over a bed of sodium fluoride pellets, then bubbles through an aqueous solution of sodium sulfite, and finally condenses in cold traps. The product obtained is rectified in a low-temperature rectification unit. Each fraction is further analyzed by gas chromatography using an activated charcoal column, 2 m long, maintained at 70°C (carrier gas is helium):

Rectification of Fluorinated Methane

Fraction No.	Boiling Point (°C)	Wt. (g)	Component
1	-162 to -159	5.6	CH_4
2	-128 to -126	6.7	CF_4
3	-80 to -76	6.1	CH_3F, CHF_3
4	-70 to hold up	2.0	CH_3F, CHF_3, CH_2F_2

Product composition (mole %): CF_4, 29.0%; CHF_3, 12.3%; CH_2F_2, 12.7%; CH_3F, 46.0%

Conversion: 43%

Total current efficiency: 44%

FLUORINATION OF ETHYL ACETATE (PREPARATION OF SODIUM TRIFLUOROACETATE) [28]

Apparatus. The apparatus consists of an electrolytic cell, hydrogen fluoride absorber, and fluorinated product trap. The electrolytic cell (made of a nickel vessel), which is not equipped with the bubbler, and the hydrogen fluoride absorber are virtually the same as those described above. The product trap consists of a pair of polyethylene bottles containing 200 and 500 ml of water.

Procedure. Ethyl acetate (34.5 g) is dissolved in anhydrous hydrogen fluoride (1 liter) that had been purified by electrolysis, then electrolysis of the solution is carried out with an anodic current density of 3.3 A/dm^2 at an average of 4.8 V at 4-6°C for 221 A hr. At the end, the voltage rises up to 5.3 V.

The outgoing gases, such as trifluoroacetyl fluoride, fluorocarbons, oxygen difluoride, hydrogen, and so on, from the cell are passed through the reflux condenser kept at -20°C into the hydrogen fluoride absorber, and are bubbled through the water in the product trap where trifluoroacetyl fluoride is converted into trifluoroacetic acid. In this system, gases that are inert to water should escape.

The aqueous solution of trifluoroacetic acid in the absorber is neutralized with sodium carbonate, and filtered. The filtrate is evaporated to dryness. Extraction of the resulting crude sodium trifluoroacetate with absolute ethanol followed by evaporation of the solvent affords sodium trifluoroacetate (54.2 g), as follows:

Yield, 51.6%; current efficiency, 40.5%.

General Scope and Limitations

Unless special provision in the electrochemical method is made, all the hydrogens of the organic substrate are replaced by fluorine. Partial fluorination may be achieved, however, by using a high concentration of the substrate [29, 30] and low current density, or in the case of gaseous substrates, bubbling these diluted with inert gas through the cell. By adjusting the feed rate, some optimization of yield toward desired products is accomplished. Thus for methane, Nagase and co-workers [26] have found a marked effect of feed rate on the yields of various partially fluorinated methanes from methane (Table 7.2).

Table 7.2 Partial Fluorination of Methane

Feed Rate (ml/min)	% Product Composition				Total % Yield	Total % Current Efficiency
	CF_4	CHF_3	CH_2F_2	CH_3F		
42	34	20	18	28	48	38
64	29	12	13	46	43	44
104	24	11	18	57	32	48

For nongaseous substrates, cell designs providing for short contact time at the electrode should likewise afford enhanced yields of partially fluorinated products.

Many compounds are fluorinated with preservation of functional groups and the original carbon skeleton. If cleavage does occur between a carbon and a functional group (C-halogen, C-O, C-N, C-S, C-Si, etc.) the parent fluorocarbon is often the resultant product.

Further general observations may be made concerning the course of reaction of specific classes of compounds:

HYDROCARBONS AND HALOCARBONS

Electrofluorinaiton of branched chains apparently occurs to give straight chain products in a radical type of isomerization (i.e., isogrpoyl → n-propyl). Highest yields of perfluorinated products are obtained from partially fluorinated starting materials:

$$\text{hexane} \xrightarrow{\text{HF}} C_6F_{14} \ (22)$$

$$2,2\text{-difluorohexane} \xrightarrow{\text{HF}} C_6F_{14} \ (62.8)$$

Better yields are likewise obtained with halocarbons bearing I, Br, or Cl substituents. Chlorine is usually retained in the product, but Br and I are replaced. (The ease of replacement is I > Br > H > Cl). Chlorofluorination can occur with chlorocarbons or with Cl_2 added to the reaction medium:

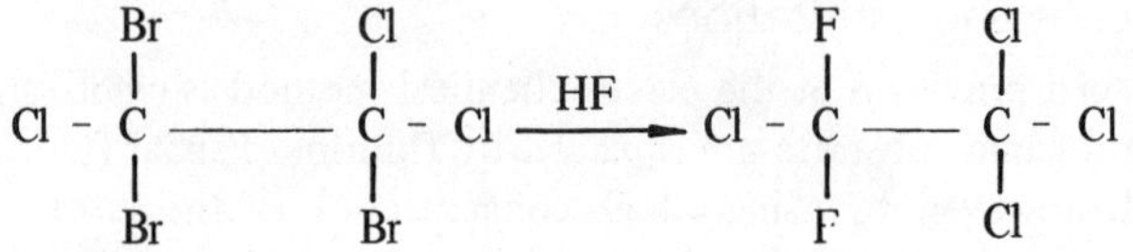

$$\underset{\underset{Br}{|}}{\overset{\overset{Br}{|}}{Cl-C}} - \underset{\underset{Br}{|}}{\overset{\overset{Cl}{|}}{C}} -Cl \xrightarrow{\text{HF}} \underset{\underset{F}{|}}{\overset{\overset{F}{|}}{Cl-C}} - \underset{\underset{Cl}{|}}{\overset{\overset{Cl}{|}}{C}} -Cl$$

$$CH_3Cl \xrightarrow{\text{HF}} CCl_2F_2, \ CCl_2HF, \ \text{etc.}$$

$$CH_3I \xrightarrow{\text{HF}} CF_4(16), \ CF_3H(16), \ CFH_3(1), \ CF_2H_2(1), \ I_2$$

$$\text{ethane} \xrightarrow[Cl_2]{\text{HF},} CHF_2CClF_2, \ CClF_2CClF_2, \ CCl_2FCClF_2,$$

$$CClF_3, \ \text{etc.}$$

Unsaturated hydrocarbons react to give the corresponding saturated perfluorocarbons. Addition of fluorine to the double bond of chlorohydrocarbons occurs first, followed by replacement of hydrogen:

$$CHCl = CCl_2 \xrightarrow{\text{HF}} CHFClCCl_2F + CClF_2CCl_2F \ (\text{in ratio } 82:12)$$

NITROGEN-, SULFUR-, AND OXYGEN-CONTAINING COMPOUNDS

Compounds containing nitrogen, sulfur, and oxygen functions are fluorinated with formation of NF_3, SF_6, and OF_2 as by-products respectively.

AMINES

Primary and secondary amines are perfluorinated to $-NF_2$ and $\backsimeq$N-F derivatives, respectively. Tertiary amines are also readily perfluorinated. Heterocyclic aromatic amines such as pyridine give the corresponding saturated perfluoro-carbon. As would be expected, aniline and cyclohexylamine afford the same product. The branched chain members of N-alkylpiperidines and N-alkylmorpholines yield a mixture of perfluorinated amines wiht normal and branched structures, an isomerization which apparently is not observed [31] in the vapor-phase fluorination of similar compounds with CoF_3. Electrochemical fluorination of amines appears to be generally superior to chemical methods (use of metal fluorides or F_2):

$$\text{aniline} \xrightarrow{\text{HF}} C_6F_{11}NF_2, \ C_6F_{12},$$

pyridine $\xrightarrow{\text{HF}}$ (7.5), (25), (5.3),

$NF_3(30)$,

CF_4

n-amylpiperidine $\xrightarrow{\text{HF}}$ perfluoro-n-amylpiperidine (42)

isoamylpiperidine $\xrightarrow{\text{HF}}$ perfluoro-n-amylpiperidine (33),
perfluoroisoamylpiperidine (11)

ETHERS

Most ethers are perfluorinated to the corresponding perfluoroether derivatives. Perfluorocarboxylic acid fluorides are formed as significant cleavage products in other studies:

diethyl ether $\xrightarrow{\text{HF}}$ $C_2F_5OC_2F_5$; CF_3COF (36.2)

di-n-propyl ether $\xrightarrow{\text{HF}}$ C_2F_5COF (20.8), CF_3COF (2.8)

tetrahydrofuran $\xrightarrow{\text{HF}}$ (42)

anisole $\xrightarrow{\text{HF}}$ $C_6F_{11}OCF_3$

$\xrightarrow{\text{HF}}$ $\begin{matrix} CF_2CF_2 \\ | \ \ \ \ \backslash \\ OCF_3 OCF_3 \end{matrix}$ (40 g/321 gsm)

ALCOHOLS, ALDEHYDES, AND KETONES

Corresponding perfluoroalcohols, aldehydes, and ketones have not been observed as products, but instead cleavage products consisting of perfluoro-carboxylic acid fluorides, fluorocarbons, and perfluoroethers. The mechanism for formation of perfluorocarboxylic acid fluorides from alcohols has been suggested as an initial replacement of hydrogen by fluorine on the α-carbon atom of the alcohol, followed by dehydrofluorination, and then exhaustive fluorination [32]:

ethanol $\xrightarrow{\text{HF}}$ CF_3COF (29.7); CF_4, C_2F_6

allyl alcohol $\xrightarrow{\text{HF}}$ CF_3COF (2.0), C_2F_5COF (1.8),

$\qquad\qquad\qquad\qquad C_3F_7COF$ (0.6), polymers.

acetaldehyde $\xrightarrow{\text{HF}}$ CF_3COF (22), $C_2F_5OC_2F_5$

acetone $\xrightarrow{\text{HF}}$ CF_3COF (59.8), CF_4, CF_3H

$CH_3CH_2COCH_2CH_3$ $\xrightarrow{\text{HF}}$ CF_3COF (5.9), C_2F_5COF (54.8)

CARBOXYLIC ACIDS, ACYL HALIDES, ANHYDRIDES

Carboxylic acids and anhydrides dissolve in liquid HF and are slowly converted to acyl fluorides:

$$RCO_2H + HF \rightleftharpoons RCOF + H_2O$$

$$(RCO)_2O + HF \rightleftharpoons RCOF + RCO_2H \rightleftharpoons 2RCOF + H_2O$$

Perfluorination of carboxylic acid derivatives is an important industrial process to perfluorocarboxylic acid, especially trifluoroacetic acid. Because of the formation of water in the above equilibra and resultant formation of OF_2 on electrolysis, side reactions are extensive. Lower yields have also been attributed to the Kolbe reaction in which CO_2 is formed. Among the principal by-products of these side reactions are fluorocarbons and perfluoroethers.

The best yields of perfluorocarboxylic acids are obtained on electrolysis of acyl fluorides. The latter are soluble in HF, but those with more than four carbon atoms require additives to increase the conductivity. Acyl chlorides, bromides [33], and iodides [33] have also been employed successfully.

An interesting side reaction, the formation of cyclic ethers, becomes increasingly important in the electrofluorination of long-chain carboxylic acids of four or more carbons:

acetic acid $\xrightarrow{\text{HF}}$ CF_3COF, CF_4, CF_3H, CF_2H_2, CO_2

acetyl chloride $\xrightarrow{\text{HF}}$ CF_3COF (20.1); CF_4, CF_3H

acetyl fluoride $\xrightarrow{\text{HF}}$ CF_3COF (85)

acetic anhydride $\xrightarrow{\text{HF}}$ CF_3COF (45)

benzoic acid $\xrightarrow{\text{HF}}$ $C_6F_{11}COF$, C_7F_{16}, C_6F_{14}

γ,γ,γ-trifluorobutyryl fluoride $\xrightarrow{\text{HF}}$ C_3F_7COF (60),

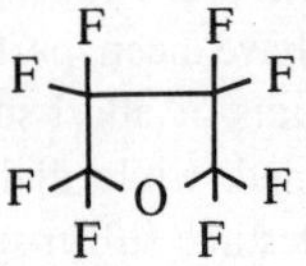

(<1)

caproyl chloride $\xrightarrow{\text{HF}}$ $CF_3(CF_2)_4COF$,

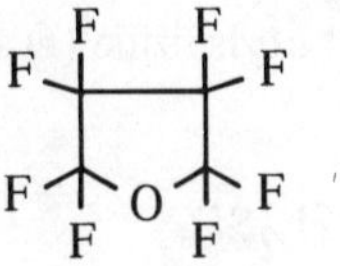

(19.8)

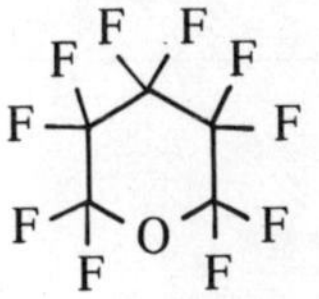

(12.5)

ESTERS

Carboxylic esters are cleaved to give perfluorocarboxylic acid fluorides in good yield. Sulfonic acid esters, on the other hand, give fair to good yields of the corresponding perfluoro derivative:

ethyl acetate $\xrightarrow{\text{HF}}$ CF_3COF (52.3)

diethyl succinate $\xrightarrow{\text{HF}}$ $\begin{array}{c} CF_2CF_2 \\ | \quad | \\ COFCOF \end{array}$ (17.0),

CF_3COF, C_2F_5COF

dimethyl sulfate $\xrightarrow{\text{HF}}$ $(CF_3)_2SO_4$ (13.4),

FSO_2OCF_2H (2.7),

FSO_2OCF_3 (16.7), etc.

AMIDES

Studies on fluorination of amides have been chiefly concerned with the production of NF_3 and CF_4. Less cleavage occurs with heavily alkylated derivatives:

$$\overset{\overset{\displaystyle O}{\parallel}}{H_2NCNH_2} \xrightarrow{\text{HF}} NF_3(38), CF_4(10), COF_2$$

$$\overset{\overset{\displaystyle O}{\parallel}}{(CH_3)_2NC\text{-}H} \xrightarrow{\text{HF}} \overset{\overset{\displaystyle O}{\parallel}}{(CF_3)_2NCF}\ (170\ g/1139\ gsm)$$

SULFUR-CONTAINING COMPOUNDS

Perfluorination products usually contain sulfur in the hexavalent state. This oxidation is believed to occur at an early stage of fluorination [7]. Few mercaptans have been perfluorinated without extensive tar formation. Perfluorinated products of alkyl sulfides, the bis(perfluoroalkyl)sulfur tetrafluorides are readily obtained, while dialkyl disulfides give the same products as those of the monosulfides in a sulfur-sulfur bond cleavage. For example, the yields of the main products from di-n-butylsulfide (**I**) and di-n-butyl disulfide (**II**) are as follows:

	n-$C_4F_9SF_5$	$(n$-$C_4F_9)_2SF_4$	SF_6	C_4F_{10}
I	18	11	–	–
II	22	3.3	–	–

The additional formation of heterocyclic sulfur compounds in this example, and from other open-chain sulfur compounds, lends support to radical processes occurring in electrochemical fluorination. Whereas cyclic sulfides appear to give a little of the corresponding perfluorinated derivative along with extensive fragmentation, the rings of cyclic sulfones are opened to yield perfluoroalkylsulfonyl fluorides. The latter are also prepared on electrofluorination of alkanesulfonyl fluorides and chlorides (the yield diminishes with increasing number of carbon atoms). In analogy to fluorination of carboxylic acids, alkylsulfonic acids afford the corresponding perfluorinated products in reduced yield:

$$\underset{O_2}{\overset{\displaystyle \square}{\underset{\displaystyle S}{\bigcirc}}} \xrightarrow{\text{HF}} CF_3(CF_2)_3SO_2F \ (40)$$

$$CH_3CH_2SO_2Cl \xrightarrow{\text{HF}} CF_3CF_2SO_2F \ (79)$$

MOLTEN INORGANIC FLUORIDES

Considerable work has been carried out in molten inorganic fluorides. The anode, consisting of carbon, generally serves as the source of carbon in the product. The inorganic melt provides the fluorine. Depending on the temperature CF_4, C_2F_6, and C_3F_8 are produced in varying yields, with CF_4 favored at higher temperatures. Figure 7.2 shows the effect of electrolyte temperature on the relative yields of CF_4, C_2F_6, and C_3F_8.

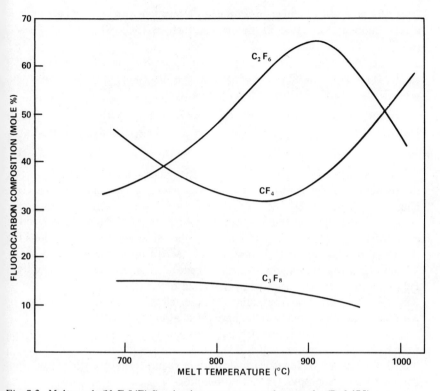

Fig. 7.2 Molten salt (NaF:LiF) fluorination at a porous carbon anode, (Ref. 175).

Proposed Mechanisms of Electrochemical Fluorination On Nickel Anodes

Almost no work on the electrochemical kinetics of fluorination has thus far appeared in the literature, necessitating postulations that are based almost entirely on the nature of the reaction products. Fortunately, the existence of a great body of electrofluorination literature covering many classes of compounds allows for reasonable insight into many aspects of the mechanism.

The following mechanisms have been considered by Burdon and Tatlow [14] in their review:

1. Elementary fluorine is generated electrochemically at the anode and reacts with the organic substrate.

2. The organic substrate undergoes direct electron transfer at the anode to give a cationic species which then reacts with the solvent-supporting electrolyte.

3. The observed products are formed in a reaction with simple high valent nickel fluorides generated at the anode such as NiF_3 of NiF_4.

4. The observed products are formed in a reaction with complex high valent nickel fluorides generated at the anode such as $(SH)_2NiF_6$ and $(SH)_3NiF_6$ (where S is the substrate).

5. The fluorinating agent is a loose complex of NiF_2 and F_2 formed at the anode; fluorination would take place between this complex and the substrate, possibly also absorbed on the nickel fluoride layer.

A furhter mechanism has been proposed by Russian workers [34]:

6. Dissociative chemisorption of the organic substrate occurs followed by reaction with solvent and further electron transfer.

Electrochemical fluorination is one of the mildest methods of introducing fluorine atoms directly into the molecules of hydrogen containing organic compounds. In contrast, chemical fluorination with F_2 is often accompanied by explosions and "burning" of substrates to HF and CF_4 unless the reaction is tamed by using an inert gas diluent or other methods. In hydrocarbons the rate of substitution of the C-H bond is tert $>$ sec $>$ prim, as would be expected.

Electrochemical fluorination, like its chemical counterparts, is apparently a radical reaction, as demonstrated by numerous observations:

a. Isolation of 2-fluoropyridine in the electrochemical fluorination of pyridine. A cationic process would have given 3-fluoropyridine.

b. The products of electrochemical fluorination are often the same as those from chemical fluorination which is known to be free-radical in nature.

c. Isomerization of branched hydrocarbon chains to normal perfluorocarbon derivatives (i.e., isopropyl $\longrightarrow$ perfluoro-n-propyl).

d. Extensive breakdown of substrates to lower molecular weight perfluorocarbons by cleavage of C-C bonds.

e. In molten salt electrochemical fluorinations the carbon anode is attacked to give the perfluorocarbon.

Considering the mechanisms proposed above, the reaction of substrate with electrochemically generated fluorine has been disputed, because an induction period is observed with new nickel anodes. Moreover, free fluorine has been observed during this induction period (even at low voltages) but not afterwards. Conversely, it is likely that fluorine atoms or free fluorine is the fluorinating agent generated on carbon anodes [25a].

The second scheme, that the organic substrate undergoes direct electron transfer followed by reaction with solvent-supporting electrolyte, has been well substantiated for electrofluorination of several compounds at Pt in CH_3CN containing fluoride supporting electrolyte [37-39a]. Here reactions are conducted at potentials below that for discharge of fluoride ion but sufficiently positive for electrooxidation of the organic compound. Typically, and in contrast to electrochemical fluorinations in HF on Ni, no perfluorinated products are formed. This mechanism, however does not account for the induction period observed in electrochemical perfluorination on Ni.

The formation of NiF_2 on the anode suggests that the third scheme (that of fluorination by a higher nickel fluoride) could be more realistic. But it has been argued that the process

$$Ni^{2+} \rightleftharpoons Ni^{3+} + e$$

occurs at more positive potentials than the liberation of fluorine from fluoride. Since reactions of organics with known high-valency metal fluorides bear little resemblance to the electrochemical method, this scheme appears also to be untenable.

The proposal [14] of fluorination by complex high valent nickel fluorides satisfies most objections. It has been shown that the compounds K_3NiF_6 and K_2NiF_6 can be formed by the action of F_2 on mixtures of potassium and nickel chlorides [39b], suggesting that the oxidation potential for the formation of these compounds is less than that required to generate fluorine. Engelbrecht, Mayer, and Pupp [39c] demonstrated that K_2NiF_6 is formed in studies of KF - HF melts. These workers also observed relatively long induction periods during which no anode gases were formed. They postulated the following mechanism:

Primary step $Ni + 2F^- \longrightarrow NiF_2 + 2e$

Induction period $NiF_2 + (2)F^- \longrightarrow NiF_{3(4)} + (2)e$
                   (formation of a black deposit on anode)

Corrosion $NiF_{3(4)} + 3(2)F^- \longrightarrow NiF_6^{3(2)-}$
                   (nickel dissolves as a complex in solution)

Nickel fluoride $NiF^{3(2)-} + substrate \longrightarrow NiF_2 + fluorinated\ substrate$
deposition

Stein, Neil, and Alms [40] have shown that nickel (III) and (IV) species are produced at a nickel anode in liquid HF, with high enough oxidation potentials to liberate oxygen from water. They suggest that these could act as vigorous fluorinating agents in reaction with organic molecules.

Several other investigations present valuable mechanistic information. Electrochemical studies on water [41, 45] and NH_4HF_2 [43] in liquid HF have been interpreted in terms of fluorination occurring on an NiF_2 layer on the surface. A "loose" complex of fluorine radicals $(F \cdot)$ on NiF_2 was proposed as the fluorinating reagent for NH_4HF_2 (see Mechanism 5). In a study [44] of the fluorination on Ni of ClF_3 in HF to give ClF_5, a mechanism was suggested in which a fluoride ion is oxidized to some form of active fluorine $(F \cdot)$ which reacts with ClF_3. Since ClF_5 was also formed at a glassy carbon anode, a mechanism involving complex nickel fluoride is not a necessary condition for successful fluorination. The three inorganic studies above would suggest that there may be several mechanisms operative in electrofluorination reactions of inorganic and perhaps organic substrates.

A study of the yield of fluorination product from isoamylpyridinium bromide on nickel anodes with different crystallographic orientations has been reported

[45]. Electrolysis on Ni with its orientation along the (001) axis shortens the induction period; but electrolysis along the (112) axis gave the best reproducibility. The decreased induction period on the (001) axis, in which the nickel
crystallites are oriented in a cubic lattice parallel to the anode surface, is
explained by higher electron density on the surface leading to more rapid formation of a nickel fluoride layer. Rolled nickel with its major orientations along
the (110) axis and lesser orientations along the (112) axis results, as would be
expected, in slower deposition of nickel fluoride and a longer induction period.
It was concluded that the electrochemical fluorination of organic compounds
can be accomplished with minimum rupture of C-C bonds only after a surface of
nickel fluoride has formed on the nickel anode.

An electrochemical study [34] of benzene at a Ni anode in liquid HF
containing NaF showed that the electrooxidation proceeds very readily at
relatively low positive potentials, compared to other organic substances. A slope
of 120-150 mV/decade was observed for the Tafel plot. A polymer with a
cross-linked structure and having a unit composition close to $C_{30}H_{20}F$ was
observed as the electrolysis product. Since the composition of the polymer is
such that dehydrogenation of benzene had occurred, a mechanism involving
dissociative chemisorption was proposed. The initial products in this reaction
are adsorbed (C_6H_{6-n}) and (nH) radicals. The basic anodic reaction is then

$$(nH) \text{ ads} \longrightarrow nH^+ + ne,$$

while the organic radicals enter polymerization reactions.

It is evident that mechanisms abound for electrochemical fluorination. To
complicate this situation, more than one mechanism may be operative. Electrochemical studies directed towards the elucidation of the nature of the anode surface both during the induction period and in the presence of an organic substrate undergoing perfluorination are needed. Moreover, the nature of the rate-
determining step, the electrochemical and chemical role of the anode material,
the effect of anode potential, and adsorption requirements of substrate and
or products on the anode require considerable research.

Electrochemical Chlorination, Bromination, and Iodination

Some characteristic features of other halogenations are considered here
according to broad classes of compounds.

Saturated Hydrocarbons

Saturated hydrocarbons undergo electrochemical chlorination and bromination at higher temperatures or on irradiation by UV light via free radical
mechanisms (*Route C* p. 2), and the products obtained appear to be similar to
those formed in homogeneous chemical reactions with these halogens. Emulsified
paraffins in aqueous HCl at 90°C give a mixture of chlorinated products [46].

Likewise cyclohexanes are converted in aqueous HBr or HCl to mono- and dihalogenated derivates on irradiation with light [47, 48].

Voltammetric studies on the chlorination of n-dodecane [49] at a carbon electrode in 20% aqueous HCl at 25-95°C showed no difference in current voltage curves from that of background. However, at 93-102°C chlorination products were produced at a porous carbon anode in 80-100% current efficiency, which was almost independent of current density, dodecane feed rate, and conversion level. The reaction mechanism was considered to involve chlorine evolution followed by a chlorination reaction taking place in a thin film adhering to the electrode. The distribution of the major products, monochlorododecanes and dichlorododecanes (higher chlorododecanes are formed also) were correlated on the basis of a calculated statistical distribution assuming a free-radical mechanism [50]:

$$R_1/R_2 = -(52 + \ln X)/25 \ln X ,$$

where R_1 is the moles of monochlorododecane per mole of dodecane feed, R_2 is the moles of dichlorododecane per mole of dodecane feed, and X is the moles of unreacted dodecane per mole of dodecane feed.

The above relationship was closely approximated at high current densities (120 mA/cm^2) or in the presence of free-radical promoters such as benzoyl peroxide in the feed.

Aromatic Hydrocarbons

In 1897, Nernst [51] expressed the opinion that it should prove possible to obtain successive stages of chlorination of organic compounds by varying the electrode potential. As early as 1913, Van Name and Maryott [52], in a study on the chlorination of benzene, reported that addition of benzene to LiCl/-HOAC solution shifted the anode (Pt) potential (measured against a Ag, Ag+ reference) to slightly more positive potentials compared to the solution without benzene. Since a similar shift was observed with CCl$_4$, it was concluded that benzene was not oxidized at the anode directly, but that chlorination proceeded by the secondary action of chlorine set free by electrolysis.

Fichter and Glanzstein [53] found that electrooxidation of benzene in a mixture of concentrated HCl and glacial acetic acid gave the products listed in Table 7.3 with increasing current density. Compared to theoretical, the overall yield of isolated products was low (11-41%), but the trend toward increasing substitution by chlorine is evident as the current density is increased, signifying higher concentrations of Cl$_2$ formed at the anode. Similar results were observed in the chlorination of toluene.

A report of successful chlorination of benzene [54] in methanolic solution is surprising, since the discharge potential of methanol is relatively low.

Table 7.3 Electrolysis of Benzene at Pt In H_2O+HOAc+HCl

Current Density A/cm^2	Products (% Yield)		
	p-Dichlorobenzene	Liquid chlorinated C_6Cl_6 hydrocarbons	Pentachlorophene and Chloranil
0.0263	17.6	– – 68.8	13.5
0.263	–	Trace 76.2	23.8
0.50	–	2.4 72.2	25.7
1.0	–	2.87 71.6	25.7
2.66	–	4.3 68.5	27.1

Although a high current density was employed, the results may indicate that chlorination occurs at the anode surface on adsorbed benzene (i.e., the concentration of methanol at the surface is low compared to chloride and benzene), in keeping with the results of other anodic reactions of aromatics in methanol.

Millington [4] has studied the anodic bromination of anthracene and naphthalene in dry acetonitrile containing Et_4NBr. Figure 7.3 shows the current potential curves for discharge of bromide ion, of anthracene, and of 9-bromoanthracene. Controlled potential electrolysis of a solution of anthracene and Et_4NBr at 0.8 V vs Ag, Ag^+, corresponding to oxidation of bromide ion, gave only a trace of 9-bromoanthracene, probably as a result of chemical bromination of anthracene. The main product at 1.2 V was 9-bromoanthracene, and at 1.6 V both 9-bromoanathracene and 9,10-dibromoanthracene were formed. Similar results were

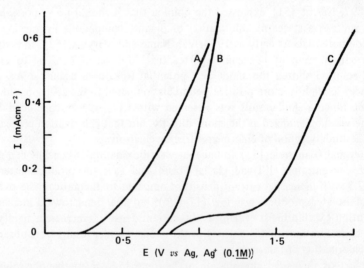

Fig. 7.3 Current-potential curves in CH_3CN at Pt. A, Et_4NBr in CH_3CN; B, anthracene in $NaClO_4/CH_3CN$; C, 9-bromoanthracene in $NaClO_4/CH_3CN$.

obtained with naphthalene. Reaction-order plots (log I versus log concentration) showed that there was no variation of current with naphthalene concentration at an electrode potential of 0.80 V, but at 1.40 V a strong dependence on naphthalene concentration was exhibited. These highly potential-dependent results suggest that the brominated products arise from an electrode process involving the aromatic compound, and not by reaction of electrochemically generated bromine with the neutral organic substrate. A mechanism involving reaction of a cationic species with bromide ion was advanced in this work; however, several alternative explanations are possible at this time. For example, it may be that the potential of discharge of the aromatic-bromine charge transfer complex must be reached before bromination occurs to any appreciable extent:

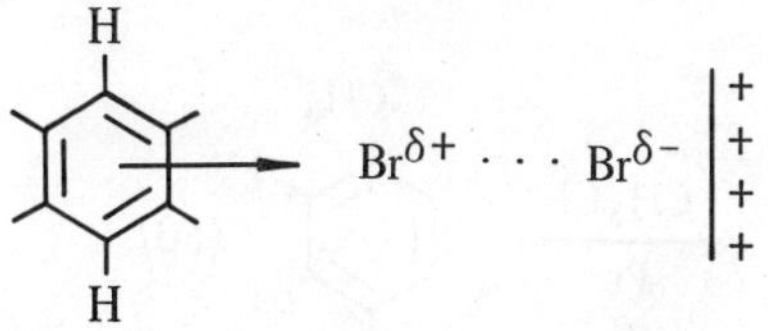

Again, the reported results may fit the proposal of Miller and co-workers [5] for iodination of aromatic compounds (p. 22).

Sasaki et al. [55] have compared the results of chemical bromination of ethylbenzene in Br_2/CH_3OH solution with electrochemical bromination using NH_4Br in CH_3OH (Table 7.4). The same products are obtained under all three

Table 7.4 Chemical and Electrochemical Bromination of Ethylbenzene in CH_3OH

Experimental Conditions	*Styrene*	*a-Methoxyethyl benzene*	*Bromoethylbenzene*	
			ortho	*para*
Chemical:				
1. Br_2/CH_3OH (dark)	3.1	Trace	12	22
2. Br_2/CH_3OH (light)	5.4	14.5	Trace	Trace
3. Electrolysis with NH_4Br/CH_3OH (dark)	20	19	2.2	6.6

conditions, but in the chemical reaction styrene and a-nethoxyethylbenzene were produced chiefly under light (daylight) conditions, while ring brominated compounds predominated under dark conditions. The product distribution obtained by electrolysis is similar to the daylight chemical reaction products. The authors suggest that the effect of electrolysis of bromide solution is equivalent to the effect of exposure of Br_2/CH_3OH to light or to the formation of bromine atoms which cause hydrogen abstraction in the side chain.

It is well recognized that, unlike chlorine and bromine, iodine does not react chemically with aromatic hydrocarbons to any appreciable extent unless special conditions [56] are employed. Miller and co-workers [5, 57] studied the possibility of electrochemical iodination by generating a positive iodine species or an aromatic cationic species. Carrying out a series of controlled potential electrolyses at 1.6-1.7 V versus Ag, Ag^+ (0.1 M) on solutions of iodine and benzenoid hydrocarbons in $CH_3CN/LiClO_4$ at Pt, they found that aromatic hydrocarbons (except anthracene and nitrobenzene) could be iodinated. For example, p-xylene was converted to a mixture of products:

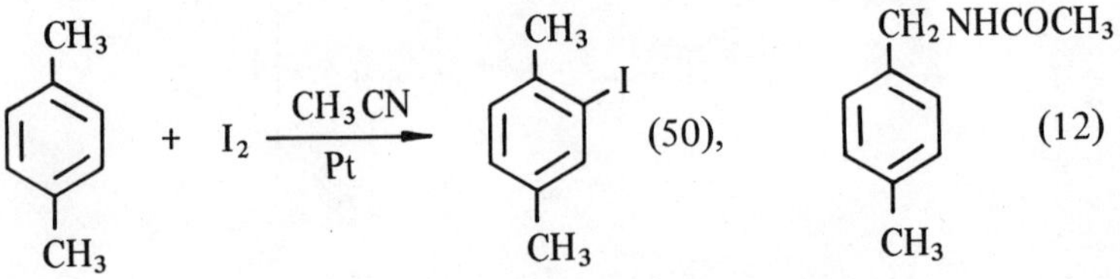

It was observed that all successful iodinations and side-chain substitutions arise from oxidations whose current-potential and current-time behavior are essentially the same as that of iodine in the absence of the aromatic. Thus anthracene, which is oxidized fairly easily compared to the iodine solution, gave no iodinated products but instead gave products characteristic of aromatic oxidation in absence of iodine. They proposed that the initial electrochemical oxidation of iodine to positive iodine controlled most of the observed chemistry.

As confirmation of this hypothesis, anodic oxidation of iodine in $CH_3CN/LiClO_4$ followed by addition of the aromatic to the oxidized solution, gave iodinated products in higher yields, with side-chain substituted products completely absent. Moreover, this procedure allowed control over the number of iodines substituted in the aromatic by varying the amount of positive iodine generated and the amount of aromatic added. Acetamide was found in appreciable yield (80% in the iodination of p-xylene) among the electrolysis products. The number of electrons transferred per iodine molecule at 1.6 V was found by coulometry to be 1.2; a pale yellow solution resulted that instantly regenerated iodine when iodide ion was added.

On the basis of this and other experiments, it was concluded that aromatic iodination probably proceeds according to the following scheme:

$$I_2 + CH_3CN \longrightarrow CH_3 - \overset{+}{C} = N - I + I^-$$

$$CH_3 - \overset{+}{C} = NI + ArH \longrightarrow ArI + CH_3CN + H^+$$

also

$$CH_3 - \overset{+}{C} = N - I + H_2O \longrightarrow CH_3\overset{O}{\overset{\|}{C}}NHI + H^+$$

$$CH_3 - \overset{O}{\overset{\|}{C}}NHI + ArH \longrightarrow ArI + CH_3\overset{O}{\overset{\|}{C}}NH_2$$

The initial steps involve electron transfer from an iodine-CH_3CN donor-acceptor complex to form the species

$$CH_3\overset{O}{\overset{\|}{C}} = NI \quad \text{and/or} \quad CH_3\overset{O}{\overset{\|}{C}}NHI.$$

Evidence for electrochemical formation of positive iodine (I^+) has previously been presented in voltammetric studies [58-61] conducted in acetonitrile and in acetic acid solutions.

Olefinic Compounds

The halofunctionalization of olefins is considered in Chapter IV, p. 368.

Other studies of halogenation of olefins have been concerned chiefly with their coulometric determination in which halogen generated electrochemically reacts with the olefin to form the dihalide:

$$2X^- \longrightarrow X_2 + 2e$$

$$\overset{\diagdown}{\diagup}C = C\overset{\diagup}{\diagdown} + X_2 \longrightarrow \overset{\diagdown}{\diagup}\underset{X}{\overset{|}{C}} - \underset{X}{\overset{|}{C}}\overset{\diagup}{\diagdown} \quad \text{where} \quad X = Br \quad \text{or} \quad Cl$$

Discrepancies from calculated values have been reported using this method which may be related to competing halofunctionalization reactions in solvent systems previously considered to be unreactive.

Amines and Amides

Hydrogen attached to nitrogen in amines and amides may be replaced by halogen electrochemically to form N-X substituted derivatives [62, 63]. The preparation of N-halo compounds is most often accomplished by the reaction of the parent nitrogen compound with the free halogen or a hypohalite in aqueous solution [64]. These reactions are usually carried out in the presence of sufficient base to neutralize the HX produced, but amides require strongly basic media.

Electrooxidation of aqueous ammonium hydroxide solutions containing KI as a halogen source and a ketone such as acetone or cyclohexanone results in formation of pyrazine derivatives. The reaction occurs by in situ generation of NI_3 which reacts with the ketone [65]. The analogous chemical reaction is known [66]:

$$\begin{array}{c} CH_3 \\ \diagdown \\ \quad\quad C = O \\ \diagup \\ CH_3 \end{array} \xrightarrow[\text{KI, Pt}]{\text{H}_2\text{O, NH}_4\text{OH}} \quad [\text{pyrazine structure}] \quad (50),\ HCl_3$$

In contrast to the above, aromatic primary amines such as aniline and 2-amino-pyridine give products of ring halogenation. It is interesting that Gilchrist [67] in 1904 found that the "decomposition voltage" of aniline in HCl solution was lower than that of the acid and on electrolysis only aniline black was formed. However, if the concentration of aniline in the region of the electrode was made sufficiently small, the "decomposition voltage" of the acid was reached, chlorine was evolved, and the aniline was chlorinated. On the other hand, no aniline black was observed in the bromination of aniline since the "decomposition voltage" of aniline in aqueous HBr was found to be slightly more positive than that of the acid.

A third mode of halogenation of amines is exemplified by the electrolysis of pyridine in CH_3CN solution containing NaI. Evidence indicates that the positive halogen compound $[I(C_5H_5N)_2]^+ClO_4^-$ is formed in solution [58, 60].

The electrochemical introduction of iodine and iodine-131 into protein material has been reported. Up to 25 iodine atoms per molecule of human albumin have been incorporated [68, 69].

Miscellaneous Classes of Compounds

Alcohols and phenols under suitable conditions may also yield positive halogen compounds in which O-H is converted to O-X. tert-Butyl alcohol is reported [62] to give tert-butyl hypochlorite [$(CH_3)_3COCl$] in 79% yield, whereas thymole is converted to a dimeric hypoiodide [70]:

$$2 \quad [\text{thymol structure, } C_3H_7, OH, CH_3] \xrightarrow[\text{KI}]{\text{H}_2\text{O, NaOH}} [\text{dimeric hypoiodide structure, } C_3H_7, OI, IO, C_3H_7, CH_3, CH_3]$$

Under other conditions phenols are ring halogenated, and alcohols are oxidized to aldehydes that are halogenated. Ethyl alcohol, for example, is converted quantitatively to chloroform in alkaline chloride solutions, whereas in acidic chloride media chloral is formed in good yield. In an interesting study on electrochemical bromination of phenol, Landsberg et al. [71] have found that a mixture of p- and o-bromophenol was formed quantitatively; however, the para-ortho ratio varied markedly with the potential of the working electrode. In the lower potential region a ratio of $p/o = 3.37$ was found. This value was somewhat larger than that found by chemical bromination under similar conditions, where $p/o = 2.96$. At more positive potentials, however, a value of $p/o = 1.96$ was observed. This remarkable activation of the ortho position in the higher potential region was ascribed to the effect of the high electric field within the double layer. High yields of ortho isomer have been observed in other anodic substitution reactions.

Ketones are halogenated in the a position to the carbonyl group in complete analogy to chemical halogenation reactions. Thus acetone may be converted quantitatively to chloroform in the high pH region or to chloroacetone at low pH. Electrochemical halogenation of ketones is a useful and convenient method of preparation of a-haloketones. The electrochemical production of iodoform was at one time an important commercial method for preparation of this material.

Reaction of electrochemically generated halogen with solvent (dimethylsulfoxide or acetonitrile) may lead to side products [1]:

$$X_2 + CH_3 \overset{O}{\overset{\|}{S}} CH_3 \longrightarrow XCH_2 \overset{O}{\overset{\|}{S}} CH_3 + HX$$

and

$$X_2 + CH_3 CN \longrightarrow XCH_2 CN + HX$$

The analogous chemical reaction of halogen with dimethylsulfoxide is known [72]. In an electrochemical study of iodine in acetic acid, good evidence was obtained for the formation of iodoacetic acid [59] in the reaction of positive halogen with the solvent:

$$I_2 \longrightarrow 2I^+ + 2e$$
$$I^+ + CH_3 CO_2 H \longrightarrow ICH_2 CO_2 H + H^+$$

In choosing a suitable solvent for electrochemical halogenation, the possibility of such side reactions should be considered, although in a great many cases, the substrate is often more reactive than the solvent.

A process of possible commercial importance for the production of melamine has been described, resulting from electrochemical formation of cyanogen halide [73]. It is known [74-76] that electrochemical oxidation of an aqueous solution of HCN containing an ammonium halide salt produces cyanogen halide according to

$$HCN + X^- \longrightarrow XCN + H^+ + e$$
$$(X = Cl, Br, I)$$

Ammonolysis of the cyanogen bromide $(X = Br)$ in a solvent such as dioxane leads to cyanamide, which may be trimerized to melamine.

In electrolysis cell: $HCN + NH_4Br \longrightarrow CNBr + H_2 + NH_3$

In CNBr reactor: $CNBr + 2NH_3 \longrightarrow H_2NCN + NH_4Br$

Trimerization: $3H_2NCN \xrightarrow[\Delta,\ pressure]{NH_3,} C_3N_6H_6 (melamine)$

Overall Reaction: $3HCN + 3NH_3 \longrightarrow C_3N_6H_6 + 3H_2$

2   LITERATURE SURVEY

The following pages contain tables surveying the literature on halogenation for fluorination, chlorination, bromination, and iodination.

Table 7.5 Fluorination: (i) Saturated and Unsaturated Hydrocarbons and Halocarbons

Compound (Medium, Experimental Conditions)	Anode	Product(s) (% Yield)	Ref.
Methane (a) (HF, NaF) CD 2.2 A/dm²	Ni	CF_4 (7.6), CF_3H (3.5), CF_2H_2 (2.7), CFH_3 (18.4)	26, 77
(b) (HF, Cl₂, NaF) CD 1.7 A/dm²	Ni	CF_4, CHF_3, $CClF_3$, CH_2F_2, $CHClF_2$, CCl_2F_2, $CHCl_2F$, CH_2ClF, CH_3Cl, CH_3F, CF_3CF_3, CH_3CF_3, (total 30.2 in proportion 12.5 : 11.2 : 6.2 : 21.6 : 3.7 : 2.5 : 0.6 : 7.2 : 22.1 : 10.9 : 0.3 : 1.2	78
(c) (KF, 2HF) CD 0.1 A/cm²; temp 74-76°C	Porous C[a]	CFH_3 (68.2), CF_2H_2 (14.2), CF_3H (6.7), CF_4 (10.9)	25a, 80a
Methyl chloride (HF, NaF) CD 2.2 A/dm²	Ni	CF_4 (10.2), $CClF_3$ (5.0), $CHClF_2$ (16.5), CH_2F_2 (13.2), CH_2ClF (5.2), CH_3F (7.7), CF_3H (17.2)	79
Methyl iodide (HF, LiF) CD 0.004 A/cm²	Ni	CF_4(16), CF_3H(16), CFH_3(1), CF_2H_2(1), I_2	80
Methylene chloride (a) (HF, KF) CD 100 mA/cm²; temp 70-90°C	Ni-Porous C[a]	CCl_3F(22.6), CCl_2FH(20.2), CCl_2F_2(37.0), $CClF_3$ (20.2)	80a
(b) (HF, LiF) CD 1.0 A/dm²; temp -10°C	Ni, Porous C	$CHFCl_2$(35.0), CF_2Cl_2(8.3), CHF_2Cl (6.4), CH_2FCl (5.0), CF_4, $CClF_3$, CF_3H, CH_2F_2, CCl_3F	25a, 81-83
Chloroform (a) (HF, KF) CD 66 mA/cm²; temp 70-90°C	Ni-Porous C[a]	CCl_3F (97.0), CCl_2F_2 (1.5)	80a
(b) HF, LiF	Ni-Porous C[a]	CCl_2F_2, CCl_3F, $CHCl_2F$, CF_4, CF_3H, CF_3Cl, CHF_2Cl	25a, 81, 83, 84
Chlorodifluoromethane (HF, NaF) CD 2.2 A/dm²	Ni	CF_4 (14.2), $CClF_3$ (22.1), CF_3H (6.2)	79
Dichlorofluoromethane (HF, NaF) CD 2.2 A/dm²	Ni	CF_4 (11.2), $CClF_3$ (7.5), CF_3H, CCl_2F_2, $CHClF_2$	79
Trichloroiodomethane (HF) CD 1-20 A/dm²	Ni	CCl_3F	85

27

28

Table 7.5 Fluorination: (i) Saturated and Unsaturated Hydrocarbons and Halocarbons (cont.)

Compound (Medium, Experimental Conditions)	Anode	Product(s) (% Yield)	Ref.
Ethane (a) (HF, NaF) CD 2.2 A/dm²	Ni	CF_3CF_3 (28.6), CF_3CHF_2 (7.0), CF_3CH_2F (6.6), CHF_2CHF_2 (9.7), CF_3CH_3 (2.2), CH_2FCH_3	77, 86
(b) (HF, Cl_2,[b] NaF) CD 1.7 A/dm²	Ni	CF_3CF_3, CHF_2CF_3, $CClF_2CF_3$, CH_2FCF_3, CHF_2CClF_2, CH_3CClF_2, CF_4, CHF_3, $CClF_3$, CH_2F_2, CCl_2FCClF_2 (total 89.7 in proportion 5.2 : 2.8 : 1.5 : 6.7 : 5.7 : 4.3 : 3.3 : 23.4 : 19.9 : 24.0 : 0 : 1.0 : 1.1 : 0 : 1.1)	78
(c) (KF.2HF) CD 0.1 A/cm²; temp 72–95°C	Porous C[a]	C_2H_5F (46.9), CH_3CHF_2 (9.8), CH_2FCH_2F (11.2), CH_3CF_3 (1.6), CF_2HCH_2F (7.8), CF_3CH_2F (2.0), CF_2HCF_2H (3.4), CF_3CHF_2 (4.1), C_2F_6 (10.1), C_4-fluorides (2.1), C_1-fluorides (1.0) (total CE 87)	25a, 80a
Ethyl chloride (KF.2HF) CD 0.1 A/cm²; temp 76°C	Porous C[a]	CH_2ClCH_2F, CH_3CHClF, CH_2ClCHF_2, $CHClFCH_2F$	25a
1,2-Dichloroethane (a) (KF.2HF) CD 0.2 A/cm²; temp 93–95°C	Porous C[a]	$C_2Cl_2F_4$, $C_2Cl_2HF_3$, $C_2Cl_2H_2F_2$, $C_2Cl_2H_3F$	25a
(b) (HF, NaF)	Ni	C_2F_6, C_2ClF_5, $CClF_2CClF_2$, Cl_2	84, 87
1,1-Difluoroethane (KF.2HF) CD 0.2 A/cm²; temp 82–84°C	Porous C[a]	CF_3CH_3 (24.0), CHF_2CH_2F (45.5), CHF_2CHF_2 (9.9), CF_3CH_2F (9.8), C_2F_5H (9.7), C_2F_6 (1.1)	25a
1,1,2-Trifluoroethane (KF.2HF) CD 0.1 A/cm²; temp 70–90°C	Porous C[a]	CHF_2CHF_2 (48.1), CF_3CH_2F (21.1), C_2F_5H (16.1), C_2F_6 (13.1), C_4-fluorides (1.1), C_1-fluorides (0.4)	25a, 80a
1,1-Dibromo-1,2,2,2-tetrachloroethane: CBr_2ClCCl_3 (HF, NaF, CaF_2) CD 1–20 A/dm²	Ni	CF_2ClCCl_3 (90)	85, 88
Ethylene (a) (HF, NaF) CD 2.2 A/dm²	Ni	CF_3CF_3 (36.0), CF_3CHF_2 (13.3), CF_3CH_2F (17.3), CHF_2CHF_2 (11.2), CF_3CH_3 (4.8), CHF_2CH_3 (3.4), CF_4 (2.0)[d]	86
(b) (HF, Cl_2,[b] NaF)	Ni	CF_3CF_3, CHF_2CF_3, $CClF_2CF_3$, CH_2FCF_3, CHF_2CClF_2, $CClF_2CClF_2$, CH_3CF_3,	78

Substrate / Conditions	Electrode	Products	Ref.
(c) (KF . 2HF) CD 0.1 A/cm²; temp 70-90°C	Porous C[a]	CH_2FCClF_2, CH_3CClF_2, CF_4, CHF_3, $CClF_3$, CH_2F_2, CH_3CHF_2, $CHClF_2$ (total 64.9 in proportion 20.3 : 6.5 : 2.7 : 5.8 : 9.2 : 5.8 : 0.7 : 23.6 : 5.0 : 7.9 : 0.7 : 2.0 : 3.5 : 1.2 : 3.7 : 1.4)	25a, 80a
Fluoroethylene (a) (HF, NaF) CD 2.2 A/dm²	Ni	C_2H_3F (10.2), C_2H_5F (3.6), CHF_2CH_3 (0.7), CH_2FCH_2F (26.6), CHF_2CH_2F (18.2), CHF_2CHF_2 (8.6), CF_3CH_2F (3.8), C_2F_5H (7.7), C_2F_6 (6.0), C_4-fluorides (13.9), C_1-fluorides (0.7) (total CE 99.6)	89
(b) (HF, Cl_2,[b] NaF) CD 1.7 A/dm²	Ni	CF_3CF_3, CHF_2CF_3, $CClF_2CF_3$, CH_2FCF_3, CH_3CF_3, CHF_2CHF_2, CHF_2CClF_2, $CClF_2CClF_2$, CH_3CF_3, CH_2FCClF_2, CH_3CClF_2, CF_4, CHF_3, $CClF_3$, CH_2F_2, $CHClF_2$, $n-C_4F_{10}$ (total 68.6 in proportion 31.4 : 3.7 : 11.3 : 8.3 : 11.5 : 20.5 : 4.2 : 0.7 : 1.1 : 0 : 0.3 : 2.8 : 1.4 : 1.2 : 0.9 : 0.7)	78
Chloroethylene (HF, NaF) CD 2.2 A/dm²	Ni	CF_3CF_3, CHF_2CF_3, $CClF_2CF_3$, CH_2FCF_3, CHF_2CClF_2, CH_2FCClF_2, CH_3CF_3, CF_4, CHF_3, $CClF_3$, $CHClF_2$, CCl_2F_2, $CClF_2CClF_2$, $CHClFCClF_2$, CCl_2FCClF_2, $n-C_4F_{10}$ (total 65.9 in proportion 39.4 : 6.9 : 14.2 : 2.6 : 1.1 : 7.7 : 4.2 : 0.6 : 7.1 : 1.0 : 7.5 : 0.7 : - : 4.6 : 0.9 : 0.9 : 0.6)[c]	89
1,2-Dibromo-1,2-dichloroethylene: $BrClC=CClBr$ (HF, NaF, CaF_2) CD 1-20 A/dm²	Ni	$F_2ClCCFCl_2$ (90)	85, 88
Acetylene (a) (HF : KF ≅ 1 : 3,N_2) CD 0.2 A/cm²; temp 120°C	Porous C[e]	$C_2F_4H_2$, C_2F_5H, C_2F_6, CF_4 (total 84 in proportion 44 : 20 : 9 : 26)	92
(b) (HF, KF) CD 133 mA/cm²; temp 88°C	Ni-Porous C[a]	1,2-Difluoroethylene, 1,1,2,2-tetrafluoroethane	25a
Propane (HF, NaF)	Ni	C_3F_8, CF_3CHFCF_3, $CF_3CF_2CHF_2$, CF_4, $C_3F_6H_2$	77

29

Table 7.5 Fluorination: (i) Saturated and Unsaturated Hydrocarbons and Halocarbons (cont.)

Compound (Medium, Experimental Conditions)	Anode	Product(s) (% Yield)	Ref.
CD 0.4-0.6 A/dm²		(30-33 current yield)	93
2,2-Difluoropropane (HF, NaF) CD 1.4 A/dm²	Ni	C_3F_8 (80)	85, 91
1,1,1,2-Tetrafluoro-3,3-dichloro-2,3-dibromopropane (HF, KF) CD 1-20 A/dm²	Ni	$CF_3CF_2CFCl_2$ (85)	80a
Cyclopropane (HF, KF) CD 167 mA/cm²; temp 75°C	Ni-Porous C[a]	Monofluorocyclopropane, 1,3-difluoropropane, 1,2,3-trifluoropropane, 1,2-difluorocyclopropane	25a
Isobutane (KF, 2HF) CD 0.066 A/cm²; temp 70-90°C	Porous C[a]	2-Fluoromethylpropane (8.4), 1-fluoromethylpropane (65.7), 1,1-difluoromethylpropane (5.8), 1,2-difluoromethylpropane (3.1), 1,3-difluoromethylpropane (15.7), trifluoromethylpropane (1.3)	
2,2-Difluorobutane (HF, NaF) CD 1.4 A/dm²	Ni	C_4F_{10} (2 isomers) (51), C_5F_{12}, C_6F_{14}	93, 94
Hexane (HF, NaF) CD 1.4 A/dm²	Ni	C_6F_{14}	93, 94
2,2-Difulorohexane (HF, NaF) CD 1.4 A/dm²	Ni	C_6F_{14} (62.8), C_4F_{10}, C_5F_{12}	93, 94
n-Octane (HF, H₂O)	Ni	C_8F_{18} (70 g), CF_4 (54 g), C_2F_6 (27 g/160 gsm), OF_2, polymer	81, 95, 96
1,1-Difluoroethylene (HF, NaF) CD 2.2 A/dm²	Ni	CF_3CF_3, CHF_2CF_3, CH_2FCF_3, CH_3CF_3, CF_4, CHF_3, $n\text{-}C_4F_{10}$ (total 66.3 in proportion 23.4 : 4.8 : 15.3 : 46.9 : 8.9 : 0.3 : 0.4)	89
Trichloroethylene (a) (HF, NaF)	Ni[e]	$CHFClCFCl_2$, $CF_2ClCFCl_2$, (in proportion 82 : 12)	81, 82
(b) (CH₃CN, AgF) CD 3-50 mA/cm²	Pt	Electrode blocked by AgCl formation	37
Tetrafluoroethylene (HF, NaF) CD 2.2 A/dm²	Ni	CF_3CF_3, CHF_2CF_3, CF_4, CHF_3, $n\text{-}C_4F_{10}$ (total 86.1 in proportion 73.5 : 2.2 : 20.4 : 0.4 : 3.5)	89

Substance (electrolyte)	Electrode	Products	Ref.
Chlorotrifluoroethylene (HF, NaF) CD 2.2 A/dm²	Ni	CF_3CF_3, $CClF_2CF_3$, CF_4, CHF_3, $CClF_3$, CCl_2F_2, $CClF_2CClF_2$, CHF_2CClF_2, n-C_4F_{10} (total 71.0 in proportion 22.0 : 31.9 : 18.6 : 2.8 : 5.9 : – : 16.7 : – : 2.1)	89
Tetrachloroethylene (HF, LiF) CD 1-20 A/dm²	Ni	CCl_2FCCl_2F (90)	91
Bromotrichloroethylene : $BrClC=CCl_2$ (HF, LiF, CaF_2) CD 1-20 A/dm²	Ni	$F_2ClCCClFCl_2$ (90)	85, 88
2,2-Difluorooctane (HF, NaF) CD 1.4 A/dm²	Ni	C_8F_{18}, (2 isomers) (40), C_7F_{16}, C_6F_{14}, C_5F_{12}, C_4F_{10} (total 500 g/350 gsm in proportion 82 : 3.8 : 6.6 : 2.6 : 5)	93, 94
Cyclohexane (a) (HF, NaF)	Ni	Little or no fluorination product	93
(b) (CH₃CN, AgF) CD 3-50 mA/cm²	Pt	1,2-Difluorocyclohexane	37
1,1-Difluorocyclohexane (HF)	Ni	Perfluorocyclohexane, perfluoromethylcyclopentane, C_6F_{14} (2 isomers), lower perfluoroalkanes (total 72.5 in proportion 40 : 55 : 3.7 : 0.4)	93
Methylcyclohexane (HF, NaF)	Ni	C_7F_{14} (18.2)	93
Trifluoromethylcyclohexane (HF, NaF) CD 1.4 A/dm²	Ni	C_7F_{14} (47)	93
Benzene (a) (HF, NaF)	Ni	Polymer ($C_{30}H_{20}F$)	93, 124
(b) (H₂O, HF)	Ni	Black polymer : $C_6H_{3.6}F_{0.1}O_{0.1}$ (0.34 g/0.4 Ahr)	90
(c) (HF, BF₃)	Pt	Polymer : $C_6H_{3.5}F_{0.06}$ (2.6 g/2.4 Ahr)	90
Iodobenzene (CH₃CN, AgF) CD 3-50 mA/cm²	Pt	Iodobenzene difluoride	37
Toluene (a) (HF, NaF) CD 1-10 mA/cm²	Ni	C_7 fluorocarbon	95, 96
(b) (H₂O, HF)	Ni	Polymer (0.5 g/0.8 Ahr)	90
Mesitylene (HF, KF)	Pt	Polymer (1.1 g/0.75 Ahr)	90
trans-Stilbene (CH₃CN, AgF) CD 3-50 mA/cm²	Pt	Fluorine containing product	37

Table 7.5 Fluorination: (i) Saturated and Unsaturated Hydrocarbons and Halocarbons (cont.)

Compound (Medium, Experimental Conditions)	Anode	Product(s) (% Yield)	Ref.
1,1-Diphenylethylene (CH$_3$CN, AgF) CD 3-50 mA/cm^2	Pt	α,α-Difluorodibenzyl	37, 97-99
Naphthalene (a) (HF, NaF)	Ni	Little or no fluorination product	93
(b) (H$_2$O, HF)	Ni	Viscous product	90
(c) (CH$_3$CN, Et$_3$N.HF) CPE	Pt	1-Fluoronaphthalene (4-5%)	39
(d) (CH$_3$CN, Et$_4$NF) CPE 1.8 V vs SCE	Pt	1,4-Difluoronaphthalene (70)	39a
Diphenyl (H$_2$O, HF)	Pt	Viscous product	90
2,2-Difluorodecahydronaphthalene (HF, NaF) CD 1.4 A/dm^2	Ni	C$_{10}$F$_{18}$ (3 cyclic isomers), C$_6$ isomers, C$_8$ isomers (total 35.5 in proportion 92.4 : 2.6 : 2.2)	93
o-Terphenyl (HF)	Pt	Polymer (3.7 g/3.5 Ahr)	90
9,10-Diphenylanthracene (CH$_3$CN, (CH$_3$)$_4$NF$_3$H$_2$) CPE 1.65 V vs SCE	Pt	9,10-Difluoro-9,10-dihydro-9,10-diphenylanthracene (43)	38
1-Fluoronaphthalene (CH$_3$CN, Et$_4$NF.3HF) CPE 1.6 V vs SCE	Pt	1,4-Difluoronaphthalene (40)	39a
1,4-Difluoronaphthalene (CH$_3$CN, Et$_4$NF.3HF) CPE 1.8 V vs SCE	Pt	(22), (0.5)	39a

Table 7.6 Fluorination: (ii) Amines

Compound (Medium, Experimental Conditions)	Anode	Product (s) (% Yield)	Ref.
Dibutylamine (HF)	Ni	C_4F_{10}	95
n-Octylamine (HF) CD 0.012 A/cm^2	Ni	C_8F_{18}	12, 95
n-Decylamine (HF)	Ni	$C_{10}F_{22}$	95
Trimethylamine (HF) CD 20 A/ft^2	Ni	$(CF_3)_3N$, CF_4, CF_3H, NF_3	95, 100
Triethylamine (HF) CD 20 A/ft^2	Ni	$(C_2F_5)_3N$, $(C_2F_5)_2NCF_3$, partially fluorinated products	95, 100, 101
Tri-n-propylamine (HF) CD 20 A/ft^2	Ni	$(C_3F_7)_3N$	95, 100
Tri-n-butylamine (HF) CD 20 A/ft^2	Ni	$(C_4F_9)_3N$, partially fluorinated products	95, 100, 101
Triamylamine (HF) CD 20 A/ft^2	Ni	$(C_5F_{11})_3N$	95, 100
Tri-n-hexylamine (HF) CD 20 A/ft^2	Ni	$(C_6F_{13})_3N$	95, 100
Di-n-propylethylamine (HF) CD 20 A/ft^2	Ni	$(C_3F_7)_2N(C_2F_5)$	95, 100
Diisopropylethylamine (HF) CD 20 A/ft^2	Ni	$(i\text{-}C_3F_7)_2N(C_2F_5)$	95, 100
Diethyl-n-propylamine (HF) CD 20 A/ft^2	Ni	$(C_2F_5)_2N(C_3F_7)$	95, 100
Diethyl-n-butylamine (HF) CD 20 A/ft^2	Ni	$(C_2F_5)_2N(C_4F_9)$	95, 100
N,N-Dimethylaniline (HF) CD 0.02 A/cm^2	Ni	$(CF_3)_2NC_6F_{11}$(34); $CF_3(CF_2)_5N(CF_3)_2$, N,N-dimethylperfluorocyclohexylamine	25, 95, 100, 101, 115
N,N-Diethylcyclohexylamine (HF) CD 20 A/ft^2	Ni	$(C_2F_5)_2N(C_6F_{11})$	95, 100

Table 7.6 Fluorination: (ii) Amines (cont.)

Compound (Medium, Experimental Conditions)	Anode	Corresponding perfluoro(N-alkylpiperidine)		Ref.
		Total Yield	% Normal Isomer	
Piperidines:				
(HF) CD 0.0125 A/cm^2	Ni			
R is H		–	–	102
methyl^g		39	–	100
isopropyl		–	35	103
n-butyl		43	–	104
isobutyl		–	67	103
n-amyl		42	96	103, 104
isoamyl		44	75	103, 104
n-hexyl		15	–	104
n-nonyl		11	–	104
3-Carbomethoxy-N-pentylpiperidine (HF)^h	Ni			105

34

N-Alkylmorpholines:

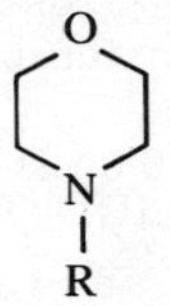

(HF)
(Ni)
CD 0.0125 A/cm²

Corresponding perfluoro(N-alkylmorpholine)

R is	Total Yield	% Normal Isomer	
H	13	–	106
methyl	14	–	106
isopropyl	–	14	103
n-butyl	40	–	104
isobutyl	–	63	103
n-amyl	44	–	104
isoamyl	–	96	103
n-heptyl	21	–	104
n-nonyl	14	–	104
n-decyl	6	–	104

N,N'-Hexamethylenebis (2-methylpyrrole):

(HF)
(Ni)
CD 0.0125 A/cm²

Perfluoro[N,N'-hexamethylenebis (2-methylpyrrole)] 104
(23)

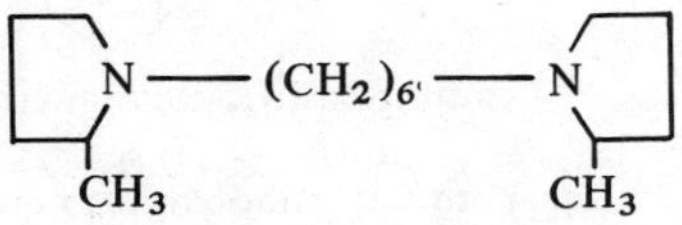

35

36

Table 7.6 Fluorination: (ii) Amines (cont.)

Compound (Medium, Experimental Conditions)	Anode	Product(s) (% Yield)	Ref.
N,N-Diethylisopentylamine (HF) CD 0.0125 A/cm²	Ni	Perfluoro(N,N-diethylisopentylamine) (22)	104
N-Butylhexamethylenimine: (HF) CD 0.0125 A/cm² [N—Bu ring structure]	Ni	Perfluoro(N-butylhexamethylenimine) (37)	104
N,N-Dialkylpiperazines: [R—N N—R ring structure] (HF) CD 0.0125 A/cm² R is n-butyl, isopentyl, n-hexyl	Ni	Corresponding perfluoro(N,N-dialkylpiperazine) (20) (18) (15)	104
Aniline (HF)	Ni	$C_6F_{11}NF_2$, C_5F_{12}, C_6F_{12}, NF_3	102
Methyl p-(diethylamino)benzoate (HF) CD 0.012 A/cm²	Ni	$(C_2F_5)_2$N-CF$\Big\langle$ $\substack{CF_2CF_2 \\ CF_2CF_2}$ $\Big\rangle$C$\substack{F \\ CO_2H}$ (14)	105
Pyridine (HF) CD 0.01 A/cm²	Ni	Perfluoropiperidine (7.5), perfluoro-1,1-dipiperidyl, 2-fluoropyridine (5.3), n-C_5F_{12} (25), CF_4, NF_3 (30)	10, 81, 95, 102, 106-109
2-Fluoropyridine (HF)	Ni	Perfluoropiperidine (13)	10, 107
Quinoline (HF)	Ni	C_9F_{18} (22 g/100 gsm), NF_3	10. 95
1,1'-Dipiperidylmethane	Ni	Perfluoropiperidine (205 g), perfluoro-1-methylpiperidine (178 g), C_5F_{10} (1100 g), NF_3 (260 g/1374 gsm)	106

4-n-Propylpyridine (HF)	Ni	Perfluoro-4-n-propylpiperidine (85.7 g), perfluoro-3-ethylhexane (754 g/1695 gsm), NF_3, C_3-C_6 fluorocarbons	106
4-Isopropylpyridine (HF)	Ni	Perfluoro-4-isopropylpiperidine (260 g)^i, C_7F_{16} isomers (74 g), C_8F_{18} isomers (582 g/1452 gsm) NF_3	106
N,N-Diethylbenzylamine (HF) CD 20 A/ft^2	Ni	$(C_2F_5)_2NCF_2C_6F_{11}$	95, 100
N,N,N',N'-Tetraethylethylenediamine (HF) CD 20 A/ft^2	Ni	$(C_2F_5)_2NCF_2CF_2N(C_2F_5)_2$	100
N,N,N',N'-n-Tetrabutylethylenediamine (HF) CD 20 A/ft^2	Ni	$(C_4F_9)_2NCF_2CF_2N(C_4F_9)_2$	100
Triethylenediamine (HF) CD 0.05 A/ft^2	Ni	Perfluorotriethylenediamine	109a
2,4,6-Tri(dimethylaminomethyl)-phenol (HF) CD 20 A/ft^2	Ni		100
N,N-Dimethylglycine methyl ester (HF)	Ni	$(CF_3)_2NCF_2COF$ (6)	110
N,N-Dimethylglycine dimethylamide (HF)	Ni	$(CF_3)_2NCOF$, $(CF_3)_3N$, C_4F_9NO, NF_3	110

37

Table 7.6 Fluorination: (ii) Amines (cont.)

Compound [Medium, Experimental Conditions]	Anode	Product(s) [% Yield]	Ref.
I(pyridine)$_2$F (CH$_3$CN, AgF) CD 3–50 mA/cm^2	Pt	I(pyridine)F$_3$	37
N-Alkylpyridinium bromides (HF) CD 0.02 A/cm^2	Ni	Products not indicated	45, 111
3-(N,N-Diethylamino)propionyl chloride hydrochloride (HF) CD 20 A/ft^2	Ni	(C$_2$F$_5$)$_2$NCF$_2$CF$_2$COF (12)	109a
N-(β-Propionyl chloride)-2,6-dimethyl-morpholine hydrochloride (HF) CD 20 A/ft^2	Ni	![structure] O S$_F$ NCF$_2$CF$_2$COF CF$_3$ / CF$_3$ (16 g/50 Ahr)	109a
N,N′-Di(β-propionyl chloride)-piperazine dihydrochloride (HF) CD 20 A/ft^2	Ni	FCO(CF$_2$)$_2$N S$_F$ N(CF$_2$)$_2$COF (14 g/50 Ahr), CF$_3$CF$_2$N S$_F$ N(CF$_2$)$_2$COF	109a
Ethyliminodipropionyl chloride hydrochloride: CH$_3$CH$_2$N(CH$_2$CH$_2$COCl)$_2$.HCl (HF)	Ni	CF$_3$CF$_2$N(CF$_2$CF$_2$COF)$_2$ (28)	109a
1-Piperidinopropionyl chloride hydro-chloride (HF) CD 20 A/ft^2	Ni	c-C$_5$F$_{10}$NC$_2$F$_4$COF (37)	109a
4-Ethyl-1-piperidinopropionyl chloride hydrochloride (HF) CD 20 A/ft^2	Ni	c-C$_2$F$_5$C$_5$F$_9$NCF$_2$CF$_2$COF (24)	109a

Table 7.7 Fluorination: (iii) Ethers

Compound (Medium, Experimental Conditions)	Anode	Product (s) (% Yield)	Ref.
Dimethyl ether (a) HF CD 20 A/ft²	Ni	CF_3OCF_3	112
(b) (KF .2HF) CD 0.1 A/cm²; temp 77°C	Porous C^a	CH_2FOCH_3, CHF_2OCH_3, CH_2FOCH_2F, CH_2FOCHF_2, CF_3OCHF_2, CF_3OCH_2F, CHF_2OCHF_2, CF_3OCHF_2, CF_3OCF_3	25a, 80a
Diethyl ether (HF) CD 3.3 A/dm²	Ni	$C_2F_5OC_2F_5$; CF_3CO_2Na (36.2)^h	13, 112, 113
Ethyl vinyl ether (HF)^h CD 3.3 A/dm²	Ni	CF_3CO_2Na (29.3), $C_2F_5CO_2Na$ (0.5), $C_3F_7CO_2Na$ (0.7)	114
Di-n-propyl ether (HF)^h CD 3.5 A/dm²	Ni	$C_2F_5CO_2Na$ (20.8), CF_3CO_2Na (2.8)	113
Diisopropyl ether (HF)	Ni	C_3F_8, C_3F_7H	95
Di-n-butyl ether (a) (HF) •CD 20 A/ft²	Ni	$C_4F_9OC_4F_9$,^{f} C_4F_{10}, C_3F_8, C_2F_6, $^bCF_4$	13, 81, 101, 112 113
(b) (HF)^h CD 3.3 A/dm²	Ni	n-C_3F_7COF (19.5), C_2F_5COF(2.4), CF_3COF(4.6	113
n-Butyl ethyl ether (HF)^h CD 3.5 A/dm²	Ni	n-$C_3F_7CO_2Na$(17.5), $C_2F_5CO_2Na$(0.7), CF_3CO_2Na(17.6)	113
n-Butyl vinyl ether (HF)^h CD 3.3 A/dm²	Ni	CF_3CO_2Na (10.5), $C_2F_5CO_2Na$(2.5), $C_3F_7CO_2Na$(16.2)	114
Di-n-amyl ether (HF) CD 20 A/ft²	Ni	$C_5F_{11}OC_5F_{11}$, C_4F_{10}	95, 112
Di-n-hexyl ether (HF) CD 20 A/ft²	Ni	$C_6F_{13}OC_6F_{13}$	112
Ethylene glycol monobutyl ether (HF) CD 20 A/ft²	Ni	$C_4F_9OCF_3$	112
Diethylene glycol diethyl ether (HF) CD 20 A/ft₂	Ni	$C_2F_5OCF_2CF_2OCF_2CF_2OC_2F_5$ (3.8)	112

39

Table 7.7 Fluorination: (iii) Ethers (cont.)

Compound (Medium, Experimental Conditions)	Anode	Products(s) (% Yield)	Ref.
Diethylene glycol dibutyl ether (HF)	Ni	$C_4F_9OCF_3$	95
Ethylene oxide (HF) CD 2.5 A/dm²	Ni	Perfluoroethylene oxide; CF_3CO_2Na	113, 116
Propylene oxide (HF) CD 3.0 A/dm²	Ni	CF_3CO_2Na (2.7), $C_2F_5CO_2Na$ (4.0)	113
Epichlorohydrin (HF)	Ni	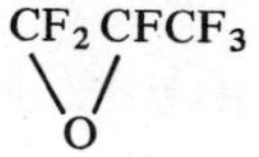 CF_2CFCF_3 (epoxide)	95

Tetrahydrofurans: (HF)

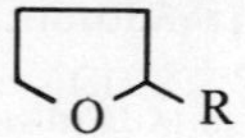

	Anode	Products(s) (% Yield)	Ref.
R is H CD 0.017–0.018 A/cm² CD 20 A/ft²	Ni	[perfluorotetrahydrofuran structure CF_2–CF_2 / CF_2 CFR_F / O] (42)	95, 113, 116, 117
methyl		(25.2)	118
ethyl		(39.8)	118
n-propyl		(40.2)	118
n-butyl		(40.8)	118, 119
ethoxymethyl		(41.8)	118, 119
n-propoxymethyl		(39.0)	118, 119
n-butoxymethyl		(42.0)	118, 119
n-pentyloxymethyl		(18.8)	118, 119
isopentyloxymethyl		(17.9)	118, 119
n-hexyloxymethyl		(8.9)	118, 119

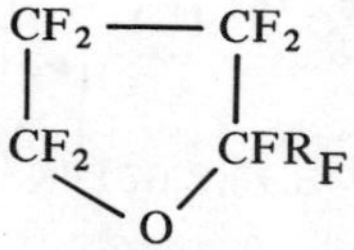

40

n-heptyloxymethyl
-CH₂-N(Et)₂ (5.4)
(16.2)

-CH₂-N (morpholine) (35.4) 119

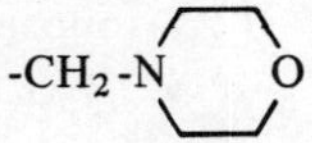

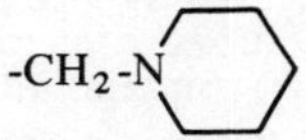

-CH₂-N (piperidine) (18.1) 119

-CH₂-N (methylpiperidine), CH₃ (3.2) 119

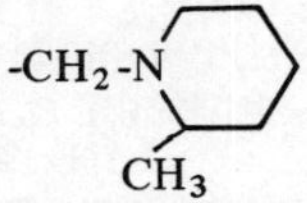

α-Butyl-β-chlorotetrahydrofuran (HF) CD 0.017 A/cm²	Ni	α-Perfluorbutyltetrahydrofuran (38.7 g/248 gsm), polymer	118
α-Octyl-β-chlorotetrahydrofuran (HF) CD 0.017 A/cm²	Ni	α-Perfluorobutyltetrahydrofuran, $C_{12}F_{23}C10$, polymer	118
Tetrahydropyran (HF) CD 20 A/ft²	Ni	Decafluoropentamethylene oxide (35)	95, 116
1,4-Dioxane (HF) CD 20 A/ft²	Ni	$CF_3OCF_2CF_2OCF_3$ (40 g/321 gsm)	95, 112
Anisole (HF)	Ni	$C_6F_{11}OCF_3$	95

Table 7.8 Fluorination: (iv) Alcohols and Phenols

Compound (Medium, Experimental Conditions)	Anode	Product(s) (% Yield)	Ref.
Methanol (HF) CD 0.04 A/cm²	Ni	CF_4 (9.6 g), CF_3H (6.4 g/35 gsm)	13, 95
Ethanol (HF)[h] CD 3.3 A/dm²	Ni	CF_3CO_2Na (29.7); CF_4, C_2F_6	13, 32, 95
n-Propyl alcohol (HF)[h] CD 2.7 A/dm²	Ni	$C_2F_5CO_2Na$ (21.7), CF_3CO_2Na (1.6); C_2F_6	11, 32, 95
Isopropyl alcohol (HF)[h] CD 3.3 A/dm²	Ni	CF_3CO_2Na (10.9); C_3F_8	32, 95
2,2,3,3-Tetrafluoropropyl alcohol (HF)[h] CD 1.6 A/dm²	Ni	CF_3CO_2Na (8.9), $C_2F_5CO_2Na$ (52.3)	120
Allyl alcohol (HF)[h] CD 3.0 A/dm²	Ni	CF_3CO_2Na (2.0), $C_2F_5CO_2Na$ (1.8), $C_3F_7CO_2Na$ (0.6), polymers	121
n-Butyl alcohol (HF)[h] CD 3.3 A/dm	Ni	$C_3F_7CO_2Na$ (17.5), $C_2F_5CO_2Na$ (1.8), CF_3CO_2Na (1.1)	32, 95

Crotyl alcohol (HF)[h] CD 3.0 A/dm²	Ni	CF_3CO_2Na (1.4), $C_2F_5CO_2Na$ (0.3), $C_3F_7CO_2Na$ (0.1), polymers	121
Amyl alcohol (HF) CD 0.012 A/cm²	Ni	C_4F_{10}, C_5F_{12}, C_2F_6, CF_4, CF_3H, CO_2	11, 12, 95
n-Hexanol (HF)	Ni	C_6F_{14}	95
Ethylene glycol (HF) CD 2.9 A/dm²	Ni	CF_3CO_2H (6.1), CF_4, CHF_3, CF_3OOCF_3, CF_3CF_3, CF_3OCF_3	117
1,3-Propanediol (HF) CD 2.9 A/dm²	Ni	$CF_2(CO_2H)_2$ (2.6), CF_3CO_2Na, $C_2F_5CO_2Na$, CF_4, CHF_3, CF_3CF_3, CHF_2CF_3, C_3F_8, $n\text{-}C_4F_{10}$, CF_3OCF_3, $C_2F_5OC_2F_5$	117
1,4-Butanediol (HF) CD 2.9 A/dm²	Ni	$(CF_2CO_2H)_2$ (1.8), CF_3CO_2Na, $C_2F_5CO_2Na$, $C_3F_7CO_2Na$, CHF_3, CF_3CF_3, C_3F_8, $n\text{-}C_4F_{10}$, CF_3OCF_3, $CF_2(CF_2)_2CF_2$—O (37.1)	117
Phenol (a) (HF) CD 0.04 A/cm²	Ni	C_6F_{12}, C_5F_{12}	13, 95
(b) (H₂O, HF)	Ni	Polymer (2.0 g/1.2 Ahr)	90
(c) (HF)	Ni	p,p'-Biphenol	90
2,6-Dimethylphenol (H₂O, HF)	Ni	3,3',5,5'-Tetramethyl-4,4'-dihydroxydiphenyl	90

43

Table 7.9 Fluorination: (v) Aldehydes and Ketones, etc.

Compound (Medium, Experimental Conditions)	Anode	Product(s) (% Yield)	Ref.
Carbon monoxide (HF, KF)[k] CD 100-167 A/cm²; temp 86°C	Porous C[a]	COF_2	7, 122
Paraldehyde (HF) CD 3.0 A/dm²	Ni	CF_3CO_2Na (1.0), $C_2F_5CO_2Na$ (1.8), $C_3F_7CO_2Na$ (1.4)	123
Acetaldehyde (HF)[h]	Ni	CF_3CO_2Na (22), $C_2F_5OC_2F_5$	123
Propionaldehyde (HF)[h] CD 3.3 A/dm²	Ni	$C_2F_5CO_2Na$ (7.1), CF_3CO_2Na (6.7)	123
Butyraldehyde (HF)[h] CD 3.3 A/dm²	Ni	$C_3F_7CO_2Na$ (1.9), $C_2F_5CO_2Na$ (1.3), CF_3CO_2Na (1.5); C_3F_8	95, 123
β-Hydroxybutyraldehyde (HF)[h] CD 3.0 A/dm²	Ni	$C_3F_7CO_2Na$ (1.4), $C_2F_5CO_2Na$ (1.8), CF_3CO_2Na (1.0)	123
Acetone (HF)[h] CD 3.3 A/dm²	Ni	CF_3CO_2Na (59.8); CF_4, CF_3H	11, 95, 125
Methyl ethyl ketone (HF)[h] CD 3.3 A/dm²	Ni	CF_3CO_2Na (41.9), $C_2F_5CO_2Na$ (7.2); $C_2F_4H_2$	95, 125
Diethyl ketone (HF)[h] CD 3.3 A/dm²	Ni	CF_3CO_2Na (5.9), $C_2F_5CO_2Na$ (54.8)	125, 126
Methyl vinyl ketone (HF)[h] CD 3.0 A/dm²	Ni	CF_3CO_2Na (12.7), $C_2F_5CO_2Na$ (1.9)	121
Methyl propyl ketone (HF)[h] CD 3.3 A/dm²	Ni	CF_3CO_2Na (35.9), $C_2F_5CO_2Na$ (2.3), $C_3F_7CO_2Na$ (5.6)	125
2,4-Butanedione (HF)[h] CD 3.5 A/dm²	Ni	CF_3CO_2Na (52.7)	125
Mesityl oxide (HF)[h] CD 3.3 A/dm²	Ni	CF_3CO_2Na (43.3), $C_2F_5CO_2Na$ (1.3), $C_3F_7CO_2Na$ (1.1)	121
Methyl pyruvate (HF)[h]	Ni	CF_3CO_2Na	125

Table 7.10 Fluorination: (vi) Carboxylic Acids, Acyl Halides, Anhydrides

Compound (Medium, Experimental Conditions)	Anode	Product(s) (% Yield)	Ref.
Phosgene: $COCl_2$ (HF)	Ni	$COClF$ (90.7), COF_2 (3.6)[l]	128
Acetic acid (a) (HF, KF) CD 0.02 A/cm^2	Ni	CF_3COF, CF_4, CF_3H, CF_2H_2, CO_2	11, 95, 129, 120
(b) (KHF$_2$), CD 0.0125 A/cm^2	Ni	CF_4, CHF_3, C_2H_6, C_2F_6, CO_2	127
Acetyl fluoride (HF, NaF)	Ni	CF_3COF (85)	33
Acetyl chloride (HF, NaF)[h]	Ni	CF_3CO_2Na (20.1); CF_4, CF_3H	11, 95, 131
Chloroacetyl fluoride (HF, NaF)[h]	Ni	CF_2ClCO_2Na (30.6), CF_3CO_2Na (6.1), CO_2, CO	131-133
Dichloroacetyl fluoride (HF, NaF)[h]	Ni	$CFCl_2CO_2Na$ (31.1), CF_2ClCO_2Na (14.1), CF_3CO_2Na (2.1), CHF_2Cl, $CHFCl_2$, $CFCl_3$, $CF_2ClCFCl_2$, CF_2Cl_2, CF_2ClCF_2Cl, CF_3CHCl_2, CO_2, CO	131, 133
Trichloroacetyl fluoride (HF, NaF)[h]	Ni	$CFCl_2CO_2Na$ (28.2), CF_2ClCO_2Na (14.0), CF_3CO_2Na (6.0), CO_2, CO	131, 133
Trifluoroacetic acid (KHF$_2$) CD 0.0125 A/cm^2	Ni	CF_4, C_2F_6, CO_2	127
Acetic anhydride (HF) CD 18 A/ft^2	Ni	CF_3COF (45)	33, 101, 129
Carboxylic acid anhydrides (HF)	Ni	Perfluorocarboxylic acid fluoride	129
Propionic acid (HF) CD 0.04 A/cm^2	Ni	CF_4, C_2F_6, C_3F_8, $C_2F_4H_2$, C_2F_5H; α-fluoropropionic acid, β-fluoropropionic acid (10), OF_2	13, 30, 95
Propionyl chloride (HF)[h]	Ni	$CF_3CF_2CO_2Na$ (<50)	134
3H-Tetrafluoropropionic acid (HF)[h]	Ni	CF_3CO_2Na, $C_2F_5CO_2Na$ (ratio 60 : 40)	120
α-Fluoropropionyl fluoride: $CH_3CFHCOF$ (HF, NaF)	Ni	C_2F_5COF (59)	135
Acrylic acid (HF)[h] CD 3.3 A/dm^2	Ni	CF_3CO_2Na (3.9), $C_2F_5CO_2Na$ (20.7)	114

45

Table 7.10 Fluorination: (vi) Carboxylic Acids, Acyl Halides, Anhydrides (cont.)

Compound (Medium, Experimental Conditions)	Anode	Product(s) (% Yield)	Ref.
n-Butyric acid (HF, NaF)[h]	Ni	$CF_3(CF_2)_2CO_2H$, $CF_3CF_2CO_2H$, CF_3CO_2H; α, β- and γ-fluorobutyric acids (proportion 1 : 2 : 12)	13, 30, 33, 95
n-Butyryl fluoride (HF, NaF)[h,m] CD 15 A/ft^2	Ni	$CF_3(CF_2)_2CO_2H$ (36), $CF_3CF_2CO_2H$ (4); C_3F_7COF (34), perfluorotetrahydrofuran (24)	33, 135
n-Butyryl chloride (HF, NaF)[n] CD 1.2 A/dm^2	Ni	Perfluorotetrahydrofuran (19); $CF_3(CF_2)_2CO_2Na$	134, 137, 138
γ-Bromobutyryl chloride: $Br(CH_2)_3COCl$ (HF, NaF)	Ni	C_3F_7COF (17)	135
γ,γ,γ-Trifluorobutyryl fluoride: $CF_3CH_2CH_2COF$ (HF, NaF)	Ni	C_3F_7COF (60), perfluorotetrahydrofuran (<1)	135
Isobutyric acid (HF) CD 0.03 A/cm^2	Ni	α-, and β-Fluoroisobutyric acids	30
Crotonic acid (HF)[h] CD 3.3 A/dm^2	Ni	CF_3CO_2Na (4.5), $C_2F_5CO_2Na$ (3.3), $C_3F_7CO_2Na$ (10.0)	114
Crotonic acid anhydride (HF)[h] CD 3.3 A/dm^2	Ni	CF_3CO_2Na (6.8), $C_2F_5CO_2Na$ (6.2), $C_3F_7CO_2Na$ (40.0)	114
2-Chloro-1,1,2-trifluorethoxyacetyl chloride: $ClCFHCF_2OCH_2COCl$ (HF, NaF)	Ni	$ClCF_2CF_2OCF_2COF$ (40)	135
3-(2',2',2'-Trifluoroethoxy)propionyl chloride: $CF_3CH_2OCH_2CH_2COCl$ (HF, NaF)	Ni	$C_2F_5OCF_2CF_2COF$ (6)	135
3-(1',1',2',2'-Tetrafluoroethoxy)-propionyl chloride: $CHF_2CF_2OCH_2CH_2COCl$ (HF, NaF)	Ni	$C_2F_5OCF_2CF_2COF$ (30)	135
3-(2'-Chloro-1',1',2'-Trifluoroethoxy) propionyl chloride: $ClCFHCF_2OCH_2CH_2COCl$ (HF, NaF)	Ni	$ClCF_2CF_2OCF_2CF_2COF$ (60)	135

46

Trimethylacetic acid (HF) CD 0.02 A/cm²	Ni	CF_4, C_3F_8	11, 95
α-Methylbutyric acid (HF)	Ni	2,3-Dimethylhexane, fluorinated α-methylbutyric acids (35.5 g/102 gsm)	30
Valeric acid (HF) CD 0.012 A/cm²	Ni	C_4F_{10}, C_5F_{12}, CF_4, CF_3H, C_2F_6	12
Caproic acid (HF)[h]	Ni	$CF_3(CF_2)_4CO_2H$; $C_5F_{11}COF$, C_6F_{14}, C_5F_{12}; $C_6F_{12}O$ (13.4)	33, 95, 139-141
Caproyl chloride (a) (HF)[h,m]	Ni	$CF_3(CF_2)_4CO_2H$	33
(b) (HF, NaF)[n] CD 0.7-1.0 A/dm²	Ni	[cyclic perfluoroether structure: CF_2–CF_2–$CFCF_2CF_3$ ring with O; CF_2–CF_2–CF_2–$CFCF_3$ ring with O]	137, 138
2-Ethylcaproyl chloride (HF, NaF)[n]	Ni	CF_3CF_2CF–CF_2–$CFCF_2CF_3$, [cyclic perfluoroether structures with $CFCF_2CF_3$ and $CFCF_3$ ring substituents] (total 32.3 in ratio 1.6 : 1)	137, 138
Octanoic acid (HF)[h]	Ni	$CF_3(CF_2)_6CO_2H$	33
Octanoyl chloride (HF)[h,m]	Ni	$CF_3(CF_2)_6CO_2H$	33

47

Table 7.10 Fluorination: (vi) Carboxylic Acids, Acyl Halides, Anhydrides (cont.)

Compound (Medium, Experimental Conditions)	Anode	Product(s) (% Yield)	Ref.
γ,γ-Difluorooctanoyl chloride: n-$C_4H_9CF_2CH_2CH_2COCl$ (HF, NaF)	Ni	n-$C_7F_{15}COF$ (31)	135
Cyclohexylacetic acid (HF)	Ni	$C_6F_{11}CF_2COF$	95
Phenylacetic acid (HF) CD 0.04 A/cm^2	Ni	$C_6F_{11}CF_2COF$	13, 95
Benzoic acid (HF)	Ni	$C_6F_{11}COF$, C_7F_{16}, C_6F_{14}	95
Benzyl chloride (HF), A/cm^2 CD 0.008-0.01	Ni	$C_6F_{11}COF$ (65-70 wt %)	136
Succinic acid (HF) CD 1.5 A/dm^2	Ni	Perfluorosuccinic acid, CF_3CO_2H, low perfluoro- and fluoroalkanes	142
Succinic anhydride (HF) CD 0.015-0.03 A/dm^2	Ni	Perfluorosuccinic acid (69), CF_3CO_2H, low perfluoro- and fluoroalkanes	101, 142, 143
Maleic acid (HF) CD 1.5 A/dm^2	Ni	Perfluorosuccinic acid, CF_3CO_2H, low perfluoro- and fluoroalkanes	142
Maleic anhydride (HF) CD 0.02 A/dm$_2^h$	Ni	Perfluorosuccinic acid (20), CF_3CO_2H, low perfluoro- and fluoroalkanes	101, 142
Adipic acid (HF)^h	Ni	$(CF_2)_4(CO_2H)_2$; C_4F_{10}	13, 33, 95, 143
Adipyl chloride (HF)^{h,m}	Ni	$(CF_2)_4(CO_2H)_2$	33
Sebacic acid (HF)	Ni	$(CF_2)_8(COF)_2$, $CF_3(CF_2)_7COF$, fluorocarbons; C_5F_{12}, C_8F_{18}, C_9F_{20}	13, 95, 144
$CH_3(CH_2)_nCO_2H$ (HF)	Ni	Cyclic fluorocarbon ethers	141

γ,γ,δ,δ-Tetrafluorosebacyl chloride: (HF, NaF)[i]	Ni	CF_2 ring structure with $(CF_2)_5CO_2H$, F, O	145
		CF_2 ring structure with $(CF_2)_4CO_2H$, F, O	
Phthalic anhydride (HF) CD 0.008-0.01	Ni	$C_6F_{11}COF$ (65-70 wt %)	136
3,3'-Oxydipropionyl fluoride: $O(CH_2CH_2COF)_2$ (HF, Ac_2O)[h] CD 0.02 A/cm^2	Ni	$O(CF_2CF_2CO_2H)_2$ (3.7), $CF_2(CO_2H)_2$ (35.2), perfluoro-(3-methoxypropionic acid), pentafluoropropionic acid (15.4), CF_3CO_2H (27)	146, 147
N,N-Dimethylcarbamoyl chloride: $(CH_3)_2NCOCl$ (HF)	Ni	$(CF_3)_2NCOCl$, $(CF_3)_2NCOF$ (37)	148, 149
N,N-Diethylcarbamoyl chloride: $(C_2H_5)_2NCOCl$ (HF)	Ni	$(CF_3)_2NCOF$ (40 g), $\overline{CF_2OCF_2CF_2NC_2F_5}$ (241g/595 gsm)	148
N,N-Dibutylcarbamoyl chloride: $(C_4H_9)_2NCOCl$ (HF)	Ni	$(CF_3)_2NCOF$, $\overline{CF_2OCF(C_2F_5)CF_2NC_4F_9}$ (211g/1176 gsm)	148
Morpholine N-carbamoyl chloride: (HF)	Ni	$(CF_3)_2NCOF$ (34. g), $O(CF_2CF_2)NCOF$ (30 g/604 gsm)	148

Morpholine structure: ring with O and NCOCl

Table 7.11 Fluorination: (vii) Esters

Compound (Medium, Experimental Conditions)	Anode	Product(s) (% Yield)	Ref.
Etlyl acetate (HF)[h] CD 3.5 A/dm$_2$	Ni	CF_3CO_2Na (52.3)	28
Vinyl acetate (HF)[h] CD 3.3 A/dm$_2$	Ni	CF_3CO_2Na (71.5)	114
Allyl acetate (HF)[h] CD 3.0 A/dm^2	Ni	CF_3CO_2Na (20.6), $C_2F_5CO_2Na$ (3.7), $C_3F_7CO_2Na$ (0.5)	121
n-Propyl acetate (HF)[h] CD 3.5 A/dm$_2$	Ni	CF_3CO_2Na, $C_2F_5CO_2Na$ (total 42.5)	28
n-Butyl acetate (HF)[h] CD 3.5 A/dm$_2$	Ni	CF_3CO_2Na, $C_2F_5CO_2Na$, $C_3F_7CO_2Na$ (total 35.8); C_4F_{10}	11, 28
Ethyl propionate (HF)[h] CD 3.5 A/dm^2	Ni	CF_3CO_2Na, $C_2F_5CO_2Na$ (total 49)	28
n-Propyl propionate (HF)[h] CD 3.5 A/dm^2	Ni	CF_3CO_2Na, $C_2F_5CO_2Na$ (total 34.8)	28, 24
n-Butyl propionate (HF)[h] CD 3.5 A/dm^2	Ni	CF_3CO_2Na, $C_2F_5CO_2Na$, $C_3F_7CO_2Na$ (total 38.6)	28
Ethyl acrylate (HF)[h] CD 3.0 A/dm^2	Ni	CF_3CO_2Na (54.8), $C_2F_5CO_2Na$	121
Butyl acrylate (HF)[h] CD 3.0 A/dm^2	Ni	CF_3CO_2Na (5.7), $C_2F_5CO_2Na$ (52.2), $C_3F_7CO_2Na$ (29.3)	121
Methyl 3-methoxypropionate (HF)[o] CD 0.02 A/cm^2	Ni	CF_3CO_2H (2.4), $C_2F_5CO_2H$ (21.1), $CF_2(CO_2H)_2$ (0.2), $CF_3OCF_2CF_2CO_2H$ (5.7), $O(CF_2CF_2CO_2H)_2$ (0.1)	152-154
Methyl n-butyrate (HF, KF) CD 0.001 A/cm^2	Ni	Methyl-α-, β-, and γ-fluorobutyrates (total 13 in proportion 25 : 40 : 35), fluorobutyric acid, γ-butyrolactone	30
Ethyl butyrate (HF)[h] CD 3.3 A/dm$_2$	Ni	CF_3CO_2Na, $C_2F_5CO_2Na$, $C_3F_7CO_2Na$ (total 32.5)	28

n-Propyl butyrate (HF)^h CD 3.0 A/dm^2	Ni	CF_3CO_2Na, $C_2F_5CO_2Na$, $C_3F_7CO_2Na$ (total 28.2)	28
n-Butyl butyrate (HF)^h CD 3.5 A/dm^2	Ni	CF_3CO_2Na, $C_2F_5CO_2Na$, $C_3F_7CO_2Na$ (total 32.5)	28
Vinyl butyrate (HF)^h CD 3.3 A/dm^2	Ni	CF_3CO_2Na (47.4), $C_2F_5CO_2Na$ (11.2), $C_3F_7CO_2Na$ (5.9)	114
Ethyl crotonate (HF)^h CD 3.3 A/dm^2	Ni	CF_3CO_2Na (52.0), $C_2F_5CO_2Na$ (3.5), $C_3F_7CO_2Na$ (28.1)	121
Dimethyl malonate (HF)^h CD 2.9 A/dm^2	Ni	$CF_2(CO_2H)_2$ (12.0), CF_3CO_2Na, $C_2F_5CO_2Na$, CF_4, CHF_3, CF_3CF_3, CHF_2CF_3, CF_3OCF_3, CH_2F_2, CF_3OOCF_3 (4.1)	117
Diethyl malonate (HF)^h CD 2.9 A/dm^2	Ni	$CF_2(CO_2H)_2$ (11.8), CF_3CO_2Na, $C_2F_5CO_2Na$	117
Dimethyl succinate (HF)^h CD 2.9 A/dm^2	Ni	$(CF_2CO_2H)_2$ (15.1), CF_3CO_2Na, $C_2F_5CO_2Na$, CF_4, CHF_3, CF_3CF_3, CHF_2CF_3, CF_3OCF_3, CH_2F_2 (1.2)	117
Diethyl succinate (HF)^h CD 2.9 A/dm$_2$	Ni	$(CF_2CO_2H)_2$ (17.0), CF_3CO_2Na, $C_2F_5CO_2Na$	117
Dimethyl maleate (HF)^h CD 2.9 A/dm^2	Ni	$(CF_2CO_2H)_2$ (8.4), CF_3CO_2Na, $C_2F_5CO_2Na$, CF_4, CHF_3, CF_3CF_3, C_3F_8, $n\text{-}C_4F_{10}$, CF_3OCF_3, $CF_2(CF_2)_2CF_2{-}O$ (0.7)	117
Diethyl maleate (HF)^h CD 2.9 A/dm^2	Ni	$(CF_2CO_2H)_2$ (9.5), CF_3CO_2Na, $C_2F_5CO_2Na$ $(CF_2CO_2H)_2$ (9.5), CF_3CO_2Na, $C_2F_5CO_2Na$	117
Methyl benzoate (HF) CD 0.015 A/cm^2	Ni	Product not indicated (48)	45
Dimethyl phthalate (HF) CD 0.02 A/cm^2	Ni	Product not indicated (68)	45
Dimethyl terephthalate (HF) CD 0.01 A/cm^2	Ni	Product not indicated (97)	45

Table 7.12 Fluorination: (viii) Amides

Compound (Medium, Experimental Conditions)	Anode	Product(s) (% Yield)	Ref.
N, N-Dimethylformamide (HF)	Ni	$(CF_3)_2NCOF$ (170 g/1139 gsm)	148
N,N-Dimethyltrifluoroacetamide (HF)	Ni	CF_3COF (65 g), $(CF_3)_2NCOF$ (137 g), $CF_3CON(CF_3)_2$ (48 g/569 gsm)	148
Tetramethylurea (HF)	Ni	$(CF_3)_2NCOF$ (46 g), $(CF_3)_2NCON(CF_3)_2$ (10 g/196 gsm)	148
Acrylamide (HF)[h] CD 3.3 A/dm^2	Ni	CF_3CO_2Na (2.2), $C_2F_5CO_2Na$ (16.5)	114
Urea (a) (HF) CD 7 mA/cm^2	Ni	NF_3 (38), CF_4 (10), COF_2	108
(b) (HF, KF) CD, 10^{-3}-10^{-2} A/cm^2; potential 5.0-5.5 V vs Pt; temp 130°C	C	$NF_3(>55)^{w'}$; F_2O, F_2; N_2, CO_2	108a
Guanidine (HF)	Ni	NF_3 (47), CF_4 (66), N_2	108, 155
Aminoguanidine (HF)	Ni	NF_3, CF_4	108
Semicarbazide (HF)	Ni	NF_3, CF_4	108

Table 7.13 Fluorination: (ix) Sulfur and Selenium Compounds

Compound (Medium, Experimental Conditions)	Anode	Product(s) (% Yield)	Ref.
n-Butyl mercaptan (HF)	Ni	C_4F_{10}	95
n-Octyl mercaptan (HF)	Ni	$CF_3(CF_2)_7SF$,[p] C_8F_{18}	156
Dimercaptoethane (HF)	Ni	Tar on electrodes	156
Dimethyl sulfide (HF) CD 0.0025 A/cm²	Ni	CF_3SF_5 (30.5 g), $(CF_3)_2SF_4$ (3.5 g/16 gsm), CF_4. SF_6	157
Dimethyl disulfide (HF)	Ni	CF_3SF_5 (149 g/282 gsm), $CH_3SF_4CH_2F$?, CH_3SF_5	158
Diethyl sulfide (HF)	Ni	$C_2F_5SF_5$ (46 g), $(C_2F_5)_2SF_4$ (12.5 g/198 gsm), SF_6	159
Di-n-propyl sulfide (HF)	Ni	$C_3F_7SF_5$ (312 g), $(C_3F_7)_2SF_4$ (316 g/826 gsm), CF_4, SF_6, $C_3H_2F_6$	159
Di-n-butyl sulfide (HF)	Ni	$C_4F_9SF_5$ (18), $(C_4F_9)_2SF_4$ (11),C_4F_{10}, C_4F_9H, $\overline{CF_2(CF_2)_3}SF_4$, SF_6	159
Di-n-butyl disulfide (HF)	Ni	$C_4F_9SF_5$ (22),[q] $(nC_4F_9)_2SF_4$ (3.3), SF_6, C_4F_{10}, $\overline{CF_2(CF_2)_3}SF_4$	159
Di-t-butyl disulfide (HF)	Ni	Liquid products (178 g), gaseous products 159 (825 g/849 gsm), SF_6	159
Di-n-hexyl sulfide (HF)	Ni	$C_6F_{13}SF_5$ (17.0), $(C_6F_{13})_2SF_4$ (2.9), C_6F_{14} (isomers) (26.2)	160
Di-(β-diethylaminoethyl)disulfide (HF)	Ni	$(C_2F_5)_2NCF_2CF_2SF_5$ (27), SF_6	159
Diphenyl sulfide (HF)	Ni	$C_6F_{11}SF_5$ (3.4), $(C_6F_{11})_2SF_4$ (0.8), C_6F_{12} (29.1), C_6F_{11}-C_6F_{11} (9.8) dodecafluoro-1, 4-thioxan	159
1,2-Di(methylthio)ethane: $\begin{array}{ccc} CH_2 & \!\!\!\!-\!\!\!\! & CH_2 \\ \mid & & \mid \\ SCH_3 & & SCH_3 \end{array}$ (HF)	Ni	CF_3SF_5 (170 g), $C_2F_5SF_4CF_3$ (90g), $C_2F_5SF_4C_2F_5$ (28 g), $CF_3SF_4C_2F_5$ (19 g), perfluoro-s-dithiane octafluoride (6 g/370 gsm)	158

53

Table 7.13 Fluorination: (ix) Sulfur and Selenium Compounds (cont.)

Compound (Medium, Experimental Conditions)	Anode	Product(s) (% Yield)	Ref.
Sulfolene: [structure] CD 0.005 A/dm² (HF)	Ni	Perfluoro-n-butylsulfonyl fluoride (45)	151
3-Methylsulfolene (HF) CD 0.005 A/cm²	Ni	$(CF_3)_2CFCF_2CF_2SO_2F$, $CF_3CF_2CF_2CF(CF_3)CF_2SO_2F$, $CF_3(CF_2)_4SO_2F$ (total 65 in proportion 41 : 24 : 35), C_5F_{12}	151
Tetrahydrothiophene (HF)	Ni	$(C_2F_5)_2SF_4$, $C_4F_9SF_5$, $C_3F_7SF_4CF_3$, $\overline{CF_2(CF_3)_2}SF_4$, SF_6	159
1,4-Thioxane: [structure] (HF)	Ni	[structure] F_4S CF_2CF_2 O CF_2CF_2 (19 g/78 gsm),	158, 159 161
s–Trithiane: [structure] CH_2-S S CH_2-S CH_2 (HF) CD 0.0008 A/cm²	Ni	SF_6; $(C_2F_5)_2O$, $C_2F_5SF_5$, $C_2F_5OCF_2CF_2SF_5$ CF_3SF_5 (90 g), $SF_5CF_2SF_5$ (55 g), perfluoro-s-trithiane (28.5 g), CS_2 (25 g/470 gsm)	158
Thioglcollic acid: $HSCH_2CO_2H$ (HF) CD 0.0008 A/cm²	Ni	$SF_3CF_2CO_2H$ (3), CF_3CO_2H, CF_3COF, CF_3SF_5, C_2F_6, CF_4, CF_3H, COF_2, SF_6, SO_2F_2, SOF_2, CO_2	80
S-Methythioglycollic acid chloride: CH_3SCH_2COCl (HF)	Ni	CF_3SF_5 (64 g), $(CF_3)_2SF_4$ (63 g), $CF_3SF_4CF_2CF_3$ (24 g), $CF_3SF_4CF_2COF$ (39 g), $CF_3SF_4SF_4CF_3$ (10 g/615 gsm)	162
Tetramethylene sulfone (HF)	Ni	$CF_3(CF_2)_3SO_2^{\cdot}F$ (40)	163

Substrate	Electrode	Products	Ref.
Hexamethylene sulfone (HF)	Ni	$CF_3(CF_2)_5SO_2F$ (high)	163
Octamethylene sulfone (HF)	Graphite	$CF_3(CF_2)_7SO_2F$ (good)	163
3-Methyltetrahydrothiophene-1,1-dioxide (HF)	Ni	$CF_3CF_2C(CF_3)FCF_2SO_2F$ (28.6), $CF_3C(CF_3)FCF_2CF_2SO_2F$ (28;0)	163
2-Butyltetrahydrothiophene-1,1-dioxide (HF)	Ni	$CF_3(CF_2)_7SO_2F$ (good), $CF_3CF_2CF_2C(C_4F_9)FSO_2F$ (good)	163
Methylsulfonyl fluoride (HF) CD 0.0014 A/cm²	Ni	CF_3SO_2F (96), CF_3H (1), CF_4	164, 165
Perfluoroethanesulfonic acid (HF)	Ni	$C_2F_5SO_2F$ (56), C_2F_6, CF_4, SO_2F_2	166
$Cl(CH_2CF_2)_2SO_2F$ (HF)	Ni	Fluorinated sulfonyl fluorides	135
$C_6H_{13}CFClCF_2SO_2F$ (HF)	Ni	$C_6F_{13}SO_2F$ (48.5), $C_6F_{13}CFClCF_2SO_2F$ (14.3), C_8F_{18} (10)	135

RSO_2F (HF), R is

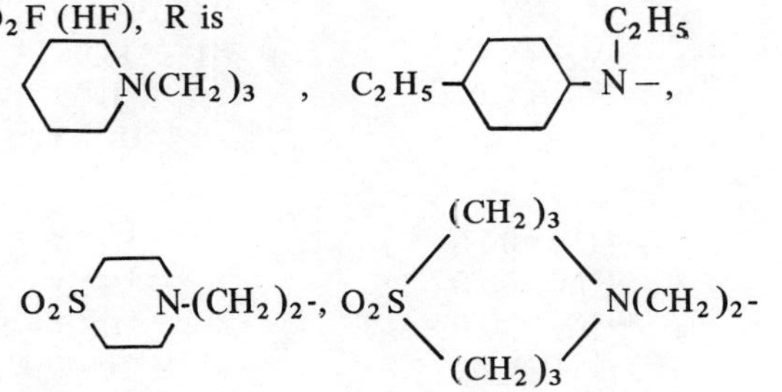

RSO_2Br (HF), R is

	Ni	R_FSO_2F	167

$C_8H_{15}N(CH_2)_3$-, or -$(CH_2)_2NCH_2CH_2$-

Sulfonyl chlorides: RSO_2Cl | Ni | R_FSO_2F (87), HCl | 164, 168

$R = CH_3$ CD = 0.0014 A/cm²

55

Table 7.13 Fluorination: (ix) Sulfur and Selenium Compounds (cont.)

Compound (Medium, Experimental Conditions)		Anode	Product(s) (% Yield)	Ref.
C_2H_5	0.004 A/cm²		(79)	166, 168
C_3H_7	0.004 A/cm²		(68)	166
C_4H_9	0.004 A/cm²		(58)	166
$n\text{-}C_5H_{11}$	0.004 A/cm²		(45)	166
$iso\text{-}C_5H_{11}$	20 A/ft²		(380 g/120 gsm)	168
$R = C_6H_{13}$	CD = 0.004 A/cm²		(36)	166, 168
C_7H_{15}	0.004 A/cm		(31)	166
C_8H_{17}	0.004 ∧/cm²		(25), C_8F_{18} (21), $CF_3(CF_2)_{0\text{-}6}SO_2F(9)$, $CF_3(CF_2)_{2\text{-}5}$ CO_2H (1), C_4F_{10} (1), C_3F_8 (1), C_2F_6 (1.5), SO_2F_2 (23), SF_6 (6)	165, 166, 168
$C_{10}H_{21}$	20 A/ft²		(—)	165
C_6H_5	20 A/ft²		(—)ˢ	168
$p\text{-}CH_3C_6H_4$	20 A/ft²		(—)ˢ	168
$o\text{-}CH_3C_6H_4$	20 A/ft²		(—)ˢ	168
$p\text{-}C_2H_5C_6H_4$	20 A/ft²		(—)ˢ	168
$p\text{-}isopropylC_6H_4$	20 A/ft²		(—)ˢ	168
$p\text{-}secbutylC_6H_4$	20 A/ft²		(—)ˢ	168
$ClO_2S(CH_2)_3$—N⟨⟩N—$(CH_2)_3$—		Ni	(3.6 g/50 Ahr)	167
N—$(CH_2)_3$— (piperidine)	20A/ft²		(5.1 g/50 Ahr)	167
N—$(CH_2)_3$— (morpholine)			(1390 g/1766 gsm)	167

CH₃ group structure:

$$CH_3-\underset{\underset{CH_3}{|}}{N}-(CH_2)_3-$$

with morpholine ring containing O and two CH₃ substituents.

Compound (conditions)	Electrode	Products	Ref.
(130 g/258 gsm)			167
Methyl chlorosulfonate (HF) CD 0.006–0.008 A/cm²	Ni	CF_3OCF_3 (28.9), CF_3OCF_2H (12.3), CF_3OSO_2F (22.4), SO_2F_2, CF_2HOCHF_2 (14.6), CF_2HOSO_2F (14.4)	169
Dimethyl sulfate (HF)[t] CD 0.004–0.006 A/cm²	Ni	CF_3OCF_3 (16.2), SO_2F_2, CF_3OCF_2H (23.0), FSO_2OCF_3 (16.7), CHF_2OCHF_2 (5.1), $(CF_3)_2SO_4$ (13.4), FSO_2OCF_2H (2.7)	170
(b) (HF)[u] CD 0.004–0.006 A/cm²	Ni	SO_2Cl_2 (9.9), CF_3OSO_2F (1.2), CF_3OCF_3 (6.9), CF_2HOSO_2 (5.4), $CH_2FOSO_2OCH_3$ (57.1), mixture of fluorinated sulfates (2.4)	170
Carbon disulfide (a) (HF) CD 0.0025 A/cm²	Ni	CF_3SF_5 (>90), CF_2 $(SF_5)_2$ (0.5), $CF_2(SF_3)_2$ (0.5), SF_6	157, 171, 172
(b) (HF, Cl₂) CD 2.7 A/dm²	Ni	CF_4 (29.3), $CClF_3$, CCl_2F_2, CCl_3F, SF_6 (46.5), CF_3SF_5 (19.9)	150
Thiourea (HF) CD 0.004 A/cm²	Ni	NF_3, CF_4, CF_3NF_3, SF_6, CF_3SF_5, CO_2, COF_2, SF_4, CSF_2	173
Sulfur dioxide (HF) CD 2.8 A/dm²	Ni	SO_2F_2 (62), SOF_4 (23), SF_6(1), OF_2 (14)	174
Thionyl chloride (HF) CD 2.8 A/dm²	Ni	SOF_4 (31.7), SOF_2, SF_6, OF_2, HCl	174
Sulfuryl chloride (HF) CD 2.8 A/dm²	Ni	SO_2F_2 (65.7), Cl_2, SO_2ClF, SF , OF_2	174
Sulfur monochloride (HF) CD 2.8 A/dm²	Ni	SF_6 (78.2), Cl_2, SOF_2, SOF_4	174
Sulfur dichloride (HF) CD 2.8 A/dm²	Ni	SF_6 (63.7), Cl_2, SF_5Cl, SOF_2, SOF_4	174
Dimethyl selenide (HF) CD 0.0025 A/cm²	Ni	No selenium compounds isolated, CH_3Cl	157
Carbon diselenide (HF) CD 0.0025 A/cm²	Ni	No selenium compounds isolated, Se[v]	157

Table 7.14 Fluorination: (x) Molten Salt Fluorinations

Compound (Medium, Experimental Conditions)	CD (A/in²)	Temp/°C	Anode	Products(s) (% Yield)					Ref.
				CF_4 :	C_2F_6 :	C_3F_8 :	C_3F_6 :	C_4F_8	
NaF:LiF (Wt % 52 : 48)									175, 176
(a)	1.25	700	Porous C	44.6	35	15.1	–	–	
(b)	13.5	700	Porous C	36.2	43.5	14.5	2.2	3.6	
(c)	2.9	800	Porous C	34.1	47.9	11.8	1.0	–	
(d)	12.1	800	Porous C	36.4	51	11.8	1.0	–	
(e)	2.9	900	Porous C	34.8	65	–	–	–	
(f)	14.3	900	Porous C	38.1	47	11.9	1.2	2.1	
(g)	4.9	1000	Porous C	55	45	–	–	–	
(h)	1	700	Coke grains	60	31	–	–	–	
(i)	3	800	Coke grains	50	50	–	–	–	
(j)	4.5	900	Coke grains	82.2	17	–	–	–	
NaF:KH; temp 1000°C			C	CF_4					9
LiF:AlF₃(wt % 38:62) temp 900°C CD 44 A/dm²			Porous Graphite	CF_4 (2.2 g), C_2F_6 (8.4 g), C_nF_{2n+2} (1.6 g/13.7 g of anode consumed)					175
CaF₂; temp 1500°C			Graphite	CF_4 (42.2)					177
CaF₂:CaCl₂(wt % 60:40) temp 985°C CD 8 A/in²			Porous C	CF_4 (18), CF_2Cl_2 (5), CF_3Cl (36), Cl_2 (40)					178
LiF:MgF₂:MgCl₂ (wt % 45:45:10), temp 850°C CD 3 A/in²			Porous C	CF_4 (16.2), CF_2Cl_2 (4.4), $CFCl_3$ (0.4), CF_3Cl (44.5)					178

Electrolyte (feed)	Electrode	Products (%)	Ref.
(LiF:MgF$_2$:CaF$_2$ (wt % 26:41.5:32.5), NaBr) temp 800°C, CD 6.25 A/in^2	Porous C	CF$_4$ (22), C$_2$F$_6$ (4), CF$_3$Br (56)	178
(NaF:CaF$_2$:MgF$_2$) (wt % 50:30:20) (a) CD 2.5 A/dm^2 at 757°C	C	CF$_4$, C$_2$F$_6$, C$_3$F$_8$, CO$_2$ (total CE 61 in proportion 20:60-65:10:5-10)	179
(b) CD 12.3-20.5 A/dm^2 at 800°C	C	CF$_4$, C$_2$F$_6$, C$_3$F$_8$ (total CE 92 in proportion 30-40:50-60:5-10)	179
(LiF:CaF$_2$:MgF$_2$ (wt % 35:30:35), NaCN) temp 850°C	Porous C	CF$_3$CN, (CN)$_2$, CF$_4$, C$_2$F$_6$ in proportion 24:32.5:11.0:3.7	180
(NaF:LiF (wt % 52:48), NaSCN) temp 725°C	Porous C	CF$_3$CN, SF$_4$	180
Hg(CN)$_2$: KF:HF (mole ratio 1:4:8 at 80°C, CD 0.02 A/dm^2	Ni	CF$_4$, C$_2$F$_6$, NF$_2$CF$_3$, (CN)$_2$, (CF$_3$)$_2$NF, (CF$_3$)$_2$NH, K$_2$NIF$_6$	39
Beryllium fluoride + alkali metal fluoride melt	C	CF$_4$	8
(Sodium fluoaluminate, cyanogen, He) temp 400-1200°C	C	C$_2$F$_6$ (66), CF$_4$ (44.1), CF$_3$CN (4.9)	181
(Sodium fluoaluminate, Cl$_2$, He) temp 400-1200°C	C	CF$_4$ (34), CF$_3$Cl (23), CF$_2$Cl$_2$ (10), C$_2$F$_6$ (6)	183
(Sodium fluoaluminate, Cl$_2$, Br$_2$, He)[b] temp 1000-1300°C	C	CF$_4$ (11.5), CF$_3$Cl (18), CF$_3$Br (29), C$_2$F$_6$ (5)	183
(Sodium fluoaluminate, Br$_2$, He) temp 400-1200°C	C	CF$_3$Br (31.4); CF$_4$, C$_2$F$_6$	182
(Sodium fluoaluminate, CCl$_4$, H$_2$) temp 400-1200°C	C	C$_2$F$_4$Cl$_2$ (1.6), C$_2$F$_3$Cl$_3$ (1.2), C$_2$F$_6$ (4.9), CF$_2$Cl$_2$ (41.8), CFCl$_3$ (4.9), CF$_3$Cl (36.9), C$_2$F$_2$Cl$_2$ (1.6)	184
(a) flow rate 263 ml/min (b) flow rate 235 ml/min		C$_2$F$_6$ (2.8), CF$_2$Cl$_2$ (66), CFCl$_3$ (2.8), CClF$_3$ (51)	184
(Sodium fluoaluminate, trichloroethylene, He) temp 400-1200°C	C	CF$_3$Cl (60), CF$_2$Cl$_2$ (6.4), C$_2$F$_2$Cl$_2$ (0.7), C$_2$F$_6$ (2.2)	184
(Sodium fluoaluminate, tetrachloroethylene, He) temp 400-1200°C	C	CF$_3$Cl (59)	184

Table 7.14 Fluorination: (x) Molten Salt Fluorinations (cont.)

Compound (Medium, Experimental Conditions)	Anode	Product(s) (% Yield)	Ref.
(Sodium fluoaluminate, acetylene, He) temp 400-1200°C	C	CF_4 (30), C_2F_6 (12)	184
$NaF:CaF_2$ (wt % 33:67) temp 1275-1340°C	C	CF $(44)^w$	185
$NaF:BaF_2:MgF_2$ (wt % 10:45:45), Br_2, He) temp 400-1200°C	C	Bromofluorocarbons	182
$NaF:BaF_2:MgF_2$ (wt % 10:45:45), Cl_2, He) temp 400-1200°C	C	Chlorofluorocarbons	183
(Potassium fluotantalate; perchloropropene: C_3Cl_6, H_2) temp 400-1200°C; CD 2 A/cm²	C	CF_2Cl_2, CF_3Cl, perfluorochloropropane	184

Table 7.15 Fluorination: (xi) Miscellaneous

Compound (Medium, Experimental Conditions)	Anode	Product(s) (% Yield)	Ref.
Acetonitrile (HF) CD 0.02 A/cm^2	Ni	CF_3CN, C_2F_6, C_2F_5H, CF_4, CF_3H	10, 95
Trifluoroacetonitrile (HF)	Ni	CF_4, C_2F_6, NF_3, $C_2F_5NF_2$ (1)	187a
Mercuric cyanide (HF) CD 0.02 A/cm^2	Ni	CF_4 (12 g/60 gsm)	10, 95
Butyl isocyanate (HF)	Ni	C_4F_{10}	95
Phenyltrichlorosilane (HF)	Ni	Phenyltrifluorosilane (83.4)	186, 187
Diphenyldichlorosilane (HF)	Ni	Diphenyldifluorosilane	186
Tetramethylsilane (HF) CD 0.005 A/cm^2	Ni	SiF_4, CHF_3 CF_4, CH_2F_2	187
Diethyldichlorosilane (HF) CD 0.005 A/cm^2	Ni	SiF_4, C_2F_6, C_2F_5H	187
Amyltrichlorosilane (HF) CD 0.005 A/cm^2	Ni	SiF_4, C_5F_{12}, $C_5F_{11}H$	187
$(Me_3Si)_2O$ (HF) CD 0.005 A/cm^2	Ni	SiF_4, CHF_3, CF_4, CH_2F_2	187

61

Table 7.16 Chlorination

Compound (Medium, Experimental Conditions)	Anode	Product(s) (% Yield)	Ref.
Benzene			
(a) (H_2O, chloride as HCl or NaCl) CD 116-120 A/cm^2	Graphite,ˣ Pt	Chlorobenzene (19 g/65 gsm)	188-192, 192a
(b) (H_2O, HCl, HOAc) CD 0.026-2.66 A/cm^2	Pt, graphite, Fe_3O_4	Chlorobenzene, p-dichlorobenzene, sym-tetrachlorobenzene, hexachlorobenzene, pentachlorophenol, chloranil	53
(c) (CH_3OH, HCl) CD 12.2 A/dm_2	Pt, graphite	Chlorobenzene (85) hexachlorobenzene, HCHO, CO_2, CO	54, 193
(d) (HOAc, LiCl) CD 0.1-0.8 A/dm^2	Pt, graphite	Chlorobenzene (50-70), hexachlorobenzene	52, 194
(e) (HClₜ ClCH₂CO₂H, cyanuric acidᵘ) CD 4.3 A/dm^2; temp 38°C	Porous C	Chlorobenzene (89)ʷ	194a
(f) (Et_2O, $ZnCl_2$) CD 0.10 A/dm^2	Pt	Chlorobenzene (65-88)	52, 194
Toluene			
(a) (H_2O, HCl)	Pt	o-Chlorotoluene (70), p-chlorotoluene (30)	195
(b) (aqu conc HCl) temp:boiling point	Pt	p-Chlorobenzoic acid,ᶠ o-Chlorobenzoic acid	196
(c) (H_2O, NaCl, H_2SO_4)	Pt, graphite	Ring substitution (55-96), chain substitution (45-4)	194
(d) (H_2O, HCl, HOAc)ʸ CD 0.005-0.05 A/cm^2	Pt	o- and p-Chlorotoluene, 2,4-dichlorotoluene, 2,4,5-trichlorotoluene, pentachlorotoluene, hexachlorobenzene,	53

Compound (conditions)	Electrode	Product (yield %)	Ref.
(e) (H$_2$O, NaCl)	Graphite^x	o- and p-Chlorotoluene (52),2,4-dichlorotoluene (14)	192
m-Xylene (H$_2$O, HCl)	Pt	2,4-Dimethyl-p-chlorobenzene	195
Ethylbenzene (CH$_3$OH, NH$_4$Br) CD 0.4–1.5mA/cm^2	Pt	Styrene (20)α-methoxyethylbenzene (19), o-bromoethylbenzene (2.2), p-bromoethylbenzene (6.6)	55
Nitrobenzene (H$_2$O, NaCl)	Graphite^x	m-Chloronitrobenzene (40)	192
Diphenyl (H$_2$O, NaCl)	Graphite^z	Chlorinated product	58
Naphthalene			
(a) (H$_2$O, Cl$^-$, MnO$_2$)	Graphite	α-Chloronaphthalene	54, 58, 197
(b) (conc. aqu. HCl, CCl$_4$, Roccal$^{v'}$) CD 0.38 A/cm^2	Pt	α-Chloronaphthalene (34)^w	192a
Anthracene (H$_2$O, NaCl)	Graphite^z	Chlorinated product	58
Phenanthrene (H$_2$O, NaCl)	Graphite^z	Chlorinated product	58
Hydrocinnamic acid (30% aqu HCl)	Pt	p-Chlorohydrocinnamic acid (99), o-chlorohydro-cinnamic acid (1)	195
Unsaturated fatty acids (HQAc, HCl)		Coulometric study	198
Methylamine (H$_2$O, KCl)^a CD 0.14 A/cm^2	Pt	N-Chloromethylamine (61)	62
Aniline			
(a) (H$_2$O, HCl)	C, Pt	Aniline black, trichloroaniline, tetrachlorobenzoquinone trichlorobenzoquinone	67, 199-201
(b) (H$_2$O, 10% HCl) CD 0.1 A/cm^2	C	Tetrachlorobenzoquinone (3 g/5 gsm)	200
(c) (H$_2$O, 20% HCl) CD 0.1 A/cm^2	C	Trichlorobenzoquinone	200
(d) (H$_2$O, conc HCl) CD 0.1 A/cm^2	C	2,4,6-Trichloroaniline (4 g/20 gsm)	200

Table 7.16 Chlorination (cont.)

Compound (Medium, Experimental Conditions)	Anode	Product(s) (%Yield)	Ref.)
Indigo (H$_2$O, HCl)	Pt	,	202
Brucine (H$_2$O, HCl)	Graphite	2,4,6-trichloroaniline C$_{21}$H$_{27}$O$_7$N$_2$Cl$_7$	203
p-Toluenesulfonamide (H$_2$O NaCl)	Pt	N-Chloro-p-toluenesulfonamide (94)	62
5,5-Dimethylhydantoin: CD 145A/ft^2	Pt	N,N'-Dichloro-5,5-dimethylhydantoin (91)	62
(H$_2$O, NaCl)[b'] CD 0.62 A/cm^2			
Cyanuric acid:	Pt[c']	(98)	62
(a) (H$_2$O, NaCl)[d',e'] CD 960 A/ft^2			

	Electrode	Product	Refs.
(b) $(H_2O, NaCl)^{d',e'}$ CD 78 A/ft²	Graphite	(structure: triazine ring with Cl, Cl, N, N, N, OH) (99)	62
(c) $(H_2O)^{d',f}$ CD 500 A/ft²	$Pt^{c'}$	Calcium bis(dichlorocyanurate) (15)	62
(d) $(H_2O, KCl)^{a',d'}$ CD 930 A/ft²	$Pt^{c'}$	Potassium dichlorocyanurate (quant)	62
Phenol (HOAc, LiCl)	Pt	Chlorophenols	52
Acetic acid (H_2O, HCl)	C, Pt	Chloroacetic acid (0.42 mole/F)	191, 194, 204
Propionic acid (H_2O, HCl)		β-Chloropropionic acid,^{f} α-chloropropionic acid	205
Acetone (a) (H_2O, HCl)	C, Pt	Chloroacetone (97)	206-209
(b) (H_2O, KCl)	Pt	Chloroform (quant)	208, 210-213
Methyl ethyl ketone (H_2O, KCl)	Pt	Chloroform (47), HOAc	211
Cyclohexanone (H_2O, HCl)	C	α-Chlorocyclohexanone	209
Acetophenone $(H_2O, HOAc, HCl)$	C	α-Chloroacetophenone (80)	209
Propiophenone $(H_2O, HOAc, HCl)$	Pt	α-Chloropropiophenone, β-chloropropiophenone	209
Benzophenone $(H_2O, HOAc, HCl)$	C	Monochlorobenzophenone (70)	209
Ethanol (a) $(H_2O, NaCl)$	Pt	Chloroform (quant)	211, 214-217
(b) $(H_2O, CaCl_2,$ cyanuric acid$^{u'})$ CD 3.0 A/dm²; temp 100-115°C	Porous C, Pt	Chloral (51),^{w} chloral alcoholate, $ClCH_2CO_2H$, monochloroacetaldehyde hydrate	218, 219, 219a, 219b

Table 7.16 Chlorination (cont.)

Compound (Medium, Experimental Conditions)	Anode	Product(s) (% Yield)	Ref.
n-Propyl alcohol			
(a) (H$_2$O, CaCl$_2$),	Pt	Chloroform	211
(b) (H$_2$O, NaCl)^g	Pt	n-Propyl hypochlorite (71)	62
CD 0.343 A/cm^2			
t-Butyl alcohol (H$_2$O, NaCl)$^{g'}$	Graphite	t-Butyl hypochlorite (79)	62
CD 0.1 A/cm^2			
Cellulosic material (H$_2$O, chloride)	Pt	Delignified product, bleached	220–223
Lignosulfonic acid (H$_2$O, HCl)	Pt	Chlorolignin; chloro- and dichloroacetaldehyde	224, 224a
Carbon disulfide	—	Carbon tetrachloride (34)	190
Benzenesulfonic acid (H$_2$O, KCl)	Pt	Potassium salts of mono-, di-, and trichlorobenzene-sulfonic acid, trichlorobenzoquinone, tetrachlorobenzoquinone	225
Potassium benzenesulfonate (H$_2$O, KCl)	Pt	Trichlorobenzoquinone, tetrachlorobenzoquinone (6), m-chlorobenzenesulfonic acid, 3,5-dichlorobenzenesulfonic acid, 3,4,5-trichloro-benzenesulfonic acid.	225
Hydrogen cyanide			
(a) (dil. aqu. HCl)	C	CNCl	74, 75
(b) (conc. aqu. HCl)	C	$\left[\overline{(NCl \cdot CHCl)_3} \right] \cdot 3HCl$	74
Ethane + HCl (NaAlCl$_4$ + AlCl$_3$ melt) temp 164–174°C	Ni	C$_2$H$_5$Cl, 1,2-dichloroethane, hexachloroethane, tetrachloroethylene, trans-dichloroethylene	224b
Ethane + HCl(FeCl$_3$ + NaCl melt) temp 170°C	Pt	1,2-Dichloroethane, ethyl chloride	224b
Ethane (HgCl$_2$ + KCl melt) temp 270°C	Pt	C$_2$H$_5$Cl, isomeric dichloroethanes, 1,1,2-trichloro-ethane, tetrachloroethylene, vinyl chloride, 1,2-dichloroethylene	224b
Cyclohexane (conc HCl, or H$_2$O, NaCl) CD 0.6 A/cm^2	Pt, C	Monochlorocyclohexane	48

n-Dodecane			
(a) (20% aqu HCl) CD 83 mA/cm²; temp 100°C	Porous C	Monochloride, dichloride (ratio 7 : 1, total CE = 73%)	224c
(b) (H_2O, HCl)^h CD 0.0172 A/cm²	Porous C	Primary monochlorides (A), secondary monochlorides (B), dichlorides (C) (current efficiency 44% in distribution: A : B : C=10.5 : 54.6 : 34.9)	6, 49
Paraffins (emulsion of melt in conc HCl at 90°C)	Pt	Mixture of chlorinated products (contains 50% Cl)	226

67

Table 7.17 Bromination

Compound (Medium, Experimental Conditions)	Anode	Product(s) (%Yield)	Ref.
Benzene			
(a) (H_2O, HBr)	Graphite^l	Bromobenzene (23)	189
(b) (H_2O, NaBr)	Graphite^l	Bromobenzene (45)	192
(c) ($AlBr_3$-benzene complex)	—	Bromobenzene	227
(d) (BrCN, $AlBr_3$)	—	Bromobenzene (45-108), benzonitrile (2-7), polymer	228
Toluene			
(a)	—	Bromotoluene, benzyl bromide	190, 229
(b) (H_2O, NaBr)	Graphite^l	o- and p-Bromotoluene (30)	192
(c) (BrCN, $AlBr_3$)	—	Bromotoluenes ($o:m:p=32:44:24$), tolunitriles ($o:m:p=35:3:62$)	228
Ethoxybenzene (H_2O, HBr)	Pt	Kinetic study	230
Naphthalene (CH_3CN, Et_4NBr)			
(a) CPE 0.80 V vs Ag, Ag^+(0.1 M)	Pt	No brominated product	4
(b) CPE 1.35 V vs Ag, Ag^+(0.1 M)	Pt	1-Bromonaphthalene (38)$^{j'}$	4
(c) CPE 2.00 V vs Ag, Ag^+(0.1 M)	Pt	1,4-Dibromonaphthalene (68)$^{k'}$	4
Anthracene (CH_3CN, Et_4NBr)			
(a) CPE 0.80 V vs Ag, Ag^+(0.1M)	Pt	Anthraquinone (2.9), 9-bromoanthracene (<0.2)	4
(b) CPE 1.20 V vs Ag, Ag^+(0.1M)	Pt	Anthraquinone (2.9), 9-bromoanthracene (18), 9,10-dibromoanthracene (<0.2)	4
(c) CPE 1.60 V vs Ag, Ag^+(0.1 M)	Pt	Anthraquinone (2.8), 9-bromoanthracene (10), 9,10-dibromoanthracene (3.6)	4
Olefins and Dienes: (HOAc, CH_3OH, KBr)	Pt	Coulometric determination—products not isolated	198, 231, 232
Phenol			
(a) (H_2O, HBr)	Pt	Polybrominated phenol; p-Bromophenol$^{l'}$; 2,4,6-tribromophenol$^{m'}$	67, 195, 233, 234
(b) (H_2O, conc HBr) CPE V vs Ag, AgBr in 1 N HBr (i) 0.621-0.710 V CD 0.00031-0.0188 A/cm^2	Pt	p- and o-Bromophenol (quant) $p/o = 3.37 \pm 0.10$	71, 235

68

(ii) 0.905–0.924 V CD 0.094–0.10 A/cm²		$p/o = 1.96 \pm 0.08$	
p-Bromophenol (H₂O, conc HBr) CPE 0.700 V vs Ag, AgBr in 1 N HBr	Pt	2,4-Dibromophenol, 2,4,6-tribromophenol	71
Aniline (H₂O, HBr) CD 0.013 A/cm²	Pt	2,4,6-Tribromoaniline	67
o-Aminobenzoic acid	Pt	Tribromoaniline: coulometric study	236
Cinnamic acid (H₂O, KBr)	—	Bromostyrene	237, 238
8-Hydroxyquinoline (H₂O, HClO₄, KBr)	Pt	Kinetic study of bromination; coulometric study	239, 240
2-Methyl-8-hydroxyquinoline (H₂O, HClO₄, KBr)	Pt	Kinetic study of bromination	239
Cyanuric acid: (H₂O, KBr)$^{d',n'}$ CD 555 A/ft²	Pt$^{c'}$	Potassium dibromocyanurate (95)	62

Indigo			
(a) (H₂O, HBr)	C	5,5,5-Dibromoindigo^f	202, 241, 242

2,4,6-tribromoaniline

(b) (pyridine 0.2HBr) temp 70°C	Pt, C	5,5'-Dibromoindigo,^{f} 2,4,6-tribromoaniline	202, 243
Succinimide (H₂O, NaBr, NaOH)	Pt	N-Bromosuccinimide (54)	63

Table 7.17 Bromination (cont.)

Compound (Medium, Experimental Conditions)	Anode	Product(s) (%Yield)	Ref.
Fluoresein: (H_2O, NaOH, KBr) [structure: fluorescein with CO_2H]	Pt	Eosine: [structure: tetrabromofluorescein with CO_2H]	244
Potassium benzenesulfonate (H_2O, KBr)	Pt	Tetrabromobenzoquinone	225
Ethanol (H_2O, NaBr, $CaBr_2$)	Pt	Bromoform	214, 245, 246
Citric acid: [structure: CH_2CO_2H / $HOCCO_2H$ / CH_2CO_2H] (H_2O, H_2SO_4, KBr) CD 1.8 A/dm^2	Pb	Pentabromoacetone (high)	247
Acetone (a) (H_2O, HBr)	Pt	Bromoacetone	206
(b) (H_2O, KBr) CD 0.066 A/cm^2	Pt	Bromoform (90.2)	248, 249
Hydrogen cyanide (H_2O_2, NH_4Br)$^{o'}$ CD 100 A/ft^2	Graphite	BrCN (98)	73, 75, 76
Cyclohexane (H_2O, 40% HBr)$^{p'}$ CD 0.4 A/cm^2	Pt	Mono- and dibromocyclohexanes	47
Methylcyclohexane (H_2O_2 40% HBr)$^{p'}$ CD 0.4 A/cm^2	Pt	Mono- and dibromomethylcyclohexanes	47

Table 7.18 Iodination

Compound (Medium, Experimental Conditions)	Anode	Product(s) (% Yield)	Ref.
Benzene			
(a) (H_2O, HI)	—	Iodobenzene	249
(b) $(I_2, CH_3CN, LiClO_4)^{q'}$ CPE 1.6–1.7 V vs Ag, Ag$^+$ (0.1 N)	Pt	Iodobenzene (96)	5
(c) $(I_2, CH_3CN, LiClO_4)$ CPE 1.6–1.7 V vs Ag, Ag$^+$ (0.1 N)	Pt	Iodobenzene	5, 57
Toluene			
(a) $(I_2, CH_3CN, LiClO_4)^{q'}$ CPE 1.6–1.7 V vs Ag, Ag$^+$ (0.1 N)	Pt	Iodotoluenes (ortho 47, para 47)	5
(b) $(I_2, CH_3CN, LiClO_4)$ CPE 1.6–1.7 V vs Ag, Ag$^+$ (0.1 N)	Pt	Iodotoluenes (ortho 16, para 16, meta 2)	5, 57
p-Xylene			
(a) $(I_2, CH_3CN, LiClO_4)^{q'}$ CPE 1.6–1.7 V vs Ag, Ag$^+$ (0.1 N)	Pt	2-Iodo-p-xylene (100); diiodo-p-xylene, tetraiodo-p-xylene	5
(b) $(I_2, CH_3CN, LiClO_4)$ CPE 1.6–1.7 V vs Ag, Ag$^+$ (0.1 N)	Pt	2-Iodo-p-xylene (50), N-p-xylylacetamide (12)	5, 57
(c) $(I_2, CH_3CN, (nPr)_4NBF_4)$ CPE 1.6–1.7 V vs Ag, Ag$^+$ (0.1 N)	Pt	2-Iodo-p-xylene (66), N-p-xylylacetamide (9)	5, 57
Mesitylene $(I_2, CH_3CN, LiClO_4)$ CPE 1.6–1.7 V vs Ag, Ag$^+$ (0.1 N)	Pt	Iodomesitylene (73)	5, 57
Anisole			
(a) $(I_2, CH_3CN, LiClO_4)^{q'}$ CPE 1.6–1.7 V vs Ag, Ag$^+$ (0.1 N)	Pt	Iodoanisoles (ortho 24, para 56)	5
(b) $(I_2, CH_3CN, LiClO_4)$ CPE 1.6–1.7 V vs Ag, Ag$^+$ (0.1 N)	Pt	Iodoanisoles (ortho 6, para 13, meta 1)	5, 57
Nitrobenzene $(I_2, CH_3CN, LiClO_4)^{q'}$ CPE 1.6–1.7 V vs Ag, Ag$^+$ (0.1 N)	Pt	No iodination product	5
Anthracene $(I_2, CH_3CN, LiClO_4)$ CPE 1.6–1.7 V vs Ag, Ag$^+$ (0.1 N)	Pt	No iodination product	5, 57

Table 7.18 Iodination (cont.)

Compound (Medium, Experimental Conditions)	Anode	Product(s) /% Yield	Ref.
Triphenylmethane (I_2, CH_3CN, $LiClO_4$) CPE 1.6–1.7 V vs Ag, Ag$^+$ (0.1 N)	Pt	Iodophenyldiphenylmethane (15), triphenylmethanol (36)	5, 57
Phenol (H_2O, HI)	–	Iodinated product	70, 249
Phenols: (H_2O, NaOH, KI)	–	ArOI	70
Thymol (H_2O, NaOH, KI) CD 1.5 A/cm^2	Pt	[structure: 2,6-substituted phenol, C_3H_7 and CH_3, OI]	70, 208, 233, 250, 251
8-Hydroxyquinoline (a) (H_2O, NaOH, KI) CD 0.5 A/dm^2	Pt	7-Iodo-8-hydroxyquinoline, 5,7-diiodo-8-hydroxyquinoline (82.4)	252
(b) (H_2O, NaOH, KI) CD 0.1 A/dm^2	Pt	7-Iodo-8-hydroxyquinoline (68)	252
Phenolphthalein (H_2O, KI, NaOH)	Pt	[structure: NaO, I, CO_2Na]	253–255
Fluorescein (H_2O, NaOH, KI)	Pt	Erythrosine	244
Pyridine^{IV} (CH_3CN, NaI, $NaClO_4$)	Pt	$[I(C_5H_5N)_2]^+ClO_4^-$	58, 60
2-Aminopyridine (H_2O, CH_3OH, KI)	–	2-Amino-5-iodopyridine	256

5,5-Dimethylhydantoin:
(H$_2$O, NaI)[8']

CH$_3$ CH$_3$ (5,5-dimethylhydantoin structure)

	Electrode	Products	References
5,5-Dimethylhydantoin: (H$_2$O, NaI)[8']	Pt	N,N'-Diiodo-5,5-dimethylhydantoin	62
Acetone (a) (H$_2$O, KI) CD 0.2 A/cm^2	Pt	Iodoform (90-95)	235, 245, 257-262
(b) (H$_2$O, NH$_4$OH, KI)	Pt	2,5-Dimethylpyrazine (50), iodoform	263
Cyclohexanone (H$_2$O, NH$_4$OH, KI) CPE 0.32 V vs SCE	Pt	dihydropyrazine C$_{12}$H$_{18}$N$_2$ (high), Iodoform (97)	263
Ethanol (H$_2$O, KI, Na$_2$CO$_3$) CD 1 A/dm^2; temp 60-70°C	Pt	Iodoform	208, 214, 245, 264, 269
Acetic acid (HOAc, HClO$_4$, I$_2$) CPE	Pt	ICH$_2$CO$_2$H	59

2,5-Dimethylpyrazine (structure)

Table Footnotes

a	A combination Ni-porous carbon anode was used.
b	CCl_4, $COCl_2$, or $HCCl_3$ were also used as Cl_2 sources; thionyl chloride, sulfuryl chloride, and sulfur mono- and dichlorides could not be used in chlorofluorination reactions.
c	Di- and trichlorides were formed in addition.
d	C_2H_6 also formed, possibly as a result of cathodic reduction of C_2H_4.
e	Rotating anode.
f	Major reaction product
g	CD 20 A/ft^2.
h	Product hydrolyzed.
i	A small amount of the 4-*n*-propyl isomer was present.
j	Product was largely perfluoro(3-ethylhexane) and a small amount of perfluoro(2-methyl-3-ethylpentane).
k	Cell temperature was 86°C.
l	Yield in weight percent.
m	The yield of $CF_3(CF_2)_xCO_2H$ was 3 to 4 times greater using the carboxylic acid fluoride compared to the carboxylic acid.
n	A markedly lower yield is obtained with the corresponding carboxylic acid.
o	Product hydrolyzed and esterified.
p	A more likely formula may be $CF_3(CH_2)_7SF_5$.
q	The product consisted of 80% normal and 20% iso-isomers.
r	Normal perfluoro derivative isolated.
s	Perfluorocyclohexyl derivative formed.
t	5-10% solution of dimethyl sulfate in HF.
u	Molar ratio of dimethyl sulfate: HF is 1:5.5.
v	Selenium coated on electrodes.
w	Current efficiency.
x	Anode consisted of graphite, NaCl, NaI, and aromatic compressed together.
y	Increasing chlorination observed with increasing current density.
z	Anode consisted of graphite, NaCl, and aromatic compressed together.
a'	Conc. HCl was added to maintain pH at 7.
b'	Conc. HCl was added to maintain pH at 4-5.
c'	Platinized titanium electrode.
d'	Cyanuric acid added slowly during electrolysis.
e'	Conc. HCl was added to maintain pH at 3-4.
f'	$CaCl_2$ and conc. HCl slowly added.
g'	Conc. HCl was added to maintain pH at 8-9.
h'	Temp 93-102°C.
i'	Anode consisted of graphite, Fe filings, NaBr, and aromatic compressed together.
j'	Current efficiency 70%.
k'	Current efficiency 71%.
l'	Obtained using an excess of phenol.

m' Obtained using high CD.

n' KBr and HBr was added to maintain pH at 7-8.

o' Temp 38°C.

p' Irradiated with UV light.

q' Aromatic added to preelectrolyzed solution of $I_2/CH_3CN/LiClO_4$.

r' Other amines studied: 2-picoline, 2,6-lutidine, 4,4'-bipyridine, 1,10-phenanthroline, 2,2',2''-tripyridine, 2,2'-bipyridine.

s' Conc. aqueous HI added during course of electrolysis.

t' Voltammetric study.

u' Catalyst.

v' Emulsifying agent.

w' The composition of the anode gas depended on the anode potential, anode CD, and conc. of urea (optimum 2-3 mole % for highest yield of NF_3).

Acknowledgements

I sincerely thank Dr. S. Nagase, Government Industrial Research Institute, Japan for his contribution to this chapter as well as for preprints of papers and useful comments. I am also indebted to Dr. J. A. Donohue of American Oil Company for his valuable comments.

References

1. M. Michlmayr and D. T. Sawyer, *J. Electroanal. Chem.*, **23**, 387 (1969).
2. L. L. Miller and A. K. Hoffmann, *J. Am. Chem. Soc.*, **89**, 593 (1967).
3. R. C. Weast, Ed., *Handbook of Chemistry and Physics*, Chemical Rubber, Cleveland, Ohio, 1969, p. D-109.
4. J. P. Millington, *J. Chem. Soc. (B)*, **1969**, 982.
5. L. L. Miller, E. P. Kujawa, and C. B. Campbell, *J. Am. Chem. Soc.*, **92**, 2821 (1970).
6. H. M. Fox and F. N. Ruehlen, U.S. Patent 3,393,249 (1968).
7. S. Nagase, in P. Tarrant, Ed., *Fluorine Chemistry Reviews*, Vol. 1, Marcel Dekker, New York, 1967, p. 77.
8. P. Lebeau and A. Damiens, *Compt. Rend.*, **182**, 1340 (1926).
9. J. A. Lyons and E. C. Broadwell, U.S. Patent 785,961 (1905).
10. J. H. Simons, *J. Electrochem. Soc.*, **95**, 47 (1949).
11. J. H. Simons, H. T. Francis, and J. A. Hogg, *J. Electrochem. Soc.*, **95**, 53 (1949).
12. J. H. Simons and W. J. Harland, *J. Electrochem. Soc.*, **95**, 55 (1949).
13. J. H. Simons, W. H. Pearlson, T. J. Brice, W. A. Wilson, and R. D. Dresdner, *J. Electrochem. Soc.*, **95**, 59 (1949).
14. J. Burdon and J. C. Tatlow, in M. Stacey, J. C. Tatlow, and A. G. Sharp, Ed., *Advances in Fluoring Chemistry*, Vol. 1, Academic Press, New York, 1960, p. 129.
15. J. H. Simons and T. J. Brice, in J. H. Simons Ed., *Fluorine Chemistry*, Vol. 1, Academic Press, New York, 1954, p. 340.
16. E. Forche, in E. Muller, Ed., *Methoden der organischen Chemie:* Houben-Weyl, Vol. 5/3, G. Thieme Stuttgart, 1962, p. 38.
17. M. Schmeisser and P. Sartori, *Chem. Ingr. Tech.*, **36**, 9 (1964).
18. For a review on handling HF, see M. Kilpatrick and J. G. Jones, in J. J. Lagowski, Ed., *Chemistry of Non-Aqueous Solvents*, Vol. II, Academic Press, New York, 1947, p. 43.

19. K. Fredenhagen and G. Cadenbach, *Z. Phys. Chem.*, **146A**, 245 (1930).
20. W. Klatt, *Z. Anorg. Chem.*, **222**, 225,285 (1935).
20.a.M. Kilpatrick and F. E. Luborsky, *J. Am. Chem. Soc.*, **76**, 5863 (1954).
21. G. G. Koerber and T. DeVries, *J. Am. Chem. Soc.*, **74**, 5008 (1952).
22. N. Hackerman, E. S. Snavely, and L. D. Fiel, *Electrochem. Acta*, **12**, 535 (1967).
23. B. Burrows and R. Jasinski, *J. Electrochem. Soc.*, **115**, 348 (1968).
24. S. Nagase, H. Baba, and R. Kojima, *Kogyo Kagaku Zasshi*, **66**, 1287 (1963); *Chem. Abstr.*, **60**, 11618 (1964).
24.a.N. Hackerman, E. S. Snavely, and L. D. Fiel, *Corrosion Sci.*, **7**, 39 (1967).
25. V. Y. Kazakov, L. A. Savel'ev, E. A. Shiskin, Y. N. Voitovich, and N. L. Gudimov, *J. Appl. Chem. USSR*, **41**, 2096 (1968).
25.a.H. M. Fox, F. N. Ruehlen, and W. V. Childs, *J. Electrochem. Soc.*, **118**, 1246 (1971); W. V. Childs and F. N. Ruehlen, U.S. Patent 3,511,761 (1970); *Chem. Abstr.*, **73**, 51709c (1970); W. V. Childs, U.S. Patent 3,511,762 (1970); W. V. Childs, U.S. Patent 3,558,449 (1971); *Chem. Abstr.*, **74**, 70920b (1971).
26. S. Nagase, K. Tanaka, and H. Baba, *Bull. Chem. Soc. Japan*, **38**, 834 (1965).
27. Oxygen difluoride may explode if condensed in a cold trap with hydrocarbons then warmed.
28. S. Nagase and R. Kojima, *Bull. Chem. Soc. Japan*, **34**, 1468 (1961).
29. G. A. Sokol'skii and M. A. Dmitriev, *J. Gen. Chem. USSR*, **31**, 1026 (1961).
30. H. Schmidt and H. D. Schmidt, *J. Prakt. Chem.*, **2**, 105 (1955).
31. V. S. Plashkin, G. P. Tataurov, and S. V. Sokolov, *J. Gen. Chem. USSR*, **36**, 1705 (1966).
32. S. Nagase and R. Kojima, *Kogyo Kagaku Zasshi*, **64**, 1397 (1961); *Chem. Abstr.*, **57**, 3281 (1962).
33. H. M. Scholberg and H. G. Bryce, U.S. Patent 2,717,871 (1955); Japan Patent 221,494 (1956).
34. G. N. Kokhanov and S. A. Per'kova, *Sov. Electrochem.*, **3**, 867 (1967).
35. J. M. Tedder, *Adv. Fluor. Chem.*, **2**, 104 (1961).
36. *Chem. Eng. News*, Jan. 12, 40 (1970).
37. H. Schmidt and H. Meinert, *Angew Chem.*, **72**, 109 (1960).
38. J. D. Domijan, C. J. Ludman, E. M. McCarbon, R. F. O'Malley, and V. J. Roman, *Inorg. Chem.*, **8**, 1534 (1969).
39. I. L. Knunyants, I. N. Rozhkov, A. V. Bukhtiarov, M. M. Gol'din, and R. V. Kudryavtsev, *Izv. Akad. Nauk SSSR Ser. Khim.*, **1970**, 1207; *Chem. Abstr.*, **73**, 65752 (1970).
39.a.I. N. Rozhkov, A. V. Bukhtiarov, N. D. Kuleshova, and I. L. Knunyants, *Dokl. Akad. Nauk SSSR*, **193**, 1322 (1970); *Chem. Abstr.*, **74**, 70878u (1971).
39.b.J. W. Mellor, *A Comprehensive Treatise On Inorganic and Theoretical Chemistry*, Vol. 15, Longmans, Green, London, 1936, p. 406.
39.c.A. Engelbrecht, E. Mayer, and C. Pupp, *Monatsh*, **95**, 633 (1964).
40. L. Stein, J. M. Neil, and G. R. Alms, *Inorg. Chem.*, **8**, 2472 (1969).
41. J. A. Donohue and A. Zietz, *J. Electrochem. Soc.*, **115**, 1039 (1968).
42. J. A. Donohue, A. Zietz, and R. J. Flannery, *J. Electrochem. Soc.*, **115**, 1042 (1968).
43. L. G. Spears and N. Hackerman, *J. Electrochem. Soc.*, **115**, 452 (1968).
44. H. H. Rogers, S. Evans, and J. H. Johnson, *J. Electrochem. Soc.*, **116**, 601 (1969).
45. Y. N. Voitovich, V. Y. Kazakov, A. N. Kozyreva, and A. I. Levin, *J. Applied Chem. USSR*, **42**, 131 (1969).
46. L. B. Lazar, *Bull. Inst. Politech. Bucuresti*, **20**, 109 (1958); *Chem. Abstr.*,

54, 5062 (1960).
47. U. K. Agaev, V. P. Smirnova, S. Z. Rizaeva, and A. F. Aliev, *Azerb. Khim. Zh.*, **1969**, 84; *Chem. Abstr.*, **71**, 87116 (1969).
48. U. K. Agaev, V. P. Smirnova, and A. F. Aliev, *Azerb. Khim. Zh.*, **1967**, 98; *Chem. Abstr.*, **69**, 102464 (1968).
49. F. N. Ruehlen, G. B. Willis, and H. M. Fox, *J. Electrochem. Soc.*, **111**, 1107 (1964).
50. A simple model was assumed in which chlorine is evolved at the electrode and substituted for hydrogen in dodecane on a statistical basis.
51. W. Nernst, *Chem. Ber.*, **30**, 1547 (1897).
52. R. G. Van Name and C. H. Maryott, *Am. J. Sci.*, **35**, 153 (1913).
53. F. Fichter and L. Glanzstein, *Chem. Ber.*, **49**, 2473 (1916).
54. P. Jayles, *Compt. Rend.*, **189**, 686 (1929).
55. K. Sasaki, H. Urata, K. Uneyama, and S. Nagaura, *Electrochim. Acta*, **12**, 137 (1967).
56. For chemical oxidation conditions see Y. Ogata and K. Nakajima, *Tetrahedron*, **20**, 43 (1964), and references therein.
57. L. L. Miller, *Tetrahedron Lett.*, **1968**, 1831.
58. I. M. Kolthoff and J. Jordan, *J. Am. Chem. Soc.*, **75**, 1571 (1953).
59. G. Durand and B. Tremillon, *Anal. Chim. Acta*, **49**, 135 (1970).
60. A. I. Popov and D. H. Geske, *J. Am. Chem. Soc.*, **80**, 1340 (1958).
61. G. Dryhurst and P. J. Elving, *Anal. Chem.*, **39**, 606 (1967).
62. E. J. Matzner, U.S. Patent 3,449,225 (1969).
63. M. Lamchen, *J Chem. Soc.*, **1950**, 747.
64. P. A. S. Smith, *The Chemistry of Open-Chain Organic Nitrogen Compounds*, Vol. I, W. A. Benjamin, New York, 1965, p. 201.
65. S. H. Wilen and A. W. Levine, *Chem. Ind.*, **1969**, 237.
66. J. H. Fellman, S. H. Wilen, and C. A. VanderWerf, *J. Org. Chem.* **21**, 713 (1965).
67. L. Gilchrist, *J. Phys. Chem.*, **8**, 539 (1904).
68. U. Rosa, F. Pennisi, R. Bianchi, G. Federighi, and L. Donato, *Biochim. Biophys. Acta*, **133**, 486 (1967).
69. R. Rzendowska and T. Majle, *Rocz. Panstw. Zakl. Hig.*, **19**, 429 (1968); *Chem. Abstr.*, **70**, 26249h (1969).
70. Farbenfabriken Vorm. Friedr. Bayer & Co, Ger. Patent 64,405 (1891); *Friedl.*, **3**, 872.
71. R. Landsberg, H. Lohse, and U. Lohse, *J. Prakt. Chem.*, **12**, 253 (1961).
72. A. V. Thomas and E. G. Rochow, *J. Am. Chem. Soc.*, **79**, 1843 (1957).
73. R. W. Foreman and J. W. Sprague, *Ind. Eng. Chem., Prod. Res. Dev.*, **2**, 303 (1963).
74. M. Sato, T. Fujisawa, and J. Sato, *Bull. Yamagata Univ. Engr.*, **5**, 55 (1958); *Chem. Abstr.*, **52**, 19639i (1958).
75. J. W. Sprague and F. Veatch, U.S. Patent 3,294,657 (1966).
76. J. W. Sprague, U.S. Patent 3,300,398 (1967).
77. P. Sartori, *Angew. Chem. Intn. Ed.*, **2**, 261 (1963).
78. S. Nagase, H. Baba, and T. Abe, *Bull. Chem. Soc. Japan*, **40**, 2358 (1967).
79. S. Nagase, H. Baba, and T. Abe, *Bull. Chem. Soc. Japan*, **39**, 2304 (1966).
80. R. N. Haszeldine and F. Nyman, *J. Chem. Soc.*, **1956**, 2684.
80.a. H. M. Fox and F. N. Ruehlen, U.S. Patent 3,511,760 (1970).
81. J. K. Wolfe, U.S. Patent 2,601,014 (1952); U.S. Patent 2,806,817 (1957).
82. General Electric Co., Brit. Patent 758,492 (1956); *Chem. Abstr.*, **51**, 9381 (1957).
83. British Thomson-Houston Co. Ltd., Brit. Patent 668,609 (1952); *Chem. Abstr.*, **46**, 6016 (1952).
84. H. Kisaki, S. Mabuchi, and T. Sakomura, *Denki Kagaku*, **34**, 29 (1966); *Chem. Abstr.*, **65**, 1048 (1966).

85. Farbenfabriken Bayer A.G., Brit. Patent 741,399 (1955).
86. S. Nagase, K. Tanaka, H. Baba, and T. Abe, *Bull. Chem. Soc. Japan*, **39**, 219 (1966).
87. G. N. Kokhanov and T. M. Izmailova, *Sov. Electrochem.*, **2**, 1204 (1966).
88. D. Goerrig, H. Jonas, and W. Moschel, Ger. Patent 1,040,009 (1958); *Chem. Abstr.*, **54**, 22358 (1960).
89. S. Nagase, T. Abe, and H. Baba, *Bull. Chem. Soc. Japan*, **40**, 584 (1967).
90. A. F. Shepard and B. F. Dannels, U.S. Patent 3,386,899 (1968).
91. Farbenfabriken Bayer A. G., Brit. Patent 740, 723 (1955).
92. P. E. Ashley and K. J. Radimer, U.S. Patent 3,298,940 (1960); *Chem. Abstr.*, **66**, 71956q (1967).
93. M. Sander and W. Bloechl, *Chem. Ingr. Tech.*, **37**, 7 (1965).
94. Saline Ludwigshalle A.-G., Brit. Patent 879,057 (1961).
95. J. H. Simons, U.S. Patent 2,519,983 (1950).
96. J. H. Simons and R. D. Dresdner, *J. Electrochem. Soc.*, **95**, 64 (1949).
97. H. Schmidt and H. D. Schmidt, *J. Prakt. Chem.*, **2**, 250 (1955).
98. H. Schmidt and H. D. Schmidt, *Chem. Tech.*, **5**, 454 (1953).
99. J. Bornstein and M. R. Borden, *Chem. Ind.*, **1958**, 441.
100. E. A. Kauck, and J. H. Simons, U.S. Patent 2,616,927 (1952); U.S. Patent 2,631,151 (1953); *Chem. Abstr.*, **47**, 5827 (1953).
101. V. Y. Kazkov, L. A. Savel'ev, R. A. Dzerzhinskaya, E. A. Shishkin, and N. L. Gudimov, *J. Appl. Chem. USSR*, **41**, 2212 (1968).
102. J. H. Simons, U.S. Patent 2,490,098 (1949); *Chem. Abstr.*, **44**, 6443 (1950).
103. V. S. Plashkin, L. N. Pushkina, V. F. Kollegov, and S. V. Sokolov, *Zh. Vses. Khim. Obshchest.*, **12**, 237 (1967); *Chem. Abstr.*, **67**, 172545 (1967).
104. S. A. Mazalov, S. I. Gerasimov, S. V. Sokolov, and V. L. Zolatavin, *J. Gen. Chem. USSR*, **35**, 485 (1965).
105. S. V. Sokolov, Z. I. Mazalova, and S. A. Mazalov, *J. Gen. Chem. USSR*, **35**, 1774 (1965).
106. T. C. Simmons, F. W. Hoffmann, R. B. Beck, H. V. Holler, T. Katz, R. J. Koshar, E. R. Larsen, J. E. Mulvaney, K. E. Paulson, F. E. Rogers, B. Singleton, and R. E. Sparks, *J. Am. Chem. Soc.*, **79**, 3429 (1957).
107. R. E. Banks, W. M. Cheng, and R. N. Haszeldine, *J. Chem. Soc.*, 3407 (1962).
108. M. Schmeisser and F. Huber, *Z. Anorg. Allg. Chem.*, **362**, 337 (1968).
108.a.N. Watanabe, A. Tasaka, and K. Nakanishi, *Denki Kagaku Oyobi Kogyo Butsuri Kagaku*, **37**, 705 (1969).
109. R. E. Banks, A. E. Ginsberg, and R. N. Haszeldine, *J. Chem. Soc.*, 1740 (1961).
109.a.W. E. Erner, U.S. Patent 3,335,143 (1967).
110. J. A. Young and R. D. Dresdner, *J. Am. Chem. Soc.*, **80**, 1889 (1958).
111. V. Y. Kazkov, L. A. Savel'ev, R. A. Dzerzhinskaya, E. A. Shiskin, and N. L. Gudimov, *J. Applied Chem. USSR*, **41**, 2089 (1968).
111.a.R. A. Guenthner, U.S. Patent 3,471,484 (1969).
112. J. H. Simons, U.S. Patent 2,500,388 (1950).
113. S. Nagase, H. Baba, and R. Kojima, *Kogyo Kagaku Zasshi*, **65**, 1183 (1962); *Chem. Abstr.*, **58**, 3098 (1963).
114. S. Nagase and H. Baba, *Kogyo Kagaku Zasshi*, **67**, 321 (1964); *Chem. Abstr.*, **61**, 2725 (1964).
115. V. S. Plashkin, L. N. Pushkina, S. L. Mertsalov, V. F. Kollegov, and S. V. Sokolov, *Zh. Org. Khim.*, **6**, 1006 (1970).
116. E. A. Kauck and J. H. Simons, U.S. Patent 2,594,272 (1952); Brit. Patent 672,720 (1952).
117. S. Nagase, T. Abe, and H. Baba, *Bull. Chem. Soc. Japan*, **41**, 1921 (1968).

118. I. P. Kolenko, N. A. Ryabinin, and B. N. Lundin, *Akad. Nauk SSSR, Ural'skii Filial Trudy Inst. Khim.*, **15**, 59 (1968).
119. N. A. Ryabinin, I. P. Kolenko, and B. N. Lundin, *J. Gen. Chem. USSR*, **37**, 1166 (1967).
120. S. Nagase, H. Baba, and R. Kojima, *Bull. Chem. Soc. Japan*, **36**, 29 (1963).
121. S. Nagase, H. Baba, and R. Kojima, *Kogyo Kagaku Zasshi*, **65**, 38 (1962); *Chem. Abstr.*, **58**, 441 (1963).
122. W. V. Childs, U.S. Patent 3,461,050 (1969); *Chem. Abstr.*, **71**, 87225q (1969).
123. S. Nagase, H. Baba, and R. Kojima, *Bull. Chem. Soc. Japan*, **35**, 1907 (1962).
124. G. N. Kokhanov and S. A. Per'kova, *Sov. Electrochem.*, **3**, 977 (1967).
125. S. Nagase, H. Baba, and R. Kojima, *Kogyo Kagaku Zasshi*, **64**, 2126 (1961); *Chem. Abstr.*, **57**, 2067 (1962).
126. S. Nagase, H. Baba, and R. Kojima, *Rept. Govt. Ind. Res. Inst. Nagoya*, **11**, 229 (1962).
127. A. V. Bukhtiarov, I. N. Rozhkov, and I. L. Knunyants, *Izv. Akd. Nauk SSSR, Ser. Khim.*, **1970** , 781; *Chem. Abstr.*, **73**, 51581e (1970).
128. Netherlands Patent 14,889 (1968); Derwent Pat. Rept., **6**, No. 16 (1969).
129. E. A. Kauck and A. R. Diesslin, *Ind. Eng. Chem.*, **43**, 2332 (1951).
130. R. Kojima, T. Hayashi, and S. Takagi, Japan Patent 17,224 (1960); *Chem. Abstr.*, **55**, 19561 (1961).
131. F. Dvorak and V. Dedek, *Coll. Czech. Chem. Commun.*, **31**, 2727 (1966).
132. F. Dvorak, Czech Patent 119,682 (1966); *Chem. Abstr.*, **67**, 17394n (1967).
133. F. Dvorak and V. Dedek, *Coll. Czech. Chem. Commun.*, **33**, 3913 (1968).
134. C. Hu and C. Ch'eb, *Chem. Abstr.*, **64**, 11077 (1966).
135. Minnesota Mining and Manufacturing Co., Brit. Patent 1,007,288 (1965); *Chem. Abstr.*, **64**, 3363 (1966).
136. G. Gambaretto, G. Troilo, and M. Napoli, *Chim. Ind. (Milan)*, **52**, 1097 (1970); *Chem Abstr.*, **74**, 41979x (1971).
137. Saline Ludwigshalle A.-G., Ger. Patent 1,069,639 (1959); *Chem. Abstr.*, **55**, 12119b (1961).
138. Saline Ludwigshalle A.-G., Brit. Patent 862,538 (1961); *Chem. Abstr.*, **57**, 16567i (1962).
139. Minnesota Mining and Manufacturing Co., Brit. Patent 718,318 (1954); *Chem. Abstr.*, **50**, 2686 (1956).
140. Saline Ludwigshalle A.-G., Ger. Patent 1,069,639 (1959); *Chem. Abstr.*, **55**, 12119b (1961).
141. E. A. Kauck and J. H. Simons, U.S. Patent 2,644,823 (1953).
142. S. Nagase, H. Baba, K. Tanaka, and T. Abe, *Kogyo Kagaku Zasshi*, **67**, 2062 (1965); *Chem. Abstr.*, **63**, 250 (1965).
143. M. A. Okatov, *Khim. Nauka i Prom.*, **4**, 675 (1959).
144. R. A. Guenthner, U.S. Patent 2,606,206 (1952).
145. Chemische Fabrik Pfersee G.M.B.H., Brit. Patent 1,077,301 (1967); *Chem. Abstr.*, **67**, 96339z (1967).
146. V. Y. Kazakov, R. A. Dzerzhinskaya, V. I. Tsimbalist, and E. A. Shiskin, *J. Gen. Chem. USSR*, **36**, 1799 (1966); *Chem. Abstr.*, **66**, 61251s (1967).
147. Minnesota Mining and Manufacturing Co., Brit. Patent 858,671 (1961); *Chem. Abstr.*, **55**, 11310 (1961).
148. J. A. Young, T. C. Simmons, and F. W. Hoffmann, *J. Am. Chem. Soc.*, **78**, 5637 (1956).
149. J. A. Young and R. D. Dresdner, *J. Org. Chem.*, **23**, 1576 (1958).
150. T. Abe, S. Nagase, K. Kodaira, and H. Baba, *Bull. Chem. Soc. Japan*, **43**, 1812 (1970).

151. V. Beyl, H. Niedersprum, and P. Voss, *Ann.*, **731**, 58 (1970).
152. R. A. Dzerzhinskaya, V. I. Tsimbalist, and V. Y. Kazakov, *J. Gen. Chem. USSR*, **37**, 1214 (1967).
153. T. J. Brice, W. H. Pearlson, and H. M. Scholberg, U.S. Patent 2,713,593 (1959); *Chem. Abstr.*, **50**, 5731 (1956).
154. I. M. Dolgopol'skii, A. V. Tumanova, B. A. Byzov, V. V. Berenblit, and G. B. Fedorova, *Authors' Certificate USSR* 119873 (1959); cited in Ref. 128.
155. A. Engelbrecht and E. Nachbaur, *Monatsch.*, **90**, 371 (1959).
156. R. D. Dresdner and J. A. Young, *J. Am. Chem. Soc.*, **81**, 574 (1959).
157. A. F. Clifford, H. K. El-Shamy, H. J. Emeleus, and R. N. Haszeldine, *J. Chem. Soc.*, **1953**, 2372.
158. J. A. Young and R. D. Dresdner, *J. Am. Chem. Soc.*, **81**, 574 (1959).
159. F. W. Hoffmann, T. C. Simmons, R. B. Beck, H. V. Holler, T. Katz, R. J. Koshar, E. R. Larsen, J. E. Mulvaney, F. E. Rogers, B. Singleton, and R. S. Sparks, *J. Am. Chem. Soc.*, **79**, 3424 (1957).
160. J. A. Young and R. D. Dresdner, *J. Org. Chem.*, **25**, 1464 (1960).
161. W. A. Severson, T. J. Brice, and R. I. Coon, 128th Meeting ACS, Minneapolis, Minn., September 11-16, 1955, Div. Ind. Eng. Chem., Fluorine Chem. Subdivision.
162. J. A. Young and R. D. Dresdner, *J. Org. Chem.*, **24**, 1021 (1959).
163. Dow Corning Corp., Brit. Patent 1,099,240 (1968); *Chem. Abstr.*, **68**, 83821w (1968).
164. T. Gramstad and R. N. Haszeldine, *J. Chem. Soc.*, **1956**, 173.
165. J. Burdon, I. Farazmand, M. Stacey, and J. C. Tatlow, *J. Chem. Soc.*, **1957**, 2574.
166. T. Gramstad and R. N. Haszeldine, *J. Chem. Soc.*, **1957**, 2640.
167. Minnesota Mining and Manuf. Co., Brit. Patent 1,146,312 (1969).
168. T. J. Brice and P. W. Trott, U.S. Patent 2,732,398 (1956); *Chem. Abstr.*, **50**, 13982h (1956).
169. G. A. Sokol'skii and M. A. Dmitriev, *J. Gen. Chem. USSR*, **31**, 648 (1961).
170. G. A. Sokol'skii and M. A. Dmitriev, *J. Gen. Chem. USSR*, **31**, 1023 (1961).
171. G. A. Silvey and G. H. Cady, U.S. Patent 2,697,726 (1954).
172. G. A. Silvey and G. H. Cady, *J. Am. Chem. Soc.*, **74**, 5792 (1952).
173. M. Schmeisser and F. Huber, *Z. Naturforsch.*, **21**, 285 (1966).
174. S. Nagase, T. Abe, and H. Baba, *Bull. Chem. Soc. Japan*, **42**, 2062 (1969).
175. F. Olstowski and J. J. Newport, U.S. Patent 3,033,767 (1962).
176. Dow Chemical Co., Brit. Patent 868, 020 (1961).
177. E. I. DuPont De Nemours & Co., Brit. Patent 885,635 (1961).
178. F. Olstowski and L. G. Dean, Brit. Patent 894,885 (1962).
179. W. R. Wolfe, Jr., Brit. Patent 863,602 (1961).
180. F. Olstowski, U.S. Patent 3,017,336 (1962).
181. K. J. Radimer, U.S. Patent 3,032,488 (1962).
182. K. J. Radimer, U.S. Patent 2,970,092 (1961); *Chem. Abstr.*, **55**, 9121 (1961).
183. W. E. Hanford and K. J. Radimer, U.S. Patent 2,970, 092, U.S. Patent 2,970,093 (1961).
184. K. J. Radimer, U.S. Patent 2,841,544 (1958).
185. L. G. Blosser, U.S. Patent 2,990,347 (1958).
186. R. Mueller and K. Bennewitz, Ger. Patent 1,139,498 (1962); *Chem. Abstr.*, **58**, 7975 (1964).
187. R. E. Seaver, *NASA Tech. Note D-1089*, 1961; *Chem. Abstr.*, **55**, 26782g (1961).
187.a. T. Abe, S. Nagase, and K. Kodaira, *Bull. Chem. Soc. Japan*, **43**, 957

(1970).
188. A. Lowy and H. S. Frank, *Trans. Am. Electrochem. Soc.*, **43**, 107 (1923).
189. C. W. Croco and A. Lowy, *Trans. Am. Electrochem. Soc.*, **50**, 315 (1926).
190. N. A. Isgarysev and W. S. Polikarpov, *Chem. Zentr.*, 885 (1941).
191. A. Hering, Ger. Patent 903,451 (1954); *Chem. Abstr.*, **52**, 7914e (1958).
192. G. Bionda and M. Civera, *Ann. Chim. (Rome)*, **41**, 814 (1951); *Chem. Abstr.*, **46**, 7908 (1952).
192.a.C. F. Hull, Ph.D. Thesis, Univ. of Cincinnati, 1958.
193. W. Jeunehomme, *Compt. Rend.*, **199**, 1027 (1934); *J. Chim. Phys.*, **32**, 173 (1935).
194. C. G. Schleuderberg, *J. Phys. Chem.*, **12**, 583 (1908).
194.a.J. C. Ghosh, S. K. Bhattacharyya, M. R. A. Rao, M. S. Muthanna, and R. B. Patnaik, *J. Sci. Indust. Res., India*, **11B**, 361 (1952).
195. H. Mühlhofer, *Diss. München Techn. Hochschule*, (1905).
196. J. B. Cohen, H. M. Dawson, and P. F. Crosland, *J. Chem. Soc.*, **87**, 1034 (1905).
197. G. Bionda, *Ann. Chim. Appl.*, **36**, 204 (1946); *Chem. Abstr.*, **41**, 3701 (1947).
198. F. Cuta and Z. Kucera, *Chem. Listy*, **47**, 1166 (1953); *Chem. Abstr.*, **48**, 3850 (1954).
199. K. Elbs and E. J. Brunschweiler, *J. Prakt. Chem. (2)*, **52**, 559 (1895).
200. J. Erdelyi, *Chem. Ber.*, **63**, 1200 (1930).
201. F. Goppelsroeder, *Compt. Rend.*, **82**, 331 (1876).
202. F. Fichter and F. Cueni, *Helv. Chim. Acta*, **14**, 658 (1931).
203. F. Fichter and H. Stenzl, *Helv. Chim. Acta*, **19**, 1173 (1936).
204. M. A. Youtz, *J. Am. Chem. Soc.*, **46**, 545 (1924).
205. F. Fichter and R. Ruegg, *Helv. Chim. Acta*, **20**, 1580 (1937).
206. A. Richard, *Compt. Rend.*, **133**, 878 (1901).
207. A. Riche, *Ann.*, **112**, 321 (1859).
208. J. W. Shipley and M. T. Rogers, *Can. J. Res.*, **17B**, 147 (1939).
209. J. Szper, *Bull. Soc. Chim. France*, **51**, 653 (1932).
210. F. Bottazzi, *Atti Accad. Lincei*, **18II**, 133 (1909). *Chem. Zentr.*, **II**, 1631 (1909).
211. J. Feyer, *Z. Elektrochem.*, **25**, 115 (1919).
212. J. E. Teeple, *J. Am. Chem. Soc.*, **26**, 536 (1904).
213. B. Waser, *Chem. Ztg.*, **34**, 141 (1910); *Chem. Abstr.*, **4**, 2064 (1910).
214. Chemische Fabrik Auf Actien vorm. E. Schering, Ger. Patent 29,711 (1884); *Friedl.*, I, 576; *Chem. Zentr.*, 447 (1885).
215. I. Shcherbakov, *Z. Elektrochem.*, **31**, 360 (1925).
216. R. Trechcinski, *Chem. Zentr.*, I, 13 (1907).
217. I. Stscherbakoff, *Z. Elektrochemie*, **31**, 360 (1925).
218. Chemische Fabrik Auf Actien vorm E. Schering, *Z. Elektrochem.*, **1**, 70 (1894).
219. J. C. Ghosh, S. K. Bhattacharyya, M. S. Muthana, and C. R. Mitra, *J. Sci. Ind. Res., India*, **11B**, 371 (1952).
219.a.Ludersdorf, *Ann. Phys. Paris*, **19**, 83 (1830).
219.b.S. Koidzumi, *Mem. Coll. Sci. Kyoto Imp. Univ.*, **8**, 155 (1925).
220. Aktiengesellschaft, A. Hering, Ger. Patent 971,953 (1959); *Chem. Abstr.*, **55**. 3979 (1961).
221. N. A. Sychev and V. N. Petrova, *Zhur. Prikkd. Khim. USSR*, **27**, 461 *Chem. Abstr.*, **48**, 8090 (1954).
222. Furdoonjee Dorabjee Pudumjee, Brit. Patent 942,958 (1963).
223. H. Kurz, U.S. Patent 2,828,253 (1958).
224. A. M. Petrova, S. A. Berezkina, L. G. Kudryavtseva, and Y. N. Sukhushin, *Tr. Tomsk. Univ.*, **170**, 73 (1964); *Chem. Abstr.*, **63**, 18446 (1965).
224.a.K. Schwabe, *Monatsh.*, **81**, 609 (1950).

224.b.G. E. Edwards, U.S. Patent 3,385,775 (1968).
224.c.H. M. Fox and F. N. Ruehlen, U.S. Patent 3,393,349 (1968).
225.   J. K. H. Inglis and F. Wootton, *J. Chem. Soc.*, **93**, 1592 (1908).
226.   L. B. Lazar, *Bull. Inst. Politechnic Bucuresti*, **20**, 109 (1958); *Chem. Abstr.*, **54**, 5062 (1960).
227.   W. Neminski and W. Plotnikow, *Zhur. Russ. Fiz.-Khim. Obshch.*, **40**, 391 (1908); *Chem. Abstr.*, **3**, 3156 (1909).
228.   L. R. Wilson, Ph.D. Thesis, University of Kansas, 1963.
229.   L. Bruner and S. Czarnecki, *Bull. Acad. Sci. Cracow*, 322 (1909); *Chem. Abstr.*, **5**, 1417 (1911).
230.   G. S. Kozak and Q. Fernando, *J. Phys. Chem.*, **67**, 811 (1965).
231.   J. W. Miller and D. D. Deford, *Anal. Chem.*, **29**, 475 (1957).
232.   F. A. Leisey and J. F. Grutsch, *Anal. Chem.*, **28**, 1553 (1956).
233.   H. Zehrlant, *Z. Elektrochem.*, **7**, 501 (1901).
234.   G. S. Kozak and Q. Fernando, *Anal. Chim. Acta*, **26**, 541 (1962).
235.   R. Landsberg, H. Lohse, and U. Lohse, *Wiss. Z. Tech. Hochsch. Chem. Leuna-Merseburg*, **2**, 461 (1959/60); *Chem. Abstr.*, **55**, 8118 (1961).
236.   L. G. Hargis and D. F. Boltz, *Talanta*, **11**, 57 (1964).
237.   Brester, *Jaresb. f. Chem.*, 87 (1866).
238.   W. Löb, *Z. Elektrochem.*, **3**, 46 (1896).
239.   G. S. Kozak, Q. Fernando, and H. Freiser, *Anal. Chem.*, **36**, 296 (1964).
240.   Q. Fernando, M. A. V. Devanathan, J. C. Rasiah, J. A. Calpin, and K. Nakulesparan, *J. Electroanal. Chem.*, **3**, 46 (1962).
241.   Farbwerke Vorm Meister, Lucius and Brüning, Ger. Patent 149,983 (1902).
242.   A. Schmidt and R. Mueller, U.S. Patent 765,996, (1904).
243.   E. Kunz, Ger. Patent 239,672 (1909); *Friedl.*, **10**, 393.
244.   Soc. Chimique des Usines du Rhône, Ger. Patent 108,838 (1899); *Chem. Zentr.*, I, 1176 (1900).
245.   K. Elbs and A. Herz, *Z. Elektrochem.*, **4**, 113 (1897).
246.   Les Etablissements Poulenc Frères, *Fr. 497,583* (1918).
246.a.I.   E. Antonova, K. N. Bil'dinov, A. P. Kharchenko, and A. A. Goncharenko, *USSR Patent 206,564* (1967).
250.   Bourgalt, *J. Pharm. Chem.*, **17**, 221 (1918).
251.   E. Moles and M. Marquina, *Anales. Real. Soc. Espan. Fis. Quim.*, **17**, 59 (1919); *Chem. Abstr.*, **13**, 3156 (1919).
252.   O. W. Brown and B. Berkowitz, *Trans. Electrochem. Soc.*, **75**, 385 (1939).
253.   A. Classen and W. Löb, *Chem. Ber.*, **28**, 1603 (1895).
254.   A. Classen, Ger. Patent 85930 (1894); *Friedl.*, **4**, 1090.
255.   A. Classen, U.S. Patent 618,168 (1908).
256.   Deutsche Gold-und Silver-Scheideanstalt vorm. Roessler, Ger. Patent 526,803 (1926); *Chem. Abstr.*, **25**, 4807 (1931).
257.   H. Abbott, *J. Phys. Chem.*, **7**, 84 (1903).
258.   G. A. Roush, *Trans. Electrochem. Soc.*, **8**, 281 (1905).
259.   J. E. Teeple, *J. Am. Chem. Soc.*, **26**, 170 (1904).
260.   R. Ramaswamy, M. S. Venketachalapathy, and H. V. K. Udupa, *J. Electrochem. Soc.*, **110**, 297 (1963).
261.   A. R. Johnson, *Trans. Wis. Acad. Sci.*, 253 (1908).
262.   J. W. Schuyl, *J. Chem. Ed.*, **46**, 518 (1969).
263.   S. H. Wilen and A. W. Levine, *Chem. Ind.*, **1969**, 237.
264.   O. Dony-Henault, *Z. Elektrochem.*, **7**, 57 (1900).
265.   F. Foerster and W. Meves, *Z. Elektrochem.*, **4**, 268 (1897).
266.   T. Kempf, U.S. Patent 372,940 (1885).
267.   G. Lazzarini, *Ind. Chim.*, **4**, 771 (1929); *Chem. Abstr.*, **24**, 334 (1930).
268.   A. Vyskočil, *Chem. Listy*, **23**, 212 (1929).
269.   E. Schering. Ger. Patent 29,771 (1884).

Chapter VIII

ELECTROCHEMICAL REDUCTION OF ORGANIC COMPOUNDS
M. R. Rifi

1 INTRODUCTION

This chapter is meant to acquaint the reader with reactions of organic compounds that take place at the cathode and to point out some of th advantages of cathodic reduction over chemical reducing agents.

Polarographically, a great many cathodic reactions have been carried out on the microscale; however, most of these studies will not be referred to in thi chapter unless the reaction has been domonstrated on a preparative scale Nevertheless, mechanisms of electrochemical reductions are broadly discussed Cathodic processes involving organometallic compounds, nitrogen heterocycles and electroinitiated polymerizations are treated, for the most part, in othe chapters.

2 GENERAL EXPERIMENTAL PROCEDURE

Electrolysis Cell. This can vary from an ordinary beaker to sophisticate equipment, depending on the nature of the reduction in question. For choice o electrolysis cells the reader should refer to Chapter II.

Cathode Material. In aqueous media, where the generation of hydrogen required, a cathode of low hydrogen over voltage (e.g., platinum) is recommenc ed. Alternatively, if it is desired that reduction take place as a result of electro transfer from the cathode to the substrate, a high overvoltage cathode (e.g mercury, lead, or zinc) should be employed. For reaction in nonaqueous medi any of the above cathode materials may be used as well as other conductiv substrates.

Supporting Electrolyte. The choice of a supporting electrolyte depends on th following criteria: (a) its solubility (and dissociation) in the medium; (b) it reduction potential, which should occur at a more negative potential than tha required to reduce the substrate; (c) its inertness (chemical and electrochemica to the substrate and its reduction product. Some of these parameters are di cussed in detail in Chapters II and III.

Solvent. The most common solvent used in electrochemical reductions is wate For compounds and supporting electrolytes that are not soluble in water, cosolvent such as ethanol, or the like, is employed. For reactions in nonaqueou media, ethanol, dioxane, formamide, and other organic liquids with hig dielectric constant and reduction potential have been used. Often, the corre choice of solvent depends on the following factors: (a) its solubilizing propertie (b) the reactivity of the solvent with the substrate and its reduction product (c) its ease of reduction; (d) its ease of separation from the product; and (c) i toxicity.

Electrode Potential. Setting the working electrode potential at a certain valu with realtion to a reference allows the selective reduction of a functional grou in a polyfunctional molecule and offers as a result advantages not easil available through conventional techniques.

3 CARBONYL COMPOUNDS

Aldehydes and Ketones

Products and Reaction Conditions

The course of reduction of aldehydes and ketones is strongly dependent on the reaction conditions, especially the pH of the medium and the nature of the cathode material. Under acidic conditions, dimeric products (pinacols) are often obtained, while under basic conditions, the corresponding alcohol is generally the main product. With lead cathodes under acidic conditions, acetone is reduced to both pinacol and isopropyl alcohol; however, with mercury as the cathode, no pinacol is formed and the product is mainly isopropyl alcohol with some propane [357]. The potential of the electrode may also affect the reduction. For example, in the reduction of benzil [452] cis-trans-isomerization takes place. This isomerization is apparently catalyzed by the electrical field in the Helmholtz double layer and is dependent on the electrode poetntial, temperature, pH, and the nature of the supporting electrolyte. The effect of current density on the reduction of acetone has been reported [460]. It was found that an increase in current density increases the formation of pinacol (probably due to the formation of a high concentration of the corresponding radical anion).

In the selection of reaction conditions, the researcher should consider the chemical effects of the variables on the reactants and products of the reaction. For example, in the reduction of aliphatic aldehydes and ketones, if the reaction solution is too basic, self-condensation of the starting material may occur. Under acidic conditions, the pinacol product may rearrange to the corresponding pinacolone. The cathode itself may, in certain cases, react with the starting material (or products) to form organometallic products. For example, in the reduction of acetone on a lead cathode in the absence of air, lead di- and tetra-isopropyl are obtained [358].

Various products may be obtained from one starting material by using different cathodes [359]. Thus for glyoxalic acid:

$$2HC-COH + 2e + \xrightarrow{\text{Fe, Ag, Ni}} \begin{array}{l} HO-CH-CO_2H \\ | \\ HO-CH-CO_2H \end{array}$$

$$\xrightarrow{\text{Pb, Cd, Hg}} \begin{array}{l} CH_2CO_2H \\ | \\ CH_2CO_2H \end{array}$$

The reduction 2-methylcyclohexanone at a mercury or lead cathode gave th
pure *trans*-2-methylcyclohexanol, while at a copper cathode, the cis form wa
obtained. Using a nickel cathode, both isomers were obtained, and on a platinur
cathode, no reduction took place [361].

In some cases the electrochemical reduction of aldehydes and ketones afford
different products than those obtained with conventional reducing agents. For
example, the cathodic reduction of *p*-aminoacetophenone on a silver catalys
under acidic conditions yields the corresponding pinacol, which could not b
easily obtained with common reducing agents [291]. The reduction of severa
steroids at a mercury cathode was studied by Kabasakalian and co-workers [447
and was found to afford the corresponding equatorial epimeric alochols wit
high degree of stereospecificity and in good yield. This may be compared wit
catalytic hydrogenation at low pH which affords mainly to the correspondin
axial alcohol [597].

The reduction of ketones can be affected by Mg^+, which is generated from
magnesium anode [448, 453]. This "anodic reduction" affords, in certain case
products different from those normally obtained in the reaction of ketones wit
magnesium-magnesium iodide reagent [454, 455]:

Chemically:

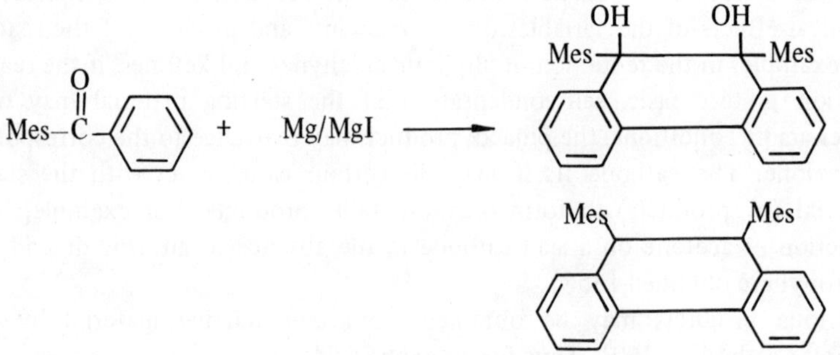

Electrochemically:

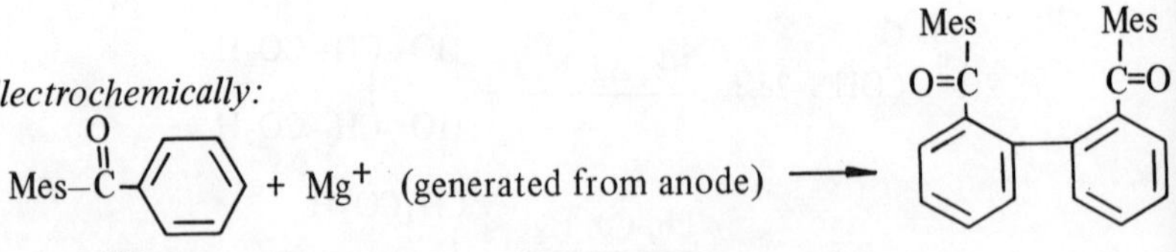

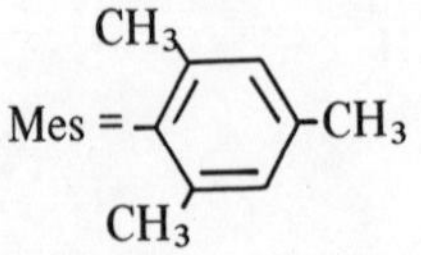

Udupa and co-workers [594] have made an interesting observation with regard to the use of stationary and rotating cathodes. At stationary cathodes (low current density) the reduction afforded the alcohol predominantly, while by using a rotating electrode (high current density) dimeric products were formed.

The reduction of carbonyl functions to methylene groups has been reported to occur at a mercury electrode in aqueous media [582].

Mechanism of Reduction

REDUCTION IN NONAQUEOUS SOLUTIONS

The reduction of aldehydes and ketones in nonaqueous solutions occurs stepwixe as follows:

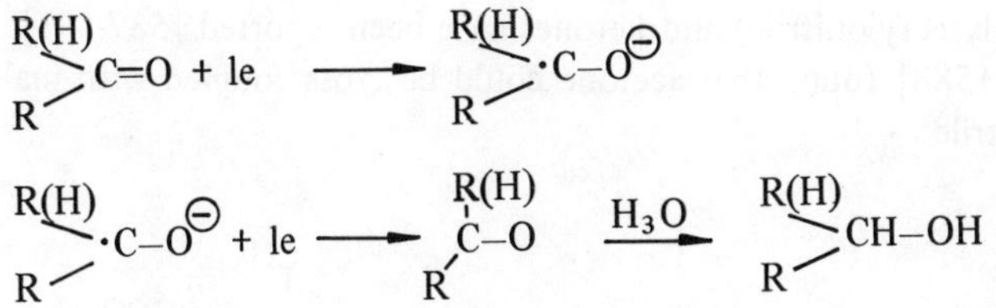

The ease of reduction will, to a large extent, depend on the stability of the anion radical. For example, it is not surprising that aldehyde generally reduce at a more negative potential than ketones. Aromatic substituents, which stabilize the radical intermediates, facilitate the reduction [265]. In certain cases, it has been possible to experimentally detect, by means of ESR measurements [360], the ketyl radical intermediate while carbonation was used to trap the anion [362]:

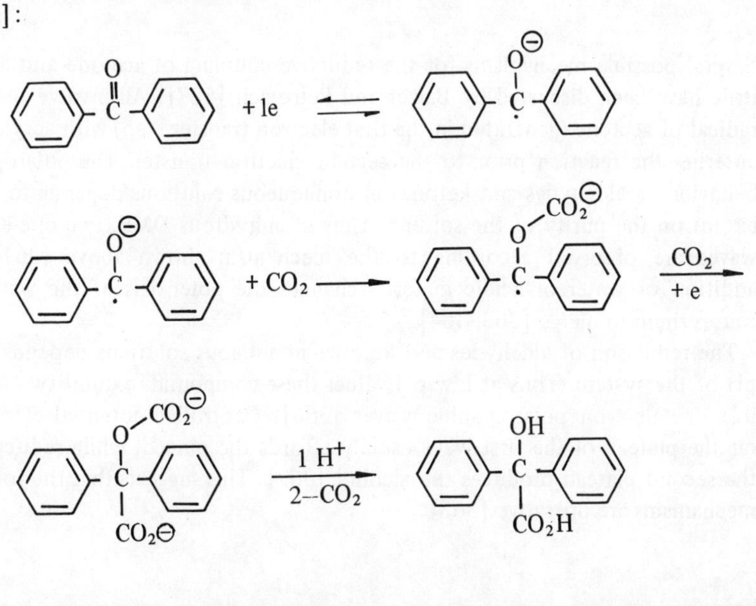

Another method successfully used to trap anions derived from carbonyl compounds it to carry out the reduction in the presence of alkyl halides, which are not reduced at the applied potential. For example, benzophenone electrolyzed in the presence of ethyl bromide gave diphenylethylcarbinol [583]:

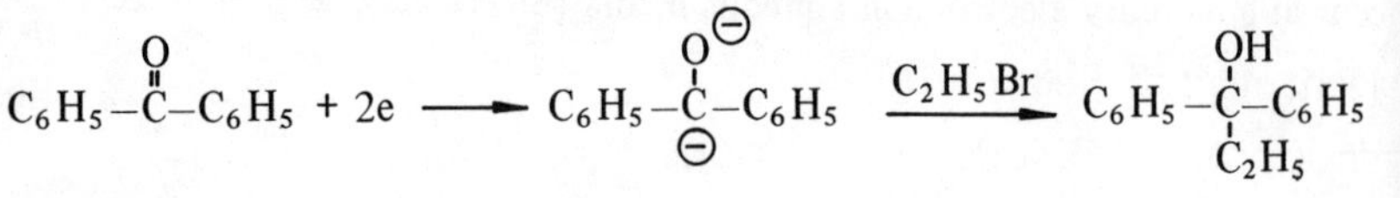

$$C_6H_5-\overset{\overset{\text{O}}{\|}}{C}-C_6H_5 + 2e \longrightarrow C_6H_5-\overset{\overset{\text{O}^{\ominus}}{|}}{\underset{\ominus}{C}}-C_6H_5 \xrightarrow{C_2H_5Br} C_6H_5-\overset{\overset{\text{OH}}{|}}{\underset{C_2H_5}{C}}-C_6H_5$$

A great many cathodic cross dimerizations between unsaturated compounds (especially acrylonitrile) and ketones have been reported [587-591]. Sugino and Nonaka [588] found that acetone could be cross coupled with maleic acid and acrylonitrile:

$$CH_3-\overset{\overset{\text{O}}{\|}}{C}-CH_3 + \overset{CHCO_2H}{\underset{CHCO_2H}{\|}} \xrightarrow[\text{Hg}]{20\% \ H_2SO_4} CH_3-O \underset{O}{\overset{\overset{CH_3}{|}}{\underset{\diagdown}{\overset{\diagup}{C}}}} \begin{matrix} CHCO_2H \\ | \\ CH_2 \end{matrix}$$

$$CH_3-\overset{\overset{\text{O}}{\|}}{C}-CH_3 + CH_2 = CHCN \xrightarrow[\text{Hg}]{20\% \ H_2SO_4} (CH_3)_2\ \overset{\overset{\text{OH}}{|}}{\underset{H}{C}}-CH_2-CH_2\,CN \quad + \\ \text{Lactone}$$

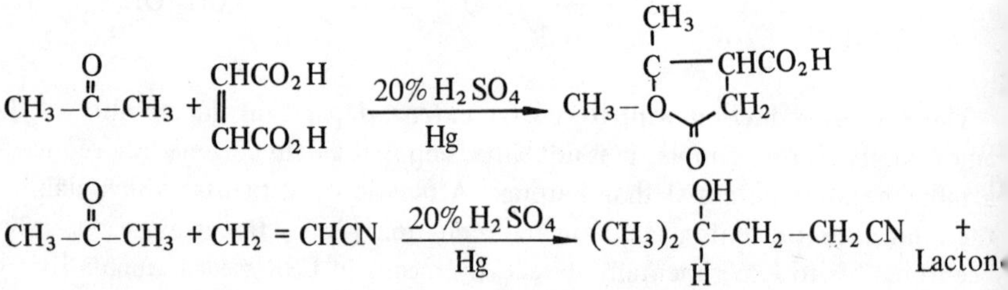

Several possible mechanisms for the reductive coupling of acetone and acrylonitrile have been discussed by Baizer and Petrovich [587]. All involve the anion radical of acetone (generated in the first electron transfer step) with acrylonitrile entering the reaction prior to the second electron transfer. The polarographic behavior of aldehydes and ketones in nonaqueous solutions depends to a great extent on the purity of the solvent. Thus in anhydrous DMF, two one-electron waves are observed according to the mechanism shown above [361]. The addition of water or acidic material changes the potentials of the waves and causes them to merge [363, 364].

The reduction of aldehydes and ketones in aqueous solutions depends on the pH of the system. Thus at low pH values these compounds exhibit two irreversible one-electron polarographic waves [366]. Controlled-potential electrolysis on the plateau of the first wave usually affords the pinacol while reduction of the second plateau produces the alcohol [367]. This suggests that the following mechanisms are operative [368]:

$$\begin{matrix} R(H) \\ \diagdown \\ \diagup \\ R \end{matrix} C = O \quad \xrightarrow{\ H_3O\ } \quad \begin{matrix} R(H) \\ \diagdown \\ \diagup \\ R \end{matrix} C = O\text{----}H$$

$$\begin{matrix} R(H) \\ \diagdown \\ \diagup \\ R \end{matrix} C = O\text{----}H \ + 1e \quad \xrightarrow{\text{wave I}} \quad \begin{matrix} R(H) \\ \diagdown \\ \diagup \\ R \end{matrix} \cdot C - O - H$$

$$2 \begin{matrix} R(H) \\ \diagdown \\ \cdot C - OH \\ | \\ R \end{matrix} \quad \longrightarrow \quad \begin{matrix} OH & OH \\ | & | \\ R - C \!\!-\!\!-\!\!-\!\! C - R \\ | & | \\ R(H) & R(H) \end{matrix}$$

$$\begin{matrix} R(H) \\ \diagdown \\ \cdot C - OH \\ | \\ R \end{matrix} + 1e \quad \xrightarrow{\text{wave II}} \quad \begin{matrix} R(H) \\ \diagdown \\ \ominus C - OH \\ | \\ R \end{matrix}$$

$$\begin{matrix} R(H) \\ \diagdown \\ \ominus C - OH \\ | \\ R \end{matrix} \quad \xrightarrow{\ \overset{\oplus}{H_3}O\ } \quad \begin{matrix} R(H) \\ \diagdown \\ CH - OH \\ | \\ R \end{matrix} + H_2O$$

With an increase in the pH of the medium, the two polarographic waves merge to form one two-electron reduction wave [368, 369]. Under these conditions the primary product is the corresponding alcohol:

$$\begin{matrix} R(H) \\ \diagdown \\ \diagup \\ R \end{matrix} C - O + 2e \quad \longrightarrow \quad \begin{matrix} R(H) \\ \diagdown \\ \ominus C - O^\ominus \\ \diagup \\ R \end{matrix}$$

$$\begin{matrix} R(H) \\ \diagdown \\ \ominus C - O^\ominus \\ | \\ R \end{matrix} + 2H_3O \quad \longrightarrow \quad R_2CHOH + 2H_2O$$

In the formation of pinacols from the electrochemical reduction of aldehydes and ketones, meso- and dl-isomers may be formed. In certain cases, the formation of one isomer over the other may depend on the reaction medium. Thus it was found by Stocker [286] that in the electrochemical reduction of acetophenone under acidic conditions, equal proportions of the dl and meso forms of the corresponding pinacols were formed. However, under basic conditions the ratio of dl to meso froms was 3:1. In order to explain these results, the author considered the stereochemistry of the meso and dl forms I and II:

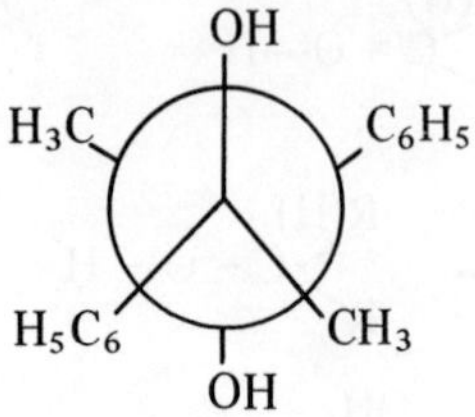

I
(yields meso)

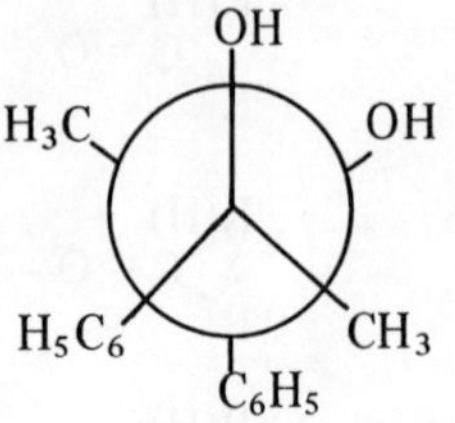

II
(yields dl)

The meso form **I** is formed because of its favored steric configuration. On the other hand, the dl form is formed because it allows for hydrogen bonding interaction. Dimerization to form compounds **I** and **II** is arrived at from a common intermediate $C_6H_5-\overset{\overset{\displaystyle OH}{|}}{C}-CH_3$. Under basic conditions, the radical anion $C_6H_5-\overset{\overset{\displaystyle O\ominus}{|}}{\underset{\displaystyle \cdot}{C}}-CH_3$ is formed, which may abstract a proton from the medium to from the corresponding radical. Thus coupling products may be formed from the interaction of the radical with the radical anion. The enhanced formation of the dl over the meso form was explained as due to the increased preference for dimerization to take place according to conformation **II** shown above.

The importance of hydrogen bonding in the formation of the dl isomer of pinacols formed in the electrochemical reduction of ketones and aldehydes, may, in some cases, be minimized. Thus it was found [599] that in the reduction of benzaldehyde, the presence of certain types of supporting electrolytes (e.g., $Et_4N^+ClO_4^-$) gave a 0.8 ratio of dl to meso isomers. The decreased effect of hydrogen bonding may be due to competitive orientation of the O–H bond in the high fields extant in the double layer.

Reduction of Diketones

The course of reduction of a-diketones is essentially the same as that of monoketones [265, 490-494]. Under acidic conditions, benzil exhibits one polarographic wave while methylphenyl diketone exhibits two waves. Reduction of the latter at the plateau of the first wave affords the corresponding dimeric diketopmacol [495]. The reduction of a-diketones approaches reversibility as the pH of the medium increases, indicating that the first step of the reduction produces the ketyl radical:

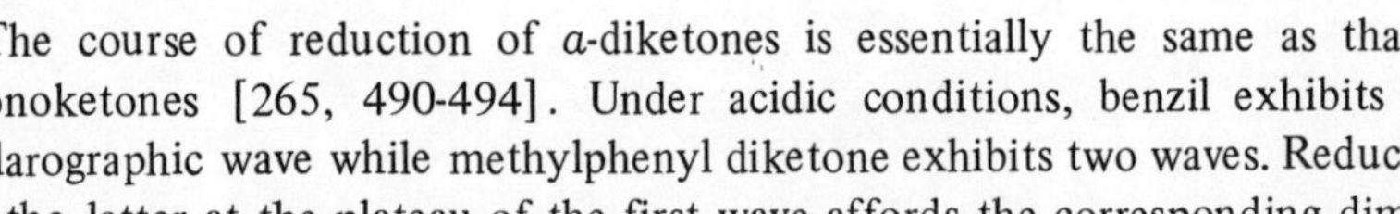

In acid solution, the anion is protonated, followed by the addition of a second electron to give the ketoalcohol (benzoin or acetoin). In contrast, under basic conditions, the initial anion survives long enough to allow its anodic oxidation to starting material. Ketyl radical anions of a-diketones have been detected by ESR spectroscopy [490].

A number of recent studies of 1,3-diketones have demonstrated reactions of possible synthetic utility. For example, the reduction of 1,1,3,-tetramethyl-cyclobutanedione in DMF exhibits two one-electron waves that merge upon the addition of water [365]. Macroscale electrolysis at the plateau of the second wave affords the keto-alcohol shown below:

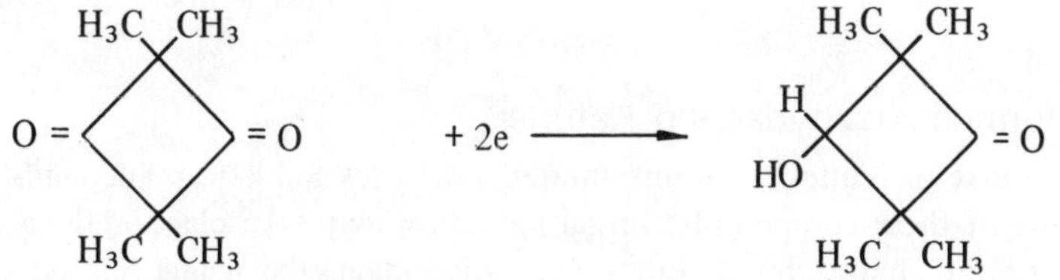

The keto-alcohol may have been formed from the partial reduction of the dione or the rearrangement of the desired bicyclic diol:

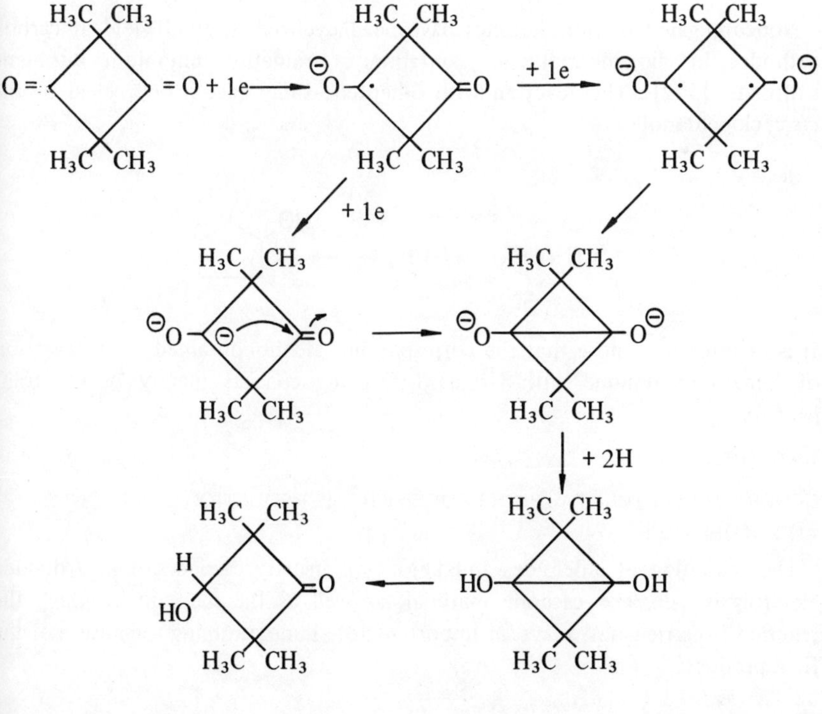

It may prove possible to isolate the bicyclic diol as the corresponding diacetate by carrying out the reduction in the presence of acetic anhydride. Curphey and co-workers [281] found that similar diacetates can be prepared from substituted 1,3-diketones:

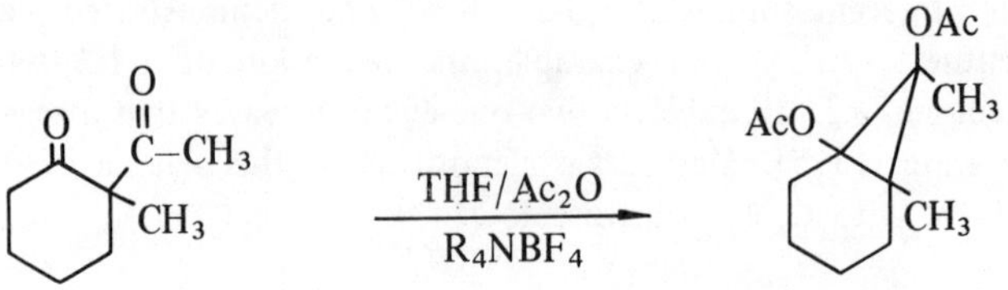

Unsaturated Aldehydes and Ketones

The course of reduction of unsaturated aldehydes and ketones depends on the structure of these compounds: Initial reduction may take place on the carbonyl group or the double bond. For further discussion, the reader should consult Refs. [265, 300, 355, 356, and 307-377]. An important reaction of a,β-unsaturated aldehydes and ketones is bimolecular reductive coupling to give the hydrodimer $O=C-\overset{|}{C}-\overset{|}{C}-\overset{|}{C}-C=O$ [587].

Nonconjugated olefinic ketones have been cyclized in good yield at carbon cathodes in dioxane/methanol containing tetramethylammonium p-toluene sulfonate [592]. The reaction with 6-heptene-2-one gave a 66% yield of the cis-cyclopentanol:

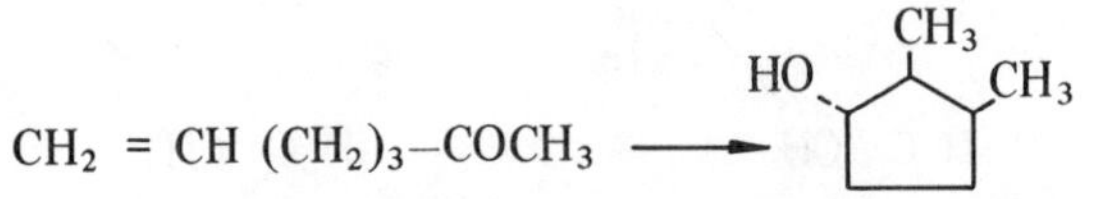

$$CH_2 = CH\ (CH_2)_3-COCH_3 \longrightarrow$$

It is of interest to note that the corresponding alcohol obtained by the reaction of 2-methylpentanone with Grignard reagents consists mainly of the trans isomers.

Experimental

GENERAL EXPERIMENTAL PROCEDURE FOR THE REDUCTION OF KETONES AND ALDEHYDES

The reduction of aldehydes and ketones is usually carried out in a divided electrolysis cell. The cathode material, as well as the medium in which the reaction is carried out, plays an important role in determining the nature of the final products:

1. *Formation of Alcohols.* This reaction is promoted by carrying out the reduction of aldehydes and ketones at a low concentration in a basic medium using a cathode of high hydrogen overvoltage.

2. *Formation of Pinacols.* Since this reaction involves the coupling of intermediates formed during electrolysis, the concentration of the aldehyde or ketone should be high. Furthermore, the medium is, in general, acidic and the cathode should have a low hydrogen overvoltage.

3. *Formation of Hydrocarbons.* It is difficult to predict to optimum conditions for the generation of hydrocarbons from the reduction of ketones and aldehydes. In general, a considerable success has been observed by carrying out the reduction in an acidic medium using cadmium as the cathode material.

Quinones

The early work on the electrochemical reduction of quinones dealt mainly with their polarographic behavior. In general, quinones (and hydroquinones) yield reversible polarographic waves, with the half-wave potential nearly equal to the standard potential. The simplicity of such reaction has made quinones the system of choice for testing new techniques in polarography.

REDUCTION IN APROTIC SOLVENTS

The polarographic reduction of benzoquinone and anthraquinone in DMF shows two reversible waves each corresponding to the addition of one electron [601, 602]:

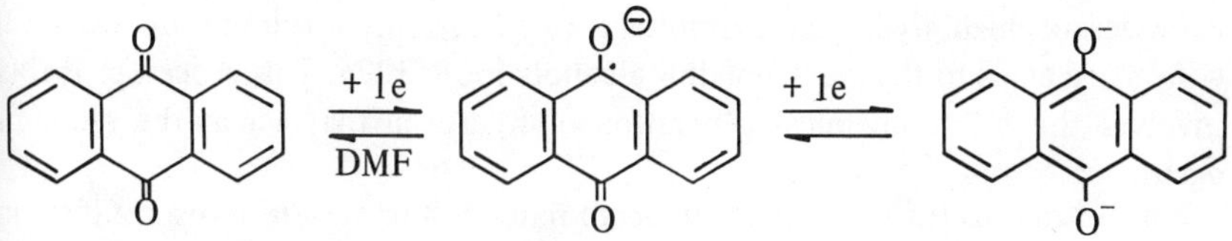

The dianion has enough stability to be trapped by chemical means. Thus if an alkyl halide is added during reduction, the corresponding di-ether is formed.

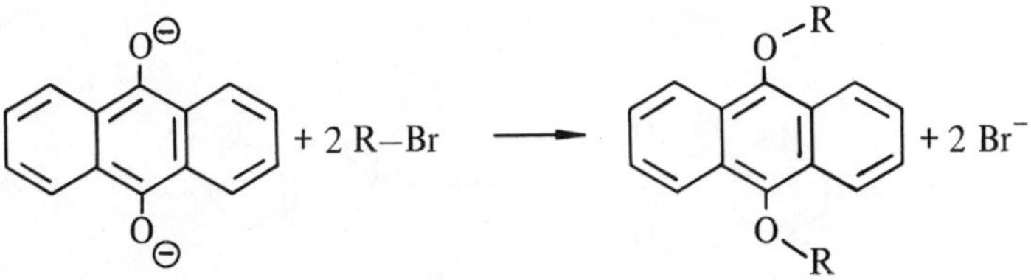

The anion radical (semiquinone anion) has been detected by ESR spectroscopy [603].

The use of lithium cations as supporting electrolytes allowed the reduction of quinones to occur at a more positive potential [605]. This was explained as due to the association of the lithium cation with the carbonyl group of the quinone. In another study [606], the use of different cations changed the polarographic behavior of quinones. This change in behavior was attributed to two kinds of phenomena: (1) adsorption, when cations such as tetraalkylammonium salt were used, and (2) kinetic, when lithium cations were used.

REDUCTION IN PROTIC SOLVENTS

In the presence of protons, the two polarographic waves, observed in aprotic solvents, approach each other and merge. This happens because the second wave appears at a more positive potential. Reduction under such conditions reduce the stability of the semiquinone anion and a two-electron reversible step is observed.

Carboxylic Acids

General Discussion

Carboxylic acids may be cathodically reduced to the corresponding alcohols, aldehydes, or even hydrocarbons [271, 272]. Iverson and Lund [581] have shown that carboxylic acids may be converted to the aldehyde stage if the carbonyl group is activated by an electron-withdrawing group. In general, cathodes of high hydrogen overvoltage (e.g., mercury or lead) are used for converting acids to the corresponding alcohols [308, 309]. This process probably involves the electrochemical generation of hydrogen that act as the reducing agent.

Formic and acetic acids do not undergo reduction under the above conditions, and while it has been reported [378] that formic acid can be reduced to the corresponding aldehyde or alcohol, the work could not be duplicated [379].

The reduction of aromatic carboxylic acid is generally more facile than aliphatic acids. The reaction may proceed through anion radical intermediates generated by direct electron transfer [458, 462, 467]. In certain cases, reduction of *ortho*-carboxyl groups results in the formation of lactones [322], for example

Wagenknecht [593] has recently described the reduction of substituted aromatic carboxylic acid to the corresponding aldehyde. The success of the reduction depended on the use of a buffer medium containing benzene. The benzene extracts the aldehyde from the vicinity of the electrode and prevents its further reduction. Thus the reduction was envisioned to proceed as follows:

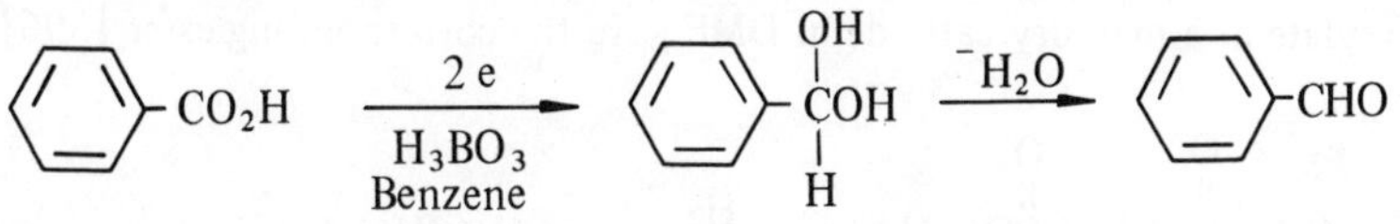

A simple experimental procedure was described for the above reaction using a practical electrolysis cell. The reader who is interested in the reduction of carboxylic acid is advised to consult Ref. [593].

General Experimental Procedure

The experimental technique used in the reduction of carboxylic acids may be straightforward and generally involves the use of a beaker, which serves as the cathode chamber, and a porous cup suspended into the beaker, which serves as the anode compartment. Common cathodes used are lead and mercury. The solvents are generally aqueous sulfuric acid. Hydrochloric acid should be avoided due to the ease of oxidation of chloride anion to chlorine.

Reduction of Carboxylic Esters

The reduction of carboxylic ester is in general similar to that of carboxylic acids. However, carboxylic esters are more difficult to reduce than the corresponding acids. For example, the esters of acetic and phenyl acetic acids cannot be reduced under the same conditions required to reduce their acid analogs. On the other hand, the esters of oxalic and phthalic acids can be reduced to the corresponding aldehydes and ketones [322, 326, 327]. In certain cases the corresponding hydrocarbons are obtained [380]:

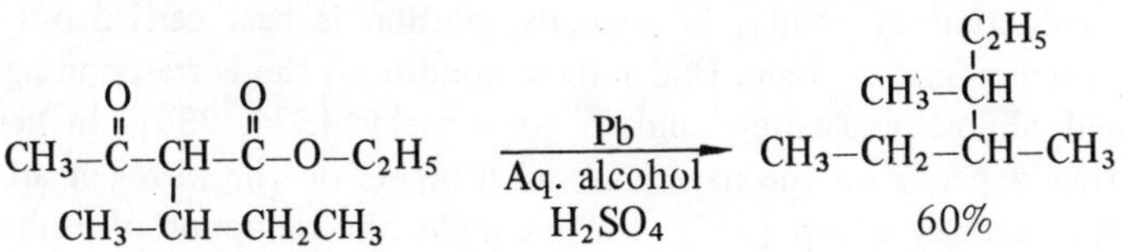

Aromatic carboxylic esters reduce, under strong acidic conditions, to ethers, which may undergo cleavage to the corresponding alcohols [381, 382].

In general, the electrochemical reduction of esters is carried out in aqueous acid media using a high hydrogen overvoltage cathode (e.g., mercury or lead).

Thus the ethyl esters of acetic, phenyl acetic acids are not reduced, while those of succinic, fumaric acetoacetic, and benzoic acids are readily reduced.

Products from keto-esters are derived from the electrochemical reduction of one or two carbonyl groups. For example, the reduction of acetoacetic ester on a lead cathode in aqueous acid at elevated temperatures afford the corresponding hydrocarbon [593]. On the other hand, the reduction of ethyl phenyl-glyoxylate at a mercury cathode in DMF gave the corresponding ester [596]:

$$CH_3-\overset{O}{\underset{\|}{C}}-CH_2-\overset{O}{\underset{\|}{C}}OC_2H_5 \quad \xrightarrow[\text{Aq. } H_2SO_4]{Hg} \quad CH_3(CH_2)_3CH_3$$

$$C_6H_5-\overset{O}{\underset{\|}{C}}-\overset{O}{\underset{\|}{C}}-OC_2H_5 \quad \xrightarrow[\substack{(C_2H_5)_4NI,\ 20^\circ \\ 0.002\ A/cm^2}]{Hg,\ DMF} \quad C_6H_5-\overset{OH}{\underset{\|}{C}}H-\overset{O}{\underset{\|}{C}}OC_2H_5$$

In some electrochemical reduction of esters, isomerization may take place as in the reduction of acetoacetic acid type esters [595]:

$$CH_3-\overset{O}{\underset{\|}{C}}-\overset{O}{\underset{\underset{C_4H_9}{|}}{C}}H\overset{O}{\underset{\|}{C}}-OC_2H_5 \quad \xrightarrow[C_2H_5OH]{Pb,\ H_2SO_4} \quad CH_3(CH_2)_6CH_3$$

Reduction of Amides

One of the most commonly used solvents in electrochemical reduction is N,N-dimethylformamide (DMF). However, several aliphatic and aromatic amides undergo reduction under acidic as well as neutral conditions. Reduction may take place via direct electron transfer to the carbonyl group or through hydrogen, which is generated from the reduction of hydrogen ions.

The reduction of amides in aqueous solution is best carried out in acidic media using a lead cathode. Under these conditions the corresponding amine is obtained almost exclusively and in good yields [334, 336]. In general, the reduction depends on the nature of substituents on the nitrogen atom rather than the carboxy group [357]. For example, the reduction of a-phenylaceta-mide yields the corresponding amine in less than 1% yield, whereas with N-methyl-a-phenyl-acetamide the yield of the corresponding amine is about 80%.

Recently, Benkeser and co-workers [333] developed a novel method for the reduction of amides to the corresponding alcohols by using methylamine as the solvent and lithium chloride as the supporting electrolyte. These authors proposed the following mechanism for the reduction:

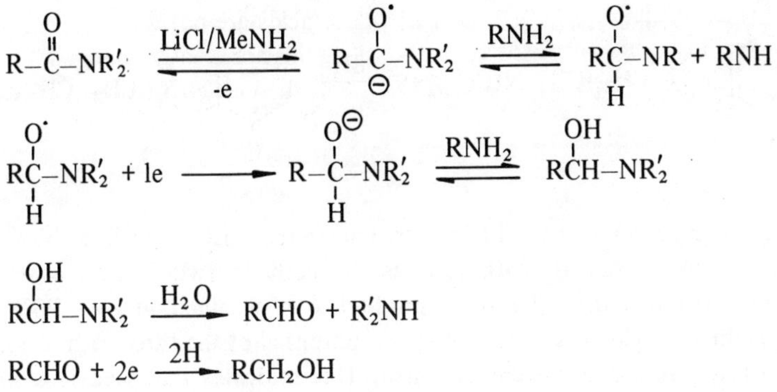

$$\underset{R-C-NR'_2}{\overset{O}{\parallel}} \xrightarrow[-e]{LiCl/MeNH_2} \underset{R-C-NR'_2}{\overset{O^{\cdot}}{\underset{\ominus}{\mid}}} \xrightarrow{RNH_2} \underset{RC-NR}{\overset{O^{\cdot}}{\underset{H}{\mid}}} + RNH$$

$$\underset{H}{\overset{O^{\cdot}}{\underset{\mid}{RC-NR'_2}}} + 1e \longrightarrow \underset{H}{\overset{O^{\ominus}}{\underset{\mid}{R-C-NR'_2}}} \xrightarrow{RNH_2} \underset{RCH-NR'_2}{\overset{OH}{\mid}}$$

$$\underset{RCH-NR'_2}{\overset{OH}{\mid}} \xrightarrow{H_2O} RCHO + R'_2NH$$

$$RCHO + 2e \xrightarrow{2H} RCH_2OH$$

The above mechanism was considered to be consistent with the observation that small amounts of aldehydes and methylimines were also isolated. The latter may be formed from the reaction of amines on the aldehyde intermediate.

Reduction of Imides

The reduction of imides is, in many respects, similar to that of amides. Unlike the reduction of amides, however, the intermediate can, in certain cases, be isolated [340]:

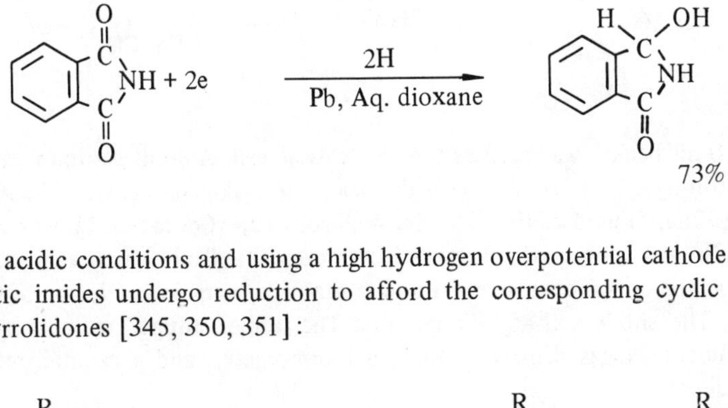

73%

Under acidic conditions and using a high hydrogen overpotential cathode, cyclic aliphatic imides undergo reduction to afford the corresponding cyclic amides and pyrrolidones [345, 350, 351]:

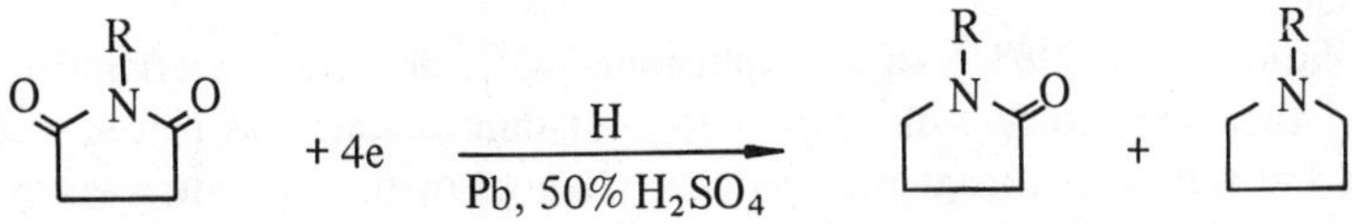

Allen [384] reported some interesting results concerning the effect of the electrode material on the reduction of N,N-dimethylaminoethyltetrachloro-phthalimide in aqueous sulfuric acid:

Using mercury, which has a high hydrogen overpotential, a yield of 80% of the isoindoline was obtained. With lead as the cathode (which has a lower overpotential than mercury) the same product was obtained in over 90% yield. Using cadmium (which has the lowest overpotential of the three cathodes), poor yields of the isoindoline were obtained. This indicates that lead is a specific cathode for this reaction and that no relationship exists here between the overpotential of the electrode and the yield of the product. Allen and coworkers also found that the reduction of the two carbonyl groups takes place in a step-wise manner as the corresponding phthalimidine could be isolated if the electrolysis was interrupted prior to completion.

General Consideration: Reduction of Acetophenone [298]

A 250-ml beaker was used as the electrolysis cell. A 1-cm platinum wire was sealed through the bottom side of the beaker to make contact with the mercury pool (150 cm^2) used as the cathode. A porous cup (Coors No. 1) was inserted into a No. 13 rubber stopper, which was then fitted into the beaker. The cup served as the andoe compartment and contained an aqueous solution of lithium nitrate. The anode was a platinum wire. The rubber stopper was also bored to accommodate a gas dispersion tube, a thermometer, and a calomel reference electrode.

A solution of 1.0088 g of acetophenone-7-C^{14}, dissolved in 60 ml of 80% ethanol that was 2.0 M with respect to potassium acetate was placed into the beaker and was stirred magnetically. The potential of the cathode was adjusted to read 1.6 V (SCE). Electrolysis proceeded at 0.2 A for 40 hr. This afforded 15.74% meso and 44.81% racemic pinacols.

Table 8.1 Carbonyl Compounds

	Cathode	Reaction Medium	Products	Yield	Ref.
I. ALDEHYDES					
A. Aliphatic					
HCHO	Hg	Aq. base	CH_3OH	—	243, 244
CH_3CHO		Aq. base	CH_3CH_2OH		245
CH_3CH_2CHO	Hg	Aq. base	$CH_3CH_2CH_2OH$		246, 247
CH_3CH_2CHO	Ni, Cd, Cu, Hg etc.		$CH_3CH_2CH_2OH$, $CH_3CH_2-CH-CH-CH_2CH_3$ with $OH\ OH$		470
$CH_3(CH_2)_2CHO$	Hg	Aq. C_2H_5OH	$CH_3(CH_2)_2CH_2OH$	—	239
$CH_3-CH=CH-CHO$	Pb	25% H_2SO_4	Butanol, and ─ OH	50	547
$CH_3-CH-CH_2CHO$ with OH	Hg	Aq. NaOH	$CH_3CH-CH_2CH_2OH$ with OH	—	240
$CH_2=CH-CHO$	Cu, Hg, Pb, Zn		$CH_2=CH-CH_2OH$	50	242
$CH_3CH=CH-CHO$ HC═CH ╲ CH─C─CHO ╲O╱			(cyclic structure)	63	249, 250
Cinnamaldehyde	Hg	Aq. buffer	$H_5C_6-CH_2-CH_2-CH_2OH$		241
OHC–CHO	Hg		$HO-CH-CH_2OH$ / $HO-CH-CH_2OH$	—	475 248

101

102

Table 8.1 Carbonyl Compounds (continued)

Fructose	Hg, Pb	—	Sorbitol, mannitol	62.5	175
Glucose	Ph, Hg, Ni-Al Pb-Hg, Zn-Hg	—	Formaldehyde	50.9	251
					251, 252
			Pentose, mannitol sorbitol		253, 261
			Polyhydric alcohols	—	262
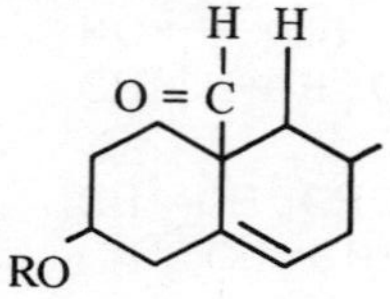 Sugars (decalin structure, O=C—H, RO)	Pb, Hg Zn	Dioxane, THF H₂SO₄	(decalin structure, CH₃, RO)	50	449
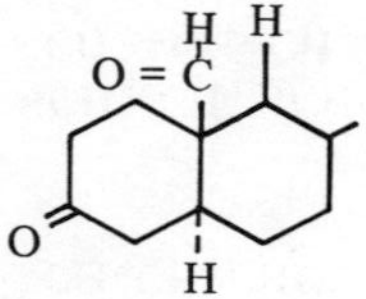 (decalin structure, O=C—H, O)	Pb, Hg Zn	Dioxane, THF H₂SO₄	(decalin structure, CH₃, H)	50	499
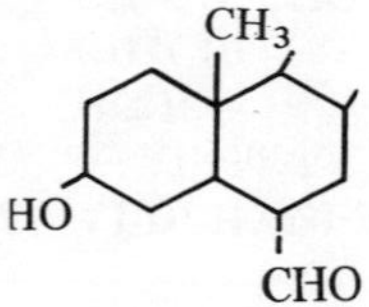 (decalin structure, CH₃, CHO, HO)	Pb, Hg	Dioxane, THF H₂SO₄	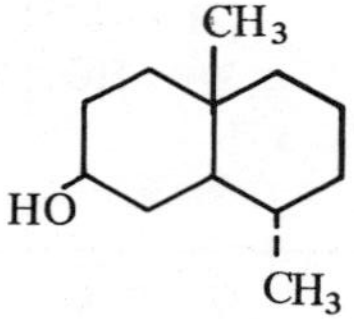 (decalin structure, CH₃, HO)	~50	449

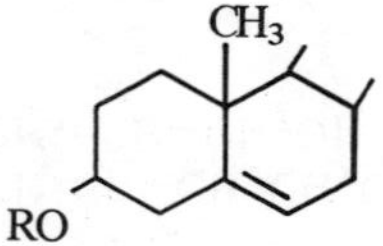

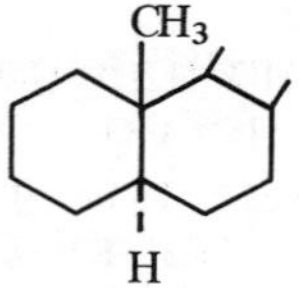

B. Aromatic

Electrode	Conditions	Starting material	Product	Yield (%)	Ref.
Hg	Aq. acid		d,l-Pinacol	36.3	264
			meso-Pinacol	38.3	264
Hg	Aq. base		d,l-Pinacol	45.6	264
			meso-Pinacol	38.3	264
Hg	C_2H_5OH, pH = 1.3		Hydrobenzoin	—	265
	pH = 8.6		Hydrobenzoin	—	265
			Benzylalcohol	—	265
Ni, Pt	Aq. Alcohol KOH, 0.5 A/dm²	3-methylbenzaldehyde (CHO, CH₃ meta)	3-methylbenzyl alcohol (CH₂OH, CH₃)	70–75	266
Ni, Pt		2,4-dimethylbenzaldehyde (CHO, CH₃, CH₃)	2,4-dimethylbenzyl alcohol (CH₂OH, CH₃, CH₃)	70–75	266
Ni, Pt		3,5-dimethylbenzaldehyde (CHO, CH₃, CH₃)	3,5-dimethylbenzyl alcohol (CH₂OH, CH₃, CH₃)		
Ni, Pt		4-isopropylbenzaldehyde (CHO, CH(CH₃)₂)	4-isopropylbenzyl alcohol (CH₂OH, CH(CH₃)₂)	70–75	266

Table 8.1 Carbonyl Compounds (continued)

Starting material	Electrode	Conditions	Product	Yield (%)	Ref.
CHO–C₆H₄–OH (4-hydroxybenzaldehyde)	Hg	2 N NaOH	(4-HO–C₆H₄)₂CH–CO–OH structure	95	267
CHO–C₆H₄–OH (3-hydroxybenzaldehyde)	—	10% Aq. NaOH 0.025 A/dm²	(3-HO–C₆H₄)₂CH–CO–OH structure	20	268
CHO–C₆H₄–N(CH₃)₂	Hg	Aq. NaOH –1.9 V (SCE)	(H₃C)₂N–C₆H₄–CH(OH)–CH(OH)–C₆H₄–N(CH₃)₂	97	269
CHO, OH, COCH₃ substituted benzaldehyde	Pb	Aq. NaOH	pinacol product with OH, COCH₃ groups	75	270

104

II. KETONES
A. Aliphatic

Substrate	Electrode	Conditions	Product		
CO_2	Pb	Aq. sol., R_4NX	Glycollic acid		456
CO_2	Hg	Aq. sol., R_4NX	Formate	30	456
(a) $CH_3\overset{O}{\overset{\|}{C}}-CH_3$	Hg, Cu, Pb		Propane, propanol, pinacol	—	271
(b) $CH_3\overset{O}{\overset{\|}{C}}-CH_3$, styrene	Hg	$Et_4N\ pTS$ $[(C_2H_5)_4 \overset{\oplus}{}\ {}^{\ominus}SO_3\text{-}C_6H_4\text{-}CH_3]$	$\underset{OH}{\overset{CH_3}{CH_3-\underset{\|}{\overset{\|}{C}}-CH_2-CH_2}}$	42	472
trans-Stilbene					
trans-Stilbene	Cu, Ni	$Et_4N\ pTS$	$H_3C-\underset{OH}{\overset{\|}{C}}-\underset{CH_3}{\overset{\|}{CH}}-\underset{C_6H_5}{\overset{\|}{CH}}-CH_2-C_6H_3$	45	472
(c) $CH_3-\overset{O}{\overset{\|}{C}}-CH_3$, $CH_2=CH-CN$	Hg	Aq. H_2SO_4	$CH_3\underset{OH}{\overset{CH_3}{\overset{\|}{C}}}-CH_2-CH_2-CN$	—	474
$R-\overset{O}{\overset{\|}{C}}-R$	Cu, Ni	Aq. H_2SO_4	Pinacol	—	552
R=alkyl, Aryl			Pinacol	—	—

Table 8.1 Carbonyl Compounds (continued)

$CH_3-\overset{O}{\overset{\|}{C}}-CH_2-CH_3$	Pb	Aq. Acid	2-Butanol Butane	12.2 15.5	272
$CH_3-\overset{O}{\overset{\|}{C}}-CH_2-\overset{CH_3}{\underset{CH_3}{CHN}}$	Pb Pb	Aq. H_2SO_4, 30% Aq. H_2SO_4, 30%	$\overset{OH}{\underset{\|}{CH_3}}-CH-CH_2CHN(CH_3)_2$	62.3	288
$CH_3-\overset{O}{\overset{\|}{C}}-(CH_2)_2-CH_3$	Cd-Bi	—	n-Pentane	71.2	273, 274
$CH_3-\overset{O}{\overset{\|}{C}}-CH=CH_2$	Hg	NaOAc/HOAC EtOH	$CH_2=C-\overset{O}{\overset{\|}{C}}-(CH_2)_4-\overset{O}{\overset{\|}{C}}-CH_3$	65	461
			$CH_2=CHC-\overset{OH}{\underset{CH_3}{\|}}-\overset{OH}{\underset{CH_3}{\|}}-CH=CH_2$ (Mixture of meso and dl)	13	461
$CH_3-\overset{O}{\overset{\|}{C}}-CH_2-\overset{OH}{\underset{CH_3}{\overset{\|}{C}}}-CH_3$	Hg	0.1 N NH_3 buffer pH > 7	$CH_3CH_2-CH_2C(CH_3)_2-OH$	47.8	275
Diacetyl	Hg	—	Acetoin		450
cyclohexanone structure	Cu, Hg,	—	2,6-Dimethylcyclohexanol (*trans*)	—	289

106

Substrate	Electrode	Solvent	Product	%	Ref.
			2,6-Dimethylcyclohexanol (*cis*)	—	289
			2,6-Dimethylcyclohexanol (*mixture*)	—	289
cyclohexanone =O, Me₃SI ($=O$, Me$_3$SI)	C	DMSO	methylenecyclohexane ($=CH_2$)	10	549
2-acetylfuran ($O=C-CH_3$, furan)	Hg	Aq. acid	furan pinacol (CH₃ CH₃, $-C-C-$, OH OH)	—	468
2-acetylthiophene ($O=C-CH_3$, thiophene)	Sn	DMF	thiophene pinacol (CH₃ CH₃, $-C-C-$, OH OH)	—	471
4-*t*-Butylcyclohexanone (*t*-Bu)	Pb, −2.4 V (SCE)	MeOH	*trans*-4-*t*-butylcyclohexanol (OH, *t*-Bu)	82	598
	Pb, −1.8 V (SCE)	MeOH, HoAc	*cis*-4-*t*-butylcyclohexanol (OH, *t*-Bu)	87	598
1-Menthane ketone (CH₃ CH₃, O, H CH₃)	Cd	Aq. C₂H₅OH H₂SO₄	1-Menthane (CH₃ CH₃, CH₃)	73	276

Table 8.1 Carbonyl Compounds (continued)

Starting material	Metal	Conditions	Product	Yield	Ref.
CH_3—C=O (decalin, H_3C)	Pb, Hg	Dioxane, THF H_2SO_4	H_3C C_2H_5 (decalin)	~50	449
R—C=O / CHOH (CH_3)	Pb, Hg	Dioxane, THF H_2SO_4	R—CH_2 / CHOH (CH_3)	50	449
R—H_2C=O (decalin)	Pg, Hg	Dioxane, THF H_2SO_4	R—H_2C (decalin)	50	449
(CH_3, CH_3) cyclohexanone H_3C H, *dl*-Isomenthane	Cd	Aq. C_2H_5OH H_2SO_4	(CH_3, CH_3, CH_3) Isomenthane	73	276
O CH_3 / CH_3 cyclohexanone	Hg	CH_3CN, Ac_2O	(AcO)HO OH(OAc) (CH_3, CH_3)	33	281

Substrate	Electrode	Electrolyte	Product	Yield	Ref
(structure)	Pb	30% H_2SO_4	(structure, OH)	1	277
				49	277
(structure, HO–OH)	Hg	Aq. HCl	(structure, OH–OH)	—	282
Croconic acid	Hg	0.5N HCl −0.35 V	(structure) ⇌ (structure)	—	459
(structure)	Platinized Pt	$(NH_4)_2SO_4$ H_2	(phenol, OH; hydroquinone, OH–OH; cyclohexanol, OH)	—	284

Table 8.1 Carbonyl Compounds (continued)

Starting compound	Metal	Electrolyte	Product	Yield (%)	Ref.
(structure)	Pb	30% H₂SO₄	(structure)	73	277
(structure)	Pb	30% H₂SO₄	(structure)	—	277
(structure)	Pb	30% H₂SO₄	(structure)	71	278
(structure)	Pb	30% H₂SO₄	(structure)	42	278
(structure)	Pb	30% H₂SO₄	(structure)	51	279, 297

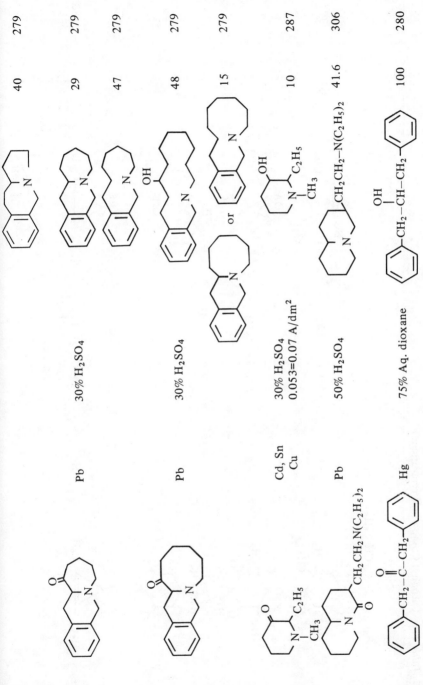

			Yield	Ref.
Pb	30% H_2SO_4		40	279
			29	279
			47	279
Pb	30% H_2SO_4		48	279
		or	15	279
Cd, Sn Cu	30% H_2SO_4 0.053=0.07 A/dm²		10	287
Pb	50% H_2SO_4		41.6	306
Hg	75% Aq. dioxane		100	280

Table 8.1 Carbonyl Compounds (continued)

112

Substrate	Metal	Conditions	Product	Yield	Ref.
$(H_5C_6)_2CH-\underset{\underset{\displaystyle O}{\|\|}}{C}-CH(C_6H_5)_2$	Hg	75% Aq. dioxane	$(H_5C_6)_2CH-\underset{\underset{\displaystyle OH}{\|}}{CH}-CH(C_6H_5)_2$	—	280
	Li	THF			387
	Hg	CH_3CH, Ac_2O		31	281
	Hg	Alcohol		—	290
	Hg	Aq. C_2H_5OH (polarography) Aq. acid		—	283

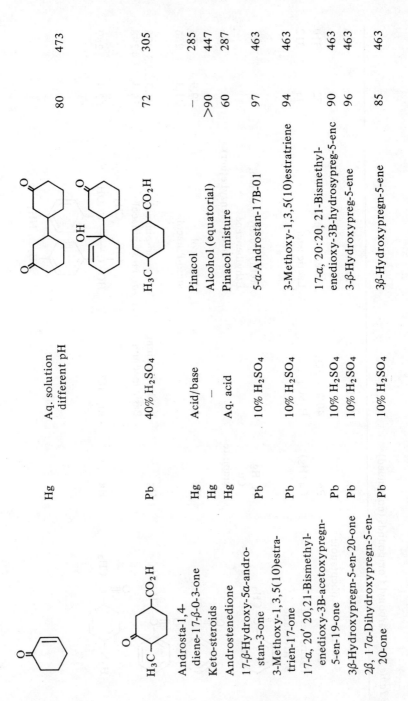

Compound	Metal	Aq. solution different pH	Product	Yield	
(cyclohex-2-enone)	Hg			80	473
H_3C—...—CO_2H	Pb	40% H_2SO_4	CO_2H	72	305
Androsta-1,4-diene-17-β-0-3-one	Hg	Acid/base	Pinacol	–	285
Keto-steroids	Hg	–	Alcohol (equatorial)	>90	447
Androstenedione	Hg	Aq. acid	Pinacol misture	60	287
17-β-Hydroxy-5α-andro-stan-3-one	Pb	10% H_2SO_4	5-α-Androstan-17B-01	97	463
3-Methoxy-1,3,5(10)estra-trien-17-one	Pb	10% H_2SO_4	3-Methoxy-1,3,5(10)estratriene	94	463
17-α, 20' 20,21-Bismethyl-enedioxy-3B-acetoxypregn-5-en-19-one	Pb	10% H_2SO_4	17-α, 20:20, 21-Bismethyl-enedioxy-3B-hydrosypreg-5-enc	90	463
3β-Hydroxypregn-5-en-20-one	Pb	10% H_2SO_4	3-β-Hydroxypreg-5-ene	96	463
2β, 17α-Dihydroxypregn-5-en-20-one	Pb	10% H_2SO_4	3β-Hydroxypregn-5-ene	85	463

Table 8.1 Carbonyl Compounds (continued)

		Acid/base			
Progesterone	Hg		Pinacol	—	285
5-Methyl-4-oxoheptanoic acid	Pb	40% H_2SO_4	5-Methylheptanoic acid	72	310
Pyruvic acid	Cu (Hg)	H_2SO_4	Lactic acid	45	312
B. Aromatic					
	Pt, Hg, Cu graphite		Methylphenyl carbinol Ethylbenzene Bis(α-methylbenzyl ether)	—	266
	Hg	Aq. acid	dl-Pinacol meso-Pinacol	39.9 31.3	286 286
	Hg	Acidic medium Basic medium	dl, meso-pinacol dl, meso-pinacol	31.3, 22.3 36.4, 13.4	298 298
	Hg	Aq. HCl		62.6	289

114

Reactant	Metal	Conditions	Product	Yield	Ref.
H_2N–C$_6$H$_4$–C(=O)–CH$_2$CH$_3$	Hg	Aq. HCl	HO–C(CH$_2$CH$_3$)(C$_6$H$_4$NH$_2$)–C(CH$_2$CH$_3$)(OH)(C$_6$H$_4$NH$_2$)	57	291
HO–C(=O)–C$_6$H$_4$–C(=O)–CH$_3$	Hg	Aq. KOH	2,3-Bis(p-Carboxyphenyl)-2,3-butanediol	96	292
R–C(=O)–C$_6$H$_4$–C(=O)–R	—	—	R–CH(OH)–C$_6$H$_4$–C(=O)–R	—	550
(H$_3$C)$_2$N–C$_6$H$_4$–C(=O)–CH$_3$	Hg	Aq. EtOH	Mixed pinacols	34	293
H$_3$C–O–C$_6$H$_4$–C(=O)–CH$_3$	Hg	KOAC	Pinacol		304
HO–C$_6$H$_4$–C(=O)–CH$_3$	Hg	2 N HaOH	HO–C(CH$_3$)(C$_6$H$_4$OH)–C(CH$_3$)(OH)(C$_6$H$_4$OH)	77	294

Table 8.1 Carbonyl Compounds (continued)

Carbonyl compound	Metal	Conditions	Product	Yield (%)	Ref.
$H_3CO-C_6H_4-C(=O)-CH_2CH_3$	Hg	80% C_2H_5 OH	$HO-C(CH_3)(C_6H_4OCH_3)-C(CH_3)(C_6H_4OCH_3)-OH$	—	295
	Hg		*dl, meso*-Pinacol *dl/meso* = 1.2	94	297
$Mes-C(=O)-C_6H_4(OCH_3)$	Mg^+	Generated at Mg anode, NaI, Pyridine	Mes–(C=O)–cyclohexyl–cyclohexyl–(O=C)–Mes diketone	38	448
$H_5C_6-C(=O)-C(=O)-(CH_2)_n-C_6H_5$ $n = 1, 4, 5, 6$	Mg^+	Generated at Mg anode, NaI, pyridene	$H_5C_6-CH(OH)-(CH_2)_n-C(OH)-C_6H_5$	—	453
$H_5C_6-C(=O)-C(=O)-(CH_2)_3-C_6H_5$	Mg^+	Generated at Mg anode, NaI, pyridene	*cis*-1,2-Diphenylcylopentane-1,2-diol	87	453
$H_5C_6-C(=O)-C(=O)-(CH_2)_3-C_6H_5$	Mg^+	Generated at Mg anode, NaI, pyridene	*cis*-1,2-Dipyenylcyclohexane-1,2-diol	—	453

Substrate	Metal	Conditions	Products	Yield (%)	Ref.
4-Cl-C6H4-C(=O)-CH3	Hg	80% C_2H_5OH	dl, $meso$-pinacol $dl/meso = 3.1$	88	297
4-Cl-C6H4-C(=O)-CH3	Hg	80% C_2H_5OH 2M KOAC	dl, $meso$-pinacol $dl/meso = 3.1$	95	297
2-Cl-C6H4-C(=O)-CH3	Hg	80% C_2H_5OH 2M KOAC	dl, $meso$-pinacol $dl/meso = 2.1$	—	297
4-(F_3C...)-C6H4-C(=O)-CH3	Hg	80% C_2H_5OH	dl, $meso$-pinacol $dl/meso = 1.0$	87	297
$CH_3-C(=O)-CH=CH-C_6H_5$	Hg	NaOH	$H_5C_6CH_2CH_2-CH_2-\overset{O}{\overset{\|}{C}}CH_3$	30-35	548
			$H_5C_6-\underset{\overset{\|}{H_5C_6-CH-CH_2-C=O}}{CH-CH_2-\overset{O}{\overset{\|}{C}}CH_3}$	18-20	548
			$H_5C_6-CH=CH-\overset{OH}{\overset{\|}{CH}}-CH_3$ $H_5C_6-CH=CH-\overset{\|}{\underset{OH}{CH}}-CH_3$	7-10	548

117

Table 8.1 Carbonyl Compounds (continued)

Compound		Conditions	Product	%	Ref.
p-Acetamido-ω-dimethyl-aminopropiophenone	Hg	50% HOAC	Pinacol	37.5	299
R = —OH, -OCH₃	Hg	Aq. C_2H_5OH pH = 1.3 pH = 4.9	Benzpinacol Benzpinacol Benzhydrol	100 25 67	265 265 265
Deoxybenzoin	Hg	Different pH	*dl, meso*-Pinacol	–	302
Acetylbenzoin	Hg		Deoxybenzoin pinacol	30	301
Benzoin methyl ether	Hg		Deoxbenzoin pinacol	100	301
R = OH R = OCH₃	Hg	Aq. C_2H_5OH	*trans*-Glycol	10	303
				50	203

Substrate	Electrode	Conditions	Product		
$H_5C_6-\overset{O}{\overset{\|}{C}}-CH_2-\overset{O}{\overset{\|}{C}}-C_5H_5$	Hg	95% Ethanol, pH 4.2 (a) 1.15 (SCE)		53	303
		(b) 1.3 V (SCE)	Enolate anion;		476, 492
					492
	Hg	0.1 M (CH₃)₄NOH, Ethanol −2.0 V (SCE)	$C_6H_5\underset{OH}{\overset{\|}{CH}}CH_2\underset{OH}{\overset{\|}{CH}}C_6H_5$		492

Table 8.1 Carbonyl Compounds (continued)

Compound		Metal	Conditions	Product	Yield	Ref
		Hg	Aq. C$_2$H$_5$OH R = OH	trans-Glycol cis-Glycol Diol	22 51 17	303 303 303
		Hg	Aq. C$_2$H$_5$OH	trans-Glycol cis-Glycol Diol	36 54 12	303 303 303
Desylamine		Hg	Basic solution	Desylamine Pinacol	—	301
Desylchloride		Hg	Acidic solution	Deoxybenzoin	50	301
Benzalacetone		Hg	C$_2$H$_5$OH, pH = 1.3	H$_3$CC—H$_2$C—CH—CH—CH$_2$—CCH$_3$	—	265
Benzalacetophenone		Hg	C$_2$H$_5$OH, pH = 1.3	H$_5$C$_6$—C—H$_2$C—CH—CH—CH$_2$—C—C$_6$H$_5$	—	265

120

Substrate	Metal	Conditions	Products	Yield (%)	Reference
R–C(=O)–(CH₂)ₙ–C(=O)–C₆H₄–R–Hg	R–Hg	n = 3, R = OH	*trans*-Glycol	47	303
			cis-Glycol	42	
		R = OCH₃		3	
			trans-Glycol	79	303
			cis-Glycol	25	
		n = 6, R = OH	Pinacolone	64	303
		R = OCH₃	*cis*-Glycol	90	303
		n = 7, R = OH	Unidentified Products	84	
		R = OCH₃	Unidentified Products		
(H₅C₆)₂C(O)–O–C(C₆H₅)₂	Hg	75% dioxane	(H₅C₆)₂C(OH)–O–C(C₆H₅)₂	—	307
			C₆H₅–CH(–CH)–C₆H₅ ; C₆H₅–CH(OH)–CH(OH)–C₆H₅	—	307
p-Acetylpyridine	Hg	Conc. HCl	2,3-bis(B-Pyridil)-2,3-butane diol	73.4	294
C₆H₅–C(=O)–CO₂H	Cu, Hg	NaOH	*meso*-Pinacol	60	312
		H₂SO₄	*dl*-Pinacol	20	
			C₆H₅–CH(OH)–CH–CO₂H	7	

Table 8.1 Carbonyl Compounds (continued)

Androstene dione	Hg	Aq. acid	Pinacol mixture	60	353
Estrone	Pb, C	Aq. C_2H_5OH	Estradiol	—	354
	Hg	Aq. acid	HO H	—	355
	Hg	Aq. acid	OH OH	—	356
	Hg	Aq. sol. pH ⩾ 12	OH	—	469

II. CARBOXYLIC ACIDS
A. Aliphatic

Formic acid	Pb	Aq. acid, low current density (CD)	Formaldehyde	—	378
	Pb	Aq. acid, high CD	Methanol	—	278
Butyric acid	—	80% H_2SO_4 0.2 A/d^2	Butyl alcohol	6.5	309

Starting material	Electrode	Electrolyte	Product	Yield (%)	Ref.
$C_6H_5CH_2CO_2H$ (phenylacetic acid)	Pb, Hg	50% H_2SO_4	Phenethyl alcohol	35.3	308
α-Hydroxypropionic acid	C	Aq. sol.	Lactaldehyde	40	311
Oxalic acid	Pb, C, Hg, Ni	Aq. H_2SO_4	Tartaric acid / Glyoxilic acid / Formaldehyde / Glycolic acid		313, 314 / 315
Maleic acid	Hg, Bi; Hg-Bi	Aq. Sol. pH = 0.3–4.0	Succinic acid	87–100	466
B. Aromatic					
$C_6H_5CO_2H$ (benzoic acid)	Pb, Hg	Aq. H_2SO_4	$C_6H_5CH_2OH$	—	316
			C_6H_5CHO	—	316
	Pb	10% H_2SO_4, 30–40 A/d^2	$C_6H_5CH_2OH$	60–92	451
2,6-diethylbenzoic acid (C_2H_5/C_2H_5/CO_2H)	Pb	Aq. H_2SO_4	ortho-ethyl CH_2OH benzene (C_2H_5, CH_2OH)	70	317
3-ethylbenzoic acid (C_2H_5/CO_2H)	Pb	Aq. H_2SO_4	meta-ethyl CH_2OH benzene (C_2H_5, CH_2OH)	67	317

Table 8.1 Carbonyl Compounds (continued)

Starting material	Metal	Conditions	Product	Yield	Ref.
2-OCH$_3$-C$_6$H$_4$-CO$_2$H	Pb	Aq. alcoholic H$_2$SO$_4$	2-OCH$_3$-C$_6$H$_4$-CH$_2$OH	70	318
2-NH$_2$-C$_6$H$_4$-CO$_2$H	Pb	50% H$_2$SO$_4$	2-NH$_2$-C$_6$H$_4$-CH$_2$OH	78	325
3-Br-C$_6$H$_4$-CO$_2$H	Pb	Aq. or alcoholic H$_2$SO$_4$	3-Br-C$_6$H$_4$-CH$_2$OH	78.5	319
2-NH$_2$-3-CH$_3$-C$_6$H$_3$-CO$_2$H	Pb	Aq. or alcoholic H$_2$SO$_4$	2-NH$_2$-3-CH$_3$-C$_6$H$_3$-CH$_2$OH	45	320, 318
4-NH$_2$-2-CH$_3$-C$_6$H$_3$-CO$_2$H	Pb	Aq. or alcoholic H$_2$SO$_4$	4-NH$_2$-2-CH$_3$-C$_6$H$_3$-CH$_2$OH	76.5	320
2-NH$_2$-4-CH$_3$-C$_6$H$_3$-CO$_2$H	Pb	Aq. or alcoholic H$_2$SO$_4$	2-NH$_2$-4-CH$_3$-C$_6$H$_3$-CH$_2$OH	65	320

Starting material	Metal	Medium	Product	Yield (%)	Ref.
H₃C—C₆H₂(NH₂)(CH₃)—CO₂H	Pb	Aq. or alcoholic H₂SO₄	H₃C—C₆H₂(NH₂)(CH₃)—CH₂OH	73	320
o-HO—C₆H₄—CO₂H	Cu, Hg	Aq. Na₂SO₄	o-HO—C₆H₄—CHO	13–46	321
(2-CO₂H—C₆H₄)—O—COCH₃	Cu, Hg	Aq. Na₂SO₄	(2-CH₂OH—C₆H₄)—OCOCH₃	30	321
o-C₆H₄(CO₂H)₂	Cd, Sn	Aq. H₂SO₄	phthalide	—	347
o-C₆H₄(CO₂H)₂	Hg	20% Dioxane 50% C₂H₅OH	2,5-Dihydrophthalic acid	—	322
m-C₆H₄(CO₂H)₂				—	323
m-C₆H₄(CO₂H)₂	Pb	Aq. or alcoholic H₂SO₄	m-C₆H₄(CH₂OH)₂	82	318

125

Table 8.1 Carbonyl Compounds (continued)

Compound	Structure (product)	Catalyst	Solvent	Yield %	Ref.
pyridine-4-carboxylic acid (CO_2H)	4-CH_3-pyridine	Hg	95% C_2H_5OH	31	324
pyridine-2-carboxylic acid (CO_2H)	2-CH_3-pyridine	Hg	95% C_2H_5OH	33	324
IV. ESTERS					
$O=C(O-CH-CH_2)$, $CH_3-CH-CH_2$	$CH_3CH=CH_2$	C	—	>90	465
phenyl-$CO_2C_2H_5$	phenyl-CH_2OH	Pb	Aq. H_2SO_4	—	387
Diethyl oxalate	Ethylglyoxalate	Hg, Pb		53	326
Methyl phthalate	Phthalide	Hg	—	322
Methyl isophthalate	m-Carboxybenzyl	Hg	—	11
Diethyl phthalate	(phthalide)	Hg	75% C_2H_5OH	51.7	327
Diethyl phthalate	CO_2H, CO_2H	Hg	—	327

126

Starting material	Electrode	Electrolyte	Product	Yield (%)	Ref.
$\overset{O}{\underset{}{C}}\!-CH_2-COC_2H_5$ (phenyl)	Cu (Hg)	Aq. NaOH	$\overset{OH}{CH}-CH_2CO_2H$ (phenyl)	44	312
D-Ribono-γ-lactone	Hg	30% H_2SO_4	*D*-Ribose	57-75	328
D-Mannono-γ-lactone	Hg	Aq. Na_2SO_4	*D*-Mannose	60	329
α-*D*-Glucoheptonelactone	Hg	Aq. Na_2SO_4	α-*D*-Glucoheptose	60	329
Lactones of monosacarides	Hg	Alkalic and alkalic acid sulfate	Monosacarides		330
Polyhydroxycarboxylic acid lactones	Hg	Aq. Na_2SO_4	Sugars	—	331

V. AMIDES
A. Aliphatic

Starting material	Electrode	Electrolyte	Product	Yield (%)	Ref.
$CH_3-CON\overset{H}{\underset{}{}}$ (phenyl)	Pb, Cd	30% H_2SO_4	(phenyl)$-NH-CH_2CH_3$	39	332
$CH_3(CH_2)_3CON(CH_3)_2$	Ni, Pb, Hg	30% H_2SO_4	$CH_3(CH_5)_4-N(CH_3)_2$	60	332
$CH_3(CH_2)_4CONH_2$	Pt	CH_3NH_2/LiCl	$CH_3(CH_2)_5OH$	65	333
$CH_3(CH_2)_6CONH_2$	Pt	CH_3NH_2/LiCl	$CH_3(CH_2)_6CHO$	22	455
$CH_3(CH_2)_8CONH_2$	Pt	CH_3NH_2/LiCl	$CH_3(CH_2)_9OH$	59	455
$CH_3(CH_2)_{10}CONH_2$	Pt	CH_3NH_2/LiCl	$CH_3(CH_2)_{11}OH$	92	455
$CH_3(CH_2)_{14}CONH_2$	Pt	CH_3N/LiCl	$CH_3(CH_2)_{15}OH$	86	455
$CH_3(CH_2)_{16}CONH_2$	Pt	CH_3N/LiCl	$CH_3(CH_2)_{17}OH$	79	455
$CH_3(CH_2)_{10}-CONHCH_3$	Pt	CH_3NH_2/LiCl	$CH_3(CH_2)_{11}OH$	84	333
$CH_3(CH_2)_{12}-CON(CH_3)_2$	Pt	CH_3NH_2/LiCl	$CH_3(CH_2)_{13}OH$	75	333

Table 8.1 Carbonyl Compounds (continued)

C$_6$H$_5$—CH$_2$—CONHCH$_3$	Hg	10% HCl	C$_6$H$_5$—CH$_2$—CH$_2$NHCH$_3$	93	334
C$_6$H$_5$—CH$_2$—CON(CH$_3$)$_2$	Au, Hg	H$_2$SO$_4$	C$_6$H$_5$—CH$_2$—CH$_2$N(CH$_3$)$_2$	100	334
2,5-Diketopiperazine	Hg	10% HCl	Piperazine	92	334
Diketopiperazine with CH$_2$C$_6$H$_5$ substituents	Pb	50% H$_2$SO$_4$	Piperazine with CH$_2$C$_6$H$_5$ substituents	70	335
Cyclic diketo compound (R substituents)	Hg	Acidic solution	2R—CH(CHO)(NH$_2$)		383
N-(2-Diethylaminoethyl)succinimide	Pb-Hg	50% H$_2$SO$_4$	N-(2-Diethylaminoethyl)pyrrolidone		306
(CH$_2$)$_2$(CONH$_2$)$_2$	Zn, Hg	50% H$_2$SO$_4$	Pyrrolidine / Pyrrolidone	14 / 58	352
(CH$_2$)$_4$(CON(CH$_3$)$_2$)$_2$	Pb	50% H$_2$SO$_4$	(H$_3$C)$_2$N(CH$_2$)$_6$N(CH$_3$)$_2$	10	336

Substrate	Metal	Acid	Product	Yield (%)	Ref.
$CH_2-CON(CH_3)_2$ | $CH_2-CON(CH_3)_2$	Pb	50% H_2SO_4	$(H_3C)_2N(CH_2)_3-CO_2H$ (hydrolysis of amide)	40	336
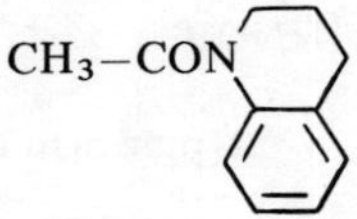 CH_3-CON- (1-acyl-1,2,3,4-tetrahydroquinoline)	Pb	Aq. H_2SO_4	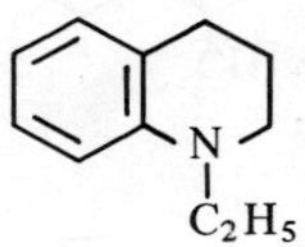 $N-C_2H_5$ (1-ethyl-1,2,3,4-tetrahydroquinoline)	76	337
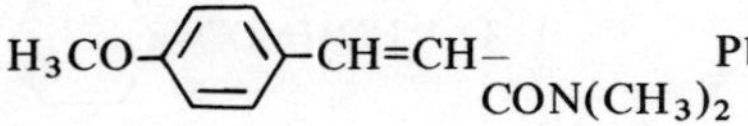 $H_3CO-C_6H_4-CH=CH-CON(CH_3)_2$	Pb	30% H_2SO_4	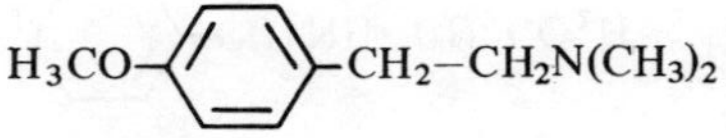 $H_3CO-C_6H_4-CH_2-CH_2N(CH_3)_2$	84.1	338

B. Aromatic

Substrate	Metal	Acid	Product	Yield (%)	Ref.
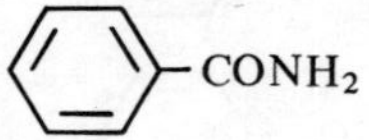 $C_6H_5-CONH_2$	Hg	10% HCl	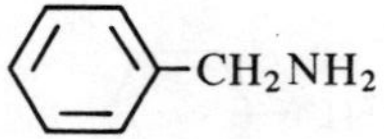 $C_6H_5-CH_2NH_2$	98	334
	Cd, Hg		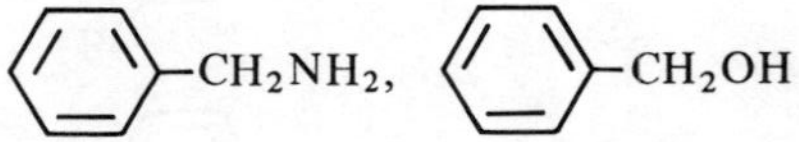 $C_6H_5-CH_2NH_2$, $C_6H_5-CH_2OH$		334
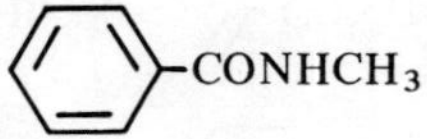 $C_6H_5-CONHCH_3$	Hg	10% HCl	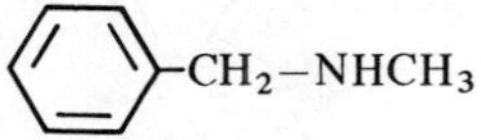 $C_6H_5-CH_2-NHCH_3$	95	334
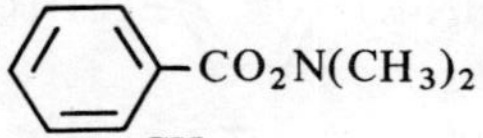 $o\text{-}CH_3\text{-}C_6H_4-CO_2N(CH_3)_2$	Hg g	10% HCl	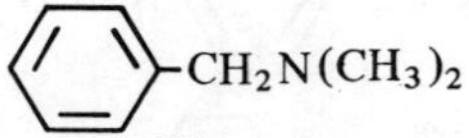 $C_6H_5-CH_2N(CH_3)_2$	100	334
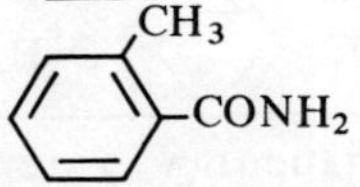 $o\text{-}CH_3\text{-}C_6H_4-CONH_2$	Pb	20% H_2SO_4	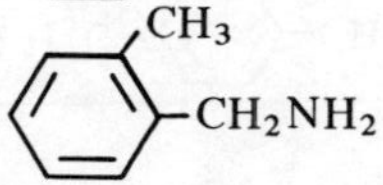 $o\text{-}C_6H_4-CH_2NH_2$	83	338

129

Table 8.1 Carbonyl Compounds (continued)

H_3C–C$_6$H$_4$–CONH$_2$	Pb	20% H_2SO_4	H_3C–C$_6$H$_4$–CH$_2$NH$_2$	79	338
H_3CO–C$_6$H$_4$–CONH$_2$	Pb	20% H_2SO_4	H_3CO–C$_6$H$_4$–CH$_2$NH$_2$	73	338
Br–C$_6$H$_4$–CONH$_2$	Pb	20% H_2SO_4	Br–C$_6$H$_4$–CH$_2$NH$_2$	67	338
Br(m)–C$_6$H$_4$–CONH$_2$	Pb	20% H_2SO_4	Br(m)–C$_6$H$_4$–CH$_2$NH$_2$	63	338
Cl–C$_6$H$_4$–CONH$_2$	Pb	20% H_2SO_4	Cl–C$_6$H$_4$–CH$_2$NH$_2$	86	338
C$_6$H$_5$–CONHCH$_2$CO$_2$H (Hippuric acid)	Pb	Aq. alcoholic H_2SO_4	C$_6$H$_5$–CH$_2$NH–CH$_2$CO$_2$H	86	339
Naphthyl–CONH$_2$	Pb	20% H_2SO_4	Naphthyl–CH$_2$NH$_2$	55	338

VI. IMIDES
A. Aliphatic

Starting material	Cathode	Electrolyte	Product	Yield	Ref.
N-CH$_3$ succinimide	Pb	50% H$_2$SO$_4$	N-CH$_3$ pyrrolidinone	80	351
N-C$_2$H$_5$ succinimide	Zn, Hg	50% H$_2$SO$_4$		18	350
(CH$_3$-substituted succinimide)	Zn, Hg	50% H$_2$SO$_4$		28	349
(C$_6$H$_5$-substituted succinimide)	Pb	50% H$_2$SO$_4$		40.3	345
(C$_6$H$_5$/H$_5$C$_6$-substituted succinimide)	Pb	50% H$_2$SO$_4$		1.2	345
(bicyclic imide)				17	344

Table 8.1 Carbonyl Compounds (continued)

B. Aromatic

Substrate	Metal	Conditions	Product	Yield	Ref.
CO–NH–CO (cyclic, benzene-fused)	Pb	Aq. dioxane	OH–CH / NH–CO (benzene-fused)	73	340
	Pb	Strong acid 60 A/d^2	NH (benzene-fused)	—	341
	Zn, Hg	12N H$_2$SO$_4$	NH (benzene-fused)	—	341
	Pb	Aq. H$_2$SO$_4$	NH (benzene-fused)	72	342
			(Isolated as the HCl salt)		
CO–NCH$_3$–CO (benzene-fused)	Pb	Strong acid	CO / NCH$_3$ (benzene-fused)	80	341
	Pb, Hg, Cd	Aq. H$_2$SO$_4$	NCH$_3$ (benzene-fused)	57.7	343
CO–N–CH$_3$ (benzene-fused)	Pb, Hg, Cl	Aq. H$_2$SO$_4$	N–CH$_3$ (benzene-fused)		343

Starting material	Electrode	Conditions	Product	Yield (%)	Ref.
(structure: CO, $N{-}C_2H_5$)	Pb	Strong acid	(structure: CO, $N{-}C_2H_5$)		341
(structure with Cl, $N{-}CH_2CH_2N(CH_3)_2$)	Pb	Aq. H_2SO_4 0.049 A/cm^2	(structure with Cl, $N{-}CH_2CH_2{-}N(CH_3)_2$)	91	384
(structure: OC, NH, OC)	Pb	50% H_2SO_4	(structure: NH)	15	345
(structure: $CH_3{-}N$, CO, OC)	Pb	50% H_2SO_4	(structure: $CH_3{-}N{-}CO$, CH_2)	49	346
			(structure: $CH_3{-}N$)	38	346

Table 8.1 Carbonyl Compounds (continued)

Structure	Electrode	Conditions	Yield (%)	Ref.
(C₂H₅–N(CH₂)– decahydronaphthalene–CO)	Pb	50% H₂SO₄	15	346
(C₂H₅–N– naphthalene dimethylene)			44	346
(HO– R– CH₃ cyclopentanol)				592

C. Nonconjugated Olefinic Ketones

$$CH_2 = CH(CH_2)_3 \overset{\overset{O}{\|}}{C}R \qquad C \qquad \text{dioxane/methanol}$$

Et₄NpTs
CPE 2.7 V (s.c.e.)

	Yield (%)
R = CH₃	66
C₂H₅	47
i-Pr	45
n-Bu	35
n-Hexyl	40

$CH_2=CH(CH_2)_n\overset{\displaystyle O}{\overset{\|}{C}}R$	C	Dioxane/methanol (or isopropanol) CPE 2.7 V (s.c.e.)		592
R = CH₃, *n* = 2				33
				50
R = CH₃, *n* = 4				10
R = CH₃, *n* = 5				25
R = C₂H₅, *n* = 9			No product	
	C	Dioxane/methanol Et₄NpTS, CPE 2.7 V (s.c.e.)		592
				12
	C	Dioxane/methanol Et₄NpTS, CPE 2.7 V (s.c.e.)		592
				40

135

Table 8.1 Carbonyl Compounds (continued)

C				
	Dioxane/methanol Et$_4$NpTS, CPE 2.7 V (s.c.e.)		64	592

4 NITRO COMPOUNDS

The reduction of nitro compounds is probably one of the most often studied reactions in electroorganic chemistry. This reduction is of value in organic synthesis because variation in reaction conditions such as the pH of the medium or the nature of the cathode can lead to a wide variety of products.

Aliphatic Nitro Compounds

Compared to aromatic nitro compounds, aliphatic nitro compounds have not been explored in great detail. This may be due to the limited information that can be obtained from their reduction. For example, in basic media, the nitro compound is transformed to the acianion which resists further reduction [222]. Under acidic conditions the nature of the electrode plays an important part in determining the nature of the product. With a high overvoltage electrode (mercury or lead) the main product is the corresponding amine. Low overvoltage cathodes, on the other hand, favor the formation hydroxylamine, with amines as the minor product [169].

Under anhydrous conditions nitroalkanes reduce first to the unstable anion radical, which cleaves to nitrite ion and the corresponding free radical [482]:

$$R-N\overset{\displaystyle O}{\underset{\displaystyle O}{<}} + 1e \longrightarrow R-\overset{..}{N}\overset{\displaystyle O^{\ominus}}{\underset{\displaystyle O}{<}}$$

$$R-\overset{..}{N}\overset{\displaystyle O^{\ominus}}{\underset{\displaystyle O}{<}} \longrightarrow R^{.} + NO_2$$

$$2R^{.} \longrightarrow R-R$$

The half-wave potentials of several nitroalkanes have been reported [236, 482]. For substituted nitroalkanes such as $R_2C(NO_2)X$, where X = $-CH_3$, $-H$, $-CH_2OH$, $-CONH_2$, $-CO_2Et$, $-CN$, a linear plot was observed between the $E_{1/2}$ of the substrate and the σ_1 of the substituent. The half-wave potential was more negative as X varied from $-CN$ to $-CH_3$.

Aromatic Nitro Compounds

Variation of Products with Reaction Conditions

One of the best examples that illustrates the use of electrolysis in organic synthesis is the electrochemical reduction of nitrobenzene under a variety of reaction conditions. Thus by simply changing the pH of the reaction medium or the nature of electrode, different products are obtained in high yields:

The cost and availability of aromatic compounds have opened the way to the synthesis of a number of important reaction intermediates. For example, benzidine, which is widely used in the dye industry can be obtained from nitrobenzene in two steps:

Mechanism of Reduction

In spite of the numerous available papers that describe the reduction of nitrobenzene and its derivatives, the mechanism of this reduction is not clearly understood. This may be due to the delicate nature of the reaction, whose course is easily altered by a slight change of electrode material as well as the pH of the solution. An attempt will be made to discuss the effects of the nature of the medium on the course of the reactions. For key papers on the effects of substituents on the ease of reduction of nitrobenzene the reader should consult Refs. [237] and [238].

REDUCTION UNDER ACIDIC CONDITIONS

The effect of the pH of the medium on the polarographic behavior of nitrobenzene is shown in Table 8.2. This table clearly demonstrates how a change in pH drastically alters the course of the reduction of nitrobenzene. This change can be, and has been, advantageous to organic chemists who seek to preferentially prepare one type of product that otherwise cannot be prepared by conventional techniques. Let us, therefore, consider what types of reactions can take place under acidic conditions.

The reduction of nitrobenzene in the pH range 4-6 requires four electrons to produce phenylhydroxylamine. This reduction proceeds through nitrosobenzene whose $E_{1/2}$ is more positive than that of nitrobenzene [224]. In fact, it was shown by Smith and co-workers [227] that the reduction of nitrosobenzene to same $E_{1/2}$ as that of nitrosobenzene (see Table 8.2).

Table 8.2 *Variation of Polarographic Behavior of Nitrobenzene with pH of Medium* [26]

pH of Medium	$E_{1/2}$ (V)	SCE
4.5	0.55	
6.5	0.65	
9.1	0.75	1.05
14	0.85	0.95

NO$_2$ + 2e + 2H $\longrightarrow$ NO + H$_2$O

NO + 2e + 2H $\rightleftharpoons$ NHOH

The isolation of phenylhydroxylamine may present a problem since this com pound often rearranges under acidic conditions to *p*-aminophenol. In general the isolation of phenylhydroxylamine may be accomplished when a low hydro gen overvoltage cathode (e.g., zinc) is used. Below a pH value of 4 and by using a high overvoltage cathode (e.g., lead) phenylhydroxylamine is further reduced to aniline [193, 223]. This reduction occurs at a more negative potential than that which produces phenylhydroxylamine. The overall reduction of nitroben zene under acidic conditions may be depicted as shown in Fig. 8.1.

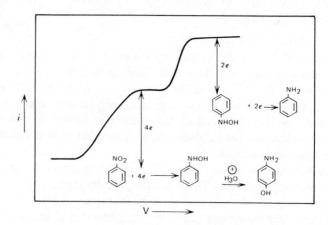

Fig. 8.1 Reduction of nitrobenzene under acidic conditions.

REDUCTION UNDER BASIC OR NEUTRAL CONDITIONS

The reduction of nitrobenzene under basic or neutral conditions occurs at a more negative potential than under acidic conditions (see Table 8.2). Aside from the pH of the medium, this reduction is sensitive to the nature of the cathode. Consider, for example, the following reactions:

$$(554)$$

$$(189)$$

$$(555)$$

The first step in the reduction of nitrobenzene probably involves the formation of the radical anion [225]. The ESR spectrum of this species has been observed

in acetonitrile [226]. Further reduction of the radical anion affords the dianion which abstracts a proton from the medium, then loses a hydroxyl ion to form nitrosobenzene:

Nitrosobenzene undergoes further reduction to give phenylhydroxylamine in similar manner as described under acidic conditions. Since under basic condition the rearrangement of phenylhydroxylamine to *p*-aminophenol is difficult condensation reaction takes place to form azoxybenzene [228] :

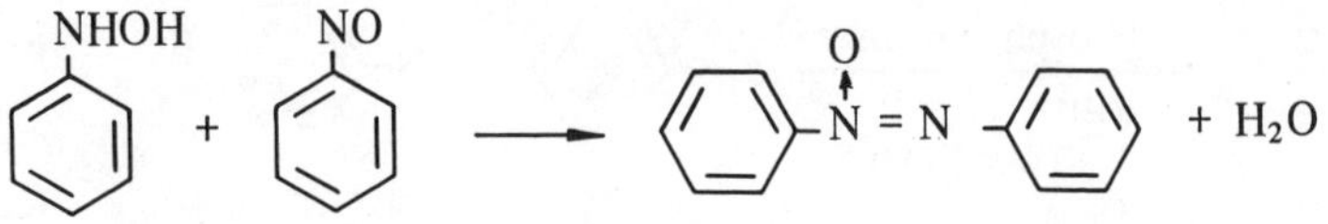

Azoxybenzene can be reduced irreversibly to give azobenzene [229, 230] :

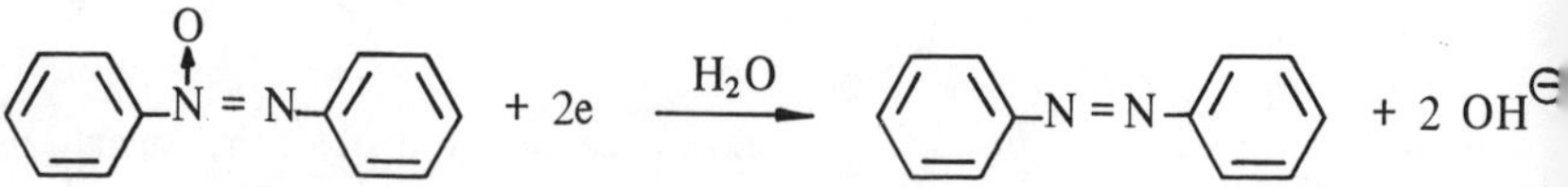

Wawzonek [230] has shown from cyclic voltammetric studies that the reduction of azobenzene to hydrazobenzene is reversible. Another indication of the reversibility of the reaction was obtained from a plot of $E_{1/2}$ versus pH which had a slope of -0.059 V:

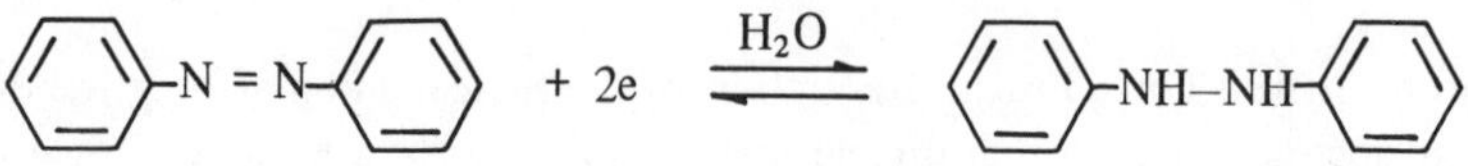

It can be seen from the above discussion that the reduction of nitrobenzene is quite complex. Indeed, one may well wonder how such a reduction can be of synthetic utility. Nevertheless, by controlling the pH of the medium as well as choosing the proper cathodes, specific compounds can be obtained in over 90% yield. For examples of such specific syntheses, the reader should consult the table at the end of this section.

The reduction of nitrobenzene derivatives to produce heterocyclic compounds has been studied by Lund and co-workers [556]. Consider the following examples:

1)

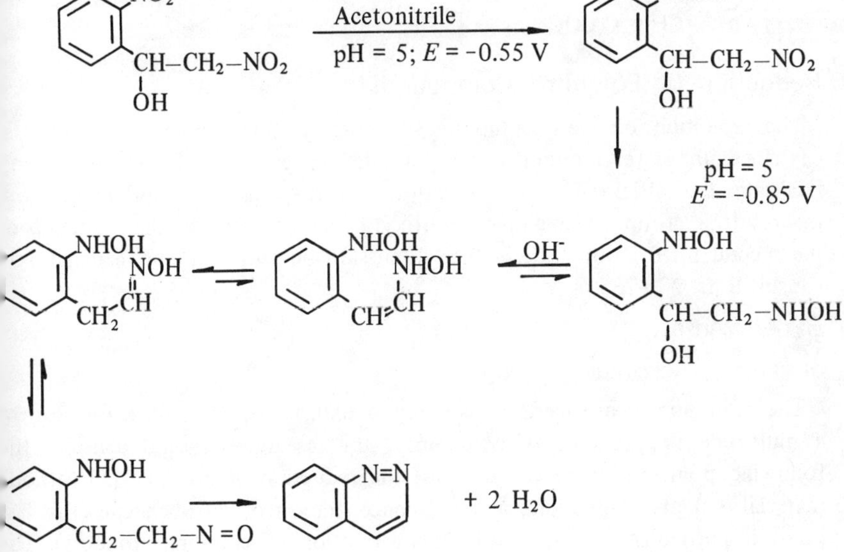
NO2
CH–CH2–NO2
OH
Acetonitrile
pH = 5; E = -0.55 V
NHOH
CH–CH2–NO2
OH
pH = 5
E = -0.85 V
NHOH
CH–CH2–NHOH
OH
OH-
NHOH
NHOH
CH=CH
NHOH
NOH
CH2 CH
NHOH
CH2–CH2–N=O
N=N
+ 2 H2O

(2)

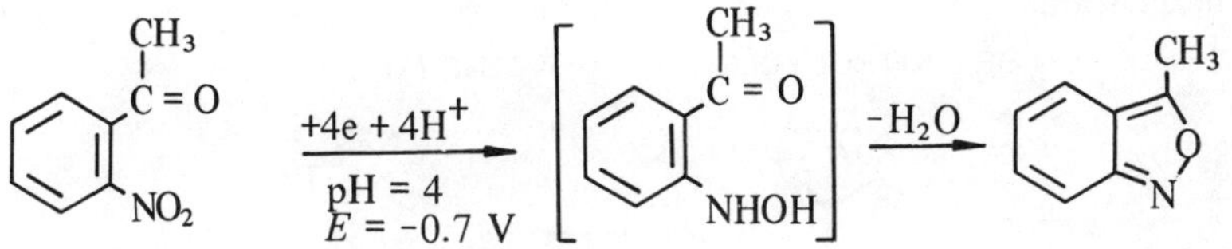
CH3
C = O
NO2
+4e + 4H+
pH = 4
E = -0.7 V
CH3
C = O
NHOH
-H2O
CH3
O
N

(3)

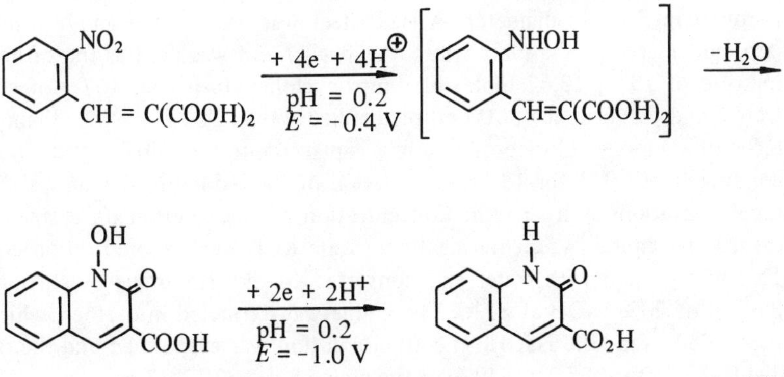
NO2
CH = C(COOH)2
+ 4e + 4H+
pH = 0.2
E = -0.4 V
NHOH
CH=C(COOH)2
-H2O
OH
N
O
COOH
+ 2e + 2H+
pH = 0.2
E = -1.0 V
H
N
O
CO2H

(4)

$$\xrightarrow[\substack{pH = 0.2 \\ E = -0.4 \text{ V}}]{+ 4e + 4H^{\oplus}}$$

Reduction of Polynitro Compounds

The reduction of these compounds occurs in stages in a similar process described above for mononitro compounds. The presence of one nitro group facilitates the initial reduction of the other. However, after the initial reduction the resulting group, for example, hydroxyl or amine, impedes the reduction of the second nitro group. For further discussion on this topic the reader should consult Refs. 233-235.

Experimental

GENERAL PROCEDURE

The reduction of nitrobenzene is complex and, if not controlled, can lead to a multitude of products. Consequently, the researcher should consider the following points prior to carrying out the reduction of interest: (1) cathode material, (2) pH of medium, and (3) concentration of nitrobenzene (this last point is particularly important in the formation of coupling products). The table at the end of this section should also be of help in allowing the researcher to choose the optimum reaction conditions for reduction of the nitrobenzene in question.

REDUCTION OF o-NITROCYCLOHEXYLBENZENE [179]

R = cyclohexyl

The electrolysis cell consisted of a 4-1 beaker, a mechanical stirrer, and a porous cup 3 in. in diameter. A lead sheet was used as the anode inside the cup, and mercury was placed into the beaker and was used as the cathode. A mixture of 137 g (0.67 mole) of o-nitrocyclohexylbenzene, 167 g of concentrated hydrochloric acid, 700 ml of distilled water, 122 g of 96% ethanol, and 6.6 g of stannous chloride dihydrate was placed into the cell. The reduction was conducted at 19 A for 13 hr. Completion of the reduction was indicated by a rapid evolution of hydrogen. Concentration of the solution on a steam bath caused the amine hydrochloride to separate as a violet colored oil on cooling. This was washed with water and converted to the free amine by stirring with 250 ml of 20% NaOH at 80°C. The amine was extracted into ether, which was then dried over solid NaOH. The ether solution was evaporated and the residue distilled to afford 98.2 g (849%) of the amine b.p. 132-134/3 mm.

Table 6.5 Nitro Compounds

Aliphatic	Cathode	Reaction Medium	Products	Yield	Ref.
Nitromethane	Pb-Hg	5% H_2SO_4	N-Methylhydroxylamine	41.5	169
1-Nitropropane	Pb-Hg	5% H_2SO_4	N-Propyldroxylamine	54.5	169
2-Nitropropane	Pb-Hg	5% H_2SO_4	N-Isopropylhydroxylamine	71.5	169
tert-Bu NO_2	Hg	Acetonitrile, R_4NX, -2.0 V (SCE)	t-Butylhydroxylamine	—	482
α-Nitroisobutyronitrile	Hg	Acetonitrile	$t-Bu-NO_2$	—	482
			Isobutyronitrile	—	482
		Et_4NBr, -1.2 V (SCE)	Tetramethylsuccinonitrile, RNO_2	—	482
α-Nitroisobutyramide	Hg	Et_4NBr, -1.2 V (SCE)	Isobutyramide, RNO_2	—	482
α-Nitroisobutyrate	Hg	Et_4NBr, -1.2 V (SCE)	Ethyl isobutyrate	—	482
			Ethyl α-nitrosoisobutyrate	—	482
			Diethyl tetramethylsuccinate	—	482
			Ethyl methacrylate	—	482
Nitrourea	Sn	Aq. H_2SO_4	Semicarbazide	69	170
1-Methyl-3-nitrourea	Sn	5% H_2SO_4	4-Methylsemicarbazide	69	196
Nitroguanidine ($HN=CNHNO_2$) $\mid$ NH_2	Sn	Aq. H_2SO_4	Aminoguanidine	80	170

145

Table 8.3 Nitro Compounds (continued)

Compound	Catalyst	Conditions	Product	Yield	Ref
Chloropicrin (Cl₃CNO₂)	Pt, Ni	Alcoholic H₂SO₄	Dichloroformoxime	—	171
	Pt, Cu	1-70% H₂SO₄	Methylhydroxylamine, Methylamine, Dichloroformoxime	—	172
H₃C–N–NO / H₃C	Pb	Aq. H₂SO₄	H₃C–N–NH₂ / H₃C	92	220
F₃CNO			$F_3C-N-N-CF_3$ (with O–N=O)		232
2-Nitro-2-methylpropanol	Sn	5% H₂SO₄	2 Hydroxylamine-2-methylpropanol	78.5	173
3-Nitro-4-heptanol	Pb	10% H₂SO₄	3-Amino-4-heptanol	75	174
NO₂ (cyclohexane)	Cu, Ni, Sn, Zn	Aq. HCl or H₂SO₄	NH₂ (cyclohexyl) , NHOH (cyclohexyl)	—	487
	Hg	6% NaHCO₃ pH = 8	NHOH (cyclohexyl)	49	488
1-(Nitromethyl)cyclohexanol (OH, CH₂NO₂)	Pb	10% H₂SO₄	OH, CH₂NH₂ (cyclohexanol)	77-83	206
			1-(Aminomethyl)-cyclohexanol	83	179

146

Substrate	Metal	Reagent	Product	Yield	Ref
3-Nitro-2-oxazolidone	Hg	10% H_2SO_4	3-Amino-2-oxazolidone	50	176
4-Nitro-1-phenyl-3-pyrazolone	Pb	Acoholic H_2SO_4	4-Amino-1-phenyl-3-pyrazolone	86	178
3-Nitro-2-oxazolidone	Hg	10% H_2SO_4	3-Amino-2-oxazolidone	50	176
5-Chloromethyl-3-nitro-2-oxazolidone	Hg	10% H_2SO_4	3-Amino-5-chloromethyl-2-oxazolidone	20	176
4,4-Dimethyl-3-nitro-2-oxazolidone	Hg	10% H_2SO_4	3-Amino-4,4-dimethyl-2-oxazolidone	45	176
o-Nitrophenylcyclohexane	Hg	Aq. HCl	o-Cyclohexylaniline	89.4	179
1,1-Dinitroethane	Hg	0.1 N NaOH	Acetamidoxime	—	179
Nitroisobutylglycerol	Pb	Aq. H_2SO_4	Tris(hydroxymethyl)methyl amine	70	180
m-Nitrotritanol	Ni	C_2H_5OH/NaOAc	m-Azotritanol		182
p-Nitrotritanol	Ni	C_2H_5OH/NaOAc	p-Azoxytritanol	—	182
			p-Azotritanol		
$CH_3CO_2-CH(-CHCH_3)$ with NO_2 group, attached to $C_6H_4-OCH_3$	Pb	C_2H_5OH, HNO_3	$CH_3CO_2CH-CH-CH_3$ with NO group, attached to $C_6H_4-OCH_3$	21	177

Table 8.3 Nitro Compounds (continued)

Substrate	Metal	Conditions	Product	Yield (%)	Ref.
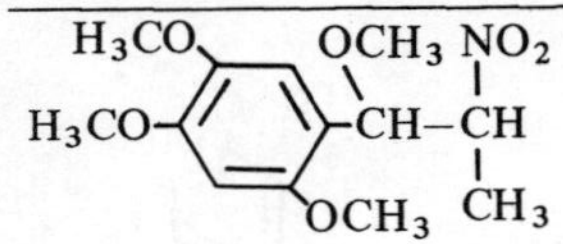 H₃CO, H₃CO ring with OCH₃, NO₂, CH–CH, CH₃, OCH₃	Pb	H_2SO_4	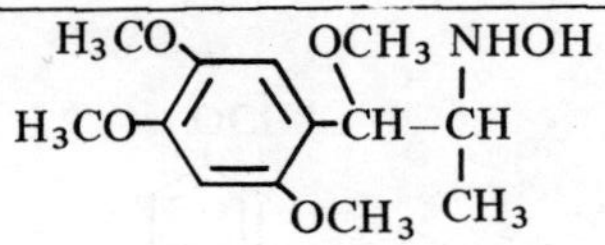 H₃CO, H₃CO ring with OCH₃, NHOH, CH–CH, CH₃, OCH₃	—	557

Aromatic Nitro Compounds

Substrate	Metal	Conditions	Product	Yield (%)	Ref.
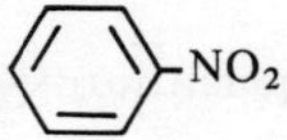 $C_6H_5NO_2$	Pt	10% HCl	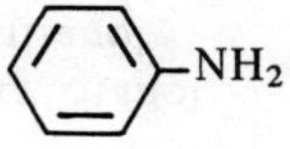 NH_2 (aniline)	91	183
	Monel	Aq. acid	 HO–C₆H₄–NH_2	72	184, 185
	Zn	Aq. acid	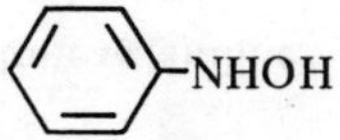 C₆H₅–NHOH	29	186
	Ni	Aq. C_2H_5OH NaOAc, reflux	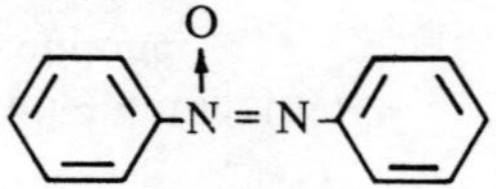 azoxybenzene (N=N→O)	95	187
	—	Sat. Na salts of organic acids	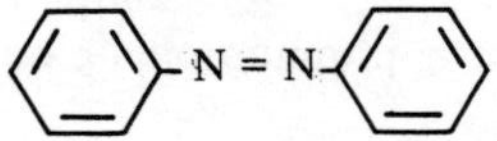 C₆H₅–N=N–C₆H₅	95	188, 189
	Zn, Sn Monel	Basic, C_2H_5OH	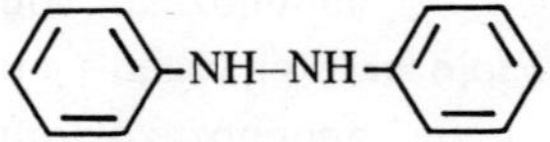 C₆H₅–NH–NH–C₆H₅	90	190, 191

148

Starting material	Catalyst	Medium	Product	Yield (%)	Ref.
p-nitrotoluene (NO₂–C₆H₄–CH₃)	Pt, Pb, Zn	Aq. acid	p-toluidine (H₃C–C₆H₄–NH₂)	—	—
			H₃C–C₆H₄–NH–NH–C₆H₄–CH₃ (p,p'-hydrazotoluene)	—	193
			H₃C–C₆H₄–N=N–C₆H₄–CH₃ (p,p'-azotoluene)	—	193
m-nitrotoluene	Hg, Monel	30% H₂SO₄	amino-cresol (NH₂, CH₃, OH)	—	194
			m,m'-azotoluene (CH₃–C₆H₄–N=N–C₆H₄–CH₃)	—	194
o-nitrotoluene	Zn, Ni		o,o'-Hydrazotoluene	—	
	Cu		o,o'-Azotoluene	—	
	Monel Fe, Pb		o-Toluidine	—	
o-nitrotoluene	—	Sat. Na-salt of org. acids	o,o'-azotoluene (CH₃–C₆H₄–N=N–C₆H₄–CH₃)	90-95	189

149

Table 8.3 Nitro Compounds (continued)

Nitro compound	Catalyst	Solvent	Product		
NO_2–C$_6$H$_3$–N(C$_2$H$_5$)$_2$ (H$_5$C$_2$) (4-nitro-N,N-diethylaniline)	Pt	Aq. H_2SO_4	HO, NH_2 substituted –N(C$_2$H$_5$)(C$_2$H$_5$) (H$_5$C$_2$)	—	558
NO_2–C$_6$H$_4$–CF$_3$ (F$_3$C)	Hg	DMF	F_3C–C$_6$H$_4$–N=N(→O)–C$_6$H$_4$–CF_3	—	477
3,5-(CH$_3$)$_2$C$_6$H$_3$–NO_2 (CH$_3$, CH$_3$)	—	80% EtOH	(CH$_3$)$_2$C$_6$H$_3$–N=N–C$_6$H$_3$(CH$_3$)$_2$ (CH$_3$, CH$_3$, CH$_3$, CH$_3$)	79	200
NO_2, C$_2$H$_5$ (2-nitroethylbenzene)	Cu	75% H_2SO_4	H_2N, OH, C$_2$H$_5$	68	195
CH(CH$_3$)$_2$, NO_2	Monel, Cu, Pb, Cd	Aq. H_2SO_4	HO, CH(CH$_3$)$_2$, NH_2, CH$_3$	54.5	204

150

Reactant	Cathode	Electrolyte	Product	Yield %	Ref.
2-nitro(cyclohexyl)benzene (NO_2)	Hg	Conc, HCl $SnCl_2$	2-amino(cyclohexyl)benzene (NH_2)	85	179
1-nitronaphthalene (NO_2)	Fe	$FeCl_2$ Sol.	1-aminonaphthalene (NH_2) / 1-naphthol (OH)	—	187
1-chloro-4-nitrobenzene (Cl–NO_2)		Sat. Na salts of org. acids	Cl–$\langle\rangle$–N=N–$\langle\rangle$–Cl	75	189
2-chloronitrobenzene (NO_2, Cl)	Fe	10% NaOH	Cl–$\langle\rangle$–NH–NH–$\langle\rangle$–Cl		
	Cu, Monel	20% C_2H_5OH	HO–$\langle\rangle$(NH_2, Cl)	76	185
4-chloronitrobenzene (Cl–NO_2)	C 2 A 1d[2]	20% H_2SO_4 25°C	Cl–$\langle\rangle$–NHOH	—	559

Table 8.3 Nitro Compounds (continued)

Substrate	Catalyst	Medium	Product		Ref.
4-Cl–C₆H₄–NO₂ (Cl, NO₂ substituted benzene)	Zn	H_2SO_4	OH, SO₃H, NH₂ substituted benzene	—	560
Dinitrobenzene (NO₂, NO₂)	Pt	Conc. H_2SO_4	NH₂, SO₃H, NH₂, OH substituted benzene	—	562
NO₂, NO₂, Br substituted benzene	Pt	H_2SO_4	NH₂, Br, OH substituted benzene	—	561
NO₂–C₆H₄–OH / NO₂ substituted benzene	Pt	H_2SO_4	NH₂, SO₃H, OH substituted benzene	—	562
Cl, NO₂ substituted benzene	Ni	NH_4Cl	Cl–C₆H₄–N=N→O–C₆H₄–Cl (azoxy compound)	—	563

152

Substrate	Catalyst	Medium	Product	Yield	Ref.
m-bromonitrobenzene	Ni	NH_4Cl	3,3'-dibromoazoxybenzene	90	564
p-bromonitrobenzene	Ni	NH_4Cl	4,4'-dibromoazoxybenzene	72	564
2,5-dichloronitrobenzene	Ni, Pb Cu Monel	Basic aq. emulsion	2,5,2',5'-tetrachlorohydrazobenzene (NH—NH)	—	197
		20% H_2SO_4	2-amino-4-chloro-...-phenol (HO, NH_2, Cl)	72	185
2-methyl-4-chloronitrobenzene (NO$_2$, CH$_3$, Cl)	Cu	20% C_2H_5OH	aminocresol (NH_2, CH_3, OH, Cl)	84	185
2-methyl-3-chloronitrobenzene (NO$_2$, CH$_3$, Cl)	Cu	20% C_2H_5OH	aminocresol (NH_2, CH_3, Cl, OH)	72	185

153

Table 8.3 Nitro Compounds (continued)

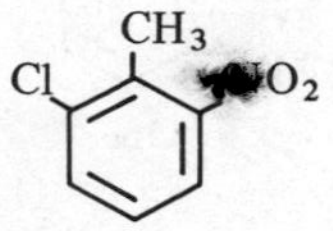	Pt	75% H_2SO_4	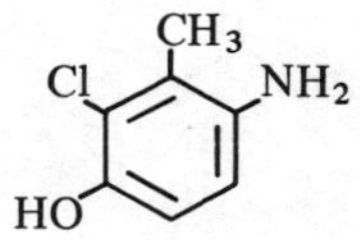	67	201
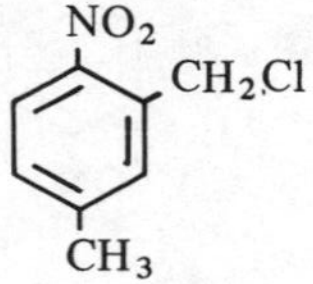	Zn	Aq. CH_3OH	Xylidine	72	205
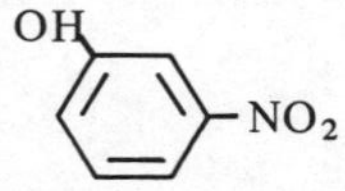	Cu	15% Caustic	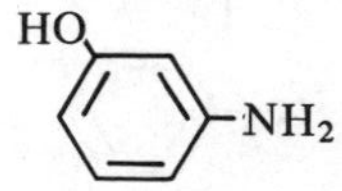	97	198
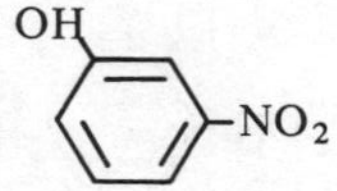	Hg	Aq. C_2H_5OH, pH = 2	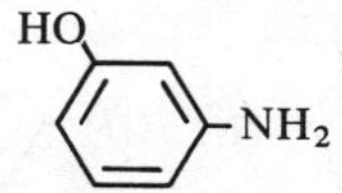	–	199
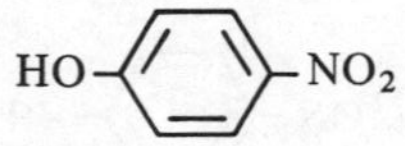	Hg	Aq. C_2H_5OH, pH = 2	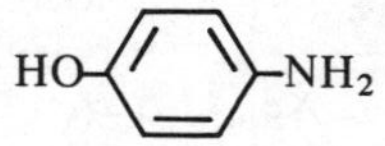	–	199
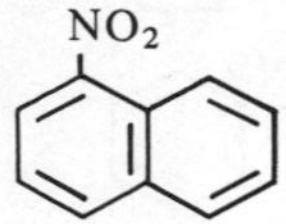	Pt	Aq. H_2SO_4	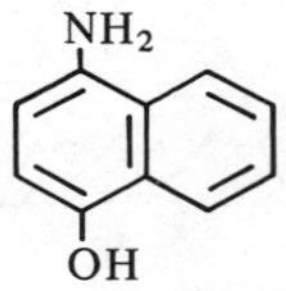	83	202

Substrate	Metal	Conditions	Product	Yield (%)	Ref.
(3-nitro-diphenylamine)	Pt	Conc. H_2SO_4	(2-amino-phenol-NH-phenyl, NH_2, OH)	65	203
(2-nitro-anisole, NO_2, OCH_3)	Fe	Basic aq. emulsion	(OCH_3–NH–NH–CH_3 biphenyl)	—	207
(RO–C_6H_4–NO_2; NCS–C_6H_4–NO_2)	Pb	NH_4NO_3	(NCS–C_6H_4–$N=N(\rightarrow O)$–C_6H_4–SCN)	—	562
(SH–C_6H_4–NO_2)	Hg	$EtOH/H_2SO_4$	(SH–C_6H_4–NH_2)	—	586
(SCH_3–C_6H_4–NO_2)	Hg	$EtOH/H_2SO_4$	(SCH_3–C_6H_4–NH_2)	100	568

Table 8.3 Nitro Compounds (continued)

	Hg	NH$_4$OAc/EtOH		

Compound	%	value
SCH$_3$—C$_6$H$_4$—NHOH	80	568
R = C$_4$H$_9$ (R–N=N–R, O)	45	208
C$_5$H$_{11}$	71	208
C$_6$H$_{13}$	86	208
C$_8$H$_{17}$	55	208
C$_9$H$_{19}$	75	208
C$_{10}$H$_{21}$	88	208
C$_{11}$H$_{23}$	78	208
C$_{12}$H$_{25}$	72	208

156

Starting material	Cathode	Electrolyte / conditions	Product	Yield / temp.	Ref.
o-dinitrobenzene (NO_2, NO_2)	Cu, An, Hg	Alc. sol. 2% NaOH	o-phenylenediamine (NH_2, NH_2)	88–90	479
m-dinitrobenzene (NO_2, NO_2)	Al, C, Sn	10 A/d^2			
	Pb, Fe, Ni				
	C	20% H_2SO_4 2 A/d^2	NO_2, NHOH	—	559
chloro-dinitrobenzene (Cl, NO_2, NO_2)	C	20% H_2SO_4 2 A/d^2	Cl, NO_2, NHOH	—	559
trinitrotoluene (CH_3, NO_2, O_2N, NO_2)	C	Aq. solution pH = 7	CH_3, NO_2, O_2N, NHOH	—	569
trinitrobenzoic acid (CO_2H, NO_2, O_2N, NO_2)	Pb	HCl	CO_2H, NH_2, H_2N, NH_2	—	570

Table 8.3 Nitro Compounds (continued)

Nitro compound	Catalyst	Solvent	Product		
H_2N—C$_6$H$_4$—NO_2	Pt	10% HCl	H_2N—C$_6$H$_4$—NH_2	95	183
NO_2, NH_2 benzene	Monel	30% H_2SO_4	HO—, NH_2, NH_2 benzene	50	183
NO_2, CH_3, H_2N benzene	Pt	Aq. H_2SO_4	NH_2, CH_3, OH benzene	—	558
CH_3—N(CH_3)—C$_6$H$_4$—NO_2	Pb	Aq. acid	$(CH_3)_2N$—C$_6$H$_4$—N=N—C$_6$H$_4$—N($CH_3)_2$; $(CH_3)_2N$—C$_6$H$_4$—NH_2	—	193
NO_2, N($CH_3)_2$ benzene	Pt	Aq. H_2SO_4	NH_2, HO—, N($CH_3)_2$ benzene	—	558

158

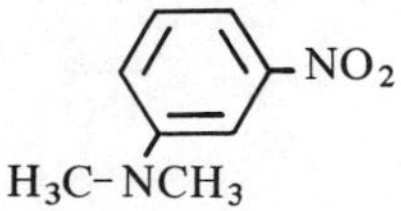	Ni	Aq. acid	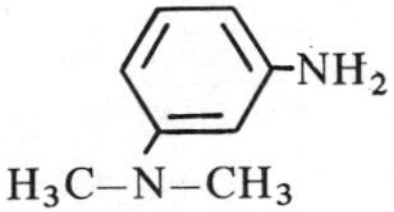		
			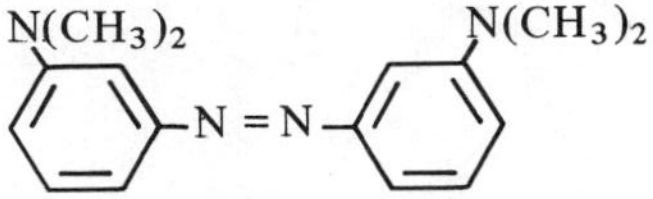	–	193
			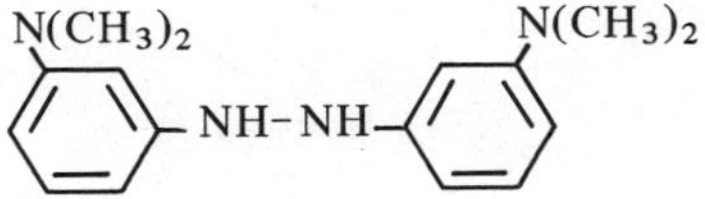	–	193
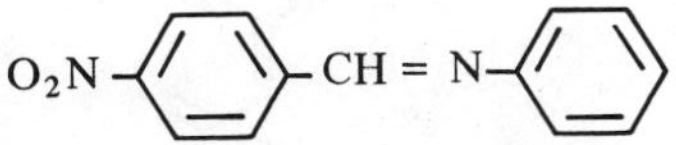	Ni	Ethanol, NaOAc, Reflux	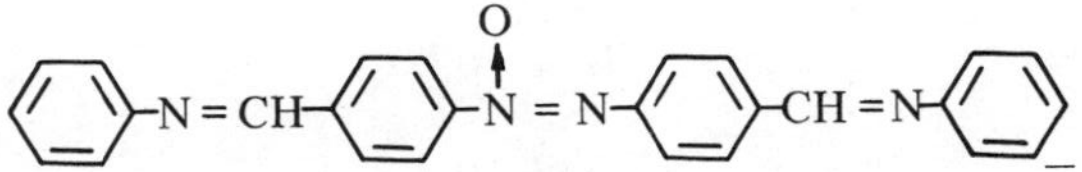	–	480
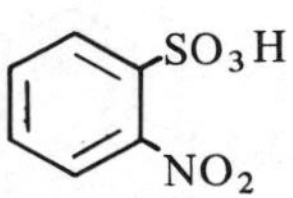			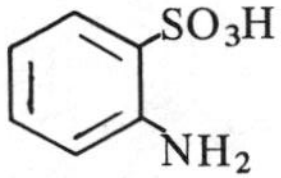	100	580
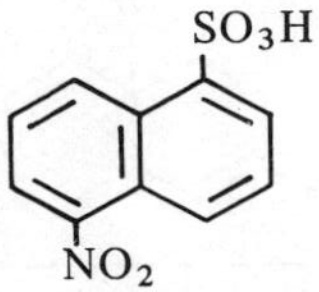	Pt	H_2SO_4	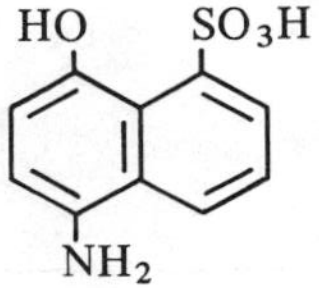	–	558

Table 8.3 Nitro Compounds (continued)

Starting material	Metal	Conditions	Product		Ref.
naphthalene with SO_3H, NO_2	Pb	Aq. H_2SO_4	naphthalene with SO_3H, NH_2	—	571
naphthalene with SO_3H, NO_2	Pb	Aq. H_2SO_4	naphthalene with NHOH, SO_3H	—	572
SO_3H–benzene–NO_2, Cl	Hg	LiCl	SO_3H–benzene–NHOH, Cl	—	573
CO_2H–benzene–NO_2	Pb, Sn	Conc. HCl	CO_2H–benzene–NH_2	—	210
CO_2H–benzene–NO_2			CO_2H–benzene–NO_2, HO	—	210
CO_2H–benzene–NO_2	Pb, Sn	Conc. HCl	CO_2H–benzene–NH_2	—	210

	Pt	Aq. H_2SO_4		574	—
	Pt	Aq. H_2SO_4		575	—
	Pt	Aq. H_2SO_4		558	—
	Pt	Aq. H_2SO_2		576	—
	Pt	Aq. H_2SO_4		576	—

Table 8.3 Nitro Compounds (continued)

Starting material	Catalyst	Conditions	Product	Yield	Ref.
NO_2 benzene + NO_2-C_6H_4-CH_3 (o)	Pt, Hg	Aq. base	C_6H_5-N=N-C_6H_4-CH_3 (azo, o-CH₃)	—	577
NO_2-C_6H_4-CO_2H + NO_2-C_6H_4-CH_3	Pt, Hg	Aq. base	H_3C-C_6H_4-N=N-C_6H_4-CO_2H	—	577
NO_2-C_6H_4-CO_2H + NO_2-C_6H_4-SO_3H	Pt, Hg	Aq. base	SO_3H-C_6H_4-N=N-C_6H_4-CO_2H	—	577
NO_2-C_6H_4-CO_2H + NO_2-C_6H_4-NH_2	Pt, Hg	Aq. base	H_2N-C_6H_4-N=N-C_6H_4-CO_2H	—	577
(CO_2H, NO_2, phenol)			HO-C_6H_3(CO_2H)(NH_2)		210
pyridine: O_2N, CN, CH_2OCH_3, H_3C, Cl	Hg	Aq. HOAC, HCl	pyridine: H_2N, CH_2NH_2, CH_3, H_3C, N	—	216

162

Starting material	Metal	Conditions	Product	Yield (%)	Ref.
HO₂C–C₆H₄–NO₂ (p)	Pb Sn	Conc. HCl	HO₂C–C₆H₄–NH₂ (p)	95	210
HO₃S–C₆H₄–NO₂ (p)	Pt, Ni	Basic, neutral and alc. solutions	p-Azobenzyldisulfonic acid	—	489
	Pt, Pb	Dilute H_2SO_4	p-Aminobenzylsulfonic acid	—	489
m-nitrobenzenesulfonic acid (NO₂, SO₃H)	Ni, Fe	H_2O	H_2N–/HO₃S– biphenyl (NH₂, SO₃H)	60	211
ethyl m-nitrobenzoate (CO₂C₂H₅, NO₂)	Pb	Aq. HCl	ethyl m-aminobenzoate (CO₂C₂H₅, NH₂)	80	212
methyl p-nitrobenzoate (H₃CO₂C, NO₂)	Pb	Aq. HCl	methyl p-aminobenzoate (H₃CO₂C, NH₂)	85	212
ClCH₂O₂C–C₆H₄–NO₂ (p)	Pb	Aq. HCl	ClCH₂O₂C–C₆H₄–NH₂ (p)	90	212
2,3-dinitrobenzoic acid (CO₂H, NO₂, NO₂)	Hg	EtOH/H_2SO_4	CO₂H, NO₂, NHOH compound	—	481

Table 8.3 Nitro Compounds (continued)

164

			Ref.
CO_2H, NO_2, NO_2 (2,4-dinitrobenzoic acid)	50:50 +20 mV (SCE)	CO_2H, NH_2, NH_2	—	481
	−400 mV (SCE)	CO_2H, NO_2, NHOH	—	481
	Hg, 0 mV (SCE)	CO_2H, NHOH, NHOH	—	481
	−400 mV (SCE)	CO_2H, NH_2, NH_2	—	481
CO_2H, NO_2, NO_2	Hg, +50 mV (SCE)	NHOH, CO_2H, O_2N	—	481
	−400 mV (SCE)	CO_2H, NH_2	—	481

	Electrode	Conditions	Product		Ref.
3,4-dinitrobenzoic acid (CO_2H, NO_2, NO_2)		0 mV (SCE)	CO_2H derivative (NO_2, NHOH)	—	481
		−400 mV (SCE)	CO_2H derivative (NH_2, NH_2)	—	481
X-dinitrobenzene (NO_2, NO_2); X = Br, CH_3, OC_2H_5	Hg	Acidic medium	X-nitro-NHOH and X-diamine (NH_2, NH_2)	—	483
(NHOH, O_2N, R)	Hg	Acidic medium	azoxy / azo NO_2 products	—	486

165

Table 8.3 Nitro Compounds (continued)

Starting compound	Electrode	Medium	Product	Yield	Ref.
O_2N–C6H3(R)–NHOH (R = H, Br, CH$_3$, OH, OMe, —CONH$_2$)	Hg	Acidic medium	azoxy compound (R, R′ substituted, O_2N–, –NO$_2$, N=N→O)	—	486
2,2′-dinitrobiphenyl (NO_2, NO_2)	Hg	80% C$_2$H$_5$OH (polarography)	biphenyl HN-NH	—	213
1,8-dinitronaphthalene (NO_2 NO_2)	Hg	Buffer sol. (polarography)	naphthalene N=N	—	478
3,3′-dinitrodiphenyl sulfone (NO_2, O=S=O, NO_2)	Ni	Alcoholic HCl 0.02–0.05 A/cm^2	3,3′-diaminodiphenyl sulfone (NH_2, O=S=O, NH_2)	94.5	214

Miscellaneous Nitro Compounds

Starting compound	Electrode	Medium	Product	Yield	Ref.
PhCH=C(Br)NO$_2$	Hg	Aq. dioxane	PhCH$_2$CN	80	215
			PhCHO	10	215

Metal	Conditions	Substrate / Product	Yield	Ref.
Hg	Aq. dioxane	C_6H_5–$C(Cl)(NO_2)$–$CHNO_2$–H $\rightarrow$ C_6H_5–CH(OH)–CH=O	—	215
Hg	Aq. dioxane	RC = C–R (NO_2, NO_2) $\rightarrow$ R–$C(O)$–$CH(NOH)$–R	—	215
Hg	Alc. solution of H_2SO_4 and HCl	CH_3O, CH_3O–C_6H_3–CH=C(NO_2)–CH_3 $\rightarrow$ CH_3O, CH_3O–C_6H_3–CH_2–CH(NH_2)–CH_3	20	218
Cu, Sn	50% HCl, $SnCl_2$	4,4'-Dinitro-2,2'-stilbenedisulfonic acid $\rightarrow$ 4,4'-Diaminostilbenedisulfonate	49–57	485
Pb	Alcoholic sol. of HCl and HOAc	phenanthrene–CH=CHNO_2 $\rightarrow$ phenanthrene–$CH_2CH_2NO_2$	50	217
Pb	Alcoholic sol. of HCl and HOAc	phenanthrene–CH=CHNO_2 $\rightarrow$ phenanthrene–$CH_2CH_2NO_2$	91	217

Table 8.3 Nitro Compounds (continued)

168

Starting material	Metal	Reducing agent	Product	Yield	Page
(structure: nitro/dimethyl biphenylmethane)	C, Sn	HCl/SnCl$_2$ · 2H$_2$O	(structure: amino/dimethyl)	96	219
(structure: CO$_2$H, dinitro)	S, Sn	HCl/SnCl$_2$·2H$_2$O	(structure: CO$_2$H, diamino)	93	219
(structure: nitrotropolone)	Hg	Aqueous acid	(structure: aminotropolone)		193
(structure: 4,4′-dinitroazo-N,N′-dioxide)	Cu	Alcoholic dil. H$_2$SO$_4$	(structure: azoxy-N,N′-dioxide)	78	221
(structure: 4-nitroquinoline N-oxide)	Pb	Alcoholic dil. H$_2$SO$_4$	Pyridinetetrahydroquinoline	—	221
	Cu, Ni	Alcoholic dil. H$_2$SO$_4$	4-Aminoquinoline		221

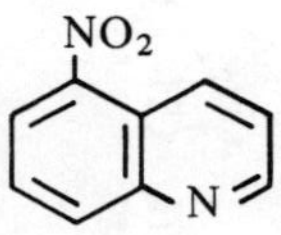

578

579

558

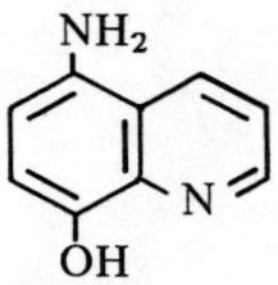

H₂SO₄

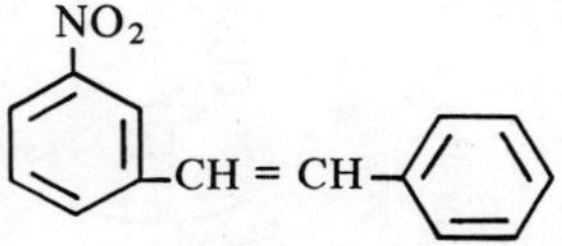

H₂SO₄

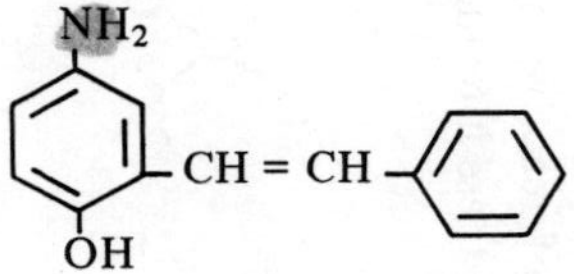

Aq. H₂SO₄

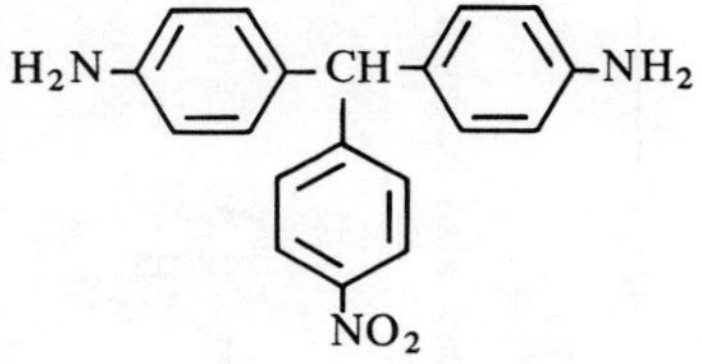

Pt

Pt

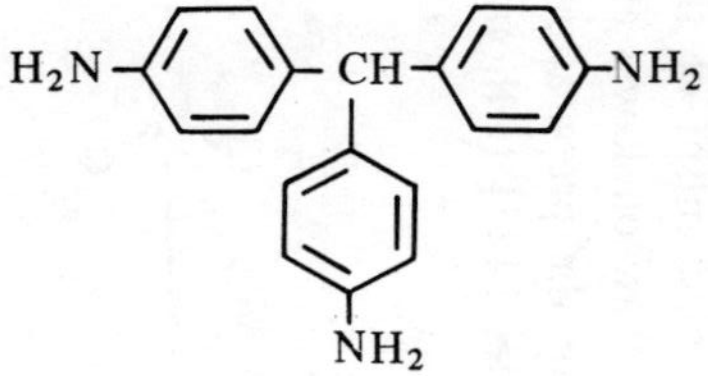

Pt

5 ORGANIC HALIDES

The electrochemical reduction of carbon-halogen bonds has been studied in detail (see Table 8.5) and their polarographic behavior has been reviewed [429-431, 500]. Unfortunately, however, a significant portion of the published work has dealt with the mechanism of the reduction rather than with it utility in organic synthesis. Wawzonek was among the first to use the reduction of carbon-halogen bonds for the preparation of reactive intermediates such a carbenes [409] and benzynes [423] (although these intermediates were formed in low yields):

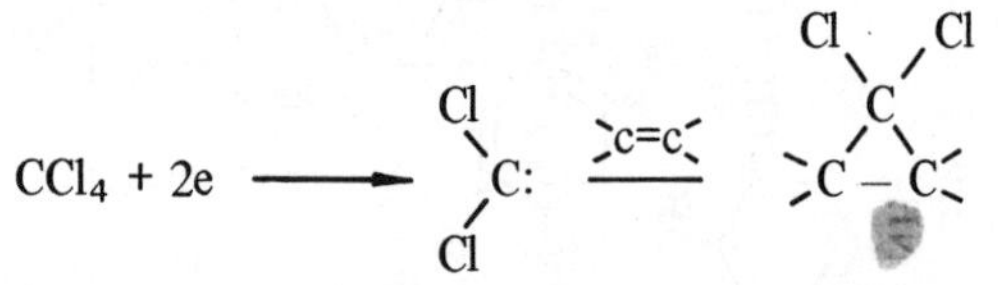

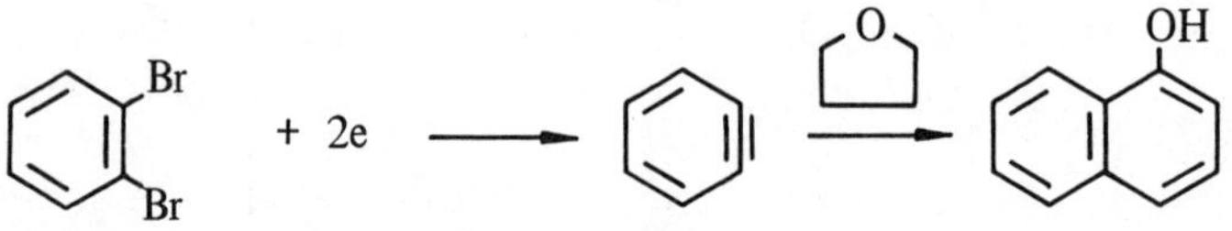

The reduction of dihalides has been explored by the author for the preparation of highly strained hydrocarbons such as bicyclobutane, bicyclopentane, spiro-pentane, and related compounds [392, 406, 408]:

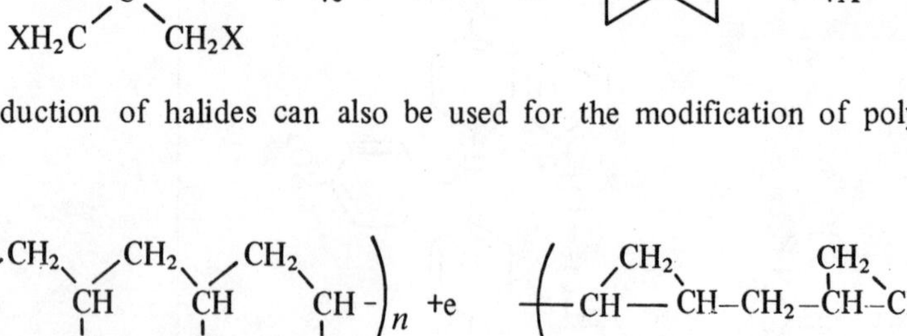

The reduction of halides can also be used for the modification of polymers [432]:

Polyvinyl chloride

Reaction Products

In general, the cathodic reduction of monohalides results in the replacement of halogen with hydrogen [389, 391, 394]. However, in certain activated halides, dimerization occurs [397]:

$$O_2N\text{—}\underset{}{\bigcirc}\text{—}CH_2Br + 2e \xrightarrow{CH_3OH} O_2N\text{—}\underset{}{\bigcirc}\text{—}CH_2\text{—}CH_2\text{—}\underset{}{\bigcirc}\text{—}NO_2$$

Vicinal dihalides, upon reduction, afford the corresponding olefins, which in general, do not undergo further reduction [389, 404, 405, 423]. 1,3-Dihalides give cyclopropanes and some olefins [392]. 1,4-Dihalides yield cyclobutanes and straight chain hydrocarbons [392]. *gem*-Halides afford mostly hydrocarbons [409-411]; however, carbene intermediates have been trapped [409]. In general the nature of the product is not related to that of the cathode material or current density [342], although there are some exceptions [398, 425].

Reduction under Controlled Potential

The case of reduction of carbon halogen bonds is in the following order: RI > RBr > RCl > RF. Table 8.4 illustrates the effect of structure on the half-wave potential of halides.

It can be seen from Table 8.4 that if a molecule contains more than one halogen, the difference in their half-wave potential will allow the selective reduction of only one, under controlled potential [428]. This is illustrated by the following examples:

$$Br\text{—}\underset{}{\bigcirc}\text{—}\overset{O}{\overset{\|}{C}}\text{—}CH_2\text{—}CH_2\text{—}CH_2Cl + 2e \xrightarrow{DMF} \underset{}{\bigcirc}\text{—}\overset{O}{\overset{\|}{C}}\text{—}CH_2\text{—}CH_2\text{—}CH_2Cl \quad 96\%$$

$$Br\text{—}\underset{}{\bigcirc}\text{—}I + 2e \xrightarrow{DMF} \underset{}{\bigcirc}\text{—}Br \quad 98\%$$

The reduction of polyhalides under controlled potential electrolysis can be used as a powerful tool in organic synthesis, as can be seen from the reactions below [392, 432]:

$$\underset{H_3C\ CH_3}{\overset{H_3C\ CH_3}{Cl_2\diamondsuit Cl_2}} + 2e \xrightarrow[DMF]{-2.0\ V\ (SCE)} \underset{H_3C\ CH_3}{\overset{H_3C\ CH_3}{Cl\text{—}\diamondsuit\text{—}Cl}}$$

$$\underset{\substack{\text{(III)}}}{\overset{\substack{\text{(I)}}}{\underset{BrH_2C}{\overset{BrH_2C}{\bigg\rangle}C\bigg\langle}\overset{CH_2Br}{\underset{CH_2Br}{}}}} \quad + \quad 2e \quad \xrightarrow[\text{DMF}]{-1.4 \text{ V (SCE)}} \quad \underset{\substack{\text{(IV)}}}{\overset{\substack{\text{(II)}}}{\triangleright}\overset{CH_2Br}{\underset{CH_2Br}{}}}$$

Table 8.4 Half-wave Potentials of Organic-halides[a,b]

Organic Halide	Solution	-E½ (SCE)	Ref
CH_3I	75% Dioxane	1.63	43
CH_3Br	75% Dioxane	2.01	43
CH_3Cl	75% Dioxane	2.23	
⟨⟩-CH_2Br	DMF	1.22	43
$CH_2 = CH-Br$	75% Dioxane	2.47	38
⟨⟩-Br	DMF	2.32	43
⟨⟩-Cl	DMF	2.54	43
$CH_2=CH-CH_2Br$	DMF	1.29	41
$Br-CH_2-CH_2Br$	DMF	1.38	39
$\underset{BrH_2C}{\overset{BrH_2C}{\bigg\rangle}C\bigg\langle}\overset{CH_2Br}{\underset{CH_2Br}{}}$	DMF	1.8, 2.3	43
$Br-CH_2-CH_2CH_2Br$	DMF	1.91	39
$Br-CH_2(CH_2)_2Br$	DMF	1.99	39
CH_2Cl_2	75% Dioxane	2.23	38
CH_2Br_2	75% Dioxane	1.48	38
$CHBr_3$	75% Dioxane	0.49, 1.09	38
CBr_4	75% Dioxane	0.30, 0.75	38
CCl_4	75% Dioxane	0.71, 0.78	38

[a]Dropping mercury was used as the cathode.
[b]For further comparison of half-wave potentials of halides, the reader should consult Refs. [445] and [446].

n any macroscale controlled potential reduction, a knowledge of the polarographic behavior of the starting material is essential. For example, compound **I** exhibits two, two-electron polarographic waves in DMF at -2.04 and -2.4 V (SCE). Its reduction at -2.0 V (SCE) allowed the formation of **II**, while reduction at -2.4 V (which is the $E_{1/2}$ of compound **II**) afforded products which could not be identified. No reduction took place between **I** and sodium or potassium metal. The reduction of compound **III** with conventional reducing agents for example, zinc, sodium, or magnesium, [437-439] gives several products which include spiropentane. However, due to subsequent reduction of **IV** under these reaction conditions, its isolation has not been possible. Compound **III** exhibits two, two-electron polarographic waves at -1.8 and -2.3 V (SCE). Its electrolysis at -1.4 V (SCE) afforded **IV** [432]; however, reduction under uncontrolled potential gave spiropentane in high yield [433]. The reduction of **IV** at -2.0 V afforded spiropentane in over 50%, thus demonstrating that the formation of spiropentane from the reduction of the tetrabromide **III** proceeds through the intermediate **IV**.

The above reactions clearly demonstrate the merit of controlled potential electrolysis in conjunction with polarography, in electroorganic synthesis. The synthesis of compound **IV** by conventional methods involves four steps [440]. While under controlled-potential electrolysis it can be prepared in one step.

Mechanism of Reduction

General Considerations

In spite of the numerous papers [396, 397, 399, 419] that have been published concerning the electrochemical cleavage of the carbon halogen bonds, no single mechanism can be used to unambiguously explain all the data on hand. In general, the electrochemical reduction of organic halides is irreversible and is independent of the pH of the medium. Elving and Pullman [430] have advanced the following general mechanism:

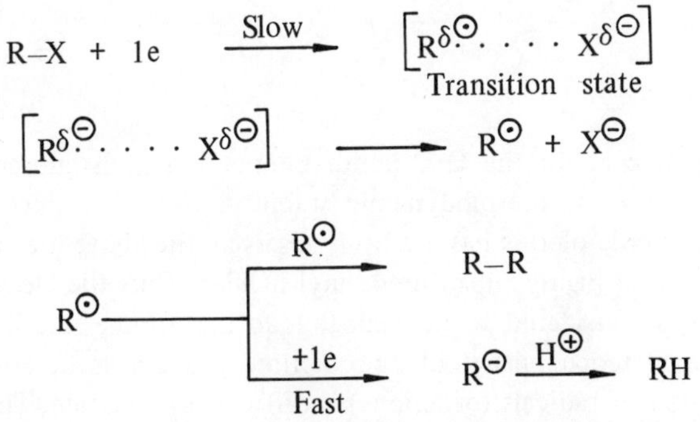

$$R{-}X + 1e \xrightarrow{\text{Slow}} \left[R^{\delta \ominus} \cdots X^{\delta \ominus} \right]$$
$$\text{Transition state}$$

$$\left[R^{\delta \ominus} \cdots X^{\delta \ominus} \right] \longrightarrow R^{\ominus} + X^{\ominus}$$

$$R^{\ominus} \begin{cases} \xrightarrow{R^{\ominus}} R{-}R \\ \xrightarrow[\text{Fast}]{+1e} R^{\ominus} \xrightarrow{H^{\oplus}} RH \end{cases}$$

Alkyl Halides

The intervention of anionic intermediates in the cleavage of monohalides is widely accepted. However, it is the path(s) that leads to these intermediates that has caused much controversy. The mechanism of Elving proposes that one electron is added, in a slow step, to form a radical intermediate. Yet it is questionable whether any of the products that have been isolated which require such intermediates as dimeric products may be formed from the attack of anion on the starting material, that is, alkyl halide. Furthermore, while the addition of one electron is quite popular, the addition of two electrons [394, 408, 429] should not be neglected as this will result in no violation of any fundamental principles. Consequently, the mechanism of the electrochemical reduction of alkyl halides may be envisioned as follows:

$$RX + 1e \xrightarrow{\text{slow}} R^{\odot} \xrightarrow[+1e]{\text{fast}} R^{\ominus}$$

or

$$RX + 2e \xrightarrow{\text{slow}} R^{\ominus} + X^{\ominus}$$

$$R^{\ominus} \begin{cases} \xrightarrow{RX} R\text{–}R \\ \xrightarrow{H^{\ominus}} RH \end{cases}$$

Vinyl Halides

The reduction of vinyl-halides may involve an initial addition of electrons to the lowest vacant π molecular orbital [436] to form the radical anion or dianion:

or

which can rotate about the C–C bond. Elimination of halide anion gives the olefin with the most thermodynamic stability. However, since the cathodic reduction of simple olefins has not been observed, the above mechanism may be applicable only to highly substituted vinyl halides. Thus the cleavage of simple vinyl halides is considered to be analogous to that of the simple monohalides. Whether the stereochemistry of the olefinic product is determined from a preequilibration of radicals (or anions) is difficult to ascertain. Thus while Skell

[443] considers the vinyl radical to be configurationally stable, others [444] have suggested that it undergoes rapid cis-trans equilibration.

Aromatic Halides

Aromatic halides reduce in an overall two-electron mechanism with the replacement of halogen by hydrogen:

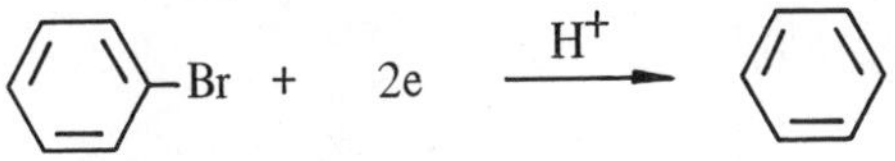

In the presence of carbon dioxide, benzoic acid can be isolated [432] together with benzene. This strongly suggests that the mechanism of reduction involves anionic intermediates.

Vicinal and a, ω-Dihalides

Vicinal dihalides reduce at a more positive potential than monohalides or 1,3-dihalides [408] and afford the corresponding olefins. Furthermore, trans dihalides reduce at a more positive potential than cis dihalides. This observation was explained as being due to the dihedral angle between the two carbon-halogen bonds [445]. A mechanism that has been recently advanced by the author [408] and which appears to accommodate vicinal as well as a, ω-dihalides in general is as follows:

Transition state

In this mechanism the degree of development of the central C–C bond in the transition state will be reflected in the half-wave potential of the dihalide. As n increases the central bond becomes weaker and the energy of the transition state (and hence the activation energy of the reaction) increases [408]. For $n = 0$ or 1, the central bond is so well developed that the reaction is considered to proceed in the concerted manner. This is supported by the finding that even affords high yield of cyclopropane [408].

In the vicinal unsaturated dihalides, trans isomers reduce at potentials which are more positive than those for the corresponding cis isomers [504]. This is an indication of a concerted mechanism for the trans isomer in which the elimination of the dihalide would be expected to proceed via a lower transition state energy than the corresponding cis analog [505].

In a recent article Fry [608] reported the electrochemical reduction of

stereoisomeric 2,4-dibromopentanes which gives trans and cis-cyclopropane:
Based on this observation, a stepwise mechanism involving rapid cyclization b
the carbanion intermediate was proposed:

$$CH_3 - \underset{\underset{Br}{|}}{CH} \overset{CH_2}{\underset{\underset{Br}{|}}{CH}} CH-CH_3 \xrightarrow{+2e} \triangle$$

dl or meso   Mixture

While this explanation is plausible, it need not be true for all 1,3-dihalides. Fo
if these cleavages do, in fact, proceed in a stepwise manner, an explanatio
(other than inductive effect) must be advance for the substantial differences i
their half-wave potentials [445]. The following example [609] will illustrat
this point:

(II)   100%

(III)   OEt

The monoethyl ether was explained as derived from the solvolysis of the mono
halide. The half-wave potential of **II** is about 0.34 V (>7 kcal/mole) mor
positive than that of compound **III**. [This datum was obtained by the autho
who kindly thanks Dr. S. Cristol for the supply of the dibromo compounds.
If the anion from compound **II** undergoes cyclization exclusively, then it mu:
cyclize at such a fast rate so as to have essentially no lifetime. In this case, :
becomes a matter of "semantics" whether the reduction proceeds via a "cor
certed" (i.e., practically no lifetime for a genuine anion) or stepwise mechanisn
The point to keep in mind here is that the geometry of 1,3-dihalides plays a ro
in determining the mechanism of reduction.

gem-Halides

In general, *gem*-halides reduce in a stepwise manner and affect the replacemen
of halogens with hydrogens [409-411]. In certain cases, carbenes have bee
trapped [409]:

$$CCl_4 + 2e \longrightarrow \overset{\ominus}{\underset{\underset{Cl}{|}}{\overset{\overset{Cl}{|}}{C}}}-Cl \longrightarrow : CCl_2 + Cl^{\ominus}$$

$$.CCl_2 + \overset{}{\underset{}{\diagdown}}C{=}C\overset{}{\underset{}{\diagup}} \longrightarrow \overset{Cl\diagdown \diagup Cl}{\underset{}{\underset{}{\diagup C \diagdown}}}\quad \overset{}{\underset{}{\diagdown}}C{-}C\overset{}{\underset{}{\diagup}}$$

Experimental

REDUCTION OF 1,3-DIBROMOPROPANE [392]

A solution of 50 g (0.2 mole) of 1,3-dibromopropane was dissolved in 1 liter 0.1 N DMF LiBr in DMF solution, which was then placed into the cathode

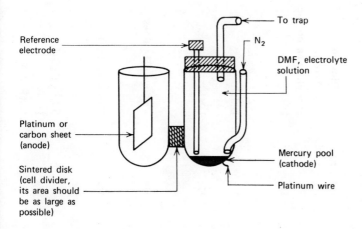

Fig. 8.2 Reduction of Organic Halides.

compartment. The anode compartment was charged with 1 liter solution of 0.1 N LiBr in DMF (see Fig. 8.2). A dry-ice-acetone trap was connected to the cathode compartment. Electrolysis proceeded at an overall cell voltage of 40-50 V and a current of 0.3-0.5 A for 12 hr. A colorless liquid was found in the trap. This was identified as cyclopropane from its NMR spectrum which exhibited a singlet (CCl_4, TMS) at 9.78 γ. A trace of DMF was also found in the trap. Based on repetitive runs the yield was about 80-85%.

Table 8.5 Organic Halides

I. ALIPHATIC

A. Monohalides

Methyl iodide	Hg	75% Dioxane	Methane	—	389
Ethyl bromide	Hg	75% Dioxane	Ethane	—	389
	Pb	Propylene carbonate	Tetraethyl lead	100	390
1-Bromohexane	Hg	DMF	Hexane	80	391
			Hexene	10	391
			Dihexylmercury	Trace	391
6-Bromo-1-hexene	Hg	DMF	1-Hexene	80	391
	Hg	DMF	Methylcyclopentane	7	391
$\overset{\oplus}{N}-CH_2-CH_2-CH_2Br$ $(C_2H_5)_3 \, X^{\ominus}$			Cyclopropane	—	392
$H_3C-\overset{\overset{\displaystyle O}{\|}}{C}-CH_2-X$ X=Br, I, Cl	Hg	Various pH values	Acetone	—	393
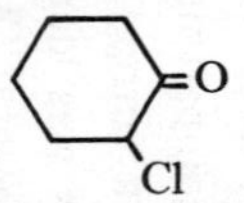 cyclohexanone (2-Cl)	Hg	Various pH values	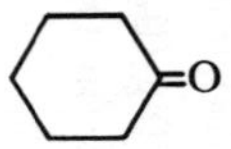 cyclohexanone	—	394
Benzylbromide	Hg	50% CH₃OH	Dibenzylmercury	—	396
Benzylchloride	Hg	DMF	Toluene	75	396
		DMF, CO₂	Phenylacetic acid	—	396

O_2N—C$_6$H$_4$—CH$_2$Cl	Hg	CH$_3$OH	(O_2N—C$_6$H$_4$—CH$_2$—)$_2$	—	397
H$_5$C$_6$, H$_5$C$_6$, Br, R (1) R = CO$_2$CH$_3$ = CO$_2$H	Hg	C$_2$H$_5$OH	H, R Partial inversion	—	398
(2) R = CH$_3$	Hg	C$_2$H$_5$OH	Partial retention	—	398
H$_5$C$_6$, H$_5$C$_6$, Br, CH$_3$	Hg	CH$_3$CN	H, CH$_3$ 25% Retention	93	399
	C		H, CH$_3$ Retention of configuration	—	403
H$_5$C$_6$, H$_5$C$_6$, Br, CH$_2$OCH$_3$	Hg	CH$_3$CN	H, CH$_2$OCH$_3$ 31% Optically pure	89.5	403
H$_5$C$_6$			=CH$_2$	6.7	403
H$_5$C$_6$, H$_5$C$_6$, I, CH$_3$	Hg	CH$_3$CN	H, CH$_3$ 3.9% Optically pure	53	403

Table 8.5 Organic Halides (continued)

Substrate	Electrode	Conditions	Product	Yield	Ref
H_5C_6, H_5C_6 cyclopropane with HgBr, CH_3; $[\alpha]^{26}_{5461} = -135°$	Hg	CH_3CN	$(H_5C_6)_2Hg$ with CH_3, H_5C_6	44	403
			H_5C_6–$C(CH_3)$ $=$ C–CH_3, CH_3	3.5	403
			H_5C_6, H_5C_6 with H, Hg, CH_3 $_2$	35	403
			H_5C_6, H_5C_6 cyclopropane $\left(CH_3 \right)_2 Hg$; $[\alpha]^{27}_{5461} = +150°$	Trace	403
Cl–$C(C_6H_5)(CH_3)$–CO_2H	Hg	96% EtOH 0.1 M Et$_4$NCl	H, H_5C_6–C–CO_2H, CH_3 (Inversion)	—	386
Cl–$CH_2CH_2CH_2CN$	Hg	Aq. solution	Adiponitrile	—	401
			Propionitrile	—	401
H_5C_2–$C(H)$ $=$ C, C_2H_5, I	Hg	DMF	H_5C_2, H C $=$ C, C_2H_5, H 90%	30	424
			H, H_5C_2 C $=$ C, C_2H_5, H	70	424

Compound	Metal	Solvent	Product	Yield (%)	Page
$\underset{H_5C_2}{\overset{H}{\big\rangle}}C=C\underset{I}{\overset{C_2H_5}{\big\langle}}$	Hg	DMF	$\underset{H}{\overset{H_5C_2}{\big\rangle}}C=C\underset{H}{\overset{C_2H_5}{\big\langle}}$ (90%) $\underset{H_5C_2}{\overset{H}{\big\rangle}}C=C\underset{H}{\overset{C_2H_5}{\big\langle}}$	6	424
$\underset{H_5C_6}{\overset{H_5C_6}{\big\rangle}}C=C\underset{Br}{\overset{C_6H_5}{\big\langle}}$	Hg	DMF	$(H_5C_6)_2CH-CH_2C_6H_5$	94	402
$\underset{HO_2C}{\overset{H}{\big\rangle}}C=C\underset{CO_2H}{\overset{Br}{\big\langle}}$	Hg	C_2H_5OH Various pH values	Maleic acid	5-37	400
			Fumaric acid	21-100	400
			$HO_2C-CH_2CH_2\underset{\overset{|}{CO_2H}}{CH}-\underset{\overset{|}{CO_2H}}{CH}-CH_2CO_2H$	—	400
$\underset{H}{\overset{HO_2C}{\big\rangle}}C=C\underset{CO_2H}{\overset{Br}{\big\langle}}$	Hg	C_2H_5OH	$\underset{H}{\overset{HO_2C}{\big\rangle}}C=C\underset{CO_2H}{\overset{H}{\big\langle}}$	84-100	400

B. Dihalides

Compound	Metal	Solvent	Product	Yield (%)	Page
$BrCH = CHBr$	Hg	75% Dioxane	Acetylene	—	389
$BrCH_2CH_2Br$	Hg	75% Dioxane	Ethylene	80	389
			Ethane	10	
$H_3C-\underset{\overset{|}{Br}}{CH}-\underset{\overset{|}{Br}}{CH}-CH_3$	Hg	75% Dioxane	Butene	—	389
$F_3CHCHClBr$	Pb	Basic solution	$F_2C = CHCl$	70	404

181

Table 8.5 Organic Halides (continued)

Compound	Metal	Solvent	Product	Yield	Ref.
$H_5C_6\diagdown$ $C=C=C=C\diagup C_6H_5$ $H_5C_6\diagup$ $\diagdown C=C\diagdown C_6H_5$ with Cl, Cl	Hg	DMF	$H_5C_6\diagdown C=C=C=C=C\diagup C_6H_5$ $H_5C_6\diagup \diagdown C_6H_5$	—	405
$Br-CH_2-CH_2CH_2Cl$	Hg	DMF	Cyclopropane	>80	406
$Br-CH_2-CH_2CH_2Br$	Hg	DMF	Cyclopropane	>80	392
	Pt	DMF	Cyclopropane	>80	392
Cl, Br (bicyclic)	Hg	DMF		—	392
				—	392
CH_3, Br, CH_3 (cis, trans)	Hg	DMF	H_3C, CH_3	58-90	392
OH, $Br-CH_2-CH-CH_2Br$	Hg	DMF	OH (cyclopropane)	—	406
Br, Br (adamantane)	Hg	DMF	Adamantane	—	406

Reactant	Electrode	Conditions	Product	Yield (%)	Ref.
$Br–CH_2(CH_2)_2–CH_2Br$	Hg	DMF	Cyclobutane	25	392
			n-Butane	75	392
	Hg	DMF	$BR–CH_2–CH_2–CH_2–CH_3$	—	407
$Br–CH_2$––[cyclobutane]––Br	Hg	DMF	[cyclobutane]–CH_3	—	408
	Hg	Hexamethylphosphoramide	$CH_2=CH–CH_2–CH=CH_2$	—	408
			[bicyclobutane]		
$HO_2C–CH–CH–CO_2H$ with $\overset{|}{Br}\ \overset{|}{Br}$	Hg	Various pH values	$CH_2=CH–CH_2–CH=CH_2$	—	
			Fumaric acid	—	418
			Maleic acid	—	418
$Br–CH_2(CH_2)_3–CH_2Br$	Hg	DMF	n-Pentane	80	392
[hexafluorocyclohexene, F_2 substituted]	Hg	Aq. C_2H_5OH	Hexafluorobenzene	50	414
[pentafluorobenzoic acid, CO_2H]	Hg	20% H_2SO_4, 01.2 V (SCE)	[pentafluorobenzoic acid with CO_2H, F substituents]	75	499

183

Table 8.5 Organic Halides (continued)

Compound	Electrode	Conditions	Products	Ratio	Ref.
Cl–$CCl=C$–CCl / Cl–$C=CCl$–C–Cl (chlorinated structure)	Hg	20% H_2SO_4 −1.30 V (SCE)	fluorinated CH_2OH benzene derivatives	2:1	499
	Hg	—	fluorinated CH_2OH benzene derivatives	1:1.3	499
	Hg	—	H–$C=C=C$–CCl–H derivatives	—	496
Cyclohexane hexachloride	Hg	Benzene	—	—	413
(chlorinated cyclopentene structure)	Hg	Aq. EtOH	(chlorinated cyclopentadienyl $Cl^{\ominus}$ structure)	—	497
Aldrin, Dieldrin, Isodrin, Endrin, Dieldrin	Hg	—	—	Polarographic studies	498
Dieldrin	Hg	—	—	—	501

184

CH₂Br, CH₂Br (ortho)	Hg	DMF	[−H₂C—⟨ ⟩—CH₂−]ₙ *o*-Xylene	51	415
Br Br | | CH–CH–NO₂ (phenyl)	Hg	Aq. alcohol	CH=CH–NO₂ (phenyl)	4	415
CH₂Br, CH₂Br (meta)	Hg	DMF	*m*-Xylene	— / 81	502 / 415
BrH₂C—⟨ ⟩—CH₂Br (para)	Hg	DMF	[−H₂C—⟨ ⟩—CH₂−]ₙ	92	415
			(fused polycyclic structure)	5	415

185

Table 8.5 Organic Halides (continued)

186

Starting material	Product	Electrode	Conditions	Yield (%)	Ref.
CH_2-CH_2 linked bis(p-bromomethylphenyl) (CH_2Br, CH_2Br)	CH_2-CH_2 linked bis(p-methylphenyl) (CH_3, CH_3)	Hg	DMF	—	415
$BrH_2C-C(CH_2Br)-CH_2Br$, BrH_2C CH_2Br	cyclopropane with CH_2Br, CH_2Br	Hg	DMF, −1.2 V (SCE)	65	432
bicyclic Cl/Br and Br/Cl	bicyclo structure, and norbornane with H, Cl	Hg	DMF	65, 35	39
bicyclic Cl^{36}, Cl					
C. Gem-halides					
CH_2Cl_2	CH_3Cl	Hg	DMF	—	409
$CHCl_3$	CH_3Cl	Hg	DMF	—	409
CCl_4	H_3C, H_3C $C-C$ Cl Cl, CH_3, CH_3	Hg	CH_3CN 2,3-Dimethyl-2-butene	—	409
cyclopropane fused to $(CH_2)_n$ ring with X, X and H	cyclopropane fused to $(CH_2)_n$ ring with X, X and H	Hg	DMF, 0°C	65–95	425

Substrate	Metal	Conditions	Product	Yield (%)	Ref.
x = Br, Cl; n = 4–6 (bicyclic cyclopropane, H, H, X, $(CH_2)_n$)	Hg	DMF 0 °C	Br-cyclopropane (bicyclic product, H, H, X, $(CH_2)_n$)	16–39	425
dibromobicyclic cyclopropane (H, H, Br, Br)					
$(Cl)C_6H_4$–$C(Cl)_2$–CCl_3	Hg	Aq. EtOH	ClH_5C_6–$C{=}CCl_2$ / ClH_5C_6	73–91	426
$ClCH_2$–CCl_2CN	Hg	EtOH	$H_2C{=}C(Cl)$–CN	—	427
Cl_3CCO_2H	Hg	KCl, NH₃, NH₄Cl	$Cl_2CH\,CO_2H$	100	410
Br_3CCO_2H	Hg	Aq. solution	Acetic acid	—	411
F_3C–CO–C₆H₅	Hg	Aq. acid	Acetophenone	14	416
$Cl(CH_2)_4CCl_3$	Hg	NH₄NO₃ 5.0 A	$Cl(CH_2)_4CHCl_2$	92	506
$AcO(CH_2)_4CCl_3$	Hg	NH₄NO₃	$AcO(CH_2)_4CHCl_2$	93	506
$NC(CH_2)_4$–CCl_3	Hg	LiNO₃	NC–$(CH_2)_4CHCl_2$	91	506

Table 8.5 Organic Halides (continued)

Substrate	Metal	Conditions	Product	Yield (%)	Ref.
$Cl{-}CH_2CH{=}CH{-}CH_2{-}CCl_3$	Hg	Me_4NCl	$NC(CH_2)_5Cl$	96	506
$HO{-}CH_2{-}CH{=}CH{-}CH_2CCl_3$	Hg	NH_4NO_3	$Cl{-}CH_2CH{=}CH{-}CH_2CHCl_2$	64	506
	Hg	NH_4NO_3	$HO{-}CH_2{-}CH{=}CH{-}CH_2CHCl_2$	94	506
		Me_4NCl	$HO{-}CH_2{-}CH{=}CH{-}CH_2CH_2Cl$	91	506
$Cl{-}(CH_2)_3{-}CH{=}CCl_2$	Hg	Me_4NCl	$Cl{-}(CH_2)_3{-}CH{=}CHCl$	63	506
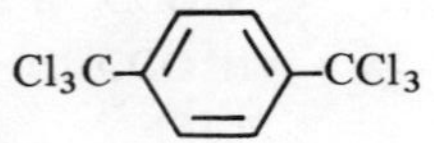 $Cl_3C{-}C_6H_4{-}CCl_3$ (1,4)	Hg	THF ($-10\,^{\circ}C$)	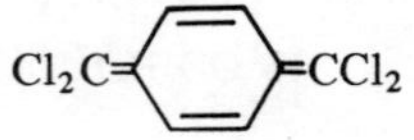 $Cl_2C{=}C_6H_4{=}CCl_2$	—	412
		HCl	$-[-\overset{Cl}{\underset{Cl}{C}}{-}C_6H_4{-}\overset{Cl}{\underset{Cl}{C}}-]_n-$	—	412
		Dioxane	$-[-\overset{Cl}{\underset{Cl}{C}}{-}C_6H_4{-}\overset{Cl}{\underset{Cl}{C}}-]_n-$	95	412
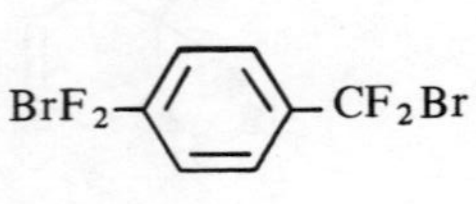 $BrF_2C{-}C_6H_4{-}CF_2Br$	Hg	Dioxane / Aq. HCl	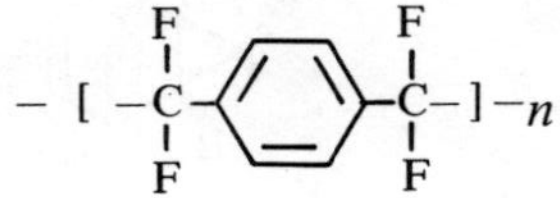 $-[-\overset{F}{\underset{F}{C}}{-}C_6H_4{-}\overset{F}{\underset{F}{C}}-]_n-$	95	412
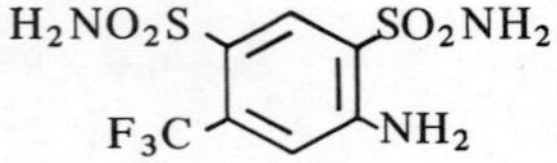 $H_2NO_2S{-}C_6H_2(CF_3)(NH_2){-}SO_2NH_2$	Hg	Aq. CH_3OH	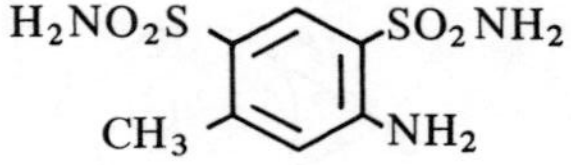 $H_2NO_2S{-}C_6H_2(CH_3)(NH_2){-}SO_2NH_2$	66.6	417

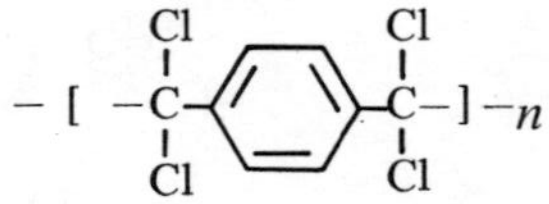

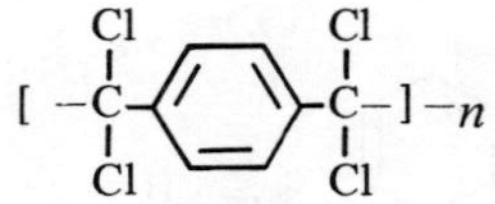

II. AROMATIC

A. Monohalides

Substrate	Metal	Solvent	Product	Yield (%)	Ref.
Bromobenzene	Hg	DMF	Benzene	100	391
2-Bromonitrobenzene (NO_2, Br ortho)	Hg	DMF, CO_2	Benzene	—	419
			Benzoic acid	~10	419
3-Bromonitrobenzene (NO_2, Br)	Hg	DMF	nitrobenzene radical ($\cdot$–NO_2), nitrobenzene cation ($^{\oplus}NO_2$)	—	420
4-Bromonitrobenzene (NO_2, Br)	Hg	DMF	O_2N–C_6H_4–CH_2–CH_2–C_6H_4–NO_2, NO_2-C_6H_4-CH_2–CH_2	—	420
4-Bromonitrobenzene (NO_2, Br)	Hg	CH_3OH	nitrobenzene radical ($\cdot$–NO_2), nitrobenzene cation ($^{\oplus}NO_2$)	—	397
3-Bromoacetophenone ($O=C$–CH_3, Br)	Hg	DMF	acetophenone (C_6H_5–CO–CH_3), Br–C_6H_4–$NO_2^{\oplus}$	—	420
	Hg	DMF	acetophenone	94	428

Table 8.5 Organic Halides (continued)

Halide	Metal	Conditions	Product		Ref.
(1-bromonaphthalene)	Hg	75% Dioxane	(naphthalene)	—	389
(3-iodobenzoic acid, CO$_2$H)	Hg	60% C$_2$H$_5$OH	Benzoic acid	—	421
(iodophthalic anhydride)	Hg	C$_2$H$_5$OH	(phthalic anhydride)	—	421
(4-chloro-2-aminopyridine)	Zn	NaOH, 80% C$_2$H$_5$OH	(2-aminopyridine, NH$_2$)	60	422
(H$_3$CO chloro-aminopyridine)	Zn	Aq. CH$_3$OH	(H$_3$CO, NH$_2$ aminopyridine)	—	422
(H$_3$C 4-chloro-6-methyl-2-aminopyridine)	Zn	NaOH, 80% C$_2$H$_5$OH	(CH$_3$, NH$_2$ pyridine)	60	422

B. Dihalides

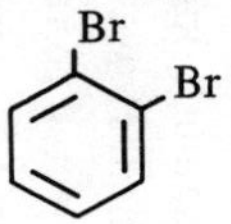

Substrate	Metal	Solvent	Product	Yield	Ref.
(1,2-dibromobenzene)	Hg	DMF, Furan		1	423
(1,3-dibromobenzene)	Hg	DMF or CH₃CN	Benzene	—	423
(1,4-dibromobenzene)	Hg	DMF	Benzene	—	423
(1-bromo-4-iodobenzene)	Hg	DMF	(bromobenzene)	98	428
CH₂CH₂CH₂Cl	Hg	DMF	CH₂CH₂CH₂Cl	99	428
O=C—CH₂CH₂CH₂Cl	Hg	DMF	O=C—CH₂CH₂CH₂Cl	96	428

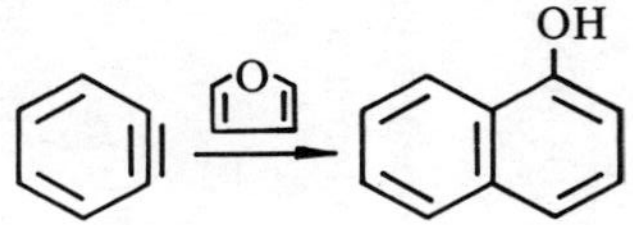

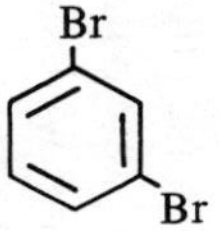

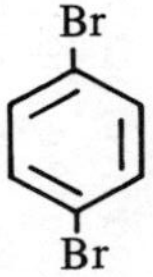

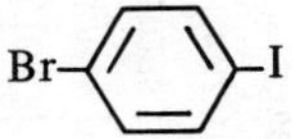

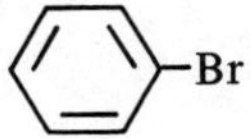

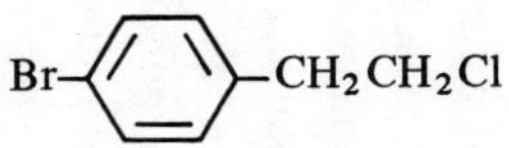

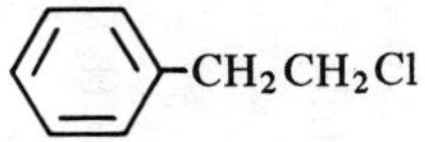

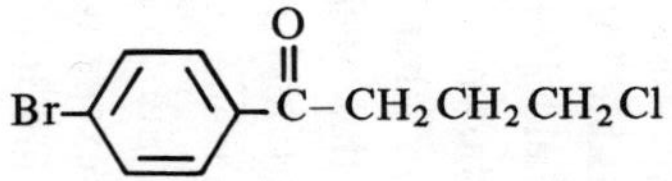

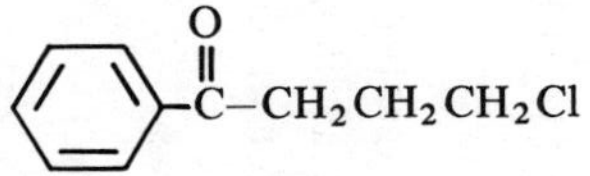

191

6 OLEFINS AND ACETYLENES

Olefins: General Considerations

The industrial preparation of adiponitrile from the electrochemical dimerization of acrylonitrile [101-104] represents a major shift of interest in electroorganic chemistry as a preparative tool in organic synthesis.

$$2\ CH_2{=}CH{-}CN + 2e + 2H^+ \longrightarrow NC{-}CH_2{-}CH_2{-}CH_2{-}CH_2{-}CN$$

This reaction, developed by Baizer and co-workers, has opened up a new
tion of acrylonitrile [101-104] represents a major shift of interest in electroreduction of olefins:

$$\underset{RO-\overset{O}{\overset{\|}{C}}-CH=CH}{}\ \overset{(CH_2)_n}{\underset{CH=CH-\overset{O}{\overset{\|}{C}}-OR}{/\ \ \ \ \ \backslash}}\longrightarrow RO-\overset{O}{\overset{\|}{C}}-CH_2-CH\underset{10\text{-}98\%}{\overset{(CH_2)_n}{\overset{/\ \ \ \ \ \backslash}{\underline{\quad\quad}}CH-CH_2-\overset{O}{\overset{\|}{C}}-OR}}$$

where n = 1-5

The reduction of activated olefins to afford polymeric products with different
properties than those obtained by conventional polymerization techniques has
also become quite popular in recent years [161].

In the reduction of simple olefins, it is generally believed that hydrogen atoms
(or molecules) generated at the surface of the electrode are the active species
responsible for reduction. For example, propylene is reduced to propane in over
90% yield [99]. However, recent studied by Sternberg and co-workers [149]
have shown that a further route to reduction may take place via the "solvated
electron" that is present in the bulk of the electrolysis medium. Thus electrolysis
using this technique was shown, in some cases, to have an advantage over
conventional reducing agents. For example, when 2,3-dimethyl-2-butene and
cyclohexene are electrochemically reduced in ethanol/hexamethylphosphor-
amide, the corresponding saturated hydrocarbons were obtained. However, only
cyclohexene was reduced when the two compounds were subjected to reduction
by lithium in ethylamine [610] (although it would be of interest to determine the

course of reduction of these compounds with lithium in ethanol/hexamethylphosphoramide).

Kuhn and Byrne [545] have recently studied the electrochemical reduction of ethylene on different cathodes and noted an increase in electrocatalytic activity with the following series of cathodes: Ru > Pt > Au > Cu. Electrochemical reduction of olefins in the presence of certain alkaloids, for example, brucine, can lead to the generation of optically active dihydro products [511]:

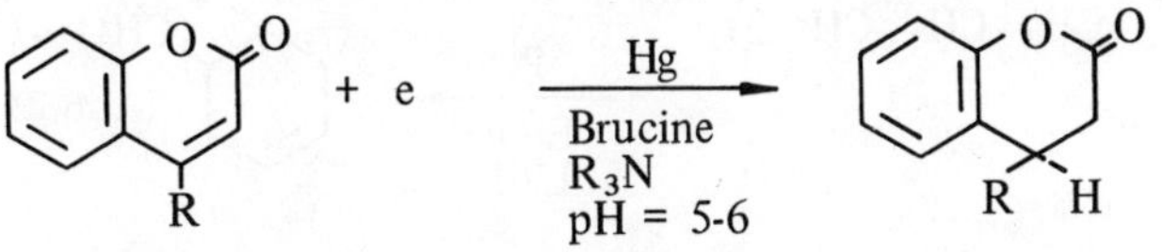

Reaction Products

The products that are formed in the electrochemical reduction of olefins may be classified in the following categories.

Saturated Hydrocarbons

These products are generally produced under protic conditions where hydrogen atoms (or molecules) generated at the cathode are responsible for the reduction:

$$CH_3{-}CH{=}CH_2 \xrightarrow[H_3PO_4]{Pt} \text{Propane} \quad 90\% \qquad [99]$$

In certain cases, reduction via the solvated electron may take place:

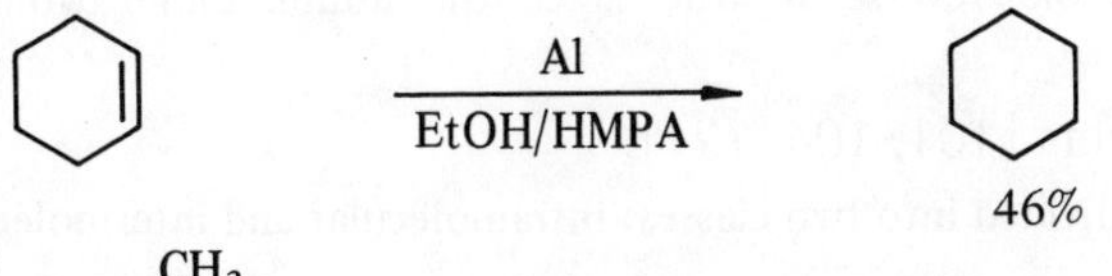

[508]

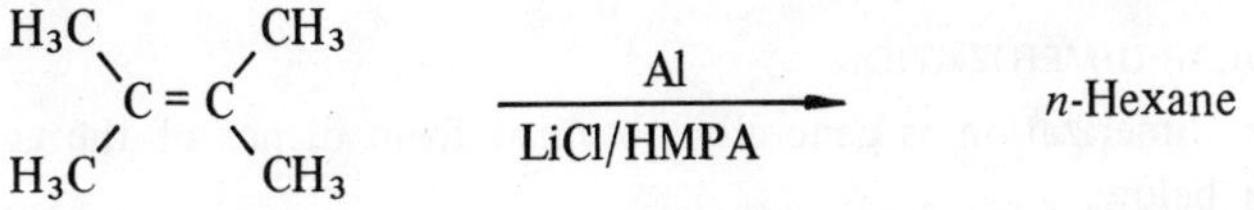

[149]

$$CH_3{-}(CH_2)_5{-}CH{=}CH_2 \xrightarrow[\text{ETOH, LiCl}]{Pt} CH_3{-}(CH_2)_6{-}CH_3 \quad 10\% \qquad [95]$$

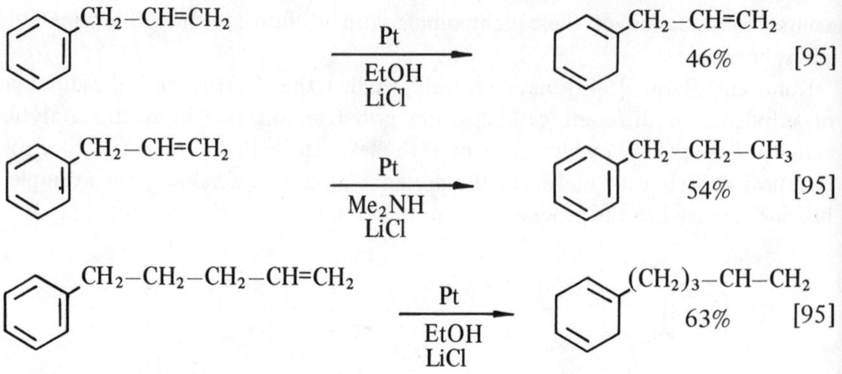

In general, the reduction of isolated double bonds is difficult to accomplish; however, Benkeser and co-workers explain the reduction of the above olefins by a preisomerization of the olefin to a conjugated system

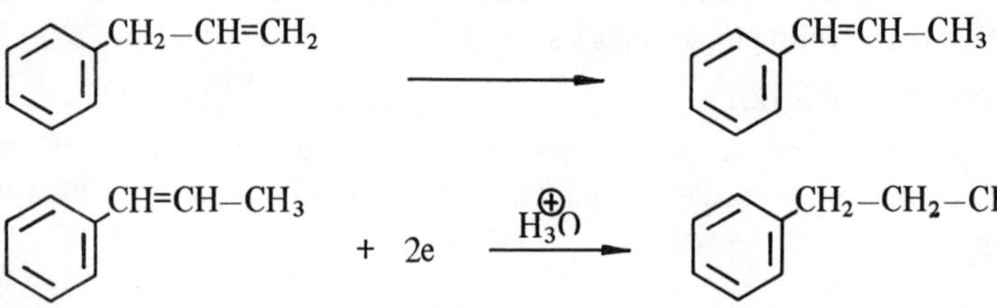

The electrochemical reduction reactions shown above demonstrate once again the importance of electrolysis over conventional reducing agents. Thus Benkeser has shown that the selective reduction of aromatic moieties in the presence of isolated double bonds is feasible. Using catalytic methods (e.g., Raney nickel and hydrogen), the reverse is true since the double bond would reduce preferentially.

Dimeric Products [101, 104, 127]

These may be divided into two classes: intramolecular and intermolecular:

INTRAMOLECULAR DIMERIZATION

This type of dimerization is generally obtained from dienes of the general formula shown below:

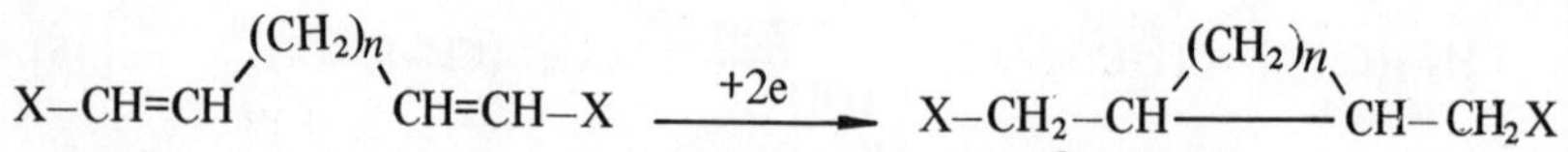

where X is an electron withdrawing group such as $-\overset{O}{\underset{\|}{C}}-OR$, CN.

INTERMOLECULAR DIMERIZATION

This includes the dimerization of an olefin with itself or with another olefin (i.e., mixed coupling):

$$2R_2C = \overset{|}{\underset{|}{C}} - X \longrightarrow X - \overset{R}{\underset{R}{C}} - \overset{R}{\underset{R}{C}} - \overset{R}{\underset{R}{C}} - \overset{|}{\underset{|}{C}} - X$$

$$R_2C = \overset{|}{\underset{|}{C}}X + R_2'C = \overset{|}{\underset{|}{C}} - Y \rightarrow X - \overset{R}{\underset{R}{C}} - \overset{R'}{\underset{R'}{C}} - \overset{|}{\underset{|}{C}} - Y$$

Polymeric Products

Activated olefins can be made to polymerize to a high molecular weight. under the proper reaction conditions:

$$[2]$$

$$CH_2 = CH - CN \xrightarrow[\substack{R_4NClO_4 \\ 25°C}]{Al} (-CH_2 - \underset{CN}{\overset{|}{CH}} -)_n \qquad [2]$$

$$[2]$$

$$CH_2 = \underset{CH_3}{\overset{|}{C}} - CH = CH_2 \xrightarrow[THF]{Pt} \text{Polyisoprene} \qquad [168]$$

Substituted Products

Intermediates produced in the reduction of olefins have been trapped by reagents such as SO_2, CO_2 [611, 131, 522, 540] to afford the following products:

Dimerization of Acrylonitrile: The "Baizer-Monsanto Process"

The electrohydrodimerization of acrylonitrile best demonstrates the important role of electrolysis in organic chemistry, particularly since this process has achieved importance on the industrial scale. For this reason, a detailed description of the process is described.

Adiponitrile is a key intermediate in the preparation of nylon, a plastic whose yearly production is close to a billion pounds. Conventionally, nylon 6.6 is prepared as follows:

$$HO-\overset{O}{\underset{\|}{C}}-(CH_2)_4-\overset{O}{\underset{\|}{C}}-OH \xrightarrow[2NH_3]{Fixed\ bed} NC(CH_2)_4CN+4H_2O$$

$$NC-(CH_2)_4-CN + 4H_2 \longrightarrow H_2N-(CH_2)_6-NH_2$$

$$HO-\overset{O}{\underset{\|}{C}}-(CH_2)_4-\overset{O}{\underset{\|}{C}}OH + H_2N-(CH_2)_6-NH_2 \rightarrow (-\overset{O}{\underset{\|}{C}}-(CH_2)_4-\overset{O}{\underset{\|}{C}}-NH-(CH_2)_6-NH-)_n$$

Nylon 6,6

Compared to the dimerization of acrylonitrile, the preparation of adiponitrile from adipic acid has two shortcomings: (1) Adipic acid is almost twice as expensive as acrylonitrile, and (2) on a weight basis, more adipic acid is required than acrylonitrile in order to prepare the same amount of adiponitrile. The reason for this is the formation of four moles of water for every mole of adiponitrile formed from adipic acid.

It can be seen from the above discussion that there was a great incentive for Baizer and co-workers to demonstrate the feasibility of the electrohydro-dimerization of acrylonitrile to adiponitrile.

The Electrolysis Cell

The design of a suitable electrolysis cell which would be practical in a large-scale operation was probably the most difficult task fo the project. Thus, because adiponitrile undergoes oxidation at the anode, a cell divider was required. Although the nature of this divider has not been revealed, it is believed to be a semipermeable cationic exchange resin membrane. The cathode used is lead while the anode is a special type alloy. The remainder of the cell is manufactured from polypropylene primarily.

The Supporting Electrolyte

The choice of a supporting electrolyte for the electrochemical preparation of a polymer-grade product is quite critical. Thus the supporting electrolyte must be inactive to electrolysis or if electrolyzed must not produce species that will react with the reactant or product. Furthermore, for economic reasons, it would be desirable to recover and reuse the supporting electrolyte.

The nature of the supporting electrolyte used in the cathode compartment has not been revealed, but it is believed to be a tetraalkylammonium salt which is prepared by the Monsanto Company. This supporting electrolyte is recovered and reused in the process. The supporting electrolyte on the anode side is a mineral acid that does not pose a corrosive problem for anode or membrane.

Concentration of Reagents

The concentration of acrylonitrile in the cathode compartment is crucial. Below 10%, propionitrile is formed in undesirable yield. On the other hand, a large excess of acrylonitrile affords polyacrylonitrile. Thus while the exact concentration of acrylonitrile used in the Baizer process has not been reported, it is probably in the neighborhood of 20-40%.

Chemical Reactions

The electrochemical reactions taking place in the electrolysis cell may be described as follows [612]:

CATHODE REACTION

$$CH_2 = CH-CN + 2e \longrightarrow [CH_2-CH-CN]^{\ominus}$$

The dianion formed probably undergoes one of the following chemical reactions:

(1) $[CH_2-CH-CN]^{\ominus} + H_2O \longrightarrow [CH_2-CH_2-CN]^{\ominus}$

$[CH_2-CH_2-CN]^{\ominus} + CH_2 = CH-CN \longrightarrow NC-CH_2-CH_2-CH_2-\overset{\ominus}{CH}-CN$

$NC-CH_2-CH_2-CH_2-\overset{\ominus}{CH}CN + H_2O \longrightarrow NC-(CH_2)_4-CN$

(2) $[CH_2-\overset{\ominus}{CH}-CN]^{\ominus} + CH_2 = CH-CN \longrightarrow NC-\overset{\ominus}{CH}-CH_2-CH_2-\overset{\ominus}{CH}-CN$

$NC-\overset{\ominus}{CH}-CH_2-CH_2-\overset{\ominus}{CH}-CN + H_2O \longrightarrow NC-(CH_2)_4-CN$

ANODE REACTION

$$H_2O-2e \longrightarrow \tfrac{1}{2}O_2 + 2H^{\oplus}$$

Large-Scale Operation

A brief description of the large-scale operation of the Baizer process is presented in the November issue of *Chemical Engineering 1965*. In general the operation is run on a continuous basis where the catholytes and anolytes are fed into their corresponding chambers and adiponitrile is isolated by means of liquid extractors and conventional distillation apparatus.

Mechanism of Reduction of Olefins and Acetylenes

The mechanistic paths for the reduction of olefins and acetylenes depend on the reaction conditions. Thus the following possibilities may be considered:

Reduction by Means of Hydrogen Atoms (or Molecules)

$$H^{\oplus} + 1e \longrightarrow H\cdot \longrightarrow \tfrac{1}{2}H_2$$

$$\overset{\diagup}{C} = \overset{\diagdown}{C} + 2H \longrightarrow \overset{\diagup}{C}H-CH_{\diagdown}$$

$$CH_3-CH = CH_2 + 2e \xrightarrow{\ H_3PO_4\ } CH_3-CH_2-CH_3 \qquad [99]$$
$$90\%$$

Alternatively,

$$\overset{\diagup}{C} = \overset{\diagdown}{C} + H\cdot \longrightarrow H-\overset{\diagup}{C}-\overset{\diagdown}{C\cdot}$$

$$2\ H\overset{\cdot}{C}-\overset{\cdot}{C}: \longrightarrow H\overset{\cdot}{C}-\overset{\cdot}{C}-\overset{\cdot}{C}-\overset{\cdot}{C}H$$

$$H\overset{\cdot}{C}-\overset{\cdot}{C}: + \overset{\diagup}{C} = \overset{\diagdown}{C} \longrightarrow \text{Polymer}$$

Direct Transfer of Electrons from Cathode to Olefin

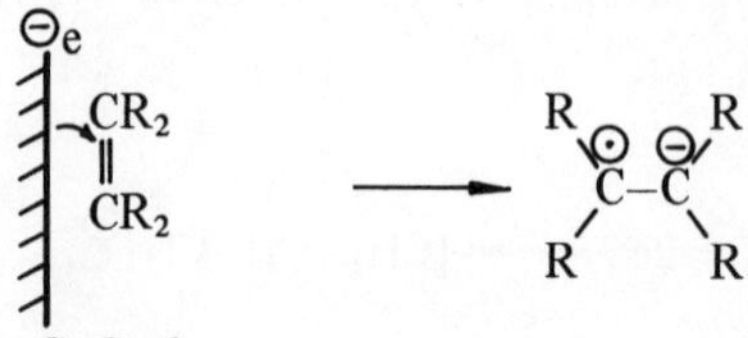

Cathode

The ease of electron transfer to olefins and acetylenes depends on their structure. Thus groups which stabilize the anion radical allow the reduction to take place at a more positive potential.

Consider the following reactions:

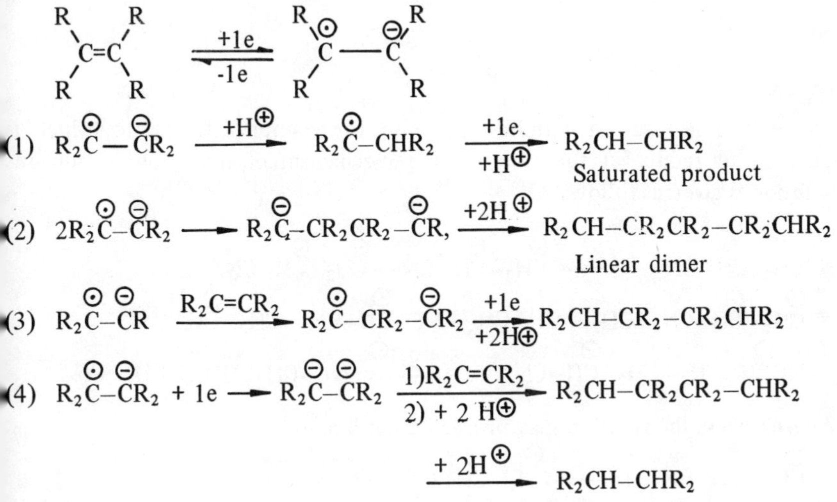

$$\underset{R}{\overset{R}{\diagdown}}C=C\underset{R}{\overset{R}{\diagup}} \quad \underset{-1e}{\overset{+1e}{\rightleftharpoons}} \quad \underset{R}{\overset{R}{\diagdown}}\overset{\odot}{C}-\overset{\ominus}{C}\underset{R}{\overset{R}{\diagup}}$$

(1) $R_2\overset{\odot}{C}-\overset{\ominus}{C}R_2 \xrightarrow{+H^{\oplus}} R_2\overset{\odot}{C}-CHR_2 \xrightarrow[+H^{\oplus}]{+1e} R_2CH-CHR_2$
Saturated product

(2) $2R_2\overset{\odot}{C}-\overset{\ominus}{C}R_2 \longrightarrow R_2\overset{\ominus}{C}-CR_2CR_2-\overset{\ominus}{C}R, \xrightarrow{+2H^{\oplus}} R_2CH-CR_2CR_2-CR_2CHR_2$
Linear dimer

(3) $R_2\overset{\odot}{C}-\overset{\ominus}{C}R \xrightarrow{R_2C=CR_2} R_2\overset{\odot}{C}-CR_2-\overset{\ominus}{C}R_2 \xrightarrow[+2H\oplus]{+1e} R_2CH-CR_2-CR_2CHR_2$

(4) $R_2\overset{\odot}{C}-\overset{\ominus}{C}R_2 + 1e \longrightarrow R_2\overset{\ominus}{C}-\overset{\ominus}{C}R_2 \xrightarrow[2) + 2\ H^{\oplus}]{1)R_2C=CR_2} R_2CH-CR_2CR_2-CHR_2$

$\xrightarrow{+ 2H^{\oplus}} R_2CH-CHR_2$

It can be seen from the above reactions 1-4 that several paths may be available for the formation of the observed products. The actual paths which are operative will depend on a number of factors: (1) the structure of the olefin, (2) the nature of the medium, that is, protic or aprotic, (3) the nature of the cathode [129], and (4) the counter ion of the supporting electrolyte [616]:

In the substituted ethylenes, for example, styrene, stilbene, and so on. Wawzonek and co-workers [613, 614] prefer the formation of dianions (even in 75% dioxane) as the reaction 4 above. The dianion is then protonated to give the observed saturated product. Other workers [615], however, have explained the two-electron pH-independent polarographic wave of such compounds according to reaction 2 above:

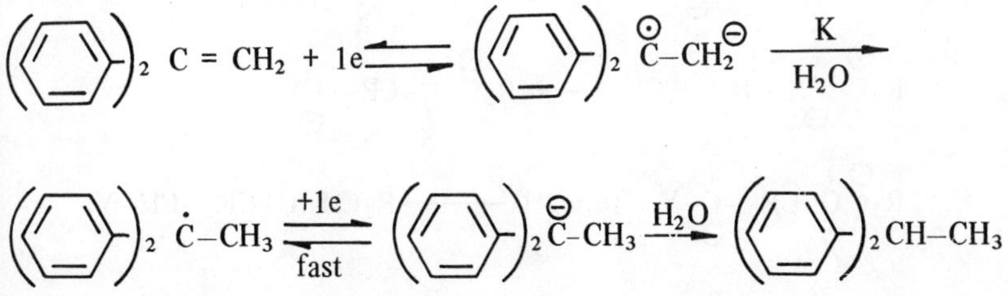

$$\left(\!\!\left(\!\!\left\langle\overline{}\right\rangle\!\!\right)\!\!\right)_2 C = CH_2 + 1e \rightleftharpoons \left(\!\!\left(\!\!\left\langle\overline{}\right\rangle\!\!\right)\!\!\right)_2 \overset{\odot}{C}-CH_2^{\ominus} \xrightarrow{K}{H_2O}$$

$$\left(\!\!\left(\!\!\left\langle\overline{}\right\rangle\!\!\right)\!\!\right)_2 \overset{\cdot}{C}-CH_3 \underset{fast}{\overset{+1e}{\rightleftharpoons}} \left(\!\!\left(\!\!\left\langle\overline{}\right\rangle\!\!\right)\!\!\right)_2\overset{\ominus}{C}-CH_3 \xrightarrow{H_2O} \left(\!\!\left(\!\!\left\langle\overline{}\right\rangle\!\!\right)\!\!\right)_2 CH-CH_3$$

This mechanism seems to be quite reasonable in view of the electron affinity of the radical as compared to the olefin.

The hydrodimerization of olefins has been studied extensively by Baizer and co-workers [101, 127, 128, 616, 617]. Thus for an olefin with the general structure:

$$\begin{array}{c} R \\ \diagdown \\ \diagup \\ R \end{array} C = \overset{\overset{\displaystyle R}{\displaystyle |}}{C} - X$$

where X is an electron withdrawing group, for example ester cyano, which is difficult to reduce at the cathode, the mechanism of reduction in aqueous solution is given as follows:

$$CH_2{=}CH{-}CN + 1e \longrightarrow \overset{\odot}{C}H_2{-}\overset{\ominus}{C}H{-}CN \longleftrightarrow \overset{\ominus}{C}H_2{-}\overset{\odot}{C}H{-}CN$$

$$\overset{\ominus}{C}H_2{-}\overset{\odot}{C}H{-}CN + CH_2 = CH{-}CN \longrightarrow NC{-}\overset{\odot}{C}H{-}CH_2{-}CH_2{-}\overset{\ominus}{C}H{-}CN$$

$$NC\overset{\odot}{C}H{-}CH_2{-}CH_2{-}\overset{\ominus}{C}H{-}CN + 1e \xrightarrow{\ +2H^{\oplus}} NC{-}CH_2CH_2CH_2CH_2{-}CN$$

Alternatively, the reaction may proceed according to

$$\overset{\odot}{C}H_2{-}\overset{\ominus}{C}H{-}CN + 1e \longrightarrow \overset{\ominus}{C}H_2{-}\overset{\ominus}{C}H{-}CN$$

or

$$CH_2{=}CH{-}CN + 2e \longrightarrow \overset{\ominus}{C}H_2{-}\overset{\ominus}{C}H{-}CN$$

$$\overset{\ominus}{C}H_2{-}\overset{\ominus}{C}H{-}CN + CH_2{=}CH{-}CN \longrightarrow NC{-}\overset{\ominus}{C}H{-}CH_2{-}CH_2{-}CH_2{-}\overset{\ominus}{C}H{-}CN \xrightarrow{H_2O}$$

$$NC{-}CH_2CH_2CH_2CH_2{-}CN$$

When more than one olefin is undergoing reduction, mixed coupling can take place [137]:

$$R_2C = \overset{|}{C}X + 1e \longrightarrow R_2\overset{\odot}{C}{-}\overset{|}{\underset{\ominus}{C}}{-}X$$

$$R_2\overset{\odot}{\underset{\ominus}{C}}{-}\overset{|}{C}X + R_2C = \overset{|}{C}X \longrightarrow R_2\overset{\odot}{C}{-}\overset{|}{\underset{X}{C}}{-}CR_2{-}\overset{|}{\underset{\ominus}{C}}{-}X$$

$$R_2\overset{\odot}{C}{-}\overset{|}{\underset{X}{C}}{-}CR_2{-}\overset{|}{\underset{\ominus}{C}}{-}X + 1e \xrightarrow{+2H^+} R_2C{-}\overset{|}{\underset{X}{C}}{-}CR_2{-}CHX$$

$$R_2\overset{\odot}{\underset{\ominus}{C}}{-}\overset{|}{C}X + R_2C = \overset{|}{C}Y \longrightarrow R_2\overset{\odot}{C}{-}\overset{|}{\underset{X}{C}}{-}CR_2{-}\overset{|}{\underset{\ominus}{C}}Y$$

$$R_2\overset{\odot}{C}\,\overset{|}{\underset{X}{C}}{-}CR_2{-}\overset{|}{\underset{\ominus}{C}}{-}Y + 1e + 2H^{\oplus} \longrightarrow R_2CH{-}\overset{|}{\underset{X}{C}}{-}CR_2{-}\overset{|}{C}H{-}Y$$

In a recent paper [128] Baizer and co-workers reported some interesting observations concerning the reduction of acrylonitrile in the presence of acetone, styrene, 1,1-diphenylethylene benzophenone, and benzaldehyde. When the reduction is carried out at potential that will reduce the compound with the more negative, half-wave potential, the current efficiency for the formation of mixed coupling products increased, the coupling which otherwise failed was found to occur, and the formation of polymeric products was decreased.

Certain dienes undergo electrochemical reduction to yield cyclic products. This method, which was also studied by Baizer and co-workers, may be of great synthethic utility:

$$
\begin{array}{c}
\diagup CH{=}CHX \\
(CH_2)n \qquad\qquad + 2e \longrightarrow \\
\diagdown CH{=}CX
\end{array}
\begin{array}{c}
\diagup CH{-}CH_2X \\
(CH_2)n \Big| \\
\diagdown CH{-}CH_2X \\
10\text{-}98\%
\end{array}
$$

where n = 1 - 5, X = CN, -CO$_2$R.

The polarographic behavior of several dienes is recorded in Table 8.6. It was found that dienes, which afforded cyclic products, reduced at a more positive potential than those which gave straight-chain hydrocarbons. Furthermore, this positive shift usually resulted in the development of two reduction waves that were difficult to discern (Fig. 8.3). The more anodic wave was considered to be

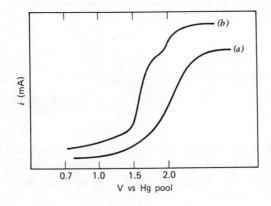

Fig. 8.3 Polarographic behavior of dienes that afford cyclic products upon reduction. (a) Diethyl 2,6-octadiene-1,8-dioate, (b) diethyl 2,7-nonadiene-1,9 dioate.

associated with the cyclization reaction and the more negative with the formation of linear hydrocarbons. This was consistent with the observation that the ratio of the wave heights correlated with the extent of cyclization. On the basis of these observations Baizer and co-workers proposed the following concerted mechanism [128]:

$$
\begin{array}{c}
\overset{(CH_2)_n}{X - CH = CH}\qquad\overset{(CH_2)_n}{CH = CH - X} \longrightarrow X - CH - CH\cdots\cdots CH\overset{\ominus}{\cdots}CH\cdots X \\
\underset{\boxed{cathode}}{+ 1e} \qquad\qquad\qquad\qquad\qquad \text{Transition state} \\
\qquad\qquad\qquad\qquad\qquad\qquad\qquad \boxed{cathode}
\end{array}
$$

$$
\overset{(CH_2)_n}{X - CH_2CH}\text{---}\overset{}{CH-CH_2} - X \xleftarrow[+ 2 H^+]{+ 1e} \overset{\odot}{X - CH} - \overset{(CH_2)}{CH}\text{---}CH - CH - X
$$

III. Reduction Via the Solvated Electron

This type of reduction is described in the section dealing with aromatic compounds. Olefins behave quite analogously.

ACETYLENES

Acetylenes are reduced at a slightly more negative potential than the corresponding olefins. Thus, the reduction of acetylenic compounds is general afford the corresponding saturated hydrocarbons. Campbell and co-workers [129] found that a good electrode for the reduction of acetylenic compounds is spongy nickel. Variations in temperature were found to have little or no effect on the current efficiency. The reduction was also found to be affected only slightly by an increase in current density:

$$
CH_3-(CH_2)_2 - C\equiv C-(CH_2)_2-CH_3 \xrightarrow[\substack{95\% \text{ EtOH} \\ H_2SO_4}]{\text{Spongy Ni}} \underset{\substack{cis\text{-4-Octene} \\ 08\%}}{\overset{\displaystyle H\diagdown\quad\diagup H}{\underset{H_3C(CH_2)_2}{C = C}\diagdown (CH_2)_2 CH_3}}
$$

The formation of the cis product was explained by invoking a mechanism that resembles catalytic hydrogenation. Of interest was the finding that when high overvoltage cathodes were used, no reduction of aliphatic acetylenic hydrocarbon could be obtained.

The reduction of several acetylenic compounds was reported by Benkeser and co-workers [95, 130]. In an attempt to determine whether reduction takes place at the suface of the electrode of in solution, the 2 and 3-octynes and 5-decyne were reduced and their products analyzed by means of VPC. In all cases, the trans isomers were isolated. It was demonstrated that under the analytical procedure the cis isomers were stable, thus indicating that all trans isomers were formed from direct reduction. Since in the reduction of such acetylenic compounds with lithium in amine solvent the same stereochemistry is obtained, Benkeser and co-workers prefer not to involve the surface of the cathode .

DIENES AND CUMELENES

For the reductions of dienes and cumelenes, the reader may consult Refs. 156 and 157.

Experimental

Reduction of Propylene and Similar Compounds

The reduction of such gaseous compounds is described in detail in Refs. [99] and [100].

Reduction of Activated Olefins [112]

$$R-CH{=}CH-CO_2Na + 2e + 2H \overset{\oplus}{\longrightarrow} R-CH_2-CH_2-CO_2Na$$

This reduction is described in *Organic Synthesis.*

Electroreductive Coupling of Olefins

The procedure for this technique as well as the apparatus are well described in Refs. 126 and 516.

Reduction of Acetylenes

A detailed reduction of acetylenes in described in Refs. 129 and 130.

Table 8.6 Olefins and Acetylenes

Reactants	Reaction Conditions		Products	Yield (%)	Ref.
	Cathode	Solvent			
Olefins					
$CH_2 = CH_2$	Pt	Aq. HCl	Ethane	—	100, 512
$F_2C = CF_2$	Ni	HF, KF	Polymer	—	545
$FCl-C = CF_2$	Ni	HF, KF	Polymer	—	94
$CH_3-CH = CH_2$	Pt	H_3PO_4	Propane	90	94
Cyclohexene	Al	Ethanol, hexamethyl phosphoramide	Cyclohexane	46	508
$CH_3(CH_2)_5CH = CH_2$	Pt	LiCl, CH_3NH_2	$CH_3(CH_2)_6CH_3$	10	95
$(H_3C)_2C = C(CH_3)_2$	Al	LiCl, HMPA	Hexane	—	149
$H_5C_6-CH = CH_2$	Pt	LiCl, CH_3NH_2	$H_5C_6CH_2CH_3$	69	95
(styrene)	Mg	Pyridine	Polystyrene	—	96
$CH = CH_2$ (phenyl), SO_2, O_2	Hg	DMF	(structure with SO_3H groups)	65	522
					12

Reactant	Catalyst	Solvent	Products		
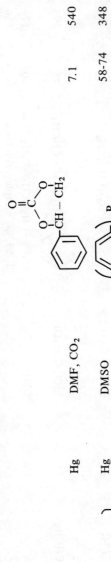 styrene (Ph–CH=CH₂)	Hg	DMF, CO_2	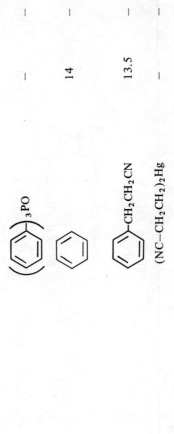	7.1	540
	Hg	DMSO	$(C_6H_5)_3P$	58-74	348
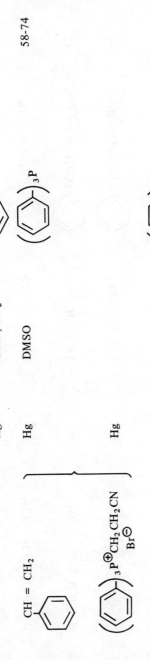 $(C_6H_5)_3\overset{\oplus}{P}CH_2CH_2CN\ \ Br^{\ominus}$	Hg		$(C_6H_5)_3PO$	—	—
			benzene	14	—
			C_6H_5–CH_2CH_2CN	13.5	—
			$(NC–CH_2CH_2)_2Hg$	—	—

Table 8.6 Olefins and Acetylenes (continued)

Substrate	Catalyst	Conditions	Product		
			$CH_2 = CH-CN$	—	—
			$NC-CH_2-CH_2CN$	—	—
			$NC-CH_2CH_2-CH$⟨C₆H₅⟩	—	—
			$NC-CH_2CH_2-CH$⟨C₆H₅⟩	—	—
$H_5C_6CH_2-CH = CH_2$	Pt	LiCl, CH_3NH_2	$H_5C_6(CH_2)_2CH_3$	54	95
$H_5C_6CH_2CH = CH_2$	Pt	LiCl, C_2H_5OH	(cyclohexadienyl)$-CH_2-CH = CH_2$	46	95
$H_5C_6(CH_2)_2CH = CH_2$	Pt	LiCl, C_2H_5OH	(cyclohexadienyl)$-(CH_2)_2-CH = CH_2$	52	95
$H_5C_6(CH_2)_3-CH = CH_2$	Pt	LiCl, C_2H_5OH	(cyclohexadienyl)$-(CH_2)_3CH = CH_2$	63	95
$H_5C_6(CH_2)_2CH = CHCH_3$	Pt	LiCl, C_2H_5OH	(cyclohexadienyl)$-(CH_2)_2CH = CHCH_3$	75	95
$H_5C_6CH = CHC_6H_5$	Hg	DMF	$H_5C_6-CH_2-CH_2-C_6H_5$	60	126
$(H_5C_6)_2C = CHC_6H_5$	Hg	DMF	1,2,3,4,-Tetraphenylbutane	—	126
			1,2,2-Triphenylethane	93	126

Substrate	Metal	Conditions	Product	Yield	Ref.
$(H_5C_6)_2C = C(C_6H_5)_2$	Hg	CH_3CN	1,1,2,2-Tetraphenylethane	48	126
			Diphenylmethane	40	126
$CH_2 = CH-CN$	Pt	DMF	Polyacrylonitrile	—	97
	Sn	NaOH	tetrabis(2-Cyanoethyl)tin	60	98
$CH_2 = CH-CN$	Pb	$R_4N\ X$, aq. sol.	Adiponitrile	—	101-104
		H_2SO_4, EtOH	$CH_3CH_2-CH_2NH_2$	—	510
			$CH_2 = CH-CH_2NH_2$	50	510
$CH_2 = \overset{\overset{\textstyle Cl}{\textstyle \vert}}{C}-CN$	Hg	40% NaOH	1,4-Dicyanobutane	68	105
$CH_3-CH = CH-CN$	Hg	40% NaOH	1,4-Dicyano-2,3-dimethylbutane	37	105
$C_6H_5-CH = CH-CN$	Hg	40% NaOH	1,4-Dicyano-2,3-diphenylbutane	60	105
$H_5C_6CH = CH-CN$	Hg	Aq. sol.	$H_5C_6CH-CH_2-CN$ $H_5C_6-CH-CH_2CN$ $H_5C_6CH-CH_2CH$ $NC-CH-CH_2C_6H_5$	16.2 41.9	127 127
$CH_2 = CH-\overset{\overset{\textstyle O}{\textstyle \|}}{C}-OCH_3$	Hg	40% NaOH	Dimethyladipate	27	105
$CH_2 = CH-\overset{\overset{\textstyle O}{\textstyle \|}}{C}-OC_2H_5$	Hg	40% NaOH	Diethyladipate	52	105
$CH_3-CH = CH-C-OC_2H_5$	Hg	40% NaOH	Diethyl 3,4-dimethyladipate	71	105

Table 8.6 Olefins and Acetylenes (continued)

Substrate	Catalyst	Conditions	Product		Ref
C_6H_5 O $CH_2 = C{-}C{-}OC_2H_5$ · (O)	Hg	40% NaOH	Diethyl 2,5-diphenyladipate	28	105
$CH_2 = CH{-}CO{-}C_3H_7$ (O)	Hg	40% NaOH	Dipropyladipate	35	105
[phenyl]$CH = CH{-}C{-}OC_2H_5$ (O)	Hg	40% NaOH	Diethyl 3,4-diphenyladipate	—	105
			Ethyl β-phenylpropionate	—	105
O_2N[phenyl]$CH = CH{-}C{-}OC_2H_5$ Hg g (O)			p-Aminophenylpropionic acid		106
Dibenzoylethylene	Hg	C_2H_5OH, pH = 4.9	Dibenzoylethane	100	300
Maleic acid	Hg	Aq. acid	Succinic acid	100	5, 107 108
Fumaric acid	Hg	Aq. acid	Succinic acid	85.2	120
[phenyl]$CH = CH{-}CO_2H$	Hg, Ni		[cyclohexyl]$-CH_2{-}CH_2{-}CO_2H$ [phenyl]$-CH_2{-}CH_2{-}CO_2H$	—	109, 110
	Cu, Pb	Aq. Na_2SO_4	2,3-Diphenyladipic acid γ-keto-β, p-diphenyl caproic acid	—	111, 112

Reactant	Metal	Conditions	Product	Yield (%)	Ref.
furyl—CH=CH—CO$_2$H	Hg	Aq. Na$_2$SO$_4$	furyl—(CH$_2$)$_2$CO$_2$H	60–70	112
o-CN—C$_6$H$_4$—CH=CH—CO$_2$H	Hg	Aq. H$_2$SO$_4$	β-(o-Carboxyphenyl)propionic acid	—	113
			β,γ-Bis(o-Cyanophenyl)-adipic acid	—	113
m-HO—C$_6$H$_4$—CH=CH—CO$_2$H	Hg	Aq. H$_2$SO$_4$	β,γ-Bis(m-Hydroxyphenyl)-adipic acid	—	113
o-Cl—C$_6$H$_4$—CH=CHCO$_2$H		25% NaOH	o-Cl—C$_6$H$_4$—CH$_2$—CH$_2$CO$_2$H		121
H$_3$CO—C$_6$H$_4$—CH=CH—CO$_2$H	Hg	Aq. H$_2$SO$_4$	β,γ-Bis(p-hydroxyphenyl)-adipic acid	55	113
o-O$_2$N—C$_6$H$_4$—CH=CHCO$_2$H	Hg	Aq. acid	Dihydrocarbostyril	—	106
p-O$_2$N—C$_6$H$_4$—CH=CHCO$_2$H	Hg	Aq. acid	β-(p-Nitrophenyl) propionic acid	—	106

Table 8.6 Olefins and Acetylenes (continued)

Substrate	Metal	Conditions	Product	Yield	Ref.
H_2N–C₆H₄–$CH=CH$–CO_2H	Hg	Aq. Acid	H_2N–C₆H₄–$CH_2CH_2CO_2H$	—	106
(3-H_3CO, 2-O–CH_3)C₆H₃–C(CH_3)=CH–CO_2H	Hg	Aq. Na_2SO_4	(2-OCH_3, 3-H_3CO)C₆H₃–$CH(CH_3)$–CH_2–CO_2H	85	114
C₆H₃$(OCH_3)_2$–$CH=CH$–CO_2H	Hg	Aq. Na_2SO_4	C₆H₃$(OCH_3)_2$–CH_2–CH_2CO_2H	80–93	114
H_3C, H_3C bicyclic furanone –C($CO_2C_2H_5$)($CO_2C_2H_5$)	Pb	Conc. H_2SO_4	H_3C, H_3C bicyclic–C($CO_2C_2H_5$)₂	85	122
CH_2CO_2H / C=C(CO_2H) / $CHCO_2H$	Hg		CH_2CO_2H / –CH–CO_2H / CH_2CO_2H	97	123
pyrroline N–R	Pb	15% H_2SO_4	pyrrolidine N–R	70–90	147

R = Alkyl

Substrate	Electrode	Conditions	Product	Yield (%)	Ref.
(coumarin structure)	Hg	Aq. MeOH HCl Brucine (alkaloid), sodium citrate	(dihydrocoumarin dimer / trimer structures)	31–89	511
$\phi-(CH=CH)_m-\overset{O}{\overset{\|}{C}}-(CH=CH)_m-\overset{O}{\overset{\|}{C}}-\phi$	Hg	60% DMF	$\phi-(CH_2-CH_2)_m-\overset{O}{\overset{\|}{C}}-(CH_2-CH_2)_m-\overset{O}{\overset{\|}{C}}-\phi$	5.5–66.9	546
$m = 0,1,2; n = 0,1,2$					
$RO_2C-CH=CH(CH_2)_n-CH=CH-CO_2R$					
$R = C_2H_5$					
$n = 1$	Hg	Aq. sol.	(cyclopropane-CO_2R)	98	128
$n = 2$			(cyclobutane-CO_2R)	41	128
$n = 3$			(cyclopentane-CO_2R)	100	128
$n = 4$			(cyclohexane-CO_2R)	81	128

Table 8.6 Olefins and Acetylenes (continued)

212

Substrate	Metal	Condition	Product	Yield	Ref
$n = 5$			![cyclohexane with two CO_2R groups]	10	128
$NC-CH=CH-CH=CH-CN$	Zn	K_3PO_4	γ,β-Di(cyanomethyl)-suberonitrile	—	509
$H_5C_6\overset{O}{\overset{\|}{C}}-CH=CH-CN$	Hg	Aq. sol.	$H_5C_6-CH-CH_2-\overset{O}{\overset{\|}{C}}-C_6H_5$ $H_5C_6-CH-CH_2-\overset{O}{\overset{\|}{C}}-C_6H_5$	45	127
			$H_5C_6\overset{O}{\overset{\|}{C}}-CH_2-CH_2CN$ $H_5C_6\overset{O}{\overset{\|}{C}}-CH-CH_2-COC_2H_5$ $H_5C_6\overset{O}{\overset{\|}{C}}-CH-CH_2C-OC_2H_5$ $H_5C_6\overset{O}{\overset{\|}{C}}-CH-CH_2-\overset{O}{\overset{\|}{C}}-OC_2H_5$ $H_5C_2-O\overset{O}{\overset{\|}{C}}-CH-CH_2-\overset{O}{\overset{\|}{C}}-C_6H_5$	55	127
$H_5C_6\overset{O}{\overset{\|}{C}}-CH=CH-COC_2H_5$	Hg	Aq. sol.			

Acetylene	Catalyst	Solvent	Product	Yield	Ref.
$CH_3(CH_2)_5C \equiv CH$	Spongy Ni	95% EtOH, H_2SO_4	1-Heptene	65	129
$CH_3(CH_2)_2{-}C \equiv C(CH_2)_2CH_3$	Spongy Ni	95% EtOH, H_2SO_4	cis-4-Octene	80	129
	Pt	LiCl, CH_3NH_2	trans-3-Octene	47	130
			cis-3-Octene	1	130
$CH_3(CH_2)_4{-}C \equiv C{-}CH_3$	Pt	LiCl, CH_3NH_2	trans-2-Octene	42.8	130
			cis-2-Octene	1	130
$CH_3(CH_2)_3C \equiv C(CH_2)_3CH_3$	Pt	LiCl, CH_3NH_2	trans-5-Decene	63	130
			cis-5-Decene	1	130
	Spongy Ni	95% EtOH, H_2SO_4	cis-5-Decene	75	129
$H_5C_6C \equiv CH$	Pt	LiCl, CH_3NH_2	Ethylbenzene	36	130
			Styrene	3.2	130
$H_5C_6C \equiv C{-}CH_2CH_3$	Pt	LiCl, CH_3NH_2	1-Phenylbutane	36.4	130
			β-Ethylstyrene	8.8	130
$H_5C_6(CH_2)_2C \equiv CH$	Pt	LiCl, CH_3NH_2	1-Phenylbutane	6.4	130
			4-Phenyl-1-butene	51.2	130
$H_5C_6(CH_2)_2C \equiv C{-}CH_3$	Pt	LiCl, CH_3NH_2	$H_5C_6(CH_2)_2CH = CHCH_3$ (trans)	—	95
$H_5C_6C \equiv C{-}C_6H_5$	Hg	DMF	Dibenzyl	68	131
			1,3,4-Tetraphenylbutane	6.1	131
Diphenylacetylene, CO_2	Hg	DMF	Diphenylfumaric acid	8	131
			Diphenylmaleic anhydride	4	131
			meso-Diphenylsuccinic acid	—	131
	Spongy Ni	95% EtOH, H_2SO_4	cis-Stilbene	80	129

213

Table 8.6 Olefins and Acetylenes (continued)

Substrate	Catalyst	Conditions	Product		Ref.
$(H_3C)_2-C-C\equiv CH$ $\quad\vert$ $OCH_2-OC_2H_5$	Ag, Cu	1% Aq. EtOH-NaOH	$(H_3C)_2-C-CH = CH_2$ $\quad\vert$ $OCH_2-O-C_2H_3$	52.2	132
$(H_3C)_2C-C\equiv CH$ $\quad\vert$ $O-CH_2-OCH(CH_3)_2$	Ag, Cu	1% Aq. EtOH-NaOH	$(H_3C)_2C-CH = CH_2$ $\quad\vert$ $O-CH_2-OCH(CH_3)_2$	54.4	132
C_2H_5 $\quad\vert$ $H_3CC-C\equiv CH$ $\quad\vert$ $O-CH_2OCH_3$		1% Aq. EtOH-NaOH	C_2H_5 $\quad\vert$ $H_3C-C-CH = CH_2$ $\quad\vert$ OCH_2CH_3	32.9	132
CH_3 $\quad\vert$ $CH_3-CH_2-C-C\equiv CH$ $\quad\vert$ C_2H_5 OH	Cu-Ag	Aq. NaOH, EtOH	CH_3 $\quad\vert$ $CH_3-CH_2-C-CH = CH_2$ $\quad\vert$ C_2H_5 OH	80	133
$\triangle C-C\equiv CH$ $\quad\vert$ OH	Cu	1% NaHCO₃	$\triangle C-CH=CH_2$ $\quad\vert$ OH	60	134
$EtO_2C-C\equiv C-CO_2Et$	Hg	HCl, KCl	a,a'-Dimethylsuccinic acid (racemic)	—	507
$CH_3\ \ OH$ $\quad\vert\quad\ \ \vert$ $H_3C-C-C-C\equiv CH$ $\quad\vert\quad\ \ \vert$ $CH_3\ \ CH_3$	Ag-Cu	Aq. NaOH, EtOH	$CH_3\ \ OH$ $\quad\vert\quad\ \ \vert$ $H_3C-C-O-CH = CH_2$ $\quad\vert\quad\ \ \vert$ $CH_3\ \ CH_3$	52	135

Dienes and higher homologs

Compound	Electrode	Conditions	Product		Ref.
Butadiene	Hg	Aq. Dioxane	$H_3C-CH = CH-CH_3$	–	380
Butadiene, SO_2, O_2	Hg	DMF	$H_2C-CH = CH-CH_3$ $\quad\vert$ SO_3H	16.9	522
$H_5C_6C = CH-CO_2H$	Hg	Aq. NaOH $0.02A/cm^2$	$H_5C_6CH-CH_2CO_2H$	81	136
$H_5C_6-C = CH-CO_2H$		THF	$H_5C_6-CH-CH_2CO_2H$	–	168
Isoprene			Polyisoprene		
(cyclopentadiene) $=CH_2$	Hg	75% Dioxane	Polarographic studies		514
$(H_5C_6)_2C = C = C = C(C_6H_5)_2$	Hg		Polarographic studies		515

7 AROMATIC HYDROCARBONS

Reaction Products

Benzene gives no polarographic wave in the accessible potential region; however, its electrochemical reduction to cyclohexadienes and cyclohexane under a variety of reaction conditions has been reported [83, 86, 91, 148, 149]. In some of these reports [148, 149], it is postulated that a direct transfer of an electron from the cathode to the benzene ring does not take place, but rather the electron is introduced into the solution as "solvated electron" which causes the reduction of benzene away from the electrode in the solution bulk:

$$e^{\ominus}(\text{cathode}) + \text{solvent} \rightleftharpoons e^{\ominus}(\text{solvents})$$

Examples of solvents used in such reactions are liquid ammonia, ethylenediamine, methylamine, ethanol/water, and hexamethylphosphoramide. The presence of electrons in such solvents was indicated by means of ESR studies [147]. Reduction under these conditions is quite analogous to that which occurs with alkali metals in liquid ammonia [150].

In some cases, the reduction of benzene with solvated electrons may involve the supporting electrolyte. Thus the following reaction was described for tetra-n-butyl-ammonium perchlorate [147]:

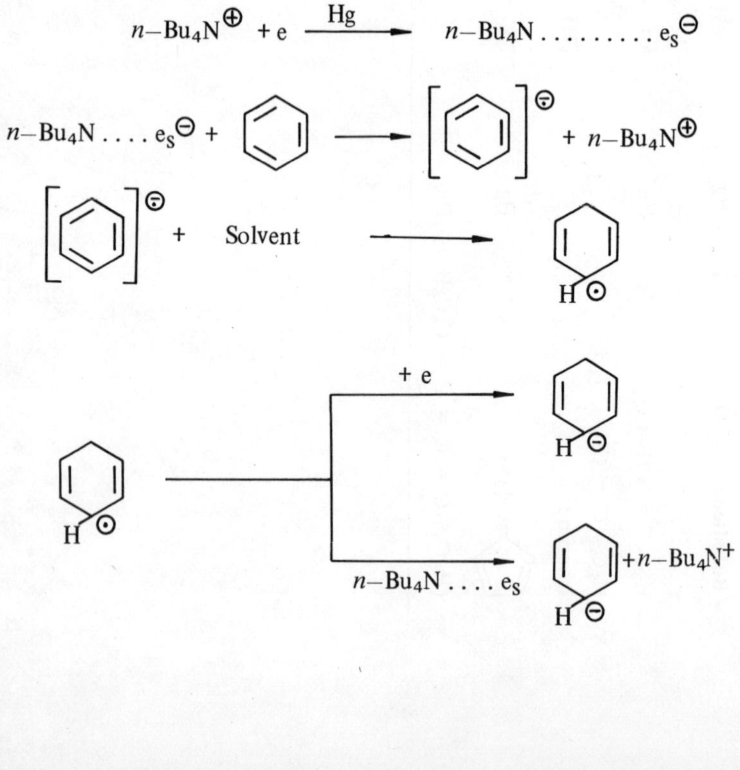

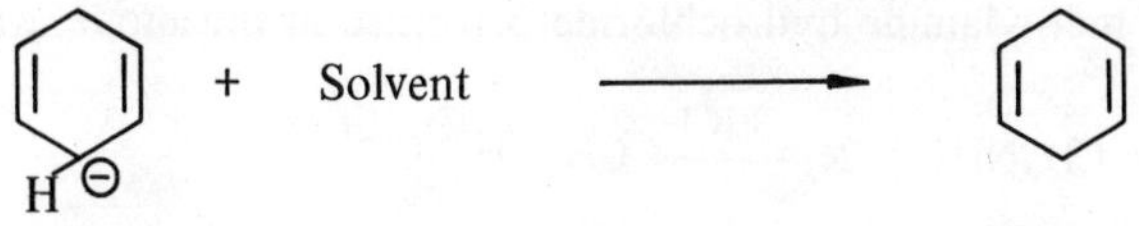

$$n-Bu_4N^{\oplus} \ldots e_S \longrightarrow n-Bu_3N + Bu^{\odot}$$

The electrochemical reduction of aromatic hydrocarbons can be of importance n organic synthesis. Thus, depending on the reaction conditions, different products may be obtained at will. Consider, for example, the reduction of benzene in a divided and in an undivided electrolysis cell [83, 84]:

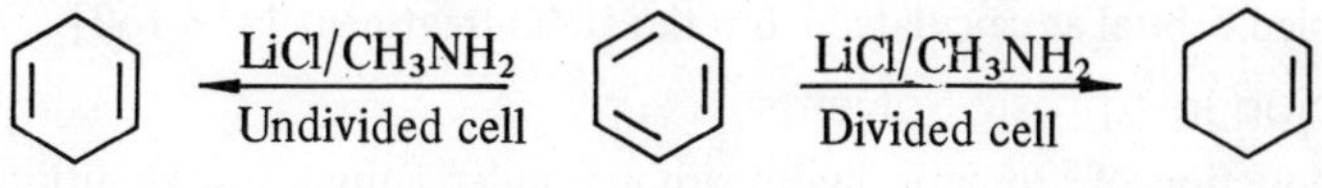

In this reaction, the reducing agent is believed to be lithium metal, which is generated electrochemically. Thus, in the overall reaction, electricity and benzene are the only materials consumed. The mechanism of this reaction may be explained as follows [83, 84]: In a divided cell, the radical anion formed is protonated by the amine to generate the amide anion, which is responsible for the formation of the conjugated cyclohexadiene which undergoes further reduction to cyclohexene:

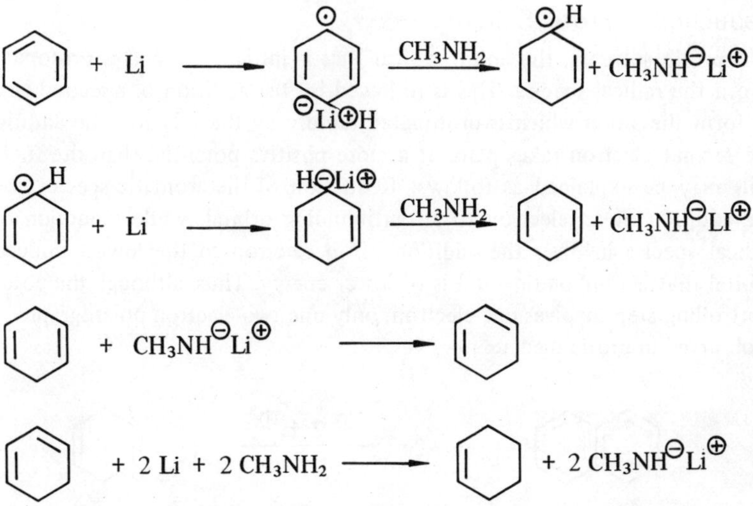

In an undivided cell, methylamine hydrochloride is formed at the anode, which

$$CH_3NH_2^- - 2e \xrightarrow{LiCl} CH_3NH^{\oplus}Cl^{\ominus}$$

neutralizes the lithium amide, formed at the cathode, and prevents it from allowing the formation of the conjugated diene. It is surprising, however, that the oxidation of methylamine would occur instead of the oxidation of chloride ion.

Mechanism of Reduction

The electrochemical reduction of aromatic hydrocarbons has been recently described by Peover [152]. In general, there is good correlation between the ease of reduction of the aromatic compound ($E_{1/2}$) and the energy of its lowest unoccupied orbital as calculated from the HMO treatment [158-160].

Reduction in Aprotic Solvents

The reduction of aromatic hydrocarbons under anhydrous conditions can be described as follows (Ar = aromatic compound):

$$Ar + e \underset{}{\overset{E_1}{\rightleftharpoons}} Ar^{\ominus} \tag{1}$$

$$Ar^{\ominus} + e \xrightarrow{E_2} Ar^{\ominus} \tag{2}$$

The first step is reversible and diffusion controlled. This was demonstrated from the cyclic voltammetry of anthracene in dimethylformamide [154]. In the absence of protons to neutralize the anion radical, another electron is added but at a more negative potential, to form the dianion species.

Reduction in Protic Solvents

In protic solvents, the anion radical that is initially formed is protonated to afford the radical species. This is followed by the addition of a second electron to form the anion which is protonated rapidly by the solvent. The addition of the second electron takes place at a more positive potential than the first one. This may be explained as follows: Reduction of the aromatic species involves the addition of an electron to an antibonding orbital, while reduction of the radical species involves the addition of an electron to the lowest unoccupied orbital that is nonbonding and is of lower energy. Thus, although the potential controlling step involves one electron, only one two-electron polarographic wave is observed in protic media:

The reduction of azulene [92], a nonalternate hydrocarbon, is of interest and differs from the reduction of alternate compounds. In protic media, azulene exhibits a one-electron reduction wave corresponding to the stable anion radical:

$$+ 1e \rightleftharpoons$$

At more negative potentials further irreversible reductions take place. In aprotic solvents, two one-electron polarographic waves are observed. If an acidic compound such as phenol is added to the solution, the first wave is unaffected while the second is doubled in intensity and a third wave appears. This behavior is explained as follows:

$$+ 1e \xrightarrow{\text{First wave}} \xrightarrow[\text{Second wave}]{+1e}$$

$$\xrightarrow{\text{Phenol}} \xrightarrow[\text{Fast}]{+e}$$

$$+ e \longrightarrow \text{Further reduction}$$

Experimental

The reduction of aromatic hydrocarbons can be carried out in simple equipment such as beakers or three-necked flasks, depending on the type of reaction in question. The most common cell is the H-type, which is divided with a sintered disk or other suitable diaphragm.

Reduction of Cumene [83]

An H-type cell whose compartments were divided with an asbestos sheet and which were fitted with dry-ice condensers was charged with reagents as follows: 12 g (0.1 mole) cumene, 17 g (0.4 mole) lithium chloride, and 450 ml of anhydrous methylamine were placed into the cathode compartment. The anode compartment was charged with 450 ml anhydrous methylamine and 17 g of lithium chloride. Platinum was used for both electrodes (2 X 5 cm each).

Electrolysis was carried out at a total cell voltage of 90 V and a current of 2.0 A for 7 hr. At this time the solution was allowed to evaporate and the mixture was hydrolyzed by the slow addition of water. After extraction into ether, drying and removal of solvent, there was obtained 9.0 g (75%) of a liquid b.p.

149-153°C. Analysis by VPC showed this liquid to be a mixture of isopropyl-cyclohexenes (89%) and 11% cumene.

The above reaction was repeated but without the cell divider. Twelve grams (0.1 mole) of cumene was reduced using 34 g (8.0 mole) lithium chloride and 900 ml methylamine. Electrolysis proceeded for 7 hr at 85 V and 2 A. Work up of the product as described above afforded 9.8 g (82%) of a liquid (b.p. 152-157°C) which was shown by VPC analysis to contain 6% isopropylcyclo-hexene, 3% of unidentified diene, 13% cumene, and 78% 2,5-dihydroisopropyl-benzene.

Table 8.1 **Aromatic Hydrocarbons**

Reactants	Reaction Conditions		Products	Yield (%)	Ref.
	Cathode	Solvent			
Benzene	Pt	LiCl, CH₃NH₂ "Undivided cell"	Cyclohexene	2	83, 84
			2,5-Cyclohexadiene	47	—
		"Divided cell"	Cyclohexane	49	83, 84
	Pt, Ru	3 N HClO₄	Cyclohexane	—	519
	Al	Ethanol, Hexamethyl-phosphoramide 28°C	Cyclohexdiene	20.2	
			Cyclohexene	8.5	
			Cyclohexane	71.3	85
	Hg	Aq. diglyme	1,4-Dihydrobenzene	—	520
	Al	Ethanol, HMPA -2°C	Cyclohexadiene	93	85
			Cyclohexene	5.8	
			Cyclohexane	1.2	
	Hg	LiCl, Ethylene-diamine	Cyclohexene	44.8	86
			C₆H₈	10.8	86
	Pb		Cyclohexane	8.3	86
		Nal, liq. NH₃, MeOH		—	91
Toluene	Hg	Aq. Diglyme	2,5-Dihydrotoluene	—	520
	Pb	LiCl, CH₃NH₂, "Divided cell"	1-Methyl-2-cyclohexene	4	83, 84
			1-Methyl-2-cyclohexane	3.2	83, 84
		"Undivided cell"	1-Methyl-2,5-cyclohexadiene		60, 83, 84
			Methylcyclohexane		

221

Table 8.7 Aromatic Hydrocarbons (continued)

Substrate	Electrode	Conditions	Product	Yield	Ref.
Azulene	Pt, Ru Ag	3N HClO$_4$ DMF	(reduced azulene ring structure)	80	519
(phenylacetic acid, —CH_2CO_2H on benzene)	Pt		CH_2—CO_2H (on cyclohexane ring)	—	92
Ethylbenzene	Pb	LiCl, CH_3NH_2	CH_2CH_3 (on cyclohexene ring)	—	93
		"Divided cell"	CH_2CH_3 (on 1,4-cyclohexadiene ring)	63	84
		"Undivided cell"	CH_2CH_3 (on 1,4-cyclohexadiene ring)	93 ⎱ 73	
			CH_2CH_3 (on cyclohexene ring)	4 ⎰	
Cumene	Pt, Ru Pb	3N HClO$_4$ LiCl, CH_3NH_2	Cyclohexane, benzene	—	519
		"Divided cell"	$CH(CH_3)_2$ (on cyclohexene ring)	77	83

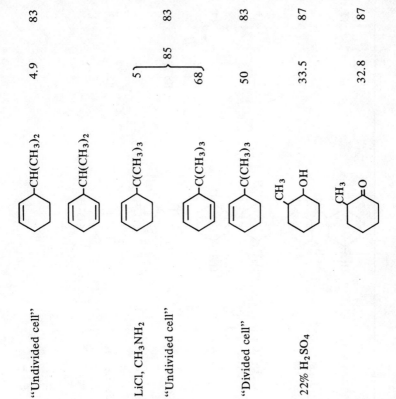

4.9   83

"Undivided cell"

5 ⎱ 85
68 ⎰   83

LiCl, CH₃NH₂
"Undivided cell"

50   83

"Divided cell"

33.5   87

22% H₂SO₄   Pt

32.8   87

t-Butylbenzene

o-Cresol

Table 8.7 Aromatic Hydrocarbons (continued)

Substrate		Catalyst	Conditions	Product	Yield	Ref.
m-Cresol		Pt	22% H_2SO_4	(HO, CH_3 cyclohexanol)	41	87
				(CH_3 ketone)	50	87
p-Cresol		Pt	22% H_2SO_4	(H_3C, OH)	24	87
				(H_3C ketone)	30	87
(CO_2H, CO_2H ortho)		Pb, Cd, Sn	5% H_2SO_4	(CO_2H, CO_2H)	88	517
(CO_2H, CO_2H para)		Pb, Cd, Sn	5% H_2SO_4	(CO_2H, CO_2H)	89	517

Substrate	Electrode	Conditions	Yield (%)	Ref.
R = NO₂, H, –C(=O)– (substituted benzene)	Hg	Polarographic study	—	518
(tetrahydronaphthalene)	—	Aq. CH₃CN	86–98, 44	521
Naphthalene derivative → CO₂H, CO₂H (tetralin dicarboxylic acid)	Hg	DMF	—	88
Naphthalene, CO₂	Hg	Aq. dioxane — Enol ether → 2-Tetralone	90 / 89	89
(OCH₂CH₃ naphthalene)	Hg	Aq. dioxane — Enol ether → 7-Methoxy-2-tetralone	80 / 93	89
(OCH₃, H₃CO naphthalene)	Hg	Aq. dioxane — Enol ether → 7,8-Dimethoxy-2-tetralone	95 / 89	89
Phenanthrene	Hg	DMF	20	88

225

Table 8.7 Aromatic Hydrocarbons (continued)

Phenanthrene, CO_2	Hg	DMF		20	88
SO_2	Hg	DMF		14.8	522

8 CARBON-NITROGEN DOUBLE BOND

A chapter on the electrochemistry of the carbon-nitrogen double bond has recently been published [142]. Consequently, only a brief discussion of this topic is described in this section.

Imines

Compared to studies on the electrochemical reduction of carbonyl compounds, few studies have been focused on the reduction of imines. In fact, interest in the electrochemical behavior of imines may have been initiated in order to determine the reduction of aliphatic ketones in the presence of amines, since the reduction of these ketones in other solvents can only be observed at high negative potentials.

$$CH_3-\overset{\overset{O}{\|}}{C}-CH_3 + RNH_2 \rightleftharpoons CH_3-\overset{\overset{NR}{\|}}{C}-CH_3 \xrightarrow[+2H^+]{+2e} CH_3-\overset{\overset{NHR}{|}}{C}H-CH_3$$

The reduction of imines which affords the corresponding amines resembles the reduction of ketones. In acid solutions imines exhibit two one-electron polarographic waves which merge at higher pH values [16]. The half-wave potential of the first wave is pH dependent while that of the second wave is not. In DMF solutions imines exhibit a single, irreversible, well-developed polarographic wave with the limiting current varying linearly with the square root of the height of the mercury column, indicating that the reduction is diffusion controlled [18].

The electrochemical reduction of imines sometimes affords products that are different from those obtained by conventional techniques. For example, the reduction of compound **I** affords the corresponding endo amine exclusively while on reduction with hydrogen over platinum the exo product is obtained exclusively [18].

Oximes

The polarographic behavior of oximes has been studied by several investigators [16, 20-22]. In general, the reduction is pH dependent and consumes four electrons to form the corresponding amine:

$$R_2C = N-OH + 4e \xrightarrow{H_2O} R_2CHNH_2$$

Reduction of oximes is easier in acid solutions, which indicates that the species initially reduced is the protonated oxime. In basic solutions, the oxime is converted to its anion, which is difficult to reduce.

At pH values above 8, a second wave at a more negative potential appears at the expense of the height of the first wave [16, 21]. Thus the overall reaction may be depicted as follows:

At Low pH

$$R_2C = \overset{\oplus}{N}OH_2 + 2e + H^+ \longrightarrow R_2C = NH$$

$$R_2C = NH + 2e \xrightarrow[\text{fast}]{2H^+} R_2CHNH_2$$

At high pH

$$R_2C = NOH \rightleftharpoons R_2C = NO^{\ominus}$$

$$\Big\downarrow \begin{matrix} +2e \\ +2H^+ \end{matrix} \quad \text{(more negative } E\frac{1}{2}\text{)}$$

$$R_2C = NH \xrightarrow[\text{fast}]{+2e,\ 2H^+} R_2CHNH_2$$

That imines are intermediates in the reduction of oximes may be seen from the reduction of benzaldehyde oxime. Reduction at a mercury cathode in buffer solution affords benzylamine. Under these conditions benzylhydroxylamine cannot be reduced; however, in slightly acidic solutions, its reduction occurs at a more negative potential than that of the oxime [16].

In certain oximes, the syn and anti forms may exhibit different reduction behaviors [16]. For example, while the reduction of *syn*-cinnamaldoxime is similar to that of the anti isomer in acid solutions, it is different in basic solutions in that it does not exhibit a polarographic wave.

Hydrazones

Like the reduction of imines, the reduction of hydrazones is similar to that of the parent ketones. The polarographic behavior of hydrazones indicates that reduction takes place in a two-electron process. The log plot slope of the polarographic wave [23] implies that the rate-determining step involves the addition of a proton and an electron. Thus the overall mechanism is depicted as follows:

$$R_2C = N-NHR' + e \xrightarrow{\ H^+\ } R_2\overset{\ominus}{C}-NH-NHR'$$

$$R_2\overset{\ominus}{C}-NH-NHR' + e \xrightarrow[\text{fast}]{\ H^+\ } R_2CH-NH-NHR'$$

Exceptions to the above mechanism are found in the reduction of benzophenone and benzaldehyde phenylhydrazones [16], which undergo a four-electron reduction to the corresponding amines.

Semicarbazones and Azines

There are only isolated examples on the reduction of semicarbazones and azines [16, 24]. In acid solutions certain semicarbazones, for example, benzalacetone, undergo hydrolysis. Those that are stable exhibit one polarographic wave whose height decreases at higher pH values. At pH values of 8 and above, another wave is observed. For example, cinnamaldehyde semicarbazone [16] exhibits a wave in mineral acid solution that is pH dependent. In alkaline solution another wave appears that is diffusion controlled and independent of the pH of the medium.

In general, the product derived from the reduction of semicarbazones is the corresponding amines:

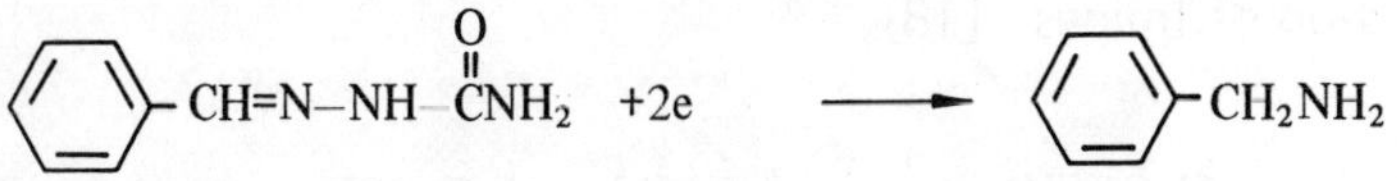

The reduction of benzalazine was studied by Lund [16], who found the compound to be unstable in strong acid. At pH values between 3.5 and 7, one polarographic wave was observed and was pH dependent. At pH 9.8 another wave was observed and the height of the first wave was substantially reduced. Controlled potential electrolysis of benzalazine in alkaline solution afforded benzaldehyde benzylhydrazone:

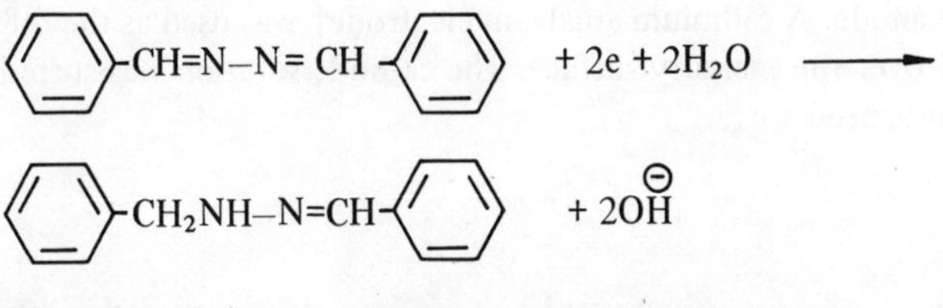

The intervention of dibenzyldiimide $C_6H_5CH_2N=N-CH_2C_6H_5$ as an intermediate in the above reduction was ruled out by Lund [26] since it would have been reduced under the reaction conditions to dibenzylhydrazine. Thus the overall reduction of benzalazine consumes six electrons to produce two moles of dibenzylamine.

Imido Esters and Amidines; $RC{-}OR$, $RC{-}NH_2$

(with $\overset{NH}{\underset{\parallel}{}}$ over each)

The imido esters are prone to hydrolysis in aqueous media. Consequently, the reduction of these compounds in nonaqueous solutions is preferred. In general, reduction affords the corresponding amines in good yield [143]. Since the esters can be produced qualitatively from nitriles [143], this may offer an alternate method for the preparation of amine from nitriles.

The reduction of amidines is similar to that of imido esters. In general, amidines of aliphatic acids are difficult to reduce while those of aromatic acids are reduced at a pH between 6 and 8 [144].

Experimential

Reduction of Imines* [18]

An H-type electrolysis cell was used with mercury as the cathode and a silver wire as the anode. A cadmium amalgam electrode† was used as the reference and was placed over the mercury surface. The cathode solution was stirred by means of a constant-speed stirrer.

---

*A general procedure was kindly supplied by Dr. A. Fry of Wesleyan Universtiy.

†Cadmium amalgam reference electrode is 0.26 V more negative than the Ag/AgBr and 0.75 V more negative than the standard calomel electrode.

In a typical macroscale electrolysis, the imine was dissolved in DMF (tetra-ethylammonium bromide) and the solution placed into the cathode chamber, The anode chamber was charged with DMF/Et$_4$NBr. The potential of the cathode was adjusted so that its value was equal to the half-wave potential of the imine (-2.14 V for the above compound). When the electrolysis was completed, the DMF solution from the cathode chamber was diluted with aqueous sodium hydroxide and the amine extracted into ether. The ether solution was dried over magnesium sulfate and the solvent evaporated. This left the amine, which was identified by standard procedures.

Table 8.8 Carbon-Nitrogen Double Bond

	Reaction Conditions				
Reactants	Cathode	Solvent	Products	Yield (%)	Ref.
Ketimines:					
Acetone, NH$_3$	Hg	NH$_3$ (NH$_4$)$_2$SO$_4$	Polarography of	—	17
cyclohexanone (O=cyclohexane)		4N Aq. HCl	cyclohexyl–N(H)(H)CH$_3$Cl	925 mg	16
H$_5$C$_6$–C(=NH)–C$_6$H$_5$ (1g)	Hg		(C$_6$H$_5$)$_2$CHNH$_3$Cl	1.05 g	16
H$_5$C$_6$–C(=N–C$_6$H$_4$CH$_3$)–C$_6$H$_5$	Hg	THF, aq. Alcohol, pH 10	H$_5$C$_6$–CH(C$_6$H$_5$)–NH–C$_6$H$_4$CH$_3$	70	79
Oximes:					
CH$_3$C(=NOH)–CO$_2$H	Hg	15% H$_2$SO$_4$	CH$_3$–C(NH$_2$)(H)–CO$_2$H	95	82
CH$_3$C(=NOH)–CH$_2$CO$_2$H	Pt	0.005 A/cm^2	CH$_3$–CH(NH$_2$)–CH$_2$–CH$_2$CO$_2$H	93.9	82

232

$CH_3CH_2-\overset{\|}{\underset{NOH}{C}}-CH_2CO_2H$	Pb	50% H_2SO_4	$CH_3CH_2-\overset{\|}{\underset{NH_2}{CH}}-CH_2CO_2H$	77	141
[Ph]$-CH=N-OH$ (1g) (anti)	Hg	THF, aq. buffer pH = g	$C_6H_5CH_2-NH_3Cl$	700 mg	16
$H_5C_6CH=CH-CH=NOH$ (anti)	Hg	Alcoholic H_2SO_4, Na_2SO_4	Cinnamylamine	225 mg	16
[Ph]$-CH=CH-\overset{\|}{\underset{}{C}}=NOH$ with CH_3 (1g)	Hg	Aq. base, benzene	Hydroxycinnamaldehyde $C_6H_5CH = CH-\overset{CH_3}{C}NH_2$	800 mg	16
	Hg	50% alcohol	↑ $C_6H_5CH=CH-\overset{H}{\underset{CH_3}{C}}-NH-\overset{O}{\overset{\|}{C}}C_6H_5$	825 mg	16
	Hg	Aq. NaOH, KCl	Benzalacetone	820 mg	16
N-Benzyl benzaldoxime (1g)	Hg	Aq. sol. pH = 7	$(C_6H_5-CH_2)_2NH_2Cl$ $CH_3-\overset{NH_2}{\underset{}{CH}}-C_6H_5$	740 mg	16
$\overset{NOH}{\overset{\|}{CH_3-C-C_6H_5}}$ Benzamide oxime (1g)	Hg	Aq. acetate buffer pH = 4.7	Benzamide ↑ Benzamide picrate	90 / 2.1 g	81 / 16
Mesoxalic acid ester oxime (1g)	Hg	$NaHCO_3$ sol.	↑ Aminomalonic acid ethylester hydrochloride	—	16

233

Table 8.8 Carbon-Nitrogen Double Bond (continued)

Substrate	Electrode	Solvent	Product	Yield	Ref
bicyclic =NC$_6$H$_5$	Hg	DMF	bicyclic —NHC$_6$H$_5$, H	20	18
H$_3$C—CH$_3$, CH$_3$ bicyclic =NC$_6$H$_5$	Hg		H, NHC$_6$H$_5$ bicyclic	80	18
			H$_3$C CH$_3$, CH$_3$ NHC$_6$H$_5$, H bicyclic	100	18
H$_5$C$_6$—C(=N—C$_6$H$_5$)—C$_6$H$_5$	Hg	DMF	Radical anion	—	18
HO—C$_6$H$_3$(OH)—C(=N—OH)—C$_6$H$_5$	Hg	C$_2$H$_5$OH, HCl	HO—C$_6$H$_3$(OH)—C(=$\overset{\oplus}{N}$H$_2$Cl$^{\ominus}$)—C$_6$H$_5$	—	19
H$_3$C—CH$_3$, CH$_3$ bicyclic =NOH	Hg	Aq. CH$_3$OH/LiCl	H$_3$C—CH$_3$, CH$_3$ bicyclic —NH$_2$	99	80

			NOH		
NOH‖ $H_5C_6-CH_2-C-CH_2C_6H_5$	Pb	50% H_2SO_4	$H_5C_6CH_2-\overset{\text{NOH}}{CH}-CH_2C_6H_5$	42	140
Azines					
$H_5C_6CH=N-N=CH-C_6H_5$ (Benzalazine, 2g)	Hg	50% alcohol, alkaline sol.	Benzaldehyde Benzylhydrazone	16	524
$H_5C_6CH=N-N=CH-C_6H_5$ (Benzalazine, 1.5 g)	Hg	1. Alkaline sol. 2. Glacial HOAc	$C_6H_5CH_2NH_3Cl$	1.08 g	16
Schiff-Bases:					
$\langle C_6H_5\rangle CH=N-NH-\langle C_6H_5\rangle$ (1g)	Hg	Dioxane, C_2H_5OH, HCl	$C_6H_5CH_2NH_3Cl$		16
$\begin{array}{c}H_5C_6\\H_5C_6\end{array}\!\!\!>\!C=N-$	Hg	Dioxane, C_2H_5OH, HCl	$(C_6H_5CH_2)_2NH_2Cl$	460 mg	16
$(CH_3)_2N-\langle C_6H_4\rangle-CH=NH-C_6H_5$ (1g)	Hg	50% Alcohol pH = 0.5	$(CH_3)_2\overset{+}{N}-\langle C_6H_4\rangle-CH_2NH_3Cl$, $\overset{\ominus}{Cl}$ (H on N)	480 mg	
$C_6H_5CH=N-\langle C_6H_5\rangle$	Hg	$Et_4Np\text{-}Ts$ (molten) 140°C CO_2, -2.0 V (Ag/Ag^+)	$C_6H_5CH-NH-\langle C_6H_5\rangle$ with CO_2H	60	

$[(C_2H_5)_4NO_3S-\langle C_6H_4\rangle-CH_3]$

Table 8.8 Carbon-Nitrogan Double Bond (continued)

Semicarbozones

Structure		Conditions	Product		
phenyl–CH=NH–NH–C(=O)–NH$_2$	Hg	Glycine buffer	1-Benzylsemicarbazide	87	524
phenyl–CH=NNHCNH$_2$ (1g), O	Hg	Dioxane, alc. HCl	C_6H_5–C(=O)Cl → phenyl–CH$_2$–NH–C(=O)–		16
phenyl–CH=CH–CH=NNHCNH$_2$ (1g), O	Hg	50% Alc. basic sol.	phenyl–CH$_2$–CH$_2$–CH=N–N–CNH$_2$–(=O)		16
phenyl–CH=CH–C(CH$_3$)=N–NCNH$_2$ (1g), O	Hg	50% Alc. pH = 13	Benzylacetone semicarbazone		
(phenyl)$_2$ C=NNHCNH$_2$, O	Hg	50% ethanolic SNHCl, O, NH$_2$NH–C–NH$_2$ HCl	(phenyl)$_2$CH–NH$_2$Cl $\xrightarrow{HCl}$	520 mg	16

9 NITRILES

The reduction of aliphatic cyano compounds can only be accomplished by using a low hydrogen overvoltage electrode such as platinum, nickel, or copper. Consequently, very little is known about their polarographic behavior. In fact, acetonitrile is often the solvent of choice in many polarographic investigations and macroscale electrolytic reductions.

Under the proper reaction conditions, the reduction of aliphatic cyano compounds may be of synthetic utility. For example, the reduction of acetonitrile [2] in aqueous acid using a platinum cathode affords ethylamine in a quantitative yield. Similarly, benzonitrile can be reduced to benzylamine [5] in nearly quantitative yield.

The mechanism of the reduction of cyano groups is generally believed to proceed as follows:

$$R\text{–}C \equiv N + 1e \rightleftharpoons R\text{–}\overset{\cdot}{C} = N^{\ominus}$$

$$R\text{–}\overset{\cdot}{C} = N^{\ominus} + H^+ \longrightarrow R\text{–}\overset{\cdot}{C} = NH \xrightarrow[+H^+]{+e} R\text{–}CH = NH$$

$$R\text{–}CH = NH + 2e + 2H^+ \longrightarrow RCH_2NH_2$$

Unsaturated nitriles reduce to give colored intermediates, some of which are long lived [11, 12, 14]. The ESR spectrum of several unsaturated nitriles in dimethylformamide has been reported [11]. The reduction of phthalonitrile and substituted benzonitrile is of interest:

The reduction of *p*-amino and *p*-fluorobenzonitrile afford dimeric products:

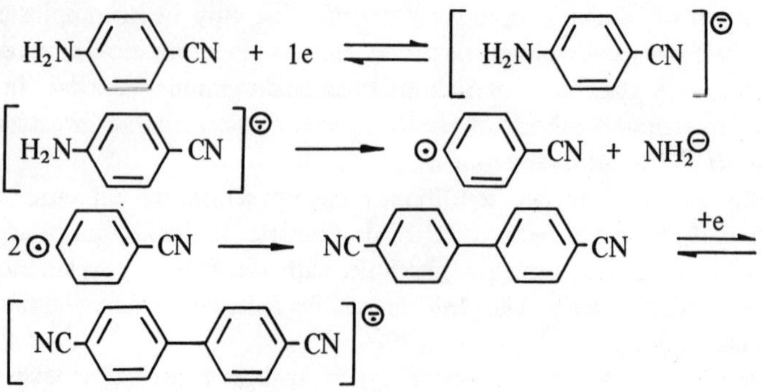

It is not clear how the precursor for the dimerization product can be obtained from the initial reduction step. Thus, an alternate mechanism is offered:

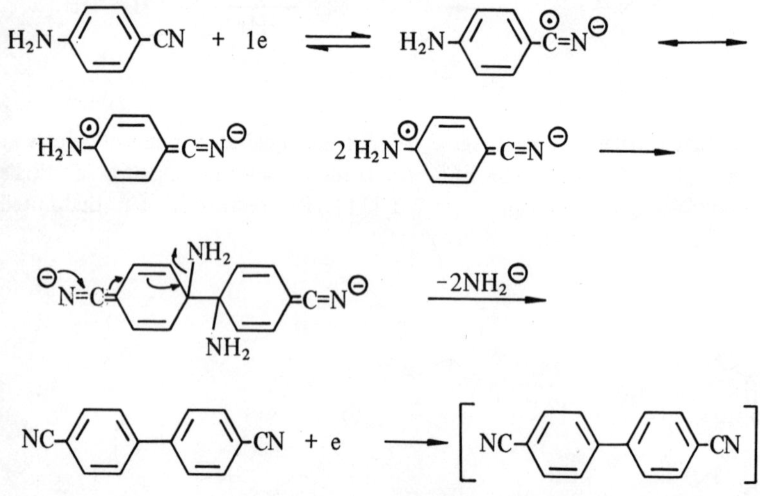

It should be remembered that the proposed mechanism does not involve bringing together two negatively charged species, since any anion produced from reduction has a counter ion from the supporting electrolyte.

The reduction of nitriles under neutral anhydrous conditions was studied in detailed by the author [12]. Thus, by using a platinum electrode, the reduction of pure acetonitrile afforded low-molecular-weight polymer whose structure was shown from spectral studies to be

$$(-\underset{\underset{CH_3}{|}}{C} = N-)_n$$

That only low-molecular-weight polymer was obtained was not surprising if one considers an anionically propagated polymerization reaction:

$$CH_3 -C \equiv N +1e \rightleftharpoons CH_3 -\overset{.}{C} = N^-$$

$$2\ CH_3\overset{.}{C} = N^- \longrightarrow {}^{\ominus}N = \underset{\underset{CH_3}{|}}{C} - \underset{\underset{CH_3}{|}}{C} = N^{\ominus}$$

polymer $\underset{\longleftarrow}{N \equiv C-CH_3}$ ${}^{\ominus}N = \underset{\underset{CH_3}{|}}{C} - \underset{\underset{CH_3}{|}}{C} = N^{\ominus}$ $\underset{\longrightarrow}{CH_3-C \equiv N}$ polymer

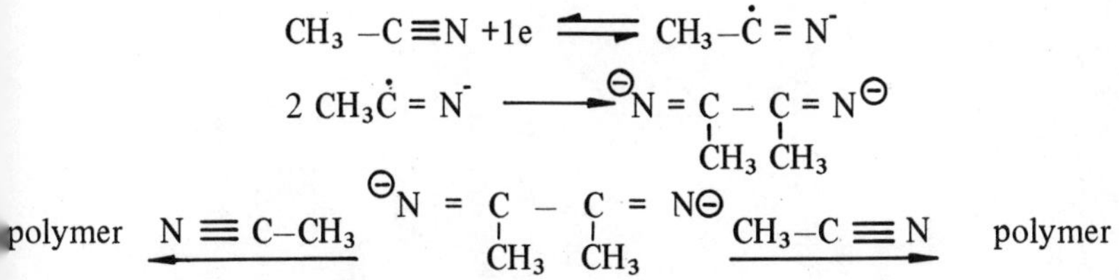

This is quite analogous to the electrochemical polymerization of styrene [96]:

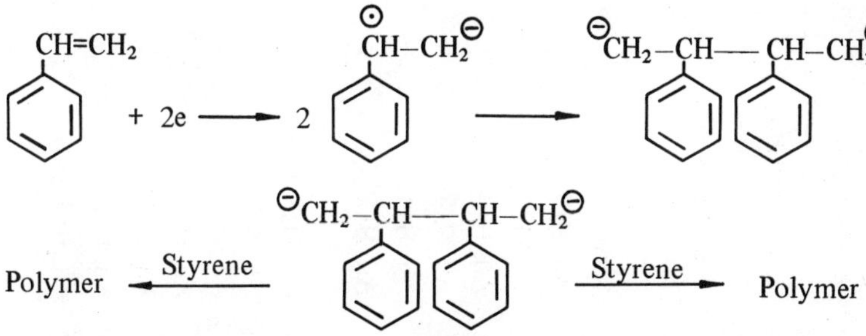

Thus a termination step in the polymerization reaction of acetonitrile is the abstraction of a proton from the monomer. To support this hypothesis, benzonitrile, which does not have acidic protons, was reduced under similar conditions and afforded high-molecular-weight polymer.

Experimental

Reduction of Acetonitrile [611]

A 200-ml beaker into which was placed a cylindrical, porous, unglased thimble (diaphragm) was used as the electrolysis cell. Platinum wires of areas about 150 cm^2 were used as electrodes. One gram (0.02 mole) of acetonitrile was dissolved in 50 ml of 8% hydrochloric acid and was placed in the beaker. The thimble was filled with dilute hydrochloric acid. Electrolysis at a current density of 3 A/cm^2 was conducted for two hours at 15°C. At this time, the catholyte was evaporated under vaccum to afford 2 g (100%) of ethylamine.

Table 8.9 Nitriles

Reactants	Reaction Conditions		Products	Yield (%)	Ref.
	Cathode	Solvent			
$H_2N-C{\equiv}N$ (Cyanamide)	Sn	Aq. $(NH_4)_2SO_4$ pH = 3–7	Formamidine	—	1
			Formic acid	—	1
			Methylamine, ammonia		1
CH_3CN		Neat	Polymer	—	12
Acetonitrile	Pt	Aq. HCl	Ethylamine	100	2
Acetonitrile	Ni–Hg	10% HCl	Ethylamine	93	5
Adiponitrile	Pt	Aq. HCl	Hexamethylenediamine	70	2
Benzyl cyanide	Ni–Pd	Aq. HCl	β-Phenylethylamine	—	3
Benzyl cyanide	Pt	Aq. HCl	β-Phenylethylamine	75	2
Cyanoacetic acid	Pt	Aq. HCl	β-Aminopropionic acid	40	2
p-Toluonitrile	Pb	Aq. H_2SO_4	CH_3—⟨benzene ring⟩—CH_2NH_2	57	10
2-Cyanoethanol	Pt	Aq. HCl	3-Amino-1-propanol	30	2
p-Cyanophenyl sulfonamide	Ni	Aq. HCl	p-Aminophenyl sulfonamide		4
Malononitrile	Pt	Aq. HCl	1,3-Diaminopropane	50	2
o-Methylphenyl cyanide	Pt	Aq. HCl	o-Methylbenzylamine	40	2
4-Amino-5-cyano-2-methyl pyridine	Ni, Hg	Aq. HCl	4-Amino-5-(aminomethyl)-2-methylpyridine		5, 6
Phenyl cyanide	Ni–Pb, Pb		Benzylamine	98	5

Starting material	Electrode	Solvent	Product		Ref.
Succinonitrile	Pt	Aq. HCl	1,4-Diaminobutane	70	2
$H_3C-C(=N)-C(NH_2)=N-...CN$	Pd	Aq. HCl	$H_3C-C(=N)-...CH_2NH_2$	—	7
Tetracyanoethylene	Hg	CH_3CN	Polarographic studies	—	8
4-Cyanopyridine	Hg	Aq. acid	4-Aminomethylpyridine	—	9
$H_2N-C_6H_4-CN$	Hg	DMF	$[NC-C_6H_4-C_6H_4-CN]$	—	11
$(NC)_2CH-C^{\ominus}-CH(CN)_2 \quad \overset{\oplus}{N}(CH_3)_4$	Hg	DMF	$\left[\begin{array}{c} CN \\ C-CH(CN)_2 \\ CH(CN)_2 \end{array}\right]$	—	11
Acrylonitrile	Pb	Aq. H_2SO_4	Allylamine	6.15	15
			Propylamine	Trace	
			Propionitrile		
[dicyclohexyl dinitrile structure]	Hg	LiCl, $EtNH_2$	[cyclohexylidenecyclohexane]	20	525
			[bicyclohexyl]	80	

241

10 OXYGEN-OXYGEN BONDS

General Considerations

The reduction of peroxides proceeds in a two-electron transfer leading to the rupture of the oxygen-oxygen bond to form hydroxyl groups:

$$R - O - O - R + 2e \xrightarrow{\quad 2H^{\oplus} \quad} 2R - OH$$

The reduction is irreversible and is, in general, not influenced greatly by the acidity of the solution [26].

Although the electrochemical reduction of peroxides is not of synthetic interest, its detection in polymers [27], oleates, linolenates, lard [28], and gasoline [29] is quite important. In most cases, detection of peroxides in these materials can be achieved with great accuracy by polarographic techniques that can measure concentrations as little as 10^{-4} molar. Most determinations are carried out in aqueous solutions under neutral or acidic conditions. Alkaline solutions should be avoided as they may hydrolyze the peroxide. For non-aqueous reduction, a 1:1 mixture of methanol-benzene is recommended [30, 31].

The ease of reduction of oxygen-oxygen bonds is in the order $RCO_3H > (RCO_2)_2 > RO_2H > RCO_3$-t-Bu > t-BuO$_2$H > RO$_2$R > (t-BuO)$_2$ [32, 33]. For a list of half-wave potentials of the oxygen-oxygen bond the reader should consult Ref. [26].

Superoxide Ion

The reduction of molecular oxygen to form superoxide ion was studied by Peover and co-workers [246, 257].

It was found that reduction of O_2 in aprotic solvents containing quaternary ammonium salts leads to the formation of a high concentration of $O_2^{\ominus}$, which is relatively stable at room temperature. Its decay in such a system may take place as follows [527]:

$$2O_2^{\ominus} + (C_2H_5)_4N^{\oplus} \longrightarrow HO_2^{\ominus} + O_2 + CH_2 = CH_2 + (C_2H_5)_3N$$

$$HO_2^{\ominus} \longrightarrow O_2 + 2OH^{\ominus}$$

In order to determine the concentration of $O_2^{\ominus}$ it is titrated with perchloric acid.

The ease of formation of $O_2^{\ominus}$ may play an important role in organic synthesis. For example, it was found that fluorene is easily oxidized on mercury in DMF solution of fluorenone in the presence of oxygen, while under the same conditions but without oxygen, no oxidation took place.

Superoxide $O_2^{\ominus}$ can also act as a nucleophile [527, 528, 539]:

$$RX + O_2^{\ominus} \longrightarrow RO_2^{\cdot} + X^{\ominus}$$

$$RO_2^{\cdot} + O_2^{\ominus} \longrightarrow RO_2^{\ominus} + O_2$$

$$RO_2^{\ominus} + RX \longrightarrow R_2O_2 + X^{\ominus}$$

11 ORGANOSULFUR

General Considerations

The polarographic behavior of organosulfur compounds has been studied in some detail [531-534]. Little emphasis was placed on the use of electrochemical reduction of such compounds in organic synthesis. The facile reaction of organosulfur compounds with mercury has somehow complicated a study of their polarographic behavior. For example, it was found [536] that in the reduction of 8,8'-diquinolydisulfide on a dropping mercury electrode, a chemical reaction precedes the electrochemical to form the following organo-mercury compound:

$$RS-S-R + Hg^{\circ} \rightleftharpoons (RS)_2Hg$$

which undergoes reduction to the mercapto product:

$$(RS)_2Hg + 2e + 2H^{\oplus} \longrightarrow 2RSH + Hg^{\circ}$$

A somewhat similar behavior was observed for the reduction of diphenyl disulfide [538]. Homocystine is similarly reduced to homocystein [145]:

$$\underset{NH_2}{\overset{O}{\underset{|}{HOC}}}-CH-CH_2-S-S-CH_2-\underset{NH_2}{\overset{O}{\underset{|}{CH-COH}}} +2e + 2H^{\oplus} \longrightarrow 2HS-CH_2\underset{NH_2}{\overset{O}{\underset{|}{CH-COH}}}$$

The reduction of sulfur-sulfur bonds on a platinum electrode was found to be irreversible as the electrode reaction is too slow [76].

The reduction of the sulfur-oxygen double bond is easier than the carbon-oxygen analog. In acid solutions, sulfones are irreversibly reduced to the sulfinic acid and mercaptan [71, 72]:

$$R-\underset{\underset{O}{\parallel}}{\overset{\overset{O}{\parallel}}{S}}-R +2e +2H^{\oplus} \longrightarrow RSO_2H + RSH$$

Sulfoxides are reduced to the corresponding sulfides [72]:

$$R-\overset{\overset{\text{O}}{\|}}{S}-R + 2e + 2H^{\oplus} \longrightarrow RSR + H_2O$$

It is of interest that the anodic oxidation of sulfides affords the corresponding sulfoxide, yet the reaction is not reversible [73, 74].

Sulfonyl chlorides undergo a two-electron reduction to give the sulfinate anion, which does not undergo further reduction [71, 75]:

$$RSO_2Cl + 2e \longrightarrow RSO_2^{\ominus} + Cl^{\ominus}$$

The reduction of carbon-sulfur bonds in desulfurization reactions may find use in organic synthesis as the method has some advantages over chemical reducing agents [529, 530].

Unlike the reduction of amides, thioamides are easily reduced in acid solution to the corresponding amines in good yield [69, 70].

The reduction of p-substituted benzenesulphonamides was studied by Zuman and co-worker [599]. This study revealed that scission of the C–S bond

occurred before that of the S–N bond. Substituents that allowed conjugation with the ring strongly facilitated the reduction of the sulfonamides. This effect was considered to be stronger than would be expressed by ρ-σ constants.

Experimental

Reduction of Homocystine [145]

A beaker was used as the electrolysis cell. A side arm extended to the bottom, through which was sealed a platinum wire that made connection with the mercury pool used as the cathode. The area of the cathode was 95.0cm². The beaker was equipped with a stirrer, thermometer, and a standard calomel electrode that touched the surface of mercury. An Alundum cup (fine porosity) 50 mm in diameter and 160 mm high served as the anode compartment. The anode consisted of a piece of platinum 10 × 10 cm². The cell was externally cooled with an ice bath.

A solution of 94 g (0.35 mole) of homocystine in 700 ml of 1.25 N sodium hydroxide was placed into the beaker. The anode compartment was charged with 1.25 N sodium hydroxide. The cathode potential was adjusted to read -1.79 V (SCE). This allowed a current of 0.059 A/cm² to pass through the cell. The temperature was maintained at 30-35°C. Towards the end of the reaction, the current density dropped to 0.014 A/cm². At this time, iodometric titration indicated that there was 100% reduction of the homocystine. The cathode solution was cooled to 15°C and rapidly adjusted to pH = 6.0 with concentrated hydriodic acid. A precipitate was formed and was quickly filtered from homocystine (formed by air oxidation) and was added to 3 liters absolute alcohol. The precipitate was allowed to settle for 1 hr in a refrigerator and was then filtered and washed with cold ethanol. This afforded 49 g (52.5%) of homocysteine m.p. (decomp.) at 234-235°C.

Table 8.10 Organosulfur

Reactants	Reaction Conditions		Products	Yield (%)	Ref.
	Cathode	Solvent			
2-Mercaptopropionic acid (5.3 g)	Cu	Conc. HCl	Propionic acid (2.3 g)	—	529
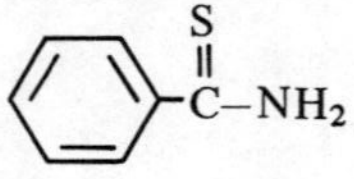 $C_6H_5-C(=S)-NH_2$	Hg	Aq. C_2H_5OH	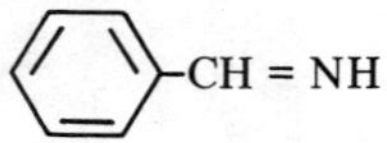 $C_6H_5-CH=NH$	—	77
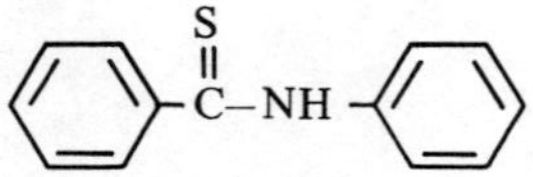 $C_6H_5-C(=S)-NH-C_6H_5$	Hg	Aq. C_2H_5OH	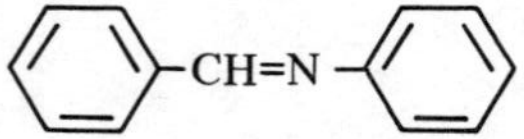 $C_6H_5-CH=N-C_6H_5$	—	77
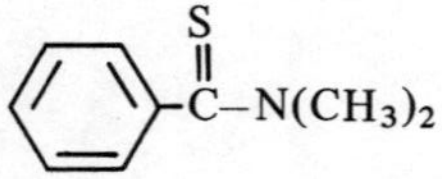 $C_6H_5-C(=S)-N(CH_3)_2$	Hg	Aq. C_2H_5OH	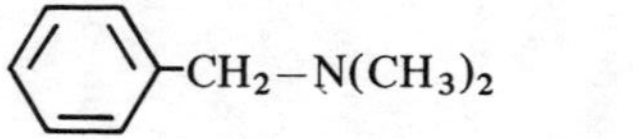 $C_6H_5-CH_2-N(CH_3)_2$	100	69, 70
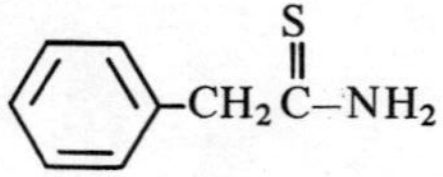 $C_6H_5-CH_2C(=S)-NH_2$	Pb	Aq. H_2SO_4	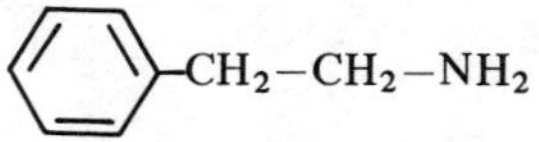 $C_6H_5-CH_2-CH_2-NH_2$	63	70
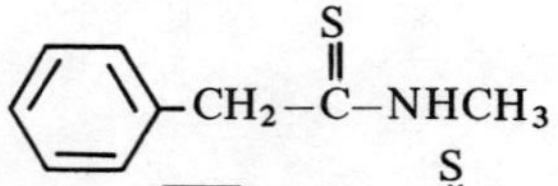 $C_6H_5-CH_2-C(=S)-NHCH_3$	Hg	Aq. H_2SO_4	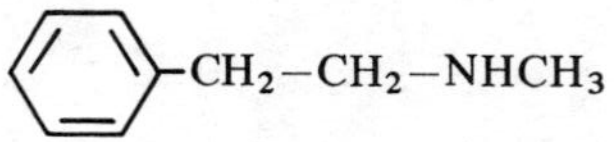 $C_6H_5-CH_2-CH_2-NHCH_3$	83	69, 70
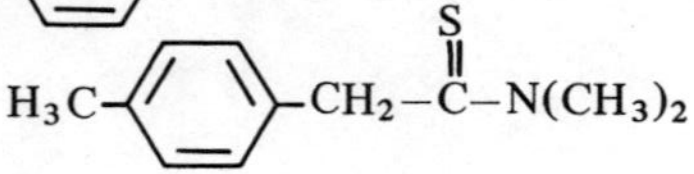 $H_3C-C_6H_4-CH_2-C(=S)-N(CH_3)_2$	Hg	Aq. H_2SO_4	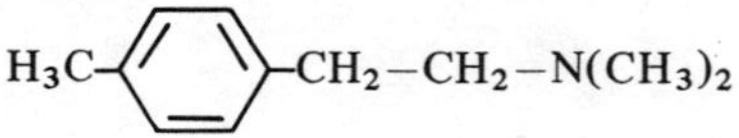 $H_3C-C_6H_4-CH_2-CH_2-N(CH_3)_2$	88	69

246

Reactant	Metal	Conditions	Product	Yield (%)	Ref.
$H_3C{-}O{-}C_6H_3(OCH_3){-}CH_2{-}C(={S}){-}N(CH_3)_2$	Hg	Aq. H_2SO_4	$H_3C{-}O{-}C_6H_4{-}CH_2{-}CH_2{-}N(CH_3)_2$	80	69
$H_3C{-}O{-}C_6H_4{-}CH_2{-}C(={S}){-}N(\text{piperidine})$	Hg	Aq. H_2SO_4	$H_3C{-}O{-}C_6H_4{-}CH_2{-}CH_2{-}N(\text{piperidine})$	69	69
$C_6H_5{-}SO_2{-}C_6H_5$	Hg	Aq. solution	$C_6H_5{-}SO_2H$ + C_6H_6	—	72
$RCO{-}SR$ (S-Benzoyl-L-cystein)	Hg	Acidic solution	L-Cystein	83	530
$C_6H_5{-}S(={O}){-}S{-}R$, R = alkyl	Hg	Acidic solution	$C_6H_5{-}S{-}R$	—	72
thioindole ($CH_2{-}C(={S}){-}N$ ring)	Hg	Strong acid	indoline ($CH_2{-}CH_2{-}NH$ ring)	63	78
$(CH_2)_7 \cdots S={C}{-}NH$ (macrocycle)	Pb	Strong acid	$NH{-}(CH_2)_8$	85	535
Homocystine	Hg	1.25 N NaOH	Homocysteine	52	145
$HO{-}(CH_2)_3{-}S{-}S{-}(CH_2)_3{-}OH$	Pb	Alc., H_2SO_4	$HO{-}(CH_2)_3SH$	70	146

Table 8.10 Organosulfur (continued)

8,8′-Diquinolyldisulfide	Hg	CH$_3$OH, NaOAc	8-Mercaptoquinoline	–	536
Methylsulfinylmethyldexylketone	Hg	Aq. DMF	2-Octanone	54	537
ω-(Methylsulfinyl)-p-methoxy-acetophenone	Hg	Aq. DMF	p-Methoxyacetophenone	74	537
Cystine	Hg	Acidic Sol.	Cysteine	–	76
2,2′-Dithiodi-(3-aminopropionic acid) (12 g)	Cu	Conc. HCl 6 A/dm^2	Benzoyl-β-alanine 7.1 g		529

12 MISCELLANEOUS REDUCTIONS

Carbon-Oxygen Bond

The reduction of carbon-oxygen bonds resembles that of carbon-halogen bonds except that halogens are better leaving groups [34-38] :

$$\text{Ph-CH-O-R'} + e \longrightarrow \text{Ph-CH·} \xrightarrow[\substack{\text{fast} \\ H^\oplus}]{+e} \text{Ph-CH}_2 + R'O^\ominus$$

The reduction appears to be independent of the pH of the medium but depends on the nature of R', $CH_3CH_2 < Ph < OAc$.

The reduction of carbon-oxygen bonds in optically active esters has recently been reported to proceed with complete loss of activity [385]:

$$H_3CO\text{—}\langle\text{ring}\rangle\text{—}\underset{H}{\overset{}{C}}\text{-CN} \;(O\text{-}\overset{O}{\underset{}{C}}\text{-Ph}) + 2e \xrightarrow[\text{Water}]{\text{Dioxane}} H_3CO\text{—}\langle\text{ring}\rangle\text{—}CH_2CN$$

This is in contrast to the reduction of some optically active carbon-halogen bonds that has been reported to proceed with a high degree of inversion [386]:

$$\langle\text{ring}\rangle\text{—}\underset{CH_3}{\overset{Cl}{C}}\text{-CO}_2H + 2e \xrightarrow{\text{EtOH}} \langle\text{ring}\rangle\text{—}\underset{CH_3}{\overset{H}{C}}\text{-CO}_2H$$

The reduction of similar carbon-oxygen bonds in esters has been reported by Wawzonek [338].

Carbon-Nitrogen Bond

The reduction of carbon-nitrogen bonds in simple amine compounds is difficult to accomplish. However, when these bonds are placed α- to carbonyl or other activating groups, their reduction becomes quite simple [227] :

The electrochemical reduction of quaternary ammonium salts (which are commonly used as supporting electrolytes in nonaqueous reactions) has been studied by several investigators [40-47]. The products from the reduction depend on the structure of the salts [40]. When R_1, R_2, and R_3 are alkyl or hydroxyalkyl groups, benzene and the corresponding amine were the products. However, when one of the R groups was allyl and the other alkyl, propylene and dialkylamine were obtained.

$$\text{Ph}-\overset{\overset{\displaystyle R_1}{|}}{\underset{\underset{\displaystyle R_3}{|}}{N}}-R_2X \longrightarrow \text{benzene} + \overset{\overset{\displaystyle R_1}{|}}{\underset{\underset{\displaystyle R_3}{|}}{N}}-R_2$$

The mechanism for the cleavage of quaternary ammonium salts may proceed via one of the following paths:

$$R_4N^{\oplus} + 1e \xrightarrow[\text{slow}]{\text{Path I}} R_3N + R^{\odot} \xrightarrow[\text{fast}]{+1e} R^{\ominus} \xrightarrow{H^{\oplus}} RH$$

$$\underset{RH}{\overset{H\cdot}{\nearrow}} \quad \underset{R-R}{\overset{R:}{\searrow}}$$

$$R_4N^{\oplus} + 2e \xrightarrow[\text{slow}]{\text{Path II}} R_3N + R^{\ominus} \xrightarrow{H^{\oplus}} RH$$

$$R^{\ominus} + R{-}\overset{\oplus}{N}R_3 \longrightarrow R{-}R + R_3N$$

Based on product analysis, both paths could be operative; however, chemical and ESR data [45, 46] make path I more attractive.

Carbon-Sulfur Bond

The reduction of carbon-sulfur single bonds is difficult to accomplish. Exceptions are the aromatic thiocyanates [62, 63]. On the other hand, carbon-sulfur double bonds reduce quite readily [64-66, 69-70].

$$\text{Ph}-\overset{\overset{\displaystyle O}{||}}{C}-CH_2-S-R + e \longrightarrow \text{Ph}-\overset{\overset{\displaystyle O}{||}}{C}-CH_2^{\odot} + RS^{\ominus}$$

$$\text{Ph}\overset{\overset{\displaystyle O}{||}}{C}-CH_2^{\odot} + e \xrightarrow{H^+} \text{Ph}-\overset{\overset{\displaystyle O}{||}}{C}-CH_3$$

$$R-\overset{\overset{\displaystyle S}{||}}{C}-NR_2 \longrightarrow RCH_2NR_2 \quad (70\text{-}100\%)$$

The reduction of trimethylsulfonium and cresyldimethylsulfonium salts has been studied [67]. These salts undergo a two-electron irreversible reduction that is preceded by a catalytic hydrogen wave.

Phenacyldialkyl sulfonium salts undergo two-electron reduction to give acetophenone and dialkyl sulfide [68]:

$$\underset{\overset{\|}{O}}{Ph-C}-CH_2S^{\oplus}R_2 \xrightarrow{+e} \underset{\overset{\|}{O}}{Ph\overset{}{C}}-CH_2{}^{\bullet} + R_2S$$

$$\underset{\overset{\|}{O}}{Ph-C}-CH_2{}^{\bullet} \xrightarrow[+H]{+e} \underset{\overset{\|}{O}}{Ph-C}-CH_3$$

In some cases, the radical intermediate undergoes dimerization [600]'

References

1. K. Odo and K. Sugino, *J. Electrochem. Soc.*, **104**, 160 (1957).
2. M. Ohta, *Bull Chem. Soc. Japan*, **17**, 485 (1942).
3. F. Kawamura and S. Suzuli, *Bull. Chem. Soc. Japan*, **55**, 476 (1952) [*Chem. Abstr.*, **48**, 3167 (1954)].
4. K. Ishifuku, H. Sakurai, and H. Okamoto, Jap. Patent, 180,563 (1949) [*Chem. Abstr.*, **46**, 3432 (1952)].
5. S. Sawa, Jap. Patent, 5019 (1951) [*Chem. Abstr.*, **47**, 2617 (1953)].
6. A. Ito, *J. Soc. Org. Syn. Chem. (Japan)*, **11**, 252 (1953) [*Chem. Abstr.*, **47**, 12056 (1953)].
7. Y. Tanaka and H. Matsuoka, *J. Agr. Chem. Soc., Japan*, **24**, 74 (1950) [*Chem. Abstr.*, **47**, 1507 (1953)].
8. P. H. Rieger, I. Bernal, and G. Fraenkel, *J. Am. Chem. Soc.*, **83**, 3918 (1961).
9. J. Volke and J. Holubek, *Coll. Czech. Chem. Commun.*, **28**, 1597 (1963).
10. K. Ogura, *Mem. Coll. Sci. Kyoto Imp. Univ.*, **12A**, 339 (1929) [*Chem. Abstr.*, **24**, 2060 (1930)].
11. P. H. Rieger, I. Bernal, W. Reinmuth, and G. Frankel, *J. Am. Chem. Soc.*, **85**, 683 (1963).
12. M. R. Rifi, unpublished results.
13. G. Sevastyanova and A. Tomilov, *J. Gen. Chem. USSR (Eng. Trans.)*, **33**, 2741 (1963).
14. M. Bargain, *Compt. Rend.*, **255**, 1948 (1962).

15. T. Nonaka and K. Sugino, *J. Electrochem. Soc.*, **114**, 1255 (1967).
16. H. Lund, *Acta Chem. Scand.*, **13**, 249 (1959).
17. P. Zuman, *Nature*, **165**, 485 (1950).
18. A. J. Fry and R. Reed, *J. Am. Chem. Soc.*, **91**, 6448 (1969).
19. H. Lund, *Acta Chem. Scand.*, **18**, 563 (1964).
20. P. Souchay and S. Ser, *J. Chim. Phys.*, **49**, C172 (1952).
21. H. J. Gardner and W. P. Georgans, *J. Chem. Soc.*, **1956**, 4180.
22. R. I. Gelb, and L. Meites, *J. Phys. Chem.*, **68**, 2599 (1964).
23. V. Prelog and O. Häfliger, *Helv. Chim. Acta*, **32**, 2088 (1949).
24 J. Tirouflet. *Advan. Polarog.*, **2**, 740 (1960).
25. G. Whitnak, J. Young, H. Sisler, and E. Gantx, *Anal. Chem.*, **28**, 833 (1956).
26. A. J. Martin, *Org. Analysis*, **4**, 1 (1960).
27. C. Barnes, R. Olofson, and G. Jones, *J. Am. Chem. Soc.*, **72**, 210 (1950).
28. C. Willits, C. Ricciuti, H. Knight, and D. Swern, *Anal. Chem.*, **24**, 785 (1952).
29. M. Whisman and B. Eccleston, *Anal. Chem.*, **30**, 1638 (1958).
30. W. Lewis and F. Quackenbush, *J. Am. Oil Chem. SoL*, **26**, 53 (1940).
31. W. Lewis, R. Quackenbush, and T. DeVries, *Anal. Chem.*, **21**, 762 (1949).
32. D. Swern and L. Silbert, *Anal. Chem.*, **35**, 880 (1963).
33. E. Kuta and F. Quackenbush, *Anal. Chem.*, **32**, 1069 (1960).
34. S. Wawzonek, H. Laitinen, and S. Kwiatkowski, *J. Am. Chem. Soc.*, **66**, 827 (1944).
35. R. Pasternak, *Helv. Chim. Acta*, **31**, 753 (1948).
36. H. Lund, *Acta Chem. Scand.*, **14**, 359 (1960).
37. H. Lund, *Acta Chem. Scand.*, **14**, 1927 (1960).
38. S. Wawezonek and J. Fredrickson, *J. Electrochem. Soc.*, **106**, 325 (1959).
39. A. J. Fry, R. G. Reed, and M. Mitnick, *J. Am. Chem. Soc.*, **94**, 8475 (1972).
40. B. Emmert, *Ber.*, **42**, 1507, 1997 (1909).
41. B. Emmert, *Ber.*, **45**, 430 (1912).
42. E. Ochiai and H. Kataoka, *J. Pharm. Soc., Japan*, **62**, 241 (1942).
43. M. Finkelstein, R. Peterson, and S. Ross, *J. Am. Chem. Soc.*, **81**, 2361 (1959).
44. B. Southworth, R. Osteryoung, K. Fleischer, and F. Nachod, *Anal. Chem.*, **33**, 208 (1961).
45. J. Mayell and A. Bard, *J. Am. Chem. Soc.*, **85**, 421 (1963).
46. S. Ross, M. Finkelstein, and R. Petersen, *J. Am. Chem. Soc.*, **92**, 6003 (1970).
47. V. Horak and P. Zuman, *Coll. Czech. Chem. Commun.*, **26**, 173 (1961).
48. J. Volke, *Acta Chim. Acad. Sci. Hung.*, **9**, 223 (1956).
49. E. Knoblock, *Coll. Czech. Chem. Commun.*, **12**, 407 (1947).
50. P. Tompkins and C. Schmidt, *Univ. Calif. Pub. Physiol.*, **8**, 221 (1943).
51. P. Tompkins and C. Schmidt, *Univ. Calif. Pub. Physiol.*, **229**, 237, 247 (1943).
52. H. Kirkpatrick, *Quart. J. Pharm. Pharmacol.*, **19**, 127, 526 (1946).
53. J. Stock, *J. Chem. Soc.*, **1944**, 427.
54. R. Kaye and H. Stonehill, *J. Chem. Soc.*, **1951**, 27.
55. S. Tang and P. Zuman, *Coll. Czech. Chem. Commun.*, **28**, 829, 1524 (1963).
56. E. Colichman and P. O'Donovan, *J. Am. Chem. Soc.*, **76**, 3588 (1954).
57. C. Chernick, H. Claassen, and B. Weinstock, *J. Am. Chem. Soc.*, **83**, 3164 (1961).

58. A. Zahlan and R. Linnell, *J. Am. Chem. Soc.*, **77**, 6207 (1955).
59. A. Balaban, C. Bratu, and C. Rentea, *Tetrahedron*, **20**, 265 (1964).
60. D. Engelkemeir, T. Geissman, W. Crowell, and S. Friess, *J. Am. Chem. Soc.*, **86**, 1229 (1947).
61. W. Struck and P. Elving, *J. Am. Chem. Soc.*, **86**, 1229 (1964).
62. K. Schwabe and J. Voight, *Z. Elektrochem.*, **56**, 44 (1952).
63. H. Lund, *Acta Chem. Scand.*, **14**, 1927 (1960).
64. I. Matzuzaki, Japan Tatent 172,748 (1946) [*Chem. Abstr.*, **43**, 6926 (1949)].
65. S. Swann Jr., *Trans. Electrochem. Sat.*, **69**, 320, 321 (1936).
66. S. Swann Jr., *Trans. Electrochem. Sat.*, **69**, 329 (1936).
67. E. Colichman and D. L. Love, *J. Org. Chem.*, **18**, 40 (1953).
68. S. Tang and P. Zuman, *Coll. Czech. Chem. Commun.*, **28**, 829, 1524 (1963).
69. K. Kindler, *Arch. Pharm.*, **265**, 390 (1927).
70. K. Kindler, *Ann.*, **431**, 187 (1923).
71. D. Barnard, M. B. Evans, G. Higgins, and F. Smith, *Chem. Ind.*, 20 (1961).
72. R. Bowers and H. Russell, *Anal. Chem.*, **32**, 405 (1960).
73. H. Drushel and J. Miller, *Anal. Chem.*, **29**, 1459 (1956).
74. M. Nicholson, *J. Am. Chem. Soc.*, **76**, 2539 (1954).
75. S. Mairanovskii and M. Neiman, *Dokl. Akad. Nauk SSSR*, **79**, 85 (1951); **87**, 805 (1952) [*Chem. Abstr.*, **46**, 28 (1953); **47**, 627 (1953)].
76. I. M. Kolthoff, W. Stricks, and N. Tanaka, *J. Am. Chem. Soc.*, **77**, 4739 (1955).
77. H. Lund, *Coll. Czech. Chem. Commun.*, **25**, 2313 (1960).
78. S. Sugasawa, I. Satoda, and J. Yamagisawa, *J. Pharm. Soc. Japan*, **58**, 139 (1938).
79. W. Hoffmann, Dissertation, Univ. of Giessen, 1914. Through S. Swann, Jr., "Electrolytic Reactions," in A. Weissberger, Ed., *Technique of Organic Chemistry*, Vol. II, Wiley, New York, 1956.
80. A. Fry and J. Newberg, *J. Am. Chem. Soc.*, **89**, 6374 (1967).
81. L. Ramberg and E. Hannerz, *Svensk Kem. Tid.*, **36**, 125 (1924) (Through Ref. 79).
82. M. Ishibashi, *M. Coll. Sci. Kyoto Imp. Univ.*, **8A**, 37 (1925) [*Chem. Abstr.*, **20**, 41 (1926)].
83. R. A. Benkeser and E. M. Kaiser, *J. Am. Chem. Soc.*, **85**, 2858 (1963).
84. R. A. Benkeser, E. M. Kaiser, and R. Lambert, *J. Am. Chem. Soc.*, **86**, 527 (1964).
85. H. W. Sternberg, R. A. Markby, I. Wender, and D. M. Mohilner, *J. Electrochem. Soc.*, **113**, 1060 (1960).
87. M. Sitaraman and V. Raman, *Current Sci.*, **16**, 23 (1947) [*Chem. Abstr.*, **41**, 4468 (1947)].
88. S. Wawzonek and D. Wearing, *J. Am. Chem. Soc.*, **81**, 2067 (1959).
89. G. B. Diamond and M. Soffer, *J. Am. Cehm. Soc.*, **74**, 4216 (1952).
90. R. Pointeau and J. Favede, *Fifth Intern. Symp. Free Radicals*, Uppsala, 1961.
91. Continental Oil Co., U.S. Patent 3,488,266 (1967).
92. P. Given and M. Peover, *Coll. Czech. Chem. Commun.*, **25**, 3195 (1960); *J. Chem. Soc.*, **1960**, 385.
93. S. Omo, *J. Electrochem. Soc. Japan*, **23**, 117 (1955) [*Chem. Abstr.*, **49**, 4910 (1955)].
94. D. Goering and H. Jonas, West Ger. Patent 937,919 (1956) [Chem. Abstr., **53**, 3950 (1959)].
95. R. A. Benkeser, Symp. Synthetic and Mechanistic Aspects of Electroorganic Chemistry, Durham, N. Carolina, 1968, p. 189.

96. J. Y. Yang, W. E. McEwen, and J. Kleinberg. *J. Am. Chem. Soc.*, **79**, 5833 (1957).
97. B. L. Funt and F. D. Williams, *J. Polymer Sci. A*, **2**, 865 (1964).
98. A. Tomilov and L. Kaabak, *Zur. Priklad. Khim.*, **32**, 2600 (1959) [*Chem. Abstr.*, **54**, 3774 (1960)].
99. H. J. Barger Jr., *J. Org. Chem.*, **34**, 1489 (1969).
100. L. D. Burke, C. Kemball, and F. A. Lewis, *Trans. Faraday Soc.*, **60**, 913 (1964).
101. Monsanto Co., U.S. Patent 553,851; Netherland Patent 6,707,472 (1967).
102. Badishe Aniline and Soda-Fabrik Aktiengesell-schaft, U.S. Patent 3,477,923 (1967).
103. A. Tomilov, Brit. Patent 1,089,707 (1967).
104. E. I. duPont d Nemours and Co., U.S. Patent 3,488,267 (1963).
105. I. Knunyants and N. Vyazankin, *Dokl. Akad. Nauk SSSR*, **113**, 112 (1957) [*Chem. Abstr.*, **51**, 14637 (1957)].
106. E. Goodings and C. Wilson, *Trans. Electrochem. Soc.*, **88**, 77 (1945).
107. P. Condit, U.S. Patent, 2,537,304 (1951) [*Chem. Abstr.*, **45**, 2341 (1951)].
108. S. Swann Jr., K. Wanderer, H. Schaffer, and W. Streaker, *J. Electrochem. Soc.*, **96**, 353 (1949).
109. C. L. Wilson and K. L. Wilson, *Trans. Electrochem. Soc.*, **80**, 151 (1941).
110. S. Omo and T. Hayaski, *Bull. Chem. Soc. Japan*, **26**, 11 (1953).
111. H. Fierz-David, L. Blangey, and M. Uhlig, *Helv. Chim. Acta*, **32**, 1414 (1949).
112. A. Ingersoll, *Organic Synthesis, Colloquium*, Vol. I, 2nd ed., Wiley, New York, 1941, p. 311.
113. C. Wilson and K. Wilson, *Trans. Electrochem. Soc.*, **84**, 153 (1943).
114. E. Woodruff, *J. Am. Chem. Soc.*, **64**, 2859 (1942).
115. M. Ferles, *Chem. Listy*, **52**, 668 (1958) [*Chem. Abstr.*, **52**, 13724 (1958)].
116. S. Szmaragd and E. Briner, *Helv. Chim. Acta*, **32**, 553 (1949) [*Chem. Abstr.*, **43**, 5677 (1949)].
117. V. Levchenko, *J. Gen. Chem. (USSR)*, **17**, 1656 (1947) [*Chem. Abstr.*, **42**, 2187 (1947)].
118. E. Ochiai, T. Teshigawara, and T. Haito, *J. Pharm. Soc. Japan*, **65**, 429 (1945); [*Chem. Abstr.*, **45**, 8376 (1951)].
119. B. Zwicker and R. Robinson, *J. Am. Chem. Soc.*, **64**, 790 (1942).
120. J. F. Norris and E. Cummings, *Ind. Eng. Chem.*, **17**, 306 (1925).
121. H. Meyer, R. Beer, and G. Lasch, *Monatsh.*, **34**, 1667 (1913) [*Chem. Abstr.*, **8**, 498 (1914)].
122. R. Lapworth and F. Royle, *J. Chem. Soc.*, **117**, 746 (1920).
123. H. Siebert, *Z. Elektrochem.*, **44**, 768 (1938).
124. L. F. Small and F. L. Cohen, *J. Am. Chem. Soc.*, **53**, 2214, 2227 (1931) **56**, 1738, 2165 (1934).
125. A. Ingersoll, in Ref. 112, p. 311.
126. S. Wawzonek, E. Blaha, R. Berkey, and M. Runner, *J. Electrochem. Soc.* **102**, 235 (1955).
127. J. P. Petrovich, M. Baizer, and M. Ort, *J. Electrochem. Soc.*, **116**, 74 (1969).
128. J. P. Petrovich, J. D. Anderson, and M. M. Baizer, *J. Org. Chem.*, **31**, 389 (1966).
129. K. Campbell and E. Young, *J. Am. Chem. Soc.*, **65**, 965 (1943).
130. R. Benkeser and C. Tincher, *J. Org. Chem.*, **33**, 2727 (1968).
131. S. Wawzonek and D. Wearing, *J. Am. Chem. Soc.*, **81**, 2067 (1959).
132. I. A. Shikhiev, *Zh. Obsh. Chim.*, **20**, 839 (1950) [*Chem. Abstr.*, **45**, 150 (1951)].

133. A. Lebedeva and T. Mishnina, *J. Gen. Chem. USSR*, **21**, 1227 (1951) [*Chem. Abstr.*, **46**, 4989 (1952)].
134. A. P. Golovchanskaya, *J. Gen. Chem. USSR*, **11**, 608 (1941) [*Chem. Abstr.*, **35**, 6931 (1941)].
135. I. A. Favorskaya, *Fh. Obsh. Khim.*, **18**, 52 (1948); [*Chem. Abstr.*, **42**, 4905 (1948)].
136. H. Fierz-David, L. Blangey, and M. Uhlig, *Helv. Chim. Acta*, **32**, 1414 (1949) [*Chem. Abstr.*, **44**, 1077 (1950)].
137. W. H. Perkin and S. Plant, *J. Chem. Soc.*, **125**, 1503 (1924).
138. M. Freund and J. Bredenberg, *Ann.*, **407**, 43 (1914).
139. C. Finziand and M. Freund, *Ber.*, **45**, 2322 (1912).
140. S. Kaplanskii, *Ber.*, **60B**, 1842 (1927) [*Chem. Abstr.*, **22**, 236 (1928)].
141. A. Anziegin and V. Bulewitsch, *Z. Physiol. Chem.*, **158**, 35 (1926); [*Chem. Abstr.*, **21**, 57 (1927)].
142. H. Lund, in S. Palai, Ed., *The Chemistry of the Carbon-Nitrogen Double Bond*, Wiley, New York, 1970, p. 505.
143. H. Wenker, *J. Am. Chem. Soc.*, **57**, 772 (1935).
144. P. D. Kane, *Z. Anal. Chem.*, **173**, 50 (1960).
145. M. J. Allen and H. G. Steinman, *J. Am. Chem. Soc.*, **74**, 3932 (1952).
146. B. Sjöberg, *Ber.*, **B75**, 13 (1942) [*Chem. Abstr.*, **36**, 6138 (1942)].
147. L. C. Craig, *J. Am. Chem. Soc.*, **55**, 2543 (1933).
148. T. Osa, T. Yamagishi, T. Kodama, and K. Misono, in Symposium on the "The Synthetic and Mechanistic Aspects of Electroorganic Chemistry" Durham, N. Carolina, Oct. 14-16, 1968, p. 157.
149. H. Sternberg, in Ref. 148, p. 179.
150. A. Krapcho and A. Bothner-By, *J. Am. Chem. Soc.*, **87**, 5799 (1965).
151. R. Benkeser, M. Burrous, J. Hazdra, and E. Kaiser, *J. Org. Chem.*, **28**, 1094 (1963).
152. M. Peover, in A. Bard, Ed., *Electroanalytical Chemistry*, Marcel Dekker, New York, 1967, p. 1.
153. G. Hoijtink, J. Van Schooten, E. DeBoer, and W. Aalbersberg, *Rec. Trav. Chim.*, **73**, 335 (1954).
154. S. Wawzonek, R. Berkey, E. Blocha, and M. Runner, *J. Electrochem. Soc.*, **102**, 235 (1955).
155. M. R. Rifi, unpublished results.
156. J. Thier and J. Weinian, *Bull. Soc. Chim. France*, 177 (1956).
157. W. Kemula and J. Kornacki, *Roczniki Chem.*, **36**, 1835, 1857 (1962) [*Chem. Abstr.*, **59**, 218 (1963)].
158. A. Maccoll, Nature, **163**, 178 (1949).
159. A. Pullman, B. Pullman, and G. Berthier, *Bull. Soc. Chim. France*, 591 (1950).
160. A. Streitwieser Jr. and I. Schwazer, *J. Phys. Chem.*, **66**, 2316 (1962).
161. B. Funt, in *Macromolecular Reviews*, Vol. 1, Wiley, New York, 1967, p. 35.
162. C. Wilson, *Rec. Chem. Prog.*, **10**, 30 (1949).
163. E. Dinneen, T. Schwan, and C. Wilson, *J. Electrochem. Soc.*, **96**, 226 (1949).
164. J. Breitenbach and H. Gabler, *Monatsch. Chem.*, **91**, 202 (1960).
165. J. Breitenbach, C. Srna, and O. Olaj, *Makromol. Chem.*, **42**, 171 (1960).
166. B. Funt and F. Williams, *J. Polymer Sci. A*, **2**, 865 (1964).
167. B. Funt, D. Richardson, and S. Bhadani, *Can. J. Chem.*, **44**, 711 (1966).
168. B. Funt, S. Bhadani, and D. Richardson, *Polymer Reprints*, **7**, 153 (1966).
169. M. W. Leeds and G. Smith, *J. Electrochem. Soc.*, **98**, 129 (1951).
170. V. Spreter and E. Briner, *Helv. Chim. Acta*, **32**, 215 (1949).
171. L. Krimen and D. Cota, *Org. Reac.*, **17**, 213 (1969).
172. H. Ziegler and E. Schneider, *Z. Electrochem.*, **53**, 109 (1949).

173. G. Smith and M. Leeds, U.S. Patent 2,589,635 (1952) [*Chem. Abstr.*, **46**, 4937 (1952)].
174. W. Gakenheimer and W. Hartung, *J. Org. Chem.*, **9**, 85 (1944).
175. G. R. Clemo, H. Perkin, Jr., and P. Robinson, *J. Chem. Soc.*, **1927**, 1589.
176. G. Gever, C. O'Keefe, G. Drake, F. Ebetino, J. Michels, and K. Hayes, *J. Am. Chem. Soc.*, **77**, 2277 (1955).
177. J. Kovacs, *Acta Univ. Szegediensis Chem. Phys.*, **2**, 56 (1948) [*Chem. Abstr.*, **44**, 6384 (1950)].
178. E. Remy and G. Kümmell, *Z. Elektrochem.*, **15**, 254 (1909).
179. T. McGuine and M. Dull, *J. Am. Chem. Soc.*, **69**, 1469 (1947).
180. M. Masui and H. Sayo, *J. Chem. Soc.*, **1961**, 5325.
181. H. Brintzinger and V. Eggers, *Z. Elektrochem.*, **56**, 158 (1952).
182. G. Wittig and B. Rartman, *Ann.*, **554**, 213 (1943).
183. M. Mizuguchi and S. Matsumota, *Yakugakn Zasshi*, **78**, 129 (1958) [*Chem. Abstr.*, **52**, 8794 (1958)].
184. J. E. Slager and J. Mirza, Fr. Patent 1,416,966 (1965) [*Chem. Abstr.*, **64**, 6561 (1966)].
185. C. L. Wilson and H. V. Udupa, *Trans Electrochem. Soc.*, **99**, 289 (1952).
186. F. M. Fredriksen, *J. Phys. Chem.*, **19**, 696 (1915).
187. R. C. Snowdon, *J. Phys. Chem.*, **15**, 797 (1911) [*Chem. Abstr.*, **6**, 617 (1912)].
188. K. Elb and D. Kopp, *Z. Elektrochem.*, **5**, 108 (1898).
189. R. H. McKee and C. J. Brockman, *Trans. Electrochem. Soc.*, **62**, 203 (1932).
190. K. Sugino and T. Sekine, *J. Electrochem. Soc.*, **104**, 497 (1957).
191. M. R. Rifi, unpublished results.
192. W. Löb, *Z. Elektrochem.*, **4**, 428 (1898).
193. I. Bergman and J. C. James, *Trans. Faraday Soc.*, **50**, 60 (1954).
194. B. B. Dey, R. K. Maller, and B. R. Pai, Ind. Patent 34, 756 (1950) [*Chem. Abstr.*, **44**, 9278 (1950)].
195. R. E. Harmon and J. Cason, *J. Org. Chem.*, **17**, 1047 (1952).
196. H. J. Backer, *Rec. Trav. Chim.*, **34**, 194 (1915).
197. B. B. Dey, T. R. Govindachari, and S. Rajagopalan, Ind. Patent 34, 756 (1948) [*Chem. Abstr.*, **44**, 6886 (1950)].
198. O. W. Brown and J. C. Warner, *Trans. Electrochem. Soc.*, **41**, 225 (1941).
199. L. N. Vertyulina and N. I. Malyugina, *Zh. Obshch. Khim.*, **28**, 304 (1958) [*Chem. Abstr.*, **52**, 10765 (1958)].
200. F. Fichter and R. Gunst, *Helv. Chim. Acta*, **22**, 554 (1939) [*Chem. Abstr.*, **33**, 3701 (1939)].
201. J. Cason, C. F. Allen, and S. Goodwin, *J. Org. Chem.* **13**, 403 (1948).
202. L. Fieser and E. L. Martin, *J. Am. Chem. Soc.*, **57**, 1840 (1935).
203. J. Piccard and L. M. Larsen, *J. Am. Chem. Soc.*, **40**, 1090 (1918).
204. C. A. Mann, M. E. Montonna, and R. Larian, *Trans. Electrochem. Soc.*, **69**, 367 (1936).
205. V. M. Berezovskii and V. S. Varkov, *Zh. Obshche. Khim.*, **23**, 100 (1953) [*Chem. Abstr.*, **47**, 5821 (1953)].
206. F. F. Blicke, N. J. Doorenbos, and R. H. Cox, *J. Am. Chem. Soc.*, **74**, 4511 (1952).
207. B. B. Deu, T. R. Govindacharic, and S. Rajogopalan, *J. Sci. Ind. Res. (Ind.)*, **4**, 637, 643 (1946) [*Chem. Abstr.*, **40**, 6347 (19460].
208. C. Weygand and R. Gabler, *J. Prakt. Chem.*, **155**, 332 (1940) [*Chem. Abstr.*, **35**, 1775 (1941)].
209. B. B. Dey, H. V. Udupa, and B. R. Pai, *Current Sci. (India)*, **16**, 186 (1947) [*Chem. Abstr.*, **42**, 41 (1948)].

210. N. A. Izgaryshev and M. Y. Fioshin, *Dokl. Akad. Nauk SSSR,* **90**, 189 (1953) [*Chem. Abstr.,* **47**, 9185 (1953)].
211. L. M. Grubina and V. V. Stender, *J. Appl. Chem. (USSR)* **13**, 1039 (1940) (in French).
212. K. Slotta and R. Kethur, *Ber.,* **71**, 335 (1938).
213. S. D. Ross, G. J. Kahan, and W. A. Leach, *J. Am. Chem. Soc.,* **74**, 4122 (1952).
214. M. J. Lacroix, *Compt. Rend.,* **178**, 483 (1924).
215. J. Armand and O. Convert. C.I.T.C.E. Meeting, Prague Czechoslovakia, Sept. 28 - Oct. 2, 1970.
216. M. Lacan, I. Tabakovic, J. Hranilovic, N. Bujas, and Z. Stunic, in Ref. 215.
217. E. Mosettig and E. L. May, *J. Am. Chem. Soc.,* **60**, 2964 (1938).
218. G. A. Alles, *J. Am. Chem. Soc.,* **54**, 271 (1932).
219. R. W. Lewis and J W. Brown, *Trans. Electrochem. Soc.,* **86**, 135 (1943).
220. D. Horvitz and E. Cerwonka, U.S. Patent 2,916,426 (1959) [*Chem. Abstr.,* **54**, 6370 (1960)].
221. E. Ochiai, T. Teshigawara, and T. Naito, *J. Pharm. Soc. Japan,* **65**, 429 (1945) [*Chem. Abstr.,* **45**, 8376 (1957)].
222. T. DeVries and R. W. Ivett, *Ind. Eng. Chem. Anal. Ed.,* **13**, 339 (1941).
223. J. Pearson, *Trans. Faraday Soc.,* **44**, 683 (1948).
224. M. Suzuki, *J. Electrochem. Soc. Japan,* **22**, 112 (1954) [*Chem. Abstr.,* **48**, 13472 (1954)].
225. B. Kastening, *Naturwiss.,* **47**, 443 (1960).
226. H. L. Piette, P. Ludwig, and R. N. Adams. *Anal. Chem.,* **34**, 916 (1962).
227. J. W. Smith and J. G. Waller, *Trans. Faraday Soc.,* **46**, 290 (1950).
228. E. Bamberger and E. Renould, *Ber.,* **30**, 2278 (1897); **33**, 271 (1960).
229. G. Costa, *Gas. Chim. Ital.,* **83**, 875 (1953).
230. S. Wawzonek and J. D. Fredrickson, *J. Am. Chem. Soc.,* **77**, 3985 (1955).
231. N. H. Garwood, U.S. Patent 3,338,806 (1963).
232. J. Gerlock, *J. Am. Chem. Soc.,* **90**, 1652 (1968).
233. J. T. Sock, *J. Chem. Soc.,* **1957**, 4532.
234. M. Masui and H. Sayo, *J. Chem. Soc.,* **1961**, 4773, 5325.
235. M. Masui and H. Sayo, *J. Chem. Soc.,* **1966**, 1733
236. N. Radin and T. DeVries, *Anal. Chem.,* **24**, 971 (1952).
237. J. Pearson, *Trans. Faraday Soc.,* **44**, 683, (1948).
238. M. Field, C. Valle, and M. Kane, *J. Am. Chem. Soc.,* **71**, 421 (1949).
239. V. T. Vodzinkii and I. A. Korshunov, *Uchenye Zapiski Gorlovsk. Gosudarsc, Unv. im. N. I. Lobachevskogo, Ser. Khim. No. 32,* 25 (1958) [*Chem. Abstr.,* **54**, 17113 (1960)].
240. I. L. Wolk, H.S. Patent, 2,419,515 (1947) [*Chem. Abstr.,* **42**, 589 (1948)].
241. W. C. Albert and A. Lowy, *Trans. Electrochem. Soc.,* **75**, 367 (1939).
242. T. G. Chambers and O. S. Slotterbeck, U.S. Patent 2,485,258 (1949) [*Chem. Abstr.,* **44**, 4807 (1950)].
243. K. Vesely and R. Brdicka, *Coll. Czech. Chem. Commun.,* **12**, 313 (1947).
244. R. Bieber and G. Trümpler, *Helv. Chim. Acta,* **30**, 706, 1109 (1947).
245. P. Valenta, *Coll. Czech. Chem. Commun.,* **25**, 853 (1960).
246. M. Peover and B. S. White, *Electrochim. Acta,* **11**, 1061 (1966).
247. R. Bieber and G. Trümpier, *Electrochim. Acta,* **31**, 5 (1948).
248. P. J. Elving and C. E. Bennett, *J. Am. Chem. Soc.,* **76**, 1412 (1954).
249. H. Adkins and F. W. Cox, *J. Am. Chem. Soc.,* **60**, 1151 (1938).
250. M. Fields and E. R. Blout, *J. Am. Chem. Soc.,* **70**, 930 (1948).
251. E. A. Parker and S. Swann Jr., *Trans. Electrochem. Soc.,* **92**, 343 (1947).
252. N. G. Belenkaya and N. A. Belozerskii, *Zh. Obshch. Khim.,* **19**, 1664 (1949) [*Chem. Abstr.,* **44**, 956 (1950)].
253. K. R. Brown, U.S. Patent 2,280,887 (1942) [*Chem. Abstr.,* **36**, 5433

(1942)].
254. R. A. Hales, U.S. Patent 2,289,189 (1943) [*Chem. Abstr.*, **37**, 42 (1943)].
255. R. A. Hales, .U.S. Patent 2,300,218 (1943) [*Chem. Abstr.*, **37**, 1660 (1943)].
256. R. A. Hales, U.S. Patent 2,303,210 (1943) [*Chem. Abstr.*, **37**, 2277 (1943)].
257. D. L. Marite and W. Hodson, *Anal. Chem.*, **31**, 1562 (1965).
258. M. L. Wolfrom, W. W. Binkley C. C. Spencer, and B. W. Lew, *Anal. Chem.*, **73**, 3357 (1951).
259. M. L. Wolfrom, M. Konigsberg, F. B. Moody, and R. M. Goepp, *Anal. Chem.* **68**, 122 (1946).
260. M. L. Wolfrom, F. B. Moody, M. Konigsberg, and R. M. Goepp, *Anal. Chem.*, **68**, 578 (1946).
261. M. L. Wolfrom, B. W. Lew, R. A. Hales, and R. M. Goepp, *Anal. Chem.*, **68**, 2342 (1946).
262. Atlas Powder Co., Brit. Patent 533,884 (1941), 533,885 (1942) [*Chem. Abstr.*, **36**, 975 (1942)].
263. T. Arai and T. Oguri, *Bull. Chem. Soc., Japan*, **33**, 1018 (1960).
264. J. Stocker and R. M. Jenevein, in Ref. 148.
265. R. A. Pasternak, *Helv. Chim. Acta*, **31**, 753 (1948).
266. H. D. Law, *J. Chem. Soc.*, **1907**, 748.
267. M. J. Allen, *J. Am. Chem. Soc.*, **72**, 3797 (1950).
268. W. Rapson and R. Robinson, *J. Chem. Soc.*, **1935**, 1537.
269. M. J. Allen, *J. Org. Chem.*, **15**, 435 (1950).
270. I. A. Pearl, *J. Am. Chem. Soc.*, **74**, 4260 (1952).
271. F. D. Popp and H. P. Schultz, *Chem. Rev.*, **62**, 19 (1962).
272. S. Swann Jr., R. W. Benoliel, L. B. Lyons, and W. H. Paht, *Trans. Electrochem. Soc.*, **79**, 83 (1940).
273. S. Swann Jr., *Trans. Electrochem. Soc.*, **62**, 177 (1932).
274. H. J. Read, *Trans. Electrochem. Soc.*, **80**, 133 (1941).
275. R. R. Read and F. A. Fletcher, *Trans. Electrochem. Soc.*, **47**, 93 (1925).
276. G. H. Keats, *J. Chem. Soc.*, **1937**, 2005.
277. N. J. Leonard, S. Swann Jr., and J. Figueras, *J. Am. Chem. Soc.*, **74**, 4620 (1952).
278. N. J. Leonard, S. Swann Jr., and E. H. Mottus, *J. Am. Chem. Soc.*, **74**, 6251 (1952).
279. N. J. Leonard, S. Swann Jr., and G. Fuller, *J. Am. Chem. Soc.*, **76**, 3193 (1954).
280. R. M. Powers and R. A. Day Jr., *J. Org. Chem.*, **24**, 722 (1959).
281. T. J. Curphey, C. W. Amelotti, T. P. Layoff, R. L. McCartney, and J. H. Williams, *J. Am. Chem. Soc.*, **91**, 2817 (1969).
282. F. Arcamone, C. Prevost, and P. Souchay, *Bull. Soc. Chim. France*, 891 (1953).
283. S. I. Zhdanov and M. I. Policvktov, *J. Gen. Chem. USSR*, **31**, 3607 (1961).
284. S. Ono, *J. Chem. Soc. Japan, Pure Chem. Sec.*, **73**, 852 (1952).
285. H. Lund, *Acta Chem. Scand.*, **11**, 283 (1957).
286. J. H. Stocker and R. M. Jenevein, *J. Org. Chem.*, **33**, 294 (1968).
287. N. J. Leonard, S. Swann Jr., and H. L. Dryden, *J. Am. Chem. Soc.*, **74**, 2871 (1952).
288. E. A. Steck and W. Boehme, *J. Am. Chem. Soc.*, **74**, 4511 (1952).
289. P. Anziani, A. Aubry, and R. Cornubert, *Compt. Rend.*, **225**, 878 (1947).
290. E. Kariv, J. Hermolin, and E. Gileadi, *J. Electrochem. Soc.*, **117**, 342 (1970).
291. M. J. Allen and A. Corwin, *J. Am. Chem. Soc.*, **72**, 117 (1950).
292. M. J. Allen, *J. Am. Chem. Soc.*, **73**, 3503 (1951).
293. M. J. Allen, J. A. Sinagusa, and W. Pierson, *J. Chem. Soc.*, **1960**, 1045.

294. M. J. Allen, *J. Org. Chem.*, **15**, 435 (1950).
295. G. I. Hobday and W. F. Short, *J. Chem. Soc.*, **1943**, 609.
296. F. D. Popp and H. P. Schultz, *Chem. Rev.*, **62**, 29 (1962).
297. J. Stocker and R. M. Jenevein, in Ref. 148.
298. J. H. Stocker and R. M. Jenevein, *J. Org. Chem.*, **33**, 2145 (1968).
299. M. J. Allen, J. E. Fearnand, and H. A. Levine, *J. Chem. Soc.*, **1952**, 2220.
300. P. Kabasakalian and J. McGlotten, *J. Am. Chem. Soc.*, **78**, 5032 (1956).
301. R. E. Juday, in Ref. 148.
302. L. Mandell, R. M. Powers, and R. A. Day, *J. Am. Chem. Soc.*, **80**, 5284 (1958).
303. R. N. Gourley and J. Grimshaw, *J. Chem. Soc. (c)*, **1968**, 2388.
304. H. A. Levine and M. J. Allen, *J. Chem. Soc. (c)*, **1952**, 254.
305. F. Sorm and J. Arient, *Coll. Czech. Chem. Commun.*, **15**, 175 (1950).
306. S. Ohkiand and Y. Noike, *J. Pharm. Soc., Japan*, **72**, 490 (1952) [*Chem. Abstr.*, **47**, 6418 (1953)].
307. R. M. Powers and R. A. Day Jr., *J. Org. Chem.*, **24**, 722 (1959).
308. S. Ono and T. Hayaski, *Bull. Chem. Soc. Japan*, **26**, 232 (1953).
309. M. Msuno, T. Asahara, S. Kuroiwa, K. Shimigu, and J. Nakano, *J. Chem. Soc. Japan, Ind. Chem. Sect.* **52**, 151 (1949) [*Chem. Abstr.*, **45**, 1884 (1951)].
310. I. L. Wolk, U.S. Patent 2,419,515 (1947) [*Chem. Abstr.*, **42**, 589 (1948)].
311. C. S. Dillon, Brit. Patent 611,674 (1948) [*Chem. Abstr.*, **43**, 4589 (1949)].
312. R. E Juday, *J. Org. Chem.*, **20**, 617 (1955), or *Trans. Electrochem. Soc.*, **84**, 173 (1943).
313. A. W. Ingersoll, in Ref. 112, p. 311.
314. T. Kuwata, Jap. Patent 9966 (1958) [*Chem. Abstr.*, **54**, 5298 (1960)].
315. V. V. Listopadov and L. I. Antropov, *Nauch. Trudy Novocherkassk. Politekh. Inst.*, **34**, 87 (1956) [*Chem. Abstr.*, **53**, 13836 (1959)].
316. S. Ono and T. Yamauchi, *Bull. Chem. Soc., Japan*, **25**, 404 (1952).
317. F. Mayer and F. A. English, *Ann.*, **417**, 60 (1918) [*Chem. Abstr.*, **13**, 1826 (1919)].
318. C. Mettler, *Ber.*, **39**, 2933 (1906).
319. C. Mettler, *Ber.*, **38**, 1747 (1905).
320. F. Mayer, W. Schäfer, and J. Rosenbach, *Arch. Pharm.*, **267**, 571 (1929) [*Chem. Abstr.*, **24**, 838 (1930)].
321. J. A. May and K. A. Kobe, *J. Electrochem. Soc.*, **97**, 183 (1950).
322. S. Ono, *Hippon Kagaku Zasshi*, **75**, 1195 (1954) [*Chem. Abstr.*, **51**, 12704 (1957)].
323. S. Ono and J. Nakaya, *J. Chem. Soc. Japan, Pure Chem. Sect.*, **74**, 907 (1953).
324. J. P. Wibaut and H. Boer, *Rec. Trav. Chim.*, **68**, 72 (1949).
325. G. H. Coleman and H. L. Johnson, *Organic Synthesis, Collequium*, Vol. III, Wiley, New York, 1955, p. 60.
326. W. Oroshnik and P. E. Spoerri, *J. Am. Chem. Soc.*, **63**, 338 (1941).
327. G. C. Whitnack, J. Reinhart, and E. St. C. Gantz, *Anal. Chem.*, **27**, 359 (1955).
328. V. Berezorskii and Y. Sobolev, *Khim. Nauka i. Prom.*, **3**, 677 (1958) [*Chem. Abstr.*, **53**, 3943 (1959)].
329. T. Sato, *J. Chem. Soc. Japan, Pure Chem. Sect.*, **71**, 310 (1950).
330. T. Hoshino and T. Sato, Jap. Patent 4359 (1950) [*Chem. Abstr.*, **47**, 3341 (1953)];
331. Hoffman-La Roche and Co., Swiss Patent 258,581 (1949) [*Chem. Abstr.*, **44**, 4352 (1950)].
332. S. Swann Jr., *Trans. Electrochem. Soc.*, **84**, 165 (1943).

333. R. A. Benkeser, in Ref. 148.
334. A. V. Koperina and N. I. Gavrilov, *J. Gen. Chem. (USSR)*, **17**, 165 (1947).
335. T. Yamazaki and M. Nagata, *Yakugaku Zasshi*, **79**, 1222 (1959) [*Chem Abstr.*, **54**, 4596 (1960)].
336. V. Prelog, *Coll. Czech. Chem. Commun.*, **2**, 716 (1930).
337. J. Tafel and T. B. Baillie, *Ber.*, **32**, 68 (1899).
338. K. Kindler, *Arch. Pharm.* **265**, 390 (1927).
339. N. I. Gavrilov and A. V. Koperina, *J. Gen. Chem. (USSR)*, **9**, 1394 (1939) [*Chem. Abstr.*, **34**, 1615 (1940)].
340. A. Dunet and A. Willemart, *Compt. Rend.*, **226**, 821 (1948) [*Chem. Abstr.*, **42**, 5011 (1948)].
341. B. Sakurai, *Bull. Chem. Soc. Japan*, **7**, 155 (1932) [*Chem. Abstr.*, **26**, 4542 (1932)]; *Bull. Chem. Soc. Japan*, **5**, 184 (1930) [*Chem. Abstr.*, **24**, 5643 (1930)].
342. A. Dunet, J. Rollet, and A. Willemart, *Bull. Cos. Chim.*, 877 (1950) [*Chem. Abstr.*, **45**, 9527 (1951)].
343. E. W. Cook and W. G. France, *J. Phys. Chem.*, **36**, 2383 (1932).
344. G. Komppa, *Ber.*, *B*, **65**, 793 (1932).
345. E. Späth and F. Breusch, *Monatsh.*, **50**, 349 (1928) [*Chem. Abstr.*, **23**, 1634 (1929)].
346. B. Sakurai, *Bull. Chem. Soc. Japan*, **14**, 173 (1939) [*Chem. Abstr.*, **33**, 6170 (1929)].
347. Badische Anilin, *Jap. Pat. Rept.*, **8**, No. 10 (1969).
348. J. H. Wagenknecht and M. M. Baizer, *J. Org. Chem.*, **31**, 3885 (1966).
349. B. Sakurai, *Bull. Soc. Chem. Japan*, **10**, 311 (1935) [*Chem. Abstr.*, **29**, 7828 (1935)].
350. B. Sakurai, *Bull. Chem. Soc. Japan*, **11**, 42 (1936) [*Chem. Abstr.*, **30**, 4100 (1936)].
351. L. C. Craig, *J. Am. Chem. Soc.*, **55**, 295 (1933).
352. B. Sakurai, *Bull. Chem. Soc. Japan*, **10**, 311 (1935) [*Chem. Abstr.*, **29**, 7828 (1935)].
353. H. Lund, *Acta Chem. Scand.*, **11**, 283 (1957).
354. H. Se, Jap. Patent 180,486 (1949) [*Chem. Abstr.*, **46**, 2432 (1952)].
355. J. Bartek, T. Mukai, T. Nozoe, and F. Santavy, *Coll. Czech. Chem. Commun.*, **19**, 885 (1954).
356. J. C. James and J. C. Speakman, *Trans. Faracay Soc.*, **48**, 474 (1952).
357. M. J. Allen, in *Organic Electrode Processes*, Reinhold, New York, 1958, p. 61.
358. J. Tafel, *Ber.*, **44**, 327 (1911).
359. E. Baur, *Z. Electrochem.*, **37**, 255 (1931).
360. N. Steinberger and G. K. Fraenkel, *J. Chem. Phys.*, **40**, 723 (1964).
361. P. Anziani, A. Aubry, and R. Cornubert, *Compt. Rend.*, **225**, 878 (1947).
362. S. Wawzonek and A. Gunderson, *J. Electrochem. Soc.*, **107**, 537 (1960).
363. P. H. Given and M. E. Peover, *Coll. Czech. Chem. Commun.*, **25**, 3195 (1960).
364. P. H. Given and M. E. Peover, *J. Chem. Soc.*, **1960**, 385.
365. M. R. Rifi, unpublished results.
366. M. Ashworth, *Coll. Czech. Chem. Commun.*, **13**, 229 (1948).
367. S. Swann Jr., *Trans. Electrochem. Soc.*, **85**, 231 (1944).
368. P. J. Elving and J. T. Leone, *J. Am. Chem. Soc.*, **80**, 1021 (1958).
369. H. J. Gardner, *Chem. Ind.*, 819 (1951).
370. H. Adkins and F. W. Cox, *J. Am. Chem. Soc.*, **60**, 1151 (1938).
371. M. Fields and E. R. Blout, *J. Am. Chem. Soc.*, **70**, 930 (1948).
372. C. Parkanyi and R. Zahradnik, *Coll. Czech. Chem. Commun.*, **27**, 1355 (1962).

373. T. A. Geissman and S. L. Friess, *J. Am. Chem. Soc.*, **71**, 3893 (1949).
374. P. Zuman, *J. Electrochem. Soc.*, **105**, 758 (1958).
375. S. Wawzonek, R. C. Reack, W. W. Vaught Jr., and J. W. Fan, *J. Am. Chem. Soc.*, **67**, 1300 (1945).
376. L. Holleck and D. Marquarding, *Naturwiss.*, **49**, 468 (1962).
377. P. Beckmann, *Australian J. Chem.*, **14**, 229 (1961).
378. C. E-lis and K. P. McElroy, U.S. Patent 867,575 (1907) [*Chem. Abstr.*, 2, 908 (1908)].
379. S. Swann Jr., *Trans. Electrochem. Soc.*, **64**, 245 (1933).
380. J. Tafel and W. Jürgens, *Ber.*, **42**, 2554 (1909).
381. J. Tafel and . Fredrichs, *Ber.*, **37**, 3187 (1904).
382. C. Mettler, *Ber.*, **37**, 3695 (1904).
383. G. W. Heimrod, *Ber.*, **17**, 338 (1914).
384. M. J. Allen and J. Ocampo, *J. Electrochem. Soc.*, **103**, 452 (1956).
385. R. E. Erickson and C. M. Fischer, *J. Org. Chem.*, **35**, 1604 (1970).
386. N. L. Weinberg, A. Kentaro Hoffman, and T. B. Reddy, *Tetrahedron Lett.*, **1971**, 2271.
387. W. Reusch and D. Priddy, *J. Am. Chem. Soc.*, **91**, 3677 (1969).
388. S. Wawzonek and J. D. Fredrickson, *J. Electrochem. Soc.*, **106**, 325 (1959).
389. M. Von Stackelberg and W. Stracke, *Z. Electrochem.*, **53**, 118 (1949).
390. H. E. Ulrey, *J. Electrochem. Soc.*, **116**, 1201 (1969).
391. J. W. Sease and R. C. Reed, Electrochemical Society Meeting, New York City, May 4-9, 1969.
392. M. R. Rifi, *J. Am. Chem. Soc.*, **89**, 4442 (1967).
393. P. J. Elving and R. E. Van Atta, *Anal. Chem.*, **27**, 1908 (1955).
394. P. J. Elving and R. E. Van Atta, *J. Electrochem. Soc.*, **103**, 676 (1956).
395. L. B. Rogers and A. J. Diefenderfer, *J. Electrochem. Soc.*, (Electrochem. Sci.) 942 (1967).
396. S. Wawzonek, R. C. Duty, and J. H. Wagenknekt, *J. Electrochem. Soc.*, **111**, 74 (1964).
397. G. Klopman, *Helv. Chim. Acta*, **44**, 1908 (1961).
398. R. Annino, R. E. Erickson, J. Michalovic, and B. McKay, *J. Am. Chem. Soc.*, **88**, 4424 (1966).
399. C. K. Mann, J. L. Webb, and H. M. Walborsky, *Tetrahedron Lett.*, **1966**, 2249.
400. P. J. Elving, I. Rosenthal, J. R. Hayes, and A. J. Martin, *Anal. Chem.*, **33**, 330 (1961).
401. S. Andreades, U.S. Patent 3,475,298 (1969).
402. L. L. Miller and E. Rickena, *J. Org. Chem.*, **34**, 3359 (1969).
403. J. L. Webb, C. K. Mann, and H. M. Walborsky, *J. Am. Chem. Soc.*, **92**, 2042 (1970).
404. L. G. Feoktistov, A. P. Tomilov, and M. M. Goldin, *Izv. Akad. Nauk Ussr Ser. Khim.*, I, 1352 (1963) [*Chem. Abstr.*, **59**, 12624 (1963)].
405. W. Kemula and J. Kornacki, *Roczniki Chem.*, **36**, 1857 (1962). [*Chem. Abstr.*, **59**, 218 (1962)].
406. M. R. Rifi, C.I.T.C.E. Meeting, Prague Czechoslovakia, Sept. 28-Oct. 2 (1970).
407. J. A. Dougherty and A. J. Diefenderfer, *J. Electroanal. Chem. Interfacial Chem.*, **21**, 531 (1969).
408. M. R. Rifi, *Tetrahedron Lett.* **1969**, 1043.
409. S. Wawzonek and R. C. Duty, *J. Electrochem. Soc.*, **108**, 1135 (1961).
410. T. Meites and L. Meites, *Anal. Chem.*, **27**, 1531 (1955).
411. P. J. Elving, I. Rosenthal, and M. K. Kramer, *J. Am. Chem. Soc.*, **73**, 1919 (1951).
412. H. Gilch, *J. Polym. Sci.*, **4**, 1351 (1966).
413. W. Kemula and A. Cisak, *Roczniki Chem.*, **36**, 1857 (1962), [*Chem.*

Abstr., **59**, 218 (1962)].
414. A. M. Doyle, A. E. Pedler, and J. C. Tatlow, *J. Chem. Soc., (c),* **1968**, 2740.
415. F. Covitz, *J. Am. Chem. Soc.,* **89**, 5403 (1967).
416. J. Stocker, *Chem. Commun.,* 934 (1968).
417. H. Lund, *Acta Chem. Scand.,* **13**, 192 (1959).
418. P. J. Elving, I. Rosenthal, and A. Martin, *J. Am. Chem. Soc.,* **77**, 5218 (1955).
419. M. R. Rifi, unpublished results.
420. T. Kitagawa, T. P. Layloff, and R. Adams, *Anal. Chem.,* **35**, 1086 (1963).
421. P. J. Elving and C. L. Hilton, *J. Chem. Soc.,* **74**, 3368 (1952).
422. K. Sigono and K. Shirai, *J. Chem. Soc. Japan (Pure Chem. Sect.),* **70**, 111 (1949); **71**, 396 (1950).
423. S. Wawzonek and J. H. Wagenknecht, *J. Electrochem..Soc.,* **110**, 420 (1963).
424. A. J. Fry and M. A. Mitnick, *J. Am. Chem. Soc.,* **91**, 6207 (1969).
425. A. J. Fry and R. H. Moore, *J. Org. Chem.,* **33**, 1283 (1968).
426. I. Rosenthal, G. Frisone, and R. J. LaCoste, *Anal. Chem.,* **29**, 1639 (1957).
427. W. H. Jura and R. J. Gaul, *J. Am. Chem. Soc.,* **80** 5402 (1958).
428. A. J. Fry, M. A. Mitnick, and R. G. Reed, *J. Org. Chem.,* **35**, 1232 (1970).
429. P. J. Elving, *Record Chem. Progr.* (Kresge-Hooker Sci. Lib.), **14**, 99 (1953).
430. P. J. Elving and B. Pullman, in I. Prigogine, Ed., *Advances in Chemical Physics,* Vol. 1, Wiley, New York, 1961, p. 1.
431. C. L. Perin, in S. G. Cohen, A. Streitwieser Jr., and R. W. Taft, Eds., *Progress in Physical Organic Chemistry,* Vol. 3, Wiley, New York, 1965, p. 256.
432. M. R. Rifi, unpublished results.
433. F. Covitz, unpublished results.
434. J. W. Sease, F. G. Burton, and S. L. Nickol, *J. Am. Chem. Soc.,* **90**, 2595 (1968).
435. F. L. Lambert, *J. Org. Chem.,* **31**, 4184 (1966).
436. L. L. Miller and E. Rickena, *J. Org. Chem.,* **34**, 3359 (1969).
437. D. E. Applequist, *J. Org. Chem.,* **23**, 1715 (1958).
438. H. O. House, R. C. Lord, and H. S. Rao, *J. Org. Chem.,* **21**, 1487 (1956).
439. V. A. Slobey, *J. Am. Chem. Soc.,* **68**, 1335 (1946).
440. M. Slobodin and I. N. Shokhor, *J. Gen. Chem. (USSR),* **21**, 2231 (1951).
441. J. W. Sease, P. Chang, and J. Groth, *J. Am. Chem. Soc.,* **86**, 3154 (1964).
442. F. Lambert and K. Kobayashi, *J. Am. Chem. Soc.,* **82**, 5324 (1960).
443. P. S. Skell and R. G. Allen, *J. Am. Chem. Soc.,* **80**, 5997 (1958).
444. A. A. Oswald, K. Griesbaum, B. B. Hudson Jr., and J. M. Bregman, *J. Am. Chem. Soc.,* **86**, 2877 (1964).
445. J. Zavada, J. Krupieka, and J. Sicher, *Coll. Czech. Chem. Commun.,* **28**, 1664 (1963).
446. P. Zuman, *Coll. Czech. Chem. Commun.,* **15**, 1107 (1950).
447. P. Kabasakalian, J. McGlotten, A. Basch, and M. D. Yudis, *J. Org. Chem.,* **26**, 1738 (1961).
448. M. D. Rausch, F. D. Popp, W. E. McEwen, and J. Kleinberg, *J. Org. Chem.,* **21**, 212 (1956).
449. L. J. Throop, U.S. Patent 3,506,549 (1970).
450. M. Fedoronko, J. Königstein, and K. Linek, *Coll. Czech. Chem. Commun.,* **32**, 3998 (1967).
451. K. Natarajan, K. S. Udupa, G. S. Subramanian, and H. V. K. Udapa, *Electrochem. Technology,* **2**, 151 (1964).
452. A. Vincenz-Chodkowska and Z. R. Grabowski, *Electrochim. Acta,* **9**, 789 (1964) [*Chem. Abstr.*, **61**, 9173 (1964)].

453. W. D. Hoffman, W. E. McEwen, and J. Kleinberg, *Tetrahedron,* 5, 293 (1959).

454. R. C. Fuxon and C. Hornberger Jr., *J. Org. Chem.,* 16, 637 (1951).

455. R. Benkeser, H. Watanabe, S. J. Mels, and M. A. Sabol, *J. Org. Chem.,* 35, 1210 (1970).

456. A. Bewick and G. P. Greener, *Tretrahedron Lett.,* 1970, 391.

457. T. S. Ivcher, E. N. Zil'berman, and E. M. Perepletchikova, *Tr. Khim. Khim. Tekhnol.,* 100 (1968).

458. A. V. Il'yasov, Y. M. Kargin, Y. A. Levin, I. D. Morozova, N. N. Sotnikova, V. K. Ivanova, and N. I. Bessolitzyna, *Izv. Akad. Nauk SSSR, Ser. Khim.,* 740 (1968).

459. B. Fleury, P. Souchay, and P. Gracian, *Bull. Soc. Chim. France,* 562 (1968).

460. V. A. Smirnov, L. I. Antropov, M. G. Smirnova, and L. A. Demchuck, *Zh. Fiz. Khim.,* 42, 1713 (1968) [*Chem. Abstr.,* 69, 82877 (1968)].

461. J. Wieman and M. L. Bouguerra, *Ann. Chim.,* 215 (1968) [*Chem. Abstr.,* 69, 112835g (1968)].

462. N. M. Przhiyalgovskaya, N. S. Yares'ko, and V. N. Belov, *Zh. Org. Khim.,* 896 (1968) [*Chem. Abstr.,* 69, 18383 (1968)].

463. L. Throop and L. Tökes, *J. Am. Chem. Soc.,* 89, 4789 (1967).

464. A. V. Il'yasov, Y. Kargin, Y. A. Levin, D. I. Morozova, and V. Kh. Ivanova, *Izv. Akad. Nauk SSSR, Ser. Khim.,* 1693 (1969).

465. A. N. Dey and B. P. Sullivan, Electrochem. Soc. Meeting, Los Angeles, Calif., 1970.

466. J. W. Johnson, S. Y. Hsieh, and W. J. James, in Ref. 465.

467. A. V. Il'yasov, Y. M. Kargin, Y. A. Levin, I. D. Morozova, and N. N. Sotnikova, *Dokl. Akad. Nauk SSSR,* 1141 (1968) [*Chem. Abstr.,* 69, 23574 (1968)].

468. M. C. Caullet, M. Salaün, and M. M. Hebert, *Compt. Rend.,* 264, 2007 (1967).

469. M. Cardinali, I. Carelli, and A. Trazza, *Ric. Sci.,* 37, 956 (1967) [*Chem. Abstr.,* 68, 101215 (1968)].

470. V. G. Khomyakov, A. P. Tomilov, and B. G. Soldatov, *Electrokhimiya,* 5, 850 (1969) [*Chem. Abstr.,* 71, 108402 (1969)].

471. E. Kryakova and A. P. Tomilov, *Elektrokhimiya,* 5, 869 (1969) [*Chem. Abstr.,* 71, 76739 (1969)].

472. M. Nicolas and R. Palland, *Compt. Rend.,* 267, 1834 (1968).

473. E. Touboul, F. Weisbuch, and J. Wiemann, *Compt. Rend.,* 268, 1170 (1969).

474. O. R. Brown and K. Lister, *Discussions Farad. Soc.,* 45, 106 (1968).

475. D. Barnes and P. Zuman, *Electroanalyt. Chem. Interfacial Electrochem.,* 16, 575 (1968).

476. R. C. Buchta and D. H. Evans, Am. Chem. Soc., Meeting, San Francisco, Calif., April 1-5, 1968.

477. W. N. Greig and J. W. Rogers, *J. Am. Chem. Soc.,* 91, 5495 (1969).

478. R. N. Boyd and A. A. Reidlinger, *J. Electrochem. Soc.,* 107, 611 (1960).

479. L. E. Ter-Minasyan, *Izvest. Akad. Nauk Armyan. SSR, Khim. Nauki,* 11, (1958) [*Chem. Abstr.,* 52, 18022 (1958)].

480. S. Wawzonek, T. Plaisance, and T. McIntyre, *J. Electrochem. Soc.,* 114, 588 (1967).

481. A. Taller, *Compt. Rend.,* 263, 722 (1966).

482. H. Sayo, Y. Tsukitani, and M. Masui, *Tetrahedron,* 24, 1717 (1968).

483. A. Tallec, *Ann. Chim.,* 3, 164 (1968).

484. H. C. Rance and J. M. Coulson, *Electrochim. Acta,* 14, 283 (1969).

485. P. N. Anantharaman, G. S. Subramanian, and H. V. K. Udupa, *Trans.*

Soc. Advan. Electrochem. Sci. Technol., **4**, 38 (1969).
486. A. Tallec, *Ann. Chim.,* **3**, 155 (1968) [*Chem. Abstr.,* **69**, 86511 (1968)].
487. V. G. Khomyakov, M. Ya. Fioshin, I. A. Avrutskava, and S. S. Sedova, *Zh. Vses. Khim. Obshchestva im. D. I. Mendeleeva,* **7**, 584 (1962) [*Chem. Abstr.,* **57**, 3193 (1962)].
488. Abbot Laboratories, Brit. Patent 1,110,184 (1968).
489. L. Weiss and K. Reiter, *Ann.,* **355**, 175 (1907) [*Chem. Abstr.,* **2**, 76 (1908)].
490. R. C. Buchta and D. H. Evans, *Anal. Chem.,* **40**, 2181 (1968).
491. H. E. Stapelfeldt and S. P. Perone, *Anal. Chem.* **41**, 623 (1969).
492. D. H. Evans and E. C. Woodbury, *J. Org. Chem.,* **32**, 2158 (1967).
493. H. E. Stapelfeldt and S. P. Perone, *Anan. Chem.,* **70**, 815 (1968).
494. N. J. Leonard, H. A. Laitinen, and E. H. Mottus, *J. Am. Chem. Soc.,* **75**, 3300 (1953); **76**, 4737 (1954).
495. T. Arai, *J. Electrochem. Soc. Japan,* **30**, E. 46 (1962).
496. L. G. Feoktistov, A. S. Solonar, and Y. S. Lyalikov, *Zh. Obshch. Khim.* **37**, 983 (1967) [*Chem. Abstr.,* **68**, 26320 (1968)].
497. L. G. Feoktistov and A. S. Solonar, *Zh. Obshch. Khim.,* **37**, 986 (1967) [*Chem. Abstr.,* **68**, 26321 (1968)].
498. C. Andrzej, *Rocz. Chem.,* **42**, 907 (1968) [*Chem. Abstr.,* **68**, 64112 (1968)].
499. P. Carrahar and F. G. Drakesmith, *Chem. Commun.,* 1562 (1968).
500. L. G. Feoktistov, *Usp. Elektrokhim. Org. Soedin., Akad. Nauk SSSR, Int. Elektrokhim,* 135 (1966).
501. O. A. Swanepoel, N. J. Van Rensburg, and G. S. Scanes, *J. S. Afr. Chem. Inst.,* **22**, 57 (1969) [*Chem. Abstr.,* **69**, 120565 (1969)].
502. O. Convert and J. Armand, *Compt. Rend.,* **265**, 1486 (1967).
503. G. J. Lawless, E. D. Bartak, and M. D. Hawley, *J. Am. Chem. Soc.,* **91**, 7121 (1969).
504. I. G. Markova and L. G. Feoktistov, *Zh. Obshch. Khim.,* **38**, 970 (1968) [*Chem. Abstr.,* **69**, 40705 (1968)].
505. S. Winstein, D. Pressman, and W. G. Young, *J. Am. Chem. Soc.,* **61**, 1645 (1939).
506. M. Nagao, N. Sato, T. Akashi, and T. Yoshida, *J. Am. Chem. Soc.,* **88**, 3447 (1966).
507. I. Rosenthal, J. R. Hayes, A. Martin, and P. Elving, *J. Am. Chem. Soc.,* **80**, 3050 (1958).
508. H. W. Sterberg, R. E. Markby, and I. Wender, Am. Chem. Soc., Meeting, April 1968.
509. A. P. Tomilov, Y. D. Smirnov, S. K. Smirnov, and S. I. Varshavskii, *Zh. Org. Khim.,* **3** (1967) [*Chem. Abstr.,* **67**, 53602 (1967)].
510. T. Nonaka and K. Sugino, *J. Electrochem. Soc.,* **114**, 1255 (1967).
511. R. N. Gourley, J. Grimshaw, and P. G. Millar, *Chem. Commun.,* 1278 (1967).
512. A. T. Kuhn, *Electrochim. Acta,* **13**, 477 (1968).
513. P. J. Elving and C. Teitelbaum, *J. Am. Chem. Scc.,* **71**, 3916 (1949).
514. J. Thiec and J. Wieman, *Bull. Chim. Soc. France* 177 (1956) [*Chem. Abstr.,* **50**, 14299 (1956)].
515. W. Kemula and J. Kornacki, *Roczniki Chem.,* **36**, 1835, 1857 (1962) [*Chem. Abstr.,* **59**, 218 (1963)].
516. M. M. Baizer, *J. Electrochem. Soc.,* **111**, 215 (1964).
517. Badische Anilin and Soda-Fabrik. Fr. Patent 1,516,599 (1968).
518. P. Zuman and O. Manousek, *J. Electroanal. Chem.,* **19**, 147 (1968).
519. S. H. Langer, *J. Electrochem. Soc.,* 1228 (1969).

20. A. Misono, T. Osa, T. Yamagishi, and T. Kodama, *J. Electrochem. Soc.*, 115, 266 (1968).
21. A. Misono, T. Osa, and T. Yamagishi, *Bull. Chem. Soc. Japan*, 41, 2921 (1968) [*Chem. Abstr.*, 70, 83573 (1969)].
22. W. C. Neikam, U.S. Patent 3,344,047 (1967).
23. J. J. Lingane, C. Gardner, . Swain, and M. Fields, *J. Am. Chem. Soc.*, 65, 1348 (1943).
24. H. Lund, *Tetrahedron Lett.* 1968, 3651.
25. P. G. Arapakas and M. K. Scott, *Tetrahedron Lett.*, 1968, 1975.
26. M. E. Peover and B. S. White, *Electrochim. Acta*, 11, 1061 (1966).
27. R. Dietz, M. E. Peover and H. P. Rothbaum, in Ref. 148.
28. A. L. Berre and Y. Berger, *Bull. Soc. Chim. (France)* 2363, 2368 (1966).
29. P. Rambacher and S. Mäke, *Angew. Chem. Internat. Ed.*, 7, 638 (1968).
30. L. Horner, M. Bretzenheim, and H. neumann, U.S. Patent 3,431,186 (1965).
31. A. Beno, M. Uher, and A. M. Kordos, *Chem. Izvest*, 23, 275 (1969) [*Chem. Abstr.*, 71, 108448 (1969)].
32. V. G. Luk'yanitsa, *Electrokhimiya*, 5, 883 (1969) [*Chem. Abstr.*, 71, 108451 (1969)].
33. R. S. Saxena, and K. C. Gupta, *Electrochim. Acta*, 13, 1749 (1968).
34. R. Elofson, F. F. Gadallah, and L. A. Gadalla, *Can. J. Chem.*, 47, 3979 (1969).
35. L. Ruzicka, M. Goldberg, M. Hübin, and H. Bockenoogen, *Helv. Chem. Acta*, 16, 1323 (1933) [*Chem. Abstr.*, 28, 1346 (1934)].
36. J. Donahue and J. W. Oliver, *Anal. Chem.*, 41, 753 (1969).
37. B. Lamm and B. Samuelsson, *Acta Chem. Scand.*, 23, 691 (1969).
38. S. Roffiz and M. Rozzi, *Ric. Sci.*, 38, 918 (1968) [*Chem. Abstr.*, 71, 118938 (1969)].
39. M. V. Merritt and D. T. Sawver, *J. Cor. Chem.*, 35, 2157 (1970).
40. W. C. Nickam, *Can. Patent* 803,992 (1969).
41. M. R. Rifi, *Tetrahedron Lett.* 1969, 5089.
42. M. R. Rifi, unpublished results.
43. J. Pinson and J. Armand, C.I.C.T.E. Meeting, Prague, Czechoslovakia, Sept. 28 - Oct. 2, 1970.
44. T. Takagi, M. Mizutani, I. Matsuda, and S. Ono, in Ref. 543.
45. A. T. Kuhn and M. Byrne, in Ref. 543.
46. E. Paspaliev and K. Batzalova, in Ref. 543.
47. G. Shima, *Mem. Coll. Sci. Kyoto Imp. Univ.*, 13, 85 (1930) [*Chem. Abstr.*, 24, 2954 (1930)].
48. G. Shima, *Mem. Coll. Sci. Kyoto Impr. Univ.*, 12A, 327 (1929) [*Chem. Abstr.*, 24, 2118 (1930)].
49. T. Shono and M. Mitanic, *Tetrahedron Lett.*, 1969, 687.
50. O. Manousek, O. Exner, and P. Zuman, C.I.C.T.E. Meeting, Prague Czechoslovakia, Sept. 28 - Oct. 2, 1970.
51. R. C. Fuson and R. O. Kerr, *J. Org. Chem.*, 19, 373 (1954).
52. T. C. Chambers, U.S. Patent 2,422,468 (1947).
53. M. J. Allen, *Organic Electrode Processes*, Reinhold, 1958.
54. R. C. Snowdon, *J. Phys. Chem.*, 15, 797 (1911).
55. M. R. Rifi, unpublished results.
56. H. Lund, in Ref. 148.
57. V. Bruckner and V. Kardos, *Ann.*, 518, 226 (1953).
58. L. Gattermann, *Ber.*, 26, 1848 (1893).
59. K. Brand, *Ber. detsch. Chem. Ges.*, 88, 3078 (1905).
60. A. Noyes and J. Dorrance, *Ber.*, 28, 235 (1895).

561. L. Gattermann, *Ber.*, **27**, 1931 (1894).
562. F. Fichter and T. Beck, *Ber.*, **44**, 3646 (1911).
563. K. Elb and C. Wipplinger, *Z. Electrochem.*, **7**, 137 (1900).
564. K. Elb and M. Henze, *Z. Electrochem.*, **7**, 138 (1900).
565. K. Elb and E. Saame, *Z. Electrochem.*, **7**, 143 (1900).
566. K. Elb and K. Gaumer, *Z. Electrochem.* **7**, 146 (1900).
567. K. Elb and A. Rohde, *Z. Electrochem.*, **7**, 136 (1900).
568. M. Leguyder and A. Darchem, *Compt. Rend.*, **267**, 1352 (1968).
569. K. Brand and T. H. Eisenmenger, *J. Prakt. Chem.* [2], **8**, 7501 (1913).
570. J. Phillops and A. Lowy, *Trans. Electrochem. Soc.*, **71**, 498 (1937).
571. V. Bruckner and E. Vinkler, *J. Prakt. Chem.*, [2], **142**, 277 (1935).
572. H. E. Fierz and P. Weissenbach, *Helv. Chim. Acta*, **8**, 305 (1920).
573. G. Palyi, *Kolor. Ert.*, **10**, 182 (1968).
574. L. Gattermann and V. Olivekrona, *Ber.*, **27**, 1937 (1894).
575. W. A. Jacobs and M. Heidelberger, *J. Am. Chem. Soc.*, **39**, 2419 (1917).
576. L. Gattermann and E. Lockhart, *Ber.*, **29**, 3034 (1896).
577. W. Löb, *Z. Elektrochem.*, **5**, 460 (1899); *Ber.*, **31**, 2201 (1898).
578. L. Gattermann, *Ber.*, **27**, 1937 (1894).
579. K. Friedländer, *Ber.*, **38**, 2839 (1905).
580. Weis and Reiter, *Ann.*, **355**, 175 (1907).
581. P. E. Iversen and H. Lund, *Acta Chem. Scand.*, **21**, 389 (1967).
582. L. Throop and L. Tokes, *J. Am. Chem. Soc.*, **89**, 4789 (1967).
583. S. Wawzonek and A. Gunderson, *J. Electrochem. Soc.*, **107**, 537 (1960).
584. S. Wawzonek, R. Berkey, E. W. Blake, and M. E. Runner, *J. Electrochem. Soc.*, **103**, 456 (1956).
585. C. S. McDowell, *Diss. Abstr. (B)*, **28**, 2348 (1967).
586. J. M. Nelson and A. M. Collins, *J. Am. Chem. Soc.*, **46**, 2250 (1924).
587. M. M. Baizer and J. P. Petrovich, in A. Streitweiser, Jr. And R. W. Taft, Ed., *Progress In Phys. Chem.*, Vol. 7, Wiley, New York 1970, p. 189.
588. K. Sugino and T. Nonaka, *Electrochim. Acta*, **13**, 613 (1968).
589. M. Nicolas and R. Palland, *Compt. Rend.*, **265C**, 1044 (1967).
590. J. Wiemann and M. L. Bouguerra, *Ann. Chim.*, **2**, 35 (1967).
591. O. R. Brown and K. Lister, Faraday Soc. Meeting, Newcastle, April 1968.
592. T. Shono and M. Mitani, *J. Am. Chem. Soc.*, **93**, 5284 (1971).
593. J. Wagenknecht, *J. Org. Chem.*, **37**, 1513 (1972).
594. H. V. K. Udupa, G. S. Subramanian, K. S. Udupa, and K. Natarajan, *Electrochem. Acta*, **9**, 313 (1964).
595. Tafel and Jürgens, *Ber.*, **42**, 2554 (1909).
596. A. P. Tomilov, E. A. Mordvintseva, and E. V. Kryukova, *Zh. Prik. Khim.*, **41**, 2524 (1968).
597. D. H. Barton, *J. Chem. Soc.*, **1953**, 1027.
598. J. P. Coleman, R. J. Kobylecki, and J. P. Utley, *Chem. Commun.*, 104 (1971).
599. O. Manovsck, D. Exner, and P. Zuman, *Coll. Czech. Comm.*, **33**, 4000 (1968).
600. C. Th. Pederson and V. D. Parker, *Tetrahedron Lett.*, **1972**, 767.
601. S. Wawzonek, R. Berkey, E. Blaha, and M. Runner, *J. Electrochem. Soc.*, **103**, 456 (1956).
602. P. Given, M. Peover, and Schoen, *J. Chem. Soc.*, **1958**, 2674.
603. R. Dehl and G. Fraenkel, *J. Chem. Phys.*, **39**, 1793 (1963).
604. R. Reieke, W. Rich, and T. Ridgway, *Tetrahedron Lett.*, **1969**, 4381.
605. B. Eggins, *Chem. Commun.*, 1267 (1969).
606. V. D. Bezuglyi, L. Ya. Kheitets, and N. Sobina, *Zh. Obshch. Khim.*, **38**, 2164 (1968).
607. H. Berg, *Naturwiss.*, **48**, 714 (1961).

608. A. J. Fry and W. E. Britton, *Tetrahedron Lett.*, **1971**, 4363.
609. S. Cristol, A. Dahl, and W. Lim, *J. Am. Chem. Soc.*, **92**, 5670 (1970).
610. A. P. Krapcho and M. E. Nadel, *J. Am. Chem. Soc.*, **86**, 1096 (1964).
611. M. R. Rifi, unpublished results.
612. M. M. Baizer, *J. Electrochem. Soc.*, **111**, 315 (1964).
613. S. Wawzonek and J. W. Fan, *J. Am. Chem. Soc.*, **68**, 2541 (1946).
614. H. A. Laitinen and S. Wawzonek, *J. Am. Chem. Soc.*, **64**, 1765 (1965).
615. G. Hoijtink, J. Van Schooten, E. DeBoer, and W. I. Aalbersberg, *Rec, Trav. Chim.*, **73**, 335 (1954).
616. J. P. Petrovich and M. Baizer, *J. Electrochem. Soc.*, **118**, 447 (1971).
617. M. M. Baizer and J. L. Chrume, *J. Electrochem. Soc.*, **118**, 450 (1971).

Chapter **IX**

PREPARATIVE ELECTROLYSES OF SYNTHETIC AND NATURALLY OCCURRING *N*-HETEROCYCLIC COMPOUNDS

Robert F. Nelson

1 INTRODUCTION

Electrochemical studies of nitrogen-containing heterocyclic compounds are almost as old as the field of electrochemistry itself. Many of these compounds were among the first investigated in the late 1800s and early 1900s; most of these initial works involved electrochemical hydrogenation of single-ring heterocyclics such as pyridines and larger molecules such as purines and alkaloids. A rather long period of relative dormancy followed, but recently the preparative electrochemistry of N-heterocyclics has been undergoing a resurgence of renewed interest, but this time also in the areas of natural product syntheses and comparisons of biochemical and electrochemical pathways of biologically important purines and pyrimidines.

Excellent review articles are available on the vast stores of polarographic and voltammetric data, as well as preparative work, that came out during the time interval (ca. 1940-1968) between these two periods of very early and very recent high activity [135, 253]. In these reviews by Volke and Lund, respectively, the authors did an excellent job of covering this middle period, a time when the general reaction pathways of N-heterocyclic molecules—both cathodic and anodic—were broadly elucidated. Because of this and the fact that the very early and recent periods have only been lightly covered in the literature, this review article will deal mainly with the latter, although an effort has been made to include all the prominent preparative work that has been reported.

It is the stated objective of this review that only preparative electrochemical investigations have been covered, and it is hoped that a rather thorough treatment has been achieved. In the large master table the electrolysis conditions and products formed, with yields wherever possible, have been tabulated; the papers cited cover all time periods, but only the early and recent works previously mentioned are treated in the narrative section. The time period covered is generally up to the end of 1972; however, a few papers from 1973 have also been included where possible. Thus, this review is intended to complement the coverages of Lund and Volke. Given these three sources, one would have a fairly comprehensive overview of the electrochemistry of N-heterocyclic compounds.

The common characteristic of all the papers cited, with only a few exceptions, is that preparative electrolytic conditions were employed and products were reported as being either isolated and characterized or identified in solution with some certainty. In cases where products are supported by some characterization data, but not definitely elucidated, a question mark follows the product cited. Voltammetric studies where products were either speculated upon or identified without supporting data have not been included in this review. This is also the case for investigations where only half-wave potential data and/or cyclic voltammetric curves are presented. In addition, a number of systems involving only electron-transfer steps and no associated chemical reactions have not been included.

2 GENERAL REACTION PATHWAYS

Although there are many compounds covered in this review the reaction pathways of *N*-heterocyclic molecules are fairly amenable to classification in a few general categories, keeping in mind that relatively little work has been done on these systems. These reactions may be enumerated as shown below:

A. Reduction
   i. Hydrogenation
   ii. Dimerization
   iii. Hydrodimerization
   iv. Halogen elimination
B. Oxidation
   i. Coupling
   ii. Fragmentation
   iii. Anodic substitution

Almost all of the early work in this area dealt with cathodic reductions carried out in aqueous H_2SO_4 at lead according to the general method of Tafel. These reactions almost invariably resulted in hydrogenation of the heterocyclic nucleus, an attached substituent group or both. In ring saturation reactions, variations of the electrolysis conditions produced some selectivity in the products obtained. For example, Sakurai has shown that in the electrolytic reduction of phthalimide derivatives the product formed could be selectively controlled via the electrolysis medium and cathode material [30, 31]:

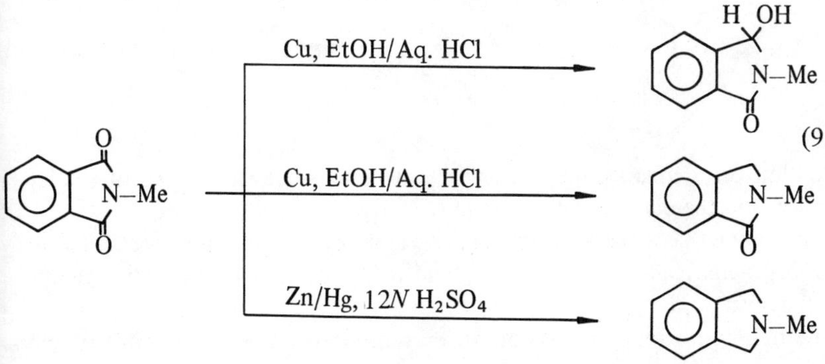

$$(9.1)$$

Similarly, Ferles and co-workers have found that in the reduction of pyridine derivatives a mixture of hydrogenated molecules results; for example, the electrolytic reduction of 3-ethylpyridine in aqueous H_2SO_4 at lead yielded a mixture of three products:

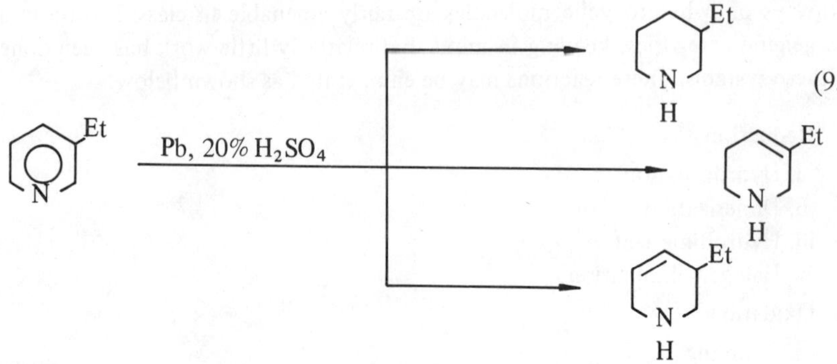

$$(9.2)$$

The majority of the papers reviewed consisted of such studies, but in some cases the hydrogenation scheme yields a single product in high yield and protonation occurs as one would anticipate based on electron distribution in the parent molecule lowest unfilled molecular orbital. The first step is usually formation of a dihydro derivative, followed by further saturation in some cases.

In a number of electrolytic hydrogenations it is not the heterocyclic nucleus that is reduced but a substituent electron-withdrawing group such as acetyl, nitroso, or nitro. In the case of the latter, this leads to the amino derivative, as shown below for N,N'-dinitropiperazine [52]:

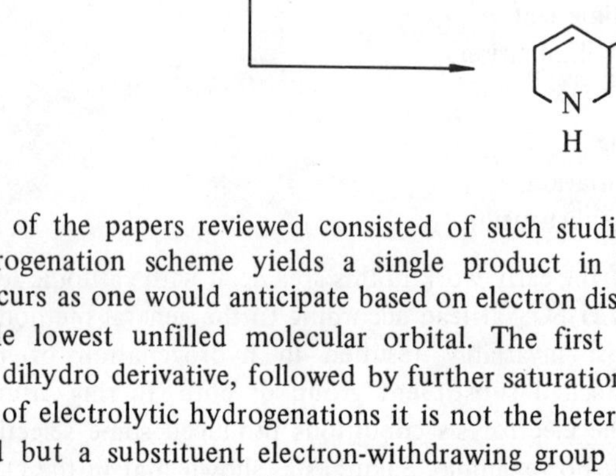

$$(9.3)$$

The dinitroso derivative may be an intermediate, since electrolysis of the latter also leads to the diaminopiperazine [51, 382].

The reductions of acetyl, aldehyde, and carboxylic acid groups occur smoothly and in general lead directly in high yield to the corresponding hydroxyalkyl or alkyl derivative.

Hydrodimerization reactions are of current interest in electrochemistry due to their potential commercial value; several examples of this type of reaction appeared in the early work on pyridine and quinoline derivatives under uncontrolled electrolytic conditions, resulting in the generation of dipiperidyls [58, 122-125] and quinoline hydrodimers [138]. More recently, Baizer and co-workers have applied this reaction route to 2- and 4-vinylpyridines in the absence and presence of additives, resulting in different products through homogeneous and cross-coupling hydrodimerization reactions [163]:

$$H_2C-CH_2-CH_2-CH_2$$

(9.4)

Halogenated N-heterocyclic molecules, when subjected to electrolytic reduction conditions, appear to undergo the same general reaction pathways as many haloaromatic and haloaliphatic molecules. This involves the addition of two electrons, followed by loss of a halide ion and addition of a proton to generate the dehalogenated parent heterocycle:

$$HET-X \quad \xrightarrow[\substack{-X^- \\ +H^+}]{+2e} \quad HET-H \tag{9.5}$$

The same reaction takes place in deuterated solvent (CH_3OD) in the case of 2-iodopyridine [161].

On the anodic side, the reaction pathway depends largely upon the degree of potential control involved in the experiments. A brief perusal of Table 9.1, p. 291, reveals that much of the early oxidation work involving the use of power supplies operating at constant applied potential (with no concomitant control of the working electrode potential) yielded product data indicating substantial fragmentation of the parent molecule. Thus the anodic oxidation of pyridine at constant applied potential yielded an n value of 20 and a variety of products [121]; under controlled-potential electrolysis conditions the oxidation process led to a coupled salt [130]:

(9.6)

This is an extreme but representative case. Much of the early work at uncontrolled working electrode potential yielded fragments, polymers, or products in low yields; under controlled-potential conditions the anodic oxidations appear to be much cleaner and often lead to coupled products consisting of two parent molecules, as reported, for example, in the oxidations of a large series of substituted carbazoles [11, 12].

As with many aromatic systems, *N*-heterocyclic molecules are susceptible to anodic substitution reactions. In light of the amount of attention being given to the introduction of substituents such as halogens, cyanide, and thiocyanate into a variety of organic molecules, it is somewhat surprising that so little attention has been devoted to *N*-heterocyclics. As shown in Table 9.1, relatively few references to anodic substitutions exist in the literature; of those presented, the introduction of cyano [131], methoxy [162], thiocyano [232], and fluoro [133] groups are noteworthy. The anodic methoxylation of *N*-methylpyrrole is of particular interest, since this pathway appears to be in marked contrast to the analogous methoxylation of furans. While dimethoxylation occurs for furan derivatives, 1-methyl-2,2,5,5-tetramethoxy-3-pyrroline is the product from *N*-methylpyrrole; no di- or trimethoxy derivatives are found, although their formation is not unreasonable "on paper" [410]. Interestingly, methoxylation of 2,6-dimethoxypyridine results in a mixture of tri-, tetra- and pentamethoxy derivatives [410]:

$$(9.7)$$

$$(9.8)$$

However, the applications have been incomplete thus far and the yields relatively low; this would appear to be an area of great potential for future electrosynthetic work in *N*-heterocyclic systems.

3 NATURAL PRODUCT SYNTHESES

Of all the areas of preparative heterocyclic electrochemistry, the one appearing to bear the greatest promise is that of natural product syntheses as developed recently by Bobbitt and his co-workers. This work is still in its infancy, but it appears that the electrolytic method leads to a much greater selectivity of product forms than chemical oxidation and platinum catalyzed oxygenation techniques. This is of particular significance in these systems since many of the compounds studied contain asymmetric carbon atoms, and thus a number of stereoisomers can be obtained in any given reaction. It should be pointed out that the role of the heterocyclic nitrogen in these reactions is not well established and at the present time it appears that the electrochemistry of these molecules is more characteristic of phenolic oxidations. A brief discussion of these studies is presented here for the sake of completeness, but these reactions are covered more fully in another section of this volume dealing with phenol electrochemistry.

Bobbitt's initial studies dealt with a very simple isoquinoline alkaloid, corypalline [327, 328]. It was found that the major electrolysis product in aqueous sodium bicarbonate solution at platinum was the C-8 dimer, 1,1'-dimethyl-6,6'-dimethoxy-7,7'-dihydroxy-1,1',2,2',3,3',4,4',-octahydro-8,8'-biisoquinoline. The same product was obtained photochemically in about the same yield [327]. A later paper was concerned with maximizing the yield of the dimer by electrolysis in several media, aqueous $Na_2B_4O_7$ being the most successful [328]. The average yield in seven runs was 55% and varied from 44 to 60%. A 2-5% yield of a C-O-C dimer was also reported.

The competitive formation of C-C and C-O-C coupled products was next studied in considerable depth by electrolytic oxidation of a number of structurally similar tetrahydroisoquinolines [329]. Previous chemical oxidation studies had shown that two possible pathways existed: oxidative coupling or oxidative dehydrogenation. In fact, under electrolytic conditions only coupling products were observed, although material balance experiments were not conducted.

In this same study, it was found that under conditions of catalytic oxygenation over platinum, C-C dimers were formed in much higher yields than by the electrolytic method (25.5%, compared to 0.7%). In addition, the Pt/O_2 method produced three discrete pairs of enantiomers from a racemic mixture of the starting material; similar results were obtained by chemical oxidation with $K_3Fe(CN)_6$. However, in a subsequent study the electrolytic conditions were modified somewhat (graphite felt anode, excess NaOMe, wet MeCN/TEAP solution) and it was found that the yield of C-C dimer was increased markedly: 77% yield as opposed to a 0.7% yield in the earlier study. More importantly, it was established that *only one* enantiomeric pair of the C-C dimer stereoisomers was obtained [331]. The C-O-C dimer was formed in 7% yield.

From an electrochemical standpoint, it is not only remarkable that the electrolytic reaction is so stereoselective but that the yields of C-C and C-O-C dimers would vary so greatly with the change in electrode/solution conditions. The authors rationalized the formation of the preferred dimer by the fact that adsorption of the radicals generated by anodic oxidation on the surface of the carbon electrode would occur in such a way that the aromatic rings would be on the surface and the bulky heterocyclic rings would be skewed away. This would lead to a single rotamer, since only those radicals with identical configurations could be close enough together to couple.

This seems reasonable in light of the data presented, and the answer may lie in the adsorptive characteristics of the two media. However, hard answers are not available and the exact nature of the interaction remains unclear. It is truly remarkable that in aqueous bicarbonate at platinum the C-O-C dimer is greatly preferred (22% as compared to 7%), whereas in the MeCN medium at a graphite cloth the C-C dimer is produced in far greater yield (77% as compared to 0.7%).

A further study was aimed toward elucidating this type of problem in that the effects of anode material, solvent conditions (including pH), cell design, and reaction time upon the relative amounts of C-C and C-O-C dimer were investigated [330]. Several differently substituted tetrahydroisoquinoline compounds were included.

It was again found that the optimum electrolysis conditions involved use of a carbon felt anode in MeCN solution. It was found that much greater success was achieved with the sodium salts of the hydroxyisoquinolines as far as product yields were concerned, presumably because the greater ease of oxidation of the phenoxide moiety led to the desired products. Three general types of coupling reactions were observed in these studies:

(9.9)

R = H, Me

(9.10)

(9.11)

Interestingly, for the latter two reactions when the nitrogen lone pair is tied up (acid solution or the N-acetyl analogs), the normal C-C coupling pathway is observed. The authors propose that in the absence of steric factors C-C coupling is preferred and that the C-O-C dimers formed above must occur via a unique route.

The two notable structural factors are that the phenolic or phenoxy groups are ortho or para to the benzylamine portion of the molecule and the ring coupling sites are ortho or para to a phenolic group. These factors would imply a resonance interaction between the benzylamine and phenolic portions of the molecules in the intermediate radicals or dications leading to coupling between an active phenoxy group and a reactive ring carbon atom. If reaction were to occur through the dication, the pathway proposed is as shown below [330]:

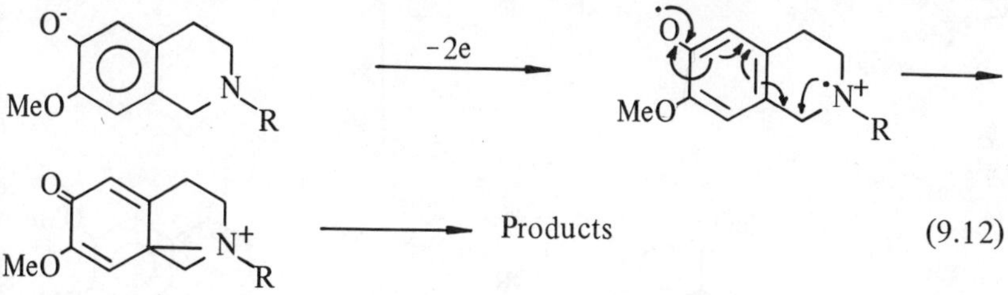

$$\tag{9.12}$$

This pathway seems reasonable, but at present it must still be considered somewhat speculative.

More recently, both Bobbitt [332] and Miller [333] have extended this basic work to the electrolytic oxidation of isoquinoline alkaloids. Bobbitt found that the electrolysis of armepavine and N-norarmepavine produced substituted 3,4-dihydroisoquinolines through oxidative fragmentation reactions. These results were of interest in comparison with enzymatic oxidations in that the electrolytic yields were much higher and no coupled products were found, in contrast with the previous work. Interestingly, with a carbethoxy group on the nitrogen, the oxidation process led to products coupled through the phenolic ring; apparently, the presence of the electron-withdrawing group on the nitrogen deactivates the isoquinoline portion of the molecule and the action is shifted to the substituent phenolic ring. The overall process is as shown below:

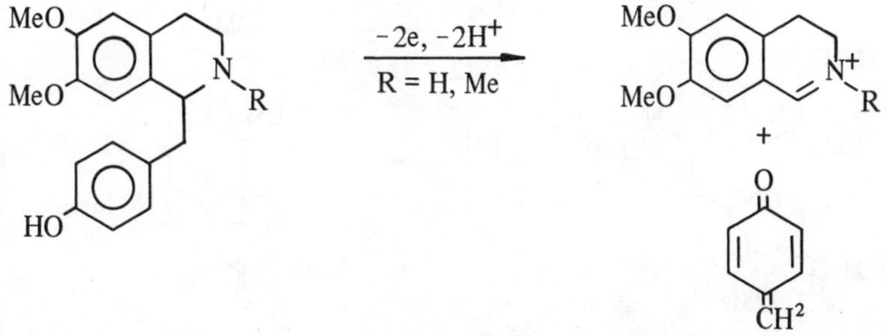

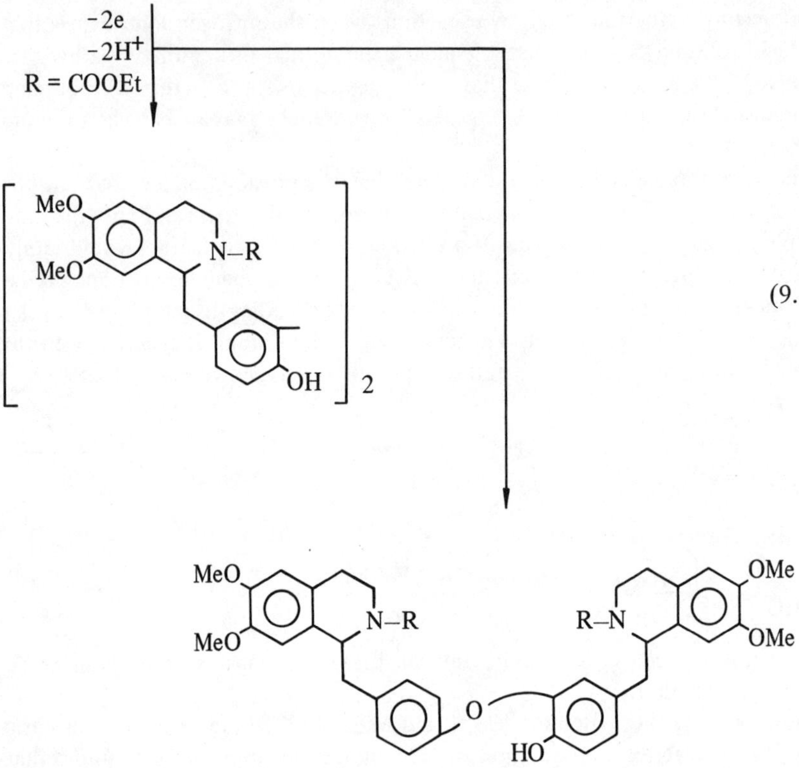

(9.13)

Fragmentation is apparently dependent upon the generation of the quinone-methide; if this species cannot be formed then intramolecular cyclization may occur. This has been shown to take place by Miller and co-workers in the electrooxidative cyclization of laudanosine in MeCN at platinum [333]. The oxidation led to O-methylflavinantine in 52% yield, compared to a 1% yield by chemical oxidation methods:

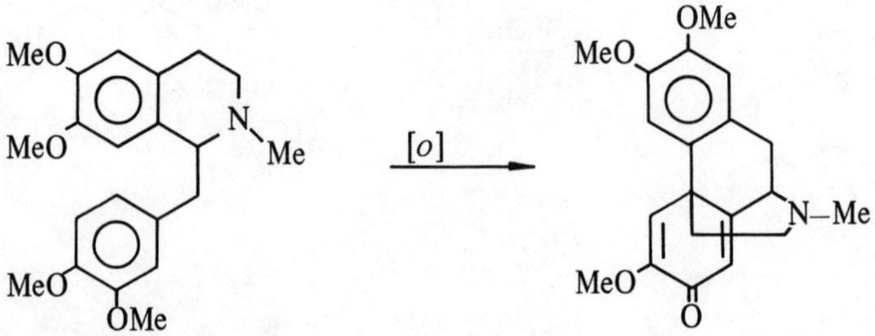

(9.14)

Whereas the electrochemistry of the molecules studied by Bobbitt is dictated largely by phenoxy groups, the oxidation of laudanosine appears to be at least partly controlled by the aromatic isoquinoline ring. Also, because of the methoxy group para to the benzylic linkage (instead of a hydroxy), the quinone-methide cannot form as with armepavine, and so a completely different pathway is followed. A great deal remains to be done concerning the mechanism here, but the high yield as compared to chemical oxidation methods is very exciting and labels this as one of the most promising areas of electrosynthesis of N-hetero-cyclics for future study. This is true of all the work cited in this section; the importance of the enhanced yields and stereoselectivity observed in these systems cannot be overemphasized.

4 PURINES AND PYRIMIDINES

In the anodic and cathodic electrolyses of purines and pyrimidines, several pathways are involved:

  i. Hydrogenation
  ii. Coupling and dimerization
  iii. Substituent elimination
  iv. Substituent oxidation
  v. Ring opening and fragmentation

Scattered early work was concerned with the anodic oxidations of some of the large purine molecules but there was little coherence. More recently, Elving and Dryhurst and their co-workers have carried out extensive electrochemical studies on these systems in hopes of correlating these data with biologically operative pathways, in particular enzymatic oxidations. Although space will not permit a thorough exposition of this work, the more pertinent highlights have, hopefully, been gleaned.

Initial studies on the parent molecules, purine and pyrimidine, showed that both are reducible at mercury and lead nominally to the dihydro derivatives. The reduction of pyrimidine has been studied in a variety of media. In aqueous solutions, the initial reduction product is 3,4-dihydropyrimidine over a wide pH range [99]. The reduction was cited as proceeding through the anion radical, which could protonate and then dimerize or be further reduced to the dihydro compound. The latter species was cited as being reduced on to the tetrahydro stage, but the structure of the tetrahydro derivative was uncertain. All of the proposed products were found to be unstable, and so the proposed mechanism must be considered somewhat speculative.

Subsequent studies probed more deeply into the aqueous electrochemistry of pyrimidine, but provided little more concrete information [371, 372]. Recently, O'Reilley and Elving investigated the electrolytic reduction of pyrimidine in MeCN and found it to be a bit more well behaved than in aqueous medium

[373]. They found the initial reduction to be a diffusion-controlled, reversible one-electron step to produce an unstable anion radical. This species was proposed to undergo the same processes reported for the aqueous reductions, but again no products were isolated. However, the overall process can be represented in this way:

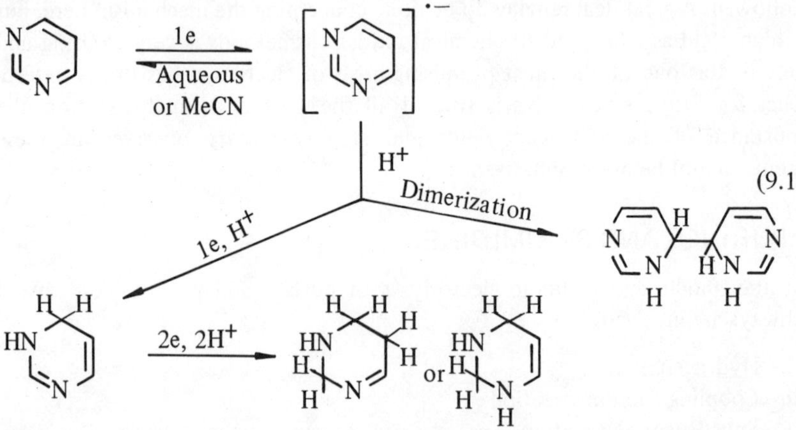

(9.15)

The reduction of purine is somewhat more complicated but basically similar [69]. The primary step leads to 1,6-dihydropurine by reductive hydrogenation; this is then carried out again in a second reduction process, producing (probably) 1,2,3,6-tetrahydropurine. The latter compound is unstable in aqueous solution and hydrolyzes to a substituted imidazole. Dimerization is blocked either by steric factors or increased delocalization (relative to pyrimidine) and so it is not observed:

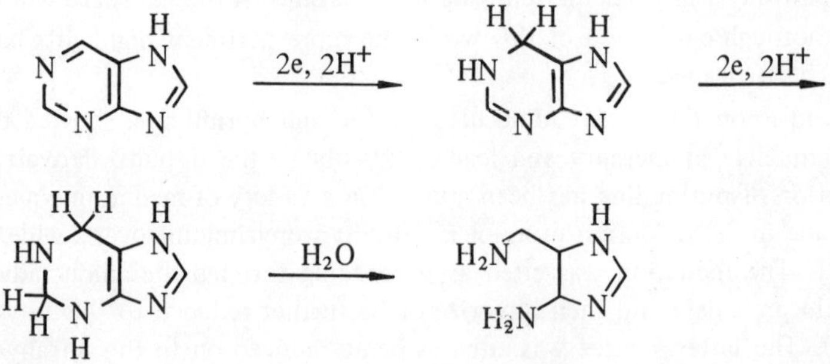

(9.16)

Since none of the intermediates or the final hydrolysis product were isolated and identified—largely due to their instability—this pathway is somewhat uncertain. However, it does fit the available data nicely and so it will serve as a basis for our looking at a variety of substituted purines.

Preparative studies of purine reductions have not been carried out in nonaqueous media at this time. Likewise, neither purine nor pyrimidine has been oxidized electrochemically.

These same early papers also covered the electrolytic reductions of adenine [69] and cytosine [99]. Although product analysis was incomplete, it was reported that cytosine (4-amino-2-hydroxypyrimidine) undergoes a three-electron reduction via a route involving formation of the 3,4-dihydrocytosine; this species then deaminates to generate 2-hydroxypyrimidine. One-electron reduction and protonation leads to a free radical that subsequently dimerizes:

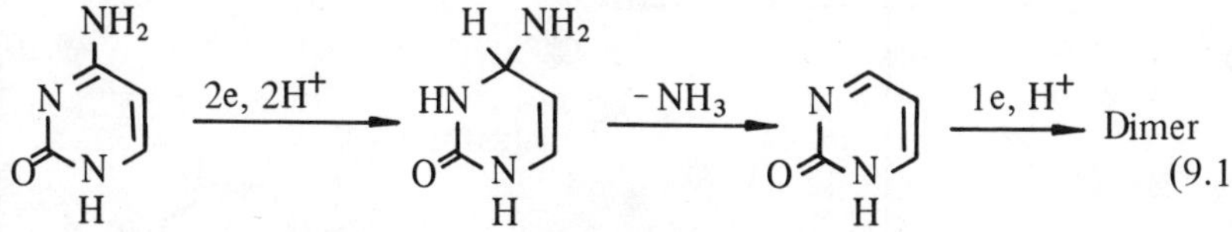

$$ (9.17) $$

Adenine (6-aminopurine) has been subjected to both electrolytic oxidation [70] and reduction [69, 71]. The reduction is reported to occur in a single six-electron step and is thought to involve a two-electron hydrogenation of the 1,6 double bond followed by a similar process at the 2,3 double bond, deamination at the 6 position and then a two-electron reduction of the regenerated double bond. Subsequent hydrolysis of the 2,3 bond then yields the same imidazole derivative as from purine. Hypoxanthine (6-hydroxypurine) was proposed to follow a similar pathway [69].

This rather unusual deamination reaction was subsequently found to be general for purines substituted with an amino or alkylamino group at the 6 position, including several adenine nucleotides [71].

The anodic oxidation of adenine at a pyrolytic graphite electrode was found to be extremely complex in aqueous acetate buffer, pH 2.3 [70]. However, the various electrolysis products were isolated and characterized, leading to the scheme shown below:

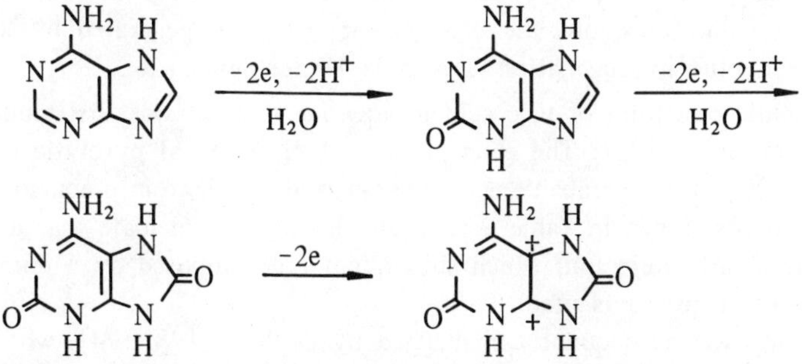

$$ (9.18) $$

The dicarbonium ion generated was proposed to decompose by a number of pathways to produce the compounds isolated:

$$\text{(9.19)}$$

4-aminopurpuric acid

Although this mechanism accounts for the products reported, it seems rather unusual that the exocyclic 6-amino group is not oxidized in preference to the ring double bonds. The alternate formation of the dicarbonium ion is exceptional, but subsequent work on uric acid, xanthine, guanine, and other purines has shown this species to be a common intermediate [374].

A subsequent paper correlated all of the electrochemical data on the compounds discussed above and centered largely on the effects of adsorption on their redox behavior [372]. Analytical applications were treated at length, but no further preparative work was presented.

The electrochemistry of uric acid (2,6,8-purinetriol) and several derivatives has been studied in considerable depth. Reduction of these compounds generally leads to saturation of one or more of the carbonyl groups of the purine rings [79-81, 83, 84]; most of this work was done under uncontrolled electrolysis conditions, but the products appear to be generated by straightforward reductive hydrogenation (see Table 9.1 for details).

The anodic oxidation of uric acid in aqueous solution was first studied by Struck and Elving [82]. The electrolysis in 1 M HOAc at pyrolytic graphite produced CO_2, urea, parabanic acid, alloxan, and an allantoin precursor in the molar amounts shown in Table 9.1. Again, the key intermediate was proposed to be the dicarbonium ion, which subsequently decomposed by a number of pathways to the products cited.

This study was subsequently challenged by Dryhurst [78, 374], who found that the same products were formed, but in vastly differing amounts. Instead of the dicarbonium ion as the key intermediate, a 4,5-diol was proposed to arise by hydrolysis of the oxidized uric acid. More recent work has suggested, however, that a more likely first intermediate would be a bis-imine that could go on to the diol by successive hydrolysis reactions [376]:

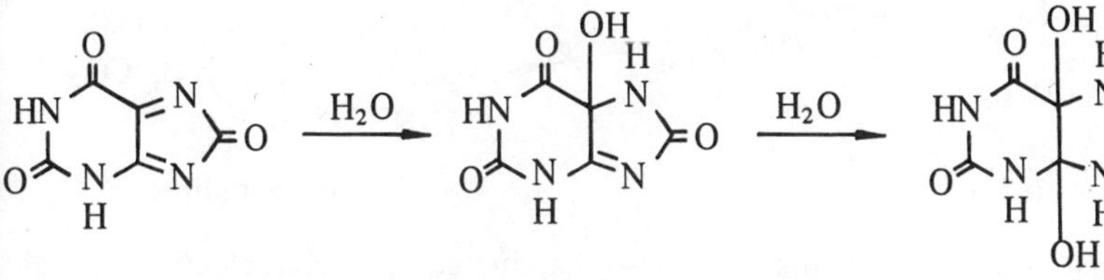

$$(9.20)$$

The overall process, then, would be as shown below:

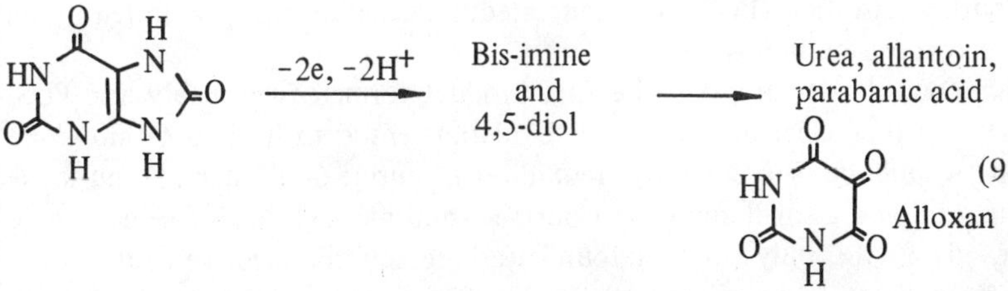

$$(9.21)$$

The nature of the products formed and their yields generally correlate quite well with enzymatic oxidation studies, and so here the electrochemical data can provide valuable information concerning the nature of intermediates such as the bis-imine and 4,5-diol and their stabilities—an expressed goal of the studies by Dryhurst and his co-workers. The electrochemical data also provided verification of the two-electron nature of the process and physical characteristics of the products formed.

Similar studies on the preparative electrolytic oxidation of guanine (2-amino-6-oxypurine) in 1 M HOAc showed a similar pathway [86]. The n value ranged from 4.2 to 4.7 and the products formed were generated from a (proposed) 4,5-diol:

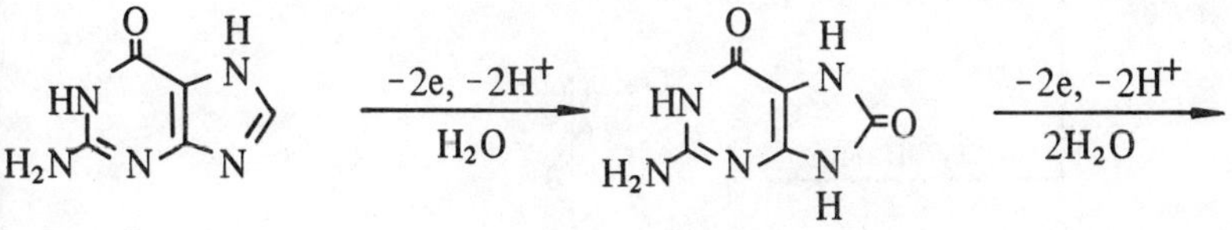

$$\text{(9.22)}$$

Oxalyl guanidine

Dryhurst and his students have also conducted extensive electrochemical studies on a series of thiopurines and purinethio acids. The work was initiated on the preparative oxidation of 6-thiopurine in 1 M HOAc and ammonia and carbonate buffers (pH 9) at a pyrolytic graphite anode [72]. In contrast to the purine derivatives previously cited, the electrolyses were very clean and relatively few products were formed. Briefly, it was found that at low pH the anodic process involved a one-electron transfer and the coupled product, bis(6-purinyl)disulfide (PDS), was generated; its isolation was prohibited by its slow decomposition in this medium.

In ammonia buffer at pH 9 the first product formed is probably also PDS, but here further electrolysis and air oxidation ensue to lead to 6-thiopurine (which would be subsequently reoxidized), purine-6-sulfonamide, purine-6-sulfinic acid, and a small amount of purine-6-sulfonic acid. In air-free carbonate buffer, pH 9, the only product found was the sulfonic acid, generated by a six-electron reaction in quantitative yields. These data can be summarized in this way:

$$\text{(9.23)}$$

The preparative electrolytic reduction of 6-thiopurine was found to proceed by an initial four-electron step at low pH [73]. The product of this reaction, 1,6-dihydropurine, is generated by reductive elimination of the thio group; from there, the second step is the same as for purine itself, that is, reduction to the tetrahydro derivative and hydrolysis to the substituted imidazole as in Eq. 9.16. At higher pH the two reductions merge into a single six-electron step, but the final result is the same.

The electrolytic reductions of purine-6-sulfinic acid and purine-6-sulfonic acid also follow the same route by initial formation of 6-thiopurine, then purine and ultimately the substituted imidazole by hydrolysis of the tetrahydropurine. The various species formed vary with pH, but the general pathway is reduction of the sulfinic or sulfonic acid groups to thio functions and subsequent reduction as for 6-thiopurine [73]; in some cases, however, the process is terminated at the 1,6-dihydropurine stage (see Table 9.1).

Some similarities to the above were found in the electrochemistry of 2-thiopurine at mercury and pyrolytic graphite [74]. Since here the thio substituent is at a relatively inert site, the reduction process does not involve it directly. As with other purines, the 1 and 6 positions are active and tend to dominate the reduction processes. Thus, reduction at the potential of the first polarographic wave produced 6,6'-bis(1,6-dihydro-2-thiopurine); electrolysis at the second wave led to 1,6-dihydro-2-thiopurine.

Anodic oxidation at pyrolytic graphite is almost completely analogous to the behavior of 6-thiopurine. Two oxidation waves are observed; preparative electrolysis at the potential of the first anodic wave generated bis(2-purinyl)-disulfide while oxidation at the second wave yielded purine-2-sulfonic acid. The overall electrochemical scheme, then, is as shown below:

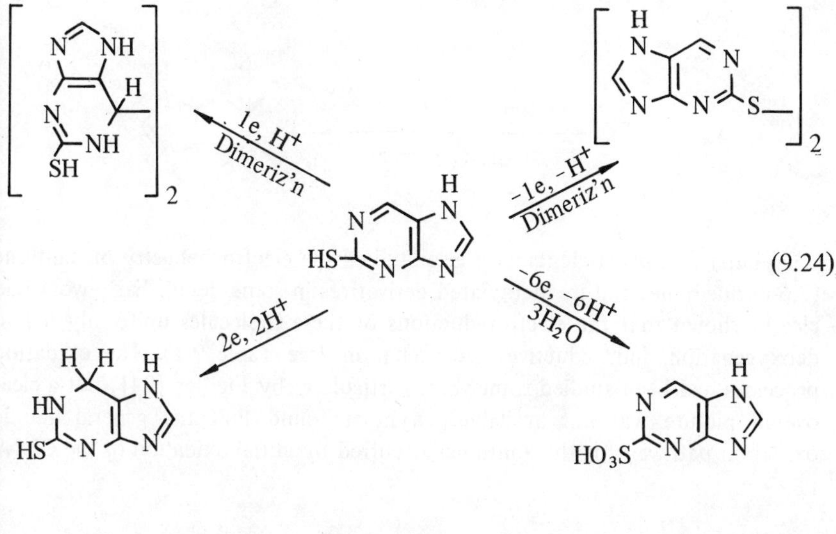

(9.24)

The electrochemistry of 2,6-dithiopurine is somewhat similar to the 2- and 6-thiopurines, but also quite different in some respects [75, 76]. The reduction process was first studied at a mercury cathode [76], but it was found that catalytic hydrogen evolution was excessive and, worse, that the dithiol was reacting rather rapidly with the mercury. Recourse to pyrolytic graphite resulted in a diminution of the hydrogen evolution, but it remained the major process. The only reduction products cited were H_2S and 1,6-dihydro-2-thiopurine; however, these must be considered as speculative since no product isolations were achieved.

The anodic process is much clearer in that three discrete products were detected at the potential of each of three voltammetric waves [75]. Preparative electrolysis past the first wave led to the usual type of coupled product, bis(6-purinyl)disulfide-2,2'-dithiol. Electrolysis at the second wave took this process one step further to yield bis(6-purinyl)disulfide-2,2'-disulfide and the third oxidation process produced purine-2,6-disulfonic acid:

$$(9.25)$$

Dryhurst and his students have also studied the electrochemistry of xanthine (2,6-purinedione) and its methylated derivatives in some detail. Early work had clearly shown that the electroreductions of these molecules uniformly led to deoxygenation and reductive hydrogenation (see Table 9.1). The oxidation processes had been studied somewhat, particularly by Fichter [94], but a clear overall picture was not available. Dryhurst found that the general anodic oxidation pathway for the xanthines occurred by initial oxidation of the C8-N9

double bond to generate the corresponding uric acids; these would then decompose along the lines described above for uric acid. Thus, xanthine was found
to yield alloxan, parabanic acid, allantoin, urea, CO_2, and NH_3 [376]; similarly,
some biologically important xanthines, including theobromine and caffeine,
yielded the same substituted products upon anodic oxidation at pyrolytic
graphite [377, 378], with methyl groups at the appropriate positions.

These same types of product distributions had been reported previously by
Fichter and Kern [94]. By variation of such experimental parameters as current
density, concentration, solution conditions (pH), and temperature they found
large variations in the product amounts, but the same basic pattern was in
evidence.

One exception to the general pathway of xanthine oxidations is theophylline
(1,3-dimethylxanthine). The initial step in the anodic oxidation of this molecule
is a one-electron transfer to generate a radical species that can be further
oxidized via the uric acid route to generate parabanic acid, 1,3-dimethylalloxan,
6,8-dimethylallantoin, dimethylurea, urea, and CO_2; alternatively, two radicals
can couple to form the C-8 dimer [375]:

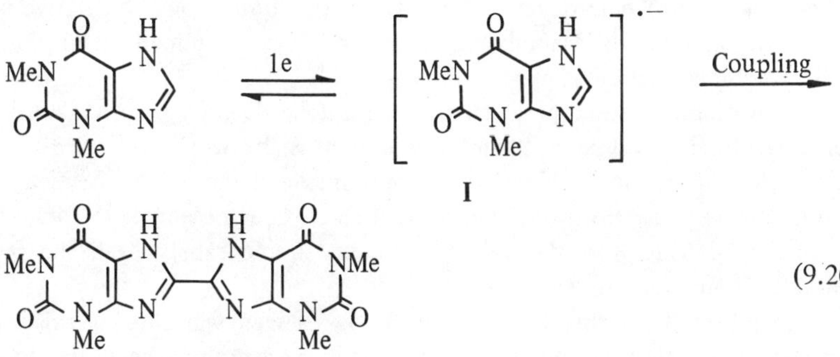

(9.26)

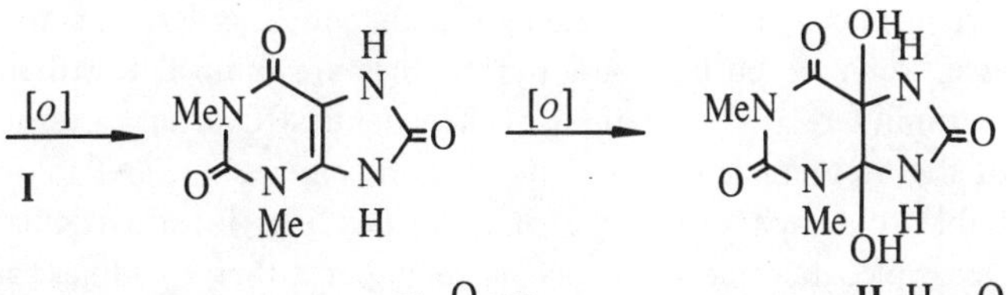

(9.27)

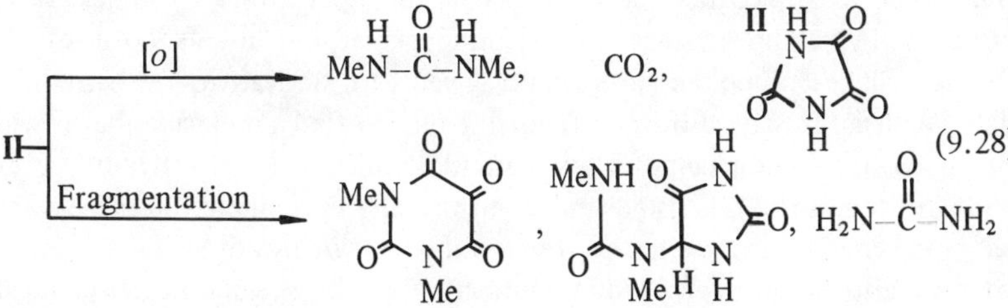

(9.28)

It is interesting to note that in the oxidations of xanthines where the 7 or 9 nitrogens are methylated the intermediate bis-imine cannot form; instead, iminium or bis-iminium ions are generated and these are hydrolyzed so rapidly that they cannot be observed on cyclic voltammograms. It is apparent that these iminium ions are the intermediates in the xanthine reactions, though, since the same product distributions are obtained as with the unmethylated analogs.

As mentioned earlier, the overall goal of this work is to determine whether the electrochemical and enzymatic oxidation pathways are similar. Once this is established—and it appears to be so for many of the compounds studied by these workers—the electrochemical data can contribute valuable information pertinent to the more inaccessible enzymatic systems.

In Dryhurst's work there is a heavy reliance on transient cyclic voltammetric peaks to establish reaction pathways and support rather involved kinetic schemes, and so these aspects of the work cited must be taken as somewhat tenuous; in fact, there have been several re-evaluations of these mechanistic pathways due to misinterpretations of transient electrochemical data. However, the products formed are sound and do compare favorably with those from enzymatic systems. Thus, this work does qualify as an area of great potential for the future in N-heterocyclic electrochemistry, both from preparative and mechanistic standpoints. Happily, it has recently been summarized in considerably more detail than that presented here [411].

It is regrettable that more discussion cannot be devoted to other areas of intense electrochemical interest and study such as the work of Ferles on the electrolytic and chemical reductions of pyridines and the electrochemistry of pyridine nucleotides. However, the pertinent data are presented in Table 9.1, and further delving into the references cited should satisfy even the most insatiable Faradaic appetite.

Several points of information regarding Table 9.1 seem warranted at this time. In general, the table is set up whereby compounds are grouped according to the basic heterocyclic nuclei, the latter being in alphabetical order. In some cases, however, classes such as purines and pyrimidines are lumped together, since they are structurally related and their electrochemical data are intermingled. The choice of the basic nucleus may sometimes be ambiguous, and so one may find it profitable to search under one or more headings for a particular compound. For example, nicotine could be listed under either Pyridines, as it is, or under Pyrrolidines, since the nicotine skeleton contains both of these moieties. Since compounds are usually given as a derivative of a basic heterocyclic nucleus, it may also be difficult for one to find a molecule that is widely known by a common name. Thus, caffeine, which is a derivative of xanthine (a substituted purine), is listed under Purines and Pyrimidines. In these cases, as well as in general, recourse to the *Merck Index* is strongly advised.

It may also be anticipated that numbering of the skeletal nuclei could be a problem, since the numbering schemes for heterocyclic molecules are variable. The currently accepted numberings for the more common *N*-heterocyclics in this review are contained in Table 9.1 as an aid to those not familiar with these molecules, but it should be stressed that referral to the original paper(s) cited is a more reliable guideline. This is advisable, since in some cases where the numbering scheme used is very unusual or confusing the original nomenclature is used so that the literature articles can be perused with greater ease.

Electrolysis conditions are ocassionally incompletely provided in the original literature, even to the electrode material employed, so there are some gaps in the data presented. However, the original papers are referenced and some useful information can be gleaned from them; the only real exception to this would be the old German patent literature that is not covered in Chemical Abstracts.

Acknowledgements

Thanks are due to the Departments of Chemistry, University of Idaho and California State University at Sacramento for financial support during the early stages of this work. The aid of Dr. Henning Lund through the donation of an early manuscript copy is also appreciated, as are similar literature manuscripts from several other individuals. The encouragement and support of Dr. R. N. Adams and Dr. D. E. Smith is gratefully acknowledged.

5 GLOSSARY OF TERMS AND GENERAL ELECTROLYSIS CONDITIONS

Unless otherwise stated, electrolyses were carried out at room temperature and the reference electrode was a saturated calomel electrode (sce). A question mark after a product denotes that it was not isolated and/or its nature could not be clearly elucidated. Since both oxidations and reductions have been covered, the table contains designations for "anodes" and "cathodes"; when only one electrode is entered for a particular process it should be understood that this is the working electrode, and thus it is defined that the system involves oxidation (anode working electrode) or reduction (cathode working electrode). In the few cases where two electrodes are given, the working electrode is underlined. The following are explanations of abbreviations included in Table 9.1.

A	amperes
Ahr	ampere-hours
C	denotes a carbon electrode; could be one of a number of forms
CD	current density, amperes per square centimeter (A/cm^2) or decimeter (A/dm^2)
CPE	controlled-potential electrolysis

DME	dropping mercury electrode
DMF	N,N-dimethylformamide
DMSO	dimethylsulfoxide
P.G.	pyrolitic graphite (electrode)
TBAP	tetra-n-butylammonium perchlorate
TEACN	tetraethylammonium cyanide
TEAP	tetraethylammonium perchlorate
TPAP	tetra-n-propylammonium perchlorate
V	volts

Table 9.1 Preparative Electrochemical Data for Nitrogen Heterocyclic Compounds

Compound	(Solvent, Electrolyte, Electrolysis Conditions)	Anode	Cathode	Product (s)	(% Yield)	Ref.
Acridines						

Compound	(Solvent, Electrolyte, Electrolysis Conditions)	Anode	Cathode	Product (s)	(% Yield)	Ref.
1. Acridine						
a.	MeCN, TEAP, CPE at 1.8 V ($n=1.02$)	Pt		Acridyl acridinium perchlorate dimer		1
b.	Strong acid, CPE		Hg (?)	C—9 dimer		2
2. N-Methylacridinium methosulfate						
a.	NaOH sol'n, $K_3Fe(CN)_6$ catalyst 3-10 A, 8-28 V applied	Fe		N-Methylacridone		3
3. 9-(o-Iodophenyl) acridine						
a.	90% EtOH/0.1 N KOH, 0.5 N KOAc CPE at −1.4 V, $n = 2.1$		Hg pool	9,10-Dihydro-9-(o-iodophenyl) acridine(90)		4
b.	90% EtOH/0.1 N KOH, 0.5 N KOAc CPE at −1.7 V, $n = 4.2$		Hg pool	9,10-Dihydro-9-phenylacridine		4
4. Lucigenin (dimethylbiacridinium ion)						
a.	MeCN, DMF, DMSO, TBAP, CPE, $n = 1$	Pt		Dimethylbiacridine		5, 6
Carbazoles						

Table 9.1 Preparative Electrochemical Data for Nitrogen Heterocyclic Compounds

1. Carbazole
   a. MeCN, TEAP, CPE at 1.25 V | Pt | 3,3'-Bicarbazyl (20-30), higher telomers | 11
   b. MeCN, TEAP, Pyridine, CPE at 1.25 V | Pt | N,N'-Bicarbazyl (100) | 11
2. N-Methylcarbazole
   a. MeCN, TEAP, CPE at 1.20 V | Pt | N,N'-Dimethyl-3,3'-bicarbazyl (100) | 11
3. N-Ethylcarbazole
   a. MeCN, TEAP, CPE at 1.20 V, $n = 2.01$ | Pt | N,N'-Diethyl-3,3'-bicarbazyl (100) | 11
4. N-Isopropylcarbazole
   a. MeCN, TEAP, CPE at 1.25 V, $n = 2.01$ | Pt | N,N'-Di-isopropyl-3,3'-bicarbazyl (100) | 11

The following N-phenyl substituted carbazoles were all electrolyzed in MeCN/TEAP at a platinum anode; in all cases coupling occurred at the 3- position of the carbazole nucleus to form a N,N'-bis(p-substituted phenyl)-3,3'-bicarbazyl.

5. N-Phenylcarbazole, CPE at 1.35 V, $n = 2.03$ | Corresponding bicarbazyl (100) | 11, 12
6. N-(p-Methoxyphenyl) carbazole, CPE at 1.25 V, $n = 1.98$ | Corresponding bicarbazyl (100) | 12
7. N-(p-Acetamidophenyl) carbazole, CPE at 1.25 V, $n = 1.93$ | Corresponding bicarbazyl (80-90) | 12
8. N-(p-Biphenyl) carbazole, CPE at 1.35 V, $n = 2.11$ | Corresponding bicarbazyl (80-90) | 12
9. N-(p-Tolyl) carbazole, CPE at 1.35 V, $n = 1.98$ | Corresponding bicarbazyl (100) | 12
10. N-(p-t-Butylphenyl) carbazole, CPE at 1.30 V, $n = 2.04$ | Corresponding bicarbazyl (100) | 12
11. N-(p-Fluorophenyl) carbazole, CPE at 1.35 V, $n = 2.08$ | Corresponding bicarbazyl (90-100) | 12
12. N-(p-Chlorophenyl) carbazole, CPE at 1.35 V, $n = 2.14$ | Corresponding bicarbazyl (80-90) | 12
13. N-(p-Bromophenyl) carbazole, CPE at 1.40 V, $n = 2.12$ | Corresponding bicarbazyl (80-90) | 12
14. N-(p-Iodophenyl) carbazole, CPE at 1.40 V, $n = 2.04$ | Corresponding bicarbazyl (90-100) | 12
15. N-(p-Carbomethoxyphenyl) carbazole, CPE at 1.40 V, $n = 1.96$ | Corresponding bicarbazyl (80-90) | 12
16. N-(p-Cyanophenyl) carbazole, CPE at 1.40 V, $n = 2.05$ | Corresponding bicarbazyl (90-100) | 12
17. N-(p-Nitrophenyl) carbazole, CPE at 1.45 V, $n = 2.36$ | Corresponding bicarbazyl (60-70) | 12

The following 3-substituted carbazoles were all electrolyzed in MeCN/TEAP at a platinum anode; in all cases coupling occured at the 6-position of the carbazole nucleus to form a 3,3'-disubstituted-6,6'-bicarbazyl. Considerable side reactions of an unspecified nature were reported. In the presence of pyridine several of these compounds formed the corresponding 3,3'-disubstituted-N,N'-bicarbazyls in near-quantitative yields (asterisked compounds).

18.	3-Methylcarbazole*, CPE at 1.20 V, n = 2.6	Corresponding bicarbazyl (20–40)	12
19.	3-Fluorocarbazole*, CPE at 1.30 V, n = 34	Corresponding bicarbazyl (20–40)	12
20.	3-Chlorocarbazole, CPE at 1.35 V, n = 3.4	Corresponding bicarbazyl (20–40)	12
21.	3-Bromocarbazole, CPE at 1.35 V, n = 3.2	Corresponding bicarbazyl (30–40)	12
22.	3-Iodocarbazole, CPE at 1.35 V, n = 3.0	Corresponding bicarbazyl (20–40)	12
23.	3-Benzoylcarbazole*, CPE at 1.45 V, n = 2.8	Corresponding bicarbazyl (30–40)	12
24.	3-Cyanocarbazole, CPE at 1.50 V, n = 4.0	Corresponding bicarbazyl (20–30)	12
25.	3-Nitrocarbazole, CPE at 1.60 V, n = 4.0	Corresponding bicarbazyl (10–20)	12
26.	3,6-Diiodo-N-phenylcarbazole		
	a. MeCN, TEAP, CPE at 1.45 V Pt	3,3'-Diiodo-N,N'-diphenyl-6,6'-bicarbazyl, I$_2$	12
27.	3,6-Diiodo-N-ethylcarbazole		
	a. MeCN, TEAP, CPE at 1.40 V Pt	3,3'-Diiodo-N,N'-diethyl-6,6'-bicarbazyl, I$_2$	12

The following 3-substituted-N-ethylcarbazoles were all electrolyzed in MeCN/TEAP at a platinum anode; in all cases coupling occurred at the 6-position of the carbazole nucleus to form a 3,3'-disubstituted-N,N'-diethyl-6,6'-bicarbazyl.

28.	3-Acetamido-N-ethylcarbazole, CPE at 1.10 V, n = 2.12	Corresponding bicarbazyl (50)	12
29.	3-Methyl-N-ethylcarbazole, CPE at 1.20 V, n = 2.08	Corresponding bicarbazyl (40–50)	12
30.	3-Fluoro-N-ethylcarbazole, CPE at 1.30 V, n = 2.03	Corresponding bicarbazyl (100)	12
31.	3-Chloro-N-ethylcarbazole, CPE at 1.35 V, n = 1.93	Corresponding bicarbazyl (60)	12
32.	3-Bromo-N-ethylcarbazole, CPE at 1.35 V, n = 2.13	Corresponding bicarbazyl (50)	12
33.	3-Iodo-N-ethylcarbazole, CPE at 1.35 V, n = 2.20	Corresponding bicarbazyl (60)	12

Table 9.1 Preparative Electrochemical Data for Nitrogen Heterocyclic Compounds

34.	3-Formyl-N-ethylcarbazole, CPE at 1.40 V, n = 2.09		Corresponding bicarbazyl (80–90)	12
35.	3-Acetyl-N-ethylcarbazole, CPE at 1.40 V, n = 2.14		Corresponding bicarbazyl (80–90)	12
36.	3-Benzoyl-N-ethylcarbazole, CPE at 1.40 V, n = 2.12		Corresponding bicarbazyl (80–90)	12
37.	N-Ethylcarbazole-3-carboxylic acid, CPE at 1.40 V, n = 2.11		Corresponding bicarbazyl (80–90)	12
38.	3-Carboethoxy-N-ethylcarbazole, CPE at 1.45 V, n = 2.22		Corresponding bicarbazyl (60–70)	12
39.	3-Cyano-N-ethylcarbazole, CPE at 1.50 V, n = 2.14		Corresponding bicarbazyl (60–70)	12
40.	3-Nitro-N-ethylcarbazole, CPE at 1.55 V, n = 2.22		Corresponding bicarbazyl (50–60)	12
41.	1,2,3,4-Tetrahydrocarbazole a. Aqueous EtOH, H_2SO_4	Pb	1,2,3,4,10,11-Hexahydrocarbazole (100)	9
42.	N-Methyl-1,2,3,4-tetrahydrocarbazole a. H_2O, H_2SO_4	Pb	N-methyl-1,2,3,4,10,11-hexahydrocarbazole	9
43.	5-Amino-1,2,3,4-tetrahydrocarbazole a. 60% H_2SO_4, 80–100°C	Pb	5-Amino-1,2,3,4,10,11-hexahydrocarbazole (13)	9
44.	6-Amino-1,2,3,4-tetrahydrocarbazole a. 60% H_2SO_4, 5 A	Pb	6-Amino-1,2,3,4,10,11-hexahydrocarbazole (10), isolated as the acetyl derivative	9
45.	7-Amino-1,2,3,4-tetrahydrocarbazole 60% H_2SO_4, 5 A current, 100°C	Pb	7-Amino-1,2,3,4,10,11-hexahydrocarbazole, isolated as the acetyl derivative	10
46.	6-Bromo-1,2,3,4-tetrahydrocarbazole a. 60% H_2SO_4, 80–100°C, CD = 0.02 A/cm^2	Pb	6-Bromo-1,2,3,4,10,11-hexahydrocarbazole	10
47.	5-Nitro-1,2,3,4,10,11-hexahydrocarbazole a. 60% H_2SO_4, 5–10°C, CD = 0.02 A/cm^2	Pb	5-Amino-1,2,3,4,10,11-hexahydrocarbazole (55)	10
48.	6-Nitro-1,2,3,4,10,11-hexahydrocarbazole a. 60% H_2SO_4, 5–10°C	Pb	6-Amino-1,2,3,4,10,11-hexahydrocarbazole (20), isolated as the acetyl derivative	10

49. 1,2,3,4,5,6,7,8-Octahydrocarbazole
   a. 60% H_2SO_4, 5.5 A current | Pb | Dodecahydrocarbazole (high yield) | 9

50. N-Methyl-1,2,3,4,5,6,7,8-octahydro-carbazole
   a. 60% H_2SO_4, 5.5 A current | Pb | N-Methyl-1,2,3,4,5,6,7,8,10,13-decahydrocarbazole (77) | 9

51. N-Ethyl-1,2,3,4,5,6,7,8-octahydro-carbazole
   a. 60% H_2SO_4, 5.5 A current | Pb | N-Ethyl-1,2,3,4,5,6,7,8,10,13 decahydrocarbazole | 9

Cinnolines

1. Cinnoline
   a. 1 N HCl, CPE at -0.4 V | Hg pool | 1,4-Dihydrocinnoline | 13

2. 3-Hydroxycinnoline
   a. pH 6.5 Phosphate buffer/butanol, CPE at -0.80 V, $n = 2$ | Hg pool | 3-Keto-1,2,3,4-tetrahydrocinnoline | 13
   b. 2 N HCl, CPE at -0.4 V, $n = 2$ | Hg pool | N-aminooxindole | 13

3. 4-Hydroxycinnoline
   a. 1 N HCl, CPE at -0.5 V, $n = 2$ | Hg pool | 4-Keto-1,2,3,4-tetrahydrocinnoline | 13
   b. 0.5 N KOH, CPE at -1.6 V, $n = 3$ | Hg pool | 4-Keto-1,2,3,4-tetrahydrocinnoline dimer | 13
   c. Acetate buffer/30% EtOH, CPE at -1.2 V, $n = 4$ | Hg pool | 4-Hydroxy-1,2,3,4-tetrahydrocinnoline | 13

Table 9.1 Preparative Electrochemical Data for Nitrogen Heterocyclic Compounds

4. 3-Methyl-4-hydroxycinnoline			
a. Acetate buffer, CPE at −0.8 V, $n = 2$	Hg pool	4-Keto-2-methyl-1,2,3,4-tetrahydrocinnoline	13
5. 4-Methylcinnoline			
a. 1 N HCl, CPE at −0.4 V, $n = 2$, 10°C	Hg pool	1,4-Dihydro-4-methylcinnoline (85)	13
b. 1 N HCl, CPE at −1.0 V, $n = 4$, 10°C	Hg pool	Skatole (3-methylindole)	13
c. Alkaline solution, CPE	Hg pool	1,4-Dihydro-4-methylcinnoline	13
6. 1,4-Dihydro-4-methylcinnoline			
a. Acid solution, CPE	Hg pool	Skatole	13
7. 2,4,-Dimethylcinnolinium iodide			
a. 1 N HCl, CPE at −0.4 V, $n = 2$	Hg pool	1,4-Dihydro-2,4-dimethylcinnonlinium iodide (?)	13
b. 1 N HCl, CPE at −1.0 V, $n = 4$	Hg pool	Skatole	13
8. 4-Chlorocinnoline			
a. Alkaline solution	DME	4,4′-Bis(cinnolinyl)	13
9. 3-Phenylcinnoline			
a. 2 N HCl/50% EtOH, CPE at −0.5 V, $n = 2$	Hg pool	1,4-Dihydro-3-phenylcinnoline (93)	13
b. 2 N HCl/50% EtOH, CPE at −1.2 V, $n = 4$	Hg pool	1-(2′-Aminophenyl)-2-phenyl-2-aminoethane (?)	13
10. 1-Methyl-3-phenylcinnolinium iodide			
a. 2 N HCl/50% EtOH, CPE at −0.5 V	Hg pool	1-Methyl-3-phenyl-1,4-dihydrocinnoline	13
11. 3-Phenyl-4-carboxycinnoline			
a. 2 N HCl/50% EtOH, CPE at −0.5 V	Hg pool	1,4-Dihydro-3-phenylcinnoline	13
b. 0.5 N NaOH/50% EtOH, CPE at −0.5 V	Hg pool	1,4-Dihydro-3-phenylcinnoline	13
12. 4-Mercaptocinnoline			
a. 1 N HCl, CPE at −0.55 V, $n = 4$	Hg pool	1,4-Dihydrocinnoline, H_2S	13

Substrate and conditions	Electrode	Product	Ref.
b. pH 9 Borate buffer/30% EtOH, CPE at -1.2 V, $n = 4$	Hg pool	1,4-Dihydrocinnoline	13
13. 4-Methoxycinnoline a. Citrate buffer, CPE, $n = 2$	Hg pool	4-Keto-1,2,3,4-tetrahydrocinnoline	13
14. Benzo[c]cinnoline a. Acetate buffer/30% EtOH, CPE at -0.5 V, $n = 2$	Hg pool	5,6-Dihydrobenzo[c]cinnoline	13
15. 5,6-Dihydrobenzo[c]cinnoline a. Acetate buffer/30% EtOH, CPE at -0.2 V	Hg pool	Benzo[c]cinnoline	13
16. 5-Methylbenzo[c]cinnolinium iodide a. Acetate buffer/30% EtOH, CPE at -0.4 V, $n = 2$	Hg pool	5,6-Dihydrobenzo[c]cinnoline	13
17. 1,4-Diphenyl-1,2,3,4-tetrahydrocinnoline a. MeCN, LiClO$_4$, Na$_2$CO$_3$, CPE at 0.4 V, $n = 2$	Pt	1,4-Diphenyl-1,4-dihydrocinnoline	14
b. MeCN/HClO$_4$, CPE at 1.4 V, $n = 4$	Pt	1,4-Diphenylcinnolinium perchlorate	14
18. 1,4-Diphenyl-1,4-dihydrocinnoline a. MeCN, LiClO$_4$, CPE at 1.0 V, $n = 2$	Pt	1,4-Diphenylcinnolinium perchlorate	14
19. 1,4-Dihydro-1-methyl-3-phenylcinnoline a. 1 N HCl/40% EtOH, CPE at -0.9 V, $n = 4$	Hg pool	2-(o-Methylaminophenyl)-1-phenylethylamine dihydrochloride	15
20. 2-Methyl-3-phenylcinnolinium iodide a. 1 N HCl/40% EtOH, CPE at -1.1 V	Hg pool	Reduced solution was diazotized and coupled with β-naphthol to give a red azo dye	15

Table 9.1 Preparative Electrochemical Data for Nitrogen Heterocyclic Compounds

Imidazoles

1. Imidazolylpropionic acid			
a. 2 N H$_2$SO$_4$, CD = 2 A/dm^2, n = 6.6	Pb, PbO$_2$	Succinic acid, urea	16
2. Histidine			
a. H$_2$O, H$_2$SO$_4$	PbO$_2$	Malonic acid, melanin, CO$_2$	17
3. Histamine			
a. H$_2$O, H$_2$SO$_4$	PbO$_2$	Urea (9), β-alanine (8.5), NH$_3$, CO$_2$	17
4. Imidazole-2-carboxylic acid			
a. 2 N HCl, CPE at -1.2 V vs Ag/AgCl, n = 2.3	Hg pool	Imidazole-2-carboxaldehyde (60), isolated as the diethylacetal derivative	18
b. 0.8 N HCl, CPE at -1.15 V, n = 2.4	Hg pool	Imidazole-2-carboxaldehyde (67), (82% yield when electrolysis is run at 0°C)	19
5. Imidazole-2-carboxamide			
a. 0.8 N HCl, CPE at -1.0 V, n = 2.1	Hg pool	Imidazole-2-carboxaldehyde (68) (85% yield when electrolysis is run at 0°C)	19
6. N-Methylimidazole-2-carboxylic acid			
a. 2 N HCl, CPE at -1.2 V vs Ag/AgCl, n = 2.3	Hg pool	N-Methylimidazole-2-carboxaldehyde (59)	18
7. N-Benzylimidazole-2-carboxylic acid			
a. 2 N HCl, CPE at -1.2 V vs Ag/AgCl, n = 2.3	Hg pool	N-Benzylimidazole-2-carboxaldehyde (78.5)	18

b. 1.3 N HCl, CPE at − 1.05 V, n = 2.4, 10°C	Hg pool	N-Benzylimidazole-2-carboxaldehyde (85) (94% yield when electrolysis is run at 0°C)	19
8. N-Benzylimidazole-2-carboxamide			
a. 0.8 N HCl/40% EtOH, CPE at − 1.05 V, n = 2.3	Hg pool	N-Benzylimidazole-2-carboxaldehyde (86) (95% yield when electrolysis is run at 0°C)	19
9. 2-(p-Dimethylaminophenyl)-4,5-diphenylimidazole			
a. Benzonitrile, TBAP, CPE at 0.15 V vs Ag/AgCl, n = 2	Pt	N-N or C-N dimer (structure undetermined)	20

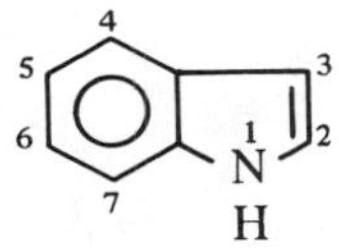

Indoles and Isoindoles

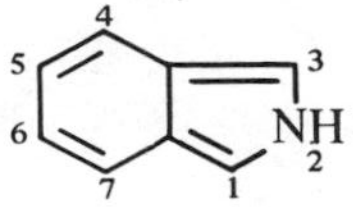

1. Indole			
a. H_2SO_4/66% EtOH	Pb	Indoline (2,3-dihydroindole), polymers	21
b. H_2O, H_2SO_4, EtOH, 100-120°C, 35 A, 36 V	Pb	Indoline (40)	22
c. 20% H_2SO_4, 60°C, CD = 0.3 A/cm²	Pb	Indoline (50)	23
d. MeCN, TEAP, CPE	Pt, C	N,N-Diethy-3-oxamylindole (24)	412
2. N-Methylindole			
a. H_2SO_4/66% EtOH	Pb	N-Methylindoline	21
3. 2-Methylidole			
a. H_2SO_4/66% EtOH	Pb	2-Methylindoline	21

299

Table 9.1 Preparative Electrochemical Data for Nitrogen Heterocyclic Compounds

4. 2,3-Dimethylindole			
a. H_2SO_4/66% EtOH	Pb	2,3-Dimethylindoline	21
5. Thiooxindole		Indoline (63), benzosulfoindoline	24
6. Dihydropentindole, $C_{11}H_{11}N$			
a. 60% H_2SO_4, CD = 0.03 A/cm^2	Pb	Tetrahydropentindole, $C_{11}H_{13}N$	25
7. N-Methylindoline methiodide			
a. Liquid NH_3, NaI, CD = 0.04 A/cm^2	Pt	N,N-dimethyl-β-phenylethylamine (75)	26
b. Liquid NH_3, 0.5% H_2O	Pt	N,N-dimethyl-β-phenylethylamine, N,N-dimethyl-2-ethylaniline	26
8. N-Nitrosoisoindoline			
a. H_2O, H_2SO_4, CD = 4.45 A/cm^2	Pb, Cd	N-aminoisoindoline (72-Pb, 92-Cd)	27
9. Isatin (indole-2,3-dione)			
a. 50% H_2SO_4, 45 A	Pb	Indoline (4)	28
10. Phthalimide (Isoindole-1,3-dione)			
a. Dioxane, Aq. HCl, 1.5 A	Pb	Hydroxyphthalimidine (3-hydroxyisoindole-1-one) (73)	29
b. 50% H_2SO_4, 45 A	Pb	1,3-Dihydroisoindole (isoindoline) (32)	29
c. MeOH, Aq. HCl	Pb	Methoxyphthalimidine (3-methoxyisoindole-1-one) (65)	29
d. EtOH, Aq. HCl	Pb	Ethoxyphthalimidine (3-ethoxyisoindole-1-one) (30-40)	29
e. H_2O, H_2SO_4, vary temp., CD	Pb	1,3-Dihydroisoindole (isoindoline) (73)	27
f. EtOH/Aq. HCl, CD = 20 A/dm^2	Pb	Hydroxyphthalimidine	30
g. EtOH/Strong HCl, 30-40°C, CD = 60 A/dm^2	Pb	Phthalimidine (isoindole-1-one)	30
h. H_2O, H_2SO_4	Pb	Indoline	28

i. $12\,N$ H$_2$SO$_4$, 80-90°C, CD = 0.6 A/cm^2	Zn amalgam	Indoline (32)	31
j. $1\,M$ HCl/50% EtOH, CPE at -1.0 V, $n = 2$	Hg	Oxyphthalimidine	408
k. $1\,M$ LiCl/50% EtOH, CPE at -1.1 V, $n = 1$	Hg	Oxyphthalimidine dimer (structure undetermined)	408
l. DMF, TEAP, CPE at -2.6 V	Hg	Phthalide (60)	427
11. Phthalimidine (isoindole-1-one)			
a. H$_2$O, H$_2$SO$_4$, vary CD, temp.	Pb, Cd, Sn	Isoindoline [0(Sn) - 71.6(Cd)]	27
12. N-Methylphthalimide			
a. H$_2$O, H$_2$SO$_4$, vary CD, temp.	Sn, Hg, Cd, Pb	N-Methylphthalimidine	27
b. EtOH/Aq. HCl	Pb	N-Methylphthalimidine	30
c. EtOH/Aq. HCl	Cu	N-Methyl-3-hydroxyisoindole-1-one	30
d. $12\,N$ H$_2$SO$_4$	Zn/Hg	N-Methylisoindoline (80)	31
13. N-Methylphthalimidine			
a. H$_2$O, H$_2$SO$_4$, vary CD, temp.	Hg, Pb, Sn, Cd	N-Methylisoindoline [0(Sn) - 68.6(Pb)]	27
14. N-Ethylphthalimide			
a. EtOH/Aq. HCl, CD = 0.02 A/cm^2	Cu, Ni	N-Ethyl-3-hydroxyisoindole-1-one	30
b. EtOH/Aq. HCl, CD = 0.06 A/cm^2	Pb	N-Ethylisoindole	30
15. N-Phenylphthalimide			
a. H$_2$O, H$_2$SO$_4$	Pb	N-Phenylisoindoline	30
b. DMF, TEAP, CPE at -2.4 V	Hg	o-Hydroxymethylbenzanilide (15), phthalide (35), aniline	427
c. DMF, TEAP, Phenol, CPE at -2.4 V	Hg	o-Hydroxymethylbenzanilide (34), phthalide (19), aniline	427

Table 9.1 Preparative Electrochemical Data for Nitrogen Heterocyclic Compounds

Conditions			Product	Ref.
d. DMF, TEAP, Ac_2O, CPE at -2.4 V		Hg	O-Acetyl-o-hydroxymethylbenzanilide (67)	427
16. N-(11-Lupinyl)phthalimide				
a. 20% H_2SO_4		Pb	N-(11-lupinyl)-1,3-dihydroisoindole	32
17. N-Anilinophthalimidine				
a. Acid solution, CPE, $n = 2$		Hg	N-Anilino-ψ-phthalazinone	33
18. Indigo ($\Delta^{2,2'}$-Biindoline)-3,3'-dione				
a. Conditions unspecified			Indigo white	34-36
b. Indigo/carbon suspension, NaOH, 105°, CD = 0.0008 A/cm^2		Hg	Indigo white (49)	37
c. Indigo/carbon electrode (compressed), Na_2CO_3	C		Indigo white (30-40)	38
d. Indigo mixed with charcoal in a linen bag as electrode, Na_2CO_3	C		Indigo white (30-40)	39
e. 51% HBr, CD = 0.03 A/cm^2	C		5,5'-Dibromoindigo (90), 5-bromoisatin, 2,4,6-tribromoaniline	40
f. 51% HBr, CD = 0.5 A/cm^2	Pt		5,5'-Dibromoindigo (90)	40
g. 51% HBr, CD = 0.03 A/cm^2, divided cell (diaphragm)	C		5,5'-Dibromoindigo (39)	40
h. 26% HBr, CD = 0.03 A/cm^2	C		5,5'-Dibromoindigo (43), 5-bromoisatin (21.6)	40
i. KBr, 78% H_2SO_4, 55°C	C		5,5'-Dibromoindigo (59)	40
j. Nitrobenzene, pyridinedihydrobromide, CD = 0.13 A/cm^2, 120-145°C	C		5,5'-Dibromoindigo (69)	40-42
k. 34% HCl, CD = 0.03 A/cm^2	C		5-Chloroisatin, 5,7-dichloroisatin (primary product), 2,4,6-trichloroaniline	40
l. 78% H_2SO_4, CD = 0.1 A/cm^2	C		Isatin (10) (isolated as the β-phenylhydrazone)	40
m. Pyridine, $LiClO_4$, Acid, CPE		Hg	Indigo white	421

No.	Substrate / Conditions	Electrode	Product	Ref.
19.	Indigo white		Indigo	43, 44
	a. Conditions unspecified			
20.	Bromoindigo + dibromoindigo		Bromoindigo white + dibromoindigo white	45
	a. Conditions unspecified			
21.	Indigosulfonic acid		Indigo white sulfonic acid	36
	a. Conditions unspecified			
22.	Bis(2-phenyl-3-indolinone)azine			
	a. DMF, TEAP, CPE at −0.40 V, Benzoic acid	Hg pool	3,3'-Azobis(2-phenylindole)	46
	b. DMF, TEAP, CPE at −1.20 V	Hg pool	3,3'-Azobis(2-phenylindole)	46
23.	2-Phenyl-3-phenyliminoindolenine			
	a. DMF, $KClO_4$, $ClCH_2COOH$, CPE at −0.40 V, $n = 2$	Hg	2-Phenyl-3-phenylaminoindole (the same process is reported for other derivatives)	385, 386
24.	Isatin-3-oxime			
	a. Aqueous solution, CPE	Hg	3-Amino-oxindole	394
25.	Isatin-3-oxime ethyl ether			
	a. pH 2-12, CPE, $n = 4$	Hg	3-Amino-oxindole	395
26.	1-Phenyl-3-iminoisoindoline (generated in situ)			
	a. pH 9 Ammonia solution, CPE at −1.70 V, $n = 4$	Hg	Dimethylamine, 1-phenylisoindoline	402
27.	1,1-Dimethylindolinium iodide			
	a. Aqueous solution, CPE	Hg	Dimethyl-(β-phenylethyl)amine (90), N,N-dimethyl-2-ethylaniline	398
28.	3-Diazo-oxindole			

Table 9.1 Preparative Electrochemical Data for Nitrogen Heterocyclic Compounds

a. Aqueous buffers, pH < 7.5, CPE, $n = 6$	Hg	3-Amino-oxindole	423
b. Aqueous buffers, pH > 8.5, CPE over first wave, $n = 2$	Hg	Isatin-3-hydrazone	423
c. Aqueous buffers, pH > 8.5, CPE over second wave, $n = 6$	Hg	3-Amino-oxindole	423
29. *N*-Nitraminoisoindoline			
a. 0.03 N H$_2$SO$_4$, CPE at -0.95 V	Hg	*N*-Hydrazinoisoindoline	426

Phthalazines

1. Phthalazine			
a. Aqueous buffers, $n = 6$	Hg	*o*-Xylene-α,α'-diamine	47
b. 8 N HCl, CPE at -0.85 V, $n = 6$	Hg pool	*o*-Xylene-α,α'-diamine (80), isoindoline (22)	48
c. 0.2 N HCl, CPE at -1.0 V, $n = 6$	Hg pool	Isoindoline (85), *o*-xylene-α,α'-diamine	48
d. 0.05 N KOH/30% EtOH, CPE at -1.65 V, $n = 2$	Hg pool	1,2-Dihydrophthalazine (83), C-1 dimer	48
e. 0.05 N KOH/30% EtOH, CPE at -1.95 V, $n = 4$	Hg pool	1,2,3,4-Tetrahydrophthalazine	48
f. 0.2 N KOH/10% EtOH, CPE, $n = 1.4$	Hg pool	C-1 dimer (40), 1,2-dihydrophthalazine (15)	48
2. 1-Methylphthalazine			
a. 2 N HCl, CPE at -0.75 V, 0-5°C, $n = 4$	Hg pool	1-Methylisoindole	48
b. 0.05 N KOH/30% EtOH, CPE at -1.65 V, $n = 2$	Hg pool	1,2-Dihydro-4-methylphthalazine	48
3. 1-Methyl-4-iodophthalazine			
a. Conditions unspecified, CPE at -0.5 V	Hg pool	1-Methylphthalazine	48

Compound / Conditions	Electrode	Product	Ref.
4. 2-Methylphthalazinium iodide 　a. Borate buffer, CPE at -1.2 V, $n = 1$	Hg pool	Two C-1 dimers	48
5. 1(2H)-Phthalazinone 　a. Phosphate buffer/20% EtOH, CPE at -1.35 V, $n = 2$	Hg pool	3,4-Dihydro-1(2H)-phthalazinone	49
6. 4-Methyl-1(2H)-phthalazinone 　a. Phosphate buffer/20% EtOH, CPE at -1.35 V, $n = 2$	Hg pool	3,4-Dihydro-4-methyl-1(2H)-phthalazinone (75)	49
7. 3,4-Dihydro-4-methyl-1(2H)-phthalazinone 　a. 2 N HCl, CPE at -1.05 V, $n = 2$ 　b. 0.2 N KOH, CPE at -0.3 V, $n = 2$	Hg pool	3-Methylphthalimidine 4-Methyl-1(2H)-phthalazinone	49 49
8. 4-(4-Dimethylaminophenyl)-1(2H)-phthalazinone 　a. 0.5 N HCl, CPE at -1.0 V, $n = 2$	Hg pool	3,4-Dihydro-4-(4-dimethylaminophenyl)-1(2H)-phthalazinone	49
9. 2,3-Dihydro-1,4-phthalazinedione 　a. Acid solution, CPE, $n = 6$	Hg pool	Phthalimidine (1,3-isoindoledione)	49
10. 2,3-Dihydro-2,3-dimethyl-1,4-phthalazinedione 　a. 4 N HCl, CPE at -1.05 V, $n = 6$ 　b. Acetate buffer, CPE, $n = 4$ 　c. pH 8.3 Borate buffer, CPE, $n = 2$	Hg pool Hg pool	2-Methylphthalimidine, one other product 3,4-Dihydro-2,3-dimethyl-1(2H)-phthalazinone (80)	49 49
11. 4-Hydroxy-3,4-dihydro-2,3-dimethyl-1-phthalazinone	Hg pool	3,4-Dihydro-2,3-dimethyl-4-hydroxy-1-phthalazinone	49

Table 9.1 Preparative Electrochemical Data for Nitrogen Heterocyclic Compounds

a. 0.5 N HCl, CPE at -0.7 V, $n = 2$	Hg pool	3,4-Dihydro-2,3-dimethyl-1-phthalazinone (85), C-4 dimer	49
b. Acetate buffer, CPE at -0.75 V, $n = 1$	Hg pool	C-4 dimer	49
12. 3,4-Dihydro-2,3-dimethyl-1-phthalazinone			
a. 3 N HCl, CPE at -1.1 V, 5°C, $n = 2$	Hg pool	N-Methyl-2-(methylaminomethyl) benzamide, 2-methyl-phthalimidine	49
13. 2-Phenyl-1-phthalazinone			
a. Various pH solutions, CPE, $n = 2$	Hg pool	3,4-Dihydro-2-phenyl-1-phthalazinone	49
14. 4-Chloro-2-phenyl-1-phthalazinone			
a. Acid solution, CPE, $n = 4$	Hg pool	3,4-Dihydro-2-phenyl-1-phthalazinone(?)	49
15. 2-(Dimethylaminoethyl)-4-benzyl-1-phthalazone hydrochloride (Ahanon)			
a. pH 0-10, $n = 2$	DME	2-(Dimethylaminoethyl)-3,4-dihydro-4-benzyl-1-phthalazinone	50
16. 4-Dimethylamino-1-phenylphthalazine			
a. 1 N HCl, CPE at -1.0 V, 0°C, $n = 4$	Hg	2-(1'-Amino-1'-phenylmethyl)-N,N-dimethyl-benzamidine (isolated as the dihydrochloride)	402
b. 0.5 N KOH/20% EtOH, CPE at -1.6 V, $n = 2$	Hg	1,2-Dihydro-4-dimethylamino-1-phenyl-phthalazine (89)	402
c. 0.5 N HCl/40% EtOH, CPE at -1.0 V, $n = 4$	Hg	1-Phenyl-3-iminoisoindoline (formed by addition of NH$_3$ to pH 9 following electrolysis)	402
17. 4-Dimethylamino-1-methylphthalazine			
a. 0.5 N KOH/20% EtOH, CPE at -1.75 V, $n = 2$	Hg	1,2-Dihydro-4-dimethylamino-1-methyl-phthalazine	402
18. 4-Methoxy-1-phenylphthalazine			

Conditions	Metal	Product	Ref.
a. 1 N HCl/30% EtOH, CPE at -0.82 V, $n = 4$, 0°C	Hg	1-Phenylpthalimidine, 3-methoxy-1-phenyl-isoindole (?), 2-(1'-amino-1'-phenylmethyl)-benzoate	402
b. 1 N HCl, CPE at -1.15 V, $n = 8$	Hg	1-Phenylisoindoline (85-100)	402
c. 0.5 N KOH/20% EtOH, CPE at -1.70 V, $n = 2$	Hg	1,2-Dihydro-4-methoxy-1-phenylphthalazine	402
19. 4-Methoxy-1-methylphthalazine			
a. 0.2 N KOH/20% EtOH, CPE at -1.80 V, $n = 2$	Hg	1,2-Dihydro-4-methoxy-1-methylphthalazine	402
20. 4-Mercapto-1-phenylphthalazine			
a. 1 N HCl/40% EtOH, CPE at -0.80 V, $n = 8$	Hg	1-Phenylisoindoline (70-80)	402
b. 1 N KOH, CPE at -1.70 V, $n = 4$	Hg	1,2-Dihydro-1-phenylphthalazine	402

Piperazines

Conditions	Metal	Product	Ref.
1. N,N'-Dinitrosopiperazine			
a. H_2O, H_2SO_4, HOAc	Cu	N,N'-Diaminopiperazine (38)	51
b. 50% HOAc, 60°C, LiCl, CPE	Hg	N,N'-Diaminopiperazine (85-95)	382
2. N,N'-Dinitropiperazine			
a. H_2O, H_2SO_4, HOAc	Pb	N,N'-diaminopiperazine (24)	52
b. 1:1 HOAC:H_2O	Cu	N,N'-diaminopiperazine (26)	52

Table 9.1 Preparative Electrochemical Data for Nitrogen Heterocyclic Compounds

3. 2,5-Piperazinedione		
a. $2 N$ HCl, CD = 0.08 A/cm^2	Hg	Aminoacetaldehyde, H_2NCH_2CHO	53
b. H_2O, H_2SO_4, EtOH, CD = 0.187 A/cm^2	Pb	Piperazine (40)	54
4. 3-Methyl-2,5-piperazinedione		
a. $2 N$ HCl, CD = 0.047 A/cm^2	Hg	Aminoacetaldehyde,β-aminopropionaldehyde	53
5. 3,6-Dimethyl-2,5-piperazinedione		
a. $2 N$ HCl, CD = 0.047 A/cm^2	Hg	β-Aminopropionaldehyde	53
6. N,N'-Dimethyl-2,5-piperazinedione		
a. $2 N$ HCl, CD = 0.047 A/cm^2	Hg	(Methylamino) acetaldehyde	53
7. N-(Phenylacetyl)-2,5-piperazinedione		
a. H_2O, H_2SO_4, EtOH, CD = 0.187 A/cm^2, 40°C	Pb	N-(β-phenylethyl)-2,5-piperazinedione (95)	54
8. 3,6-Di-isobutylpiperazine-2,5-dione		
a. MeCN, NaClO$_4$, CPE at 2.2 V vs Ag$^+$/AgNO$_3$	Pt	1,6-Di-isopropyl-3,8-dimethyl-5H,1OH-di-imidazo(1,5-a: 1',5'-d) pyrazine-5, 10-dione	278

Piperidines

1. Piperidine		
a. $2 N$ H_2SO_4, CD = 0.05 A/cm^2	Pb	δ-Aminovaleric acid, succinic acid, NH_3, CO_2, CO, glutaric acid, formic acid	55
2. N-Nitrosopiperidine		
a. 30% H_2SO_4, CD = 0.15 A/cm^2,	Pt	Dipiperidine base, $C_{10}H_{18}N_2$	56, 57

Substrate / conditions	Electrode	Product	Ref.
b. 10% H_2SO_4	Cu	N-Aminopiperidine (piperylhydrazine) (81), 10% yield using Pb cathode	51, 58, 59
3. *N*-Nitropiperidine			
a. 10% NaOAc	Ni	Piperylhydrazine (52), *N,N*-pentamethylene-hydrazine	52
b. 10% H_2SO_4	Cu	Piperylhydrazine (52), *N,N*-pentamethylene-hydrazine	52
c. 0.5 *N* NaOH, CPE at -1.15 V, $n = 2$	Hg	*N*-Nitrosopiperidine	60
4. 2-Methyl-*N*-nitrosopiperidine			
a. 30% H_2SO_4/EtOH, 5-6 V, CD = 14 A/dm^2	Pt	Aminocaproic acid (structure undetermined), base $C_{12}H_{24}N_2$	61
b. H_2O, H_2SO_4	Pb	2-Methylpiperylhydrazine	58
5. 3-Methyl-*N*-nitrosopiperidine			
a. H_2O, H_2SO_4	Pb	3-Methylpiperylhydrazine	59
6. 4-Methyl-*N*-nitrosopiperidine			
a. H_2O, H_2SO_4	Pb	4-Methylpiperylhydrazine	59
7. 2,6-Dimethyl-*N*-nitrosopiperidine			
a. H_2O, H_2SO_4	Pb	2,6-Dimethylpiperylhydrazine	59
8. 5-Ethyl-2-methyl-*N*-nitrosopiperidine			
a. H_2O, H_2SO_4	Pb	5-Ethyl-2-methylpiperylhydrazine	58
9. 2,4,6-Trimethyl-*N*-nitrosopiperidine			
a. H_2O, H_2SO_4	Pb	2,4,6-Trimethylpiperylhydrazine	59
10. 2,2,6,6-Tetramethylpiperidine-4-one			
a. Aqueous $(NH_4)_2SO_4$	Pb(?)	4-Hydroxy-2,2,6,6-tetramethylpiperidine	62
11. 4-Ketopiperidine-2,6-dicarboxylic acid			

Table 9.1 Preparative Electrochemical Data for Nitrogen Heterocyclic Compounds

a. 1 N NaOH, CD = 11 A/dm^2	Pb	4-Hydroxypiperidine-2,6-dicarboxylic acid	63
12. N-Benzoylpiperidine			
a. 70% H$_2$SO$_4$, 40°C	Pb	N-Benzylpiperidine (77)	64, 65
13. N-(Phenylacetyl)piperidine			
a. H$_2$O, H$_2$SO$_4$	Pb	N-(β-Phenylethyl)piperidine (63)	65
b. H$_2$O, H$_2$SO$_4$, EtOH, CD = 0.187 A/cm^2	Pb	N-(β-Phenylethyl)piperidine (96–100)	54
14. N-(p-Anisylthioacetyl)piperidine			
a. H$_2$O, H$_2$SO$_4$	Pb	N-(β-p-Anisylethyl)piperidine	65
15. 2,6-Piperidinedione (glutarimide)			
a. 20% H$_2$SO$_4$, CD = 0.05 A/cm^2	Pb	Piperidine-2-one (15)	66
b. 50% H$_2$SO$_4$	Zn amalgam	Piperidine (47)	66
16. Piperidine-2-one			
a. 20% H$_2$SO$_4$	Zn amalgam	Piperidine (52)	66
17. N-Methyl-2,6-piperidinedione			
a. 50% H$_2$SO$_4$, CD = 10–15 A/dm^2, 293 Ahr	Pb	Piperidine-2-one	67
b. 50% H$_2$SO$_4$, 600 Ahr	Pb	Piperidine	67
18. N-Ethyl-2,6-piperidinedione			
a. 50% H$_2$SO$_4$, CD = 0.05 A/cm^2	Pb	N-Ethylpiperidine-2-one (72)	66
b. 50% H$_2$SO$_4$	Zn amalgam	N-Ethylpiperidine (75)	66
19. N-Ethylpiperidine-2-one			
a. 50% H$_2$SO$_4$	Zn amalgam	N-Ethylpiperidine (52)	66

20.	N-Phenyl-2,6-piperidinedione			
	a. 80% H_2SO_4, CD = 0.05 A/cm^2	Pb	N-Phenylpiperidine-2-one (75)	66
	b. 90% H_2SO_4	Zn amalgam	N-Phenylpiperidine (76)	66
21.	N-Phenylpiperidine-2-one			
	a. 50% H_2SO_4, 18°C	Zn amalgam	N-Phenylpiperidine	66
22.	3,5-Dicyano-4,4-dimethyl-2,6-piperidinedione			
	a. H_2O, H_2SO_4, EtOH, CD = 0.04 A/cm^2	Rot. Pb	3,5-Dicyano-4,4-dimethylpiperidine-2-one	68
23.	N-Methylpiperidine-3-one			
	a. 20% H_2SO_4	Pb	1,2-Dimethylpyrrolidine, methylpentylamine	181
24.	4,4'-Dipiperidyl			
	a. H_2O, H_2SO_4, CPE	Pb	Hydrogenated derivatives (14-40) (yield varies with %H_2SO_4, applied potential)	414
25.	4-Piperidyl-4'-pyridyl			
	a. H_2O, H_2SO_4, CPE	Pb	Hydrogenated derivatives (6-68) (yield varies with %H_2SO_4, applied potential)	414
26.	2-Piperidyl-2'-pyridyl			
	a. H_2O, H_2SO_4, CPE	Pb	Hydrogenated derivatives (19-32) (yield varies with %H_2SO_4, applied potential)	414

Purines, Pyrimidines, Related Compounds

Table 9.1 Preparative Electrochemical Data for Nitrogen Heterocyclic Compounds

1. Purine (7-imidazo [4.5-d] pyrimidine)				
a. pH 1-5, CPE at -0.8 to -1.1 V, $n = 2$		Hg	1,6-Dihydropurine	69
b. pH 4.7, CPE at -1.25 V, $n = 4$		Hg	1,2,3,6-Tetrahydropurine (?)	69
c. pH 7.5, CPE		Hg	1,2,3,6-Tetrahydropurine	419
2. Adenine (6-aminopurine)				
a. 1 M HOAc, CPE at 1.2 V, $n = 6$	P.G.		Parabanic acid, oxaluric acid, allantoin, 4-aminopurpuric acid, urea, NH_3, CO_2	70
b. Aqueous buffers, CPE, $n = 6$		Hg	1,6-Dihydropurine	71
c. pH 1.3-2.3, CPE at -1.2 to -1.3 V, $n = 6$		Hg	1,2,3,6-Tetrahydropurine	69
d. 1 N HCl, CPE at -0.95 V, $n = 6$		Hg	NH_3, purine, 1,2,3,6-tetrahydropurine	391
3. 6-Thiopurine				
a. 1 M HOAc, CPE at 0.7 V, $n = 1$	P.G.		Bis(6-purinyl)disulfide	72
b. pH 9.1 Ammonia buffer, CPE at 0.4 V, $n = 4$	P.G.		Purine-6-sulfinic acid, purine-6-sulfonamide	72
c. pH 9.1 Ammonia buffer, CPE at 0.7 V, $n = 4.3$	P.G.		Purine-6-sulfinic acid (37), purine-6-sulfonamide (53)	72
d. pH 9 Carbonate buffer, CPE at 0.8 V, $n = 6$	P.G.		Purine-6-sulfonic acid (100)	72
e. pH 1-3, CPE at -1.0 V		Hg	1,2,3,6-Tetrahydropurine (hydrolyzes to a substituted 4-aminoimidazole), polymeric material	73
f. pH 7 McIlvaine buffer, CPE at -1.5 V		Hg	1,2,3,6-Tetrahydropurine (hydrolyzes as above)	73
4. 2-Thiopurine				
a. Acetate buffer, CPE at 1.5 V, $n = 5.8$	P.G.		Purine-2-sulfonic acid	74
b. Acetate buffer, CPE at 0.3 V, $n = 0.8$	P.G.		Bis(2-purinyl)disulfide	74
c. pH 1-8, CPE (first wave), $n = 0.95$		Hg	6,6'-Bis(1,6-dihydro-2-thiopurine)	74
d. pH 1.5, CPE (second wave)		Hg	1,6-Dihydro-2-thiopurine	74

5. 2,6-Dithiopurine			
a. 1 M HOAc, CPE at 1.4 V	P.G.	Purine-2,6-disulfonic acid (20)	75
b. pH 1-7, CPE	P.G.	1,6-Dihydro-6-thiopurine	76
6. Purine-6-sulfinic acid			
a. pH 3.6-5.4, CPE at -0.8 V	Hg	6-Thiopurine, purine	73
b. pH 7, CPE at -1.0 V, $n = 2$	Hg	Purine	73
c. pH 9.1 Ammonia buffer, CPE, $n = 4$	Hg	1,6-Dihydropurine	73
d. pH 11.6, CPE, $n = 4$	Hg	1,6-Dihydropurine	73
7. Purine-6-sulfonic acid			
a. pH 3-4, CPE at -0.7 V, $n = 4$	Hg	6-Thiopurine, purine (above pH 4, only purine forms)	73
b. pH 3.6-5.4, CPE at -1.1 V	Hg	1,6-Dihydropurine (above pH 8, only purine forms)	73
8. Purine-2,6-disulfonic acid			
a. pH 0-2, CPE at -0.7 V, $n = 5-6$	Hg	Purine-2-sulfonic acid	77
b. pH 0-2, CPE at -0.9 V, $n = 13-14$	Hg	1,6-Dihydropurine, 1,2,3,6-tetrahydropurine (hydrolyzes)	77
c. pH 4, CPE at -1.1 V, $n = 11$	Hg	1,6-Dihydropurine	77
d. pH 7-13, CPE at -1.4 V, $n = 2$	Hg	Purine-2-sulfonic acid	77
e. pH 6, CPE at -1.3 V, $n = 12$	Hg	1,2,3,6-Tetrahydropurine (hydrolyzes)	77
f. pH 7-9, CPE at -1.6 V, $n = 8$	Hg	1,2,3,6-Tetrahydropurine (hydrolyzes)	77
9. Uric acid (2,6,8-purinetriol)			
a. 1:1 pH 3.7/MeOH, CPE, $n = 2$	P.G.	Allantoin (20), 4,5-dimethoxyuric acid (?)	78
b. pH 3.7 Acetate buffer, CPE, $n = 2$	P.G.	Allantoin (90), parabanic acid	78
c. 70% H_2SO_4, CD = 0.082 A/dm^2, 0-5°C	Pb	Purone, isopurone, tetrahydrouric acid	79, 81
d. Dilute HOAc, CPE at 0.8 V, $n = 2.2$	C	Alloxan (0.3), parabanic acid (0.3), urea (0.75), CO_2(0.25), allantoin precursor (0.25); molar yield per mole of uric acid.	82

313

Table 9.1 Preparative Electrochemical Data for Nitrogen Heterocyclic Compounds

10. 3-Methyluric acid a. 70% H_2SO_4, CD = 12 A/dm^2, 13-15°C	Pb	3-Methylpurone(3-methyldesoxyuric acid) (35)	83
11. 1,3-Dimethyluric acid a. 75% H_2SO_4, CD = 12 A/dm^2	Pb	1,3-Dimethylpurone (65)	83
12. 3,9-Dimethyluric acid a. 75% H_2SO_4, CD = 12 A/dm^2	Pb	3,9-Dimethylpurone (45)	83
13. 7,9-Dimethyluric acid a. 75% H_2SO_4, CD = 12 A/dm^2	Pb	7,9-Dimethylpurone	83
14. 1,3,7-Trimethyluric acid a. 60% H_2SO_4, CD = 1.5 A/dm^2	Pb	1,3,7-Trimethylpurone (50)	83
15. 1,3,7,9-Tetramethyluric acid a. 50% H_2SO_4, CD = 1 A/dm^2	Pb	1,3,7,9-Tetramethylpurone (28)	83
16. Xanthine (2,6-Purinedione) a. 50% H_2SO_4, CD = 12 A/dm^2	Pb	Desoxyxanthine (70)	84

17. Guanine (2-Aminohypoxanthine)
  a. 60% H_2SO_4, CD = 12 A/dm^2, 12-22°C
  b. 1 M HOAc, CPE
	Pb
	
	P.G.
		Desoxyguanine (2-amino-1,6-dihydropurine) (75)	85

		Parabanic acid (0.35), oxalyl guanidine (0.55), guanidine (0.35); molar yields per mole of guanine	86
18. 3-Methylxanthine a. 50% H_2SO_4, CD = 12 A/dm^2, 1-7°C	Pb	3-Methyldesoxyxanthine (3-methyl-1,6-dihydro-purine-2-one)	87
19. 7-Methylxanthine a. 50% H_2SO_4, CD = 12 A/dm^2, 9-11°C	Pb	7-Methyldesoxyxanthine (66)	87
20. 1,3-Dimethylxanthine (Theophylline) a. 30% H_2SO_4, CD = 12 A/dm^2	Pb	Desoxytheophylline (60)	88

b. 1 M HOAc, CPE	P.G.		C-8 dimer, parabanic acid, 1,3-dimethylalloxan, 6,8-dimethylallantoin, dimethylurea, urea, CO_2	375
21. 8-Chlorotheophylline				
a. 50% H_2SO_4, 1.8 A		Hg	Desoxytheophylline	89, 90
b. 50% H_2SO_4, 1.5 A, 31-39°C		Pb	Theophylline	91
22. 1,7-Dimethylxanthine (Paraxanthine)				
30% H_2SO_4, CD = 12 A/dm^2		Pb	Desoxyparaxanthine (76)	88
23. 3,7-Dimethylxanthine (Theobromine)				
a. 50% H_2SO_4, CD = 0.12 A/cm^2	PbO_2	Pb	Desoxytheobromine (60-70)	92, 93
b. 4 N H_2SO_4, CD = 0.012 A/cm^2, $n = 4$			Methylalloxan, methylparabanic acid, urea, CH_3NH_2, HCOOH, NH_3	94, 95
c. 1 M HOAc, CPE, $n = 4$	P.G.		Methylparabanic acid, methylurea, urea, NH_3, CO_2, 6-methylallantoin, methylalloxan	377
24. Desoxytheobromine				
1 N H_2SO_4, CD = 0.007 A/cm^2, $n = 8$	PbO_2		Methylparabanic acid (51.6), CH_3NH_2, NH_3, CO_2	94
25. 1,3,7-Trimethylxanthine (Caffeine)				
a. 50% H_2SO_4, CD = 0.12 A/cm^2, 18°C	PbO_2	Pb	Desoxycaffeine (80)	93, 96, 97
b. 2 N H_2SO_4, CD = 0.002 A/cm^2, 18°C			Dimethylalloxan, parabanic acid, CH_3NH_2, NH_3, HCOOH	94, 98
c. 50% HOAc, CD = 0.01 A/cm^2, 45-50°C, $n = 2$	Pt		Ammonium tetramethylpurpurate (murexoin)	94
d. 1 M HOAc, CPE, $n = 4$	P.G.		Methylparabanic acid, dimethylurea, urea, CO_2, dimethylalloxan, 6,8-dimethylallantoin, NH_3	377
26. Desoxycaffeine				
a. 1 N H_2SO_4, $n = 10$	PbO_2		Dimethylparabanic acid (11.6)	94

315

Table 9.1 Preparative Electrochemical Data for Nitrogen Heterocyclic Compounds

27. 7,8-Dichlorocaffeine			
a. 50% H_2SO_4, 4 hr	Pb	Caffeine, 8-chlorocaffeine	91
b. 50% H_2SO_4, 2 hr	Pb	8-Chlorocaffeine	91
c. 50% H_2SO_4, 1% HOAc	Pb	Caffeine	91
d. 50% H_2SO_4	Hg	Desoxycaffeine	91
28. Pyrimidine			
a. pH 0.4-9.2, CPE, $n = 6$	Hg	Tetrahydropyrimidine (hydrolyzes), dimer (C-4?)	99
29. 2-Aminopyrimidine			
a. pH 5.7, CPE at -1.45 V, $n = 2$	Hg	2-Amino-3,4-dihydropyrimidine, dimer	99
b. pH 6-8, CD = 3 A/dm^2, 10-12°C	Pb	2-Amino-1,2-dihydropyrimidine (83), isolated as the picrate	106
30. 2-Amino-4-methylpyrimidine			
a. pH 8, CPE at -1.65 V, $n = 2$	Hg	2-Amino-3,4-dihydropyrimidine, dimer	99
31. 4-Amino-2,6-dimethylpyrimidine			
a. pH 4.7, CPE, $n = 4$	Hg	2,6-Dimethyl-3,4-dihydropyrimidine	99
32. 2-Hydroxypyrimidine			
a. pH 1.3-9.2, CPE, $n = 1$	Hg	Dimer (?)	99
33. 2-Hydroxy-4-aminopyrimidine (Cytosine)			
a. pH 4.7, CPE at -1.45 V, $n = 3$	Hg	2-Hydroxypyrimidine (deamination), then dimer	99, 422
34. 2,4,5,6-Pyrimidinetetrone (Alloxan)			
a. 50% H_2SO_4	Pb	Alloxantin, dihydrouracil	100
b. pH 4 acetate buffer, CPE, $n = 2$	Hg	Dialuric acid sodium salt	101
c. 1 M HOAc, CPE	Hg	Dialuric acid	381
35. Alloxantin dihydrate			
a. pH 4 acetate buffer, CPE at -0.4 V	Hg	Dialuric acid sodium salt	101

36. Dialuric acid			
a. 70% H_2SO_4, CD = 12 A/dm^2, 18-20°C	Pb	Dihydrouracil, tetrahydropyrimidine-2-one	100
37. 2,4,6(1H, 3H, 5H)-Pyrimidinetrione (barbituric acid)			
a. 50% H_2SO_4, 0-18°C	Pb	Dihydrouracil (major), tetrahydropyrimidine-2-one	102
b. 50% H_2SO_4 · 40-50°C	Pb	Tetrahydropyrimidine-2-one (major), dihydrouracil	102
38. 5-Ethylbarbituric acid			
a. 75% H_2SO_4	Pb	5-Ethyluracil, 5-ethyldihydrouracil	103
b. 75% H_2SO_4, low CD, low temp.	Pb	5-Ethyldihydrouracil	103
39. 5-Isopropylbarbituric acid			
a. 75% H_2SO_4, 12-15°C	Pb	5-Isopropyluracil (55), 5-isopropylpyrimidine-2,4-dione	103
40. 5,5-Diethylbarbituric acid (Veronal)			
a. 75% H_2SO_4	Pb	5,5-Diethyl-1,2-dihydropyrimidine-4,6-dione (desoxyveronal)	104
b. 75% H_2SO_4, 48-53°C, 30 A	Pb	Polymeric material	104
41. 4-Methyl-2,4(1H, 3H)-pyrimidinedione (4-Methyluracil)			
a. 50% H_2SO_4, CD = 12 A/dm^2, 7-9°C	Pb	4-Methyltetrahydropyrimidine-2-one, 1,3-diaminobutane	105
42. 5-Aminobarbituric acid (Uramil)			
a. 68% H_2SO_4, CD = 12 A/dm^2, 5-7°C	Pb	Dihydrouracil	100
43. 2-Amino-4-chloropyrimidine			
a. pH 7.2, CD = 2.0 A/dm^2, 72-3°C	Cd	2-Aminopyrimidine (91)	106

318

Table 9.1 Preparative Electrochemical Data for Nitrogen Heterocyclic Compounds

b. H_2O, MeOH, $(NH_4)_2SO_4$, 60-5°C	Cd, Zn	2-Aminopyrimidine (90-100)	106
c. H_2O, NaOH, EtOH, CD = 2.5-5.0 A/dm²	Pb	2-Aminopyrimidine (60)	106
d. 80% EtOH, NaOH	Pd	2-Aminopyrimidine (60)	107
e. MeOH, H_2O	Pd	2-Amino-4-methoxypyrimidine	107
f. pH 7.2-7.4, 72-3°C	Spongy Cd	2-Aminopyrimidine (95-100)	108
44. 2-Amino-4-chloro-6-methylpyrimidine			
a. H_2O, MeOH, $(NH_4)_2SO_4$, 60-5°C	Cd, Zn	2-Amino-6-methylpyrimidine (90-100)	106
b. H_2O, NaOH, EtOH, CD = 2.5-5.0 A/dm²	Pb	2-Amino-6-methylpyrimidine (60)	106
c. H_2O, MeOH, $(NH_4)_2SO_4$, 60-5°C, 2.5 A/dm²	Zn	2-Amino-6-methylpyrimidine (90)	106, 109
d. 80% EtOH, NaOH	Pd	2-Amino-6-methylpyrimidine (60)	107
e. MeOH, H_2O	Pd	2-Amino-4-methoxy-6-methylpyrimidine	107
f. EtOH, H_2O	Cu	2-Amino-4-ethoxy-6-methylpyrimidine	107
45. Pyrimidones, thiopyrimidones			
a. H_2O, acid, 20% MeOH	Hg	Corresponding dimers (structures undetermined)	110
46. Parabanic acid			
a. 70% H_2SO_4, CD = 12 A/dm²	Pb	Hydantoin (2,4-(3H, 5H)-imidazoledione), imidazole-2-one	100
b. 1 *M* HOAc, CPE	Hg	Urea, glycolic acid, 5-hydroxyhydantoin	111
47. 1-Methylpurinium iodide			
a. pH 10.1, CPE, *n* = 6	Hg	1,2,3,6-Tetrahydro-1-methylpurine	391
48. 6-Amino-1-methylpurinium iodide			
a. 1 *N* HCl, CPE at −0.98 V, 0°C, *n* = 6	Hg	NH_3, 1-methyl-1,2,3,6-tetrahydropurine	391, 419
49. 6-Amino-9-ribosyl-1-methylpurinium iodide			
a. 2 *N* HCl/40% EtOH, CPE, −35°C, *n* = 6	Hg	NH_3, 9-ribosyl-1-methyl-1,2,3,6-tetrahydro-	391

50. 6-Methylaminopurine
   a. 1 N HCl, CPE at -1.15 V | Hg | MeNH$_2$, 1,2,3,6-tetrahydropurine | 391

Pyrazines, Pyrazoles

1. Pyrazine
   a. Aqueous solution | DME | 1,4-Dihydropyrazine (inferred from polarographic data) | 112, 113
   b. Aqueous acid, CPE at -0.26 V | Hg | 1,4-Dihydropyrazine cation radical (verified by EPR) | 403
   c. pH 3 buffer, CPE | Hg | 1,4-Dihydropyrazine quaternary salt | 403
2. 2,3-Dimethyl-5,6-dihydropyrazine
   a. pH 7/40% MeOH, CPE at -0.9 V, $n=2$ | Hg | 2,3-Dimethyl-1,4,5,6-tetrahydropyrazine | 114
3. 2,3-Diphenyl-5,6-dihydropyrazine
   a. pH 7/40% MeOH, CPE at -0.9 V, $n=2$ | Hg | 2,3-Diphenyl-1,4,5,6-tetrahydropyrazine | 114
4. 2-Phenyl-3-methyl-5,6-dihydropyrazine
   a. pH 7/40% MeOH, CPE at -0.9 V, $n=2$ | Hg | 2-Phenyl-3-methyl-1,4,5,6-tetrahydropyrazine | 114
5. 3,5-Dimethylpyrazole
   a. 10% Na$_2$SO$_4$, CD $=0.05$ A/cm^2 | PbO$_2$ | Pyrazole-3-carboxylic acid | 115
6. 1-Phenyl-3-methylpyrazole
   a. K$_2$CO$_3$ solution, CD $=0.02$ A/cm^2 | Pt | Pyrazole-3-carboxylic acid | 115
   b. 2 N H$_2$SO$_4$, CD $=0.04$ A/cm^2 | PbO$_2$ | p-Benzoquinone, hydroquinone, oxalic acid | 115

Table 9.1 Preparative Electrochemical Data for Nitrogen Heterocyclic Compounds

	Electrode	Product	Ref.
7. 1-Phenyl-3-methyl-5-pyrazolone	Pt	Methylenebis(1-phenyl-3-methyl-5-pyrazolone), 4,4'-bis(1-phenyl-3-methyl-5-pyrazolone) (10)	115
a. 2 N H_2SO_4			
8. 1-Phenyl-3,4-dimethyl-5-pyrazolone	Pt	4,4'-Bis(1-phenyl-3,4-dimethyl-5-pyrazolone) (5), fumaric acid	115
a. 2 N H_2SO_4, CD = 0.4 A/cm^2			
9. 1-Phenyl-2,3-dimethyl-3-pyrazolin-5-one (Antipyrine)	Pt	Polymeric products	115
a. H_2O, H_2SO_4			
10. 1-Phenyl-3-methylpyrazole-4,5-dione	Pt	Polymeric material	115
a. 2 N H_2SO_4		1-Phenyl-3-methyl-4-hydroxy-5-pyrazolone (46)	115
b. NaHSO$_3$, HOAc, CD = 0.1 A/cm^2			
11. 4-Amino-1-phenyl-3-methyl-5-pyrazolone	PbO$_2$	Rubazonic acid	115
a. 2 N H_2SO_4, CD = 0.04 A/cm^2			
12. 4-Nitro-1-phenyl-3-pyrazolone	Pb	4-Amino-1-phenyl-3-pyrazolone	116
a. H_2O, H_2SO_4, EtOH, 80°C			
13. N-Nitropyrazole	Hg	Pyrazole	60
a. 0.5 N HClO$_4$, CPE at −0.6 V			
14. 4-Isonitroso-1-phenyl-3-methyl-5-pyrazolone	PbO$_2$	4-Amino-1-phenyl-3-methyl-5-pyrazolone	115
a. 2 N H_2SO_4, CD = 0.02 A/cm^2			
15. 4-Nitroantipyrine	Hg	4-Amino-1-phenyl-3-methyl-5-pyrazolone	383
a. 1 N HCl/MeOH, CPE			
16. 2,3-Dimethyl-4-nitro-1-phenylpyrazolidine-5-one	Hg	4-Aminoantipyrine, 4-amino-2,3-dimethyl-1-	383
a. Aqueous solution, CPE			

17. 2,5-Diphenylpyrazine			
a. 0.2 N NaOH/80% MeOH, CPE at -1.4 V	Hg	1,2-Dihydro-2,5-diphenylpyrazine (formed by rearrangement of the initially formed 1,4-dihydro derivative)	399
18. 1,2-Dihydro-2,5-diphenylpyrazine			
a. 0.5 N NaOH/50% MeOH, pH 13.4, CPE at -1.8 V	Hg	2,5-Diphenylpiperazine (61)	399
b. pH 12.6/60% MeOH, CPE at -1.55 V	Hg	1,2,3,4-Tetrahydro-5-phenyl-6-methylpyrazine	399
19. 2,3-Diphenylpyrazine			
a. 0.5 N NaOH/50% MeOH, pH 13.4, CPE at -1.4 V, $n = 2$	Hg	1,2,-Dihydro-2,3-diphenylpyrazine (formed by rearrangement of the initially formed 1,4-dihydro derivative)	399
b. 0.5 N NaOH/50% MeOH, CPE at -1.75 V, $n = 6$	Hg	2,3-Diphenylpiperazine (14)	399
20. 2,3,5,6-Tetramethylpyrazine			
a. pH 9.2/20% MeOH, CPE at -1.3 V, $n = 2$	Hg	1,2-Dihydro-2,3,5,6-tetramethylpyrazine (formed by rearrangement of the initially formed 1,4-dihydro derivative)	399
b. pH 9.2, CPE, $n = 6$	Hg	2,3,5,6-Tetramethylpiperazine	399
21. 1,2-Dimethyl-3-phenylpyrazolium perchlorate			
a. DMF, TEAP, CPE	Hg	1,2-Dimethyl-3-phenylpyrazoline	420

Pyridazines

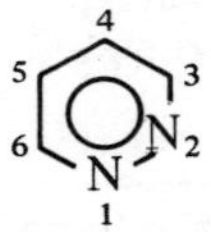

321

Table 9.1 Preparative Electrochemical Data for Nitrogen Heterocyclic Compounds

1. 3-Phenyl-6-dimethylaminopyridazine			
a. Acetate buffer, CPE	Hg	4,5-Dihydro-3-phenyl-6-dimethylaminopyri-dazine	117
2. 6-Chloro-3-methylpyridazine			
a. Aqueous buffer, CPE, $n = 2$	Hg	3-Methyl-6-pyridazinone	117
3. 3-Phenyl-6-methoxypyridazine			
a. Acetate buffer, CPE	Hg	3-Phenyl-6-pyridazinone	117
4. 1-Methyl-3,6-diphenylpyridazinium iodide			
a. Acid solution, CPE, $n = 2$	Hg	1,4-Dihydro-1-methyl-3,6-diphenylpyridazine	117
5. 3,6-Diphenylpyridazine			
a. Alkaline solution, CPE	Hg	3,6-Diphenyl-2,3,4,5-tetrahydropyridazine	117
6. 1-Methyl-3,6-diphenyl-5-t-butyl-pyridazinium iodide			
a. Acid solution, CPE	Hg	1-Methyl-2,5-diphenyl-3-t-butylpyrrole, 2,5-diphenyl-3-t-butylpyrrole (formed in a 3:1 ratio)	118
7. 6-Methylpyridaz-3-one			
a. Aqueous buffer, CPE	Hg	4,5-Dihydro-6-methylpyridaz-3-one	119
8. 3-Carbomethoxy-4-oxo-5-methyl-(2,3-d)furopyridazine			
a. pH 4.56, CPE at −1.35 V	Hg	6,7-Dihydro derivative (60) (pH 9, 20% yield)	392

Pyridines

1. Pyridine				
a. H_2O, H_2SO_4	Pt		1-(2-Pyridyl)pyridinium sulfate, glutacondialdehyde	120
b. $2 N$ H_2SO_4, CD = 0.05 A/cm^2, 68°C, $n = 20$	PbO_2		CO, CO_2, HNO_3, HCOOH, HCHO, NH_3	121
c. 10% H_2SO_4, CD = 17.1 A/dm^2		Pb	Piperidine, 2,2'- and 4,4'-dipiperidyls	58, 122, 123
d. H_2O, H_2SO_4, 50–80°C		Pb	Piperidine (75), 2,2'- and 4,4'-dipiperidyls	124, 124, 379
e. H_2O, H_2SO_4		Pb	Piperidine, di- and polypiperidyls (7)	126
f. $2 N$ H_2SO_4, catalyst poisons		Ni	Piperidine	127
g. 10% H_2SO_4		Cd, Pb, Tl, Hg	Piperidine (58–88; yield varies with cathode material)	128
h. H_2O, H_2SO_4, 30–40°C		Pb	Piperidine (65), tetrahydropyridine	129
i. MeCN, $LiClO_4$, CPE at 1.6 V vs Ag/Ag$^+$	C		1-(2-Pyridyl)pyridinium perchlorate	130
j. MeCN, TEACN, CPE	Pt		2-Cyanopyridine (15)	131
k. Pyridine (solvent), $AlCl_3$, CPE		Hg, Pt, Cu, C	Polyamine and polyamide copolymers (formed by air oxidation and hydrolysis of electrolysis product)	132
l. Anhydrous HF, 50 A	Ni	Fe	Perfluoropiperidine, perfluorodipiperidyl, perfluoro-n-pentane	133
m. 35% H_2SO_4		Pb	Piperidine (43), 1,2,3,6-tetrahydropyridine (13)	145
2. N-Methylpyridinium methosulfate				
a. NaOH, H_2O, $K_3[Fe(CN)_6]$	Fe		N-Methyl-α-pyridone (80–95)	3, 134, 136
b. 10% H_2SO_4		Pb	1-Methylpiperidine, 1,1'-dimethyl-4,4'-dipiperidyl	137
c. H_2O, H_2SO_4		Pb	N-Methylpiperidine (72), N-methyl-1,2,3,6-tetrahydropyridine (17)	142

324

Table 9.1 Preparative Electrochemical Data for Nitrogen Heterocyclic Compounds

3. *N*-Benzylpyridinium chloride			
a. 10% H_2SO_4	Pb	1-Benzylpiperidine, 1,1'-dibenzyl-4,4'-dipiperidyl	137
b. H_2O, H_2SO_4	Pb	1,1'-Dibenzyl-4,4'-dihydrodipyridyl	138-140
c. H_2O, H_2SO_4, 50-60°C	Pt	1,1'-Dibenzyl-4,4'-dihydrodipyridyl	138
d. H_2O, H_2SO_4, CPE	Hg	1-Benzylpiperidine (98)	398
4. *N*-Methylpyridinium methosulfate + *N*-Benzylpyridinium chloride			
a. 10% H_2SO_4	Pb	1-Methyl-1'-benzyl-4,4'-dipiperidyl	141
5. *N*, 2-Dimethylpyridinium methosulfate			
a. H_2O, H_2SO_4	Pb	1,2-Dimethylpiperidine (65), 1,2-dimethyl-1,2,3,6-Tetrahydropyridine (26.5)	142, 183
6. *N*, 3-Dimethylpyridinium methosulfate			
a. H_2O, H_2SO_4	Pb	1,3-Dimethylpiperidine, 1,3-dimethyl-1,2,5,6 tetrahydropyridine (53)	142, 182
7. *N*, 4-Dimethylpyridinium methosulfate			
a. H_2O, H_2SO_4	Pb	1,4-dimethylpiperidine (17), 1,4-dimethyl-1,2,3,6-tetrahydropyridine (8)	142
8. *N*-Methyl-3-carbethoxypyridinium methosulfate			
a. H_2O, H_2SO_4	Pb	1,3-Dimethylpiperidine, 1,3-dimethyl-1,2,3,6-tetrahydropyridine, 1-methyl-1,2,5,6-tetra-hydropyridine-3-carboxylic acid	142, 143
9. *N*-Methyl-4-carbethoxypyridinium methosulfate			

Substrate / Conditions	Electrode	Products	Ref.
a. H_2O, H_2SO_4	Pb	1,4-Dimethylpiperidine, 1,4-dimethyl-1,2,3,6-tetrahydropyridine, 1-methylpiperidine-4-carboxylic acid	142
10. *N*-Methyl-4-carbomethoxypyridinium methosulfate a. H_2O, H_2SO_4	Pb	1,4-Dimethylpiperidine, 1,4-dimethyl-1,2,3,6-tetrahydropyridine, 1-methylpiperidine-4-carboxylic acid	142
11. *N*-Ethylpyridinium bromide-4-carboxylic acid a. pH 0.15-9.15 buffers, CPE	Hg	1-Ethyl-4-formylpyridinium bromide (0-75, yield varies with pH)	144
12. 2-Methylpyridine (*α*-Picoline) a. 35% H_2SO_4	Pb	2-Methylpiperidine (50), 2-methyl-1,2,3,6-tetrahydropyridine (19)	145
b. $2\,N$ H_2SO_4, CD = 0.05 A/cm²	PbO₂	Picolinic acid (major), CH_3COOH, HCOOH, HNO_3, CO, CO_2, NH_3, pyridine-2-aldehyde	146
13. 3-Methylpyridine (*β*-Picoline) a. 35% H_2SO_4	Pb	3-Methylpiperidine (58), 3-methyl-1,2,5,6-tetrahydropyridine (20)	145
b. 20% H_2SO_4	Pb	3-Methylpiperidine, 3-methyl-1,2,5,6-tetrahydropyridine	147
c. $7\,N$ H_2SO_4, CD = 10 A/dm², 30-40°C	PbO₂	Nicotinic acid (65)	148-150
14. 4-Methylpyridine (*γ*-Picoline) a. 35% H_2SO_4	Pb	4-Methylpiperidine (39), 4-methyl-1,2,3,6-tetrahydropyridine (26)	145

Table 9.1 Preparative Electrochemical Data for Nitrogen Heterocyclic Compounds

b. H_2O, H_2SO_4, CD = 0.1 A/cm²	PbO_2	Isonicotinic acid (42.5), pyridine-4-aldehyde (16.8)	151
c. 20% HNO_3, CD = 0.1 A/cm²	Pt	Isonicotinic acid (27)	152
d. Aqueous acid, CD = 0.00026 A/cm²	Pt	Isonicotinic acid (15)	153
15. 2,3-Dimethylpyridine			
a. H_2O, H_2SO_4	Pb	2,3-Dimethylpiperidine, 2,3-dimethyl-1,2,5,6-tetrahydropyridine	154
b. 20% H_2SO_4	Pb	cis- and trans-2,3-Dimethylpiperidine	189
16. 2,4-Dimethylpyridine			
a. H_2O, H_2SO_4, vary CD, temp.	Pb	Lutidinic acid (pyridine-2,4-dicarboxylic acid) (22)	155
17. 3,4-Dimethylpyridine			
a. H_2O, H_2SO_4	Pb	3,4-dimethylpiperidine, 3,4-dimethyl-1,2,5,6-tetrahydropyridine	154
b. 20% H_2SO_4	Pb	cis- and trans-3,4-Dimethylpiperidine	189
18. 2,5-Dimethylpyridine			
a. H_2O, H_2SO_4	Pb	2,5-Dimethylpiperidine, 2,5-dimethyl-1,2,5,6-tetrahydropyridine	154
19. 2-Methyl-5-ethylpyridine			
a. H_2O, H_2SO_4	Pb	Isocinchomeronic acid (pyridine-2,5-dicarboxylic acid)	156
b. 20% H_2SO_4	Pb	2-Methyl-5-ethylpiperidine, 2-methyl-5-ethyl-1,2,3,6-tetrahydropyridine	147
20. 3-Methyl-4-ethylpyridine			
a. H_2O, H_2SO_4	Pb	3-Methyl-4-ethylpiperidine, 3-methyl-4-ethyl-1,2,5,6-tetrahydropyridine	147, 154

	Substrate / conditions	Electrode	Products	Ref.
21.	2,6-Dimethylpyridine			
a.	35% H_2SO_4	Pb	2,6-Dimethylpiperidine (72), 2,6-dimethyl-1,2,5,6-tetrahydropyridine (19)	145
22.	3-Ethylpyridine			
a.	20% H_2SO_4	Pb	3-Ethylpiperidine, 3-ethyl-1,2,5,6-tetrahydropyridine, 3-ethyl-1,2,3,6-tetrahydropyridine	147
23.	4-Ethylpyridine			
a.	20% H_2SO_4, CD = 0.026 A/cm^2	Pb	4-Ethylpiperidine, 4-ethyl-1,2,5,6-tetrahydropyridine	147
b.	20% HNO_3, CD = 0.1 A/cm^2, 30°C	Pt	Isonicotinic acid (75)	152
c.	Aqueous acid, CD = 0.00026 A/cm^2	Pt	Isonicotinic acid (15)	153
24.	3-Isopropylpyridine			
a.	20% H_2SO_4, CD = 0.026 A/cm^2	Pb	3-Isopropyl-1,2,3,6-tetrahydropyridine, 3-Isopropylpiperidine	147
25.	4-Isopropylpyridine			
a.	20% H_2SO_4, CD = 0.026 A/cm^2	Pb	4-Isopropyl-1,2,5,6-tetrahydropyridine, 4-Isopropylpiperidine	147
b.	Anhydrous HF, 50 A	Ni	Perfluoro-isopropylpiperidine, C_8F_{18} isomers	133
26.	4-n-Propylpyridine			
a.	Anhydrous HF, 50 A	Ni	Perfluoro-4-n-propylpiperidine, perfluoro-3-ethylhexane	133
27.	1-Hydroxypyridine-2-thione			
a.	pH 2.9-8.1, CPE, $n = 1$	Hg	Corresponding mercuric dimercaptide	157
28.	2-Aminopyridine			
a.	Aqueous MeOH, KI, 0-5° C, low CD	Pt(?)	2-Amino-5-iodopyridine	158

Table 9.1 Preparative Electrochemical Data for Nitrogen Heterocyclic Compounds

Compound / Conditions	Electrode	Products	Ref.
29. 4-Aminopyridine			
a. H_2O, H_2SO_4	Pb	4-Aminopiperidine (20), piperidine, NH_3	159
30. 4-Anilinomethylpyridine			
a. 0.2 N HCl, CPE at -0.85 V	Hg	Aniline (95), γ-picoline	160
31. 2-Iodopyridine			
a. D_2O buffer, pH 13/50% CH_3OD, CPE at -1.38 V	Hg	Pyridine-2-d	161
32. 2-Vinylpyridine			
a. Aqueous solution	Pt	Pyridine-2-aldehyde (3), 2-(β-hydroxyethyl)-pyridine (6)	151
b. H_2O, tetraethylammonium p-toluenesulfonate	Hg	1,4-Bis (2-pyridyl)butane (68.8)	163
c. H_2O, dibutylmaleate, tetraethylammonium p-toluenesulfonate, DMF	Hg	Mixed reductive coupling product of 2-vinylpyridine and dibutylmaleate	163
33. 4-Vinylpyridine			
a. H_2O, DMF, Methyltriethylammonium p-toluenesulfonate	Hg	1,4-Bis(4-pyridyl)butane (15)	163
b. H_2O, DMF, methyl vinyl ketone, tetraethylammonium p-toluenesulfonate	Hg	Methyl ω-(4-pyridyl)butyl ketone	163
34. 3-Nitro-6-chloropyridine			
a. 20% H_2SO_4, CD = 2.5 A/cm^2	Cu	3-Amino-6-chloropyridine (53), 3-amino-pyridine (trace)	164
35. 6-Chloropyridine-3-arsinic acid			
a. 5% H_2SO_4, CD = 0.1 A/cm^2	Pb	Pyridine-3-arsine, 3,3'-arsenopyridine	164
36. 2-[γ,γ,γ-Trichloro-β-hydroxypropyl]-pyridine			

Compound / Conditions	Electrode	Products (yields %)	Ref.
a. H_2O, HCl, CD = 1.1-2.3 A/dm²	Pb, Zn amalgam	2-[γ,γ-Dichloro-β-hydroxypropyl] pyridine (50), trace indolizine, pyrrocoline	165
b. H_2O, HCl, CD = 6 A/dm²	Cu	2-[γ,γ-Dichloro-β-hydroxypropyl] pyridine, 2-(2-propenyl)pyridine	165
37. 2,6-Dimethoxypyridine	Pt		
a. KOH, MeOH, H_2O		3,6-Dihydro-2,3,3,6-pentamethoxypyridine (26), 2,3,5,6-tetramethoxypyridine, 2,3,4,6-tetramethoxypyridine	162
38. 2-Cyanopyridine			
a. Phosphate buffer, CPE	Hg	2-Picolylamine (2-aminomethylpyridine) (90)	166
39. 4-Cyanopyridine			
a. Phosphate buffer, CPE	Hg	Pyridine, 4-picolylamine, CN⁻	166, 167
b. Strong acid, CPE	Hg	4-picolylamine	166, 167
c. 0.1 M KOH, CPE	Hg	Pyridine, CN⁻	167
d. 0.1 M KOH, CPE at −1.40 V, 5°C, $n = 2$	Hg	Pyridine (90-95)	160
40. 2-Acetylpyridine			
a. H_2O, HCl, CPE at −1.8 V	Hg	2,3-Bis(2-pyridyl)-2,3-butanediol (73.4)	168
b. Aqueous buffer/80% EtOH, CPE	Hg, Cu	dl/meso-Pinacol, carbinol (yields of each var. with pH, applied voltage)	169
c. 20% H_2SO_4, CD = 0.013 A/cm²	Pb	2-Ethylpiperidine (58), 2-ethylpyridine (3), 2-ethyl-1,2,3,6-tetrahydropyridine (24)	170
41. 3-Acetylpyridine			
a. 20% H_2SO_4, CD = 0.013 A/cm²	Pb	3-Ethylpiperidine (44), 3-ethyl-1,2,5,6-tetrahydropyridine (17), 3-ethyl-1,2,3,6-tetrahydropyridine (14)	170

Table 9.1 Preparative Electrochemical Data for Nitrogen Heterocyclic Compounds

42. 4-Acetylpyridine a. 20% H_2SO_4, CD = 0.013 A/cm²	Pb	4-Ethylpiperidine (53), 4-ethylpyridine (25), 4-ethyl-1,2,3,6-tetrahydropyridine (4)	170
43. 2-Hydroxymethylpyridine a. 20% H_2SO_4, CD = 0.013 A/cm²	Pb	2-Methylpyridine (25), 2-methylpiperidine (37), 2-methyl-1,2,3,6-tetrahydropyridine (16)	170
44. 3-Hydroxymethylpyridine a. 20% H_2SO_4, CD = 0.013 A/cm²	Pb	3-Methylpyridine (2), 3-methylpiperidine (34), 3-methylenepiperidine (26)	170
45. 3-(α-Hydroxyethyl)pyridine a. 20% H_2SO_4, CD = 0.013 A/cm²	Pb	3-Ethylpyridine(2.3), 3-ethylpiperidine (36), 3-ethyl-1,2,5,6-tetrahydropyridine (22), 3-ethyl-1,2,3,6-tetrahydropyridine (5)	170
46. 3-Dimethylhydroxymethylpyridine a. 20% H_2SO_4, CD = 0.013 A/cm²	Pb	3-Isopropylpiperidine (52), 3-isopropylidine-piperidine (21), 3-isopropyl-1,2,3,6-tetrahydro-pyridine (7)	170
47. 4-Hydroxymethylpyridine a. 20% H_2SO_4, CD = 0.013 A/cm²	Pb	4-Methylpyridine (7), 4-methylpiperidine (20), 4-methyl-1,2,3,6-tetrahydropyridine (63)	170
48. 4-Dimethylhydroxymethylpyridine a. 20% H_2SO_4, CD = 0.013 A/cm²	Pb	4-Isopropylpyridine (33), 4-isopropylpiperidine (18), 4-isopropyl-1,2,3,6-tetrahydropyridine (38)	170, 175
49. Pyridine-2-aldehyde (Picolinaldehyde) a. 20% H_2SO_4, CD = 0.013 A/cm²	Pb	2-Methylpyridine (41), 2-methylpiperidine (25), 2-methyl-1,2,3,6-tetrahydropyridine (11)	170

No.	Conditions	Electrode	Products	Ref.
50.	Pyridine-3-aldehyde (Nicotinaldehyde)			
	a. 20% H_2SO_4, CD = 0.013 A/cm²	Pb	3-Methylpyridine (10), 3-methylpiperidine (15), 3-methylenepiperidine (10)	170
51.	Pyridine-4-aldehyde (Isonicotinaldehyde)			
	a. 20% H_2SO_4, CD = 0.013 A/cm²	Pb	4-Methylpyridine (29), 4-methylpiperidine (20), 4-methyl-1,2,3,6-tetrahydropyridine (42)	170
	b. pH 6 Phosphate buffer, CPE	Hg	Corresponding pinacol	171
52.	Pyridine-2-carboxylic acid (Picolinic acid)			
	a. 2 N H_2SO_4, CD = 0.05 A/cm² PbO_2		HCOOH, NH_3, HCHO	146
	b. H_2O, H_2SO_4, EtOH, 10-14°C, CD = 0.17 A/cm²	Hg	α-Picoline (33)	143, 173
	c. 35% H_2SO_4	Pb	α-Picoline, 2-methylpiperidine, 2-methyl-1,2,3,6-tetrahydropyridine, traces of pipecolinic acid, 1,2,3,6-tetrahydropicolinic acid	174
	d. Acid solution, CPE, 0-5°C, $n = 2$	Hg	Pyridine-2-aldehyde	144
53.	Pyridine-3-carboxylic acid (Nicotinic acid)			
	a. 2 N H_2SO_4, CD = 0.04 A/cm² PbO_2	Hg	HCOOH, NH_3, CO, CO_2, alkylamines (trace)	172
	b. H_2O, H_2SO_4	Hg	β-Picoline (?)	143
54.	Pyridine-4-carboxylic acid (Isonicotinic acid)			
	a. pH 2.85 Citrate buffer/1 M KCl, 0-5°C, CPE at −1.0 V, $n = 2$	Hg	Pyridine-4-aldehyde (12-69, yield varies with pH)	144
	b. Aqueous solution	Sn	Propionaldehyde (23.4), NH_3	151
	c. Aqueous solution	Pt black	Me_2CHCHO (10.7), NH_3	151
	d. Aqueous solution	Hg	Me_2CHCHO (25), NH_3	151
	e. H_2O, H_2SO_4, EtOH, 10-14°C, CD = 0.17 A/cm²	Hg	γ-Picoline (31)	143, 173

Table 9.1 Preparative Electrochemical Data for Nitrogen Heterocyclic Compounds

55. Pyridine-2-carboxamide (Pincolinamide)			
a. 2 N HCl, 10-15°C, CPE at −0.8 V vs Ag/AgCl, n = 2.3	Hg	Pyridine-2-aldehyde (72) (poorer yields at higher temperatures)	18
b. H$_2$O, HCl, CPE	Hg	Pyridine-4-aldehyde	160
c. Acetate buffer, CPE	Hg	2-Hydroxymethylpyridine	160
56. Pyridine-3-carboxamide (Nicotinamide)			
a. H$_2$O, HCl, CPE	Hg	Pyridine-3-aldehyde	160
b. Acetate buffer, CPE	Hg	3-Hydroxymethylpyridine	160
57. Pyridine-4-carboxamide (Isonicotinamide)			
a. 2 N HCl, CPE at −0.8 V vs Ag/AgCl, n = 2.3	Hg	Pyridine-4-aldehyde (66) (poorer yields at higher temperatures,	18
b. 0.8 N HCl, CPE at −0.8 V, 5°C, n = 2	Hg	Pyridine-4-aldehyde (87)	160
c. pH 3.5 Citrate buffer/30% EtOH, CPE, n = 4	Hg	4-Hydroxymethylpyridine	160
58. Isonicotinic anilide			
a. 2 N HCl, CPE at −0.75 V, n = 4	Hg	4-Anilinomethylpyridine (40)	160
b. 0.4 N KOH/10% EtOH, CPE at −1.40 V, n = 2	Hg	1,4-Dihydroisonicotinic anilide (?)	160
59. Isonicotinic thiamide			
a. 1 N HCl, CPE, 5°C, n = 2	Hg	Pyridine-4-aldehyde	160
b. 1 N HCl, CPE, 5°C, n = 4	Hg	4-Aminomethylpyridine	160
c. pH 4.8 acetate buffer, CPE at −0.85 V, n = 4	Hg	4-Mercaptomethylpyridine, 4-aminomethyl-pyridine	160
d. 0.05 N KOH/0.5 N KCl, CPE at −0.2 V, 5°C, n = 2	Hg	4-Cyanopyridine	160

	Electrode	Products	Ref.
60. N-Benzylisonicotinic amide			
a. H_2O, HCl, CPE	Hg	Pyridine-4-aldehyde, benzylamine	160
b. pH 5 buffer, CPE	Hg	4-Hydroxymethylpyridine, benzylamine	160
61. N-Cyclohexylisonicotinic amide			
a. H_2O, HCl, CPE	Hg	Pyridine-4-aldehyde, cyclohexylamine	160
b. pH 5 buffer, CPE	Hg	4-Hydroxymethylpyridine, cyclohexylamine	160
62. N-Methyl-N-phenylisonicotinic amide			
a. 1 N HCl, CPE, $n = 4$	Hg	4-Hydroxymethylpyridine, N-methylaniline	160
b. pH 5 buffer, CPE	Hg	4-Hydroxymethylpyridine, N-methylaniline	160
63. Isonicotinic hydrazide (Isoniazid)			
a. 0.4 N HCl, CPE at -0.5 V, $n = 2$	Hg	Isonicotinic amide (74)	171
b. pH 11 phosphate buffer, CPE, $n = 4$	Hg	NH_3, Dihydroisonicotinic amide (?)	171
c. pH 11 phosphate buffer, CPE at 0.0 V, $0-5°C$, $n = 2$, CD = 0.002 A/cm^2	Hg	1,2-Diisonicotinoyl hydrazine (90), pyridine-4-aldehyde (5)	171
d. 0.2 M KOH/ 1 M KCl, CPE at 0.0 V, $n = 2.9$, CD = 0.01 A/cm^2	Hg	Isonicotinic acid (45), 1,2-diisonicotinoyl hydrazine (55)	171
64. 1-Isonicotinoyl-1-phenylhydrazine			
a. 0.4 N HCl, CPE at -0.5 V	Hg	Isonicotinic amide, aniline	171
b. 0.2 N KOH, CPE, $n = 4$	Hg	Aniline (90)	171
c. pH 11.9 phosphate buffer, CPE at 0.0 V, $n = 2.5$	Hg	Isonicotinic acid (90), aniline (trace)	171

Table 9.1 Preparative Electrochemical Data for Nitrogen Heterocyclic Compounds

65. 1,2-Diisonicotinoyl hydrazine			
a. H_2O, HCl, CPE	Hg	Pyridine-4-aldehyde	171
66. Ethylpyridine-2-carboxylate			
a. 35% H_2SO_4	Pb	2-Methylpyridine, 2-methylpiperidine, 2-methyl-1,2,3,6-tetrahydropyridine, traces of pipecolinic acid, 1,2,3,6-tetrahydropicolinic acid	174
67. Ethylpyridine-3-carboxylate			
a. 50% H_2SO_4	Pb	3-Methylpiperidine, 3-methyl-1,2,3,6-tetrahydropyridine, 3-methyl-1,2,5,6-tetrahydropyridine, 1,2,5,6-tetrahydronicotinic acid	174
68. Ethylpyridine-4-carboxylate			
a. Aqueous solutions, CPE	Hg	Pyridine-4-aldehyde (yield varies with pH)	160
69. 1,2-Bis(2-pyridyl)ethylene			
a. EtOH/1 M KCl, CPE at −1.6 V	Hg	1,2-Bis(2-pyridyl)ethane (94)	176
70. 1,2-Bis(4-pyridyl)ethylene			
a. 1 N HCl/50% EtOH, CPE at −1.0 V	Hg	1,2-Bis(4-pyridyl)ethane (95)	176
71. 1-(2-Pyridyl)-2-(4-pyridyl)ethylene			
a. EtOH/1 M LiCl, CPE at −1.6 V	Hg	1-(2-Pyridyl)-2-(4-pyridyl)ethane (92.5)	176
72. *syn*-Phenyl-2-pyridylketone oxime			
a. 0.1 M HCl, CPE at −0.8 V, $n = 4$	Hg	α-(2-Pyridyl)benzylamine	177
73. Isonicotinic acid aldoxime			
a. Aqueous buffer, CPE, $n = 4$	Hg	4-Aminomethylpyridine	178
74. 3-Hydroxy-5-hydroxymethyl-2-isonicotinaldehyde oxime			
a. 0.1 N HCl, CPE, $n = 4$	Hg	Pyridoxamine dihydrochloride (70)	179
b. Aqueous acid, CPE at −0.8 V	Hg	Pyridoxamine dihydrochloride (100)	180

334

c. Aqueous acid, CPE at -1.4 V	Hg	4-Deoxypyridoxine hydrochloride (100)	180
75. Pyridoxal pyridoxime			
a. Aqueous acid, constant current electrolysis	Hg	4-Deoxypyridoxine hydrochloride	180
76. 1-Ethyl-4-carbamoylpyridinium bromide			
a. 1 N HCl, CPE	Hg	Pyridine-4-aldehyde	160
b. pH 4.5 acetate buffer, CPE	Hg	4-Hydroxymethylpyridine, 1-ethyl-4-formyl-pyridinium bromide (trace) (?)	160
77. 1-Methyl-2-carbethoxypyridinium methosulfate			
a. 39% H_2SO_4	Pb	1,2-Dimethylpiperidine (major), traces of N-methylpipecolinic acid, 1-methyl-1,2,3,6-tetrahydropicolinic acid	174
78. 1-Methyl-3-hydroxypyridinium methosulfate			
a. 20% H_2SO_4	Pb	1-Methyl-3-hydroxypiperidine, methylpentyl-amine, 1,2-dimethylpyrrolidine	181
79. 1-Ethyl-3-hydroxypyridinium methosulfate			
a. 20% H_2SO_4	Pb	1-Ethylpiperidine, ethylpentylamine, 1-ethyl-2-methylpyrrolidine	181
80. 1-Phenyl-3-hydroxypyridinium chloride			
a. 20% H_2SO_4	Pb	N-Pentylaniline, 1-phenyl-2-methylpyrrolidine	181

Table 9.1 Preparative Electrochemical Data for Nitrogen Heterocyclic Compounds

81.	1-Methyl-3-carbomethoxypyridinium methosulfate		
	a. 20% H_2SO_4	Pb	1,3-Dimethylpiperidine (17), 1,3-dimethyl-1,2,5,6-tetrahydropyridine (2), 1,3-dimethyl-1,2,3,6-tetrahydropyridine (3), 1-methyl-3-methylenepiperidine 181
82.	1-Methyl-3-hydroxymethylpyridinium methosulfate		
	a. 20% H_2SO_4	Pb	Same products as in (81); yields, respectively: (16), (5), (3), (8). 182
83.	1-Methyl-3-formylpyridinium methosulfate		
	a. 20% H_2SO_4	Pb	Same products as in (81); yields, respectively: (13), (1), (2), (8). 182
84.	1,2,4-Trimethylpyridinium methosulfate		
	a. 20% H_2SO_4	Pb	1,2,4-Trimethylpiperidine, two stereoisomers (16), 1,2,4-trimethyl-1,2,5,6-tetrahydropyridine (1), 1,2,4,-trimethyl-1,2,3,6-tetrahydropyridine (50) 183
85.	1-Methoxypyridinium methosulfate		
	a. 20% H_2SO_4, CD = 0.04 A/cm^2	Pb	Piperidine (27), pyridine (15), 1,2,5,6-tetrahydropyridine (17) 184
86.	1-Methoxy-3-methylpyridinium methosulfate		
	a. 20% H_2SO_4, CD = 0.04 A/cm^2	Pb	3-Methylpiperidine (32), 3-methylpyridine (4), 3-methyl-1,2,3,6-tetrahydropyridine (17), 3-methyl-1,2,5,6-tetrahydropyridine (37) 184

No.	Compound	Cathode	Products	Ref.
87.	1-Methoxy-4-methylpyridinium methosulfate			
	a. 20% H_2SO_4, CD = 0.04 A/cm²	Pb	4-Methylpiperidine (20), 4-methylpyridine (29), 4-methyl-1,2,3,6-tetrahydropyridine (37)	184
88.	Pyridine + acetone			
	a. 20% H_2SO_4	Pb	2-(2-Hydroxy-2-propyl)-1,2,5,6-tetrahydropyridine, 4-(2-hydroxy-2-propyl)piperidine	185
89.	2-Methylpyridine + acetone			
	a. 20% H_2SO_4	Pb	2-Methyl-6-(2-hydroxy-2-propyl)-1,2,3,6-tetrahydropyridine, 2-methyl-4-(2-hydroxy-2-propyl)piperidine	185
90.	4-Methylpyridine + acetone			
	a. 20% H_2SO_4	Pb	2-(2-Hydroxy-2-propyl)-4-methyl-1,2,5,6-tetrahydropyridine	185
91.	4-Methylpyridine + methylethyl ketone			
	a. 20% H_2SO_4	Pb	2-(2-Hydroxy-2-butyl)-4-methyl-1,2,5,6-tetrahydropyridine	185
92.	2,4-Dimethylpyridine + acetone			
	a. 20% H_2SO_4	Pb	4,6-Dimethyl-1-2-(2-hydroxy-2-propyl)-1,2,5,6-tetrahydropyridine	185
93.	2,4-Dimethylpyridine + methylethyl ketone			
	a. 20% H_2SO_4	Pb	4,6-Dimethyl-2-(2-hydroxy-2-butyl)-1,2,5,6-tetrahydropyridine	185
94.	2,6-Dimethylpyridine + acetone			
	a. 20% H_2SO_4	Pb	2,6-Dimethyl-4-(2-hydroxy-2-propyl)piperidine	185

Table 9.1 Preparative Electrochemical Data for Nitrogen Heterocyclic Compounds

95.	2,6-Dimethylpyridine + methylethyl ketone			
	a. 20% H_2SO_4	Pb	2,6-Dimethyl-4-(2-hydroxy-2-butyl)piperidine	185
96.	1,4-Dimethylpyridinium methosulfate + acetone			
	a. 20% H_2SO_4	Pb	1,4-Dimethyl-2-(2-hydroxy-2-propyl)-1,2,5,6-tetrahydropyridine	185
97.	1-Methyl-2(1H)pyridone			
	a. H_2O, H_2SO_4	Pb	1-Methylpiperidine, 1-methyl-1,2,5,6-tetrahydropyridine	186
98.	1-Ethyl-2(1H)pyridone			
	a. H_2O, H_2SO_4	Pb	1-Ethylpiperidine, 1-ethyl-1,2,5,6-tetrahydropyridine	186
99.	1,3-Dimethyl-2(1H)pyridone			
	a. H_2O, H_2SO_4	Pb	1,3-Dimethylpiperidine, 1,3-dimethyl-1,2,5,6-tetrahydropyridine, 1,3-dimethyl-1,2,3,6-tetrahydropyridine	186
100.	1,6-Dimethyl-2(1H)pyridone			
	a. H_2O, H_2SO_4	Pb	1,2-Dimethylpiperidine, 1,2-dimethyl-1,2,3,6-tetrahydropyridine, 1,2-dimethyl-1,2,5,6-tetrahydropyridine	186
101.	2,4,6-Trimethylpyridine			
	a. 20% H_2SO_4	Pb	2,4,6-Trimethylpiperidine (three stereoisomers), 2,4,6-trimethyl-1,2,5,6-tetrahydropyridine	187
102.	1,2,4,6-Tetramethylpyridinium methosulfate			
	a. 20% H_2SO_4	Pb	cis,cis,cis-1,2,4,6-Tetramethylpiperidine (9), cis-1,2,4,6-tetramethyl-...	187

	Electrode	Product	Ref.
103. 1-Methyl-3-carbamidopyridinium chloride			
a. pH 7-8.1 phosphate buffer/0.1 M KCl, CPE at -1.2 V, $n = 1$	Hg	C-6 dimer (?)	190
b. pH 7-8.1 phosphate buffer/0.1 M KCl, CPE at -1.2 V, then -1.8 V, $n = 3$	Hg	Tetrahydro form of C-6 dimer	190
c. pH 7.8-9.2 phosphate buffer, CPE at -1.85 V, $n = 2$	Hg	1-Methyl-3-carbamido-1,4-dihydropyridine	190
d. 0.1 N NaOH, CPE at -1.7 to -1.8 V, $n = 2$	Hg	C-6 dimer (?)	188
104. Nicotinamide adenine dinucleotide (NAD, DPN)			
a. pH 7.6 buffer, CPE	Hg	Dihydro derivative, DPNH (partial activity)	192
b. pH 7.5 buffer, CPE	Pd, Pt, Ag, Ni, Pb	DPNH (partial activity)	193
c. Aqueous buffer, CPE at -1.7 V	Hg	DPNH (3-80% activity)	194
d. Aqueous buffer, CPE at -1.75 V	Hg	DPNH (partial activity)	195
e. pH 7 buffer, CPE at -2.0 V	Pt	DPNH (65-76% activity)	196
f. pH 7-9, CPE at -1.2 V	Hg	4,4'-dimer (?)	390
g. pH 7-9, CPE at -1.85 V	Hg	DPNH	390
105. Nicotinamide adenine trinucleotide (TPN)			
a. Aqueous buffer, CPE at -1.7 V	Hg	TPNH (94% activity)	194
b. pH7 buffer, CPE at -1.8 V	Pt	TPNH (95% activity)	196
c. pH 7-9 buffer, CPE at -1.3 V, $n = 1$	Hg	C-4 dimer (?)	191

Table 9.1 Preparative Electrochemical Data for Nitrogen Heterocyclic Compounds

d. pH 7-9 buffer, CPE at −1.6 to −1.7 V		Hg	C-4 dimer, TPNH	191
106. Nicotine [1-methyl-2-(3-pyridyl)-pyrrolidine]				
a. 2 N H$_2$SO$_4$, CD = 0.04 A/cm^2, n = 16	PbO$_2$		Nicotinic acid, malonic acid, methylamine, oxalic acid, CO, CO$_2$, NH$_3$, HCOOH	172
b. 2 N H$_2$SO$_4$, CD = 0.25 A/cm^2	Pt		Nicotinic acid (18)	197
c. H$_2$O, KMnO$_4$, 80°C	Ni, Fe		Nicotinic acid (77[Fe]-82[Ni])	198
107. 1,1′-Ethylenebis(3-carbamidopyridinium bromide)				
a. Aqueous solutions, pH 5-9, CPE, n = 2		Hg	Intramolecular cyclized 6,6′-coupled product (?)	388
108. 4-Styrylpyridine				
a. LiCl, NaOAc, HOAc, MeOH, H$_2$O(pH 3.7), CPE at −1.4 V		Hg	4-(2-Phenethyl)pyridine, *meso*-2,3-diphenyl-1,4-di-(4-pyridyl)butane, (±)-2,3-diphenyl-1,4-di-(4-pyridyl)butane	400
109. 4-(4′-Methylstyryl)pyridine				
a. KCl, MeOH, pH 3.0 buffer, CPE at −1.4 V		Hg	4-[2-(*p*-Tolyl)ethyl] pyridine, *meso*-2,3-di-(*p*-tolyl)-1,4-di-(4-pyridyl)butane, (±)2,3-di-(*p*-tolyl)-1,4-di-(4-pyridyl)butane	400
110. 4-(4′-Methoxystyryl)pyridine				
a. KCl, MeOH, pH 3.0 buffer, CPE at −1.4 V		Hg	4-[2-(*p*-Anisyl)ethyl] pyridine, *meso*-2,3-di-(*p*-anisyl)-1,4-di-(4-pyridyl)butane, (±)-2,3-di-(*p*-anisyl)-1,4-di-(4-pyridyl)butane	400
111. 1-Methylpyridinium chloride				
a. H$_2$O, H$_2$SO$_4$, CPE		Hg	1-Methylpiperidine (87)	398

No.	Compound / Conditions	Electrode	Products	Ref.
112.	1-Methyl-1,4-dihydronicotinamide			
	a. MeCN, TEAP, t-Butylamine, CPE at 0.30 V vs Ag/Ag$^+$, $n = 2$	Pt	1-Methyl-3-carbamidopyridinium perchlorate (100)	389
	b. MeCN, TEAP, CPE	Pt	1-Methyl-3-carbamidopyridinium perchlorate (40)	389
113.	1-(2,6-Dichlorobenzyl)-1,4-dihydronicotinamide			
	a. MeCN, TEAP, t-butylamine, CPE	Pt	1-(2,6-Dichlorobenzyl)-3-carbamidopyridinium perchlorate (100)	389
	b. MeCN, TEAP, CPE	Pt	1-(2,6-Dichlorobenzyl)-3-carbamidopyridinium perchlorate (40)	389
114.	2-Hydroxy-5-acetoxy-8-t-butyl-1-(2-pyridyl)naphthalene			
	a. MeCN, NaClO$_4$, CPE at 1.15 V	Pt	Zwitterion of intramolecularly cyclized compound (40)	405
115.	2-Hydroxy-1-(2-pyridyl)naphthalene			
	a. MeCN, NaClO$_4$, CPE at 1.15 V	Pt	Zwitterion of intramolecularly cyclized compound (40), C—O—C dimer (40) (?)	405
116.	2-Hydroxy-5-acetoxy-6-bromo-8-t-butyl-1-(2-pyridyl)naphthalene			
	a. MeCN, NaClO$_4$, CPE at 1.1 V	Pt	Zwitterion of intramolecularly cyclized compound (45)	405
117.	2,2'-Dipyridyl			
	a. H$_2$O, H$_2$SO$_4$, CPE	Pb	Hydrogenated derivatives (11-40) (yield varies with % H$_2$SO$_4$, applied potential.)	414

Table 9.1 Preparative Electrochemical Data for Nitrogen Heterocyclic Compounds

118. 2-Nitraminopyridine			
a. 1 N HCl/40% EtOH, CPE at −0.45 V, n = 5.4-5.6, 5°C and 25°C	Hg	2-Hydrazinopyridinium chloride (40), 2-amino-pyridinium chloride (45), N_2	418
b. 2 N HCl, CPE	Hg	2-Hydrazinopyridinium chloride (50-60), 2-aminopyridinium chloride (40-50), N_2	418
c. 6 N HCl, CPE	Hg	2-Hydrazinopyridinium chloride (75), 2-Amino-pyridinium chloride (25), N_2	418
d. pH 5 acetate buffer, n = 6.2	Hg	2-Hydrazinopyridinium salt (95-100)	418
e. 0.2 N KOH, CPE at −1.7 V, n = 3.95	Hg	2-Aminopyridine (major product), pyridine, N_2	418
119. 3-Nitraminopyridine			
a. 6 N HCl/40% EtOH, CPE at −0.45 V, n = 6.2	Hg	3-Hydrazinopyridinium chloride (90-100)	418
b. 1 N HCl, CPE, n = 5.8	Hg	3-Hydrazinopyridinium chloride (55), 3-Amino-pyridinium chloride (35)	418
c. 0.2 N KOH/40% EtOH, CPE at −1.7 V, n = 4.5	Hg	3-Aminopyridine	418
120. 4-Nitraminopyridine			
a. 1 N HCl/40% EtOH, CPE at −1.7 V, n = 6.1-6.2	Hg	4-Hydrazinopyridinium chloride (45), 4-amino-pyridinium chloride (25)	418
b. 6 N HCl, CPE	Hg	4-Hydrazinopyridinium chloride (100)	418
c. 0.2 N KOH/40% EtOH, CPE	Hg	4-Aminopyridine	418
121. 1,2-Dihydro-1-methyl-2-nitroiminopyridine			
a. 1 N HCl/40% EtOH, CPE at −0.45 V, n = 5.6-5.7	Hg	2-Hydrazino-1-methylpyridinium chloride (75), 2-amino-1-methylpyridinium chloride (25)	418
b. pH 12.5 phosphate buffer, CPE at −1.2 V, n = 3.9	Hg	2-Amino-1-methylpyridine	418

122. 1-Aminopyridinium iodide

a. 0.5 N HCl, CPE at -0.90 V, $n = 2$	Hg	Pyridine hydrochloride	418
b. 0.2 N KOH, CPE at -1.35 V, $n = 2$	Hg	Pyridine	418

123. Pyridine-1-nitroimide

a. 0.2 N HCl/30% EtOH, CPE at -0.75 V, $n = 2.2$	Hg	Pyridinium chloride	418
b. 0.1 N KOH, CPE at -1.10 V, $n = 1.7$	Hg	Pyridine	418

124. 2-Hydrazino-1-methylpyridinium chloride

a. 0.16 N HCl, CPE at -1.2 V, $n = 1.7$	Hg	2-Amino-1-methylpyridinium chloride	418

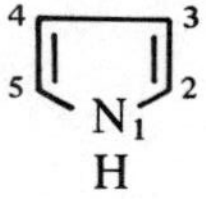

Pyrroles, Pyrrolidines

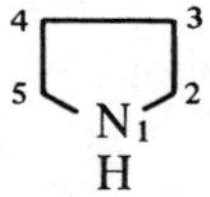

1. Pyrrole

a. 10% H_2SO_4, EtOH, 30-34°C, Ni catalyst	Pb, Cu	Pyrrolidine (15.6) (17.6% yield with a platinum black cathode)	199, 200
b. MeCN, CPE	Pt	Polymeric material	2
c. MeCN, LiClO$_4$, benzaldehyde, CPE at 1.3 V, $n = 3.5$	Pt	Tetraphenylporphin	201

2. N-Methylpyrrole

a. MeOH, KOH, CPE	Pt	N-Methyl-2,2,5,5-tetramethoxy-3-pyrroline (57)	162

3. 1,2-Dimethyl-2-pyrroline

a. 15% H_2SO_4	Pb	1,2-Dimethylpyrrolidine (45), dimer (structure undetermined)	202

343

Table 9.1 Preparative Electrochemical Data for Nitrogen Heterocyclic Compounds

4. N-Methyl-2-ethyl-2-pyrroline			
a. 15% H_2SO_4	Pb	N-Methyl-2-ethylpyrrolidine	202
5. N-Methyl-2-n-propyl-2-pyrroline			
a. 15% H_2SO_4	Pb	N-Methyl-2-n-propylpyrrolidine	202
6. N-Methyl-2-n-butyl-2-pyrroline			
a. 15% H_2SO_4	Pb	N-Methyl-2-n-butylpyrrolidine	202
7. Pyrrolidine-2-carboxylic acid			
a. 2 N H_2SO_4, CD = 2 A/dm²	PbO_2	Succinimide, pyrrolidine	203–205
8. 2-Pyrrolidone-5-carboxylic acid			
a. 2 N H_2SO_4, CD = 2 A/dm²	PbO_2	Succinimide, CO_2, succinic acid (trace)	206, 207
9. N-Acetyl-2-pyrrolidone			
a. 50% H_2SO_4, CD = 0.12 A/cm²	Pb	N-Ethyl-2-pyrrolidone, pyrrolidone	208
10. 2,5-Pyrrolidinedione (Succinimide)			
a. 50% H_2SO_4, CD = 0.12 A/cm²	Pb	2-Pyrrolidone (60), pyrrolidine (trace)	28, 208, 209
b. 50% H_2SO_4, CD = 1 A/cm², 38-40°C, 6 hr	Zn/Hg	Pyrrolidine (14), 2-pyrrolidone (58)	210
c. 50% H_2SO_4, CD = 1 A/cm², 38-40°C, 12 hr	Zn/Hg	Pyrrolidine (28), 2-pyrrolidone	210
11. N-Methylsuccinimide			
a. 50% H_2SO_4, CD = 0.03 A/cm²	Pb	N-Methyl-2-pyrrolidone (80)	211,212
b. 50% H_2SO_4, CD = 0.03 A/cm²	Zn/Hg	N-Methyl-2-pyrrolidone	212
c. 50% H_2SO_4, CD = 1 A/cm²	Zn/Hg	N-Methylpyrrolidine	212
12. N-Ethylsuccinimide			
a. 10% H_2SO_4, CD = 0.03 A/cm²	Pb	N-Ethyl-2-pyrrolidone	212
b. 50% H_2SO_4, CD = 1 A/cm²	Zn/Hg	N-Ethylpyrrolidine	212

13. *N*-Isopropylsuccinimide			
a. 50% H_2SO_4, CD = 0.12 A/cm²	Pb	*N*-Isopropyl-2-pyrrolidone (78)	208
14. *N*-Phenylsuccinimide			
a. 90% H_2SO_4, 50°C	Pb	*N*-Phenyl-2-pyrrolidone (65)	64
b. 50% H_2SO_4	Zn/Hg	*N*-Phenylpyrrolidine	213
c. 90% H_2SO_4, CD = 0.05 A/cm², 55°C	Pb	*N*-Phenyl-2-pyrrolidone	213
15. *N*-Phenyl-2-pyrrolidone			
a. 50% H_2SO_4, CD = 0.95 A/cm²	Zn/Hg	*N*-Phenylpyrrolidine (30)	213
16. *N*-(*p*-Tolyl)succinimide			
a. 95% H_2SO_4, CD = 0.12 A/cm², 40-50°C	Pb	*N*-(*p*-Tolyl)-2-pyrrolidone	208
17. *N*-(11-Lupinyl)succinimide			
a. 20% H_2SO_4	Pb	*N*-(11-Lupinyl)pyrrolidine (50)	32
18. *N*-Chlorosuccinimide			
a. MeCN, Pyridine	Ag	Succinimide	214
19. 3-Methylsuccinimide			
a. 50% H_2SO_4	Pb	3-Methylpyrrolidine (32)	28
20. 3-Phenylsuccinimide			
a. 50% H_2SO_4	Pb	3-Phenylpyrrolidine (40)	28
21. 3,4-Diphenylsuccinimide			
a. 50% H_2SO_4	Pb	3,4-Diphenylpyrrolidine (1.2)	28
22. 1-Methyl-2-ethyl-2-pyrroline-5-one			
a. 50% H_2SO_4, CD = 5 A/cm⁴	Pt	1-Methyl-2-ethyl-5-pyrrolidone (50), 1-methyl-2-ethylpyrrolidine (10)	215, 216
23. 1-Methyl-2-ethyl-2-pyrroline			
a. 50% H_2SO_4	Pt	1-Methyl-2-ethylpyrrolidine	215, 216

345

Table 9.1 Preparative Electrochemical Data for Nitrogen Heterocyclic Compounds

24.	1-Methyl-2-phenyl-2-hydroxy-5-pyrrolidone			
	a. H_2O, H_2SO_4, HOAc, CD = 2 A/cm^2	Pt	1-Methyl-2-phenylpyrrolidine	215, 216
25.	1-Methyl-2-phenyl-2-pyrroline			
	a. 50% H_2SO_4	Pt	1-Methyl-2-phenylpyrrolidine	215
26.	1-Methyl-2-benzyl-2-pyrroline-5-one			
	a. 30% H_2SO_4	Pb	1-Methyl-2-benzylpyrrolidine, 1-methyl-2-benzylpyrrolidine-5-one	215
27.	1-Methyl-2-benzyl-2-pyrroline			
	H_2O, H_2SO_4	Pb	1-Methyl-2-benzylpyrrolidine	215
28.	3-Pyrroline-2,5-dione (Maleimide)			
	a. H_2O, H_2SO_4, Ni catalyst	Pb, Cu	Succinimide (100)	217
	b. H_2O, H_2SO_4	Zn/Hg	Succinic acid, NH_3	217
29.	3-Pyrroline			
	a. H_2O, H_2SO_4	Pb	Pyrrolidine	217
30.	2,3,4,5-Tetraphenylpyrrole			
	a. MeCN, $LiClO_4$, CPE	Pt	2-Hydroxy-2,3,4,5-tetraphenylpyrrolenine (60)	404
	b. MeCN, $LiClO_4$, Na_2CO_3, CPE	Pt	2-Hydroxy-2,3,4,5-tetraphenylpyrrolenine (10), 2,2'-bis (2,3,4,5-tetraphenylpyrrolenine)-peroxide (20), lactam isomer of the alcohol (10)	404
	c. MeOH, $LiClO_4$, CPE	Pt	2-Hydroxy-2,3,4,5-tetraphenylpyrrolenine (1.5), 2-methoxy-2,3,4,5-tetraphenylpyrrolenine (80)	404
	d. EtOH, $LiClO_4$, CPE	Pt	2-Ethoxy-2,3,4,5-tetraphenylpyrrolenine (35)	404
	e. MeOH, H_2O (10^{-2} M), $LiClO_4$, CPE	Pt	2-Hydroxy derivative (28), 2-methoxy derivative (40)	404
	f. MeOH, H_2O (3.5 M), $LiClO_4$, CPE	Pt	2-Hydroxy derivative (51), 2-methoxy derivative (40)	404

	Electrode	Product(s) (yield %)	Page
g. MeNO$_2$, TEAP, CPE, 0°C or less, $n = 2$	Pt	2-Hydroxy-2,3,4,5-tetraphenylpyrrole (60)	407
h. MeNO$_2$, TEAP, CPE, 40°C or greater, $n = 1$	Pt	2,3,4,6-Tetraphenylpyridine (30), dibenzoylstilbene (5), lactam derivative (10)	407
31. 2,3,4,5-Tetraanisylpyrrole			
a. MeNO$_2$, TEAP, CPE, 15°C, $n = 2$	Pt	2-Hydroxy-2,3,4,5,-tetraanisyl-1-2-pyrrolenine (90)	415
b. MeNO$_2$, TEAP, CPE	Pt	2-Hydroxy-2,3,4,5-tetraanisyl-2-pyrrolenine (40), 2,3,4,6-tetraanisylpyridine (10)	415
c. MeNO$_2$, TEAP, Na$_2$CO$_3$, CPE, 90°C	Pt	2,3,4,4-Tetraanisyl-2-pyrroline-5-one (10), 2,3,4,6-tetraanisylpyridine (45)	415
d. MeCN, TEAP, CPE, 15°C	Pt	2,3,4,4-Tetraanisyl-2-pyrroline-5-one (20), 2-hydroxy-2,3,4,5-tetraanisyl-2-pyrrolenine (65)	415
e. MeOH, LiClO$_4$, CPE, $n = 2$	Pt	2-Hydroxy-2,3,4,5-tetraanisyl-2-pyrrolenine perchlorate salt	415
32. 2,3,4,5-Tetra-p-tolylpyrrole			
a. MeNO$_2$, TEAP, Na$_2$CO$_3$, CPE, 16°C	Pt	2-Hydroxy-2,3,4,5-tetra-p-tolyl-2-pyrrolenine (40), 2,3,4,4-tetra-p-tolyl-2-pyrroline-5-one (10), 2,3,4,6-tetra-p-tolylpyridine (5), 1,2,3,4-tetra-p-tolyl-2-butene-1,4-dione(5)	416
b. MeNO$_2$, TEAP, Na$_2$CO$_3$, 90°C, CPE	Pt	2,3,4,4-Tetra-p-tolyl-2-pyrroline-5-one (15), 2,3,4,6-tetra-p-tolylpyridine (40), 1,2,3,4-tetra-p-tolyl-2-butene-1,4-dione (30)	416
c. MeOH, LiClO$_4$, CPE	Pt	2-Hydroxy-2,3,4,5-tetra-p-tolyl-2-pyrrolenine perchlorate salt	416
33. 2-(N-Pyrrolidino)propane immonium salt			

Table 9.1 Preparative Electrochemical Data for Nitrogen Heterocyclic Compounds

a. MeCN, TEAP, CPE	Pt	1,2-Bis(N-pyrrolidino)-1,1,2,2-tetramethylethane	417
34. Phenyl-1-N-pyrrolidinomethane immonium salt			
a. MeCN, TEAP, CPE	Pt	1,2-Bis(N-pyrrolidino)-1,2-diphenylethane	417

Quinazolines

1. Quinazoline			
a. 0.2 N KOH, $n = 1.23$	Hg	C-4 dimer(?)	218
b. Aqueous solution, low conc.	Hg	3,4-Dihydroquinazoline	218
2. 3,4-Dihydroquinazoline			
a. pH 8.5 borate buffer, CPE at -1.5 V, $n = 2.2$	Hg	1,2,3,4-Tetrahydroquinazoline	218
3. 4-Aminoquinazoline			
a. Citrate buffer, pH 2.9, CPE at -1.20 V, $n = 4$	Hg	3,4-Dihydroquinazoline	409, 419
b. 4 N HCl, CPE at -0.80 V, $n = 4$	Hg	3,4-Dihydroquinazoline	409, 419
c. LiCl, LiOH, CPE at -1.80 V, $n = 6$	Hg	1,2,3,4-Tetrahydroquinazoline	409, 419
4. 4-Diethylaminoquinazoline			
a. Citrate buffer, pH 2.9, CPE, $n = 4$	Hg	3,4-Dihydroquinazoline	409, 419
5. 4-Mercaptoquinazoline			
a. Citrate buffer, pH 2.9, CPE, $n = 4$	Hg	3,4-Dihydroquinazoline	409, 419
6. 4-Methylthioquinazoline			
a. Citrate buffer, pH 2.9, CPE, $n = 4$	Hg	3,4-Dihydroquinazoline	409, 419

Compound and conditions	Electrode	Product	References
7. 4-Methoxyquinazoline			
a. MeOH/2 N HCl, CPE at −0.80 V, n = 2	Hg	3,4-Dihydro-4-methoxyquinazoline (?)	409, 419
b. Citrate buffer, pH 2.9, CPE	Hg	3,4-Dihydro-4-methoxyquinazoline(?)	409, 419
c. MeOH/LiOMe, CPE at −1.70 V	Hg	Quinazolinium salt, 3,4-dihydroquinazolinium salt, 4,4′-biquinazoline	409, 419
8. 4-Chloroquinazoline			
a. pH 6.5 Phosphate buffer/30% EtOH, CPE at −0.85 V, 0-5°C, n = 2	Hg	Quinazoline (95-100)	218
9. 3-Phenyl-3,4-dihydroquinazoline-4-one			
a. H_2O, EtOH, NaOAc, 50-60°C	Pt	4-Hydroxy-3-phenyl-1,2,3,4,-tetrahydroquinazoline (17)	219
10. 4-Hydroxy-3-phenyl-1,2,3,4-tetrahydroquinazoline			
a. H_2O, EtOH, NaOAc	Cu/Pt Black	3-Phenyl-1,2,3,4-tetrahydroquinazoline (30)	219
b. H_2O, Na_2CO_3, 50-60°C	Pt	3-Phenyl-1,2,3,4-tetrahydroquinazoline	219
11. 2-Methyl-3-(o-tolyl)-3,4-dihydro-4-quinazolinone (Methaqualone)			
a. pH 3.5 acetate buffer, CPE, n = 2	Hg	2-Methyl-3-(o-tolyl)-1,2,3,4-tetrahydro-4-quinazolinone, C-1 dimer (?)	220, 222
12. 4-Quinazolinone			
a. Aqueous solution, CPE	Hg	1,2-Dihydro-4-quinazolinone, C-1 dimer	222, 387
13. 3-Phenyl-4-quinazolinone			
a. Aqueous solution, CPE	Hg	1,2-Dihydro-3-phenyl-4-quinazolinone	222, 387
14. 2,3-Diphenyl-4-quinazolinone			
a. Aqueous solution, CPE	Hg	1,2-Dihydro-2,3-diphenyl-4-quinazolinone	222, 387

Table 9.1 Preparative Electrochemical Data for Nitrogen Heterocyclic Compounds

15. 4,4'-Biquinazoline

a. 70% Aqueous DMF, NaClO$_4$, acetate buffer, pH 5, CPE at -0.80 V, $n = 2$	Hg	Dihydro derivative (reoxidizes quantitatively to 4,4'-biquinazoline)	409

Quinolines and Isoquinolines

1. Quinoline

a. H$_2$O, H$_2$SO$_4$	Pt, PbO$_2$	Quinolinic acid (77)	148, 149, 223
b. H$_2$O, H$_2$SO$_4$, (NH$_4$)$_2$ SO$_4$, V$_2$O$_5$, CD = 0.05 A/cm^2	Pb	Quinolinic acid	225
c. 1 N H$_2$SO$_4$, CD = 0.05 A/cm^2	PbO$_2$	Quinolinic acid, 3-carboxy-2-pyridylacrylic acid, 8-hydroxyquinoline, HCOOH, HNO$_3$, CO, CO$_2$, NH$_3$	226
d. H$_2$O, H$_2$SO$_4$, CD = 0.15-0.50 A/cm^2	Pb	1,2,3,4-Tetrahydroquinoline (70), polymers	225
e. 10% H$_2$SO$_4$	Pb	1,2,3,4-Tetrahydroquinoline, trimolecular dihydroquinoline (structure undetermined)	58
f. H$_2$O, H$_2$SO$_4$	Pb	Dihydroquinoline (structure undetermined)	227, 228
g. 9% KOH	Hg	1,4-Dihydroquinoline	229, 250, 380
h. 10% KOH, CD = 0.1 A/cm^2	Ni	1,2,3,4-Tetrahydroquinoline, polymers	230
i. 10% H$_2$SO$_4$, CD = 0.1 A/cm^2	Pb	1,2,3,4-Tetrahydroquinoline (8), polymers	230
j. H$_2$O, H$_2$SO$_4$	Pb, Tl, In, Cd, Al, Sn, Pt, C	Tetrahydroquinoline, dihydroquinoline, polymers	250, 251

2. 1-Methylquinolinium methosulfate

Substrate / Conditions	Catalyst	Product (yield)	Ref.
a. H_2O, NaOH, $K_3[Fe(CN)_6]$ catalyst	Fe	1-Methyl-1,2-dihydroquinoline-2-one (95-100)	134
3. 1-Methylquinolinium iodide			
a. H_2O, Na_2CO_3	Pt	1,1'-Dimethyl-1,1',4,4'-tetrahydro-4,4'-bi-quinolinyl (20)	138
4. 1-Ethylquinolinium iodide			
a. H_2O, Na_2CO_3	Pt	1,1'Diethyl-1,1',4,4'-tetrahydro-4,4'-bi-quinolinyl (20)	138
5. 1,6-Dimethylquinolinium methosulfate			
a. H_2O, NaOH, $K_3[Fe(CN)_6]$ catalyst	Fe	1,6-Dimethyl-1,2-dihydroquinoline-2-one	3
6. 1,8-Dimethylquinolinium methosulfate			
a. H_2O, NaOH, $K_3[Fe(CN)_6]$ catalyst	Fe	1,8-Dimethyl-1,2-dihydroquinoline-2-one	3
7. 1-Methyl-6-methoxyquinolinium methosulfate			
a. H_2O, NaOH, $K_3[Fe(CN)_6]$ catalyst	Fe	1-Methyl-6-methoxy-1,2-dihydroquinoline-2-one (high yield)	136
8. 1-Methyl-8-methoxyquinolinium methosulfate			
a. H_2O, NaOH, $K_3[Fe(CN)_6]$ catalyst	Fe	1-Methyl-8-methoxy-1,2-dihydroquinoline-2-one (high yield)	136
9. 1-Methylbenzo[f]quinolinium methosulfate			
a. H_2O, NaOH, $K_3[Fe(CN)_6]$ catalyst	Fe	1-Methyl-1,2-dihydrobenzo[f]quinoline-2-one	134
10. 2-Methylquinoline (Quinaldine)			
a. 10% H_2SO_4	Pb	2-Methyl-1,2,3,4-tetrahydroquinoline, polymer	58
11. 2-Hydroxyquinoline			
a. 10% KOH, CD = 0.1 A/cm²	Ni	2-Hydroxy-3,4-dihydroquinoline (49)	230

Table 9.1 Preparative Electrochemical Data for Nitrogen Heterocyclic Compounds

Compound / conditions	Electrode	Product (yield)	Ref.
12. 3-Hydroxyquinoline			
a. 10% H_2SO_4, CD = 0.1 A/cm²	Pb	3-Hydroxy-1,2,3,4-tetrahydroquinoline (40)	230
13. 5-Hydroxyquinoline			
a. 10% H_2SO_4, CD = 0.1 A/cm²	Pb	5-Hydroxy-1,2,3,4-tetrahydroquinoline (92)	230
b. 10% KOH, CD = 0.1 A/cm²	Ni	5-Hydroxy-1,2,3,4-tetrahydroquinoline (80)	230
14. 6-Hydroxyquinoline			
a. 10% H_2SO_4, CD = 0.1 A/cm²	Pb	6-Hydroxy-1,2,3,4-tetrahydroquinoline (46)	230
b. 10% KOH, CD = 0.1 A/cm²	Ni	6-Hydroxy-1,2,3,4-tetrahydroquinoline (74)	230
15. 7-Hydroxyquinoline			
a. 10% H_2SO_4, CD = 0.1 A/cm²	Pb	7-Hydroxy-1,2,3,4-tetrahydroquinoline (46)	230
b. 10% KOH, CD = 0.1 A/cm²	Ni	7-Hydroxy-1,2,3,4-tetrahydroquinoline (76)	230
16. 8-Hydroxyquinoline			
a. 10% H_2SO_4, CD = 0.1 A/cm²	Pb	8-Hydroxy-1,2,3,4-tetrahydroquinoline (79)	230
b. 10% KOH, CD = 0.1 A/cm²	Ni	8-Hydroxy-1,2,3,4-tetrahydroquinoline (95)	230
c. H_2O, NaOH, KI, CD = 0.005 A/cm²	Pt	8-Hydroxy-7-iodoquinoline, 8-hydroxy-5,7-diiodoquinoline (relative yields vary with CD, temperature, electrolysis time)	231
d. H_2O, NaOH, KI, CD = 0.001 A/cm²	Pt	8-hydroxy-7-iodoquinoline (68)	231
e. H_2O, EtOH, NH_4SCN, 0°C, CD = 0.02-0.03 A/cm²	Pt	8-Hydroxy-5-thiocyanoquinoline	232
17. 3-Aminoquinoline			
a. H_2O, H_2SO_4, 70°C	Pt	Polymeric material	223
18. 2-Phenylquinoline			
a. H_2O, H_2SO_4, 75-80°C	Pt	Polymeric material	223
19. 3-Fluoroquinoline			
a. 80% H_2SO_4, 85-90°C	Pt	5-Fluoroquinolinic acid (26)	223

20. 2-Chloroquinoline a. H_2O, H_2SO_4, 65°C	PbO_2	Polymeric material	223
21. 3-Chloroquinoline a. H_2O, H_2SO_4, 80-90°C	Pt	5-Chloroquinolinic acid (54)	223
22. 4-Chloroquinoline a. H_2O, H_2SO_4, 60°C	PbO_2	Polymeric material	223
23. 3-Bromoquinoline a. H_2O, H_2SO_4, 55-65°C	PbO_2	5-Bromoquinolinic acid (73)	223
24. 3-Iodoquinoline a. H_2O, H_2SO_4, 80°C	Pt	5-Hydroxyquinolinic acid hemihydrate (10)	223
25. 2-Phenylquinoline-4-carboxylic acid a. Aqueous solution	Hg	2-Phenyl-1,2,3,4-tetrahydroquinoline-4-carboxylic acid	233
26. Quinolinesulfonic acid a. $6 N$ H_2SO_4, 65-75°C, CD = 0.015 A/cm^2	PbO_2	Quinolinic acid	234
27. 5-Nitroquinoline a. H_2O, H_2SO_4	Pb	5-Amino-8-hydroxyquinoline	235, 236
28. 8-Nitroquinoline a. H_2O, H_2SO_4	Pb	5-Hydroxy-8-nitroquinoline	235, 236
29. 5-Nitro-6-methylquinoline a. H_2O, H_2SO_4 b. H_2O, H_2SO_4	Pb	5-Nitro-6-methyl-8-hydroxyquinoline 5-Amino-6-methylquinoline (90)	235, 236 237
30. 5-Nitro-8-methylquinoline a. H_2O, H_2SO_4	Pb	8,8'-Dimethyl-5-azoxyquinoline (30)	237

Table 9.1 Preparative Electrochemical Data for Nitrogen Heterocyclic Compounds

Compound / Conditions	Electrode	Product	Ref.
b. H_2O, H_2SO_4	Pb	8,8'-Dimethyl-5-azoquinoline (60)	237
c. H_2O, H_2SO_4	Pb	5-Amino-8-methylquinoline (85)	237
31. 5-Nitro-6,8-dimethylquinoline			
a. H_2O, H_2SO_4	Pb	5-Amino-6,8-dimethylquinoline (90)	237
32. 1,2,3,4-Tetrahydroquinoline			
a. 1 N H_2SO_4, 15°C, CD = 0.01 A/cm^2	PbO_2	Gentisic acid, propionic acid, maleic acid, oxalic acid, HCOOH, CO, CO_2, NH_3	238
33. N-Nitroso-1,2,3,4-tetrahydroquinoline			
a. H_2O, H_2SO_4, EtOH, CD = 0.14 A/cm^2	Pt	1,2,3,4-Tetrahydroquinoline	61
34. 1,2,3,4-Tetrahydroisoquinoline-1,3-dione (Homophthalimidine)			
a. 50% H_2SO_4	Pb	1,2,3,4-Tetrahydroisoquinoline (15)	28
35. 5-Nitroisoquinoline			
a. Conc. H_2SO_4, 70–80°C	Pb	5-Amino-8-hydroxyisoquinoline	239
36. 3-Methyl-5-nitroisoquinoline			
a. Conc. H_2SO_4, 60–70°C	Pt	3-Methyl-5-amino-8-hydroxyisoquinoline disulfate (65)	240
37. 1-Methyl-2-cyanoquinolinium perchlorate			
a. pH 5 buffer, CPE at −0.7 V, $n = 1.1$	Hg	4,4'-Bis(1-methyl-2-cyano-1,4-dihydroquinol-inyl)	241
38. 1-Methyl-3-cyanoquinolinium perchlorate			
a. pH 5 buffer, CPE at −0.80 V, $n = 1.0$	Hg	4,4'-Bis(1-methyl-3-cyano-1,4-dihydroquinol-inyl) (59)	241
b. DMF, CPE at −0.80 V, $n = 0.91$	Hg	4,4'-Bis(1-methyl-3-cyano-1,4-dihydroquinol-inyl) (64)	241

39. 1-Methyl-4-cyanoquinolinium perchlorate			
a. pH 7 buffer, EtOH, O_2, CPE at -0.70 V, $n = 1.0$	Hg	1-Methyl-4-cyano-2-quinolone (24)	241
b. pH 7 buffer, EtOH, O_2, CPE at -1.50 V	Hg	1-Methyl-4-cyano-2-quinolone (12)	241
40. 1,1-Dimethyl-1,2,3,4-tetrahydroquinolinium iodide			
a. MeOH, Me_4NCl, CPE	Hg	Dimethyl-(γ-phenylpropyl)amine (88)	398
41. 1,1-Dimethyl-homotetrahydroquinolinium iodide			
a. MeOH, Me_4NCl, CPE, 45°C	Hg	Dimethyl-(δ-phenyl-n-butyl)amine (90)	398

Quinoxalines

1. Quinoxaline			
a. 5% NaOH/5% $KMnO_4$, 40°C, CD = 0.15 A/cm²	Cu	Pyrazine-2,3-dicarboxylic acid (91.7)	224
b. Aqueous alkaline sol'n	Hg	1,4-Dihydroquinoxaline, dimer	243
c. pH 13.5/5% MeOH, CPE at -1.3 V, $n = 2$	Hg	1,4-Dihydroquinoxaline (20)	114, 244
d. 0.2 M HCl, CPE (first wave)	Hg	1,2-Dihydroquinoxaline	396
e. 0.2 M HCl, CPE (second wave)	Hg	Acetaldehyde, o-phenylenediamine	396
2. 2-Methylquinoxaline			
a. pH 7.4/80% MeOH, CPE at -0.90 V, $n = 2.2$	Hg	3,4-Dihydro-2-methylquinoxaline (15) (unstable)	244, 245

Table 9.1 Preparative Electrochemical Data for Nitrogen Heterocyclic Compounds

b. pH 13.7/80% MeOH, CPE at −1.15 V, $n = 2.1$	Hg	3,4-Dihydro-2-methylquinoxaline (15)	244, 245
c. pH 9.2/80% EtOH, CPE at −1.35 V, $n = 4.5$	Hg	1,2,3,4-Tetrahydro-2-methylquinoxaline (5)	244
3. 2,3-Dimethylquinoxaline			
a. pH 7.4/80% MeOH, CPE at −0.80 V, $n = 2.6$	Hg	1,2-Dihydro-2,3-dimethylquinoxaline (70)	244, 245
b. pH 13.7/80% MeOH, CPE at −1.50 V	Hg	1,2-Dihydro-2,3-dimethylquinoxaline (78)	244, 245
c. pH 9.2/40% EtOH, CPE at −1.70 V, $n = 4.9$	Hg	cis-1,2,3,4-Tetrahydro-2,3-dimethylquinoxaline (45)	244
d. 0.2 M HCl, CPE	Hg	1,2-Dihydro-2,3-dimethylquinoxaline	396
e. pH 8, CPE at −1.4 V	Hg	1,2-Dihydro-2,3-dimethylquinoxaline	396
4. 1,2-Dihydro-2,3-dimethylquinoxaline			
a. pH 10, CPE at −2.2 V	Hg	1,2,3,4-Tetrahydro-2,3-dimethylquinoxaline	396
5. 2-Phenylquinoxaline			
a. pH 7-10, CPE at −1.0 V	Hg	1,4-Dihydro-2-phenylquinoxaline (rearranges to 1,2-dihydro-2-phenylquinoxaline)	246
b. pH 7.4/80% MeOH, CPE at −0.80 V, $n = 2.3$	Hg	1,2-Dihydro-2-phenylquinoxaline (75)	242, 244 245
c. pH 13.4/80% MeOH, CPE at −1.30 V, $n = 2.3$	Hg	1,2-Dihydro-2-phenylquinoxaline (50)	244, 245
d. pH 11.9/80% DMF, CPE at −1.30 V, 0°C	Hg	1,4-Dihydro-2-phenylquinoxaline (10)	242, 244
e. pH 7.5/40% EtOH, CPE at −1.60 V	Hg	1,2,3,4-Tetrahydro-2-phenylquinoxaline	244
6. 2,3-Diphenylquinoxaline			
a. pH 7.4/80% MeOH, CPE at −1.00 V, $n = 2$	Hg	1,4-Dihydro-2,3-diphenylquinoxaline (50) (70% yield at pH 9.2)	244, 245

Substrate / Conditions	Electrode	Product	Ref.
b. pH 13.7/80% MeOH, CPE at −1.50 V	Hg	cis-1,2,3,4-Tetrahydro-2,3-diphenylquinoxaline (50)	244, 245
7. 1,2-Dihydro-2,3-diphenylquinoxaline			
a. pH 13.7/50% MeOH, CPE at −1.50 V	Hg	cis-1,2,3,4-Tetrahydro-2,3-diphenylquinoxaline (25)	244, 245
8. 2-Phenyl-3-methylquinoxaline			
a. pH 7.7/80% MeOH, CPE at −0.90 V, $n = 2.6$	Hg	1,2-Dihydro-2-phenyl-3-methylquinoxaline (30)	244
b. pH 7.5/80% MeOH, CPE at −1.35 V	Hg	1,2-Dihydro-2-phenyl-3-methylquinoxaline (60)	244
c. pH 14/80% MeOH, CPE at −1.35 V	Hg	1,2-Dihydro-2-phenyl-3-methylquinoxaline (20)	244, 245
9. 3,4-Dihydro-2-phenyl-3-methylquinoxaline			
a. pH 13.8/80% MeOH, CPE at −1.80 V	Hg	cis-1,2,3,4-Tetrahydro-2-phenyl-3-methyl-quinoxaline (47)	224, 245
10. Quinoxalone			
a. Aqueous solution, CPE	Hg	3,4-Dihydroquinoxalone	425
11. 3-Methylquinoxalone			
a. Aqueous solution, CPE	Hg	3,4-Dihydro-3-methylquinoxalone, C-4 dimer	247
12. 1,3-Dimethylquinoxalone			
a. Aqueous solution, CPE	Hg	3,4-Dihydro-1,3-dimethylquinoxalone, C-4 dimer	247
13. 1,3-Bis (p-methoxybenzyl)quinoxalone			
a. Aqueous solution, CPE	Hg	4,4'-Dimer	247
14. 3-Phenyl-2(1H) quinoxalinone			
a. pH 4.5 acetate buffer/EtOH, CPE at −1.3 V	Hg	3,4-Dihydro-3-phenyl-2(1H)quinoxalinone	248, 249
15. 1-Methyl-3-phenyl-2(1H)-quinoxalinone			
a. pH 4.5 acetate buffer/EtOH, CPE at −1.3 V	Hg	3,4-Dihydro-1-methyl-3-phenyl-2(1H)quin-oxalinone	248, 249

Table 9.1 Preparative Electrochemical Data for Nitrogen Heterocyclic Compounds

16. 3-Aryl-2(1H)quinoxalinones			
a. Aqueous solution, CPE	Hg	3,4-Dihydro-3-aryl-2(1H)quinoxalinones	249
17. 1-Alkyl-3-aryl-2(1H)quinoxalinones			
a. Aqueous solution, CPE	Hg	3,4-Dihydro-1-alkyl-3-aryl-2(1H)quinoxalinones	249
18. 1,3-Diaryl-2(1H)quinoxalinones			
a. Aqueous solution, CPE	Hg	3,4-Dihydro-1,3-diaryl-2(1H)quinoxalinones, C-4 dimers	249
Tetra- and Triazines			
1. 1,4-Dihydro-3-phenylbenzo-1,2,4-triazine			
a. pH 4 buffer, CPE at -1.1 V	Hg	2-Phenylbenzimidazole	253
2. 1,2,3-Benzotriazole (Benzotriazole)			
a. 4 N HCl, CPE at -1.0 V, 0-5°C, n = 4	Hg	o-Aminophenylhydrazine (72.5)	254
3. 1-Methylbenzotriazole			
a. 4 NHCl, CPE at -1.0 V, 0-5°C, n = 4	Hg	1-(2'-Aminophenyl)-1-methylhydrazine (?)	254
4. 2-Methylbenzotriazole			
a. 4 N HCl, CPE at -1.0 V, 0-5°C, n = 4	Hg	1-(2'-Aminophenyl)-2-methylhydrazine	254
b. 0.2 N KOH, CPE at -1.80 V, 0-5°C, n = 2	Hg	1,3-Dihydro-2-methylbenzotriazole (?)	254
5. 1,3-Dimethylbenzotriazolium iodide			
a. 4 N HCl, CPE at -1.0V, 0-5°C, n = 4	Hg	1-(2'-Methylaminophenyl)-1-methylhydrazine (?)	254
6. 1-Methoxybenzotriazole			
a. 1 N HCl, CPE at -0.75 V, n = 2	Hg	Benzotriazole	254
b. Alkaline solution, CPE at -1.75 V	Hg	Benzotriazole	254

7. 1-Acetylbenzotriazole a. 0.5 N HCl/20% EtOH, CPE at -0.95 V, 0°C, $n = 4$	Hg	2-Acetamidophenylhydrazine	254
8. Triphenylformazan a. MeCN, TEAP, CPE	Pt	Triphenyltetrazolium perchlorate	2
9. Triphenyltetrazolium salts a. Acid solution, CPE	Hg	Phenylbenzamidrazone, aniline, triphenylformazan	2
10. Pophyrin derivatives a. MeOH, benzene, NH$_4$Cl, boric acid b. MeOD, benzene, LiCl	Hg Hg	Corresponding dihydro compounds Corresponding dideutero compounds	255-258 255-258
11. Cyanuric fluoride, (CNF)$_3$ a. H$_2$O, NaF, 13°C, CD $= 0.02$ A/cm^2	Monel	CF$_4$(48), NF$_3$(47), CF$_3$NF$_2$(5), (CF$_3$)$_2$NF (trace) (relative yields)	277
12. Thiocyanuric acid, (CNSH)$_3$ a. H$_2$O, NaF, 13°C, CD $= 0.02$ A/cm^2	Monel	CF$_4$(24), NF$_3$(30), CF$_3$NF$_2$(7), (CF$_3$)$_2$NF(6), SF$_6$(33) (relative yields)	277
13. 1,3,5-Triamino-s-triazine a. H$_2$O, NaF, 13°C, CD $= 0.02$ A/cm^2	Monel	CF$_4$(26), NF$_3$(64), CF$_3$NF$_2$(5), (CF$_3$)$_2$NF(4), CF$_2$(NF$_2$)$_2$(1) (relative yields)	277
14. 5,6-Diphenyl-as-triazine-3-one a. pH 1.0, CPE at -0.35 V, $n = 2$ b. pH 6.3, CPE at -0.95 V, $n = 2$ c. pH 11.0, CPE at -1.10 V, $n = 2$	Hg Hg Hg	4,5-Dihydro-5,6-diphenyl-as-triazine-3-one (67) 1,4-Dihydro-5,6-diphenyl-as-triazine-3-one (75) 4,5-Dihydro-5,6-diphenyl-as-triazine-3-one	401 401 401
15. 5,6-Diphenyl-2-methyl-as-triazine-3-one a. pH 1.0, CPE at -0.35 V, $n = 2$	Hg	1,4-Dihydro-5,6-diphenyl-2-methyl-as-triazine-3-one (slowly rearranges to the 4,5-dihydro derivative)	401

Table 9.1 Preparative Electrochemical Data for Nitrogen Heterocyclic Compounds

b. pH 11.0, CPE at −1.05 V, n = 2.1	Hg	4,5-Dihydro-5,6-diphenyl-2-methyl-as-triazine-3-one (70)	401
16. 5,6-Diphenyl-as-triazine-3-thione			
a. pH 7.5, CPE at −0.90 V, n = 1.8	Hg	1,4-Dihydro-5,6-diphenyl-as-triazine-3-thione (70)	401
b. pH 11.0, CPE at −1.10 V, n = 1.9	Hg	4,5-Dihydro-5,6-diphenyl-as-triazine-3-thione (70)	401
17. 4,5-Dihydro-5,6-diphenyl-as-triazine-3-one			
a. pH 0.93, CPE at −1.00 V	Hg	4,5-Diphenylimidazole-2-one, *trans*-1,4,5,6-tetrahydro-5,6-diphenyl-as-triazine-3-one	401
b. pH 6, CPE at −1.50 V	Hg	*cis*-1,4,5,6-Tetrahydro-5,6-diphenyl-as-triazine-3-one	401
18. 4,5-Dihydro-5,6-diphenyl-2-methyl-as-triazine-3-one			
a. acid solution, CPE at −1.00 V	Hg	4,5-Diphenyl-3-methylimidazole-2-one (50)	401
19. 4,5-Dihydro-5,6-diphenyl-as-triazine-3-thione			
a. pH 10.67, CPE at −2.00 V	Hg	*cis*-1,4,5,6-Tetrahydro-5,6-diphenyl-as-triazine-3-thione (60)	401
20. 3,5,6-Triphenyl-as-triazine			
a. pH 3.60, CPE at −0.70 V, 10°C	Hg	1,2-Dihydro-3,5,6-triphenyl-as-triazine (23), 4,5-dihydro-3,5,6-triphenyl-as-triazine (63)	401
b. 0.1 N NaOH/MeCN (pH 13.47), CPE at −1.30 V	Hg	4,5-Dihydro-3,5,6-triphenyl-as-triazine (80)	401
c. pH 3.60, CPE at −1.30 V, n = 4	Hg	2,4,5-Triphenylimidazole (80), *cis*-1,4,5,6-tetrahydro-3,5,6-triphenyl-as-triazine (20)	401

360

Conditions	Electrode	Product	Ref.
d. pH 6.20, CPE at -1.40 V, $n = 4$	Hg	2,4,5-Triphenylimidazole (55)	401
21. 4,5-Dihydro-3,5,6-triphenyl-*as*-triazine			
a. pH 3.60, CPE at -1.20 V, $n = 2.2$	Hg	*cis*-1,4,5,6-Tetrahydro-3,5,6-triphenyl-*as*-triazine (45)	401

Miscellaneous N-*Heterocyclics*

Conditions	Electrode	Product	Ref.
1. 3,3-Pentamethylenediaziridine			
a. 1 N NaOH, CPE, $n = 2$	Hg	3,3-Pentamethylenediazirine (81)	260
2. 3,3-Pentamethylenediazirine			
a. 0.2 N KOH, CPE at -1.50 V, $n = 2$	Hg	3,3-Pentamethylenediaziridine (79)	260
b. pH 3/25% EtOH, CPE at -1.30 V, $n = 4.2$	Hg	Cyclohexanone, NH_3	260
3. Phenazine			
a. MeOH, CPE	Pt	Dihydrophenazine	261
b. DMSO, DMF, MeCN, $HClO_4$ added, CPE, $n = 2$	Pt	Dihydrophenazine	262
4. 5,10-Dihydro-5,10-dimethylphenazine			
a. pH 1.5/acetone, CPE at 1.1 V	Pt	5,10-Dihydro-5-methylphenazine	263
5. 1-Azabicyclo [2.2.2]-octane-2-carboxylic acid			
a. MeOH, NaOMe	Pt	2-Methoxy-1-azabicyclo[2.2.2]-octane (43)	264
6. Cyclobutane-1,2-dicarboxylimide			
a. 50% H_2SO_4, 0-5°C	Pb	2-Aminomethylcyclobutane-1-carboxylic acid γ-lactam (15)	265

Table 9.1 Preparative Electrochemical Data for Nitrogen Heterocyclic Compounds

Substrate / Conditions	Metal	Product	Ref.
7. Cyclopentane-1,2-dicarboxylimide a. 50% H_2SO_4, 0-5°C	Pb	2-Aminomethylcyclopentane-1-carboxylic acid γ-lactam (25)	265
8. Cyclohexane-1,2-dicarboxylimide a. 65% H_2SO_4, 0-5°C	Pb	Isogranatanin (17)	265
9. 3,3-Dimethylcyclopropane-1,2-dicarboxylimide (Caronimide) a. 50% H_2SO_4, 0-5°C	Pb	4,4-Dimethylpiperidine-2-one (5)	265
10. Tropinone a. H_2O, NH_3, $(NH_4)_2SO_4$	Pb	Tropine	266
11. 2-Carbethoxytropinone a. H_2O, H_2SO_4	Pb, Hg	2-Carbethoxytropine	267, 268
12. *N*-Methylgranatonin a. H_2O, H_2SO_4, CD = 0.12 A/cm^2	Pb	*N*-Methylgranatanin (69), two stereoisomers of *N*-methylgranatoline	269, 270
13. Cyclooctanonethioisoxime a. 55% H_2SO_4, CD = 0.067 A/cm^2	Pb	Cyclooctamethyleneimine (85)	271
14. Cyclohexadecanonethioisoxime a. H_2O, H_2SO_4, EtOH, CD = 0.067 A/cm^2, 50°C	Pb	Cyclohexadecamethyleneimine	271
15. Dihydroquinacridone a. H_2O, NaOH, MeOH, 55°C, CD = 0.12 A/cm^2	Ni	Quinacridone (100)	272
16. *m*-Nitro-γ-stilbazole a. Conc. H_2SO_4, 2.5-3 A	Pb	1-Amino-4-hydroxy-3-γ-stilbazole	273, 274

No.	Compound / Conditions	Electrode	Product	Ref.
17.	3-Nitrophthalic acid anil			
	a. H_2O, H_2SO_4	Pb	3,3'-Azophthalic acid (91)	275
18.	4-Nitrophthalic acid anil			
	a. H_2O, H_2SO_4	Pb	4,4'-Azophthalic acid (91)	275
19.	3-Methyl-4-phenylazo-5-phenylisoxazole			
	a. Conditions unspecified	?	3-Methyl-4-(β-phenylhydrazino)-5-phenyl-isoxazole	276
20.	2,3-Dihydro-5,6-diphenyl-1,4-diazepinium perchlorate			
	a. 0.1 M $HClO_4$/EtOH	Cu	Chalcone, benzylacetophenone (40% total yield)	406
21.	2,3-Dihydro-1,4-diphenyl-1,4-diazepinium perchlorate			
	a. 0.1 M $HClO_4$/EtOH	Cu	Dianilinoethane (15), perhydro derivative (40)	406
22.	7-Chloro-1-methyl-5-phenyl-3H-1,4-benzodiazepine-2-one (Valium)			
	a. 70% DMF/1 N LiCl (aqueous), CPE at -1.6 V, $n = 2$	Hg	4,5-Dihydro derivative	7
23.	7-Chloro-2-methylamino-5-phenyl-3H-1,4-benzodiazepine 4-oxide (Librium)			
	a. pH 4.6 buffer, no CPE	DME	7-Chloro-2-methylamino-5-phenyl-3H-1,4-benzodiazepine and 7-chloro-2-methylamino-5-phenyl-3H-4,5-dihydro-1,4-benzodiazepine	8
24.	7-Chloro-2-methylamino-5-phenyl-3H-4,5-dihydro-1,4-benzodiazepine			
	a. Aqueous solution, CPE	Hg	6-Chloro-2-methyl-4-phenyl-3,4-dihydroquinazoline	384

Table 9.1 Preparative Electrochemical Data for Nitrogen Heterocyclic Compounds

Alkaloids, Natural Product Syntheses

1. Peganine, $C_{11}H_{14}N_2$ (Vasicine)			
a. 20% H_2SO_4, CD = 0.055 A/cm²	Pb	N-(o-aminobenzyl)pyrrolidine (85)	279
b. 25% H_2SO_4, CD = 0.1 A/cm²	Pb	N-(o-aminobenzyl)pyrrolidine	280, 281
2. 9-Pegen-1-one, $C_{11}H_{10}N_2O$			
a. 25% H_2SO_4, CD = 0.056 A/cm²	Pb	N-(o-aminobenzyl)pyrrolidine (55)	282
3. 9-Pegen-8-one, $C_{11}H_{10}N_2O$			
a. 30% H_2SO_4, CD = 0.1 A/cm²	Pb	N-(o-aminobenzyl)pyrrolidine (65)	283
4. α-Matrinidin, $C_{12}H_{20}N_2$			
a. 50% H_2SO_4, CD = 3-4 A/dm²	Pb	Dihydro-α-matrinidin, $C_{12}H_{22}N_2$ (72)	284
5. Anagyrine, $C_{15}H_{20}N_2O$ (Monolupine, Rhombinin)			
a. 50% H_2SO_4	Pb	Hexahydrodeoxyanagyrine (d-spartein), $C_{15}H_{26}N_2$	285
6. Thermopsin, $C_{15}H_{20}N_2O$			
a. 50% H_2SO_4, 40-45°C, 10 A	Pb	Base, $C_{15}H_{28}N_2$, ring opening? (15)	286
7. Aphyllidine, $C_{15}H_{22}N_2O$			
a. 50% H_2SO_4, 10 A, 8 V, 40-45°C	Pb	Pachycarpine (d-spartein), $C_{15}H_{26}N_2$ (30)	287
8. Sophocarpine, $C_{15}H_{24}N_2O$			
a. 50% H_2SO_4, 10 A, 8 V, 40-45°C	Pb	1-Rotatory base, $C_{15}H_{26}N_2$ (30)	288
9. Sophoridine, $C_{15}H_{26}N_2O$			
a. 50% H_2SO_4	Pb	Base, $C_{15}H_{26}N_2$ (31)	288
10. Quinine, $C_{20}H_{24}N_2O_2$			
a. 50% H_2SO_4, n = 4	Pb	Tetrahydroquinine, $C_{20}H_{28}N_2O_2$	289

No. / Starting material	Cathode	Product	Reference
11. Cinchonine, $C_{19}H_{22}N_2O$ a. 50% H_2SO_4, CD = 0.023 A/cm^2, 30–35°C	Pb	Dihydrodeoxycinchonine, $C_{19}H_{24}N_2O$ (40)—two stereoisomers	289, 290
12. Cinchonidine, $C_{19}H_{22}N_2O$ (stereoisomer with cinchonine) a. 50% H_2SO_4.	Pb	Tetrahydrocinchonidine, $C_{19}H_{26}N_2O$	290
13. Cinchotine, $C_{19}H_{24}N_2O$ a. 50% H_2SO_4	Pb	Dihydrodeoxycinchotine, $C_{19}H_{26}N_2$	290
14. Papaverine, $C_{20}H_{21}NO_4$ a. 35% H_2SO_4, CD = 0.13 A/cm^2, 45°C	Pb	Tetrahydropapaverine, $C_{20}H_{25}NO_4$	291, 292
15. N-Methylpapaverinium methosulfate a. H_2O, H_2SO_4	Pb	Racemic laudanosine	293
16. Narcotine, $C_{22}H_{23}NO_7$ a. 30% H_2SO_4, 8–10 A, 35°C	Pb	Tetrahydronarcotine, $C_{22}H_{27}NO_7$ (20–25)	294
17. Cotarnine, $C_{12}H_{15}NO_4$ a. 20% H_2SO_4, 3.5 A, 5.3 V	Pb	Hydrocotarnine, $C_{12}H_{15}NO_3$	295, 296
18. Hydrastinine, $C_{11}H_{13}NO_3$ a. 20% H_2SO_4, 3.5 A, 5.3 V	Pb	Hydrohydrastinine, $C_{11}H_{13}NO_2$	295, 296
19. 1-Veratryl-6,7-methylenedioxydihydroisoquinoline chloromethylate, $C_{20}H_{22}NO_4Cl$ a. H_2O, H_2SO_4, 5 A	Pb	1-Veratrylhydrohydrastinine, $C_{20}H_{23}NO_4$	297
20. Berberine(sulfate), $C_{20}H_{19}NO_5$ a. 30% H_2SO_4/EtOH	Pb	Tetrahydroberberine, $C_{20}H_{21}NO_4$ (44.1)	298
21. 4-Methyldihydroberberine, $C_{21}H_{21}NO_4$			

365

Table 9.1 Preparative Electrochemical Data for Nitrogen Heterocyclic Compounds

a. 30% H_2SO_4/EtOH	Pb	Two stereoisomers of 4-methyltetrahydroberberine, $C_{21}H_{23}NO_4$ (66)	298
22. Dihydroberberine methiodide, $C_{20}H_{19}NO_4$-CH_3I a. H_2O, H_2SO_4	Pb	Tetrahydroberberine methiodide, $C_{20}H_{21}NO_4$-CH_3I	298
23. Des-*N*-methyl-4-methyldihydroberberine, $C_{22}H_{23}NO_4$ a. 30% H_2SO_4/EtOH	Pb	Tetrahydro-des-*N*-methyl-4-methyldihydroberberine, $C_{22}H_{27}NO_4$, isolated as the hydrocloride	298
24. *A*-Des-*N*-methyltetrahydroberberine, $C_{21}H_{23}NO_4$ a. 30% H_2SO_4/EtOH	Pb	Hydro-*A*-des-*N*-methyltetrahydroberberine, $C_{21}H_{25}NO_4$	298
25. Benzyldihydroberberine a. H_2O, H_2SO_4, EtOH, 15 A, 50-60°C	Pb	Two stereoisomers of benzyltetrahydroberberine	299
26. Des-*N*-methylbenzyldihydroberberine a. H_2O, H_2SO_4, EtOH	Pb	*β*-Hydrodes-*N*-methylbenzyldihydroberberine	299
27. Methyldihydroberberine methiodide a. H_2O, H_2SO_4, EtOH	Pb	Two stereoisomers of hydrodes-*N*-methyl-dihydroberberine	300
28. Des-*N*-methylethyldihydroberberine a. H_2O, H_2SO_4, EtOH	Pb	Two stereoisomers of hydrodes-*N*-methylethyl-dihydroberberine	301
29. Isopropyldihydroberberine hydriodide a. H_2O, H_2SO_4, EtOH	Pb	Two stereoisomers of isopropyltetrahydrober-berine (60 and 20%)	302

	Substrate / Conditions	Electrode	Product	Ref.
30.	β-Des-N-methylisopropyltetrahydroberberine			
	a. H_2O, H_2SO_4, EtOH	Pb	Hydro-β-des-N-methylisopropyltetrahydroberberine	302
31.	Des-N-methylisopropyldihydroberberine			
	a. H_2O, H_2SO_4, EtOH	Pb	Two stereoisomers of hydrodes-N-methylisopropyldihydroberberine	302
32.	Des-N-methylisobutyldihydroberberine			
	a. H_2O, H_2SO_4, EtOH	Pb	Two stereoisomers of hydrodes-N-methylisobutyldihydroberberine	303
33.	Isobutyldihydroberberine			
	a. H_2O, H_2SO_4, 10 A	Pb	Two stereoisomers of isobutyltetrahydroberberine	303
34.	Isoamyldihydroberberine			
	a. H_2O, H_2SO_4, 2.5 A, 40-50°C	Pb	Two stereoisomers of isoamyltetrahydroberberine	304
35.	Des-N-methylisoamyldihydroberberine			
	a. H_2O, H_2SO_4	Pb	Two stereoisomers of hydrodes-N-methylisoamyldihydroberberine	304
36.	Phenyldihydroberberine			
	a. H_2O, H_2SO_4, EtOH, 15 A, 24 V	Pb	Two stereoisomers of phenyltetrahydroberberine	305
37.	Des-N-methylphenyltetrahydroberberine			
	a. H_2O, H_2SO_4. EtOH	Pb	Hydrodes-N-methylphenyltetrahydroberberine	305
38.	1-Benzyl-4-phenylazodihydroberberine, $C_{33}H_{29}N_3O_4$			

Table 9.1 Preparative Electrochemical Data for Nitrogen Heterocyclic Compounds

a. H_2O, H_2SO_4, ÉtOH, 10 A, 30-40°C	Pb	Two stereoisomers of 1-benzyltetrahydrober-berine, $C_{27}H_{27}NO_4$ (36 and 14%)	306
39. Morphine hydrochloride, $C_{17}H_{19}NO_3 \cdot HCl$ a. 10% HCl, CD = 4.0 A/dm²	Pt	Dihydromorphine hydrochloride, $C_{17}H_{21}NO_3 \cdot$-HCl (89)	307
40. Codeine phosphate, $C_{18}H_{21}NO_3 \cdot H_2PO_4$ a. 10% H_3PO_4, CD = 4.0 A/dm²	Pt	Dihydrocodeine phosphate, $C_{18}H_{23}NO_3 \cdot H_2PO_4$ (97)	307
41. Pseudocodeine, $C_{18}H_{21}NO_3$ a. H_2O, H_2SO_4	Pb	Dihydropseudocodeine, $C_{18}H_{23}NO_3$	308
42. Desoxycodeine-A, $C_{18}H_{21}NO_2$ a. 20% H_2SO_4, CD = 0.15 A/cm²	Pb	Dihydrodesoxycodeine-A, $C_{18}H_{23}NO_3 \cdot \frac{1}{2}H_2O$	309
43. Desoxycodeine-C, $C_{18}H_{21}NO_2$ a. 20% H_2SO_4, CD = 0.15 A/cm², 10-20°C	Pb	Dihydrodesoxycodeine-B, $C_{18}H_{23}NO_3 \cdot \frac{1}{2}H_2O$ (80)	310
44. α-Chlorocodide, $C_{18}H_{20}NO_2Cl$ a. 20% H_2SO_4, CD = 0.15 A/cm², 10-20°C	Pb	Desoxycodeine-B, $C_{18}H_{21}NO_2$ (100)	311, 312
45. Dihydrodesoxycodeine-D, $C_{18}N_{23}NO_2$ a. 20% H_2SO_4, CD = 0.1 A/cm²	Pb	Tetrahydrodesoxycodeine, $C_{18}H_{25}NO_2$ (100)	312
46. Chlorodihydrocodide, $C_{18}H_{22}NO_2Cl$ a. 20% H_2SO_4, CD = 0.15 A/cm², 10-20°C	Pb	Dihydrodesoxycodeine-C, $C_{18}H_{23}NO_2$ (65)	309, 310
47. δ-Ethylthiocodide, $C_{20}H_{25}NO_2S$ a. 20% H_2SO_4, CD = 0.1 A/cm², 12-18°C	Pb	Dihydro-δ-ethylthiocodide-A, $C_{20}H_{27}NO_2S$, some dihydrodesoxycodeine	312
48. Bromocodeinone a. 20% H_2SO_4, EtOH, CD = 7-8 A/dm²	Pb	Dihydrodesoxycodeine-E (55)	312, 314

No.	Substrate and conditions	Electrode	Product	Ref.
49.	Ozodihydrocodeine, $C_{18}H_{23}NO_5$			
	a. 30% H_2SO_4, 10 A	Pb	5-Desoxydihydromorphine carboxylic acid, $C_{17}H_{23}NO_5$ (70)	315
50.	Strychnine, $C_{21}H_{22}N_2O_2$			
	a. 60% H_2SO_4, CD = 12 A/dm^2, 18°C	Pb	Tetrahydrostrychnine, $C_{21}H_{26}N_2O_2$ (20), strychnidine, $C_{21}H_{24}N_2O$ (70)	289, 316-318
51.	Methoxymethyldihydrostrychnidine, $C_{23}H_{30}N_2O_2$			
	a. 20% H_2SO_4, 5 A	Pb	Methoxymethyltetrahydrostrychnidine, $C_{23}H_{32}N_2O_2$	318
52.	Brucine, $C_{23}H_{26}N_2O_4$			
	a. 50% H_2SO_4, CD = 12 A/cm^2, 15°C	Pb	Tetrahydrobrucine, $C_{23}H_{30}N_2O_4$	289
	b. Conc. H_2SO_4	PbO$_2$	Dehydrobis (apomethyl)brucine, $C_{21}H_{20}N_2O_4$	197
	c. H_2O, H_2SO_4, 15-20°C	Pb	Brucidine (50), tetrahydrobrucine (30)	319
53.	Brucidine, $C_{23}H_{28}N_2O_3$			
	a. 50% H_2SO_4	Pb	Tetrahydrobrucine	319, 320
54.	N-Methylbrucine			
	a. 60% H_2SO_4, 5 A, 18°C	Pb	Brucidine methohydrogen carbonate	320
55.	N-Methoxymethyldihydrobrucidine, $C_{25}H_{34}N_2O_4$			
	a. H_2O, H_2SO_4	Pb	Methoxymethyltetrahydrobrucidine, $C_{25}H_{36}N_2O_4$	320
56.	Vomicine, $C_{22}H_{24}N_2O_4$			
	a. H_2O, H_2SO_4	Pb	Vomicidine, $C_{22}H_{26}N_2O_3$	321

Table 9.1 Preparative Electrochemical Data for Nitrogen Heterocyclic Compounds

57. Vomicine methosulfate			
a. 40% H_2SO_4, CD = 0.025 A/cm²	Pb	Methylvomicidinium iodide (after treatment with KI)	322
58. Isodesoxyvomicine, $C_{22}H_{24}N_2O_3$			
a. H_2O, H_2SO_4, 5-6 A	Pb	Desoxyvomicidine, $C_{22}H_{26}N_2O_2$	323
59. Dihydrodesoxyvomicine, $C_{22}H_{26}N_2O_3$			
a. 60% H_2SO_4, 7 A	Pb	Dihydrodesoxyvomicidine, $C_{22}H_{28}N_2O_2$ (60)	324
60. Isodihydrovomicidine			
a. 60% H_2SO_4, 7.5 A	Pb	Isodihydrodesoxyvomicidine	324
61. Alkaloid base, $C_{22}H_{33}NO_4$ (from Stemona Sessilifolia)			
a. 20% H_2SO_4, 17 A, 20 V	Pb	Bisdesoxybase, $C_{22}H_{37}NO_2$ (35)	325
62. 6,7-Dimethoxyisoquinoline quaternary base + β-Phenylethyl chloride			
a. Conditions unspecified	?	N-(β-Phenylethyl)-6,7-dimethoxy-1,2,3,4-tetrahydroisoquinoline	326
63. 7-Hydroxy-6-methoxy-N-methyl-1,2,3,4-tetrahydroisoquinoline (Corypalline)			
a. 0.1 M NaHCO₃, EtOH, CPE at 0.3 V	Pt	C–C and C–O–C dimers (see text)	327-329
b. 0.1 N HCl, 5-10°C, CPE at 0.72 V	Pt	C–C dimer, C–O–C dimer (trace) (see text) (same products obtained with the sodium salt of corypalline in basic media)	330
64. 1,2-Dimethyl-7-hydroxy-6-methoxy-1,2,3,4-tetrahydroisoquinoline			
a. 0.1 M NaHCO₃, EtOH, CPE at 0.3 V	Pt	C–C and C–O–C dimers, C–O–C trimer (see text)	329

	Electrode	Products	Ref.
b. MeCN, TEAP, NaOMe, CPE at 0.04 V	C	C–C dimer, enantiomeric pair (77), C–O–C dimer (7)	331
65. 1-Ethyl-7-hydroxy-6-methoxy-2-methyl-1,2,3,4-tetrahydroisoquinoline			
a. 0.1 M $Na_2B_4O_7$/30% MeCN, CPE at 0.4 V	Pt	C–O–C dimer, enantiomeric pair (see text)	329
66. 7-Hydroxy-6-methoxy-1,2,3,4-tetrahydroisoquinoline			
a. 0.1 N HCl, CPE at 0.78 V	Pt	C–8 dimer (48)	330
b. 0.1 M $Na_2B_4O_7$/MeCN, CPE at 0.2 V	C	C–8 dimer (37.5)	330
c. 0.1 M $Na_2B_4O_7$, CPE at 0.3-0.4 V	Pt	C–8 dimer, isolated as acetate derivative (39)	330
67. 2-Acetyl-7-hydroxy-6-methoxy-1,2,3,4-tetrahydroisoquinoline			
a. 0.1 M $Na_2B_4O_7$/MeCN, CPE at 0.1 V	C	C–8 dimer (68)	330
68. 2-Methyl-6-hydroxy-7-methoxy-1,2,3,4-tetrahydroisoquinoline			
a. 0.1 N HCl, CPE at 0.72 V	Pt	C–8 dimer, isolated at the hydrochloride salt (15)	330
b. 0.1 M $Na_2B_4O_7$/MeCN, CPE at −0.03 to 0.06 V, 1- and 2-compartment cells	C	C–O–C dimer between a 6-hydroxy group and a C–8 ring position (1-comp, cell:62.1; 2-comp. cell:80)	330
69. 2-Acetyl-6-hydroxy-7-methoxy-1,2,3,4-tetrahydroisoquinoline			
a. 0.1 M $Na_2B_4O_7$/MeCN, CPE at 0.04-0.1 V	C	C–8 dimer (84)	330

Table 9.1 Preparative Electrochemical Data for Nitrogen Heterocyclic Compounds

	Electrode	Product	Ref.
70. 8-Hydroxy-7-methoxy-2-methyl-1,2,3,4-tetrahydroisoquinoline			
a. 0.1 N HCl, CPE at 0.72 V	Pt	C–8 dimer, isolated as the acetate derivative (24)	330
b. 0.1 M Na$_2$B$_4$O$_7$/MeCN, 1-comp. cell, CPE at 0.16 V	C	C–O–C dimer between an 8-hydroxy group and a C–5 ring position (51.7)	330
c. 0.1 M Na$_2$B$_4$O$_7$/MeCN, 2-comp. cell, CPE at 0.0 V	C	C–O–C dimer between an 8-hydroxy group and a C–5 ring position (71)	330
71. 2-Acetyl-8-hydroxy-7-methoxy-1,2,3,4-tetrahydroisoquinoline			
a. 0.1 M Na$_2$B$_4$O$_7$, CPE at 0.9 V	Pt	C–5 dimer (40)	330
b. 0.05 M Na$_2$B$_4$O$_7$/MeCN, CPE at 0.2 V	C	C–O–C dimer between an 8-hydroxy group and a C–5 ring position (56)	330
c. MeCN, TEAP, CPE at 0.16 to 0.2 V	C	C–5 dimer (41)	330
72. 6,7-Dimethoxy-1-(p-hydroxybenzyl)-2-methyl-1,2,3,4-tetrahydroisoquinoline (Armepavine)			
a. MeCN, TEAP, H$_2$O (10%), NaOMe, CPE at 0.1 V	C	6,7-Dimethoxy-2-methyl-3,4-dihydroiso-quinolinium salt (86)	332
73. 6,7-Dimethoxy-1-(p-hydroxybenzyl)-1,2,3,4-tetrahydroisoquinoline (N-Norarmepavine)			
a. MeCN, TEAP, H$_2$O (10%), NaOMe, CPE at 0.1 V	C	6,7-Dimethoxy-3,4-dihydroisoquinolinium salt (32)	332
74. 6,7-Dimethoxy-1-(p-hydroxybenzyl)-2-carbethoxy-1,2,3,4-tetrahydroisoquinoline			

Conditions	Electrode	Products	Ref.
a. MeCN, TEAP, H₂O (10%), NaOMe, CPE at 0.3 V	C	C–C dimer coupled through the meta positions of the benzylic rings (45)	332
75. 6,7-Dimethoxy-1-(3,4-dimethoxybenzyl)-2-methyl-1,2,3,4-tetrahydroisoquinoline (Laudanosine)			
a. MeCN, NMe$_4$BF$_4$, Na$_2$CO$_3$, 0°C, CPE at 1.1 V vs Ag/AgNO$_3$	Pt	O-Methylflavinantine (52)	333

Heterocyclic N-Oxides

Conditions	Electrode	Products	Ref.
1. 4-Methyl-1,2,3-benzotriazine-N-oxide			
a. pH 4 buffer, CPE, $n = 4$	Hg	3-Methylindazole, hydroxylamine	243
2. 1-Methylbenzotriazole-3-oxide			
a. Acetate buffer/30% EtOH, CPE at -1.15 V, $n = 2$	Hg	1-Methylbenzotriazole	254
3. 2-(4'Hydroxyphenyl)benzotriazole-1-oxide			
a. 0.1 N KOH, CPE at -1.2 V, $n = 2$	Hg	2-(4'-Hydroxypheny)benzotriazole	254
4. 3,8-Bis(dimethylamino)phenazone-N-oxide			
a. H$_2$O, NaOAc, EtOH	Ni	3,8-Bis(dimethylamino)phenazone (60)	334
5. Adenine 1-N-oxide			
a. 0.05 N HClO$_4$	Pt	Adenine	335
b. pH 1.4 chloride buffer, CPE at -0.9 V	Hg	Adenine	336
6. 2-Azaadenine 1-N-oxide			
a. 0.05 N HClO$_4$	Pt	2-Azaadenine	335
7. 4-Methylcinnoline-N-1-oxide			
a. Acid solution, CPE, $n = 2$	Hg	4-Methylcinnoline	13

Table 9.1 Preparative Electrochemical Data for Nitrogen Heterocyclic Compounds

8. 4-Methylcinnoline-N-2-oxide			
a. acid solution, CPE, $n = 2$	Hg	4-Methylcinnoline	13
9. Benzo[c]cinnoline-N-oxide			
a. Acid solution, CPE, $n = 2$	Hg	Benzo[c]cinnoline	13
10. 3-Acetylpyridine-N-oxide			
a. 0.1 N H$_2$SO$_4$, CPE at -1.0 V	Hg	3-(α-hydroxyethyl)pyridine-N-oxide	337
11. 4-Acetylpyridine-N-oxide			
a. pH 13, CPE at -1.1 V	Hg	2,3-Bis(4-pyridyl)-2,3-butanediol,4-(α-hydroxyethyl) pyridine	337
12. 3-Formylpyridine-N-oxide			
a. pH 5.1, CPE at -1.05 V	Hg	3-Hydroxymethylpyridine-N-oxide	338
b. pH 5.1, CPE at -1.45 V	Hg	3-Hydroxymethylpyridine	338
13. 4-Formylpyridine-N-oxide			
a. pH 1.0, CPE at -0.60 V	Hg	4-Hydroxymethylpyridine chlorohydrate	338
14. 3-Chloro-2,5-dimethylpyrazine-1-oxide			
a. Aqueous solution, CPE	Hg	1,4-Dihydro-2,5-dimethylpyrazine	339
15. Benzofuroxan-N-oxide			
a. Aqueous buffer/30% EtOH, CPE	Hg	o-Phenylenediamine	348, 351
16. Pyridine-N-oxide			
a. MeOH, Me$_4$NCl, CPE	Hg	pyridine (81)	398
17. 2-Methylpyridine-N-oxide			
a. MeOH, Me$_4$NCl, CPE	Hg	2-Methylpyridine (96)	398
18. 4-Methylpyridine-N-oxide			
a. MeOH, Me$_4$NCl, CPE	Hg	4-Methylpyridine (80)	398

Substrate / Conditions	Product	Electrode	Ref.
19. Quinoline-N-oxide			
a. MeOH, Me$_4$NCl, CPE	Quinoline (78)	Hg	398
20. 4,5,9,10-Tetraazapyrene-N-oxides (crude mixture)			
a. Water/DMF, pH 6.8, CPE at −0.90 V	Tetraazapyrene (72)	Hg	397
b. Water/acetone, pH 6.1, CPE at −0.45 V	Tetraazapyrene mono-N-oxide (65)	Hg	397
21. 2-Phenyl-3-aryliminoindolenine-N-oxide (several derivatives)			
a. DMF, KClO$_4$, benzoic acid, CPE over first wave	1-Hydroxy-2-phenyl-3-aryliminoindoles	Hg	424
b. DMF, KClO$_4$, benzoic acid, CPE over second wave	2-Phenyl-3-arylaminoindoles	Hg	424

N–O *and* N–S *Heterocyclics*

Substrate / Conditions	Product	Electrode	Ref.
1. N-Nitrosomorpholine			
a. H$_2$O, NaOH, KCl, CPE at −1.3 V	Morpholine	Hg	340
2. Gallocyanine (Phenoxazine dye)			
a. H$_2$O, H$_2$SO$_4$, 70-80°C	Leucogallocyanine	Pb	341
3. Gallamine Blue (Phenoxazine dye)			
a. H$_2$O, HCl, 70-80°C	Leucogallamine blue	Pt	341
4. 2-t-Butyl-3-phenyloxaziridine			
a. 0.3 N HCl/30% EtOH, CPE at −0.3 V, 0 to 5°C, $n = 2$	Benzylidene-t-butylamine (not isolated)	Hg	342
b. 0.3 N HCl/30% EtOH, CPE at −1.1 V, $n = 3.8$	Benzyl alcohol, t-butylbenzylamine	Hg	342

Table 9.1 Preparative Electrochemical Data for Nitrogen Heterocyclic Compounds

5. 3-Methylanthranil			
a. Acid solution, CPE	Hg	o-Aminoacetophenone	343
b. Alkaline solution, CPE, $n = 3$	Hg	Several dimers	343
6. 4-(p-Anisyl)-2,3-benzoxazin-1-one			
a. 0.5 N HCl/60% EtOH, CPE, $n = 2$	Hg	2-(4'-Methoxybenzoyl)benzoic acid ketimine	344
7. Substituted oxazoles			
a. Aqueous solution	Hg	Ring-opened derivatives (?)	345
8. Substituted Oxadiazoles			
a. Aqueous solution	Hg	Ring-opened derivatives (?)	345
9. 2,5-Diphenyloxazole			
a. DMF/TPAP, CPE	Hg	1,2-Dihydro-2,5-diphenyloxazole (?)	346
10. 2-(1-Naphthyl)-5-phenyloxazole			
a. DMF/TPAP, CPE	Hg	1,2-Dihydro-2-(1-naphthyl)-5-phenyloxazole (?)	346
11. 2,5-Diphenyloxadiazole			
a. DMF/TPAP, CPE, $n = 6$	Hg	Ring-opened products (?)	347
12. 2-(m-Tolyl)-5-(1-naphthyl)oxadiazole			
a. DMF/TPAP, CPE, $n = 6$	Hg	Ring-opened products (?)	347
13. Benzofurazan			
a. Aqueous solutions, CPE	Hg	o-Phenylenediamine	348, 380
14. Benzofuroxan			
a. pH 2.1 buffer, CPE at -0.4 V	Hg	o-Benzoquinone dioxime, I (14), 2,3-diamino-phenazine, II (22), o-phenylenediamine, III (42)	349
b. pH 6.5 buffer, CPE at -0.4 V	Hg	I (33), II (28), III (12)	349
c. pH 9.9 buffer, CPE at -0.4 V	Hg	I (28), II (7), III (61)	349

15. 4,4'-Dithiodimorpholine			
a. pH 5.6 buffer, CPE	Hg	Morpholine, S	243
16. Sydnone			
a. Aqueous acid/EtOH, CPE, $n = 4$	Hg	Corresponding hydrazine derivative	350
17. 3-Methyl-, 3-benzyl-, and 3-phenyl-sydnones			
a. Aqueous acid/EtOH, CPE, $n = 4$	Hg	Corresponding hydrazine derivatives	350
18. Phenolphthalein oxime			
a. Acid solution, CPE, $n = 2$	Hg	2-(4'-hydroxyphenyl)-3-(4''-hydroxyphenyl)-2,3-dihydro-2-isobenzazol-1-one	352
19. Phenolphthalein oxime dimethyl ether			
a. Aqueous solution, CPE	Hg	2-(4'-anisyl)-3-(4''-anisyl)-2,3-dihydro-2-isobenzazol-1-one	352
20. Phenolphthalein oxime trimethyl ether			
a. pH 3-5, CPE, $n = 2$	Hg	2-(4'-anisyl)-3-(4''-anisyl)-2,3-dihydro-2-isobenzazol-1-one	352
21. 2-Mercapto-4-hydroxythiazole (Rhodanine)			
a. Aqueous solution, CPE at -1.8 V, $n = 4$	Hg	4-Hydroxythiazole	353
22. 2-Thiazolecarboxamide			
a. 2 N HCl, CPE at -0.75 V, $10°C$, $n = 2.1$	Hg	2-Thiazolecarbaldehyde (57)	18, 354
b. pH 4.3 acetate buffer/20% EtOH, CPE at -1.05 V, $n = 4$	Hg	2-Hydroxymethylthiazole (59)	354
c. 0.5 N HCl, CPE at -0.75 V, $n = 2.2$	Hg	2-Thiazolecarbaldehyde (86) (93% yield at 0°C)	354
23. 2-Thiazolecarbohydrazide			
a. 0.8 N HCl, CPE at -0.55 V, $n = 3$	Hg	2-Thiazolecarboxamide (45), 2-thiazolecarbaldehyde (50)	354

Table 9.1 Preparative Electrochemical Data for Nitrogen Heterocyclic Compounds

b. 0.8 N HCl, CPE at -0.65 V, $n = 4.9$	Hg	2-Thiazolecarbaldehyde (21)	354
24. 2-Thiazolecarboxylic acid			
a. 0.8 N HCl, CPE at -0.85 V, $n = 3.9$	Hg	2-Thiazolecarbaldehyde (17) (25% yield at 0°C)	354
25. 2-Thiazolecarboxanilide			
a. 0.8 N HCl/60% EtOH, CPE at -0.7 V, $n = 4.5$	Hg	2-Anilinomethylthiazole (50), 2-thiazolecarbaldehyde (3), 2-hydroxymethylthiazole, aniline	354
26. 2-Thiazole-N-methylcarboxanilide			
a. 1 N HCl/20% EtOH, CPE at -0.65 V, $n = 4.0$	Hg	2-Hydroxymethylthiazole, N-methylaniline	354
27. 2-Anilinomethylthiazole			
a. 0.8 N HCl, CPE at -1.0 V	Hg	2-Methylthiazole (50), aniline	354
28. 2-Thiazole-N-butylcarboxamide			
a. 0.8 N HCl, CPE at -0.85 V, $n = 2.7$	Hg	2-Thiazolecarbaldehyde (58)	354
29. 2-Thiazole-N-benzylcarboxamide			
a. 0.8 N HCl/40% EtOH, CPE at -0.75 V, $n = 3.7$	Hg	2-Thiazolecarbaldehyde (47) (60% yield at 0°C)	354
30. Ethyl 2-thiazolecarboxylate			
a. 0.8 N HCl/40% EtOH, CPE at -0.85 V, $n = 3.0$	Hg	2-Thiazolecarbaldehyde (59) (72% yield at 0°C)	354
31. 4-Methyl-2-thiazolecarboxamide			
a. 2 N HCl/40% EtOH, CPE at -0.80 V vs Ag/AgCl, $n = 3.0$	Hg	4-Methyl-2-thiazolecarbaldehyde (62) (determined polarographically; 31% isolated, 52.5% isolated at 10-15°C)	18
32. Phenothiazine			
a. MeCN, LiClO$_4$, CPE at 0.90 V vs Ag/Ag$^+$	Pt	Phenothiazinium perchlorate (radical salt)	355-360

33.	10-[3-(4-β-Hydroxyethyl-1-piperazinyl)-propyl]-2-chlorophenothiazine		
	a. H_2O, H_2SO_4, EtOH	Au	Corresponding sulfoxide (80)	361
34.	Chlorpromazine		
	a. $1 N$ H_2SO_4, CPE at 0.95 V, $n = 2$	Pt	Chlorpromazine-5-oxide (100)	362-364

All of the following six compounds were electrolyzed in $1 N$ H_2SO_4 at a Pt anode, yielding an n value of two and the corresponding sulfoxide in 90-100% yields.

35.	Promethazine, CPE at 0.75 V			364
36.	Promazine, CPE at 1.05 V			364
37.	Trifluoperazine, CPE at 1.00 V			364
38.	Prochlorperazine, CPE at 1.00 V			364
39.	Thioridazine, CPE at 0.75 V			364
40.	Trifluopromazine, CPE at 1.05 V			364
41.	2-Amino-7-methoxyphenothiazine-9,9-dioxide			
	a. Aqueous solution, CPE	Hg	Stable diimine form	365
42.	N–S Heterocyclic sulfonamides			
	a. pH 9.3 borate buffer, CPE	Hg	SO_2, NH_3, corresponding N–S heterocyclic nucleus	366
43.	Saccharin			
	a. H_2O, H_2SO_4, EtOH, 50°C	Pb	Benzylsultam	367
	b. H_2O, NaOH (3%)	Pb	Benzamide (65), benzaldehyde	367
44.	6-Trifluoromethyl-7-sulfamyl-3,4-dihydro-1,2,4-benzothiadiazine-1,1-dioxide			

Table 9.1 Preparative Electrochemical Data for Nitrogen Heterocyclic Compounds

a. H_2O, $KHCO_3$/30% MeOH, CPE at -1.70 V, $n = 6.3$	Hg	6-Methyl-7-sulfamyl-3,4-dihydro-1,2,4-benzo-thiadiazine-1,1-dioxide (55)	368
45. 1,2-Naphthofuroxan-4-sulfonic acid			
a. pH 5.4-10.1/30% EtOH, CPE	Hg	1,2-Naphthoquinone dioxime-4-sulfonic acid	351
46. 3-Methylbenzothiazolium iodide			
a. Aqueous solution, CPE, $n = 1$-2	Hg	2,3-Dihydro-3-methylbenzothiazole, dimer	243
47. 2-(2-Methyl-3-hydroxy-5-hydroxymethyl-4-pyridyl)thiazolidine-4-carboxylic acid			
a. pH 5.8 phosphate buffer, CPE at -0.95 V	Hg	*N*-Pyridoxylcysteine (ring-opened product)	369
48. 3-Amino-1,2-benzisoxazole			
a. 1 *N* H_2SO_4/EtOH, CPE at -0.9 V	Hg	*o*-Aminobenzamide (100)	393
49. 3-Amino-6-methoxy-1,2-benzisoxazole			
a. 1 *N* H_2SO_4/EtOH, CPE at -0.9 V	Hg	2-Amino-4-methoxybenzamide (100)	393
50. 3-Amino-6-chloro-1,2-benzisoxazole			
a. 1 *M* NH_4OAc/EtOH, CPE at -1.7 V	Hg	2-Amino-4-chlorobenzamide (100)	393

REFERENCES

1.. L. S. Marcoux, Ph.D. Thesis, Kanşas University, 1967.
2. H. Lund, *Elektrodereaktioner i organisk polarografi og voltammetri.*, Aarhus Stifsbogtrykkerie, Aarhus, 1961.
3. K. Neudlinger and M. Chur, *J. Prakt. Chem.* [2], **89**, 466 (1912) [*Chem. Abstr.*, **8**, 2695, (1914)].
4. J. J. Lingane, C. G. Swain, and M. Fields, *J. Amer. Chem. Soc.*, **65**, 1348 (1943).
5. K. D. Legg and D. M. Hercules, *J. Amer. Chem. Soc.*, **91**, 1902 (1969).
6. K. D. Legg, D. W. Shive, and D. M. Hercules, *Anal. Chem.*, **44**, 1650 (1972).
7. H. Oelschläger, J. Volke, and H. Hoffmann, *Coll. Czech. Chem. Commun.*, **31**, 1264 (1966).
8. H. Oelschläger, J. Volke, H. Hoffmann, and E. Kurek, *Arch. Pharm.*, **300**, 250 (1967) [*Chem. Abstr.*, **67**, 5730 (1967)].
9. W. H. Perkin, Jr., and S. G. Plant, *J. Chem. Soc.*, **1924**, 1503.
10. J. Gurney and S. G. Plant, *J. Chem. Soc.*, **1927**, 1314.
11. J. F. Ambrose and R. F. Nelson, *J. Electrochem. Soc.*, **115**, 1159 (1968).
12. J. F. Ambrose, L. L. Carpenter, and R. F. Nelson, *J. Electrochem. Soc.*, **121**, 000 (1974).
13. H. Lund, *Acta Chem. Scand.*, **21**, 2525 (1967).
14. G. Cauquis and M. Genies, *Tetrahedron Lett.*, **1970**, 3403.
15. D. E. Ames, B. Novitt, D. Waite, and H. Lund, *J. Chem. Soc. (C)*, **1969**, 796.
16. Y. Takayama, *Bull. Chem. Soc. Japan*, **8**, 189 (1933) [*Chem. Abstr.*, **27**, 5253 (1933)].
17. Y. Takayama and H. Oeda, *Bull. Chem. Soc. Japan*, **9**, 535 (1934) [*Chem. Abstr.*, **29**, 1392 (1935).
18. P. E. Iversen, *Acta Chem. Scand.*, **24**, 2459 (1970).
19. P. E. Iversen and H. Lund, *Acta Chem. Scand.*, **21**, 279 (1967).
20. W. Sümmermann and H. Baumgärtel, *Coll. Czech. Chem. Commun.*, **36**, 575 (1971).
21. O. Carrasco, *Gazz. Chim. Ital.*, **38**, 301 (1908) [*Chem. Abstr.*, **2**, 3229 (1908)].
22. J. von Braun and W. Sobecki, *Ber.*, **44**, 2158 (1911) [*Chem. Abstr.*, **5**, 3432 (1911)].
23. J. T. Wrobel and K. M. Pazdro, *Rocz. Chem.*, **41**, 637 (1967) [*Chem. Abstr.*, **67**, 39542m (1967)].
24. S. Sugasawa, I. Satoda, and J. Yanagisawa, *J. Pharm. Soc. Japan*, **58**, 29 (1938) [*Chem. Abstr.*, **32**, 4161 (1938)].
25. S. G. Plant and D. M. L. Rippon, *J. Chem. Soc.*, **1928**, 1906.
26. J. T. Wrobel, K. M. Pazdro, and A. S. Bien, *Chem. Ind. (London)*, 1760 (1966).
27. E. W. Cook and W. G. France, *J. Phys. Chem.*, **36**, 2383 (1932).
28. E. Späth and F. Breusch, *Monatsh. Chem.*, **50**, 349 (1928) [*Chem. Abstr.*, **23**, 1634 (1929)].

29. A. Dunet and A. Willemart, *Bull. Soc. Chim. Fr.*, 887 (1948).
30. B. Sakurai, *Bull. Chem. Soc. Japan*, 5, 184 (1930) [*Chem. Abstr.*, 24, 5643 (1930)].
31. B. Sakurai, *Bull. Chem. Soc. Japan*, 7, 155 (1932) [*Chem. Abstr.*, 26, 4542 (1932)].
32. G. R. Clemo, R. Raper, and C. R. S. Tenniswood, *J. Chem. Soc.*, 1931, 429.
33. H. Lund, *Tetrahedron Lett.*, 1965, 3973.
34. H. Grotthus, *Ann. Chim. Phys. [1]*, 63, 18 (1807).
35. F. Goppelsroeder, *Bull. Soc. Ind. Mulhouse*, 54, 343 (1884).
36. J. Mullerus, *Chem. Ztg.*, 17, 1454 (1893).
37. J. Nevyas and A. Lowy, *Trans. Am. Electrochem. Soc.*, 50, 349 (1926) [*Chem. Abstr.*, 20, 3395 (1926)].
38. H. Chaumat, *Compt. Rendu.*, 146, 231 (1907) [*Chem. Abstr.*, 2, 1389 (1908)].
39. H. Chaumat, *Compt. Rendu.*, 145, 1419 (1907) [*Chem. Abstr.*, 2, 1002 (1908)].
40. F. Fichter and F. Cueni, *Helv. Chim. Acta*, 14, 651 (1931) [*Chem. Abstr.*, 25, 3650 (1931)].
41. E. Kunz, Ger. Patent 239,672 (1909) [*Chem. Abstr.*, 6, 2177 (1912)].
42. E. Kunz, Ger. Patent 248,262 (1911) [*Chem. Abstr.*, 6, 2689 (1912)].
43. A. Binz, *Z. Elektrochem.*, 5, 5, 103 (1898).
44. A. Binz and A. Hagenbach, *Z. Elektrochem.*, 6, 261 (1899).
45. Farbwerke vorm. Meister, Lucius and Brüning, Ger. Patent 145,602(1902).
46. R. Andruzzi, M. E. Cardinali, I. Carelli, and A. Trazza, *J. Electroanal. Chem.*, 26, 211 (1970).
47. C. Furlani, S. Bertola, and G. Morpurgo, *Ann. Chim. (Rome)*, 50, 858 (1960) [*Chem. Abstr.*, 55, 3242 (1961)].
48. H. Lund and E. T. Jensen, *Acta Chem. Scand.*, 24, 1867 (1970).
49. H. Lund, *Coll. Czech. Chem. Commun.*, 30, 4237 (1965).
50. P. Pflegel and G. Wagner, *Pharmazie*, 22, 147 (1967) [*Chem. Abstr.*, 67, 102850c (1967)].
51. H. J. Backer, *Rec. Trav. Chim. Pays-Bas*, 32, 39 (1913) [*Chem. Abstr.*, 7, 1703 (1913)].
52 H. J. Backer, *Rec. Trav. Chim. Pays-Bas*, 31, 142 (1912) [*Chem. Abstr.*, 6, 2744 (1912)].
53. G. W. Heimrod, *Ber.*, 47, 338 (1914) [*Chem. Abstr.*, 8, 1424 (1914)].
54. N. I. Gavrilov and A. V. Koperina, *Zh. Obshch. Khim.*, 9, 1394 (1939) [*Chem. Abstr.*, 34, 1615 (1940)].
55. M. Yokoyama and K. Yamamoto, *Bull. Chem. Soc. Japan*, 8, 306 (1933) [*Chem. Abstr.*, 28, 1702 (1934)].
56. F. B. Ahrens, *Ber.*, 30, 533 (1897).
57. F. B. Ahrens, *Ber.*, 31, 2272 (1898).
58. F. B. Ahrens, *Z. Elektrochem.*, 2, 578 (1896).
59. F. B. Ahrens and L. Sollmann, *Chem. Ztschr.*, 2, 414 (1903).
60. E. Laviron and P. Fournari, *Bull. Soc. Chim. Fr.*, 518 (1966).

61. R. Widera, *Ber.*, **31**, 2276 (1898).
62. E. Schering, Ger. Patent 95,623 (1896).
63. B. Emmert and A. Herterich, *Ber.*, **45**, 661 (1912) [*Chem. Abstr.*, **6**, 1439 (1912)].
64. Th. B. Baille and J. Tafel, *Ber.*, **32**, 68 (1899).
65. K. Kindler, *Arch. Pharm.*, **265**, 389 (1927) [*Chem. Abstr.*, **21**, 2668 (1927)].
66. B. Sakurai, *Bull. Chem. Soc. Japan*, **13**, 482 (1938) [*Chem. Abstr.*, **32**, 8281 (1938)].
67. R. Lukeš and V. Smetačková, *Coll. Czech. Chem. Commun.* **5**, 61 (1933) [*Chem. Abstr.*, **27**, 2629 (1933)].
68. G. Gratton and G. R. Ramage, *J. Chem. Soc.*, **1935**, 539.
69. D. L. Smith and P. J. Elving, *J. Am. Chem. Soc.*, **84**, 1412 (1962).
70. G. Dryhurst and P. J. Elving, *J. Electrochem. Soc.*, **115**, 1014 (1968).
71. B. Janik and P. J. Elving, *J. Electrochem. Soc.*, **117**, 457 (1970).
72. G. Dryhurst, *J. Electrochem. Soc.*, **116**, 1097 (1969).
73. G. Dryhurst, *J. Electrochem. Soc.*, **116**, 1357 (1969).
74. G. Dryhurst, *J. Electroanal. Chem.*, **28**, 33 (1970).
75. G. Dryhurst, *J. Electrochem. Soc.*, **117**, 1113 (1970).
76. G. Dryhurst, *J. Electrochem. Soc.*, **117**, 1118 (1970).
77. D. L. McAllister and G. Dryhurst, *J. Electroanal. Chem.*, **32**, 387 (1971).
78. G. Dryhurst, *J. Electrochem. Soc.*, **118**, 699 (1971).
79. J. Tafel, *Ber.*, **34**, 258 (1901).
80. J. Tafel, *Ber.*, **34**, 1181 (1901).
81. J. Tafel and P. A. Houseman, *Ber.*, **40**, 3743 (1907) [*Chem. Abstr.*, **2**, 128 (1908)].
82. W. A. Struck and P. J. Elving, *Biochem. J.*, **4**, 1343 (1965).
83. J. Tafel, *Ber.*, **34**, 279 (1901).
84. J. Tafel and B. Ach, *Ber.*, **34**, 1165 (1901).
85. J. Tafel and B. Ach, *Ber.*, **34**, 1170 (1901).
86. G. Dryhurst and G. F. Pace, *J. Electrochem. Soc.*, **117**, 1259 (1970).
87. J. Tafel and A. Weinschenk, *Ber.*, **33**, 3369 (1900).
88. J. Tafel and J. Dodt, *Ber.*, **40**, 3752 (1907) [*Chem. Abstr.*, **2**, 128 (1908)].
89. Y. Yoshitomi, *J. Pharm. Soc. Japan*, **512**, 839 (1924) [*Chem. Abstr.*, **19**, 2345 (1925)].
90. Y. Yoshitomi, *J. Pharm. Soc. Japan*, **508**, 460 (1924) [*Chem. Abstr.*, **18**, 3174 (1924)].
91. Y. Yoshitomi, *J. Pharm. Soc. Japan*, **510**, 649 (1924) [*Chem. Abstr.*, **19**, 2303 (1925)].
92. J. Tafel, *Ber.*, **32**, 3194 (1899).
93. C. F. Böhringer and Söhne, Ger. Patent 108,577 (1898).
94. F. Fichter and W. Kern, *Helv. Chim. Acta*, **9**, 429 (1926) [*Chem. Abstr.*, **21**, 3185 (1927)].
95. F. Rochleder, *Justus Liebigs Ann. Chem.*, **79**, 124 (1851).
96. Th. B. Baillie and J. Tafel, *Ber.*, **32**, 75 (1899).
97. Th. B. Baillie and J. Tafel, *Ber.*, **32**, 3206 (1899).

98. L. Pommerehne, *Arch. Pharm. (Weinheim)*, **235**, 365 (1897).
99. D. L. Smith and P. J. Elving, *J. Am. Chem. Soc.*, **84**, 2741 (1962).
100. J. Tafel and L. Reindl, *Ber.*, **34**, 3286 (1901).
101. W. A. Struck and P. J. Elving, *J. Am. Chem. Soc.*, **86**, 1229 (1964).
102. J. Tafel and A. Weinschenk, *Ber.*, **33**, 3383 (1900).
103. F. Fichter and H. Stenzl, *Helv. Chim. Acta*, **17**, 665 (1934) [*Chem. Abstr.*, **28**, 6711 (1934)].
104. J. Tafel and A. B. Thompson, *Ber.*, **40**, 4480 (1907) [*Chem. Abstr.*, **2**, 534 (1908)].
105. J. Tafel and A. Weinschenk, *Ber.*, **33**, 3378 (1900).
106. K. Sugino, K. Shirai, T. Sekine, and K. Odo, *J. Electrochem. Soc.*, **104**, 667 (1957).
107. K. Sugino and K. Shirai, *J. Chem. Soc. Japan*, **70**, 111 (1949) [*Chem. Abstr.*, **45**, 6641 (1951)].
108. K. Shirai, *J. Electrochem. Soc. Japan*, **21**, 387 (1953) [*Chem. Abstr.*, **48**, 13484 (1954)].
109. K. Sugino, K. Odo, and K. Shirai, *J. Chem. Soc. Japan*, **71**, 396 (1950) [*Chem. Abstr.*, **45**, 6512 (1951)].
110. G. K. Budnikov, *J. Gen. Chem. USSR*, **38**, 2350 (1968) [*Chem. Abstr.*, **70**, 53408 (1969)].
111. G. Dryhurst, B. H. Hansen, and E. B. Harkins, *J. Electroanal. Chem.*, **27**, 375 (1970).
112. L. F. Wiggins and W. S. Wise, *J. Chem. Soc.*, **1956**, 4780.
113. J. Volke, D. Dumanović and V. Volková, *Coll. Czech. Chem. Commun.*, **30**, 246 (1965).
114. J. Pinson and J. Armand, *Compt. Rendu., Ser. C*, **266**, 1081 (1968).
115. F. Fichter and H. de Montmollin, *Helv. Chim. Acta*, **5**, 256 (1922) [*Chem. Abstr.*, **16**, 1774 (1922)].
116. E. Remy and G. Kummell, *Z. Elektrochem.*, **15**, 254 (1909) [*Chem. Abstr.*, **3**, 1863 (1909)].
117. H. Lund, *Oesterr. Chem. -Ztg.*, **68**, 43 (1967) [*Chem. Abstr.*, **67**, 32075 (1967)].
118. H. Lund and P. Lunde, *Acta Chem. Scand.*, **21**, 1067 (1967).
119. P. Pflegel, G. Wagner, and O. Manoušek, *Z. Chem.*, **6**, 263 (1966) [*Chem. Abstr.*, **65**, 13205 (1966)].
120. P. Baumgarten and E. Dammann, *Ber.*, **66B**, 1633 (1933) [*Chem. Abstr.*, **28**, 476 (1934)].
121. M. Yokoyama and K. Yamamoto, *Bull. Chem. Soc. Japan*, **7**, 28 (1932) [*Chem. Abstr.*, **26**, 2194 (1932)].
122. B. Emmert, *Ber.*, **46**, 1716 (1913) [*Chem. Abstr.*, **7**, 2753 (1913)].
123. L. Pincussohn, *Z. Anorg. Allg. Chem.*, **14**, 379 (1897).
124. Farbenfabriken vorm. F. Bayer & Co., Ger. Patent 310,023 (1916) [*Chem. Abstr.*, **15**, 2038 (1921)].
125. Robinson Brothers, Ltd., D. W. Parkes, Br. Patent 395,741 (1931) [*Chem. Abstr.*, **28**, 421 (1934)].

126. N. S. Drosdov, *J. Gen. Chem. USSR*, **3**, 351 (1933) [*Chem. Abstr.*, **28**, 2277 (1934)].

127. N. I. Kobozev and V. V. Monblanova, *Acta Physicochem. USSR*, **4**, 395 (1936) [*Chem. Abstr.*, **30**, 8040 (1936)].

128. S. Szmaragd and E. Briner, *Helv. Chim. Acta*, **32**, 553 (1949) [*Chem. Abstr.*, **43**, 5677 (1949)].

129. C. Marie and G. Lejeune, *J. Chim. Phys.*, **22**, 59 (1925) [*Chem. Abstr.*, **19**, 1708 (1925)].

130. W. R. Turner and P. J. Elving, *Anal. Chem.*, **37**, 467 (1965).

131. S. Andreades and E. W. Zahnow, *J. Am. Chem. Soc.*, **91**, 4181 (1969).

132. A. Cisak and P. J. Elving, *Electrochim. Acta*, **10**, 935 (1965).

133. T. C. Simmons and F. W. Hoffmann, *J. Am. Chem. Soc.*, **79**, 3429 (1957).

134. O. Fischer and K. Neudlinger, *Ber.*, **46**, 2544 (1913) [*Chem. Abstr.*, **7**, 3974 (1913)].

135. J. Volke, in A. R. Katritzky, Ed., *Physical Methods in Heterocyclic Chemistry*, Vol. 1, Academic, New York, 1963.

136. O. Fischer and M. Chur, *J. Prakt. Chem. [2]*, **93**, 363 (1916) [*Chem. Abstr.*, **11**, 1637 (1917)].

137. E. Ochiai and H. Kataoka, *J. Pharm. Soc., Japan*, **62**, 241 (1942) [*Chem. Abstr.*, **45**, 5150 (1951)].

138. B. Emmert, *Ber.*, **42**, 1997 (1909) [*Chem. Abstr.*, **3**, 2436 (1909)].

139. B. Emmert, *Ber.*, **52B**, 1351 (1919) [*Chem. Abstr.*, **14**, 56 (1920)].

140. B. Emmert, *Ber.*, **53**, 370 (1920 [*Chem. Abstr.*, **14**, 2794 (1920)].

141. E. Ochiai and N. Kawagoye, *J. Pharm. Soc. Japan*, **63**, 313 (1943) [*Chem. Abstr.*, **45**, 5153 (1951)].

142. M. Ferles, *Coll. Czech. Chem. Commun.*, **24**, 2221 (1959).

143. F. Šorm, *Coll. Czech. Chem. Commun.*, **13**, 57 (1948) [*Chem. Abstr.*, **42**, 7298 (1948)]

144. H. Lund, *Acta Chem. Scand.*, **17**, 972 (1963).

145. M. Ferles, *Coll. Czech. Chem. Commun.*, **24**, 1029 (1959).

146. M. Yokoyama, *Bull. Chem. Soc. Japan*, **7**, 69 (1932) [*Chem. Abstr.*, **26**, 3256 (1932)].

147. M. Ferles, M. Havel, and A. Tesarova, *Coll. Czech. Chem. Commun.*, **31**, 4121 (1966).

148. M. Kulka, *J. Am. Chem. Soc.*, **68**, 2472 (1946).

149. V. G. Khomyakov, S. S. Kruglikov, and V. M. Berezovskii, *Zh. Obshch. Khim.*, **28**, 2898 (1958) [*Chem. Abstr.*, **53**, 5916 (1959)].

150. V. G. Khomyakov, S. S. Kruglikov, and N. A. Izgaryshev, *Dokl. Akad. Nauk SSSR*, **115**, 557 (1957) [*Chem. Abstr.*, **52**, 8794 (1958)].

151. L. N. Ferguson and A. J. Levant, *Nature*, **167**, 817 (1951).

152. A. Ito and K. Kawada, *Ann. Rept. Takamine Lab.*, **5**, 14 (1953) [*Chem. Abstr.*, **49**, 14516 (1955)].

153. T. Mutavchiev and A. Marinov, *Godishnik Khim. -Tekhnol. Inst.*, **2**, 193 (1956) [*Chem. Abstr.*, **52**, 11631 (1958)].

154. M. Ferles and A. Šilhánková, *Z. Chem.*, **8**, 175 (1968) [*Chem. Abstr.*, **69**,

32450 (1968)].
155. V. G. Khomyakov, S. S. Kruglikov, and L. I. Kazakova, *Tr. Mosk. Khim. -Tekhnol. Inst.*, 189, (1961) [*Chem. Abstr.*, **57**, 15065 (1962)].
156. L. D. Borhki and V. G. Khomyakov, USSR Patent 187,024 (1966) [*Chem. Abstr.*, **67**, 17395 (1967)].
157. A. F. Krivis and E. S. Gazda, *Anal. Chem.*, **41**, 212 (1969).
158. Deutsche Gold- und Silber-Scheideanstalt vormals Roessler, Ger. Patent 526,803 (1926) [*Chem. Abstr.*, **25**, 4807 (1931)].
159. B. Emmert and W. Dorn, *Ber.*, **48**, 687 (1915) [*Chem. Abstr.*, **9**, 2093 (1915)].
160. H. Lund, *Acta Chem. Scand.*, **17**, 2325 (1963).
161. R. F. Evilia and A. J. Diefenderfer, *J. Electroanal. Chem.*, **22**, 407 (1969).
162. N. L. Weinberg and E. A. Brown, *J. Org. Chem.*, **31**, 4054 (1966).
163. J. D. Anderson, M. M. Baizer, and E. J. Prill, *J. Org. Chem.*, **30**, 1645 (1965).
164. A. Binz and O. v. Schickh, *Ber.*, **68B**, 315 (1935) [*Chem. Abstr.*, **29**, 3339 (1935)].
165. K. Brand and K. Reuter, *Ber.*, **72**, 1668 (1939) [*Chem. Abstr.*, **33**, 9304 (1939)].
166. J. Volke and A. M. Kardos, *Coll. Czech. Chem. Commun.*, **33**, 2560 (1968).
167. J. Volke and J. Holubek, *Coll. Czech. Chem. Commun.*, **28**, 1597 (1963).
168. M. J. Allen, *J. Org. Chem.*, **15**, 435 (1950).
169. J. H. Stocker and R. M. Jenevein, *J. Org. Chem.*, **34**, 2807 (1969).
170. M. Ferles and A. Tesařová, *Coll. Czech. Chem. Commun.*, **32**, 1631 (1967).
171. H. Lund, *Acta Chem. Scand.*, **17**, 1077 (1963).
172. M. Yokoyama, *Bull. Chem. Soc. Japan*, **7**, 103 (1932) [*Chem. Abstr.*, **26**, 4050 (1932)].
173. J. P. Wibaut and H. Boer, *Rec. Trav. Chim.* Pays-Bas, **68**, 72 (1949) [*Chem. Abstr.*, **43**, 4266 (1932)].
174. M. Ferles and M. Prystaš, *Coll. Czech. Chem. Commun.*, **24**, 3326 (1959).
175. J. Lakomy, A. Silhánková, M. Ferles, and O. Exner, *Coll. Czech. Chem. Commun.*, **33**, 1700 (1968).
176. J. Volke and J. Holubek, *Coll. Czech. Chem. Commun.*, **27**, 1777 (1962).
177. L. W. Harrison and G. E. Cheney, *Talanta*, **15**, 1413 (1968).
178. J. Volke, R. Kubiček, and F. Šantavý, *Coll. Czech. Chem. Commun.*, **25**, 871 (1960).
179. O. Manoušek, *Coll. Czech. Chem. Commun.*, **25**, 2250 (1960).
180. M. Matsumoto, M. Miyazaki, and M. Ishii, *Yakugaku Zasshi*, **88**, 1093 (1968) [*Chem. Abstr.*, **70**, 16525 (1969)].
181. M. Ferles, A. H. Attia, and H. Hruba, *Coll. Czech. Chem. Commun.*, **36**, 2057 (1971)].
182. M. Ferles and M. Holík, *Coll. Czech. Chem. Commun.*, **31**, 2416 (1966).
183. M. Holík, A. Tesařová, and M. Ferles, *Coll. Czech. Chem. Commun.*, **32**, 1730 (1967).
184. M. Jankovský and M. Ferles, *Coll. Czech. Chem. Commun.*, **35**, 2802 (1970).

185. M. Ferles, M. Vanka, and A. Šilhánková, *Coll. Czech. Chem. Commun.*, **34**, 2108 (1969).

186. M. Ferles and H. Hrubá, *Z. Chem.*, **9**, 450 (1969) [*Chem. Abstr.*, **72**, 55176 (1970)].

187. A. Šilhánková, D. Doskočilová, and M. Ferles, *Coll. Czech. Chem. Commun.*, **34**, 1985 (1969).

188. S. J. Leach, J. H. Baxendale, and M. G. Evans, *Austr. J. Chem.*, **6**, 395 (1953).

189. A. Šilhánková, D. Doskočilová, and M. Ferles, *Coll. Czech. Chem. Commun.*, **34**, 1976 (1969).

190. J. N. Burnett and A. L. Underwood, *J. Org. Chem.*, **30**, 1154 (1965).

191. A. J. Cunningham and A. L. Underwood, *Arch. Biochem. Biophys.*, **117**, 88 (1966).

192. B. Ke, *Arch. Biochem. Biophys.*, **60**, 505 (1956).

193. B. Ke, *J. Am. Chem. Soc.*, **78**, 3649 (1956).

194. R. F. Powning and C. C. Kratzing, *Arch. Biochem. Biophys.*, **66**, 249 (1957).

195. T. Kono, *Bull. Agr. Chem. Soc. Japan*, **21**, 115 (1957) [*Chem. Abstr.*, **51**, 10626 (1957)].

196. T. Kono and S. Nakamura, *Bull. Agr. Chem. Soc. Japan*, **22**, 399 (1958) [*Chem. Abstr.*, **53**, 18120 (1959)].

197. F. Fichter and H. Stenzl, *Helv. Chim. Acta*, **19**, 1171 (1936) [*Chem. Abstr.*, **31**, 324 (1937)].

198. B. Lovreček, *Radovi Jugoslav. Akad. Znanosti i Umjetnosti*, **296**, 65, (1953) [*Chem. Abstr.*, **48**, 6882 (1954)].

199. M. Dennstedt, Ger. Patent 127,086 (1901).

200. B. Sakurai, *Bull. Chem. Soc. Japan*, **11**, 374 (1936) [*Chem. Abstr.*, **30**, 6288 (1936).

201. A. Stanienda, *Z. Naturforsch.*, **22b**, 1107 (1967).

202. L. C. Craig, *J. Am. Chem. Soc.*, **55**, 2543 (1933).

203. Y. Takayama, *Bull. Chem. Soc. Japan*, **8**, 213 (1933) [*Chem. Abstr.*, **28**, 105 (1934)].

204. Y. Takayama, *J. Chem. Soc. Japan*, **56**, 781 (1935) [*Chem. Abstr.*, **29**, 7973 (1935)].

205. Y. Takayama, *Bull. Chem. Soc. Japan*, **11**, 138 (1936) [*Chem. Abstr.*, **30**, 5184 (1936)].

206. Y. Takayama, *Bull. Chem. Soc. Japan*, **8**, 137 (1933) [*Chem. Abstr.*, **27**, 4177 (1933)].

207. Y. Takayama, *J. Chem. Soc. Japan*, **52**, 544 (1931) [*Chem. Abstr.*, **26**, 5018 (1932)].

208. J. Tafel and M. Stern, *Ber.*, **33**, 2224 (1900).

209. J. Tafel and B. Emmert, *Z. Phys. Chem.*, **54**, 433 (1906).

210. B. Sakurai, *Bull. Chem. Soc. Japan*, **10**, 311 (1935) [*Chem. Abstr.*, **29**, 7828 (1935)].

211. L. C. Craig, *J. Am. Chem. Soc.*, **55**, 295 (1933).

212. B. Sakurai, *Bull. Chem. Soc. Japan*, 11, 41 (1936) [*Chem. Abstr.*, 30, 4100 (1936)].
213. B. Sakurai, *Bull. Chem. Soc. Japan*, 13, 350 (1938) [*Chem. Abstr.*, 32, 6555 (1938)].
214. H. R. L. Streight and E. G. Hallonquist, *Trans. Am. Electrochem. Soc.*, 56, 485 (1929) [*Chem. Abstr.*, 23, 4624 (1929)].
215. R. Lukes, *Coll. Czech. Chem. Commun.*, 4, 351 (1932) [*Chem. Abstr.*, 27, 290 (1933)].
216. R. Lukes, *Chem. Listy*, 27, 392, 409 (1933) [*Chem. Abstr.*, 29, 1720 (1935)].
217. B. Sakurai, *Bull. Chem. Soc., Japan*, 12, 8 (1937) [*Chem. Abstr.*, 31, 3393 (1937)].
218. H. Lund, *Acta Chem. Scand.*, 18, 1984 (1964).
219. H. Itomi, *Mem. Coll. Sci. Kyoto Univ. [A]*, 13, 311 (1930) [*Chem. Abstr.*, 25, 2057 (1931)].
220. P. Pflegel and G. Wagner, *Pharmazie*, 22, 60 (1967) [*Chem. Abstr.*, 67, 7499 (1967)].
221. P. Pflegel and G. Wagner, *Pharmazie*, 22, 643 (1967) [*Chem. Abstr.*, 69, 2345 (1968)].
222. P. Pflegel and G. Wagner, *Z. Chem.*, 9, 151 (1969) [*Chem. Abstr.*, 71, 9063 (1969)].
223. J. C. Cochran and W. F. Little, *J. Org. Chem.*, 26, 808 (1961).
224. T. Kimura, S. Yamada, K. Yoshizue, and T. Nagoya, *Yakugaku Zasshi*, 77, 891 (1957) [*Chem. Abstr.*, 52, 1181 (1958)].
225. V. V. Tsodikov, L. D. Borkhi, V. G. Brudz, N. E. Khomutov, and G. Khomyakov, *Khim Geterotsikl. Soedin.*, 112 (1967) [*Chem. Abstr.*, 67, 17256u (1967)].
226. M. Yokoyama and K. Yamamoto, *Bull. Chem. Soc. Japan*, 18, 121 (1943) [*Chem. Abstr.*, 41, 4494 (1947)].
227. E. Merck, Ger. Patent 90,308 (1896).
228. E. Merck, Ger. Patent 104,664 (1898).
229. V. V. Levtushchenko, *Zh. Obshch. Khim.*, 11, 686 (1941) [*Chem. Abstr.*, 36, 39 (1942)].
230. S. Szmaragd and E. Briner, *Helv. Chim. Acta*, 32, 1278 (1949) [*Chem. Abstr.*, 43, 7355 (1949)].
231. O. W. Brown and B. Berkowitz, *Trans. Am. Electrochem. Soc.*, 75, 385 (1939) [*Chem. Abstr.*, 33, 2420 (1939)].
232. N. N. Mel'nikov, S. I. Sklyarenko, and E. M. Cherkasova, *Zh. Obshch. Khim.*, 9, 1819 (1939) [*Chem. Abstr.*, 34, 3699 (1940)].
233. F. Zuckmayer, Ger. Patent 342,048 (1916) [*Chem. Abstr.*, 16, 3528 (1922)].
234. J. B. Conn and J. Van de Kamp, U.S. Patent 2,453,701 (1948) [*Chem. Abstr.*, 43, 1443 (1949)].
235. L. Gattermann, *Ber.*, 27, 1927 (1894).
236. Farbenfabriken vorm. J. Bayer & Co., Ger. Patent 80,978 (1894).
237. K. Elbs, *Z. Elektrochem.*, 10, 579 (1904).

238. M. Yokoyama and K. Yamamoto, *Bull. Chem. Soc. Japan*, **18**, 126 (1943) [*Chem. Abstr.*, **41**, 4494 (1947)].

239. L. F. Fieser and E. L. Martin, *J. Am. Chem. Soc.*, **57**, 1840 (1935).

240. M. M. Joullie and J. K. Puthenpurayil, *J. Heterocycl. Chem.*, **6**, 697 (1969).

241. S. Kato, J. Nakaya, and E. Imoto, *.Bull. Chem. Soc. Japan*, **44**, 1928 (1971).

242. J. Pinson, J. -P. Launay, and J. Armand, *Compt. Rendu. Ser. C*, **270**, 1881 (1970).

243. H. Lund, in A. R. Katritzky and A. J. Boulton, Eds., *Advances in Heterocyclic Chemistry*, Vol. 12, Academic, New York, 1970.

244. J. Pinson and J. Armand, *Coll. Czech. Chem. Commun.*, **36**, 585 (1971).

245. J. Pinson and J. Armand, *Compt. Rendu. Ser. C*, **268**, 629 (1969).

246. M. Schellenberg, *Helv. Chim. Acta*, **53**, 1151 (1970).

247. P. Pflegel and G. Wagner, *Pharmazie*, **24**, 308 (1969) [*Chem. Abstr.*, **71**, 108422 (1969)].

248. P. Pflegel and G. Wagner, *Pharmazie*, **24**, 384 (1969) [*Chem. Abstr.*, **71**, 108423 (1969)].

249. P. Pflegel and G. Wagner, *Z. Chem.*, **8**, 179 (1968) [*Chem. Abstr.*, **69**, 32478 (1968)].

250. V. V. Levtushchenko, *Zh. Obshch. Khim.*, **18**, 1245 (1948) [*Chem. Abstr.*, **43**, 955 (1949)].

251. N. E. Khomutov and V. V. Tsodikov, *Elektrokhim.*, **1**, 482 (1965) [*Chem. Abstr.*, **63**, 14375 (1965)].

252. N. E. Khomutov and V. V. Tsodikov, *Elektrokhim.*, **2**, 722 (1966) [*Chem. Abstr.*, **65**, 6721 (1966)].

253. S. Kwee and H. Lund, *Acta Chem. Scand.*, **23**, 2711 (1969).

254. H. Lund and S. Kwee, *Acta Chem. Scand.*, **22**, 2879 (1968).

255. H. H. Inhoffen and P. Jäger, *Tetrahedron Lett.*, **1964**, 1317.

256. H. H. Inhoffen and P. Jäger, *Tetrahedron Lett.*, **1965**, 3387.

257. H. H. Inhoffen and R. Mählhop, *Tetrahedron Lett.*, **1966**, 4283.

258. H. H. Inhoffen, P. Jäger, R. Mählhop, and C. -D. Mengler, *Justus Liebigs Ann. Chem.*, **704**, 188 (1967).

259. H. Lund and S. Gruhn, *Acta Chem. Scand.*, **20**, 2637 (1966).

260. H. Lund, *Coll. Czech. Chem. Commun.*, **31**, 4175 (1966).

261. D. N. Bailey, D. M. Hercules, and D. K. Roe, *J. Electrochem. Soc.*, **116**, 190 (1969).

262. D. T. Sawyer and R. Y. Komai, *Anal. Chem.*, **44**, 715 (1972).

263. R. F. Nelson, D. W. Leedy, E. T. Seo, and R. N. Adams, *Fresnius' Z. Anal. Chem.*, **224**, 184 (1967) [*Chem. Abstr.*, **66**, 51580 (1967)].

264. P. G. Gassman and B. L. Fox, *J. Org. Chem.*, **32**, 480 (1967).

265. K. N. Menon and J. L. Simonsen, *J. Chem. Soc.*, **1929**, 302.

266. E. Schering, Ger. Patent 96,362 (1897).

267. E. Merck, Ger. Patents 406,215 (1922); 408,869 (1923).

268. R. Willstätter, Ger. Patent 302,401 (1917).

269. R. Willstätter and H. Veraguth, *Ber.*, **38**, 1984 (1905).

270. A. Piccinini, *Gazz. Chim. Ital.*, **32**, 260 (1902).
271. L. Ruzička, M. W. Goldberg, M. Hürbin, and H. A. Boekenoogen, *Helv. Chim. Acta*, **16**, 1323 (1933) [*Chem. Abstr.*, **28**, 1346 (1934)].
272. F. Akiba and T. Tobita, Jap. Patent 9,419 (1965) [*Chem. Abstr.*, **63**, 13226 (1965)].
273. K. Friedlander, *Ber.*, **38**, 2837 (1905).
274. K. Friedlander, *Z. Elektrochem.*, **12**, 685 (1906).
275. K. Elbs and H. Koch, *Z. Elektrochem.*, **7**, 142 (1900).
276. G. Wittig, H. Kleiner, and J. Conrad, *Justus Liebigs Ann. Chem.*, **469**, 1 (1929) [*Chem. Abstr.*, **23**, 2974 (1929)].
277. T. Abe, S. Nagase, and K. Kodaira, *Bull. Chem. Soc. Japan*, **43**, 957 (1970).
278. L. A. Simonson and C. K. Mann, *Tetrahedron Lett.*, **1970**, 3303.
279. E. Späth, F. Kuffner, and J. Lintner, *Ber.*, **69B**, 2052 (1936) [*Chem. Abstr.*, **30**, 8227 (1936)].
280. K. S. Narang and I. N. Ray, *J. Chem. Soc.*, **1936**, 686.
281. H. R. Juneja, K. S. Narang, and I. N. Ray, *J. Chem. Soc.*, **1935**, 1277.
282. E. Späth and N. Platzer, *Ber.*, **69B**, 387 (1936) [*Chem. Abstr.*, **30**, 3826 (1936)].
283. E. Späth and N. Platzer, *Ber.*, **68B**, 2221 (1935) [*Chem. Abstr.*, **30**, 2196 (1936)].
284. K. Tsuda and K. Murakami, *J. Pharm. Soc. Japan*, **57**, 307 (1937) [*Chem. Abstr.*, **31**, 5365 (1937)].
285. H. R. Ing, *J. Chem. Soc.*, **1933**, 504.
286. A. Orekhov, H. Gurevich, and T. Okolskaya, *Ber.*, **68A**, 820 (1935) [*Chem. Abstr.*, **29**, 4767 (1935)].
287. A. Orekhov and S. Norkina, *Ber.*, **67B**, 1845 (1934) [*Chem. Abstr.*, **29**, 1093 (1935)].
288. A. Orekhov, M. Rabinovitsch, and R. Konovalova, *Ber.*, **67B**, 1850 (1934) [*Chem. Abstr.*, **29**, 1093 (1935)].
289. J. Tafel and K. Naumann, *Ber.*, **34**, 3291 (1901).
290. M. Freund and J. A. W. Bredenberg, *Justus Liebigs Ann. Chem.*, **407**, 43 (1914) [*Chem. Abstr.*, **9**, 457 (1915)].
291. E. Späth and A. Burger, *Ber.*, **60B**, 704 (1927) [*Chem. Abstr.*, **21**, 1817 (1927)].
292. E. Späth and W. Gruber, *Ber.*, **70**, 1538 (1937) [*Chem. Abstr.*, **31**, 6665 (1937)].
293. E. Späth and A. Burger, *Monatsh. Chem.*, **47**, 733 (1926) [*Chem. Abstr.*, **21**, 3367 (1927)].
294. C. Finzi and M. Freund, *Ber.*, **45**, 2322 (1912) [*Chem. Abstr.*, **7**, 80 (1913)].
295. R. Wolffenstein and E. Bandow, Ger. Patent 94,949 (1897).
296. E. Bandow and R. Wolffenstein, *Ber.*, **31**, 1577 (1897).
297. R. D. Haworth, W. H. Perkin, Jr., and J. Rankin, *J. Chem. Soc.*, **1925**, 2018.

298. M. Freund and K. Fleischer, *Justus Liebigs Ann. Chem.*, **409**, 188 (1915) [*Chem. Abstr.*, **9**, 2872 (1915)].

299. M. Freund and K. Fleischer, *Justus Liebigs Ann. Chem.*, **397**, 30 (1913) [*Chem. Abstr.*, **7**, 2197 (1913)].

300. M. Freund and H. Commessmann, *Justus Liebigs Ann. Chem.*, **397**, 52 (1913) [*Chem. Abstr.*, **7**, 2198 (1913)].

301. M. Freund and H. Commessmann, *Justus Liebigs Ann. Chem.*, **397**, 57 (1913) [*Chem. Abstr.*, **7**, 2199 (1913)].

302. M. Freund and R. Lachmann, *Justus Liebigs Ann. Chem.*, **397**, 70 (1913) [*Chem. Abstr.*, **7**, 2200 (1913)].

303. M. Freund and H. Hammel, *Justus Liebigs Ann. Chem.*, **397**, 85 (1913) [*Chem. Abstr.*, **7**, 2200 (1913)].

304. M. Freund and D. Steinberger, *Justus Liebigs Ann. Chem.*, **397**, 94 (1913) [*Chem. Abstr.*, **7**, 2201 (1913)].

305. M. Freund and E. Zorn, *Justus Liebigs Ann. Chem.*, **397**, 107 (1913) [*Chem. Abstr.*, **7**, 2202 (1913)].

306. M. Freund and K. Fleischer, *Justus Liebigs Ann. Chem.*, **411**, 1 (1916) [*Chem. Abstr.*, **10**, 1329 (1916)].

307. S. Takagi and T. Ueda, *J. Pharm. Soc. Japan*, **56**, 44 (1936) [*Chem. Abstr.*, **30**, 4099 (1936].

308. E. Speyer and W. Krauss, *Justus Liebigs Ann. Chem.*, **432**, 233 (1923) [*Chem. Abstr.*, **17**, 3508 (1923)].

309. L. F. Small and F. L. Cohen, *J. Am. Chem. Soc.*, **53**, 2227 (1931).

310. L. F. Small and R. E. Lutz, *J. Am. Chem. Soc.*, **56**, 1738 (1934).

311. M. Freund, W. W. Melber, and E. Schlesinger, *J. Prakt. Chem.*, **101**, 1 (1920) [*Chem. Abstr.*, **15**, 834 (1921)].

312. L. F. Small and F. L. Cohen, *J. Am. Chem. Soc.*, **53**, 2214 (1931).

313. D. E. Morris and L. F. Small, *J. Am. Chem. Soc.*, **56**, 2159 (1934).

314. E. Speyer and K. Sarre, *Ber.*, **57B**, 1404 (1924) [*Chem. Abstr.*, **19**, 297 (1925)].

315. E. Speyer and A. Popp, *Ber.*, **59B**, 390 (1926) [*Chem. Abstr.*, **20**, 2164 (1926)].

316. J. Tafel, *Justus Liebigs Ann. Chem.*, **301**, 291 (1898).

317. H. Leuchs and W. Wegener. *Ber.*, **63B**, 2215 (1930) [*Chem. Abstr.*, **25**, 705 (1931)].

318. G. R. Clemo, W. H. Perkin, Jr., and R. Robinson, *J. Chem. Soc.*, **1927**, 1589.

319. H. Leuchs and H. Beyer, *Ber.*, **64B**, 2156 (1931) [*Chem. Abstr.*, **26**, 467 (1932)].

320. J. M. Gulland, W. H. Perkin, Jr., and R. Robinson, *J. Chem. Soc.*, **1927**, 1627.

321. H. Wieland and W. W. Moyer, *Justus Liebigs Ann. Chem.*, **491**, 129 (1931) [*Chem. Abstr.*, **26**, 1290 (1932)].

322. H. Wieland and O. Müller, *Justus Liebigs Ann. Chem.*, **545**, 59 (1940) [*Chem. Abstr.*, **35**, 124 (1941)].

323. H. Wieland and J. Kimmig, *Justus Liebigs Ann. Chem.*, **527**, 151 (1937) [*Chem. Abstr.*, **31**, 1817 (1937)].

324. H. Wieland and R. G. Jennen, *Justus Liebigs Ann. Chem.*, **545**, 86 (1940) [*Chem. Abstr.*, **35**, 126 (1941)].

325. H. Schild, *Ber.*, **69B**, 74 (1936) [*Chem. Abstr.*, **30**, 2973 (1936)].

326. Troponwerke Dinklage & Co., Ger. Patent 707,705 (1941) [*Chem. Abstr.*, **37**, 2888 (1943)].

327. J. M. Bobbitt, J. T. Stock, A. Marchand, and K. H. Weisgraber, *Chem. Ind. (London)*, 2127 (1966).

328. G. F. Kirkbright, J. T. Stock, R. D. Pugliese, and J. M. Bobbitt, *J. Electrochem. Soc.*, **116**, 219 (1969).

329. J. M. Bobbitt, K. H. Weisgraber, A. S. Steinfeld, and S. G. Weiss, *J. Org. Chem.*, **35**, 2884 (1970).

330. J. M. Bobbitt, H. Yagi, S. Shibuya, and J. T. Stock, *J. Org. Chem.*, **36**, 3006 (1971).

331. J. M. Bobbitt, I. Noguchi, H. Yagi, and K. H. Weisgraber, *J. Am. Chem. Soc.*, **93**, 3551 (1971).

332. J. M. Bobbitt and R. C. Hallcher, *Chem. Commun.*, 543 (1971).

333. L. L. Miller, F. R. Stermitz, and J. R. Falck, *J. Am. Chem. Soc.*, **93**, 5941 (1971).

334. F. Ullmann and P. Dieterle, *Ber.*, **37**, 23 (1904).

335. F. A. McGinn and G. B. Brown, *J. Am. Chem. Soc.*, **82**, 3193 (1960).

336. C. R. Warner and P. J. Elving, *Coll. Czech. Chem. Commun.*, **30**, 4210 (1965).

337. E. Laviron, R. Gavasso, and M. Pay, *Talanta*, **17**, 747 (1970).

338. E. Laviron and R. Gavasso, *Talanta*, **16**, 293 (1969).

339. T. Okano and K. Ohira, *Yakugaku Zasshi*, **88**, 1170 (1968) [*Chem. Abstr.*, **70**, 16609 (1969)].

340. H. Lund, *Acta Chem. Scand.*, **11**, 990 (1957).

341. Farbwerke vorm. L. Durand, Huguenin & Co., Ger. Patent 164,320 (1905).

342. H. Lund, *Acta Chem. Scand.*, **23**, 563 (1969).

343. H. Lund and A. D. Thomsen, *Acta Chem. Scand.*, **23**, 3567 (1969).

344. H. Lund, *Acta Chem. Scand.*, **18**, 563 (1964).

345. V. D. Bezuglyi, N. P. Shimanskaya, and E. M. Peresleni, *Zh. Obshch. Khim.*, **34**, 3540 (1964) [*Chem. Abstr.*, **62**, 8983 (1965)].

346. W. N. Grieg and J. W. Rogers, *J. Electrochem. Soc.*, **117**, 1141 (1970).

347. G. L. Smith and J. W. Rogers, *J. Electrochem. Soc.*, **118**, 1089 (1971).

348. R. Schindler, H. Will, and L. Holleck, *Z. Elektrochem.*, **63**, 596 (1959) [*Chem. Abstr.*, **54**, 105 (1960)].

349. C. D. Thompson and R. T. Foley, *J. Electrochem. Soc.*, **119**, 177 (1972).

350. P. Zuman, *Coll. Czech. Chem. Commun.*, **25**, 3245 (1960).

351. E. S. Levin, Z. I. Fodiman, and Z. V. Todres, *Elektrokhimiya*, **2**, 175 (1966) [*Chem. Abstr.*, **65**, 5022 (1966)].

352. H. Lund, *Acta Chem. Scand.*, **14**, 359 (1960).

353. C. Dreux, M.-L. Girard, and P. Souchay, *Compt. Rendu.*, **262**, 1565 (1966).

354. P. E. Iversen and H. Lund, *Acta Chem. Scand.*, **21**, 389 (1967).

355. J. P. Billon, G. Cauquis, J. Combrisson, and M. Li, *Bull. Soc. Chim. Fr.*, 2062 (1960).
356. J. P. Billon, *Bull. Soc. Chim. Fr.*, 1784 (1960).
357. J. P. Billon, *Bull. Soc. Chim. Fr.*, 1923 (1961).
358. J. P. Billon, G. Cauquis, and J. Combrisson, *Compt. Rendu.*, **253**, 1593 (1961).
359. J. P. Billon, *Ann. Chim. (Paris)*, **7**, 183 (1962).
360. J. P. Billon, G. Cauquis, and J. Combrisson, *J. Chim. Phys.*, **61**, 374 (1964).
361. P. Kabasakalian and J. McGlotten, *Anal. Chem.*, **31**, 431 (1959).
362. F. H. Merkle, *Diss. Abstr.*, **25**, 2198 (1964) [*Chem. Abstr.*, **62**, 5.863 (1965)].
363. F. H. Merkle and C. A. Discher, *J. Pharm. Sci.*, **53**, 620 (1964) [*Chem. Abstr.*, **61**, 5462 (1964)].
364. F. H. Merkle and C. A. Discher, *Anal. Chem.*, **36**, 1639 (1964).
365. L. Erdey, E. Banyai, and E. B. Gere, *Talanta*, **3**, 54 (1959).
366. O. Manoušek, O. Exner, and P. Zuman, *Coll. Czech. Chem. Commun.*, **33**, 4000 (1968).
367. M. Matsui, T. Sawamura, and T. Adachi, *Mem. Coll. Sci. Kyoto Univ. [A]*, **15**, 151 (1932) [*Chem. Abstr.*, **26**, 5264 (1932)].
368. H. Lund, *Acta Chem. Scand.*, **13**, 192 (1959).
369. O. Manoušek and P. Zuman, *Coll. Czech. Chem. Commun.*, **29**, 1718 (1964).
370. V. Sh. Tsveniashvili, S. I. Zhdanov, and Z. V. Todres, *Fresnius' Z. Anal. Chem.*, **224**, 389 (1967) [*Chem. Abstr.*, **66**, 52057 (1967)].
371. J. E. O'Reilley and P. J. Elving, *J. Electroanal. Chem.*, **21**, 169 (1969).
372. G. Dryhurst and P. J. Elving, *Talanta*, **16**, 855 (1969).
373. J. E. O'Reilley and P. J. Elving, *J. Am. Chem. Soc.*, **93**, 1871 (1971).
374. G. Dryhurst, *J. Electrochem. Soc.*, **116**, 1411 (1969).
375. B. H. Hansen and G. Dryhurst, *J. Electroanal. Chem.*, **32**, 405 (1971).
376. G. Dryhurst, *J. Electrochem. Soc.*, **119**, 1659 (1972).
377. G. Dryhurst and B. H. Hansen, *J. Electroanal. Chem.*, **30**, 407 (1971).
378. G. Dryhurst and B. H. Hansen, *J. Electroanal. Chem.*, **30**, 417 (1971).
379. D. W. Parkes, U.S. Patent 1,947,732 (1934) [*Chem. Abstr.*, **28**, 2628 (1934)].
380. V. V. Levtushchenko, *Zh. Obshch. Khim.*, **17**, 1656 (1947) [*Chem. Abstr.*, **42**, 2187 (1948)].
381. B. H. Hansen and G. Dryhurst, *J. Electrochem. Soc.*, **118**, 1747 (1971).
382. P. E. Iversen, *Chem. Ber.*, **105**, 358 (1972).
383. D. M. Hamel and H. Oelschläger, *J. Electroanal. Chem.*, **28**, 197 (1970).
384. H. Oelschläger and H. Hoffmann, *Arch. Pharm. (Weinheim)*, **300**, 817 (1967).
385. R. Andruzzi, M. E. Cardinali, and A. Trazza, *Ann. Chim. (Rome)*, **61**, 66 (1971).
386. R. Andruzzi, M. E. Cardinali, I. Carelli, and A. Trazza, *J. Electroanal. Chem.*, **36**, 147 (1972).

387. P. Pflegel and G. Wagner, *Pharmazie*, **27**, 24 (1972).
388. D. J. McClemens, A. K. Garrison, and A. L. Underwood, *J. Org. Chem.*, **34**, 1867 (1969).
389. W. J. Blaedel and R. G. Haas, *Anal. Chem.*, **42**, 918 (1970).
390. J. N. Burnett and A. L. Underwood, *Biochem.*, **4**. 2060 (1965).
391. S. Kwee and H. Lund, *Acta Chem. Scand.*, **26**, 1195 (1972).
392. A. Daver, *Compt. Rendu., Ser. C*, **274**, 244 (1972).
393. M. Jubault and D. Peltier, *Bull. Soc. Chim. Fr.*, 2365 (1972).
394. R. Andruzzi, M. E. Cardinali, I. Carelli, and A. Trazza, *Ann. Chim. (Rome)*, **61**, 415 (1971).
395. R. Andruzzi, M. E. Cardinali, and A. Trazza, *Electrochim. Acta*, **17**, 1524 (1972).
396. M. Fedoronko and I. Jezo, *Coll. Czech. Chem. Commun.*, **37**, 1781 (1972).
397. E. Laviron, D. Bernard, and G. Tainturier, *Tetrahedron Lett.*, **1972**, 3643.
398. L. Horner and H. Röder, *Chem. Ber.*, **101**, 4179 (1968).
399. J. Armand, P. Bassinet, K. Chekir, J. Pinson, and P. Souchay, *Compt. Rendu. Ser. C*, **275**, 279 (1972).
400. K. Alwair, J. F. Archer, and J. Grimshaw, *J. Chem. Soc., Perkin Trans.*, **II**, 1663 (1972).
401. J. Pinson, J. P. M'Packo, N. Vinot, J. Armand, and P. Bassinet, *Can. J. Chem.*, **50**, 1581 (1972).
402. H. Lund and E. T. Jensen, *Acta Chem. Scand.*, **25**, 2727 (1971).
403. L. N. Klatt and R. L. Rouseff, *J. Am. Chem. Soc.*, **94**, 7295 (1972).
404. M. Libert, C. Caullet, and S. Longchamp, *Bull. Soc. Chim. Fr.*, 2367 (1971).
405. G. Popp, *J. Org. Chem.*, **37**, 3058 (1972).
406. H. P. Cleghorn, J. E. Gaskin, and D. Lloyd, *J. Chem. Soc. (B)*, **1971**, 1615.
407. M. Libert and C. Caullet, *Bull. Soc. Chim. Fr.*, 1947 (1971).
408. A. Ryvolová-Kejharová and P. Zuman, *Coll. Czech. Chem. Commun.*, **36**, 1019 (1971).
409. S. Kwee and H. Lund, *Acta Chem. Scand.*, **25**, 1813 (1971).
410. N. L. Weinberg and H. R. Weinberg, *Chem. Rev.*, **68**, 449 (1968).
411. G. Dryhurst, *Fortsch. Chem. Forsch.*, **34**, 47 (1972).
412. C. J. Nielsen and R. F. Nelson, unpublished data.
413. M. Ito and T. Kuwana, *J. Electroanal. Chem.*, **32**, 415 (1971).
414. Yu N. Forostyan, A. P. Oleinik, and V. M. Artemova, *Elektrokhim.*, **7**, 715 (1971).
415. M. Libert, C. Caullet, and J. Huguet, *Bull. Soc. Chim. Fr.*, 3639 (1972).
416. M. Libert, C. Caullet, and G. Barbey, *Bull. Soc. Chim. Fr.*, 536 (1973).
417. C. P. Andrieux and J. M. Savéant, *J. Electroanal. Chem.*, **26**, 223 (1970).
418. H. Lund and S. K. Sharma, *Acta Chem. Scand.*, **26**, 2329 (1972).
419. S. Kwee and H. Lund, *Experientia Suppl.*, **18**, 387 (1971).
420. Z. N. Timofeeva, N. M. Omar, L. S. Tikhonova, and A. V. El'tsov, *Zh. Obshch. Khim.*, **40**, 2072 (1970).
421. D. A. Hall and P. J. Elving, *Israel J. Chem.*, **8**, 839 (1970).
422. J. W. Webb, B. Janik, and P. J. Elving, *J. Am. Chem. Soc.*, **95**, 991 (1973).

423. M. E. Cardinali, I. Carelli, and A. Trazza, *J. Electroanal. Chem.*, **34**, 543 (1972).
424. R. Andruzzi, M. E. Cardinali, and A. Trazza, *J. Electroanal. Chem.*, **41**, 67 (1973).
425. P. Pflegel and G. Wagner, *Pharmazie*, **22**, 643 (1967).
426. P. Pflegel, J. Petzold, and W. Horsch, *Pharmazie*, **26**, 603 (1971).
427. D. W. Leedy and D. L. Muck, *J. Am. Chem. Soc.*, **93**, 4264 (1971).

Chapter **X**

ELECTROLYTIC SYNTHESIS AND REACTIONS OF ORGANOMETALLIC COMPOUNDS

William J. Settineri and L. Dennis McKeever

397

1   INTRODUCTION

Most electrolyses of organic compounds are conducted at metallic electrodes, and it is often found that organometallic species of a fleeting nature are formed. In certain cases stable products are produced and electroorganic techniques provide a novel and facile method for their synthesis. Commercial interest in the electrochemical synthesis of organometallic compounds developed only after Ziegler reported [249] convenient electrolytic techniques for the synthesis of alkylaluminum and alkyllead compounds. In fact, one of the few commercial electroorganic processes involves the synthesis of an organometallic compound. Tetraethyllead, an important material of commerce, at least at present, is made by the Nalco process [296].

Several reviews have appeared, some of which are specific to the area of electroorganometallic chemistry [249, 356, 363], while other more general articles [18, 19, 204, 205, 47, 294, 299, 338, 357, 139, 148] contain significant portions dealing with the material of this chapter. Bibliographies by Swann [339, 340, 341] also include electroorganometallic entries. A series of articles by R. E. Dessy and associates constitutes a significant review of this subject matter, and these papers are referenced at appropriate points in the text.

As the chapter title states, reactions discussed herein could be placed into two categories: (1) reactions of organometallic compounds to form either new organometallic compounds or simple organic compounds, and (2) reactions of simple organic compounds to form organometallic compounds. This was undesirable and unworkable as was a systematic grouping of the material following the periodic table. The more logical presentation follows the actual organization within the literature and breaks the material into cathodic and anodic processes. The cathodic section contains various preparations of organometallics from nonorganometallic compounds and also a systematic presentation of electroreductions of organometallics that follows a periodic table classification. Much work of a very similar nature for organomagnesium, organoboron, and organoaluminum compounds falls into an anodic section. Breakdown of the material on organotransition metal compounds into anodic and cathodic reactions did not seem to add any clarity, so this material is discussed together in a single section. A miscellaneous section is used to collect loose ends and is followed finally by two tested electrochemical preparations of organometallic compounds. In general, only reactions where significant quantities of materials are isolated and characterized are discussed. Electroanalytical studies are included only where they elucidate a particular reaction mechanism.

Organometallic compounds may be formed in either cathodic or anodic processes. Dissolution of the cathode material often occurs, for example, in the reduction of aldehydes, ketones, alkylhalides and certain olefinic compounds.

Diisopropylmercury, for example, is formed during the reduction of acetone at the mercury cathode:

$$Me_2C = O + 6H^+ + 6e \longrightarrow Me_2CHHgCHMe_2 + 2H_2O$$

In general, these reactions grossly appear to proceed by a radical mechanism, that is, the organic radical initially formed at the cathode reacts with the cathode metal:

$$R^+ + e \longrightarrow R\cdot$$
$$R\cdot + M \longrightarrow RM$$

Relative to cathodic processes, direct metallation of organic radicals via anodic processes is rare. Anodic oxidation of organometallics (indirect metallations) to yield new organometallics (e.g., Grignard reagents → tetraalkyllead compounds) is, however, an important field that has been studied extensively. In these processes, free radicals are formed at the anode upon electrolysis of an organometallic electrolyte. The resultant radical then attacks the anode metal in a manner that, outwardly at least, seems analogous to the reactions of cathodically generated radicals with cathode metals:

$$R^- \longrightarrow R\cdot + e$$
$$R\cdot + M \longrightarrow RM$$

In either cathodic or anodic processes, the generated radicals may alternatively dimerize or split out hydrogen to yield an unsaturated species.

Organometallics of the form R_2M typically are not electroconductive in the pure state. Even in donor solvents (e.g., ethers), their conductivity is very low, decreasing in the order $R_2Mg > R_2Be > R_2Zn > R_2Cd > R_2Hg$ [335, 337]. The addition of certain salts, metal hydrides, or other organometallic reagents to these compounds tends to significantly increase conductivity. For example, admixing ethylsodium and diethylzinc yields a highly conductive salt [180, 185]:

$$NaZnEt_3 \longrightarrow Na^+ + ZnEt_3^-$$

Application of the complexation technique has instigated a significant number of studies of the electroorganic reactions of complex organometallics. These studies will be detailed in subsequent sections.

2   CATHODIC PROCESSES

Organoalkalimetal Compounds

Reduction of Vinyl Monomers

Much of the recent work on the electroorganic synthesis of organoalkalimetal compounds has been stimulated by electrolytic initiation of anionic polymerization [233, 414]. Vinyl monomers that yield relatively stable carbanionic species upon reduction have been studied most extensively.

$$2R_2C = CH_2 + 2e \longrightarrow R_2\bar{C}-CH_2CH_2-\bar{C}R_2$$
$$R_2\bar{C}-CH_2CH_2\bar{C}R + M \longrightarrow {}^-M(M)_nR_2C-CH_2 CH_2-CR_2(M)_nM^-$$

Reaction of the initially formed dicarbanionic species with more monomer, M, yields a "living polymer" whose molecular weight is proportional to the amount of monomer and the quantity of current passed. Funt has reviewed this in detail (see Chapter XI).

McKeever and Waack have similarly found that nonpolymerizable 1,1-disubstituted and 1,1,2,2-tetrasubstituted olefins undergo facile electrolytic reduction in polar aprotic solvents, yielding stable carbanionic species [263]. Reductions were conducted in hexamethylphosphoramide using alkali metal halides as supporting electrolytes. For example, reduction (potential not controlled) of 1,1-diphenylethylene results in quantitative yields of the 1,1,4,4-tetraphenyl-

$$2(C_6H_5)_2C=CH_2 + 2M^+ + 2e \longrightarrow$$
$$[(C_6H_5)_2\bar{C}-CH_2 CH_2-\bar{C}-(C_6H_5)_2]M_2^+$$

butane dianion. Comparison of the quantity of current passed with the moles of active organometallic species formed indicated that one mole of base was formed for every equivalent of electricity passed. It was not established whether direct reduction of the olefin or indirect reduction by alkali metal took place.

Analogous reduction of 1,1-diphenylethylene in hexamethylphosphoramide (HMPA) solution containing calcium iodide electrolyte yielded the dimeric 1,1,4,4-tetraphenylbutane dianion:

$$2(C_6H_5)_2C{=}CH_2 + Ca^{++} + 2e \longrightarrow [(C_6H_5)_2C{-}CH_2CH_2{-}C(C_6H_5)_2]^{=}Ca^{++}$$

Electroreduction of aryl halides, such as triphenylmethylchloride, in HMPA solution containing alkali metal halide supporting electrolyte yields the corresponding carbanionic species [263]:

$$(C_6H_5)_3CCl + M^+ \xrightarrow{\ e\ } (C_6H_5)_3C^-M^+ + Cl^-$$

It is likely that the dimer is an intermediate that is easily reduced further to the carbanion.

Reduction of Aromatic Compounds

Electroreduction of solutions of difficultly reducible aromatic compounds containing alkali metal salts as supporting electrolytes usually occurs by indirect electron transfer. However, in some cases where the aromatic species is easily reducible, direct electron transfer occurs [334, 47].

Birch reported the electrolytic reduction of m-tolyl methyl ether to the di-hydro adduct in ammonia solutions containing sodium ethoxide electrolyte [31]. A smooth copper cathode was employed, and the reduction occurred with low current efficiency. In subsequent studies, Benkeser and co-workers studied the selective reduction of various aromatic compounds to the dihydro and tetrahydro products [26, 27]. Electrolysis of benzene, for example, was carried out in methylamine solvent using lithium chloride electrolyte. Interestingly, the nature of the reduction is dependent on whether the cell is divided or not.

In electrolyses conducted in a divided cell, the catholyte typically assumed the deep blue color of the solvated electron. Product formation in the divided cell was explained by reaction of the dihydro products with the lithium methylamide produced in the cathode compartment. In the undivided cell, the lithium alkyl-amide was suggested to be neutralized by reaction with the methylamine hydrohalide formed at the anode. Under these conditions, the dihydro adducts are stable to further reduction.

Sternberg and co-workers [331, 334] have similarly studied the electro-chemical reduction of aromatic hydrocarbons in ethylenediamine solutions containing lithium chloride electrolyte. As illustrated by the current voltage data in Fig. 10.1, reduction of the polycyclic hydrocarbons, naphthalene, anthracene, and biphenyl proceeds via direct electron transfer. Benzene, however, has a more negative reduction potential (-1.1 V versus Zn(Hg)-ZnCl$_2$), and

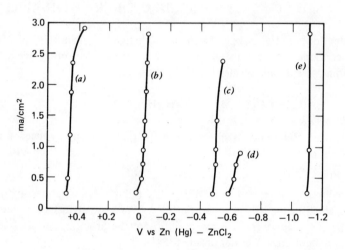

Fig. 10.1 Current-voltage curves in ethylenediamine, 0.3 M in LiCl. (*a*) [H$_2$NCH$_2$CH$_2$NH$_3$]$^+$; (*b*) anthracene; (*c*) naphthalene; (*d*) biphenyl (all 0.017 M); (*e*) Li$^+$ (0.3 M). (From Ref. 334.)

reduction therefore occurs by indirect transfer from the metal. The overall current efficiency for benzene reduction is about 64%, resulting in cyclohexadiene (17%); cyclohexene (70%) and cyclohexane (13%). Hydrogen was the main by-product.

$$\text{Li}^+_{(s)} \cdots \text{e}^-_{(s)} + \text{B}_{(s)} \rightleftharpoons \text{Li}^+_{(s)} + \text{B}^-_{(s)}$$

Sternberg, Markby, Wender, and Mohilner later found that benzene and tetralin could be reduced in ethanol/hexamethylphosphoramide solution at high current efficiency (cf. Table 10.1) [332, 333]. It is especially notable that HMPA suppresses hydrogen evolution from ethanol at the cathode. It was also found that olefins such as hexene-1, hexene-2, and 2,3-dimethyl-2-butene were easily reduced electrochemically [332]. Reduction in ethanol-HMPA appeared to occur via electrochemically generated solvated electrons. The transient organolithium reagent (e.g., benzene anion) is protonated by solvent, further reduced and protonated to complete the hydrogenation of the double bond.

Table 10.1. Electrolytic Reduction of Benzene

| | | Product Yield (mole %) | | |
Solvent	Current Efficiency	Cyclo-hexadiene	Cyclo-hexene	Cyclo-hexane
Ethylenediamine	64	17	70	13
Ethanol-HMPA ($\frac{2}{3}$ ethanol)	95	22.8	10.0	67.2

Analogous studies by Asahara, Senō, and Kaneko [9] revealed that electro-reduction of naphthalene in hexamethylphosphoramide/alcohol solutions proceeded via electrochemically generated solvated electrons. Lithium chloride was used as the electrolyte. The product composition (dihydronaphthalene, tetralin, hexalin, octalin, and decalin) was found to be dependent on reduction potential, alcohol type, and concentration of hexamethylphosphoramide.

Electroreduction of Onium Ions

This section is divided as follows: (1) the formation of organic amalgams via onium ion electroreductions, (2) electrolysis of onium ions at sacrificial cathodes to yield organometallics, and (3) the electrolysis of organometallic onium ions. Onium ion types treated here are sulfonium, quaternary ammonium, arsonium, iodonium, and phosphonium.

Organic Amalgams

Organic amalgams are formed by the electroreduction of certain onium ions at mercury cathodes. The overall process is similar to that for alkali metal amalgams:

$$Na^+ + e + XHg \longrightarrow Na\cdot/XHg$$
$$R_4N^+ + e + XHg \longrightarrow R_4N\cdot/XHg$$

We will denote organic amalgams as above, by placing an electron near the organic portion of the formula. No implications are intended regarding the association between the cation and the electron.

The study of organic amalgams seems to have been inspired, in part, by the formation of an ammonium amalgam [310, 323] that was obtained by electrolysis of ammonium carbonate at a mercury cathode. Early [260] interest in obtaining organic amalgams was also sparked by a desire to obtain substances of a metallic nature composed in part of nonmetallic elements. The ability of metal cations to accept electric charge and become free metals was equated with

Table 10.2. Organic Amalgams

Depolarizer	(Supporting Electrolyte, Solvent)	Cathode (Potential, Other)	Products (% Yield)	Ref.
		Alkylammonium Amalgams		
$(CH_3)_4NBF_4$	(Depolarizer, CH_3CN or N-methyl-pyrollidone)	Hg (20 mA/cm^2, 0°C)	Tetramethylammonium amalgam $(CH_3)_4N^{\cdot}$/XHg	203, 242
$(CH_3)_4NCl$	(Depolarizer, absolute EtOH)	Hg ($-$, -10°C)	As above–white silvery crystals	260
$(CH_3)_4NCl$	(Depolarizer, absolute EtOH or CH_3CN n-PrOH)	Hg ($-$,-34°C)	As above	261
$(CH_3)_nNH_{4-n}Cl$ $n = 1,2,3,4$	(Depolarizer, H_2O)	Hg ($-$, $-$)	Supposed quaternary ammonium amalgam, no isolation, decomposed in H_2O	236
$(Et)_nNH_{4-n}Cl$	(Depolarizer, H_2O)	Hg ($-$, $-$)	As above	236
$(Et)_4NI$	(Depolarizer, CH_3CN)	Hg (20 mA/cm^2, 0°C)	$(Et)_4N^{\cdot}$/XHg	203
$(Bu)_4NI$	(Depolarizer, dimethylformamide)	Hg ($-$, -20°C)	$(Bu)_4N^{\cdot}$/XHg	203
$CH_3N(Et)_3Cl$	(Depolarizer, hexamethylphosphoramide)	Hg (50 mA/cm^2,-10°C)	$CH_3N(Bu)_3^{\cdot}$/XHg	203
n-$C_{16}H_{33}N(CH_3)_3Br$	(Depolarizer, dimethylformamide)	Hg (20 mA/cm^2,-25°C)	n-$C_{16}H_{33}N(CH_3)_3^{\cdot}$/XHg	203
R_4NA^a R = Me, Et, n-Pr, n-Bu	(Depolarizer, CH_3CN or dimethylformamide)	Hg ($-$, low temp.)	$R_4N^{\cdot}$/12Hg	241
n-$C_{12}H_{25}N(CH_3)_3A^a$	(Depolarizer, CH_3CN or dimethylformamide)	Hg ($-$, low temp.)	n-$C_{12}H_{25}N(CH_3)_3^{\cdot}$/12Hg	241
n-$C_{16}H_{33}N(CH_3)_3A^a$	(Depolarizer, as above)	Hg ($-$, low temp.)	n-$C_{16}H_{33}N(CH_3)_3^{\cdot}$/12Hg	241
n-$C_{16}H_{33}N(CH_3)_3Br$	(Depolarizer, CH_3CN)	Hg (60 mA/cm^2,-15°C)	n-$C_{16}H_{33}N(CH_3)_3^{\cdot}$/7 Hg	37
$PhCH_2N(CH_3)_3I$	(Depolarizer, CH_3CN)	Hg ($-$, -30°C)	$PhCH_2N(CH_3)_3^{\cdot}$/XHg	242

404

Sulfonium and Phosphonium Amalgams

	(Depolarizer, CH$_3$CN or dimethylformamide)	Hg (−, low temp.)		65, 412
R$_4$PI R$_1$=R$_2$=R$_3$=R$_4$ R=Me, n-Bu, Ph	(Depolarizer, CH$_3$CN or dimethylformamide)	Hg (−, low temp.)	R$_4$P'/XHg (not isolated)	65, 412
$(n\text{-Bu})_3\overset{\text{I}}{P}-(CH_2)_n-\overset{\text{I}}{P}(n\text{-Bu})_3$ $n = 1,2,3,4$	(Depolarizer, CH$_3$CN or dimethylformamide)	Hg (−, low temp.)	$(n\text{-Bu})_3P-(CH_2)_nP(n\text{-Bu})_3$/XHg	65, 412
$(Ph)_nP(CH_3)_{4-n}\overset{\text{I}}{}$ $n = 1,2,3$	(Depolarizer, CH$_3$CN or dimethylformamide)	Hg (−, low temp.)	$(Ph)_nP(CH_3)_{4-n}$/XHg (not isolated)	65, 412
$(R)_3SI$ or Br R = Me, Et, n-Bu, Ph	(Depolarizer, CH$_3$CN or dimethylformamide)	Hg (−, low temp.)	$(R)_3S'$/XHg (not isolated)	65, 412
$(Ph)_2SCH_3PF_6$	(Depolarizer, CH$_3$CN or dimethylformamide)	Hg (−, low temp.)	$(Ph)_2SCH_3'$/XHg (not isolated)	65, 412
$(CH_3)_3SI$	(Depolarizer, CH$_3$CN or dimethylformamide)	Hg (−, low temp.)	(Amalgam)?	65, 412

^{a} A, anion unspecified.

the tendency of onium ions to accept electrons, hence the search for metallic properties in this latter species.

Conditions for the electrochemical preparation and isolation of organic amalgams are fairly restrictive regarding temperature and solvent systems. Generally their successful preparation has been carried out below $0°C$ in solvents more inert to reduction than water (alcohols, polar aprotics). Thus, early experiments [236] involving electrolysis of mono, di, tri, and tetramethylammonium ions in water did not yield the corresponding amalgams. Similar observations were also made [236] for mono, di, and triethylammonium ions in water.

Tetramethylammonium amalgam was isolated [260] in the reduction of the corresponding ammonium chloride in absolute alcohol at $-10°C$.

$$(CH_3)_4N^+ + e + XHg \longrightarrow (CH_3)_4N \cdot /XHg$$

The amalgam was described as crystalline, silvery white, and violently reactive, evolving $N(CH_3)_3$ when warmed to room temperature. Emphasizing the effect of temperature on the process, a more efficient preparation was subsequently described [261] when the electroreduction was carried out in a manner similar to the above preparation while maintaining the temperature at $-34°C$.

In addition to the alkylammonium amalgams, sulfonium and phosphonium amalgams have been reported [65]. By analogy with the quaternary ammonium amalgams [260], the formation of sulfonium or phosphonium amalgams could be represented as follows:

$$(R)_3S^+ + e + XHg \longrightarrow (R)_3S \cdot /XHg$$

Many properties are common to all the amalgams mentioned above, for example, their violent decomposition in water, thermal decomposition, ability to transfer electrons to certain other molecules, and the observed residual potential of the amalgams measured against a standard electrode when the driving voltage is removed during their preparation by electrolysis.

Two routes have been mentioned [65, 241, 260, 280] for decomposition of these amalgams: (1) the clean dry amalgams can be thermally ($25°C$) decomposed to parent amines [241, 260], sulfides [65], phosphines [65], and hydrocarbon radicals [295],

$$(CH_3)_4N \cdot /XHg \xrightarrow{20°C} (CH_3)_3N + CH_3 \cdot + XHg$$
$$(R)_3S \cdot /XHg + H \cdot \xrightarrow{20°C} (R)_2S + RH^* + XHg$$

*RH can be either saturated or unsaturated hydrocarbon, but no dimeric (R-R) products were observed for either sulfonium or phosphonium amalgams [65].

and (2) electrons can be transferred to acceptor species such as water, alcohols, aromatics, and metal cations, generating a reduced species plus the original onium ion.

The above modes of decomposition can be used to make inferences regarding amalgam composition and reduction potential. Decomposition of $(CH_3)_4N^{\cdot}/XHg$ in water (violent evolution of hydrogen) and quantitative observation of onium ion generated led to a value of 12 or 13 Hg atoms per nitrogen in the amalgam [241]. The displacement of cations (Cu^{++}, Na^+, K^+, Rb^+, Cs^+) from solution yielded either the free metal or metal amalgam [260]. This fact was used to judge that the amalgam reducing potentials were more than for the above cations. Myatt and Todd [280] reported $(R)_4N^{\cdot}/XHg$ amalgams ($R= CH_3$, Et, n-Pr, n-Bu, dodecyl) showed no ESR signal but were able to transfer electrons to naphthalene, anthracene, terephthalonitrile, nitrobenzene, and biphenyl, giving paramagnetic solutions. Of these, biphenyl has the most negative reducing potential (-2.7 V versus SCE), and it was concluded that the reduction potential of the amalgams was more negative than this.

The crystalline nature of the $(CH_3)_4N^{\cdot}/12Hg$ was confirmed [241] by means of an x-ray powder diagram of the dendritic crystals that were visible to the naked eye.

Tetramethylammonium amalgam has a density less than that of mercury [260], but the amalgam does not show the huge volume expansion noted for ammonium amalgam ($NH_4^{\cdot}/XHg$). A volume expansion more like that of ammonium amalgam was noted [260] for $CH_3NH_3^{\cdot}/XHg$.

Electroreductions to form organic amalgams contrast sharply with many other organic electrochemical reactions, especially those of this section in which radicals (free or otherwise) are probable intermediate species. Positively charged depolarizers make the generation of neutral uncharged adducts possible, and at the low temperatures employed for organic amalgam preparation, the lifetime of these adducts is far in excess of corresponding species generated by electron transfer in the systems to be subsequently described. In the latter situations the adduct moves on to final products in a relatively short interval of time. Furthermore, in the amalgam systems, we have seen that the original onium ion can even be regenerated by discharging the amalgams with suitable electron acceptors.

Electroreduction of Onium Ions—Sacrificial Cathodes

Transfer of electrons from metal cathodes such as mercury or lead to many onium ions, near room temperature and above, causes rapid discharge of the cation, scission of a carbon-heteroatom bond and generation of organometallic compounds:

$$(R)_n A^+ + e + Metal \rightarrow (R)Metal + R_{(n-1)}A$$

$$A = \text{heteroatom S, P, I, Sb, N}$$

This is true for some sulfonium, phosphonium, stibonium, iodonium, and quaternary ammonium compounds.

Many aspects of the present reductions are similar to the generation of organic amalgams via onium ion reductions, but the isolable electron adducts that were stable at the low temperatures employed for amalgam preparation are apparently quickly decomposed at room or higher temperatures, giving very different isolable products. By raising the temperature of the organic amalgams, many of the products generated in the present syntheses are observed. This correspondence is shown in Fig. 10.2 for trialkylsulfonium ions.

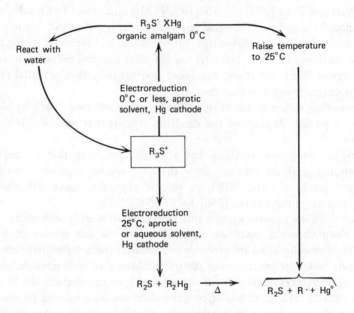

Fig. 10.2 Electroreduction of trialkylsulfonium ions under different conditions; mercury cathode.

Complete electrochemical studies of both triphenylsulfonium [264] and tetraphenylstibonium [276] ions on mercury cathodes have shown that a two-step reduction process is operative and that the mercury cathode was involved in the first reduction step. Morris, McKinney, and Woodbury [276] have proposed a reduction scheme for the tetraphenylstibonium ion.

$$(C_6H_5)_4Sb^+\text{.....}Hg + e \rightarrow (C_6H_5)_3Sb + [C_6H_5Hg\cdot]$$
$$2\,[C_6H_5Hg\cdot] \rightarrow (C_6H_5)_2Hg + Hg^\circ$$
$$[C_6H_5Hg\cdot] + H^+ + e \rightarrow C_6H_6 + Hg^\circ$$

It bears many features in common with schemes for other onium ion reductions, particularly that for the triphenylsulfonium ion. These common features are (1) appearance of two reduction steps, each consuming one electron per reduced species, (2) required association of the cathode metal with an intermediate in the process to account for observed products, and (3) the existence of a competitive situation between disproportionation of $[C_6H_5Hg\cdot]$ and further reduction to $C_6H_5:^-$.

The relative amounts of the two products, diphenylmercury and benzene, shown in the above scheme depend on reduction potential, current density and no doubt many other factors. For example, by reducing the tetraphenyl-stibonium ion at a first reduction stage (-.75 V versus SCE), only triphenyl-stibine and diphenylmercury were isolated, but reduction at a second stage (-1.4 V versus SCE) gave only benzene and triphenylstibine [276].

Behavior similar to the above was noted [264] for the triphenylsulfonium ion in aqueous systems. In this case the products were diphenylmercury, diphenyl-sulfide, and benzene. Reduction at potentials beyond a second polarographic wave gave benzene, diphenylsulfide, and some diphenylmercury, while at a first (more positive) wave diphenylmercury and sulfide were produced exclusively. In addition to the voltage variable the authors noted a further effect, that of depolarizer concentration, on the product mix (diphenylmercury versus benzene) and coulometric n value. Even at reduction potentials beyond the second polarographic wave, increasing amounts of diphenylmercury were formed if the concentration of triphenylsulfonium ion was increased in the millimolar range. This phenomenon was interpreted as an effect of the competition between the disproportionation process (see above stibonium reduction scheme) and the further reduction of the $[C_6H_5Hg\cdot]$ species. Higher concentrations of depolarizer gave greater reduction rates of the triphenylsulfonium ion, hence greater cathode surface concentrations of $[C_6H_5Hg\cdot]$, which stood a better chance of disproportionating at increased concentrations, even though the reduction potential was sufficiently negative for reduction of $[C_6H_5Hg\cdot]$.

Many benzylic sulfonium ions when electroreduced at mercury or lead cathodes also yield corresponding organometallics [20, 325a]. Again, the mechanistic thinking is similar to that described above. The reactions at mercury cathodes are conveniently carried out in water and are high-yield processes for obtaining symmetrical organomercurials. The reduction of the benzyldimethyl-sulfonium ion is a representative example [325a].

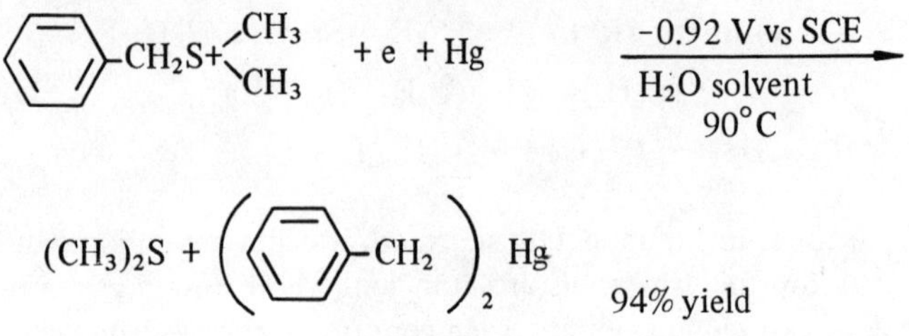

$$\text{Ph–}CH_2S^+\!\!\begin{array}{c}CH_3\\CH_3\end{array} + e + Hg \quad \xrightarrow[\substack{H_2O \text{ solvent}\\90^\circ C}]{-0.92 \text{ V vs SCE}}$$

$$(CH_3)_2S + \left(\text{Ph–}CH_2\right)_2 Hg \qquad 94\% \text{ yield}$$

The product mercurial is crystalline and in a relatively pure form as generated in the catholyte solution. Similar products are formed when various alkyl groups (Et, *n*-Bu, CH_2CH_2OH) and ring substituents (halogen, alkyl, MeS, Ph) are employed [325a]. By employing a lead cathode, nitrogen purge and dimethylformamide solvent the isolated product is tetrabenzyllead [325a].

In view of the many benzylic sulfoniums (Table 10.3) that form organometallic compounds at mercury or lead cathodes, it is interesting to note three structural changes in these molecules which prevent the formation of organometallic products when the compounds are reduced at either mercury or lead cathodes. These structures are shown below. All three variations yield direct carbon coupling in contrast to inclusion of mercury or lead atoms in the products for other benzylic sulfoniums.

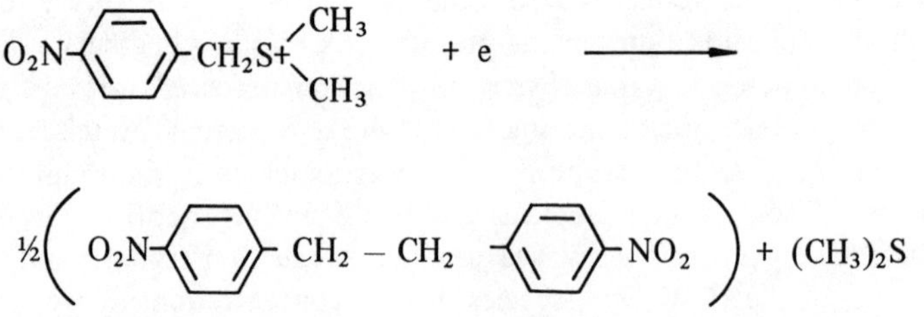

$$O_2N\text{–}C_6H_4\text{–}CH_2S^+\!\!\begin{array}{c}CH_3\\CH_3\end{array} + e \quad \longrightarrow$$

$$\tfrac{1}{2}\left(O_2N\text{–}C_6H_4\text{–}CH_2 - CH_2\text{–}C_6H_4\text{–}NO_2 \right) + (CH_3)_2S \quad [406]$$

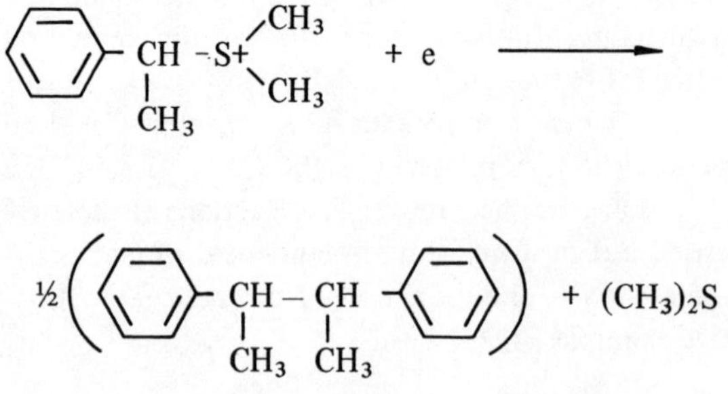

$$Ph\text{–}\underset{CH_3}{\underset{|}{CH}}\text{–}S^+\!\!\begin{array}{c}CH_3\\CH_3\end{array} + e \quad \longrightarrow$$

$$\tfrac{1}{2}\left(Ph\text{–}\underset{CH_3}{\underset{|}{CH}} - \underset{CH_3}{\underset{|}{CH}}\text{–}Ph \right) + (CH_3)_2S \qquad\qquad [325b]$$

Table 10.3. Electroreduction of Onium Compounds—Sacrificial Cathodes

Depolarizer	(Supporting Electrolyte, Solvent)	Cathode (Potential, Other)	Products (% Yield)	Ref.
		Sulfonium Compounds		
$(Ph)_3SBr$	(KCl, H_2O)	Hg (-1.25 V vs SCE)	$(Ph)_2Hg$ $(Ph)_2S$ (quantitative conversion)	264
$(Ph)_3SBr$ 3.5×10^{-3}	$(KCl, H_2O$ with gelatin)	Hg (-1.6 V vs SCE)	$C_6H_6 + (Ph)_2S$	264
$(Ph)_3SBr$ 3.5×10^{-3}	$(KCl, H_2O + $ Triton X 100)	Hg (-1.6 V vs SCE)	Some $(Ph)_2Hg + (Ph)_2S + C_6H_6$	264
$(Et)_3SBr$	(Depolarizer, CH_3CN)	Pb (0.02-0.03 A/cm^2)	$(Et)_4Pb$ (33) based on lead consumption	327
$(Et)_3SI$	(Et_4NBr, DMF)	Pb (-1.8 V vs SCE)	$(Et)_4Pb$ (50) based on sulfonium consumed. Trace of ethane.	325a
$(CH_3)_3SI$	(Et_4NBr, DMF)	Pb (-1.8 V vs SCE)	$(CH_3)_4Pb$	325a
$NC(CH_2)_3S(CH_3)_2OTos^a$	$(Et_4N \ OTos, DMSO)$	Hg (-1.90-2.05 V vs SCE)	$(CH_3)_2Hg$, $CH_3Hg(CH_2)_3CN$, $[NC(CH_2)_3]_2Hg$	396
$NC(CH_2)_2S(CH_3)_2OTos$	$(Et_4N \ OTos, DMSO)$	Hg (-1.6 V vs SCE)	$H_3CHgCH_2CH_2CN$ (5.3), $(NCCH_2CH_2)_2Hg$ (5)	396
$(PhCH_2)_3SHSO_4$	(2 M LiOAC-HOAC, 50% EtOH-H_2O)	Hg (-0.80 V vs SCE)	$(PhCH_2)_2Hg$, $(PhCH_2)_2S$	20
$PhCH_2S(CH_3)_2OTos$	(Et_4NBr, DMF)	Pb (-1.3 V vs SCE, N_2 pressure)	$(PhCH_2)_4Pb$ (9) yellow O_2 sensitive crystals	325a
$PhCH_2S(CH_3)_2A$ $A = Cl, NO_3^-, OTos^-$	(KCl, H_2O)	Hg (V vs SCE, 90°C) -1.08 -1.05 -0.91	$(PhCH_2)_2Hg + PhCH_3$ (54) (67) (94)	325a

411

Table 10.3. Electroreduction of Onium Compounds—Sacrificial Cathodes (cont.)

Depolarizer	(Supporting Electrolyte, Solvent)	Cathode (Potential, Other)	Products (% Yield)	Ref.
PhCH$_2$S(CH$_3$)$_2$OTos	(KOTos, DMF)	Hg (-1.16 V vs SCE, 25°C)	(PhCH$_2$)$_2$Hg (92)	325a
PhCH$_2$S(CH$_3$)$_2$Cl	(KCl, H$_2$O)	Hg (-1.30 V vs SCE, 25°C)	(PhCH$_2$)$_2$Hg (12)	325a
R—⟨C$_6$H$_4$⟩—CH$_2$S(CH$_3$)$_2$Cl	(KCl, H$_2$O)	Hg (V vs SCE, temp., °C)	(R—⟨C$_6$H$_4$⟩—CH$_2$)$_2$Hg + (CH$_3$)$_2$S	325a

	Cathode (V vs SCE, temp., °C)	R (% Yield)
R = p—CH$_3$	-1.22 / 25	(38)
R = m—CH$_3$	-1.0 / 50	(—)
R = o—CH$_3$	-1.1 / 25	(30)
R = p-t-amyl	-1.38 / 50	(83)
R = p-Ph	-1.18 / 25	(—)
R = p-C$_{12}$H$_{25}$	-1.10 / 38	(55)
R = p-CH$_3$S	-1.1 / 25	(—)

Depolarizer	Supporting Electrolyte, Solvent	Cathode (V vs SCE, temp., °C)	Products	X Yield
(X)—⟨C$_6$H$_4$⟩—CH$_2$S(CH$_3$)$_2$A		Hg (V vs SCE, temp., °C)	((X)—⟨C$_6$H$_4$⟩—CH$_2$)$_2$Hg	
X = o,p-2Cl A = Cl⁻	KCl, H$_2$O	-0.95 / 40		o,p-2Cl (95)
p-F OTos⁻	Et$_4$NOTos, DMF	-0.74 / 40		p-F (57)
p-Cl OTos⁻	Et$_4$NBr, DMF	-0.87 / 25		p-Cl (66)
p-Br Br⁻	KCl, H$_2$O	-1.05 / 25		p-Br (—)
o—⟨C$_6$H$_4$⟩—CH$_2$S(CH$_3$)$_2$ 2Cl	(KCl, H$_2$O)	Hg (-1.5 V vs SCE, 25°C)	(CH$_3$—⟨C$_6$H$_4$⟩—O—⟨C$_6$H$_4$⟩—CH$_2$)$_2$Hg	325a

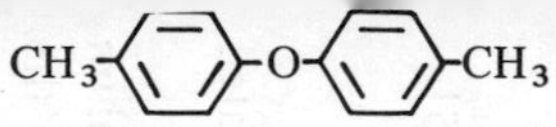

Quaternary Ammonium Compounds

(Et)$_4$NBr $\quad$ CH$_3$—〈benzene〉—O—〈benzene〉—CH$_3$ $\quad$ Pb (0.02-0.03 A/cm^2) (Depolarizer, CH$_3$CN) $\quad$ (Et)$_4$Pb (12) based on lead consumption $\quad$ 327

Phosphonium Compounds

(Ph)$_3$P(CH$_2$)$_n$CN
$n = 2,3,4$ $\quad$ Hg (-1.50 to -1.75 V vs SCE) (Depolarizer, DMF) $\quad$ [NC(CH$_2$)$_n$]$_2$Hg, $n = 2,3,4$ $\quad$ 395

Iodonium Compounds

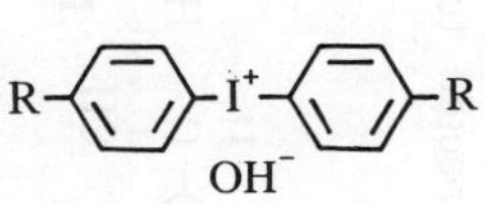

R—〈benzene〉—I$^+$—〈benzene〉—R OH$^-$

R = H
$\quad$ CH$_3$
$\quad$ OCH$_3$

Pt (-1.6 V vs SCE, CPE)^b (2N NaOH, H$_2$O) $\quad$ R—〈benzene〉—Hg—〈benzene〉—R $\quad$ 12

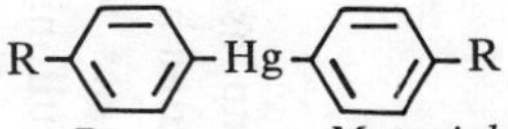

R	Material Yield
H	(51)
CH$_3$	(41)
OCH$_3$	(58)

Stibonium Compounds

(Ph)$_4$SbBr $\quad$ Hg (-0.75 V vs SCE) (KCl, H$_2$O) $\quad$ (Ph)$_2$Hg, (Ph)$_3$Sb $\quad$ Quantitative conversion 276

$\quad$ Hg (-1.4 V vs SCE) $\quad$ C$_6$H$_6$ (Ph)$_3$Sb

---

^{a}OTos is p-toluenesolfonate anion.
^{b}CPE is controlled-potential electrolysis.

413

$$\underset{CH_3}{\overset{CH_3}{>}} {}^+SCH_2 - \left\langle \underset{}{\bigcirc} \right\rangle - CH_2S^+ \underset{CH_3}{\overset{CH_3}{<}} \quad + \; 2e \longrightarrow$$

$$\left[CH_2 - \left\langle \underset{}{\bigcirc} \right\rangle - CH_2 - \right]_n \quad + \; 2(CH_3)_2S \qquad\qquad [405]$$

The effect of the para-nitro group, as we will see in a later section, is duplicated in the electrolysis of ring-substituted benqyl bromides [168, 217].

The contrast in reaction courses for the a-methylbenzyldimethylsulfonium cation and the benzyldimethylsulfonium cation is somewhat analogous to the situation prevailing when ketones having a carbonyl group adjacent to a phenyl group are electroreduced. Although electroreductions of benzaldehyde [6] **I**, phenylacetone [5] **II**, and benzylic sulfoniums **III** all yield the symmetrical mercurials **IV**, **V**, and **VI**, respectively, the compounds **VII** and **VIII** do not generate mercurials when similarly reduced at mercury cathodes.

$$\left\langle \bigcirc \right\rangle - \overset{O}{\underset{H}{\overset{\|}{C}}} \qquad\qquad \left\langle \bigcirc \right\rangle - \overset{O}{\underset{}{\overset{\|}{CH_2C}}} - CH_3 \qquad\qquad \left\langle \bigcirc \right\rangle - CH_2S^+ \underset{CH_3}{\overset{CH_3}{<}}$$

$$\textbf{I} \qquad\qquad\qquad\qquad \textbf{II} \qquad\qquad\qquad\qquad \textbf{III}$$

$$\left(\left\langle \bigcirc \right\rangle - CH_2 \right)_2 Hg \qquad \left(\left\langle \bigcirc \right\rangle - \underset{CH_3}{\overset{}{CH_2CH}} \right)_2 Hg \qquad \left(\left\langle \bigcirc \right\rangle - CH_2 \right)_2 Hg$$

$$\textbf{IV} \qquad\qquad\qquad\qquad \textbf{V} \qquad\qquad\qquad\qquad \textbf{VI}$$

$$\left\langle \bigcirc \right\rangle - \overset{O}{\underset{(CH_3)}{\overset{\|}{C}}} \qquad\qquad \left\langle \bigcirc \right\rangle - \underset{(CH_3)}{\overset{H}{\underset{}{C \, S^+}}} \overset{CH_3}{\underset{CH_3}{<}}$$

$$\textbf{VII} \qquad\qquad\qquad\qquad\qquad \textbf{VIII}$$

The circled methyl groups of compounds **VII** and **VIII** may both be acting in a similar manner to prevent formation of organomercurial since the same mercurial, **IX**, would be expected to be generated from both compounds **VII** and **VIII**.

$$\left\langle \bigcirc \right\rangle - \underset{CH_3}{\overset{H}{\underset{}{C}}} - Hg - \underset{CH_3}{\overset{H}{\underset{}{C}}} - \left\langle \bigcirc \right\rangle$$

$$\textbf{IX}$$

If the formation of organometallics by electrolysis of the benzylic sulfonium compounds follows the above proposed scheme for stibonium compounds, we should observe a dependence in the relative proportions of observed products (e.g., dibenzylmercury and toluene) with electrolysis potential. Two polarographic waves at -0.65 and -1.15 V versus SCE were observed [20] for the tribenzylsulfonium ion. These are well-separated waves and controlled-potential electrolysis using a mercury cathode at -0.8 V versus SCE gave only dibenzylmercury and benzylsulfide [20]. Any further or second-step reduction of an intermediate radical, for example, $[PhCH_2Hg\cdot]$, in the aqueous system would require the formation of a third product, probably toluene.* Now, in the present case the first reduction stage lies at the relatively positive value of -0.65 V versus SCE compared to benzyldialkylsulfonium ions, where the first half-wave value is approximately -1.3 V versus SCE [325a]. In the latter case an electrolysis carried out at -1.30 V versus SCE at 25°C in water on a mercury cathode yielded only 12% of dibenzylmercury while yielding 83% of toluene. At this more negative setting we are likely seeing a large amount of a second-stage reduction resulting in toluene:

$$[PhCH_2 - - - - - Hg\cdot] + e \rightarrow PhCH_2{:}^- + Hg^\circ$$
$$PhCH_2{:}^- + H^+ \rightarrow PhCH_3$$

Bar [20] has observed a shift of the first reduction wave for the tribenzylsulfonium ion toward positive values with increasing temperatures. Working at 90°C, it is possible to reduce the benzyldimethylsulfonium ion at -0.92 V versus SCE and obtain a 94% yield of dibenzylmercury [325a]. We have favored the first reduction step and subsequent steps leading to mercurial, relative to the competing second reduction step that generates toluene.

Cathodic reductions of trialkylsulfonium ions [325a, 327, 396] and tetraalkylammonium ions [327] at mercury or lead cathodes lead to organomercury or lead compounds. The following reactions are illustrative:

$$NCCH_2CH_2\overset{\overset{\displaystyle CH_3}{\displaystyle |}}{\underset{\underset{\displaystyle CH_3}{\displaystyle |}}{S}} + \text{+e} \xrightarrow[\text{Hg cathode}]{\text{DMSO}} CH_3HgCH_2CH_2CN + (NCCH_2CH_2)_2Hg + (CH_3)_2S$$

$$(Et)_3S^+ + e \xrightarrow[\text{Pb cathode}]{\text{DMF}} (Et)_4Pb + (CH_3)_2S$$

---

*A coulmetric n value of 0.98 [20] also argues for a high yield of product generated in a first stage reduction since generation of toluene would require two electrons per sulfonium ion.

The proposed [396] mechanism for trialkylsulfonium compounds is similar to that discussed above for benzylic and aryl sulfonium compounds. The initial reduction product of an overall two-step reduction process was thought to be a reactive radical that formed an organometallic product at the active cathode. Further reduction generated a carbanion from the radical. In support of the above thinking the cyanomethyldimethylsulfonium ion was observed to undergo only a single two-electron reduction step and no organomercurials were observed on macroelectrolysis of this onium ion [396]. Only carbanion products were formed:

$$\text{NCCH}_2\overset{\overset{\displaystyle CH_3}{|}}{\underset{\underset{\displaystyle CH_3}{|}}{S}}\text{+} +2e \rightarrow (CH_3)_2S + NCCH_2^-$$

$$NCCH_2^- + H^+ \rightarrow CH_3CN$$

Azoo, Coll, and Grimshaw found that diaryliodonium ions, like sulfonium ions, will form organomercurials when reduced at a mercury cathode [12]. Corresponding diarylmercury compounds were isolated when diphenyl- and ring-substituted diphenyliodonium compounds were reduced in aqueous 2 N NaOH at -1.6 V versus SCE:

R=H, CH₃, OCH₃

The authors of Ref. 12 pointed out that competing nucleophilic attack on the iodonium cation by hydroxide ion limited the preparative value of the electrolysis. This was especially true for rings bearing electron-withdrawing substituents. In view of a report [298] showing that diphenyliodonium salts could react with finely divided mercury at 56°C, the authors of Ref. 12 showed that diphenyliodonium hydroxide did not enter into a significant chemical reaction with the mercury pool cathode [12].

A good deal of mechanistic background for the above syntheses is provided by a number of investigations into the polarography and macroscale electro-reduction of iodonium ions [13-15, 29, 58, 59, 423]. Three major reduction waves were observed [13-15, 58] for diphenyl- and ring-substituted diphenyl-iodonium ions. The reduction stages were interpreted [13, 14] as follows:

$$\text{I} \quad \phi I^+ \phi \; + e \; \xrightleftharpoons{\; -0.2\ \text{V vs SCE}\;} \; \phi\, I^{\cdot}\; \phi \longrightarrow \tfrac{1}{2}\, \phi{-}\phi + \phi\, I$$

$$\text{II} \quad \phi\, I^{\cdot}\; \phi + e + H^+ \; \xrightarrow{\; -0.7\ \text{to}\ -1.3\ \text{V vs SCE}\;} \; \phi\, H + \phi\, I$$

$$\text{III} \quad \phi\, I + \; 2e \; + H^+ \; \xrightarrow{\; -1.7\ \text{V vs SCE}\;} \; \phi\, H + I^-$$

A similar interpretation by Colichman and Maffei included mercury atoms in the scheme [58]. We can also see continuing similarity with previous onium ions discussed. Preparative electrolysis carried out at the first reduction stage (−0.2 V versus SCE) gave benzene, biphenyl, and phenyl iodide, but no mercurials [14]. The isolation of mercurial at −1.6 V versus SCE argues strongly for inclusion of mercury atoms at some point in the overall reduction scheme [12].

Electroreduction of Onium Ions—Nonsacrificial Cathodes

Some onium ions, although they did not form organometallics on electro-reduction, are themselves organometallics and as such their electroreduction reactions are logged in Table 10.4.

Horner and Haufe [196] have investigated the cathodic cleavage of arsonium salts, $Ph_3\overset{+}{As}RX^-$, where R was alkyl, aralkyl, or aryl. The reduction was found to cleave groups at the arsenic atom in a manner that depended on the nature of R:

$$Ph_3\overset{+}{As}R + 2e + H^+ \begin{cases} Ph_2AsR + PhH \\[4pt] Ph_3As + RH \end{cases}$$

The ease of cleavage of the leaving groups increased in the order Me < Et < p-CH$_3$C$_6$H$_4$ < n-Bu < Ph < CH$_2$=CH–CH$_2$, PhCH$_2$, EtO$_2$CCH$_2$, PhCOCH$_2$.

Horner and Fuchs [194] demonstrated that cathodic cleavage of optically active Me, Ph, PhCH$_2\overset{+}{As}$CH$_2$CH=CH$_2$ occurred with retention of configuration in the tertiary arsine. In the present case the benzyl group was cleaved as was expected from the above list.

Continuing with studies of preferential cleavage, Horner and Röttger [199] showed that the most stable anion or radical was cleaved during cathodic reduction of a large number of arsonium ions. The same authors obtained a

Table 10.4. Electroreduction of Onium Ions—Nonsacrificial Cathodes; Arsonium Ions

Depolarizer	(Supporting Electrolyte, Solvent)	Cathode (Potential, Other)	Products (% Yield)	Ref.
$[(Ph)_3AsR]Br$	(Depolarizer, H_2O)	Hg (0.02-0.03 A/cm^2, 0-10°C)	Benzene + RH + Ph_3As + Ph_2AsR	196
R			Benzene RH	
p-Tolyl			(92) (8)	
p-Isopropyl-phenyl			(91) (9)	
o-Tolyl			(84) (16)	
Me-			(81) (19)	
m-Tolyl			(80) (20)	
Ph			(75) (25)	
m-Biphenylyl			(65) (35)	
Et			(52) (48)	
Benzyl (depolarizer, 50% MeOH-H_2O)			(0) (100)	
$(Ph)CH_3As(CH_2Ph)_2$ Br	(Depolarizer, H_2O)	Hg (−, −)	$(Ph)PhCH_2AsCH_3$(99)	199
$(Ph)_3AsCH_2COOC_2H_5$ Br	(Depolarizer, H_2O)	Hg (−, −)	$(Ph)_3As$ (97)	199
$(PhCH_2)_3AsCH_3$ Cl	(Depolarizer, H_2O)	Hg (−, −)	$(PhCH_2)_2AsCH_3$ (95)	199
$(Ph)_3AsCH_2COPh$ Br	(Depolarizer, H_2O)	Hg (−, −)	$(Ph)_3As$ (96)	199
$(Ph)_2AsCH_3(CH_2CH=CH_2)$ Br	(Depolarizer, H_2O)	Hg (−,−)	$(Ph)_2AsCH_3$ (93)	199

number of polarographic reduction potentials of arsoniums in aqueous and alcoholic solutions [199].

Cathodic Reactions of Neutral Organics—Sacrificial Cathodes

"Oxidative and reductive coupling is one of the most common reactions in the field of organic electrochemistry..."[3]. Most of these reactions, with the notable exception of the reductive coupling work by M. Baizer et al. [18a], are thought to involve radical species at some intermediate point. It is the interaction of these electrolytically generated radicals with metal electrodes, rather than their simple coupling or dimerization, that leads to many electrochemical syntheses of organometallic compounds and provides not only the material of this section, but a large portion of the material of this chapter.

In this section we consider electrochemical reduction of uncharged organic molecules carried out at sacrificial cathodes to form organometallic compounds. Organic molecules dealt with here possess electroactive sites such as carbonyl groups, unsaturated carbon atoms, and halogen atoms. Reduction of these sites generates an intermediate (most often considered to be radical in nature) which is capable of interacting with the cathode metal to finally yield an organometallic molecule. Cathode metals which are attacked (Hg, Pb, Sb, Bi, Sn, Zn, and others) are also those metals that react with radicals in the classical Paneth reactions.

Carbonyl and Unsaturated Compounds

J. Tafel seems to have been the first to report [345] that organometallics, specifically bisorganomercurials, could be formed by electrolysis of carbonyl compounds. The supposed product when acetone was reduced at a mercury cathode was mercury diisopropyl [345]. Immediately following this, Tafel obtained a 20% yield of di-*sec*-butylmercury from the reduction of methyl ethyl ketone at a mercury cathode in 30% sulfuric acid solution [342]. A current density of 0.1-0.2 A/cm^2 was used at a temperature of 45-50°C. The following equation represents the latter reaction:

$$\underset{\substack{\| \\ }}{\overset{O}{CH_3C}}\!-\!Et + Hg + 6H^+ + 6e \rightarrow \underset{\substack{| \\ HC\!-\!Hg\!-\!CH \\ | \qquad | \\ Et \qquad Et}}{CH_3 \quad CH_3} + 2\ H_2O$$

In addition to organomercurials, Tafel also synthesized and identified organolead compounds in the electrolysis of acetone in 20% aqueous H_2SO_4 at a lead cathode [343]. Tetraisopropyllead was identified and in addition a small quantity of a red-brown oil was thought to be lead diisopropyl.

As a rule, strong acid electrolytes [5, 6, 170, 178, 235] and high temperatures

[5, 6, 317, 342] both favor formation of organometallic compounds in the electrolysis of carbonyl compounds. It is also generally thought that ketones that have a carbonyl radical adjacent to a phenyl radical are not reduced to mercury compounds [5, 281]. For example, acetophenone was reduced at a mercury cathode in 1 N HCl at 60°C by controlled-potential electrolysis (CPE) at −1.05 V versus SCE with only pinacol formation noted [281].

Reasonable agreement [48, 170, 324] exists on the mechanism of formation of the organometallic compounds discussed above and many others (Table 10.5) since reported. Just as with other competitive schemes noted throughout this chapter, an intermediate radical is first generated. Thus, for acetone [324]:

$$\underset{CH_3}{\overset{CH_3}{\diagdown}}C{=}O \xrightarrow{\;+H^+,\; +e\;} \underset{CH_3}{\overset{CH_3}{\diagdown}}\cdot C{-}OH$$

This radical may couple to form pinacol,

$$2\;\underset{CH_3}{\overset{CH_3}{\diagdown}}\cdot C{-}OH \rightarrow \underset{CH_3\;\;OH}{\overset{CH_3}{\diagdown}}C{-}\underset{OH\;CH_3}{\overset{CH_3}{C}}$$

or be reduced further in a second reduction step to yield either alcohol or organometallic compound,

$$\underset{CH_3}{\overset{CH_3}{\diagdown}}\cdot C{-}OH \xrightarrow{\;+H^+,\; +e\;} \underset{CH_3}{\overset{CH_3}{\diagdown}}HC{-}OH$$

$$\xrightarrow[\;Hg\;]{-H_2O\,+\,H^+\,+e} \underset{CH_3}{\overset{CH_3}{\diagdown}}HC{-}Hg{-}CH\overset{CH_3}{\underset{CH_3}{\diagup}}$$

The observation of propane [324] both at mercury and lead cathodes argues for a third reduction step, possibly involving sorbed R· or the organometallic compound. This third step and the normal second reduction observed for alkyl halides, and so on, appear to be very similar:

$$\underset{CH_3}{\overset{CH_3}{\diagdown}}HC\cdot \;\;+e\;+H^+ \rightarrow \underset{CH_3}{\overset{CH_3}{\diagdown}}HCH$$

Table 10.5. Electroreduction of Carbonyl and Unsaturated Compounds to Yield Organometallic Compounds

Depolarizer	(Supporting Electrolyte, Solvent)	Cathode (Potential, Other)	Products (% Yield)	Ref.
Me–C=O (on naphthalene)	(45% Aqueous H_2SO_4)	Hg (-1.5 V vs SCE, –)	Et–(naphthalene), Unstable oil of proposed structure $[\text{Me–HC–(naphthalene)}]_2\text{Hg}$	170
CH_3CCH_3 (=O)	(20% H_2SO_4, H_2O)	Pb–Cu alloys, Pb–Sn alloys (0.04 A/cm², –)	Metal organics and cathode destruction observed	361, 362
CH_3CCH_3 (=O)	(H_2SO_4, H_2O)	Hg (– –)	Supposed $CH_3\!-\!\overset{\displaystyle CH_3}{\underset{\displaystyle CH_3}{\text{HC–Hg–CH}}}\!-\!CH_3 \;+\; \overset{\displaystyle CH_3}{\underset{\displaystyle CH_3}{\text{HC–OH}}}$	345
CH_3C–Et (=O)	(30% H_2SO_4, H_2O)	Hg (–, 45–50°C)	$\left[CH_3\!-\!CH_2\!-\!\underset{\displaystyle CH_3}{\text{CH}}\right]_2\!Hg$ (20)	342
CH_3C–Et (=O)	(30% H_2SO_4, H_2O)	Pb (– –)	Red organolead oil	342

421

Table 10.5. Electroreduction of Carbonyl and Unsaturated Compounds to Yield Organometallic Compounds (cont.)

Depolarizer	(Supporting Electrolyte, Solvent)	Cathode (Potential, Other)	Products (% Yield)	Ref.
$CH_3\overset{O}{\overset{\|}{C}}-CH_3$	(20% H_2SO_4, H_2O)	Hg ⎱ (-1.37 V vs SCE, CPE) Pb ⎰	Isopropanol (95), pinacol (3), propane (2), diisopropylmercury (tr) Isopropanol (68), pinacol (7), propane (25), diisopropyllead (tr)	324
$CH_3\overset{O}{\overset{\|}{C}}-CH_3$	(0.5 M H_2SO_4, H_2O)	Hg (V vs SCE, mA/cm², divided cell) -1.27 1.6 -1.30 5.0 -1.37 8.4	Current Efficiencies Isopropanol / Pinacol / Diisopropyl-mercury 22 2.5 17 21 2.1 28 27 1.1 11	48
$CH_3\overset{O}{\overset{\|}{C}}-CH_3$	(40% H_2SO_4, H_2O)	Hg (2.8 A/dm²)	$\underset{CH_3}{\overset{CH_3}{>}}HC-Hg-CH\underset{CH_3}{\overset{CH_3}{<}}$ (12) Based on starting material	178
$CH_3CH_2C\overset{O}{\underset{H}{<}}$	(K_2PO_4, H_2O)	Sn (4 A/dm², 20°C)	$CH_3CH_2CH_2OH$ (19) 3,4-hexanediol (11) Unstable organotin cmpd. white amorphous ppt. 7%C, 2.6%H, 50% Sn	214

Substrate	Conditions	Electrode (metal)	Products / Observations	Ref.
Butraldehyde Heptylaldehyde Acetaldehyde } Same observations and conditions as for propionaldehyde above.				215
$CH_3C{-}CH_3$ ($=O$)	(H_2SO_4 or KOH, H_2O)	Ge ($\sim 10^{-2}$ A/cm^2, 20°C) n and p types	Isopropanol H_2 Resin containing Ge; decomp. 300°C	110
$CH_3C{-}CH_3$ ($=O$)	(NaOH or buffer, H_2O)	Zn (0.05–0.1 A/cm^2, 6–20°C)	Isopropanol, pinacol, spongy zinc deposit thought due to unstable organozinc compound.	213
$CH_3C{-}CH_3$ ($=O$)	(20% H_2SO_4, H_2O)	Pb (— —)	Lead diisopropyl (supposed Tetraisopropyllead	343
Isoamyl methyl ketone $\frac{CH_3}{CH_3}{>}CH{-}CH_2CH_2CH_2{-}C(=O){-}CH_3$		Pb (— —)	Isoheptane Organolead compounds	344
Methyl ethyl ketone	(20% H_2SO_4, —)	Pb (— —)	Tetra-(*sec*-butyl)lead	305
Diethyl ketone	(20% H_2SO_4, —)	Pb (— —)	Tetra-(3-pentyl)lead	305
$\frac{CH_3}{CH_3}{>}C{=}CH{-}C(=O){-}CH_3$ $\frac{CH_3}{CH_3}{>}C{=}CHCH_2CH_2{-}C(=O){-}CH_3$ Citral }	(H_2SO_4, H_2O-alcohol)	Pb (1.2 A/150 cm^2, —)	Red organolead unidentified	235

Table 10.5. Electroreduction of Carbonyl and Unsaturated Compounds to Yield Organometallic Compounds (cont.)

Depolarizer	(Supporting Electrolyte, Solvent)	Cathode (Potential, Other)	Products (% Yield)	Ref.
Menthone CH_3—CH—CH_2 CH_2—CH_2—$C=O$ CH_2—CH—$CH(CH_3)_2$	(H_2SO_4, H_2O-methanol)	Hg (10 A/50 cm^2, 70–77°C) Same electrolysis with Pb or Cd cathodes gave indication of unstable organoleads and organocadmium compounds.	Dimenthylmercury $\left[\begin{array}{c} CH_3 \\ CH \\ CH_2 \\ CH_2 \\ CH_2—CH_2 \\ CH—CH(CH_3)_2 \end{array}\right.$—Hg$\Big]_2$	317
Phenylacetone $C_6H_5CH_2CCH_3$ $\underset{\parallel}{O}$	$\left[\begin{array}{l} 5\%\ H_2SO_4,\ H_2O \\ 30\%\ H_2SO_4,\ H_2O \\ 5\%\ H_2SO_4,\ H_2O \\ HCl + HOAc,\ H_2O \end{array}\right.$	$Hg\left[\begin{array}{l} 131\ mA/cm^2,\ 18°C \\ 131\ mA/cm^2,\ 20°C \\ 197\ mA/cm^2,\ 55°C \\ 131\ mA/cm^2,\ 16°C \end{array}\right.$	No mercury cmpd. Yes mercury cmpd. $\left[C_6H_5CH_2CH\underset{CH_3}{-}Hg\right]_2$ (25-30) Yes mercury cmpd. Yes mercury cmpd.	5
Cyclopentanone	$\left[\begin{array}{l} 5\%\ H_2SO_4,\ H_2O \\ 5\%\ H_2SO_4,\ H_2O \\ HCl + H_2SO_4,\ H_2O \\ HCl + HOAc,\ H_2O \end{array}\right.$	$Hg\left[\begin{array}{l} 131\ mA/cm^2,\ 16°C \\ 197\ mA/cm^2,\ 50°C \\ 197\ mA/cm^2,\ 20°C \\ 197\ mA/cm^2,\ 21°C \end{array}\right.$	(25-30)	5
Cyclohexanone	$\left[\begin{array}{l} 5\%\ H_2SO_4,\ H_2O \\ 30\%\ H_2SO_4,\ H_2O \end{array}\right.$	$Hg\left[\begin{array}{l} 260\ mA/cm^2,\ 55°C \\ 230\ mA/cm^2,\ 16°C \end{array}\right.$	(25-30)	5

2-Methylcyclohexanone 3-Methylcyclohexanone 4-Methylcyclohexanone	(5% H_2SO_4, H_2O)	Hg (197 mA/cm², 55°C)	[cyclic thioether S–Hg–S structure, CH_3] (25-30)	5
Benzaldehyde [C_6H_5–C(=O)H]	(50% H_2SO_4, H_2O)	Hg (19.7 A/dm², <25°C)	$[C_6H_5CH_2\text{–}]_2Hg$ Based on starting material (8.5)	6
$H_2C{=}CHC(O){-}CH_3$	(pH < 5, – –)	Hg (–, –)	$CH_3C(O){-}CH_2CH_2)_2Hg$	191
$H_2C{=}CHC(O){-}CH_3$	HOAC, NaOAc, EtOH	Hg (–1.35 V vs ?, –)	$(CH_3C(O){-}CH_2CH_2)_2Hg$	407
$(CH_3)_2C{=}CHCH_2CH_2C(O){-}CH_3$ $(CH_3)_2C{=}CHCH_2CH_2C(CH_3){=}CHCHO$	(H_2SO_4, H_2O-alcohol)	Pb (1.2 A/150 cm, –)	Red organolead oils, unidentified	235
$H_2C{=}CH{-}C{\equiv}N$	(0.7 N NaOH, initially saturated with acrylonitrile, which was added constantly throughout exp.)	Sn $\left(\begin{array}{l}200 \text{ A/m}^2, 15°C\\2000 \text{ A/m}^2, 15°C\end{array}\right)$	$(NCCH_2CH_2)_4Sn$ $\begin{bmatrix}\text{Current yield}\\-36\\-\text{nil}\end{bmatrix}$ Tetra(β-cyanotehyl)stannane	358 208

Table 10.5. Electroreduction of Carbonyl and Unsaturated Compounds to Yield Organometallic Compounds (cont.)

Depolarizer	Supporting Electrolyte, Solvent	Cathode (Potential, Other)	Products (% Yield)	Ref.
$H_2C=C-C\equiv N$ $\quad\quad\overset{\displaystyle CH_3}{\vert}$	(0.7 N NaOH 135 ml, containing 15 ml of depolarizer)	Sn (170 A/m², 27°C)	$(N\equiv C-CHCH_2)_4Sn$ $\quad\quad\quad\overset{\displaystyle CH_3}{\vert}$ (14) Tetra(β-cyanopropyl)tin	208
$H_2C=C-C\equiv N$ $\quad\quad\overset{\displaystyle CH}{\vert}\ \overset{\displaystyle CH_2}{\vert}$	(Aqueous alkaline medium, probably as above)	Sn (−, −)	Organotin compound, yellow grey powder	208
S, acrylonitrile	(1 N Na₂SO₄, H₂O) pH 7.5-9	Gra- ⎡10 A/625 cm², 40°C	$(NCCH_2CH_2)_2S$ (15)	360
Se, acrylonitrile		phite ⎢0.011 A/cm², 46-50°C	$(NCCH_2CH_2)_2Se$ (23)	207
Te, acrylonitrile		⎣0.011 A/cm², 46-50°C	$(NCCH_2CH_2)_2Te$ —	207
Dimethylethynyl-carbinol $HOC(CH_3)_2$ $\quad\quad\overset{\displaystyle CH}{\underset{\displaystyle \vert\vert}{\ }}$ $\quad\quad CH$	—	Copper (−, −)	Organocopper compound C_5H_7OCu	235a
$\overset{O}{\underset{\vert\vert}{S}}(OCH_2CH_3)_2$	((Et)₄NBr, acetonitrile)	Pb (0.02-0.03 A/cm², 25-60°C) 4 hr	$(CH_3CH_2)_4Pb$ (81) Lead yield	327
$CH_3-\overset{O}{\underset{\vert\vert}{C}}-O-CH_2CH_3$	((Et)₄NBr, acetonitrile)	Pb (0.02-0.03 A/cm², 25-60°C)	$(CH_3CH_2)_4Pb$ (10) Lead yield	327
Acrylonitrile	(Ca₃(PO₄)₂-H₃PO₄, H₂O)	Sn (3 A/389 cm², 20-22°C)	$(NCCH_2CH_2)_4Sn$ $[(NCCH_2CH_2)_3Sn]_2$ (26)	35

The scheme for reduction of carbonyl compounds is distinguished from most others mentioned in this chapter by the greater number of competing reactions which it contains. These branch points all tend to lower the yield of organometallic product.

When fast potential sweeps (960 V/sec) were employed in the electrochemical reduction of acetone, all of a species postulated to be

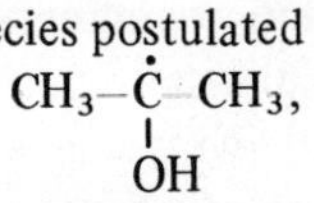

$$CH_3-\overset{\displaystyle \cdot}{\underset{\displaystyle \underset{\textstyle OH}{|}}{C}} CH_3,$$

from a first reduction was detected by reoxidation [48]. This, together with products isolated in many reductions (Table 10.5), lends support to the above scheme for carbonyl reduction.

Tomilov et al. used lead-tin [362] and lead-copper [361] alloy cathodes for the reduction of acetone in 20% H_2SO_4 solutions. Organometallics were not identified but were visually observed as dark brown or dark red oils, and cathode destruction by weight loss was noted. These studies revealed a correlation between degree of cathode destruction (formation of organometallics) and the proportion of acetone reduced to isopropanol. This is consistent with the scheme for acetone reduction proposed by Sekine et al. [324], which was noted above. Both isopropanol and organometallic were formed in a common second reduction step at the expense of pinacol formation. The scheme for acetone reduction proposed by Tomilov et al. [361] explained the correlation between isopropanol formation and cathode weight loss by invoking a common organometallic intermediate that finally led to both isopropanol and stable organometallic. Protonation of acetone was followed by a double electron reduction:

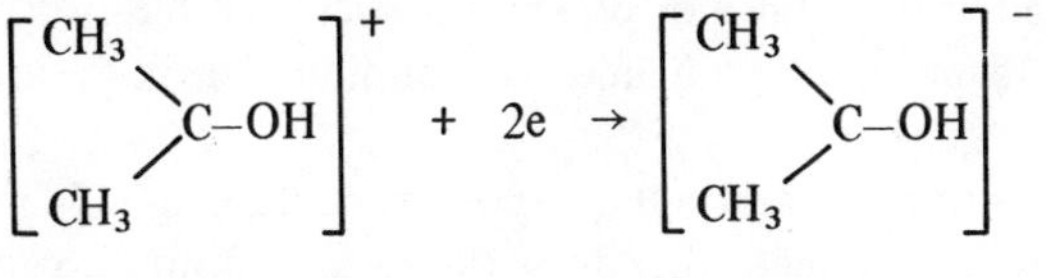

$$\left[\begin{array}{c} CH_3 \\ \diagdown \\ \diagup \\ CH_3 \end{array} C{-}OH \right]^{+} + 2e \rightarrow \left[\begin{array}{c} CH_3 \\ \diagdown \\ \diagup \\ CH_3 \end{array} C{-}OH \right]^{-}$$

The anion was to react with lead cations thought to be present in the double layer to generate a weakly bonded organometallic that required further reduction to obtain either isopropanol or tetraorganoleads:

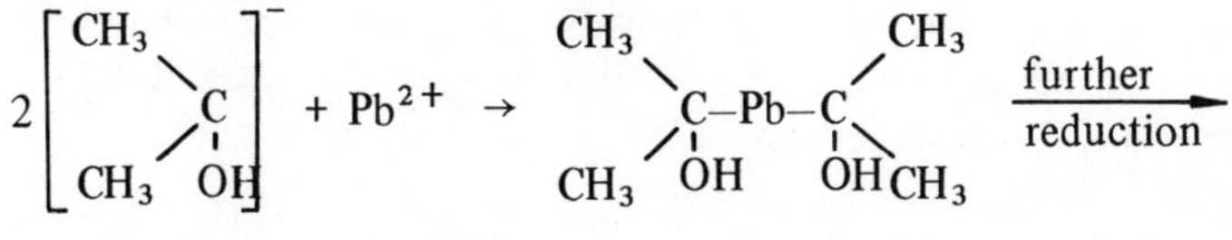

$$2\left[\begin{array}{c} CH_3 \\ \diagdown \\ \diagup \\ CH_3 \;\; OH \end{array} C \right]^{-} + Pb^{2+} \rightarrow \begin{array}{c} CH_3 \qquad\quad CH_3 \\ \diagdown \qquad\quad \diagup \\ C{-}Pb{-}C \\ \diagup \;\; | \qquad | \;\; \diagdown \\ CH_3 \;\; OH \quad OH CH_3 \end{array} \xrightarrow[\text{reduction}]{\text{further}}$$

With radicals as tenable intermediates in these carbonyl reductions, and by analogy to the classical Paneth reactions of metals and radicals, it is not surprising that a number of metals can be sacrificial cathodes when carbonyl compounds are reduced. Suitable cathodes include mercury, lead, tin [214, 215], zinc [213], cadmium [317, 324], and germanium [110]. In the latter four cases, the organometallics were not isolated and identified. They were postulated on the basis of metal loss from the respective cathodes during reduction and additionally in two cases [110, 214] by inclusion of cathode metal (Sn, Ge) in the isolated reduction products.

No great success has been had in attempts.to form organomercurials by reduction of aldehydes [6, 319] with the exception of experiments with benzaldehyde [6]. Electroreduction of benzaldehyde at a mercury cathode gave an 8.5% yield of dibenzylmercury [6]. The best yield was obtained by working at or below room temperature, while at 50°C no mercurial was obtained. Attempts to form mercurials with other aldehydes such as vanillin, piperonal, anisaldehyde, furfural, and salicylaldehyde gave only resinous materials [6].

Electrolysis of propionaldehyde on a mercury cathode in 30% H_2SO_4 gave no mercurial but resulted in a catholyte containing grey finely divided mercury [319]. It is tempting to speculate that an unstable mercury-containing intermediate is forming and decomposing in this case, with its decomposition giving rise to colloidal mercury. This same observation of a grey or black suspension in catholyte solutions has been made by others who studied electrolyses of ketones at mercury cathodes to form mercurials [5, 178], and of quaternary ammonium ions to form organic amalgams [260, 330]. In the case of quaternary ammonium ions, the turbidity was associated with a decomposing onium ion amalgam at the cathode-water interface. In the case of ketones [5, 178], although mercurials were formed, disproportionation of RHg· radicals to yield $R_2Hg + Hg°$ probably occurs.

In view of the low yields of organometallics in Table 10.5, it is suggested that depolarizers such as alkyl halides or sulfonium salts be used for preparative purposes.

Electroreduction of carbonyl compounds containing α,β-unsaturation leads to formation of carbon-metal bonds at the carbon atom β to the carbonyl group. This is an interesting result when compared to the placement of the carbon-metal bond in saturated carbonyl compounds discussed above. For example, methyl vinyl ketone has two possible sites for bond formation to a metal atom, the carbonyl group or the conjugated β-carbon atom. It is the β-carbon atom that participates in formation of a carbon-mercury bond when the unsaturated ketone is reduced at a mercury cathode at a pH less than 5 [191, 407]. The situation is viewed as follows:

$$\begin{array}{c} H\ H\ O \\ |\ \ |\ \ \| \\ C=C-C \\ |\ \ \ \ \ | \\ H\ \ \ \ CH_3 \end{array} \xrightarrow{+H^+} \begin{array}{c} H\ \ \ \ \ \ \ H\ \ \ OH \\ \diagdown\ {}^+\ |\ \diagup \\ C-C=C \\ \diagup\ \ \ \ \ \ \ \diagdown \\ H\ \ \ \ \ \ \ \ CH_3 \end{array} \xrightarrow{e}$$

$$\left[\begin{array}{c} \underset{H}{\overset{H}{\diagdown}} \cdot C - CH = C \underset{CH_3}{\overset{OH}{\diagup}} \\ \updownarrow \\ \underset{H}{\overset{H}{\diagdown}} C - CH_2 - C \underset{CH_3}{\overset{O}{\diagup}} \end{array} \right]$$

$$Hg(CH_2CH_2\overset{O}{\overset{\|}{C}}CH_3)_2 \longleftarrow \cdot CH_2CH_2\overset{O}{\overset{\|}{C}}CH_3 \quad \dot{H}gCH_2CH_2\overset{O}{\overset{\|}{C}}CH_3 \overset{Hg}{\longleftarrow} \bigg]$$

a,β-Unsaturated nitriles form organometallic compounds only with a tin cathode [358]. Thus, electrolysis of acrylonitrile in an aqueous alkaline solution gave tetra-(β-cyanoethyl)stannane [358]:

$$4 \ H_2C=CHCN \xrightarrow{\ 4e, \ 4H^+, \ Sn \ } Sn(CH_2CH_2CN)_4$$

In similar experiments with lead and mercury cathodes, the formation of an organometallic compound was not observed [354, 358]. Subsequent work with methacrylonitrile, I, and vinyl acrylonitrile, II, also gave organotin compounds while crotonic, III, and vinyl acetic, IV, acid nitriles did not form organo-metallics [208]:

$$\underset{\underset{CH_3}{|}}{H_2C=C-CN} \qquad \underset{\underset{\underset{CH_2}{\|}}{\overset{|}{CH}}}{H_2C=C-CN} \qquad \underset{\underset{CH_3}{|}}{HC=CHCN} \qquad H_2C=CHCH_2CN$$

$$\textbf{I} \qquad\qquad \textbf{II} \qquad\qquad \textbf{III} \qquad\qquad \textbf{IV}$$

Although crotononitrile, III, would not form an organometallic compound at tin cathodes, it would form a hydrodimer (β,β'-dimethyladiponitrile) when reduced at a graphite cathode [359]:

$$2 \ \underset{\underset{CH_3}{|}}{NCCH=CH} \xrightarrow{\ 2e, \ 2 \ H^+ \ } \underset{\underset{CH_3}{|}}{(NCCH_2CH)_2}$$

The β-methyl group of crotononitrile is the only structural difference from acrylonitrile which did reduce to form an organometallic compound. The ability of a methyl group so placed as to prevent organometallic formation in reduction of sulfonium and carbonyl compounds was discussed on page 414. The differing behavior of acrylonitrile and crotononitrile presents another example of such an effect.

The yield and formation of organotin compounds from α,β-unsaturated nitriles is very sensitive to electrolysis conditions such as current density, temperature, and catholyte pH. Low current densities favor high yields of organometallic [208, 358], for example, decreasing the current density from 2,000 to 200 A/m^2 increased the current yield of tetra-(β-cyanoethyl)tin from zero to 50%. An increase in temperature has a favorable influence [208], since at 26-28°C the yield of tetra-(β-cyanopropyl)tin is 14%, but at 0°C only traces are formed. Neutral to slightly alkaline pH values improved yields of organotin compounds [35].

Organoselenium, tellurium, and sulfur compounds are obtained when powdered Se, Te, or S are reduced in the presence of acrylonitrile [207, 360]. Electroreduction of powdered Se at a graphite cathode in 1 N Na$_2$SO$_4$, upon acidification, gives SeH$_2$. If the same electrolysis is carried out in the presence of acrylonitrile, Se(CH$_2$CH$_2$CN)$_2$ is formed in 23% yield. Similar results hold for S and Te. Some authors [363] view this as formation of MH$_2$, which then enters into a cyanoethylation reaction with acrylonitrile:

$$2 \ CH_2{=}CHCN + MH_2 \rightarrow (CNCH_2CH_2)_2M$$

There are differing opinions on the mechanism of formation of organometallics when acrylonitrile is reduced at tin cathodes. Some [358] consider that a radical attack on the metal cathode takes place,

$$NCCH_2CH_2\cdot + M \rightarrow NCCH_2CH_2M\cdot \ \text{and so on}$$

while others [354, 392] believe, on the basis of polarographic and preparative work [354] that a doubly charged anion $(CH_2CHCN)^{-2}$ is first protonated by water and this then reacts with the cathode metal and withdraws a cation from the crystal lattice of the metal:

$$Sn^{+4} + 4(CH_2CH_2CN)^- \rightarrow Sn(CH_2CH_2CN)_4.$$

Organic Halides

This section deals with electrochemical synthesis of organometallics by reduction of alkyl, allyl, and benzylic halides at sacrificial cathodes. The reactions here are all classed as direct metalations since organometallics are formed from nonorganometallic compounds.

There is a great deal of common ground between the areas of metal alloy reductions and electrochemical reductions, and this similarity is very effectively demonstrated by metal-alloy and electrochemical reductions of organic halides, both to yield organometallic compounds. As an interesting aside the reader is referred to the review by Jones and Gilman [205], specifically the section entitled "Reactions of Alloys with Organic Halides." Familiarity with the very many metal-alloy (Na, K) reductions of organic halides to yield organometallics and the syntheses presented in Table 10.7 makes a reasonable case for similarity in these two areas. The similarity of products is striking.

Mentioned in the foregoing section was the fact that acrylonitrile could be reduced at tin cathodes to form tetra-(β-cyanoethyl)tin, but would not similarly form organolead or mercury compounds at the respective cathodes. In contrast, β-iodopropionitrile is reduced at tin, lead, mercury, and thallium cathodes, and good yields of organotin, I, lead, II, mercury, III, and thallium compounds are obtained [365].

$$(NCCH_2CH_2)_4Sn \quad (NCCH_2CH_2)_4Pb \quad (NCCH_2CH_2)_2Hg \quad I\ Tl(CH_2CH_2CN)_2$$

$$\textbf{I} \qquad\qquad \textbf{II} \qquad\qquad \textbf{III} \qquad\qquad \textbf{IV}$$

In reductions involving acrylonitrile, failure to produce organolead or organomercury compounds was explained by Tomilov and Kaabak [354, 358] in terms of the more negative reduction potential of acrylonitrile on lead and mercury cathodes relative to the tin cathode.

The importance of potential has been considered [47, 127, 128, 354, 365, 401] also in the reduction of alkyl halides:

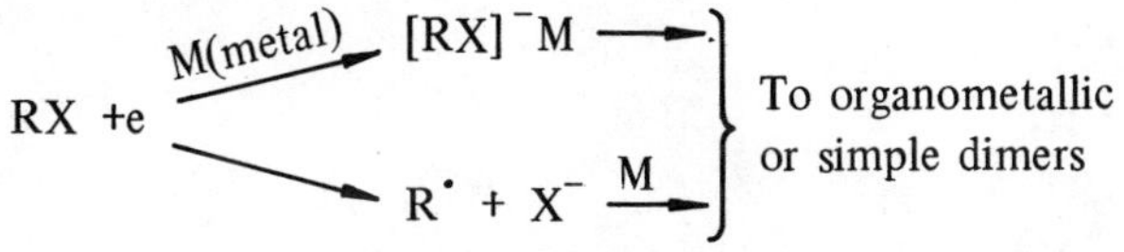

If reduction of RX to some initial species can occur at a sufficiently positive potential, then the chances for further reduction of intermediates such as [RX]⁻M, R·, or RM· become less and the likelihood increases for forming either dimers or organometallic compounds. At more negative potentials the radical R· (or some form of it) is itself rapidly reduced to form a carbanion which may be protonated:

$$R\cdot + e \rightarrow R^-$$
$$R^- + H^+ \rightarrow RH$$

Success in forming organometallics with β-iodopropionitrile relative to the situation with β-bromopropionitrile where carbanion formation is noted is consistent with the above thinking. β-Iodopropionitrile is reduced polaro-graphically in two one-electron steps (−1.2 and −1.5 V versus SCE) [127, 128]. The intermediate radical has an opportunity to form organometallics or be reduced further. The corresponding bromonitrile is reduced at sufficiently negative potentials (−1.8 to −2.0 V versus SCE) that the reduction rate of the intermediate radical overwhelms any other route, and only a single two-electron wave is observed.

There is tremendous variability in reducibility among benzylic, allylic, and alkyl halides, but within a given group the process is more facile as the halogen atom becomes less electronegative. It is therefore generally easier to form dimers and organometallics with iodo and bromo compounds than chloro and fluoro derivatives. These relative reduction potentials are based on polarography at a dropping mercury cathode and since shifts in reduction potentials do occur on different metals, we should note that the same relative positions are maintained when alkyl halides are examined on stationary lead cathodes [371]. Polaro-graphy of alkyl halides on dropping gallium cathodes [16, 17] provides an example in which the reduction of alkyl halides (CH_3I, C_2H_5I, C_3H_7I, and C_4H_9I) proceeds at more positive potentials than on mercury cathodes [16]. It was claimed that organogallium compounds were formed although isolation was not attempted.

The electrochemical reduction of cyclopropyl halides provides a good example of the effect of reduction potential on the formation of organometallics. Studies [247, 401] of optically active 1-halo-1-methyl-2,2-diphenylcyclopro-panes, I, have resulted in mercurial synthesis, and also to a degree, have illumin-ated the area of reductive cleavage of the carbon-halogen bond. Both the bromo

and iodo derivatives gave the cyclopropane, II, with net retention of configura-tion, and in addition, the iodo compound yielded the optically pure mercurial, III [401]. Again the more negative reduction potential required for the bromo derivative facilitated formation of cyclopropane, whereas the more easily reduced iodo compound gave a mercurial. Significantly, the yield of mercurial in the iodo case was shown to be dependent on reduction potential (tabulation below). In the two-step process, reduction of the intermediate radical versus dimerization was judged to be important.

Although a number of patents [54, 265, 327, 328, 418] and publications [148, 149, 150, 292, 371] exist describing cathodic conversion of alkyl halides

Variation in Yield of R_2Hg with Potential in the Reduction of RI

Cathode Potential (V vs Ag/Ag$^+$)	% Conversion of RI to R_2Hg
-2.1	39.9
-2.5	29.9
-2.9	18.9

to tetraethyllead and tetramethyllead, no commercial cathodic processes have surfaced to rival anodic routes to these materials. Calingaert [54] first disclosed formation of tetraalkylleads by electrolyzing a caustic alcoholic catholyte containing alkyl iodide, and Mead [265] followed showing that aqueous caustic systems containing casein could be used as electrolytes. The yield from these processes was less than 10%. When ethyl bromide was used the yield was found to be negligible.

Silversmith and Sloan [327] used nonhydroxylic solvents such as acetonitrile or dimethylformamide with an alkyl halide and claimed improved yields of tetraalkylleads. A very large number of solvent-electrolyte systems were examined, but acetonitrile with various supporting electrolytes gave consistently higher (80-100% based on lead consumption) yields of tetraethyllead than a number of other nonhydroxylic solvents. Smeltz [328] showed that additions of small amounts of hydroxylic solvents (water, alcohols) to the acetonitrile system resulted in better yields of tetraalkylleads, lower power consumption, less gassing, and higher rates of tetraalkyllead production.

Yang et al. [418] electrolyzed ethyl bromide in a special catholyte solvent system in which current yields of tetraalkyllead were nearly 100%. The system employed tetrabutylammonium bromide or tetrabutylphosphonium bromide in aqueous acetone or an aqueous solution of a water soluble ether such as tetrahydrofuran. The aqueous system was used with a diaphragm cell and the authors asserted there were fewer problems of diaphragm-solvent interactions relative to aprotic solvent systems, and also that anodically generated bromine could be conveniently distilled from the aqueous anolyte. A dramatic increase in the efficiency of lead conversion was observed with variation in the water-supporting electrolyte ratio [418]. This is shown in the accompanying plot (Fig. 10.3). These authors claim that commercially attractive yields only occur above a water-supporting electrolyte ratio of about 30 while an earlier patent [328] teaches the ratio should not exceed 20.

It was pointed out [418] that the nature of the solubilizers and current carriers was important to high yields of tetraethyllead. Acetone or tetrahydrofuran as solubilizers gave the desired results, but the use of ethanol gave very low yields. Quaternary ammonium salts such as tetraethylammonium bromide were vastly inferior to tetrabutylammonium bromide in promoting TEL yields.

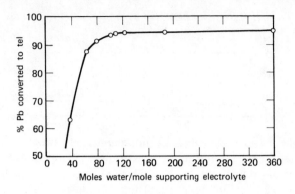

Fig. 10.3 Dependence of TEL yield on the moles of water per mole of supporting electrolyte, cd = 180 mA/cm^2, grams acetone/grams water = 0.675, excess ethyl bromide [418].

Other authors [150] have concurred on the importance of onium ion supporting electrolytes and their concentrations. Galli [150] has shown that supporting electrolytes such as R$_4$NX and R$_3$SX in propylene carbonate solutions of EtBr were beneficial to good yields of TEL while inorganic supporting electrolytes such as LiClO$_4$ actually prevented TEL formation.

TEL can also be formed by the electrolysis of EtBr alone, without cosolvent, using onium salts (e.g., n-Bu$_4$NX) as supporting electrolytes [150, 418].

Yang et al. [418] noted an interesting aspect in the synthesis of tetramethyllead from methyl bromide in their above described aqueous systems. The TML was formed in an extremely pure form and only a water wash was necessary to render the TML useful directly as an antiknock additive.

Electroreduction of benzylic halides maintains some general features displayed in the electrolysis of allyl halides, benzylic and other onium compounds, and even some more weakly bonded alkyl halides, all of which can lead to organometallic compounds. The bulk of these products in the cases of benzyl and ring-substituted benzyl halides are organomercurials.

Continuing what seems to be a common feature in the above list of halides, the benzyl halides reduce in a two-step process that involves the generation of an intermediate radical. This radical (possibly PhCH$_2$· or some absorbed form, PhCH$_2$M·) lies at a branch point in the reduction scheme, that is, the radical, as we have seen previously, can go to different products via a coupling reaction with or without metal atoms, or be further reduced at the cathode. This broad statement is made recognizing that different viewpoints exist in this area. For example, the electroreducibility of the resulting organomercurial product, usually dibenzylmercury, has been claimed [400] and disclaimed [202], electroreducibility of the intermediate radicals has been contested [168] and proposed [250], and evidence has been presented both for [202, 216, 250, 258, 308, 400] and against [250] a spontaneous chemical reduction involving an electron transfer from the mercury cathode to the halide molecule.

Benzyl bromide has received the most attention as a depolarizer and the majority of these experiments have involved sacrificial mercury cathodes (Table 10.6). Not surprisingly, though, Silversmith and Sloan found arsenic to be a sacrificial cathode material when benzyl bromide was reduced in an acetonitrile catholyte [327]. Tetrabenzylarsonium bromide was formed in the catholyte and was thought to arise by interaction of cathodically formed tribenzylarsene with unreduced benzyl bromide. These authors also stated that cathodes of bismuth and antimony were attacked, but products were not identified.

Electrolysis of benzyl bromide at a mercury cathode had earlier [250] been reported by Marple et al. to yield bibenzyl, but the work has since been corrected [308] and dibenzylmercury identified as a product of the electrolysis. Grimshaw and Ramsey [168] later showed that a large number of ring-substituted (halo, alkyl) benzyl bromides would yield corresponding dibenzyl-mercury compounds when electrolyzed in methanol under controlled-potential conditions.

Coproduced along with the dibenzylmercury compounds I in the above-mentioned work were the corresponding bibenzyl compounds [168] II.

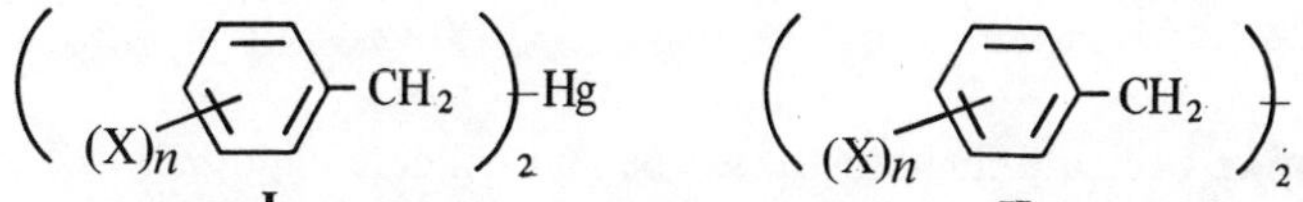

The authors ascribed both products to coupling of the intermediate radical and concluded it was impossible to further reduce the radical:

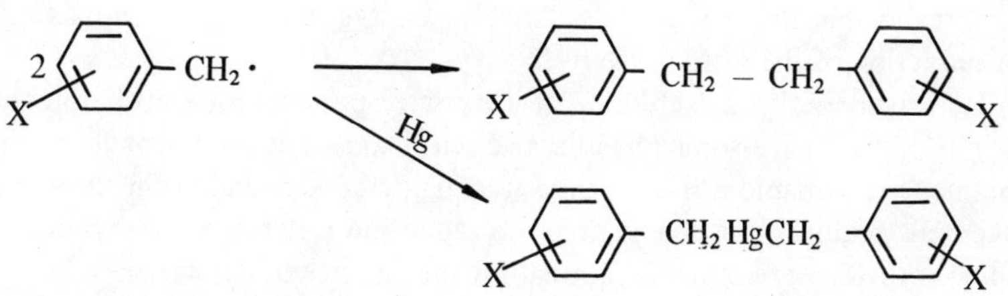

Table 10.6 shows the effect of reduction potential on yields of two types of compounds. The general increase in bibenzyl and decrease in mercurial with increasing negative potentials can be used to argue for the more usual reduction

process, that is, generation of radicals that can either interact with the cathode metal or be further reduced to carbanions. The carbanion would have been generated in greater amounts at more negative voltages and displace bromide from unreacted benzyl bromide:

Table 10.6. Products from Electrolytic Reduction of 3,4-Dichlorobenzyl Bromide in Methanolic 2 N Lithium Bromide (Data from Ref. 168)

Cathode Potential[a]	$\left(Cl_2C_6H_3\text{-}CH_2\right)_2$	$\left(Cl_2C_6H_3\text{-}CH_2\right)_2 Hg$
-0.94	2.1	39
-1.05	25.0	26.6
-1.30	21.0	4.6

[a]Versus Ag/AgCl wire in aqueous 0.1M KCl.

$$ Cl_2C_6H_3\text{-}CH_2^- \ + \ Cl_2C_6H_3\text{-}CH_2Br \ \longrightarrow \ \left(Cl_2C_6H_3\text{-}CH_2\right)_2 \ + \ Br^- $$

Contrasting with the ring-substituted benzyl halide compounds in Table 10.7, which did give corresponding dibenzylmercury compounds, 4-nitrobenzyl bromide was reported [168] * to yield 4,4′-dinitrobibenzyl when reduced at a mercury cathode. This sounds like the same effect of the para-nitro substituent mentioned earlier for 4-nitrobenzyldimethylsulfonium salts at mercury and lead cathodes. A radical mechanism was suggested [168, 217] in this case and is more tenable in view of the relatively positive reduction potentials required for reduction of the nitro derivative.

Electrolysis of benzyl chloride at a mercury pool cathode in the presence of carbon dioxide gave some phenylacetic acid [400], but no dibenzylmercury was obtained. A carbanion intermediate was suggested here and at the more negative potentials required for benzyl chloride reduction (relative to benzyl bromide or iodide) provides a reasonable rational for the absence of mercurial.

* Another account of this synthesis is reported [217], but it is not clear if a mercury cathode was employed.

Table 10.7. Electroreduction of Organic Halides

Depolarizer	(Supporting Electrolyte, Solvent)	Cathode (Potential, Other)	Products (% Yield)	Ref.
ICH$_2$CH$_2$CN 40 g/300 ml of electrolyte	(0.5 N H$_2$SO$_4$ or 1 N Na$_2$SO$_4$) H$_2$O	Sn	[(NCCH$_2$CH$_2$)$_3$Sn]$_2$ (12)	365
			(NCCH$_2$CH$_2$)$_4$Sn (—)	365
		Pb (0.035 A/cm^2, — —)	(NCCH$_2$CH$_2$)$_4$Pb (13)	365
		Hg	(NCCH$_2$CH$_2$)$_2$Hg (46)	364
			NCCH$_2$CH$_2$HgI CH$_3$CH$_2$CN (4.1)	
		Tl	(NCCH$_2$CH$_2$)$_2$TlI (4.7)	365
ClCH$_2$CH$_2$CN	(— —, — —)	Sn (— —, — —)	[(NCCH$_2$CH$_2$)$_3$Sn]$_2$ (12)	354
Ph Ph $\overset{CH_3}{\triangle}$ I	(Tetraethylammonium bromide, acetonitrile)		Optically pure $\left[Ph\ Ph\ \overset{CH_3}{\triangle}Hg \right]_2$ (40) (19)	401 247
1-Iodo-1-methyl-2,2'-diphenylcyclopropane Optically active		Hg $\left[\begin{array}{l}-2.1 \text{ V vs Ag/AgCl, CPE}\\ -2.9 \text{ V vs Ag/AgCl, CPE}\end{array}\right]$	Optically pure $Ph\ Ph\ \overset{CH_3}{\underset{H}{\triangle}}$ (60) (81)	
CH$_3$CH$_2$I	(1 N H$_2$SO$_4$:95% EtOH = 50:50)	Pb (-1.2 V vs SCE) CPE; Cd (-1.25 V vs SCE); Zn (-0.99 V vs SCE) $\left\}\begin{array}{l}\text{All potentials read at 2 mA/cm}^2\end{array}\right.$	[In each case cathode lost weight and metals detected in catholyte solutions]	292
CH$_3$CH$_2$I	(NaOH, alcohol)	Pb (— —, — —)	(CH$_3$CH$_2$)$_4$Pb	54
CH$_3$Br	((Et)$_4$NBr, acetonitrile-H$_2$O)	Pb (—)	(CH$_3$)$_4$Pb (100) Yield based on current and metal consumption	371

437

Table 10.7. Electroreduction of Organic Halides (cont.)

Depolarizer	(Supporting Electrolyte, Solvent)	Cathode (Potential, Other)	Products (% Yield)	Ref.
CH_3CH_2Br	((Et)$_4$NBr, acetonitrile-H_2O)	Pb (—)	(CH$_3$CH$_2$)$_4$Pb (~60) Yield based on current and metal consumption [(CH$_3$CH$_2$)$_3$Pb]$_2$	371
CH_3CH_2Br	(0.1 M (Et)$_4$NBr, propylene carbonate)	Pb^a [−500 mV / −500 to −1400 mV / More negative than −1400 mV]	(CH$_3$CH$_2$)$_4$Pb (0) / (70-100) / (0) Efficiency of metal consumption	149, 148
CH_3CH_2Br	(Aqueous alkaline emulsion of EtBr)	Pb^a (−500 to −1400 mV)	(CH$_3$CH$_2$)$_4$Pb CH$_3$CH$_3$ Current Yield Efficiency of lead consumption	148
CH_3CH_2Br ~1 mole	(~0.1 moles (Et)$_4$NBr/kg of acetonitrile N,N-dimethylformamide Dimethyl sulfoxide 1,2-Dimethoxyethane Methylene chloride Propylene carbonate Ethyl thiocyanate N,N-dimethylcyanamide)	Pb (0.02-0.03 A/cm^2, 25-60°C) 4 hr Separate compartments	(Et$_4$Pb Current Yield (77) Efficiency of lead consumption (81) (46) (16) (47) (71) (40) (84) (44)	148

438

				Current Yield	Efficiency of lead consumption	
CH_3CH_2Br ~1 mole	(LiBr 0.524 moles/Kg acetonitrile	Pb (0.02-0.03 A/cm², 25 25-60 C) ~4 hr Separate compartments	$(Et)_4Pb$		(100)	327
	$(Et)_3SBr$ 0.121 moles/Kg acetonitrile				(81)	
	NaI 0.159 moles/Kg acetonitrile				(76)	
	KSCN 0.243 moles/Kg acetonitrile				(100)	
	$LiClO_4$ 0.149 moles/Kg acetonitrile				(100)	
	KI <0.145 moles/Kg acetonitrile				(84)	
	$(Ph)_3EtPI$ 0.091 moles/Kg acetonitrile				(86)	
	$CaBr_2$ <0.012 moles/Kg acetonitrile				(100)	
	$(Ph)_3MeBr$ 0.094 moles/Kg acetonitrile				(72)	
CH_3CH_2I CH_3CH_2Cl	$((Et)_4NBr$, acetonitrile)	Pb (0.02-0.03 A/cm², 25-60°C) Separate compartments	$(Et)_4Pb$		Lead Yield (72) (93)	327
CH_3CH_2Br	(10 g Dioctadecyldimethyl-ammonium chloride in 290 g EtBr, 8 g acetonitrile)	Pb (−, −) Separate compartments	$(Et)_4Pb$		Lead Yield (26)	327

439

Table 10.7. Electroreduction of Organic Halides (cont.)

Depolarizer	(Supporting Electrolyte, Solvent)	Cathode (Potential, Other)	Products (% Yield)	Ref.
CH₃Cl CH₃I CH₃Br	((Et)₄NBr, acetonitrile)	Pb (0.02-0.03 A/cm², Separate compartments	(CH₃)₄Pb (93) (—) (—)	327
CH₂=CH–CH₂Cl	(LiCl, *N,N*-dimethylformamide)	Pb (0.012 A/cm², ~30°C)	(CH₂=CH–CH₂)₄Pb (65) Supposed, but no identification given	327
C₆H₅CH₂Br	((Et)₄NBr, acetonitrile)	As (0.001 A/cm², 25-30°C under N₂ atm)	(C₆H₅CH₂)₄AsBr (9) Based on As Rx thought to form (C₆H₅CH₂)₃As which interacts with C₆H₅CH₂Br to give observed onium product.	327
C₆H₅CH₂Br	((Et)₄NBr, acetonitrile)	Bi or Sb	Both give soluble organometallics	327.
CH₃CH₂Br	((Et)₄NBr, acetonitrile)	Sn (0.02-0.03 A/cm², 25-60°C)	(Et)₄Sn (72) metal yield	327
CH₃Br	((Bu)₄NBr, acetonitrile containing 0.5-10% by wt. of H₂O or MeOH)	Pb (up to 0.15 A/cm², — —)	(CH₃)₄Pb (80-93) metal yield	328
CH₃Br	((Bu)₄NBr, acetone:H₂O = 13 g : 41 g)	Pb (200 mA/3 cm², 25°C)	(CH₃)₄Pb (98) current yield	418
CH₃CH₂Br	((Bu)₄PBr, acetone 0.7 moles – H₂O 27 moles)	Pb (150 mA/3 cm², —)	(Et)₄Pb (100) current yield	418

Substrate	Conditions	Product	Yield	Ref
CH_3CH_2Br	((Bu)$_4$PBr, acetone 0.7 moles Pb (150 mA/3 cm², —) – H$_2$O 13 moles	(Et)$_4$Pb	(88) current yield	418
CH_3CH_2Br	((Bu)$_4$NBr, THF 0.4 to 0.9 moles – H$_2$O 1.4 moles) Pb (150 mA/3 cm², —)	(Et)$_4$Pb	(85) current yield	418
CH_3CH_2Br	((Bu)$_4$NBr, CH$_3$CN:H$_2$O −50 g:150 g) Pb (38 mA/cm², —)	(Et)$_4$Pb	(91) metal yield	418
CH_3CH_2Br	(0.1 M (Et)$_4$NClO$_4$, propylene carbonate) Pb (1150-1200 mV)[a]	(Et)$_4$Pb	(100) current yield	150
CH_3CH_2Br	(1 M Me EtSBr, propylene carbonate) Pb (1000-1100 mV)[a]	(Et)$_4$Pb	(77) current yield	150
CH_3CH_2Br	(0.1 M NH$_4$PF$_6$, propylene carbonate) Pb (1000-1100 mV)[a]	(Et)$_4$Pb	(0) current yield	150
CH_3Br 47 g	(25 g (Et)$_4$NBr in 260 g CH$_3$CN + 36 g H$_2$O) Pb (0.06 A/cm², 45°C)	(Me)$_4$Pb	(84) current yield	108
⟨Ph⟩–CH$_2$Br	Hg (−0.50 V vs, CPE)	[⟨Ph⟩–CH$_2$]$_2$Hg		308
(CH$_3$)$_3$C–⟨C$_6$H$_4$⟩–CH$_2$Br (LiBr, MeOH)	Hg (−1.50 V vs Ag/Ag⁺, CPE)	[(CH$_3$)$_3$C–⟨C$_6$H$_4$⟩–CH$_2$]$_2$Hg (50)		168

441

Table 10.7 Electroreduction of Organic Halides (cont.)

Depolarizer	(Supporting Electrolyte, Solvent)	Cathode (Potential, Other)	Products (% Yield)	Ref.
C_6H_5–CH_2Br	(LiBr, MeOH)	Hg (−1.40 V vs Ag/$\overset{+}{Ag}$, CPE)	$[C_6H_5$–$CH_2]_2Hg$ (41)	168
CH_3–C_6H_4–CH_2Br	(LiBr, MeOH)	Hg (−1.20 V vs Ag/$\overset{+}{Ag}$, CPE)	$[CH_3$–C_6H_4–$CH_2]_2Hg$ (64)	168
Br–C_6H_4–CH_2Br	(LiCl, MeOH)	Hg (−1.20 V vs Ag/$\overset{+}{Ag}$, CPE)	$[Br$–C_6H_4–$CH_2]_2Hg$ (6.5)	168
Cl,Cl–C_6H_3–CH_2Br	(LiBr, MeOH)	Hg (−1.05 V vs Ag/$\overset{+}{Ag}$, CPE)	$[Cl,Cl$–C_6H_3–$CH_2]_2Hg$ (27)	168
O_2N–C_6H_4–CH_2Br	(LiBr, MeOH)	Hg (−0.50 V vs Ag/$\overset{+}{Ag}$, CPE)	O_2N–C_6H_4–CH_2–CH_2–C_6H_4–NO_2 (74)	168
C_6H_5–CH_2I	(−, H_2O-EtOH)	Hg (− −, − −)	C_6H_5–CH_2HgI	202

[a] Values are experimental cathode polarization obtained by measuring electrode potential with current flow and subtracting the static potential.

442

Macroscale electrolysis of benzyl iodide was reported [202] to yield benzylmercuric iodide, but conditions were not given in detail.

Electroreduction of Organomercurials

Organomercury Salts (RHgX)

Early interest [231] in the electrochemistry of these compounds, just as in the case of organic amalgams, stemmed from a desire to isolate the adduct of an electropositive material and an electron. Would the adduct have metallic properties? Along with compounds such as $(CH_3)_4Sb^+I^-$, $(CH_3)_3S^+I^-$, and $(C_6H_5)_2I^+I^-$, RHgX salts were electrolyzed at platinum electrodes in liquid ammonia [231]. Only the RHgX compounds yielded isolable electron adducts $RHg^\cdot$. Kraus [231] considered that he had prepared $CH_3Hg^\cdot$, $C_2H_5Hg^\cdot$, and n-$C_3H_7Hg^\cdot$ (as black conductive deposits on the cathode) by electrolysis of the corresponding alkylmercury chlorides:

$$RHgCl + e \rightarrow RH^\cdot + Cl^-$$

He further established that the materials decomposed on warming according to $2\ RHg^\cdot \rightarrow R_2Hg + Hg^\circ$. The isolated $RHg^\cdot$ materials exhibited metallic luster when compressed or tamped. Kraus concluded that the materials possessed matallic properties. The work was later repeated for $CH_3Hg^\cdot$ by Rice and Evering [303], who found the same behavior.

Similar attempts [231] by Kraus to prepare free "organometals" from electrolysis of n-pentyl, n-octyl, and phenylmercuric iodides gave little or no indication of the desired products.

In a more recent study by Gowenlock and Trotman [166] RHgCl compounds (R = Me, Et, n-Pr, i-Pr, n-Bu, Ph, PhCH$_2$) were electrolyzed at Pt electrodes in liquid ammonia ($-78°C$) and were reported to form $RHg^\cdot$ compounds. These all formed R_2Hg compounds when warmed to reported decomposition temperatures. $trans$-ClCH=CH–HgCL was an exception as it formed mercury metal and colorless crystals at a Pt cathode on electrolysis as above. The white crystals were considered to be $(ClCH=CH)_2Hg$.

Several authors [165, 166] have argued that the dark $RHg^\cdot$ cathode deposits of Kraus were more likely a mercurous compound RHgHgR instead of the free radical $RHg^\cdot$. They were unable to trap any radicals with styrene or diphenylpicrylhydrazyl during electrolysis or on thermal decomposition to R_2Hg. Moreover [165], examination of these substances with electron-spin resonance spectroscopy failed to demonstrate the presence of radicals, and attempts to measure the conductance of these "organometals" were likewise negative.

The symmetrical organomercurials R_2Hg are the main products resulting from the electroreduction of RHgX. From Table 10.8 it may be seen that mercury, platinum, or copper amalgam cathodes can be used for this conversion. Evidently the mercury atoms in RHgX are sufficient for the generation of R_2Hg and cathode mercury is not necessary. It would be interesting, however, to compare yields of R_2Hg for electrolyses at mercury pool cathodes relative to platinum cathodes.

Electroreduction of *trans*-2-chloromercuri-1-acetamidocyclohexane, I, at a mercury pool cathode yielded either the mercurial II or the hydrogen-substituted demercuration product III, depending on the solvent system and reduction potential [402]. When I was reduced chemically with either sodium borohydride or hydrazine hydrate, the mercurial was also formed, and reduction by Na-Hg in water led again to the demercuration product [56]:

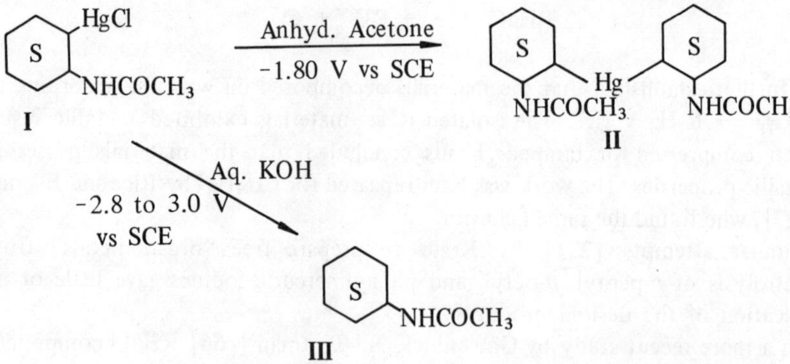

One example of cyclization via electroreduction of an organomercurial has been reported [83]. 1-Phenylethynyl-8-chloromercurinaphthalene can be converted in dimethoxyethane to the bis(organo)mercurial at a first reduction wave, or at a second more negative potential setting, to 1-phenylacenaphthalene. In the ring closure, a carbanion is thought to be generated in the *a*-position which then attacks the triple bond:

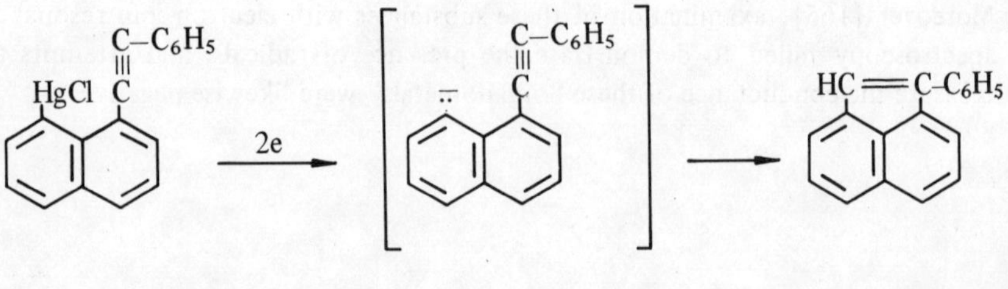

In addition to the above preparative work, a large number of papers [21, 22, 24, 25, 51, 53, 71, 72, 86, 133, 202, 216, 228, 247, 284-286, 311, 366, 367, 368, 393, 400, 401, 413] have dealt with the polarography of a tremendous number of RHgX compounds. (Some polarographic work is still omitted in the above references.) Impetus for this type of study, in part at least, seems due to the wide use of organic mercury compounds in the study of biologically occurring thiols, and also in pharmaceutical preparations and diuretics [24].

Recognizing that exceptions exist (see below), the bulk of RHgX compounds examined polarographically display two single electron reduction waves. This is true for a vast number of different R and X groups with X ranging from halogens, hydroxyl, nitrate, and acetate to amide and imide groups. RHgX compounds are reduced much more easily than corresponding R_2Hg compounds.

The mechanistic interpretations of these polarographic data are in agreement with results of preparative electrolyses discussed above, and furthermore, there is nearly a unanimous choice for the general course of reduction of RHgX compounds. The first reduction step (range of 0 to -0.5 V vs SCE) is interpreted to be generation of an organomercury radical,

$$RHgX + e \longrightarrow RHg^{\cdot} + X^-$$

which is stable or unstable, depending on temperature. If it is unstable the radical disproportionates to give the symmetrical organomercury compound (R_2Hg) and mercury metal. If the reduction potential is sufficiently negative, however, these radical species can be reduced to carbanion intermediates:

$$RHg^{\cdot} + e \longrightarrow R^- + Hg^{\circ}$$
$$R^- + H^+ \longrightarrow RH$$

The first reduction step has been considered reversible by some [21, 53, 72, 202, 311, 366, 393]; however, an explanation was offered by others [367, 368] for the apparent nonreversibility of the first polarographic wave. Dimerization or disproportionation of $RHg^{\cdot}$ was considered as causing the apparent lack of reversibility of the wave. When complexing agents for RHg^+ such as SCN^- or thiourea were present in sufficient concentration the first wave was reversible, but without them it was judged irreversible. Toropova et al. concluded that the complexing agents hindered dimerization. A simple macroscale electrolysis might provide interesting data here.

In view of the isolation of the radical product ($RHg^{\cdot}$) from the first reduction at low temperatures, it is not unreasonable to expect such a species to have enough lifetime at ambient temperatures to be detected by cyclic voltammetric methods (polarographic reversibility). This has been the case at mercury electrodes [72, 133, 393], but no anodic peak was observed with a platinum electrode [133].

Table 10.8. Electroreduction of RHgX Compounds

Depolarizer	(Supporting Electrolyte, Solvent)	Cathode (Potential, Other)	Products (% Yield)		Ref.
			Primary Isolated	And on Warming	
CH₃HgCl CH₃CH₂HgCl CH₃CH₂CH₂HgCl	(Depolarizer, liquid NH₃)	Pt (−, −60°C)	CH₃Hg· CH₃CH₂Hg· CH₃CH₂CH₂Hg·	(CH₃)₂Hg (CH₃CH₂)₂Hg (CH₃CH₂CH₂)₂Hg	231, 303 231 231
RHgCl R = CH₃, CH₃CH₂, n-Pr, i-Pr, n-Bu, Ph, PhCH₂	(Depolarizer, liquid NH₃ or EtOH)	Pt (−78°C)	RHg·	(R)₂Hg	166
CH₃CH₂$\overset{*}{C}$HHgBr CH₃ Optically active	(Depolarizer, liquid NH₃ or EtOH)	Pt (−78°C)	CH₃CH₂CHHg· CH₃ Black deposit, decomp. ~−37°C	(CH₃CH₂)₂Hg CH₃ Optically inactive	30
n-C₅H₁₁HgCl	(Depolarizer, liquid NH₃ or EtOH)	Pt (−78°C)	Little indication of n-C₅H₁₁Hg· deposit		30
i-BuCH₂HgCl	(Depolarizer, liquid NH₃ or EtOH)	Pt (−78°C)	Little indication of n-C₅H₁₁Hg· deposit		30
n-C₆H₁₃HgCl	(Depolarizer, liquid NH₃ or EtOH)	Pt (−78°C)	Light brown deposit Decomp. ~−30°C Supposed n-C₆H₁₃Hg·		30
n-C₇H₁₅HgCl	(Depolarizer, liquid NH₃ or EtOH)	Pt (−78°C)	None reported		30

446

Compound	Conditions	Electrode	Product / Notes	Ref.
Cyclohexyl-HgCl	(Depolarizer, liquid NH_3 or EtOH)	Pt (-78°C)	Black deposit, Decomp. -20°C, Supposed cyclohexyl-Hg·	30
CH_3—⟨C₆H₄⟩—HgCl	(Depolarizer, liquid NH_3 or EtOH)	Pt (-78°C)	Black flaky deposit, Supposed CH_3—⟨C₆H₄⟩—Hg·	30
ClCH=CHHgCl	(Depolarizer, liquid NH_3)	Pt (-78°C)	Supposed $(ClCH=CH)_2Hg$	166
CH_3HgOAc	(25% depolarizer + 2X molar X^S of pyridine, H_2O)	Pt (1.8 A/dm², 30-40°C)	$(CH_3)_2Hg$ (92)	259
Same as above for CH_3CH_2HgCl, $PrHgCl$, but no yields				
$(RHg)_2SO_4$ R = Et, Pr, Bu, Amyl	(−, −)	(-0.01-0.05 A/cm², 20-80°C)	R_2Hg (quant.)	267
$HOCH_2CH_2HgCl$	(NaOH + $4M$ in depolarizer, H_2O)	Hg (-0.8 V vs ?, CPE)	$(HOCH_2CH_2)_2Hg$	114
$HOCH_2CH_2HgOAc$	(pH ⩽5, H_2O)	Cu/Hg (-0.45 to -0.64 V vs ?, 0.02 A/cm²)	$H_2C=CH_2$ (~100)	366
$HOCH_2CH_2HgOAc$	(pH 12.1, H_2O)	Cu/Hg (-1.1 to -1.3 V vs ?, 0.02 A/cm²)	EtOH (30), $H_2C=CH_2$	366
$HOCH_2CH_2HgOAc$	(NaOH pH 8-14, H_2O)	Cu/Hg (-1.7 to -1.9 V vs ?, 25°C, 0.01 A/cm²)	$H_2C=CH_2$ + $HOCH_2CH_3$	113

447

Table 10.8. Electroreduction of RHgX Compounds (cont.)

Depolarizer	(Supporting Electrolyte, Solvent)	Cathode (Potential, Other)	Products (% Yield) Primary Isolated	And on Warming	Ref.
$CH_3-\overset{H}{\underset{HO}{C}}-CH_2HgOAc$	NaOH pH 8-14, H_2O	Cu/Hg (-1.7 to -1.9 V vs ?, 25°C, 0.01 A/cm²)	$CH_3-CH=CH_2$ + $CH_3-\overset{H}{\underset{OH}{C}}-CH_3$		113
$CH_3-\overset{OH}{\underset{CH_3}{C}}-CH_2HgOAc$	NaOH pH 8-14, H_2O	Cu/Hg (-1.7 to -1.9 V vs ?, 25°C, 0.01 A/cm²)	$CH_3-\overset{}{\underset{CH_3}{C}}=CH_2$ + $CH_3-\overset{OH}{\underset{CH_3}{C}}-CH_3$		113
C₆H₅HgBr	(-, -)	Hg (-)	(C₆H₅)₂Hg (quantitative)		86
C₆H₅HgOAc	(Depolarizer, CCl₄-MeOH) 30-60 ml	Pt (0.003 A/cm², 10-15°C)	C₆H₅HgCl (40) + (C₆H₅)₂Hg (20)		422
C₆H₅CH₂HgOAc	(Depolarizer, CCl₄-MeOH) 30-60 ml	Pt (0.003 A/cm², 10-15°C)	C₆H₅CH₂HgCl (1) + (C₆H₅CH₂)₂Hg (40)		422
Cyclohexyl-HgOAc	(Depolarizer, CCl₄-MeOH) 30-60 ml	Pt (0.003 A/cm², 10-15°C)	Cyclohexyl-HgCl (30) + (Cyclohexyl)₂Hg (6)		422
C₆H₅-Hg-O-C(=O)-CH₃	0.1 M KNO₃, H₂O	Hg (-1.4 V vs SCE, CPE)	(C₆H₅)₂Hg		24

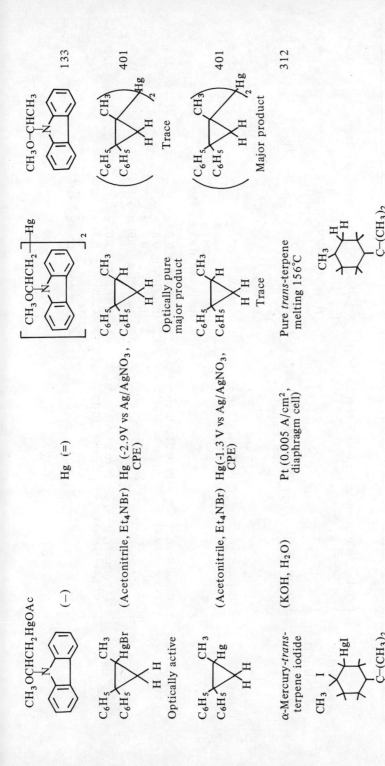

449

Table 10.8. Electroreduction of RHgX Compounds (cont.)

Depolarizer	(Supporting Electrolyte, Solvent)	Cathode (Potential, Other)	Products (% Yield)	Ref.

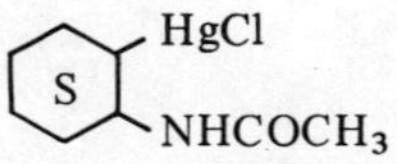

Row 1:

HgCl / NHCOCH₃ depolarizer

(n-Bu)₄NClO₄, anhyd. Hg (-1.8 V vs SCE) acetone

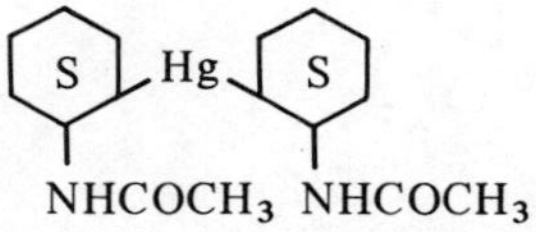

(42)

402

Row 2:

(KOH, H₂O)

Hg (-2.8 to -3.0 V vs SCE, const. current)

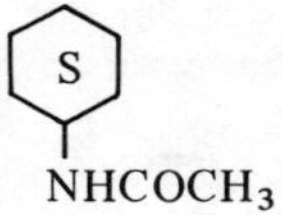

NHCOCH₃

(71)

402

Row 3:

PhHgOAc
3.37 g

(Depolarizer, CCl₄ + MeOH)

Pt (0.01 A/3 cm², N₂ atm., 10-15°C)

1. PhHgCl   1.21 g
2. Ph₂Hg   0.46 g
3. Hg   0.41 g
4. Recovered   0.85 g SM[a]

232

Row 4:

PhCH₂HgOAc
3.51 g

(Depolarizer, CCl₄ + MeOH)

Pt (0.01 A/3 cm², N₂ atm., 10-15°C)

1. PhCH₂HgCl   0.05 g
2. (PhCH₂)₂Hg   0.87 g
3. Hg   0.5 g
4. Recovered   1.02 g SM[a]

232

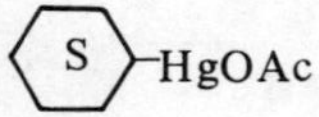

Row 5:

—HgOAc
3.4 g

(Depolarizer, CCl₄ + MeOH)

Pt (0.01 A/3 cm², N₂ atm., 10-15°C)

1. C₆H₁₁HgCl   1.05 g
2. (C₆H₁₁)₂Hg   0.11 g
3. Hg   0.4 g
4. Recovered   0.5 g SM[a]

232

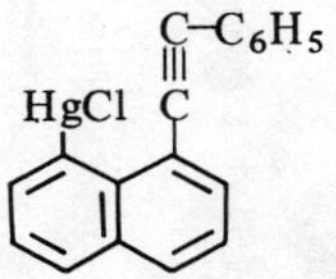

83

C≡C-C₆H₅

HgCl

(Bu)₄NClO₄, dimethoxyethane

Hg

-1.38 V vs Ag/AgClO₄, CPE

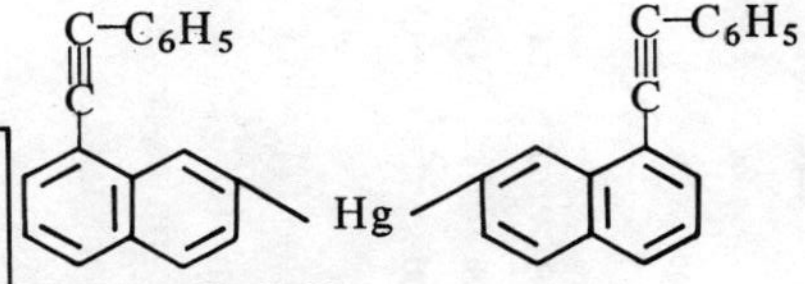

-2.76 V vs Ag/AgClO₄, CPE

C≡C-C₆H₅

Hg

C≡C-C₆H₅

HC=C-C₆H₅

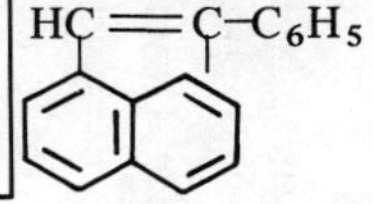

[a] SM, starting material.

451

Some insight into the nature of the RHg˙ species generated near the potential of the first reduction wave of RHgX has been obtained by the macroscale electrolysis of optically active secondary butylmercuric bromide at platinum [30] and mercury cathodes [86]. In the sequence

$$sec\text{-}C_4H_9HgBr \xrightarrow{\ e\ } [sec\text{-}C_4H_9Hg^{\cdot}] \longrightarrow (sec\text{-}C_4H_9)_2Hg \xrightarrow{\ HgBr_2\ } sec\text{-}C_4H_9HgBr$$

optically active $a = 0$

the addition of $HgBr_2$ to the symmetrical mercurial is found to proceed with retention of configuration. It was concluded that either a route to racemization of R in RHg˙ is available, or the conversion to R_2Hg goes with the inversion of one R group.

Dessy et al. [86] have carried out electrolyses using labeled (Hg^{203}) $C_6H_5Hg^{*}Br$. When $C_6H_5Hg^{*}Br$ was electrolyzed in glyme, less than 10% of the original activity was detected in the product $(C_6H_5)_2Hg$. Other routes for activity loss were ruled out by experiment, leaving the conclusion that phenyl group migration took place sometime prior to $(C_6H_5)_2Hg$ formation:

$$C_6H_5Hg^{203} + Hg \rightleftharpoons C_6H_5Hg + Hg^{203}$$

It was suggested [86] that this process may be related to the racemization of the sec-butyl group involving a species such as Hg— —R— —Hg ("a skittering radical on the electrode surface").

An interesting preparative electrolysis [113, 114, 366] serves to further illustrate the overall reduction scheme for RHgX compounds. Electrolysis of β-hydroxyethylmercuric acetate at more negative potentials than its second reduction wave in an alkaline medium gave ethanol and some ethylene:

$$HOCH_2CH_2HgOAc + e \longrightarrow HOCH_2CH_2Hg^{\cdot} + {}^{-}OAc$$

$$HOCH_2CH_2Hg^{\cdot} + e + H^{+} \longrightarrow HOCH_2CH_3 \ (30\% \text{ yield})$$

At high pH and at the first reduction wave the expected mercurial $(HOCH_2CH_2)_2Hg$ and some ethylene are formed [114]. The ethylene was attributed [366] to the reaction of the intermediate species with protons since acidic conditions lead to ~100% ethylene formation [366].

$$HOCH_2CH_2Hg^{\cdot} + H^{+} \longrightarrow H_2^{+}OCH_2CH_2Hg^{\cdot} \longrightarrow \tfrac{1}{2}Hg_2^{+} + C_2H_4 + H_2O$$

The following is an exception to the general reaction route. The first reduction step for (α-carbethoxy)-benzylmercuric bromide or nitrate proceeds in a manner similar to other RHgX compounds [21]:

$$
\underset{\substack{\text{C}_6\text{H}_5\\ \text{CHg}^+\ \text{NO}_3^- \\ \text{COOEt}}}{} + \ e \ \rightleftharpoons \ \underset{\substack{\text{C}_6\text{H}_5\\ \text{CHg}^{\cdot} \\ \text{COOEt}}}{} + \ \text{NO}_3^-
$$

Possibilities for the second reduction step were considered as either a one-electron reduction of RHg$^{\cdot}$, or a two-electron reduction of a symmetrical organomercurial formed by disproportionation of the organomercury radical:

$$
2\ \underset{\substack{\text{C}_6\text{H}_5\\ \text{CHg}^{\cdot} \\ \text{COOEt}}}{} \longrightarrow \left(\underset{\substack{\text{C}_6\text{H}_5\\ \text{C} \\ \text{COOEt}}}{} \right)_2 \text{Hg} \ + \ \text{Hg}^\circ
$$

$$
\left(\underset{\substack{\text{C}_6\text{H}_5\\ \text{C} \\ \text{COOEt}}}{} \right)_2 \text{Hg} \ + \ 2e \ \xrightarrow{\ 2\,\text{H}^+\ } \ 2\ \underset{\substack{\text{C}_6\text{H}_5\\ \text{CH} \\ \text{COOEt}}}{} \ + \ \text{Hg}^\circ
$$

It was pointed out that the height of the second wave would be the same in either case. The latter route was preferred [21], in part, since the characteristics of the second reduction wave for (a-carbethoxy)-benzylmercuric salts were close to those measured [22] for the symmetrical organomercurial, bis-[(a-carbethoxy)benzyl]-mercury. The bis(organo)mercurial also gave a single two-electron reduction wave, and addition of it to a solution of a-(carbethoxy)-benzylmercuric bromide resulted in proportionate increases in the height of the second wave of the bromide. Finally, the pH-dependent shift in the wave of the R$_2$Hg compound was the same as for the second wave of the bromide.

Bis(organo)mercury Compounds

References concerning the polarographic reduction of the symmetrical bis(organo)mercurials R$_2$Hg are reasonably abundant, but reports dealing with actual macroscale electrolysis followed by product isolation are limited. We can, based on polarographic and some macroscale work, have a fairly firm picture of the electrolytic reduction process for many R$_2$Hg compounds as an irreversible two-electron process yielding mercury and a carbanion:

$$
\text{R}_2\text{Hg} \ + \ 2e \ \rightarrow \ \text{Hg}^\circ \ + \ 2\text{R}^- \tag{1}
$$

A number of divinyl-, dialkenyl- and diphenylmercury compounds were reduced by polarographic and controlled-potential electrolysis techniques in dimethoxyethane [86] and carbanions were reported as products (Table 10.9). The method employed for their detection was not given.

Table 10.9. Electroreduction of *Bis*(Organo)Mercury Compounds

Depolarizer (Supporting Electrolyte, Solvent)	Cathode (Potential, Other)	Products (% Yield)	Ref.
$(R)_2Hg$ (Bu$_4$NClO$_4$, dimethoxyethane)	Hg (V vs Ag/AgClO$_4$, CPE)	$R^- + Hg$ Corresponding carbanion unless otherwise noted; method of detection unclear	86
$R = CH_2=CH$	−3.34 V		
C_6H_5	−3.32 V		
$C_3H_7C\equiv C$	−2.90 V		
$C_6H_5CH=CH$	−2.75 V		
C_6Cl_5	−2.63 V		
$C_6H_5C\equiv C$	−2.25 V		
C_6F_5	−1.81 V		
CCl_3	−1.43 V		
$C_6H_5HgCCl_3$ (Bu$_4$NClO$_4$, dimethoxyethane)	Hg (−2.8 V vs Ag/AgClO$_4$, CPE)	HCCl$_3$ and/or CCl$_2$=CCl$_2$ + Hg HCCl$_3$ and/or CCl$_2$=CCl$_2$ + (C$_6$H$_5$)$_2$Hg	86
O$_2$N–C$_6$H$_4$–Hg–C$_6$H$_4$–NO$_2$ (Bu$_4$NClO$_4$, dimethoxyethane); O$_2$N–C$_6$H$_4$–Hg–C$_6$H$_5$; *m*-O$_2$N-C$_6$H$_4$–Hg–C$_6$H$_4$-NO$_2$-*m*	Hg (−, CPE)	Radical anions and dianions, no C–Hg bond cleavage, species are reoxidizable to parent molecule	87

$(Bu_4NClO_4,$ Hg $(-,$ CPE$)$ dimethoxyethane$)$

Radical anions and dianions, no C–Hg bond cleavage, species are reoxidizable to parent molecule

87

In addition to reductions in which C-Hg bond cleavage occurs, Dessy et al. [87] found that a number of mono- and bis(nitroaryl)mercury compounds yielded radical anion and dianion species which could be reoxidized to the parent molecule. For bis(p-nitrophenyl)mercury the scheme was represented as

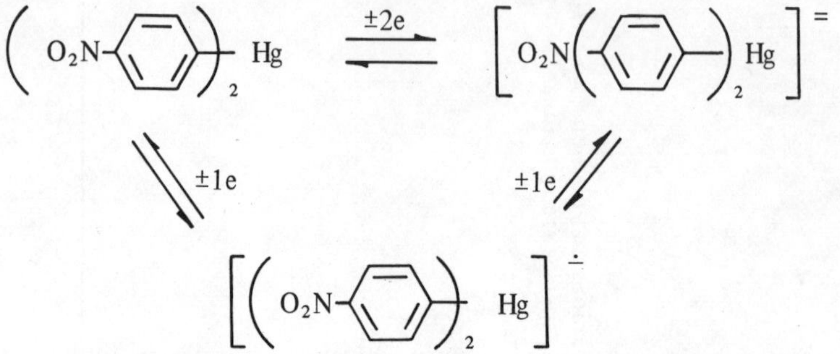

Electroreduction on a macroscale of bis(β-hydroxyethyl)mercury was encountered when the mercurial was generated at an intermediate stage in the electrolysis of $HOCH_2CH_2HgX$ [113]. When β-hydroxyethylmercuric salts were electrolyzed in an alkaline medium at a large copper amalgam cathode using a controlled potential of -0.8 V versus ?,* bis(β-hydroxyethyl)mercury and ethylene were isolated from the reaction mixture [113, 114] (Table 10.8). The above electrolysis was repeated again at -0.8 V versus SCE, but instead of actually isolating the known product, bis(β-hydroxyethyl)mercury, the system was electrolyzed further at -1.7 to -1.9 V versus ?.* The isolated products were now ethanol and ethylene. This last potential setting is sufficient to reduce bis(β-hydroxyethyl)mercury, which was reported to reduce at -1.87 V versus ?.* The authors interpreted the reduction at -1.7 to -1.9 V versus ?* as reduction of the bis(organo)mercurial formed in the initial reaction stage at -0.8 V versus ?.* Their scheme is illustrated below; the sum total is a two electron reduction of R_2Hg:

$$Hg(CH_2CH_2OH)_2 + e \xrightarrow[\text{Slow}]{H_2O} HgCH_2CH_2OH + CH_3CH_2OH$$

$$HgCH_2CH_2OH + e \xrightarrow{\text{Rapid}} [HgC_2H_4OH]^- \longrightarrow Hg + C_2H_4 + OH^-$$

The same qualitative and quantitative product mix as above was observed when $(HOCH_2CH_2)_2Hg$ was reduced chemically by sodium amalgam [114].

Bis(4-benzoyl-1,2,3-triazoyl)mercury **I** can be formed by the ionization of metallic mercury in contact with the sodium or thallium salts of the T^- anion:

*Reference electrodes were probably all the same, but the type was not given [113, 114].

$$Hg + 2T^- \longrightarrow HgT_2 + 2e$$

$$I$$

$$T^- = HC \overset{\displaystyle }{\underset{\displaystyle N}{\diagdown}} \overset{\displaystyle}{\underset{\displaystyle \ominus}{N}} \overset{\displaystyle O}{\underset{\displaystyle}{\overset{\parallel}{C}}} \overset{}{\underset{}{C}} - \phi$$

Gavrilova and Zhdanov [152] employed polarography and controlled-potential electrolysis to study **I**. The reduction presents an extreme example of an easily reduced R_2Hg compound (polar C-Hg bonds), while the products still correspond to the simple Eq. 1 for reduction of R_2Hg compounds.

Three two-electron cathodic diffusion waves were recorded for **I** at +0.075, -1.75 and -1.96 V versus SCE. The first wave was attributed to the process

$$HgT_2 + 2e \rightleftharpoons Hg^\circ + 2T^- \tag{2}$$

In support of this, it was shown that electrolysis of a solution of HgT_2 at the potential of the limiting current of the first cathodic wave resulted in solutions that when analyzed polarographically yielded an anodic reaction (oxidation process) exactly at the potential of the first cathodic wave (+0.075 V versus SCE). The anodic process was reformation of HgT_2. This, in addition to analysis of the wave shape, showed the reduction to be reversible and given by Eq. 2. The last two reduction steps at -1.75 and -1.96 V versus SCE were attributed to stepwise reduction of the carbonyl group.

Most symmetrical compounds R_2Hg are difficult to reduce [22, 401]. Indeed, it was claimed [202] that dibenzylmercury could not be reduced at a dropping mercury electrode, but others have since shown that dibenzylmercury can be reduced at -1.6 V versus a mercury pool reference in anhydrous acetonitrile. The many other reports [22, 51, 52, 228, 286] concerning polarography of these compounds demonstrate they are reducible in a variety of solvents.

Butin et al. [52] observed a dependence of $E_{1/2}$ on ring substitution for the series of compounds $(XC_6H_4CH_2)_2Hg$. This is shown in Fig. 10.4. The Hammett σ value (a measure of electron-donating or -withdrawing power exerted at the benzylic carbon) for the X group is plotted against the half-wave potential for the corresponding X-substituted compound. The more powerful withdrawing groups (positive σ values) apparently polarize the C-Hg bond and thus ease the reduction process.

Mercury(II)carboxylates are symmetrical organomercurials, but these are salts that behave very differently from the type of mercurials considered above. A large number of these are reduced polarographically in the range -0.5 to 0

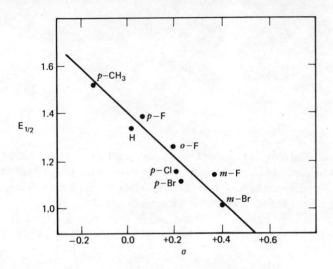

Fig. 10.4 Influence of ring substitution on $E_{1/2}$ for the series of compounds $(XC_6H_5CH_2)_2Hg$. $E_{1/2}$ plotted against the Hammett σ value for X substituents. (From Ref. 52.)

V versus SCE [50]. The suggested mode of reduction is entirely different than for R_2Hg compounds:

$$(RCOO)_2Hg \;\rightleftharpoons\; 2RCOO^- + Hg^{+2}$$

$$Hg^{+2} + 2e \;\rightleftharpoons\; Hg^\circ$$

Electroreduction of Organothallium Compounds

Organomercury products usually result when organothallium compounds are subjected to bulk electrolysis at mercury cathodes [105, 124, 151] (Table 10.10). This holds true unless reaction conditions are such that an intermediate species that is generated in an initial electron transfer is further reduced and thus diverted from the mercurial synthesis. The important variables in this regard are reduction potential and depolarizer concentration. For example, electrolysis of organothallium(III)cations such as I or II gave diphenylmercury, benzene and Tl° in proportions depending on the potential and concentration of depolarizer employed. Potentials in the range of -1.0 to -1.4 V versus SCE and low depolarizer concentrations (less than 10^{-3} M) favor total benzene formation, while very high yields of diphenylmercury are obtained at more positive potentials.

I $(E_{1/2} = 0.01$ V vs SCE [Ref. 124]   II $(E_{1/2} = 0.55$ V vs SCE)
[Ref. 105]

Electrolyses of both I and II respond similarly to the above variables, but the monophenylthallium cation was reduced much more easily than II. Electrolysis of I at -0.1 V versus SCE also gave diphenylmercury, but instead of Tl° the Tl^+ cation was observed in solution.

In discussing these mercurial syntheses it is interesting to note that heating diphenylthallium bromide with pyridine in the presence of Hg° gave good yields of diphenylmercury [160]. Also, triphenylthallium was reacted with mercury in ether solution to yield diphenylmercury and thallium amalgam [108]. Mercury was used to reduce diphenylthallium bromide to thallous bromide plus a high yield of diphenylmercury [108].

Polarographic and electrocapillary measurements yield schemes that are in accord with the above synthetic work and in addition provide evidence for adsorption phenomena. Adsorption at a dropping mercury electrode of both depolarizer and reduction product was observed for cations I and II and for dialkylthallium halides [60]. This adsorption phenomenon was included in a scheme proposed [124] for the electroreduction of monophenylthallium perchlorate which reduced polarographically in three main stages (Fig. 10.5).

The first stage of reduction is thought to occur stepwise with prior adsorption,

$$\underset{\substack{| \\ \text{Hg cathode}}}{C_6H_5Tl^{+2}} + e \longrightarrow \underset{\substack{| \\ \text{Hg cathode}}}{C_6H_5Tl^{+}} \tag{1}$$

$$2\ C_6H_5-Tl^{+} + Hg \longrightarrow (C_6H_5)_2Hg + 2\ Tl^{+} \tag{2}$$

the second stage also two reactions,

$$\underset{\substack{| \\ \text{Hg cathode}}}{C_6H_5Tl^{+}} + e \longrightarrow \underset{\substack{| \\ \text{Hg cathode}}}{C_6H_5Tl} \tag{3}$$

$$2\ \underset{\substack{| \\ \text{Hg cathode}}}{C_6H_5Tl} + Hg \longrightarrow (C_6H_5)_2Hg + 2\ Tl^{\circ}(Hg) \tag{4}$$

and the third stage one reaction.

Table 10.10. Electroreduction of Organothallium Compounds

Depolarizer	(Supporting Electrolyte, Solvent)	Cathode (Potential, Other)	Products (% Yield)	Ref.
Thallium compounds				
—Tl(ClO₄)₂		Hg (−0.10 V vs SCE, CPE)	(93) Tl⁺ in solution (99.8)	124
Monophenylthallium (III) Perchlorate	(NaClO₄ -0.1 *M*, H₂O)	Hg (∼0.7 V vs SCE, CPE)	Diphenyl mercury (94.2) no Tl ion in solution observe Tl/Hg	124
[—Tl⁺²] > 10⁻³ *M*		Hg (∼−1.0 V vs SCE, CPE)	Diphenyl mercury	124
[—Tl⁺²] = 10⁻³ *M*		Hg (∼−1.0 V vs SCE, CPE)	No diphenyl mercury, Benzene detected	124

Compound	Medium	Electrode (V vs SCE, CPE)	Products: $(Ph)_2Hg$ ( ) $Tl°$ ( )	Ref.
$\left(\right)_2 Tl^+ F^-$ Diphenylthallium (III) Fluoride	$(H_2PO_4 \cdot KH_2PO_4,\ H_2O,$ pH 6.2)	Hg (V vs SCE, CPE)		
		-0.80	(83), (~100)	105
		-1.10	(48), (~100) (Some benzene)	105
		-1.40	(34), (~100) (Some benzene)	105
$[(Ph)_2Tl^+] < 10^{-4}\ M$		-1.40	(0), (~100) + benzene	105
Cyclopentadienylthallium (I)	(LiClO₄, dimethyl-formamide)	Hg (-0.76 V vs SCE, 25°C)	Tl/Hg amalgam and cyclopentadiene	151
	(LiClO₄, dimethyl-formamide)	Hg (-0.17 V vs SCE, 25°C) anodic process	$(C_5H_5)_2Hg + Tl^+$	151
$(Et)_2TlCl$	(−, −)	Pt (−)	Tl° + unsaturated flamable gas	326

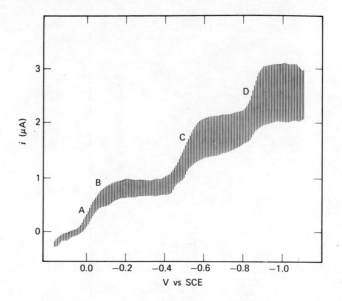

Fig. 10.5 Polarogram of $C_6H_5Tl^{+2}$ showing an adsorption prewave at A, and three main reduction stages B, C, and D. (From Ref. 124.)

$$C_6H_5Tl + H^+ + e \longrightarrow C_6H_6 + Tl^°(Hg) \qquad (5)$$
$$\overset{|}{Hg} \text{ cathode}$$

These are in accord with products reported in Table 10.10, and the whole process is very similar to the situation for cation **II**, $(C_6H_5)_4Sb^+$ [276], and RHgX compounds [24].

Electrolysis of cyclopentadienylthallium [151], formed by a neutralization reaction of cyclopentadiene and thallium hydroxide,

$$C_5H_6 + TlOH \rightarrow TlC_5H_5 + H_2O \qquad (6)$$

gives cyclopentadiene, dicyclopentadienyl mercury, and $Tl^°/Hg$. Cyclopentadienylthallium (CPT) is a neutral entity and yields three polarographic waves at -0.17, -0.42 and -0.76 V versus SCE. The first and third are due to undissociated CPT while the second was due to Tl^+. The first wave was a reversible wave, and electrolysis of CPT at -0.17 V versus SCE resulted in oxidation of CPT to Tl^+ and the cyclopentadienyl radical, which interacted with mercury to form a mercurial:

$$C_5H_5Tl \; -e \longrightarrow C_5H_5{}^{\textbf{·}} \; + \; Tl^+$$

$$2 \; C_5H_5{}^{\textbf{·}} \; + \; Hg \longrightarrow (C_5H_5)_2Hg$$

The mercurial could interact with Tl^+ to reform CPT and Hg^{+2}. Electrolysis at the third wave was similar to Eq. 5 above for cation I and produced cyclopentadiene and Tl/Hg via a carbanion route.

Electroreduction of Group IV-A Oranometallic Compounds: Si, Ge, Sn, Pb

The electroreductions of organometallic compounds of Si, Ge, Sn, and Pb have all been dealt with, but the greater emphasis has been placed on Sn and Pb organometallics. Discussed here are symmetrical and unsymmetrical derivatives, $(R)_nM$ and $(R)_nMX_{4-n}$, and organodimetals both homo- and heterodimetallic, R_nMMR_n and $R_nMM'R_n$. All may be studied in organic solvents such as dimethylformamide or dimethoxyethane and some in aqueous systems. Distinct differences will be observed in aqueous versus the organic systems.

A report [85] that broadly covers the electrochemistry of Group IV-A organometallics is available and covers polarography, controlled-potential electrolysis, and cyclic voltammetry of the aryl organometallic compounds of this group.*

Authors of the above report have set forth the conceptual framework for the possible results of electron addition to such organometallic assemblies as encountered here. Realizing that the valency for m can be 4, it is expedient to quote the report [85]:

"Conceptually, one can envisage, for the reductive mode only, the addition of an electron or electrons to an organometallic assembly RmQ, where R is a σ- or π-bonded organic moiety, m is a metal and two of its valency positions, and Q is another R group, halo or oxyfunction. The fate of the affected assembly may be to exist for long time periods as the radical anion, or to lose the ligand Q as an anion, generating the subvalent organometal Rm. The fate of Rm might then take one of several courses: (1) ultimate decomposition to R$^{\textbf{·}}$ and m; (2) disproportionation, RmR + m, which may be concomitant with 1; (3) abstraction of H from the surroundings, yielding RmH; or (4) coupling to yield RmmR. Further reduction may yield Rm, which could decompose."

---

*This report is an initial communication and part of a larger series dealing with organometallic electrochemistry authored by R. E. Dessy and others. Subsequent reports are referenced throughout this chapter.

Results for Group IV-A organometallics will show this conceptual framework to be supported by facts.

Tetraorganometals of Group IV-A

Tetraorgano compounds of silicon, germanium, tin, and lead have been reduced. Tetraethyllead was reported [390] to undergo electrochemical reduction at -0.65 to -0.7 V versus SCE, but no reduction products were reported. Likewise, tetrabutyl- and tetraphenyltin were reported [1] to reduce polarographically in a single wave. I and II gave reversible waves at a dropping mercury electrode [66], and uninegative radical anions were thought [66] to result.

$(CH_3)_3$—Sn⟨⟩⟨⟩ $(CH_3)_3Sn$⟨⟩⟨⟩—$Sn(CH_3)_3$

I   II

Employment of a cell divider between the anolyte and catholyte was critical for the electroreduction of phenyltrimethylsilane [28]. In the presence of LiCl and anhydrous methylamine, use of an undivided cell resulted in a good yield of (1,4-dihydrophenyl)trimethylsilane III, while use of a divided cell gave a good deal of carbon-silicon bond cleavage and the major products hexamethyl-

⟨⟩—$Si(Me)_3$ $(Me)_3SiOSi(Me)_3$

III   IV

disiloxane IV and cyclohexene. The authors point out the major function of the cell divider here is to prevent the immediate neutralization of lithium methylamide formed in the cathode compartment during the reduction. They speculate that nucleophilic attack on the silicon atom of cathodically formed III by lithium methylamide could give the cyclohexadiene carbanion. Protonation of this could yield 1,3-cyclohexadiene which could be reduced further to observed cyclohexene and finally to cyclohexane.

In the same undivided cell and solvent-electrolyte system as above (phenylethynyl)trimethylsilane V was reduced both with and without addition of isopropanol. In the absence of isopropanol ethylbenzene was a major product

$$\langle\!\!\!\text{—}\!\!\!\rangle\text{—}C\!\equiv\!C\text{—}Si(Me)_3 \qquad\qquad \langle\!\!\!\text{—}\!\!\!\rangle\text{—}CH_2CH_2Si(Me)_3$$

$$\text{V} \qquad\qquad\qquad\qquad\qquad \text{VI}$$

but in the presence of isopropanol ethylbenzene was nearly eliminated and a considerable amount of (2-phenylethyl)trimethylsilane VI was formed. Phenylacetylene was a major product in both cases indicating that considerable cleavage of the carbon silicon bond had occurred in each case. Regarding this cleavage in the presence of isopropanol, it was shown that extensive chemical cleavage of the carbon-silicon bond in V occurred in the presence of LiCl, methylamine, and isopropanol.

All of the above organosilicon reductions leave open the question of whether or not any cleavage of the carbon-silicon bond actually occurs in or as the result of a primary electrode process.

The electrochemical fluorination of organosilicon compounds has been attempted [322]. Me_4Si, $(Me_3Si)_2O$, Et_2SiCl_2, amyltrichlorosilane, and $PhSiCl_3$ were all electrolyzed in an Inconel cell with Ni electrodes. No fluoroorganosilicon compounds were identified. The main products were SiF_4, fluorinated carbon compounds, and various organofluorosilanes. These were not products of an electrode process but due to nucleophilic attack of the fluoride ion on the carbon-silicon bond.

Other electrochemical reductions of tetraorganosilanes and germanes have been reported [2, 211, 243]. Anion radicals of p- and m-(trimethylsilyl)nitrobenzene were obtained [211]. A number of silyl and germyl compounds such as VII and VIII, with the exception of 1,4-bis(trimethylgermyl)benzene

$$(CH_3)_3M\text{—}\langle\!\!\!\text{—}\!\!\!\rangle_n\text{—}M(CH_3)_3$$

$$M = Si, Ge$$
$$n = 1,2,3,4$$

$$\text{VII}$$

$$M(CH_3)_3$$
$$M(CH_3)_3$$

$$\text{VIII}$$

showed a one-electron reversible wave in dimethylformamide. Numbers reported were peak voltages from AC polarography and were in the range of -2.7 to -3.4 V versus $Ag/AgNO_3$.

Chemical reductions of the above class of tetraorganometals are well known and for example, stable negative ions were observed by ESR when tetraphenyl and other similarly substituted silanes were reduced by alkali metal [369].

Hexaorganodimetals and R_nMX_{4-n} Organometallics of Group IV-A

This large group of compounds is profitably discussed together in a single section because of the very interesting similarities and contrasts within the group and also because of a degree of convenience derived from following this same organization within the literature.

Aqueous and organic solvent systems are employed here depending on depolarizer stability. For example, all silicon and germanium compounds are electrolyzed in anhydrous organics, while tin and lead compounds may be examined in either type of system. Aside from symmetrical organoderivatives of tin and lead, these two metals form a number of partially alkylated derivatives in which the central metal atoms may be bonded to one, two, or three organic groups. These are usually prepared as halides, covalent in the solid state and in some organic solvent systems [350, 351], but ionized to varying extents in aqueous systems. These compounds belong to a larger class of σ-bonded organometallic cations that are stable in aqueous systems. The class also includes the metals Tl, Hg, Au, and Pt [353]. All are capable of acidic hydrolysis [353]. Due to the very different characteristics of aqueous and organic systems, the different electrochemical behavior sometimes observed in such solvents is not unexpected.

The hexaorganodimetals, both homodimetals and heterodimetals are represented as $R_3MM'R_3$, where M and M' may or may not be the same metal. Dessy et al. [94] have also considered a class of "polymetallics" $R_3MM'M''R_3$ where various combinations of different metals are possible. A tremendous number of examples that involved not only group IV-A metals but also group V and transition metals, were exhaustively electrolyzed [85, 94] in dimethoxyethane at mercury electrodes. The authors could summarize their results by pointing out that scission of metal-metal bonds always occurred, and in the following manner:

$$M-M' \quad \xrightarrow{2e} \quad M^- + M'^-$$
$$\xrightarrow{1e} \quad M'\cdot + M^-$$

A two-electron addition gave two anion fragments, while a one-electron addition gave a radical and anion. It was further pointed out that homodimetallic compounds always gave two anions (two-electron route), while heterodimetallic compounds could reduce by either route.

The following collection of reactions for the hexaorganodimetals of group IV-A is illustrative of the above generalizations [85]:

$Ph_3SiSiPh_3$ $\longrightarrow$ no reduction observed

$Ph_3GeGePh_3$ $\xrightarrow[-3.5 \text{ V}]{2e}$ $2 \, Ph_3Ge{:}^-$

$Ph_3SnSnPh_3$ $\xrightarrow[-2.9 \text{ V}]{2e}$ $2 \, Ph_3Sn{:}^-$

$Ph_3SiGePh_3$ $\longrightarrow$ no reduction observed

$Ph_3SiSnPh_3$ $\xrightarrow[-3.1 \text{ V}]{e}$ $Ph_3Sn{:}^-$ + Ph_3SiH

$Ph_3PbPbPh_3$ $\xrightarrow[-2.0 \text{ V}]{2e}$ $2 \, Ph_3Pb{:}^-$

(all V versus Ag/Ag^+ reference electrode).

The final fate of the anions or radicals depends upon the environment they are produced in. The anions can transfer electrons to other assemblies, act as nucleophiles, or participate in redistribution reactions, whereas the radicals can couple, abstract, or undergo further reduction as the coupled product. In the heterometallic assemblies the most easily reduced moiety became the anion and the remaining portion formed the radical species. Reference 94 should be consulted for a more thorough discussion of the effects of structure and conditions on the course of events. Notice that no reduction of the Si-Si bond was achieved even though cleavage of hexaphenyldisilane by lithium metal is well known.

The same triphenyllead anion that was formed by electroreduction of $Ph_3PbPbPh_3$ at -2.0 V is also observed when triphenyllead acetate [85] is reduced at -2.2 V versus the same Ag^+/Ag reference electrode. The acetate may also be reduced at -1.4 V, leading to entirely different products (radical migration to yield diphenylmercury). Aside from the phenyl compounds discussed here, only a polarographic entry [390] was located for an alkyldilead, $Et_3PbPbEt_3$. This reduced at -1.8 to -2.0 V versus SCE in absolute ethanol.

Electroreduction of polymetallic compounds such as $Ph_3SnSeSnPh_3$ and $Ph_3SnSeGePh_3$ give results that are very similar to the above for dimetallics. For example, the Sn-Se-Sn polymetallic compound was reduced in either a one- or two-electron process depending on reduction potential [85]. Anions and radicals result as above:

$$Ph_3SnSeSnPh_3 \begin{cases} \xrightarrow[-2.4 \text{ V}]{e} & Ph_3Sn^{\bullet} + Ph_3SnSe^- \\ \\ \xrightarrow[-2.9 \text{ V}]{2e} & Ph_3SnSe^- + Ph_3Sn^- \end{cases} \quad (\text{V versus } Ag/Ag^+)$$

Table 10.11. Electroreduction of Group IV-A Organometallic Compounds

Depolarizer (Supporting Electrolyte, Solvent)	Cathode (Potential, Other)	Products (% Yield)	Ref.
	Electroreduction of Organolead Compounds		
$(Et)_2Pb^{+2}(OAc)_2^{-2}$ (0.5 M $HClO_4$, H_2O)	Hg (-0.7 V vs SCE, CPE)	$(Et)_4Pb$ Major prod. / $(Et)_2Hg$ Minor 8-40, depending on concentration of depolarizer / Pb°	273
$(R)_3Pb^+OH^-$ (2 N KOH, H_2O) R = Et, Me	Pb (, diaphragm cell)	$(R)_3Pb^{\cdot}$ reported, probably $(R)_3Pb\text{-}Pb(R)_3$ (80-90)	183
$(R)_3Pb^+OH^-$ (2 N KOH, H_2O) R = Et, Me	Zn (, diaphragm cell)	Above compound not isolated - spongy lead observed	183
$(Et)_3Pb^+OH^-$ (Depolarizer, 95% EtOH)	Pb (0.01 A/cm², no diaphragm)	$(Et)_3PbPb(Et)_3$	268
(C₆H₅)₃-PbOAc (Bu_4NClO_4, dimethoxyethane (DME))	Hg (-1.4 V vs 0.001 M $AgClO_4$/Ag, CPE)	C₆H₅-Hg-C₆H₅ (quantitative)	85
(C₆H₅)₃-PbOAc (Bu_4NClO_4, DME)	Hg (-2.2 V vs 0.001 M $AgClO_4$/Ag, CPE)	(C₆H₅)₃Pb⁻	85
(C₆H₅)₂-Pb(OAc)₂ (Bu_4NClO_4, DME)	Hg (2.1 V vs 0.001 M $AgClO_4$/Ag, CPE)	C₆H₅-Hg-C₆H₅ (quantitative)	85
(C₆H₅)₂-Pb(OAc)₂ (Bu_4NClO_4, DME)	Hg (-1.1 V vs 0.001 M $AgClO_4$/Ag, CPE)	some diphenylmercury plus suspected ((C₆H₅)₂Pb-Pb(C₆H₅)₂)	85

Cp = cyclopentadienyl

Compound	Electrolyte/Solvent	Electrode (conditions)	Products	Ref.
(triphenyl-Pb-Pb-triphenyl) structure	(Bu$_4$NClO$_4$, DME)	Hg (-2.0 V vs 0.001 M AgClO$_4$/Ag, CPE)	-Pb:$^{-}$ structure	85
CpFe(CO)$_2$Pb(Ph)$_3$	(Bu$_4$NClO$_4$ 0.1 M, DME)	Hg (-2.1 V vs 10^{-3} M Ag^{+}/Ag)	(Ph)$_3$Pb:$^{-}$ + CpFe(CO)$_2$:$^{-}$	94
(OC)$_5$Mn–Pb(Ph)$_3$	(Bu$_4$NClO$_4$ 0.1 M, DME)	Hg (-2.1 V vs 10^{-3} M Ag^{+}/Ag)	(OC)$_5$Mn:$^{-}$ + (Ph)$_3$Pb:$^{-}$	94
(OC)$_5$Mn–Pb(Et)$_3$	(Bu$_4$NClO$_4$ 0.1 M, DME)	Hg (-1.8 V vs 10^{-3} M Ag^{+}/Ag)	(OC)$_5$Mn:$^{-}$ + (Et)$_3$Pb·	94
CpMo(CO)$_3$–Pb(Ph)$_3$	(Bu$_4$NClO$_4$ 0.1 M, DME)	Hg (-2.2 V vs 10^{-3} M Ag^{+}/Ag)	CpMo(CO)$_3$:$^{-}$ + (Ph)$_3$Pb:$^{-}$	94
Lead tetraphenylporphyrin (structure)	(n-Pr$_4$NClO$_4$, THF-DMSO)	Hg or Pt (-1.30 V vs SCE, CPE)	Lead tetraphenylporphyrin radical anion	126

Table 10.11. Electroreduction of Group IV-A Organometallic Compounds

Depolarizer (Supporting Electrolyte, Solvent)	Cathode (Potential, Other)	Products (% Yield)	Ref.
	Electroreduction of Organotin Compounds		
$(Et)_2Sn^{+2}(H_2O)_4$ (0.1 M KCl + 0.1 M HCl, H_2O) Aquodiethyltin cation	Hg (-0.8 V vs SCE, CPE)	Greenish yellow oil that could form thin film at cathode. Material was polymeric and soluble in benzene, CCl_4 and cyclohexane.	275
	Hg (some doubt but assume ~0.9 V vs SCE, CPE)	Hexaphenylditin	388
$BuSnCl_3$ (0.1 M $HClO_4$, H_2O)	Hg (-1.2 V vs SCE, CPE)	Brown polymeric skin on Hg surface (proposed) $(BuSn)n$	266
$EtSnCl_3$ (—, H_2O)	Hg (-1.6 V vs SCE, CPE)	$(EtSn)n$ red, insoluble, polymeric	104
Bu_2SnCl_2 (0.1 M $HClO_4$, H_2O)	Hg (-1.1 V vs SCE, CPE)	Bright yellow layer on cathode. Examination of catholyte by extraction and GLC gave no evidence for organotin cmp.	266
Bu_3SnCl (—, H_2O)	Hg (-1.5 V vs SCE, CPE)	Bu_4Sn, $Bu_3SnSnBu_3$	266
Bu_3SnCl (—, H_2O)	Hg (-0.9 V vs SCE, CPE)	$Bu_3SnSnBu_3$	266

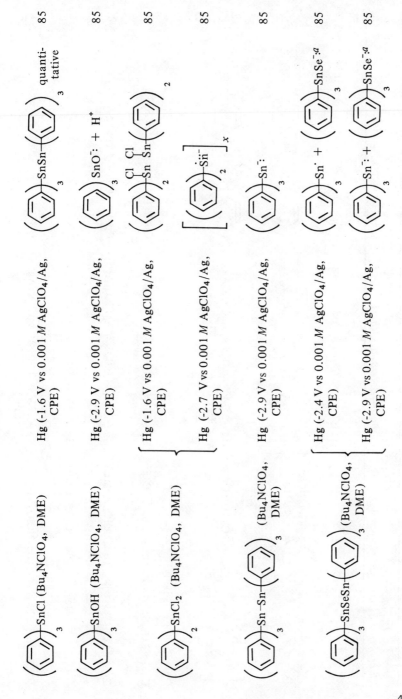

Table 10.11. Electroreduction of Group IV-A Organometallic Compounds (cont.)

Depolarizer (Supporting Electrolyte, Solvent)	Cathode (Potential, Other)	Products (% Yield)	Ref.
$(Ph)_3SnSeGe(Ph)_3$ (Bu₄NClO₄, DME)	Hg (-2.4 V vs 0.001 M AgClO₄/Ag, CPE)	$(Ph)_3GeH + (Ph)_3SnSe^{-}$:[a]	85
	Hg (-2.9 V vs 0.001 M AgClO₄/Ag, CPE)	$(Ph)_3Ge^{-}$: + $(Ph)_3SnSe^{-}$:[a]	85
$[(Ph)_3Sn]_2$ (Bu₄NClO₄, DME)	Hg (-2.9 V vs 0.001 M AgClO₄/Ag, CPE)	$(Ph)_3Sn$:⁻ + $(Ph)_3SnO$:⁻[b]	85
Cp = Cyclopentadienyl			
CpMo(CO)₂Sn(Me)₃ (Bu₄NClO₄, DME)	Hg (-1.9 V vs 0.001 M AgClO₄/Ag, CPE)	CpMo (CO)₂:⁻ + (Me)₃Sn·	94
CpMo(CO)₃Sn(Ph)₃ (Bu₄NClO₄, DME)	Hg (-2.4 V vs 0.001 M AgClO₄/Ag, CPE)	CpMo(CO)₃:⁻ + (Ph)₃Sn·	94
(OC)₅Mn-Sn(Me)₃ (Bu₄NClO₄, DME)	Hg (-1.9 V vs 0.001 M AgClO₄/Ag, CPE)	(OC)₅Mn:⁻ + (Me)₃Sn·	94
(OC)₅Mn-Sn(Ph)₃ (Bu₄NClO₄, DME)	Hg (-2.5 V vs 0.001 M AgClO₄/Ag, CPE)	(OC)₅Mn:⁻ + (Ph)₃Sn·	94

472

CpFe(CO)₂-Sn(Ph)₃ (Bu₄NClO₄, DME)	Hg (-2.6 V vs 0.001 M AgClO₄/Ag, CPE)	CpFe(CO)₂:⁻ + (Ph)₃Sn·	94
(OC)₄Co-Sn(Ph)₃ (Bu₄NClO₄, DME)	Hg (-1.6 V vs 0.001 M AgClO₄/Ag, CPE)	(OC)₄Co:⁻ + (Ph)₃Sn·	94
CpMo(CO)₂Sn(Me)₂Mo(CO)₃Cp (Bu₄NClO₄, DME)	Hg (-1.8 V vs 0.001 M AgClO₄/Ag, CPE)	2 CpMo(CO)₂:⁻ + Me₂Sn	94
CpFe(CO)₂Sn(Me)₂Fe(CO)₂Cp (Bu₄NClO₄, DME)	Hg (-2.7 V vs 0.001 M AgClO₄/Ag, CPE)	2 CpFe(CO)₂:⁻ + Me₂Sn	94
RSnCl₃ (2 N NaOH-alcohol 1:1) R = C₆H₅, Et	Hg (-1.6 to -1.7 V vs ?, CPE)	RSn (10-20)	103
C₆H₅SnCl₃ (2 N HClO₄-methanol 1:1)	Hg (-0.8 V vs ?, CPE)	(C₆H₅)₂Sn	103
Tin tetraphenylporphyrin diacetate (OAc)₂ (n-Pr₄NClO₄, DMSO)	Hg or Pt (-1.01 V vs SCE)	Tin tetraphenylporphyrin diacetate radical anion	126

473

Table 10.11. Electroreduction of Group IV-A Organometallic Compounds (cont.)

Depolarizer (Supporting Electrolyte, Solvent)	Cathode (Potential, Other)	Products (% Yield)	Ref.
GeCl ((Bu)$_4$NClO$_4$, DME) with 3 phenyl groups	Hg (-2.8 V vs 0.001 M Ag$^+$/Ag, CPE)	(phenyl)$_3$GeH	85
GeCl ((Bu)$_4$NClO$_4$, DME) with 2 phenyl groups	Hg (-2.6 V vs 0.001 M Ag$^+$/Ag, CPE)	(phenyl)$_2$GeH$_2$	85
(phenyl)$_3$GeGe(phenyl)$_3$ ((Bu)$_4$NClO$_4$, DME)	Hg (-3.5 V vs 0.001 M Ag$^+$/Ag, CPE)	(phenyl)$_3$Ge$^-$	85

Electroreduction of Organosilicon Compounds

phenyl-Si(Me)$_3$ (LiCl, Anhyd. CH$_3$NH$_2$)	Pt (—, undivided cell / divided cell)	cyclohexadienyl-Si(Me)$_3$	(76)	28
		(Me)$_3$SiOSi(Me)$_3$	(37)	28
		cyclohexene	(36)	36

 (5)

Depolarizer (7)

CH_2CH_3 (38) 28

$C{\equiv}CH$ (47)

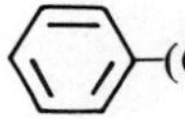

$(CH_2)_2Si(Me)_3$ (13)

CH_2CH_3 (1) 28

$C{\equiv}CH$ (34)

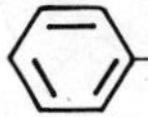

$(CH_2)_2Si(Me)_3$ (31)

Depolarizer (23)

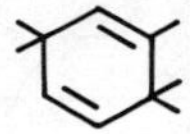

$(CH_2)_2Si(Me)_3$ (7)

$C{\equiv}CSi(Me)_3$ (LiCl, Anhyd. CH_3NH_2) Pt (−, undivided cell)

$C{\equiv}CSi(Me)_3$ (LiCl, 12 g isopropanol in 400 ml of CH_3NH_2)

475

Table 10.11. Electroreduction of Group IV-A Organometallic Compounds (cont.)

Depolarizer (Supporting Electrolyte, Solvent)	Cathode (Potential, Other)	Products (% Yield)	Ref.
$(C_6H_5)_3SiCl$ (Bu$_4$NClO$_4$, DME)	Hg (-3.1 V vs 0.001 M AgClO$_4$/Ag, CPE)	$(C_6H_5)_3SiH$	85
$(C_6H_5)_2SiCl_2$ (Bu$_4$NClO$_4$, DME)	Hg (-1.9 V vs 0.001 M AgClO$_4$/Ag, CPE)	$(C_6H_5)_2SiH_2$	85

[a] The species $(C_6H_5)_2SnSe^-$ is presumed on the basis of its reaction with $(C_6H_5)_2SnCl$.

[b] $(C_6H_5)_2Sn:^-$ and $(C_6H_5)_2SnO^-$ were presumed on the basis of their reaction with $(C_6H_5)_2SnCl$.

For a discussion of specificity of bond cleavage, see Refs. 85 and 94.

Electrochemical studies for the group of organometallics denoted by R_nMX_{4-n}, where X is mainly halogen, hydroxyl, or acetate, have dealt primarily with M = Sn and Pb. Work is also reported for Si and Ge compounds of this class.

What was probably the earliest report [268] for a member of this class of compounds was the electroreduction of triethyllead hydroxide in 95% ethanol at a lead cathode to yield hexaethyldilead, $[(Et)_3Pb]_2$. Others [183] have since claimed a nice yield improvement (80-90%) of hexaethyldilead by operating with a diaphragm cell and aqueous alkali solvent system.

The earliest electroreduction work located for an organotin compound was reported by Riccoboni and Popoff [302], who reduced diethyltin dichloride polarographically and obtained a single reversible two-electron wave. Diethyltin was the postulated product of the reduction:

$$(Et)_2Sn^{+2} + 2e \rightleftharpoons (Et)_2\overset{\cdot}{\underset{\cdot}{Sn}}$$

Within group IV-A the element (Si, Ge, Sn, Pb) involved in the organometallic compound is a key variable affecting electrolysis results. Very interesting and wide-ranging comparisons were made for a number of di- and triphenylmetal halides and acetates (ϕMX, $\phi_2 MX_2$) reduced at mercury cathodes in dimethoxyethane [85]. An intermediate radical, $\phi_3 M\cdot$, is generated for each element,

$$\phi_3 MX + e \rightarrow \phi_3 M\cdot + X^-$$

but depending on the type, can abstract hydrogen (Si, Ge), couple (Sn), or arylate the mercury cathode (Pb). These different results, all in dimethoxyethane, were summarized by Dessy et al. as follows (all V versus Ag/Ag^+):

$$Ph_3SiCl \xrightarrow[-3.1\ V]{e} Ph_3SiH$$

$$Ph_3GeCl \xrightarrow[-2.8\ V]{e} Ph_3GeH$$

$$Ph_3SnCl \xrightarrow[-1.6\ V]{e} 0.5\ Ph_3SnSnPh_3 \xrightarrow[-2.9\ V]{e} Ph_3Sn\text{:}^-$$

$$Ph_3PbOAc \begin{cases} \xrightarrow[-1.4\ V]{e} Ph_3Pb\cdot \xrightarrow{Hg} Ph_2Hg \\ \\ \xrightarrow[-2.2\ V]{2e} Ph_3Pb\text{:}^- \end{cases}$$

Very similar results were obtained [85] with Ph_2MX_2 compounds. Considerable difference exists in the reduction potential required for electrolysis of the above Ph_3MX compounds, the lead derivatives being most easily reduced. It is also true that spontaneous nonelectrolytic reductions are observed for some triphenyl- [85] and triethyllead [183] compounds. The observed products are the same as the electrolysis products. In the case of triphenyllead chloride, mercury was the reducing agent and diphenylmercury a product, while the triethyllead cation interacted with lead metal and yielded hexaethyldilead.

There is similarity in the main course of events for electrolysis of triorganolead and tin compounds. This holds for aqueous as well as organic solvent systems

$$R_3M^+ + e \xrightarrow{\text{aqueous}} R_3M\cdot \longrightarrow \tfrac{1}{2}[R_3M]_2$$

$$R_3M-X + e \xrightarrow{\text{aprotics}} R_3M\cdot + X^- \longrightarrow \tfrac{1}{2}[R_3M]_2$$

$$R = \phi, \text{ alkyl}$$

$$M = Pb, Sn$$

in spite of the fact that very different states of aggregation exist in the two solvent systems [353]. This general view is supported both by polarographic work [1, 61, 101, 102, 220, 227, 301] and macroscale electrolyses [85, 183, 268, 388]. Significant differences do exist, however, in aqueous compared to organic solvent systems. These involve adsorption of reduction products in aqueous systems [101, 102] but not in organic systems [85], and the ability to generate R_3M^-* products at the more negative potentials available in the organic systems [85].

Some early work on the electrolysis of the triethyllead cation to form hexaethyldilead was referred to on p. 468. Increased yields of hexaethyldilead were obtained in aqueous systems with diaphragms [183] as opposed to ethanolic systems without diaphragms [268]. In the aqueous system the product adsorbed and dripped from the lead cathode, while in the ethanolic medium the solvated hexaethyldilead was subject to destruction at the anode. This provides a simple example of the influence of solvent on yield even though the major reaction is probably similar in each solvent system.

A number of triphenyl- and trialkyllead and tin compounds were shown to be part of an electrochemically reversible system that depended on the ability of the hexaorganodimetal to dissociate into free radicals at platinized platinum

---

*Triphenyltin anions $(\phi_3 Sn^-)$ so generated can reduce triphenylsilicon chloride $(\phi_3 SiCl)$ to yield the same silane as detected in controlled potential electrolysis of $\phi_3 SiCl$ [90]:

$$\phi_3 Sn^- + \phi_3 SiCl + H\cdot \rightarrow \phi_3 Sn\cdot + \phi_3 SiH$$
$$\searrow$$
$$[\phi_3 Sn]_2$$

electrodes [106, 107, 346, 347]. The equations are illustrative of the sequence.

$$Me_6Sn_2 \rightleftharpoons 2 \ Me_3Sn \cdot \rightleftharpoons 2 \ Me_3Sn^+ + 2e$$

$$Ph_6Pb_2 \rightleftharpoons 2 \ Ph_3Pb \cdot \rightleftharpoons 2 \ Ph_3Pb^+ + 2e$$

The formation of radicals by homolytic dissociation of the metal-metal bond was rate determining and would occur at platinized platinum but not smooth platinum or mercury electrodes. Isotopic exchange was observed between tagged hexaphenyldilead and triphenyllead ions in the presence of platinum black. This was taken as evidence for the homolytic dissociation of hexaphenyldilead in the presence of the catalyst. We can note here also that Dessy et al. have shown the sequence

$$Ph_3Sn^+ \xrightleftharpoons[- e]{+ e} Ph_3Sn \cdot$$

to be polarographically reversible in dimethoxyethane [85].

Aquodiethyltin and lead cations are generated by dissociation of the corresponding chlorides and acetates in water:

$$Et_2Pb(OAc)_2 \xrightarrow{H_2O} Et_2Pb^{+2} + 2 \ ^-OAc$$

$$Et_2SnCl_2 \xrightarrow{H_2O} Et_2Sn^{+2} + 2 \ Cl^-$$

Both reduce initially at mercury cathodes to yield diradical species [99, 136, 273-275, 302] but there the similarity ends:

$$Et_2Pb^{+2} + 2e \rightarrow Et_2\overset{\cdot\cdot}{P}b$$

$$Et_2Sn^{+2} + 2e \rightarrow Et_2\overset{\cdot\cdot}{S}n$$

In the case of aquodiethyltin, a greenish-yellow insoluble oil, polydiethyltin (II), $[(Et)_2Sn]_n$, was isolated [274, 275]. Cyclic voltammetric methods showed a product of the electroreduction, probably $(Et)_2\overset{\cdot}{S}n$, could be reoxidized but attempts to electrochemically reoxidize the oil from the macroscale electrolysis were unsuccessful [274].

Aquodiethyllead reduces with ease at -0.37 V versus SCE, and the initially formed diradical, $Et_2\overset{\cdot\cdot}{P}b$, primarily disproportionates to $(Et)_4Pb$, but also undergoes some transmetallation to yield $(Et)_2Hg$ [273]:

$$2(Et)_2\overset{..}{Pb} \rightarrow (Et)_4Pb + Pb^\circ$$

$$(Et)_2\overset{..}{Pb} + Hg^\bullet \rightarrow (Et)_2Hg + Pb^\circ$$

These observed differences between lead and tin were explained in terms of the bond energies for lead-carbon (31 kcal/mole), versus the tin-carbon bond (54 kcal/mole). Transmetallation and disproportionation were both observed for lead, since the lead-carbon bond energy was comparable to the mercury-carbon bond energy (27 kcal/mole) [353].

Finally, as examples of monoorganometal compounds, ethyl- and phenyltin-trichloride have been electroreduced in aqueous alcoholic systems [103]. When the system was very basic, 10-20% of a red amorphous RSn product was formed, but in acid, diphenyltin was isolated.

Electroreduction of Group V-A Organometallics: As, Sb, Bi

Dessy et al. [79] have studied electroreduction of organometallics derived from Group V-A elements. These were electrolyses carried out at stirred mercury cathodes in anhydrous dimethoxyethane. Systems of the type R_2MX were reduced in order to form and study the subvalent species $R_2M\cdot$.

$$R_2MX + e \rightarrow R_2M\cdot + X^-$$

$$M = As, Sb, Bi$$

$$X = halide \text{ or } acetate$$

It was postulated that $R_2M\cdot$ could follow a number of pathways such as (a) stable species, (b) coupling to R_2MMR_2, (c) abstraction of $H\cdot$, (d) arylation of the mercury pool, and (e) disproportionation. Coupling was the major path for the $R_2M\cdot$ species of As, Sb, and Bi.

Coupling of the intermediate species $R_2M\cdot$ was observed when diphenylarsenic bromide was electroreduced:

$$(Ph)_2AsBr + e \xrightarrow[\text{Ag}^+/\text{Ag}]{-0.9 \text{ V vs}} 0.5(Ph)_2AsAs(Ph)_2 + Br^-$$

This was followed, at sufficiently negative potentials, by a second electron step (reduction of the coupled product). These products were identified by

$$(Ph)_2AsAs(Ph)_2 + e \xrightarrow[\text{Ag}^+/\text{Ag}]{-2.7 \text{ V vs}} (Ph)_2As^- + ?$$

observation of their polarographic and cyclic voltammetric behavior in comparison to known solutions. Dessy et al. noted the same diphenylarsenic anion was formed in tetrahydrofuran by cleavage of triphenylarsenic with lithium metal [409].

Along the same lines, Dehn [73] showed that the final cathodic reduction product of both cacodyl chloride and cacodyl was dimethylarsine. Successive overall reactions were shown to be

$$(CH_3)_2AsCl \; + \; 2 \; H \; \rightarrow \; [(CH_3)_2As]_2 \; + \; 2 \; HCl$$

$$[(CH_3)_2As]_2 \; + \; 2 \; H \; \rightarrow \; 2(CH_3)_2AsH$$

Electrolysis of diphenylantimony acetate at -1.9 V versus Ag/Ag^+ gave a colorless solution of tetraphenyldistibene [79]. Rapid coupling of diphenylantimony radicals was presumed here.

$$(Ph)_2SbOAc \; + \; e \; \rightarrow \; OAc^- \; + \; (Ph)_2Sb\cdot$$

$$2(Ph)_2Sb\cdot \; \xrightarrow{\text{fast}} \; (Ph)_2SbSb(Ph)_2$$

The tetraphenyldistibene was reduced at -2.45 V versus Ag/Ag^+ to the diphenylantimony anion just as was the analogous arsenic compound above. Ultraviolet spectra were used to identify these antimony compounds.

Diphenylbismuth chloride, on electrolysis as above for arsenic and antimony compounds, gave results similar to the above compounds. Conclusions regarding the products, $[Ph_2Bi]_2$ and $(Ph)_2Bi^-$, were not based on any analytical methods.

A later communication [94] compared the reduction potentials for well-characterized tetraphenyldi-arsenic, antimony, and bismuth compounds and found the reduction to be more facile as one proceeded from arsenic to antimony to bismuth.

Pentacovalent species of arsenic, antimony, and bismuth all behaved similarly, yielding chloride ion and a triorganometal compound:

$$R_3MCl_2 \; + \; 2e \; \rightarrow \; R_3M \; + \; 2 \; Cl^-$$

$$M \; = \; As, \; Sb, \; Bi$$

$$R \; = \; Me, \; Ph$$

A carbon arsenic bond was generated when o-nitrobenzenearsonic acid was prepared by electrolytic diazotization of o-nitroaniline in the presence of $NaNO_2$, $Cu(NO_3)_2$, and $NaAsO_2$ [421].

Table 10.12. Electroreduction of Group V-A Elements

Depolarizer	Supporting Electrolyte, Solvent	Cathode (Potential, Other)	Products (% Yield)	Ref.
Arsenic Compounds				
$(C_6H_5)_2AsBr$	(Bu_4NClO_4, DME)	Hg $\{$ $(-0.9$ V vs Ag/Ag$^+)$; $(-2.7$ V vs Ag/Ag$^+)$ $\}$	$(C_6H_5)_2AsAs(C_6H_5)_2$; $(C_6H_5)_2As$:$^-$	79 ; 79
Di(α-naphthyl)arsenic chloride	(Bu_4NClO_4, DME)	Hg $(-2.7$ V vs Ag/Ag$^+)$	$(\alpha$-Naphthyl$)_2As$:$^-$	79
$(C_6H_5)_2AsOAs(C_6H_5)_2$	(Bu_4NClO_4, DME)	Hg $(-2.8$ V vs Ag/Ag$^+)$	$(C_6H_5)_2As$:$^-$ + $(C_6H_5)_2AsO$:$^-$	79
$(C_6H_5)_2AsAs(C_6H_5)_2$	(Bu_4NClO_4, DME)	Hg $(-2.7$ V vs Ag/Ag$^+)$	$(C_6H_5)_2As$:$^-$	94, 79
$(CH_3)_2AsCl$ Cacodyl chloride	(Formic acid-alcohol)	Pt (—, separate compartments)	$[(CH_3)_2As]_2$, $(CH_3)_2AsH$	73
$[(CH_3)_2As]_2$	(HCl, alcohol)	Pt (—, separate compartments)	$(CH_3)_2AsH$	73
[benzene ring with NO_2 and NH_2]	(—, —)	Pt (—, —) ; C (—, —)	[benzene ring with NO_2 and $\overset{O}{\underset{}{As(OH)_2}}$] (73) ; (69)	421
[benzene ring with $O{:}As(OH)_2$ and O_2N]	(1 N NaCl, aqueous) ; (3 N NaOH, aqueous)	Pb (0.5 A/dm^2, 65°C) Iron catalyst ; Cu (4.76 ma/cm^2, 25°C)	[benzene ring with $O{:}As(OH)_2$ and H_2N] (96) ; (57)	420 ; 41_

482

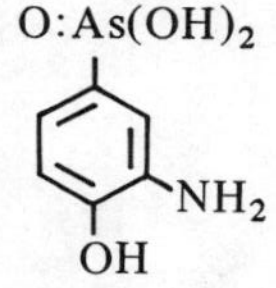

Hg (3 A/dm², 30-40°C)
Divided cell
(3 N HCl, H₂O)

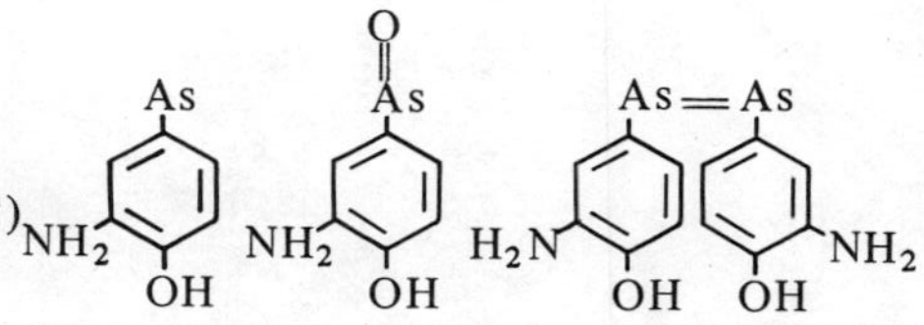

$O{:}As(OH)_2$ — NH_2, OH

As — OH, NH_2 1

$O{=}As$ — OH, NH_2 2

$As{=}As$ — OH, H_2N / NH_2, OH 3   389

Products 1, 2 and 3 at successive stages of reduction.

Hg (−, −)

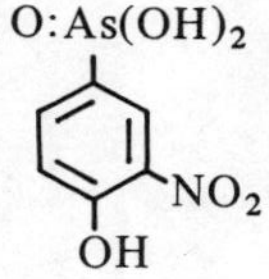

HCl above 4.7 N
HCl below 4.7 N or H₂SO₄

$O{:}As(OH)_2$ — NO_2, OH

HO—$As{=}As$—$NH_2{\cdot}HCl$ / HO—AsH_2, $NH_2{\cdot}HCl$

OH, $NH_2{\cdot}HCl$ 256

Pt, Ni, Cu (−)
(HCl, H₂O)

$O{:}As(OH)_2$ — NO_2, OH

$O{:}As(OH)_2$—$N{=}N$—$As(OH)_2{:}O$, OH
$O{:}As(OH)_2$—$N{\rightarrow}N{\rightarrow}O$, OH
+
$O{:}As(OH)_2$ — NH_2, OH 255   277

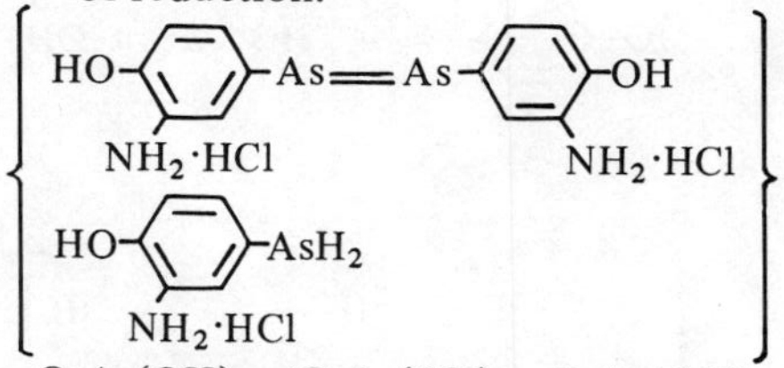

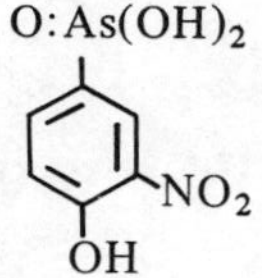

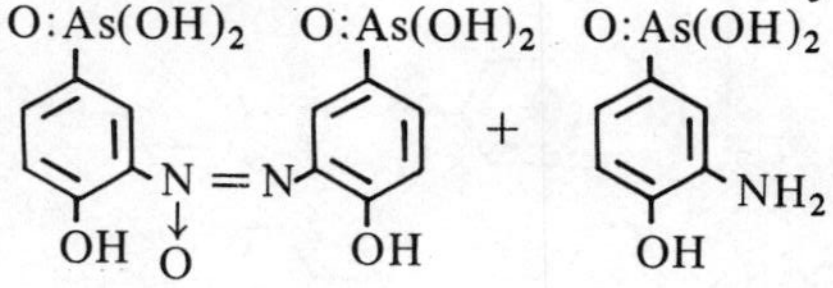

Cu (−)
(HCl, H₂O)

$O{:}As(OH)_2$ — NO_2, NO_2

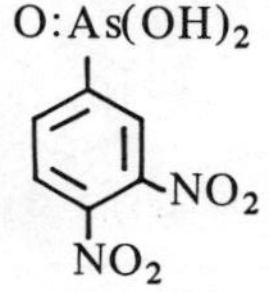

$O{:}As(OH)_2$ — NH_2, NH_2 255

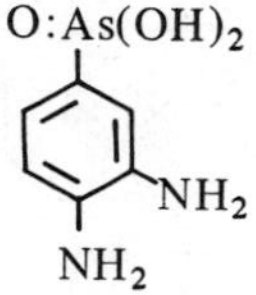

Table 10.12. Electroreduction of Group V-A Elements (cont.)

Depolarizer	Supporting Electrolyte, Solvent	Cathode (Potential, Other)	Products (% Yield)	Ref.
$O{:}As(OH)_2$ — (C₆H₄) — NH_2	HCl above 8 N HCl below 8 N	Hg (−)	$Cl{\cdot}H_3N$—〈C₆H₄〉—AsH_2 $Cl{\cdot}H_3N$—〈C₆H₄〉—$As{=}As$—〈C₆H₄〉—$NH_3{\cdot}Cl$	255
$O{:}As(OH)_2$ — (C₆H₄) — OH	HCl above 4 N HCl below 4 N	Hg (−)	HO—〈C₆H₄〉—$As{=}As$—〈C₆H₄〉—OH HO—〈C₆H₄〉—AsH_2	255
$O{:}AS(OH)_2$ — (C₆H₃) — NH_2, OH	(3 N H_2SO_4 + KI, H_2O)	cd, A/dm² Pb {(a) 1, (b) 2, (c) 4, (d) 8} divided cell 50-55°C	$HO{-}(AsR)_n OH$ $R =$ 〈C₆H₃〉 with OH, NH_2 n (a) 7.8 (b) 10 (c) 13.4 (d) 14.5	230

Substrate	Conditions	Metals	Product	Yields	Ref.
$O{:}As(OH)_2$ — C_6H_4 — NH_2	$(2\,N\ HCl,\ H_2O)$ Roughly similar yields were also obtained in H_2O-alcohol	Cu (a) 0.08 A/cm² Pb (b) 0.08 A/cm² Pb/Hg (c) 0.08 A/cm² Zn/Hg (d) 0.08 A/cm² Hg (e) 0.18 A/cm² Pb (0.019, 0.038, 0.076, 0.114 A/cm²)	H_2N — C_6H_4 — AsH_2	(a) (32.9) (b) (37) (c) (72) (d) (82) (e) (73)	130
$O{:}As(OH)_2$ — C_6H_4 — NH_2	$(2\,N\ HCl,\ H_2O)$	Above metals (above conditions, high temp. relative to above R_x)	$AsH_3,\ PhNH_2,\ H_3AsO_3$ Current yields (24, 22, 35 and 54)		130
$O{:}As(OH)_2$ — C_6H_5	$(2\,N\ HCl,\ H_2O)$	Above metals (above conditions)	C_6H_5 — AsH_2		130
AsO — C_6H_4 — NH_2	$(HCl\ or\ 2\,N\ NaOH,\ H_2O)$	(a) Cu (—) (b) Pb (c) Pb/Hg (d) Hg (e) Zn/Hg	$(H_2NC_6H_4As{:})_2$	(a) 29 current yield (b) 156 current yield (c) 129 current yield (d) 200 current yield (e) 129 current yield	130

Table 10.12. Electroreduction of Group V-A Elements (cont.)

Depolarizer	Supporting Electrolyte, Solvent	Cathode (Potential, Other)	Products (% Yield)	Ref.
O:As(OH)₂ O₂N— (benzene ring)	$\left\{\begin{array}{l}(2\,N\ \text{NaOAc}, -) \\ (\text{HCl}, H_2O)\end{array}\right.$	Pt (−) Cooled Pb (−)	$(H_2O_3AsC_6H_4NH)_2$ $(o\text{-ClH}\cdot H_2NC_6H_4As:)_2$ (83)	130 130
Me_2AsO_2H	$(2\,N\ H_2SO_4, H_2O)$	Pb or Zn/Hg (−)	$(Me_2As)_2 + (Me)_2AsH$	130
$[W(CO)_4AsMe_2]_2$	$((Bu)_4NClO_4, DME)$	Hg (−1.8 V vs Ag/Ag⁺)		88
$[Cr_2(CO)_{10}As_2Me_4]n$	$((Bu)_4NClO_4, DME)$	Hg (−2.1 V vs Ag/Ag⁺)	Dianions	94
Ph—Ph / Ph—Ph As—Ph (ring)	$((Bu)_4NClO_4, DME)$	Hg (−2.6 V vs Ag/Ag⁺)	Blue radical anion subject to further reduction at -3.0 V vs Ag/Ag⁺	89
Antimony Compounds				
$[(OC)_3FeAs(CH_3)_2]_2$	$((Bu)_4NClO_4, DME)$	Hg (−1.9 V vs Ag/Ag⁺)	Dianion (stable)	88 94
$(C_6H_5)_2SbOAc$	$((Bu)_4NClO_4, DME)$	Hg (−1.9 V vs Ag/Ag⁺)	$(C_6H_5)_2SbSb(C_6H_5)_2$	79
$(C_6H_5)_2SbSb(C_6H_5)_2$	$((Bu)_4NClO_4, DME)$	Hg (−2.4 V vs Ag/Ag⁺)	$(C_6H_5)_2Sb:^-$	79 94

Bismuth Compounds

$(Ph)_2BiCl$	$((Bu)_4NClO_4,\ DME)$	Hg (-1.2 V vs Ag/Ag$^+$)	$(Ph)_2BiBi(Ph)_2$ Supposed products by analogy with work on As and Sb	79
$(Ph)_2BiBi(Ph)_2$	$((Bu)_4NClO_4,\ DME)$	Hg (-2.3 V vs Ag/Ag$^+$)	$(Ph)\ Bi:^-$	94

Some cases involve reduction of sites in molecules other than the carbon-metal bond. For example, o-arsanilic acid is prepared from o-nitrobenzenearsonic acid [419, 420]:

$$O_2N \underset{}{\overset{O:As(OH)_2}{\bigcirc}} \longrightarrow H_2N \underset{}{\overset{O:As(OH)_2}{\bigcirc}}$$

The product yield was 57% in 3 N NaOH when a copper cathode and copper powder catalyst were used at 25°C, and 96% when in 1 N NaCl a lead cathode and iron catalyst were employed at 65°C. In another case [255] electroreduction of 3-nitro-4-hydroxyphenylarsonic acid at Pt, Ni, and copper cathodes did not affect the AsO_3H_2 group, but when the same molecule was reduced at a mercury cathode [256] the group (AsO_3H_2) was affected. Thus:

$$O:As(OH)_2 \text{ with } NO_2, HO \begin{cases} \xrightarrow[\text{HCl} >4.7\ N]{\text{Hg cathode}} & HO-\bigcirc-As=As-\bigcirc-OH \\ & \quad HCl \cdot H_2N \qquad HCl \cdot H_2N \\ \\ \xrightarrow[\text{HCl} <4.7\ N]{\text{Hg cathode}} & HO-\bigcirc-AsH_2 \\ & \quad HCl \cdot H_2N \end{cases}$$

Polymers I have been observed [230] when 3-amino-4-hydroxyphenylarsonic acid was reduced at lead cathodes and the degree of polymerization was regulated by careful controll of the current density.

$$HO(AsR)_nOH \qquad (n \text{ varies from 7 to 14})$$
$$\textbf{I}$$

Electrolytic reduction of arsonic acids is believed [389] to take the following course:

$$RAsO_3H_2 \rightarrow RAsO \rightarrow (RAsOH)_2 \rightarrow (RAs)_2O \rightarrow (RAs)_2 \rightarrow (RAsH)_2$$
$$\rightarrow RAsH_2$$

Fichter and Elkind [130] reported an extensive study of the electroreduction of arsonic acids. Generally, arsines were obtained by electrolysis of arsonic acids and yields of arsine varied with cathode metal and current density. The best yield of p-aminophenylarsine was obtained by electrolysis of p-amino-benzenearsonic acid at an amalgamated zinc cathode in $2\,N$ HCl using a current density of 0.08 A/cm^2. If the low temperature was not maintained, only PhNH$_2$, H$_3$AsO$_3$, and AsH$_3$ were isolated from the electrolysis.

A number of polarographic studies concerned with arsonic and arsinic acids have been reported. Arsonic acid with its unsubstituted ring was reduced more easily than ring-substituted arsonic acids regardless of whether the substitution was electron-withdrawing or -donating [41]. Others [304, 398] have concluded, based on polarographic work, that the As=O group is very reducible but AsOH is not. Diphenylarsinic acid [38] and ring-substituted [253, 254] bis(phenyl)-arsinic acids have also been studied.

Dessy and co-workers were able to generate radical anions when pentaphenyl-arsole II [89] and a number of bridged bimetallic species [88, 94] such as III were reduced at mercury pool cathodes. On reduction at -2.6 V versus Ag/Ag$^+$, II was converted to a blue radical anion with a half-life of about 1 min. This could be destroyed on addition of another electron at -3.0 V versus Ag/Ag$^+$. III could also be reduced to a radical anion [88] or dianion [94].

$$[(OC)_3FeAs(CH_3)_2]_2$$

II   III

3 ANODIC PROCESSES

Electrooxidation of Organoalkalimetal Compounds

Electrolysis of organoalkali metal reagents (RM) frequently results in the formation of dimeric products (R-R). Summarized in Table 10.13 are representative data. Morgat and Pallaud [272] have extensively studied the electrolytic oxidation of aliphatic and alicyclic lithium compounds at the mercury electrode. Mechanistically the reactions appear to proceed as follows:

$$2R^- - e \xrightarrow{\text{Hg}} R-R$$

$$Li^+ + e \xrightarrow{\text{Pt}} Li$$

In studies of the thermodynamic stability of organic anions, McKeever and Taft [262] found that electrooxidation of triarylmethyl alkali metal reagents yield the corresponding dimeric hexaarylethanes.

Table 10.13. Electrolysis of Organoalkali Metal Compounds

Compound (Solvent)	Electrode	Product(s) (% Yield)	Ref.
Dimethyldithiocarbamate: $$Me_2N-\overset{\text{S}}{\underset{\parallel}{C}}-SNa$$	Ni	Tetramethylthiurame disulfide: $$Me_2N\overset{\text{S}}{\underset{\parallel}{C}}-S-S-\overset{\text{S}}{\underset{\parallel}{C}}-NMe_2$$	435
Phenyllithium (ether)[a]	Hg	Diphenyl (50-60)	272
α-Naphthyllithium (ether)[a]	Hg	1,1′-Dinaphthanene (48)	272
Menthyllithium (ether)[a]	Hg	1,1′-Dimethyl-4,4′-diisopropyldicyclohexyl (48-50)	272
Butyllithium (ether)[a]	Hg	Octane (44)	272
Amyllithium (ether)[a]	Hg	Decane (49)	272
Hexyllithium (ether)[a]	Hg	Dodecane (47)	272
Heptyllithium (ether)[a]	Hg	Tetradecane (42)	272
Octyllithium (ether)[a]	Hg	Hexadecane (55)	272
Nonyllithium (ether)[a]	Hg	Octadecane (52)	272
Decyllithium (ether)[a]	Hg	Eicosane (50-55)	272
Undecyllithium (ether)[a]	Hg	Docosane (10)	272
Dodecyllithium (ether)[a]	Hg	Tetracosane (38)	272
Tetradecyllithium (ether)[a]	Hg	Octacosane (40)	272

Hexadecyllithium (ether)[a]	Hg	Dotriacontane (45)	272
Octadecyllithium (ether)[a]	Hg	Hexatriacontane (40)	272
Cyclopentyllithium (ether)[a]	Hg	Dicyclopentyl (44)	272
Cyclohexyllithium (ether)[a]	Hg	Dicyclohexyl (50)	272
Bornyllithium (ether)[a]	Hg	Diboronyl (40)	272
Farnesyllithium (ether)[a]	Hg	Squalene (30)	272
Fenchyllithium (ether)[a]	Hg	Difenchyl (25)	272
Geranyllithium (ether)[a]	Hg	Digeranyl (33)	272
Dimethyl Sodiomalonate (hexamethylphosporamide)	Pt	Tetramethylethane-1,1,2,2-tetracarboxylate, plus others	39
Ethyl Sodiomethylacetoacetonate (N,N,-dimethylacetamide)	Pt	Diethyl-α,α'-diacetyl-α,α'-dimethyl succinate, plus others	40

[a]Corresponding Cd, Zn, and Mg derivatives yield similar dimeric products upon electrolysis of the metal alkyl.

491

The classic Kolbe synthesis similarly yields dimeric products. Electrolytic oxidation of alkali metal carboxylates yields dimer and carbon dioxide.

$$2R\overset{\overset{\displaystyle O}{\|}}{C}-O^- \ M^+ \ \xrightarrow{\ -2e\ } \ R-R \ + \ 2CO_2$$

This area is reviewed in detail in Chapter VI. Analogous electrolytic oxidations of alkali metal salts of malonic acid esters yield dimeric adducts. Fichter has summarized early developments in this area [129]. Gelin and co-workers similarly observed dimer formation in the oxidation of sodium-1,3-butadien-olates in dimethylformamide [153].

Schäfer has recently reported electrochemical studies of the oxidative addition of anions to olefins [315]. For example, electrooxidation of sodium acetylacetonate in the presence of 1,3-butadiene yields *trans*-5,9-tetradecadiene-2,13-dione at 40% current efficiency:

$$(CH_3-\overset{\overset{\displaystyle O}{\|}}{C})_2CH^- \ \xrightarrow{\qquad} \ CH_3-\overset{\overset{\displaystyle O}{\|}}{C}-CH-CH_2-CH\cdots CH\cdots CH_2 \cdot$$

$$(CH_3-\overset{\overset{\displaystyle O}{\|}}{C}-CH_2-CH_2CH{=}CH \ CH_2)_2$$

Other representative examples are summarized in Table 10.14.

Schmidt has reported a novel method for the preparation of anhydrous alkali and alkaline earth metal acetates [321]. For example, anhydrous lithium acetate was formed in 98% yield by the electrolysis of lithium chloride at graphite electrodes in anhydrous acetic acid containing minor quantities of cyclohexene. The chlorine liberated at the anode reacted with the cyclohexene scavenger forming 1,2-dichlorohexane.

Electrooxidation of Organoalkaline Earth Compounds

Synthesis of Organomagnesium Compounds

In general, organomagnesium compounds are most conveniently synthesized via direct chemical routes. Most typically, the organomagnesium reagent is prepared by direct reaction of an organohalide with magnesium metal in a donor solvent such as ether. In a number of cases, however, the synthesis of organo-magnesiums has been achieved by electrolysis. Pertinent examples are summar-ized in Table 10.15. A patent by Kobetz and Pinkerton [219], in particular, details the electrolytic synthesis of organomagnesium compounds. The primary reaction scheme involves the electrolysis of molten alkylaluminum salts at a magnesium anode:

Table 10.14. Oxidative Addition of Anions to Olefins

Compound	Reactant (Solvent)	Electrode (Potential,[a] Temp.)	Product (% Yield)	Ref.
$(CH_3-\overset{O}{\overset{\|}{C}})_2CH_2{}^-Na^+$	1,3-Butadiene (methanol)	Pt (+1.0 V, 0°C)	$(CH_3-\overset{O}{\overset{\|}{C}})_2CH_2CH_2-CH=CH-CH_2)_2$ (40)	315
$(CH_3O-\overset{O}{\overset{\|}{C}})_2CH^-Na^+$	Vinylethylether (methanol)	Pt (+0.8 V, 20°C)	$(CH_3-\overset{O}{\overset{\|}{C}})_2CHCH_2CH(OCH_3)(OC_2H_5)$ (37)	315
$(CH_3)_2C-NO_2{}^-Na^+$	Styrene (methanol)	Pt (+0.4 V -0.87 V)	$[(CH_3)_2\overset{NO_2}{\overset{\|}{C}}-CH_2-CH(C_6H_5)\overline{]}_2$ + $(CH_3)_2-\overset{NO_2}{\overset{\|}{C}}-CH_2CH(C_6H_5)(OCH_3)$	315
$(CH_3)_2C-NO_2{}^-Na^+$	Vinylethylether (methanol)	Pt (+0.4 V -0.87 V)	$(CH_3)_2\overset{NO_2}{\overset{\|}{C}}-CH_2CH(OCH_3)(OC_2H_5)$	315

[a] Potential V versus Ag/AgCl.

493

$$2MAIR_4 \ + \ Mg \ \xrightarrow{\text{Electrolysis}} \ MgR_2 \ + \ AIR_3 \ + \ 2M$$

where M = alkalimetal salt and R = alkyl or aryl.

Table 10.15. Electrolytic Synthesis of Organomagnesium Compounds

Compound	Electrode	Product(s) (% Yield)	Ref.
$AIR_3 + MX$ M=Na; X=halide	Mg	MgR_2	307, 430
$MAI(C_2H_5)_4$ (diethylether solvent) M≡Na or K	Mg	$Mg(C_2H_5)_2$ (95)	239
Alkyls of 2-6 carbons		$Mg(C_{2-6})_2$	425
$NaOCH_3$ (CH_3OH)	Mg	$Mg(OCH_3)_2$	341a
$NaB(C_2H_5)_4(H_2O)$	Mg	$Mg(C_2H_5)_2$ (73)	428
$NaF \cdot 2AI(C_2H_5)_3$	Mg	$Mg(C_2H_5)_2$	429
$NaZn(C_2H_5)_3$	Mg	$Mg(C_2H_5)_2$	187
$NaAI(CH_3)_4 + NaAI(C_2H_5)_4$	Mg	$Mg(CH_3)_2$; $Mg(C_2H_5)_2$; $Mg(CH_3)(C_2H_5)$(95 overall)	219
$KAI(CH_3)_4 + KAI(n\text{-}C_3H_7)_4$	Mg	$Mg(n\text{-}C_3H_7)_2$	219
$NaAI(CH_3)_4 + NaAI(C_6H_5)_4$	Mg	$MgR_2 \cdot 2AIR_3$ R≡CH_3 or C_6H_5	219
$LiAI(CH_3)_4 + NaAI(C_4H_9)$	Mg	$MgR_2 \cdot AIR_3$ R≡CH_3 or C_4H_9	219
$RbAI(CH_3)_4 + NaAI(CH_3)_4$	Mg	$Mg(CH_3)_2 \cdot AI(CH_3)_3$	219

Electrolysis of Grignard Reagents at Inert Electrodes

Organomagnesium compounds, particularly Grignard reagents, have long been useful intermediates in organic synthesis. Of the alkali and alkaline earth metals, the electrolytic reactions and synthesis of organomagnesium compounds have received most study.

Although usually denoted RMgX, the constitution of Grignard reagents in solution (usually ether solvents) is quite complex [397]. Equilibria of the type illustrated below have been studied by Duval [109] and Martinot [252]. The existence of such ions was inferred on the basis of electrical discharge of magnesium at both the anode and cathode.

$$RMgX \ \rightleftharpoons \ R^- \ + \ MgX^+$$

$$R_2Mg \ \rightleftharpoons \ R^- \ + \ RMg^+$$

$$R_2Mg \ + \ R^- \ \rightleftharpoons \ R_3Mg^-$$

$$R^- + MgX_2 \rightleftharpoons RMgX_2^-$$

$$MgX_2 \rightleftharpoons MgX^+ + X^-$$

$$R_2Mg + X^- \rightleftharpoons R_2MgX^-$$

$$MgX_2 + X^- \rightleftharpoons MgX_3^-$$

Schlenk type equilibria [80-82, 320] are also recognized,

$$2RMgX \rightleftharpoons [RMgX]_2 \rightleftharpoons R_2Mg + MgX_2$$

The nature of the complex Grignard equilibria is determined by the solvent environment, temperature, concentration, purity, and so on. Consequently, it is not surprising that electrolysis products vary significantly with variations in the above conditions. Pioneering studies by Kondyrev [221-226] and Evans [115-123, 283] established the following mechanisms for cathodic and anodic discharge of magnesium:

At Cathode
$$\begin{cases} 2RMg^+ + 2e \longrightarrow Mg + R_2Mg \\ \\ 2MgX^+ + 2e \longrightarrow Mg + MgX_2 \end{cases}$$

At Anode
$$\begin{cases} R_3Mg^- \longrightarrow R\cdot + R_2Mg + e \\ \\ R_2MgX^- \longrightarrow R\cdot + RMgX + e \\ \\ RMgX_2^- \longrightarrow R\cdot + MgX_2 + e \end{cases}$$

Kondyrev found that magnesium was formed at the platinum cathode as a finely dispersed black solid, while the amalgam was formed at the mercury cathode [223, 225].

Detailed electroanalytical studies by Martinot [252] have shown that the true electroactive species at the anode is a solvated complex resulting from the spontaneous ionization of the organomagnesium compound in solution:

$$R_2Mg\cdot(Et_2O)_2 \rightleftharpoons RMg^+\cdot Et_2) + R^- + Et_2O$$

Upon subsequent reaction, R^- forms new complexes:

$$R^- + R_2Mg\cdot(Et_2O)_2 \rightleftharpoons R_3Mg^-\cdot Et_2O + Et_2O$$

and

$$R^- + MgBr_2(Et_2O)_2 \rightleftharpoons RMgBr_2^-\cdot Et_2O + Et_2O$$

The electroactive species were shown to be $R_3Mg^-\cdot Et_2O$ and $RMgBr_2^-\cdot Et_2O$. In an earlier electroanalytical study, Martinot [251] suggested that the rate-determining step in the electrolysis of simple aliphatic Grignard reagents was electron transfer at the anode and electrocrystallization of the magnesium deposit at the cathode. Gillet [158, 159] accounted for dimer formation by suggesting that during the electrolysis of RMgBr, ionization of the reagent is induced by the electric field yielding R^+:

$$R^+ + RMgBr \longrightarrow R-R + MgBr^+$$

The anodically formed radicals then enter into a number of subsequent reactions, depending on conditions. For example, they may:

(1) react with certain anode metals to yield another organometallic species,
(2) dimerize $(2R\cdot \rightarrow R_2)$,
(3) disproportionate $(2R\cdot \rightarrow RH + R'CH=CH_2)$, or
(4) react with solvent or another reagent.

Reaction with other anode metals is of particular importance in the synthesis of organoaluminum and organolead compounds. This is discussed in detail under the respective metal. The formation of dimeric adducts during the oxidation of Grignard reagents has been studied in detail, and pertinent data are summarized in Table 10.16. In Grignard reagents of four or more carbons, Evans and co-workers [115, 116] observed dimer formation in very high yield. Although the tabulated data indicate less than 100% yield, Evans [116] noted circumstantial evidence pointing to quantitative dimer formation during oxidation at the platinum anode. The lack of rigorous analytical tools (e.g., VPC) at the time of the study precluded precise and quantitative product analysis. Evans, however, generalized that linear and branched free radicals of four or more carbon atoms tend to quantitatively couple [116]. Evans and Lee [119, 121] further found that the dimerization reaction proceeds in low yield at low current density. Increased concentration of the reagent as well as increased current density was found to increase the yield of dimer. It was also found that organomagnesium chlorides yielded more dimer than corresponding bromides and iodides [117].

Reaction of anodically formed radicals with ether solvents is common, particularly for Grignard reagents containing less than four carbons. The primary reaction is hydrogen abstraction with concomitant formation of a new radical: Hydrogen, either α or β to oxygen, can undergo abstraction. Evans and Field [118] observed both types of extraction in their study of the electrolysis of methylmagnesium iodide in n-butylether. Anodic discharge of methyl radical led to the following transformations:

Table 10.16. Electrolysis of Grignard Reagents

Compound (Solvent)	Electrode (cd)	Product(s) (% Yield)	Ref.
CH_3MgBr (ether)	Pt	Methane (21-79); ethane (20-71); ethylene (2-5); isobutylene (5-17)	119, 121
CH_3MgI (ether)	Pt	Methane (57-64); ethane (13-28); ethylene (6-7); isobutylene (8-17)	119, 121
$C_2H_5Mg(Cl)(Br)(I)$ (ether)	Pt	Ethane (51); ethylene (48); hydrogen (1)	119, 121
C_3H_7MgBr (ether)	Pt	Propane (50); propylene (49); hydrogen (1)	119, 121
CH_3MgI (n-butylether)	Pt (0.2-1.6 A/dm^2)	Methane, ethane, butane, butylene, carbon dioxide, butanol-1, and pentanol-1	118
n-C_3H_7MgBr (ether)	Pt (0.8-1.6 A/dm^2)	Propane, propylene, hexane, n-propanol	115
i-C_3H_7MgBr (ether)	Pt (0.8-1.6 A/dm^2)	Propane, propylene, i-propanol, ethanol, and 2,3-dimethylbutane	115
n-C_4H_9MgBr (ether)	Pt (0.4-2.0 A/dm^2)	Octane (85); butane; butene-1	116
n-$C_5H_{11}MgBr$ (ether)	Hg	Octane (41)	272
i-C_4H_9MgBr (ether)	Pt (0.4-2.0 A/dm^2)	2,5-Dimethylhexane (96)	116
s-C_4H_9MgBr (ether)	Pt (0.4-2.0 A/dm^2)	3,4-Dimethylhexane (49)	116
t-C_4H_9MgBr (ether)	Pt (0.4-2.0 A/dm^2)	Tetramethylbutane	116
n-Amyl MgBr (ether)	Hg	Decane (55-60)	272
n-Hexyl MgBr (ether)	Pt (0.4-2.0 A/dm^2)	Dodecane (83)	116

497

Table 10.16. Electrolysis of Grignard Reagents (cont.)

Compound (Solvent)	Electrode (cd)	Product(s) (% Yield)	Ref.
n-Hexyl MgBr (ether)	Hg	Dodecane (45)	272
n-Heptyl MgBr (ether)	Hg	Tetradecane (50)	272
n-Octyl MgBr (ether)	Hg	Hexadecane (47)	272
n-Nonyl MgBr (ether)	Hg	Octadecane (55-60)	272
n-Decyl MgBr (ether)	Hg	Eicosane (35)	272
n-Undecyl MgBr (ether)	Hg	Docosane (5-6)	272
n-Dodecyl MgBr (ether)	Hg	Tetracosane (40)	272
n-Tetradecyl MgBr (ether)	Hg	Octacosane (40)	272
n-Hexadecyl MgBr (ether)	Hg	Dotriacontane (52)	272
n-Octadecyl MgBr (ether)	Hg	Hexatriacontane (54)	272
Cyclopentyl MgBr (ether)	Hg	Dicyclopentyl (35)	272
Cyclohexyl MgBr (ether)	Hg	Dicyclohexyl (53)	272
Bornyl MgBr (ether)	Hg	Dibornyl (40)	272
Farnesyl MgBr (ether)	Hg	Squalene (33)	272
Fenchyl MgBr (ether)	Hg	Difenchyl (35)	272
Geranyl MgBr (ether)	Hg	Digeranyl (60)	272

Menthyl MgBr (ether)	Hg	Dimenthyl (48)	272
C_6H_5MgBr (ether)	Hg	Biphenyl (55)	272
C_6H_5MgBr (ether)	Pt (0.24-0.48 A/dm²)	Benzene; diphenyl; p-terphenyl; styrene; ethanol and polymer	123
p-Tolyl MgBr (ether)	Pt (0.16 A/dm²)	p-Methylstyrene; polymer	123
p-chlorophenyl-MgBr (ether)	Pt (0.24 A/dm²)	p-chlorostyrene; polymer	123
1-Naphthyl-MgBr (ether)	Hg	1,1'-Binaphthyl (43)	272
Benzyl-MgCl	Pt	Bibenzyl (50)	147
Benzyl-MgBr (ether)	Pt (0.24 A/dm²)	Bibenzyl (~80)	123

$$R\cdot + R'CH_2 - O - CH_2R'' \longrightarrow RH + R' - \overset{\cdot}{C}H - O - CH_2 - R''$$

a Abstraction:

$$CH_3\cdot + C_4H_9 - O - CH_2C_3H_7 \longrightarrow$$

$$CH_4 + C_4H_9 - O - \overset{\cdot}{C}HC_3H_7 \longrightarrow$$

$$C_4H_9 + C_3H_7C \overset{\displaystyle\nearrow O}{\underset{\displaystyle\diagdown H}{}}$$

$\downarrow$ CH_3MgI

C_4H_8 and C_4H_{10} C_3H_7 $CHOHCH_3$

(Pentanol-2)

β Abstraction:

$$CH_3\cdot + \quad C_4H_9 - O - CH_2CH_2C_2H_5 \longrightarrow$$

$$CH_4 + C_4H_9 - O - CH_2-\overset{\cdot}{C}H-C_2H_5 \longrightarrow$$

$$C_4H_9O\cdot \quad + \quad CH_2 = CH-C_2H_5$$

$\Big\downarrow CH_3MgI/_{H_2O}$

$$C_4H_9OMgI + CH_3\cdot$$

$\downarrow$

C_4H_9OH
(Butanol-1)

Similarly, electrolysis of n-propylmagnesium bromide in ethyl ether is accompanied by reaction of the propyl radical with solvent [115]:

a Abstraction:

$$C_3H_7\cdot + CH_3CH_2 - O - C_2H_5 \longrightarrow C_3H_8 + CH_3 - \overset{\cdot}{C}H - O - C_2H_5$$

$$CH_3\overset{\cdot}{C}H - O - C_2H_5 \longrightarrow C_2H_5 + CH_3CHO$$

$\downarrow$

$C_2H_4 + C_2H_6$ $\Big\downarrow C_3H_7MgBr/_{H_2O}$

$\downarrow$

s-amyl alcohol

β Abstraction:

$$C_3H_7 + CH_3CH_2 - O - C_2H_5 \longrightarrow C_3H_8 + \overset{\bullet}{C}H_2-CH_2-O-C_2H_5$$

$$\overset{\bullet}{C}H_2CH_2 - O - C_2H_5 \longrightarrow C_2H_4 + C_2H_5O$$

$$\downarrow C_3H_7MgBr/_{H_2O}$$

ethanol

Electrolysis of aryl Grignard reagents likewise led to substantial reaction of the anodically generated radicals with ether solvent [123]. The formation of significant quantities of styrene derivatives was accounted for by a-hydrogen abstraction from ether:

$$C_6H_5\cdot + CH_3CH_2 - O - C_2H_5 \longrightarrow C_6H_6 + CH_3\overset{\bullet}{C}H - O - C_2H_5$$

$$\downarrow CH_3CHO + C_2H_5\cdot$$

$$\downarrow C_6H_5MgBr$$

$$C_6H_5CH=CH_2 + MgBrOH \longleftarrow C_6H_5CH-CH_3$$
$$| $$
$$OMgBr$$

Examination of the current-voltage curves for a variety of Grignard Reagents in ether solution (Table 10.17) relates structure with decomposition potential in an expected manner [356, 363].

Table 10.17. Decomposition Potentials of Grignard Reagents in Ether Solution

Compound	Decomposition Potential (V)
C_6H_5MgBr	2.17
CH_3MgBr	1.94
C_3H_7MgBr	1.42
C_4H_9MgBr	1.32
C_2H_5MgBr	1.28
$C_2H_5(CH_3)CHMgBr$	1.24
$(CH_3)_2CHMgBr$	0.97
$(CH_3)_3CMgBr$	0.97
$CH_2=CHCH_2MgBr$	0.86

The differences in these potentials were attributed to the differences in discharge overvoltage of the corresponding anions. Thus, aryl and alkyl radicals require a more negative potential for their formation than the easily formed allyl radical.

In analogous studies on the oxidative dimerization of olefins, Schäfer [315] found that anodic oxidation of Grignard reagents in the presence of vinyl monomers yielded dimeric addition products. For example, oxidation of an ether solution of n-butylmagnesium bromide in the presence of styrene yielded 6,7-diphenyl-dodecane in 30% yield:

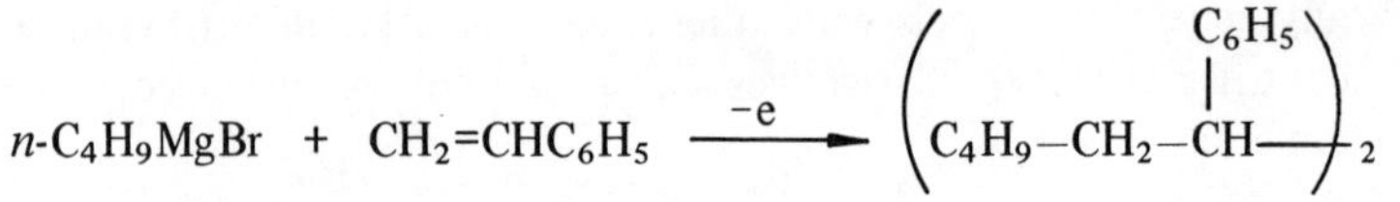

$$n\text{-}C_4H_9MgBr \ + \ CH_2{=}CHC_6H_5 \ \xrightarrow{-e} \ \left(C_4H_9{-}CH_2{-}\overset{\displaystyle C_6H_5}{\underset{\displaystyle |}{CH}}{-} \right)_2$$

Similarly oxidation of t-butylmagnesium bromide yielded 2,2,7,7-tetramethyl-4,5-diphenyloctane in 15% yield:

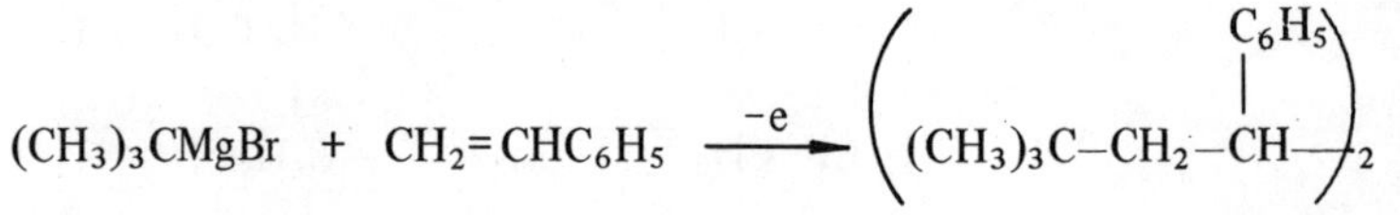

$$(CH_3)_3CMgBr \ + \ CH_2{=}CHC_6H_5 \ \xrightarrow{-e} \ \left((CH_3)_3C{-}CH_2{-}\overset{\displaystyle C_6H_5}{\underset{\displaystyle |}{CH}}{-} \right)_2$$

The addition of 1,3-butadiene to the oxidation product of n-butylmagnesium bromide yielded mixed dimeric isomers in approximately 20% yield:

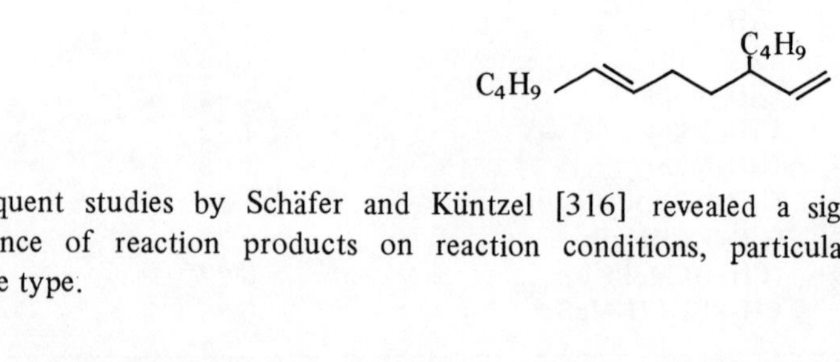

Subsequent studies by Schäfer and Küntzel [316] revealed a significant dependence of reaction products on reaction conditions, particularly the electrode type.

Formation of Organometallics at Sacrificial Anodes

Electrolysis of Grignard Reagents

The electrolysis of Grignard solutions provides a facile route to a number of other organometallic reagents, particularly organolead compounds. The overall reaction of these anodically formed reagents is believed to proceed as follows:

$$4 \ RMgX \ + \ M \ \xrightarrow{-4e} \ R_4M \ + \ 2MgX_2 \ + \ 2Mg$$

$$RX \ + \ Mg \ \longrightarrow RMgX$$

R is typically an alkyl such as methyl or ethyl, and M is lead. Organoboron, aluminum, and phosphorous compounds are similarly formed. Tomilov and Brago [356, 363] have extensively reviewed this important area, so only key reactions are discussed herein.

The Nalco process for the formation of tetraethyllead achieved commercial significance some years ago [296]. Key patents by Braithwaite [374] report the electrolysis of ethylmagnesium halides in high-boiling ethers such as the dibutyl ether of ethylene glycol [381]. The electrolysis is conducted in a specially designed cell and the tetraethyllead is distilled off as it is formed. Additional Grignard reagent is formed during the electrolysis by reaction of added alkyl halide with the magnesium liberated at the cathode. Magnesium halide is precipitated by the addition of dioxane. The schematic flow diagram for the process is illustrated in Fig. 10.6. The Grignard reagent is prepared in 8,000-gallon kettles and is then forwarded (along with excess alkyl halide) to the electrolysis cells. The lead pellets fed to the top of the cell serve as the anode while the steel cell walls function as the cathode. A nonconducting, permeable membrane separates the cathode from the anode. Detailed surveys of the significant developments in this important area are found in the reviews of Marlett [249] and Tomilov [356, 363].

Summarized in Table 10.18 are representative examples on the synthesis of organolead and organoaluminum, boron, and phosphorous compounds from Grignard reagents.

Electrolysis of Organoaluminum Compounds

The development of facile preparative methods for the synthesis of trialkylaluminum compounds lead to extensive investigations of the electrolytic synthesis of a number of other organometallics using R_3Al as the precursor. The aluminum alkyls are typically complexed with alkali metal halides or alkylalkali metal compounds that give rise to highly conductive solutions. Electrolysis of these solutions at anodes of lead, zinc, tin antimony, cadmium, or mercury, for example, lead to the formation of the corresponding metal alkyls, usually in high yield.

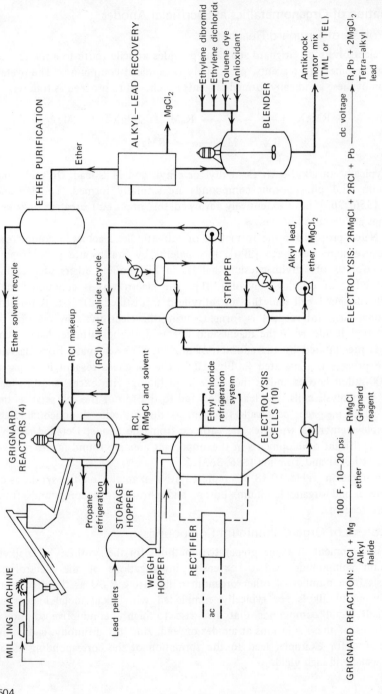

Fig. 10.6 Flow relations for the manufacture of alkyl leads by electrolysis. (From Charles L. Mantell, *Electro-Organic Chemical Processing*, Noyes Development Corporation, Park Ridge, N. J., 1968, P. 168).

MILLING MACHINE

GRIGNARD REACTORS (4)

ETHER PURIFICATION

ALKYL—LEAD RECOVERY

Ether solvent recycle

Ether

MgCl₂

Ethylene dibromid
Ethylene dichlorid
Toluene dye
Antioxidant

BLENDER

Antiknock
motor mix
(TML or TEL)

R₄Pb + 2MgCl₂
Tetra–alkyl
lead

dc voltage

ELECTROLYSIS: 2RMgCl + 2RCl + Pb → R₄Pb + 2MgCl₂

RCl makeup

(RCl) Alkyl halide recycle

STRIPPER

Alkyl lead,
ether, MgCl₂

RCl,
RMgCl and solvent

Propane
refrigeration

STORAGE
HOPPER

Lead pellets

WEIGH
HOPPER

Ethyl chloride
refrigeration
system

ELECTROLYSIS
CELLS (10)

RECTIFIER

ac

RMgCl
Grignard
reagent

100 F, 10–20 psi

ether

GRIGNARD REACTION: RCl + Mg → RMgCl

Alkyl
halide

504

Compound	(Solvent)	Electrode (CD-A/cm²)		Products (% Yield)		Ref.
CH₃MgCl	Diethylene glycol in dibutyl ether	Pb	(0.026)	Pb(CH₃)₄	(97)	384
C₂H₅MgBr	Diethyl ether	Pb	(0.002)	Pb(C₂H₅)₄	High	374
C₂H₅MgBr	Diethyl ether	Pb	(0.01)	Pb(C₂H₅)₄	(86)	43
C₂H₅MgCl	Dibutyl ether of diethylene glycol	Pb	(0.003)	Pb(C₂H₅)₄	(100)	374
C₂H₅MgCl	Dimethyl ether of diethylene glycol	Pb	(0.018)	Pb(C₂H₅)₄	–	45
Ethylmagnesium glycol	Dibutyl ether of diethylene glycol		–	Pb(C₂H₅)₄	(98)	381
C₂H₅MgI	Diethyl ether	Al		Al(C₂H₅)₃		122
CH₃MgCl	Diethyl ether	B		B(CH₃)₃		380
C₂H₅MgCl	Diethyl ether	B		B(C₂H₅)₃		380
C₃H₇MgCl	Diethyl ether	B		B(C₃H₇)₃		380
i-C₃H₇MgCl	Diethyl ether	B		B(i-C₃H₇)₃		380
C₄H₃MgCl	Diethyl ether	B		B(C₄H₉)₃		380
C₆H₅MgCl	Diethyl ether	B		B(C₆H₅)₃		380
C₂H₅MgCl	Diethyl ether	Black phosphorous	(0.1)	P(C₂H₅)₃		379
Alkyl-MgCl R is C₁–C₈	Diethyl ether	P		P(Alk)₃		379
C₆H₅MgCl	Diethyl ether	P		P(C₆H₅)₃		379
C₆H₅CH₂MgCl	Diethyl ether	P		P(C₆H₅CH₂)₃		379

As illustrated below, even alkylaluminums themselves can be formed by electrolysis at the aluminum anode:

$$3\ NaAl(C_2H_5)_4\ +\ Al\ \xrightarrow{-e}\ 4\ Al(C_2H_5)_3\ +\ 3\ Na$$

Ziegler [426] found that by recycling three of the four moles of triethylaluminum and regenerating the electrolyte by reaction with sodium, ethylene, and hydrogen, the total synthesis of triethylaluminum resulted:

$$3\ Na\ +\ 3C_2H_4\ +\ {}^3/_2\,H_2\ +\ Al(C_2H_5)_3\ \longrightarrow\ 3NaAl(C_2H_5)_4$$

Exemplary of the synthesis of other metal alkyls, is the formation of tetraethyllead [427]:

$$4\ NaAl(C_2H_5)_4\ +\ Pb\ \xrightarrow{-e}\ Pb(C_2H_5)_4\ +\ 4Na$$
$$+\ Al(C_2H_5)_3$$

Tetraethyllead can similarly be formed upon electrolysis of $NaF \cdot 2Al(C_2H_5)_3$ [431]. One of the improved methods for the electrosynthesis of tetraethyllead involves the use of a mercury cathode and a potassium aluminumalkyl electrolyte. The potassium electrolyte exhibits improved conductivity [44, 154, 155, 156, 375]:

$$KAl(C_2H_5)_4\ +\ Pb\ \xrightarrow{-e}\ Pb(C_2H_5)_4\ +\ 4Al(C_2H_5)_3$$
$$+\ 4K(Hg)_x$$

$NaAl(C_2H_5)_4$ is reformed as indicated above and sodium is then exchanged for potassium:

$$NaAl(C_2H_5)_4\ +\ K(Hg)_x\ \longrightarrow\ KAl(C_2H_5)_4\ +\ Na(Hg)_x$$

and then the potassium amalgam is regenerated:

$$Na(Hg)_x\ +\ KOH\ \longrightarrow\ K(Hg)_x\ +\ NaOH$$

Summarized in Table 10.19 are representative examples of the synthesis of various metal alkyls from aluminum alkyl precursors. Since this important area is adequately reviewed by Marlett [249] and Tomilov and co-workers [356, 363] only limited examples are provided herein. Additional key references are cited in the bibliographies of Swann [341, 339, 340].

Table 10.19. Electrochemical Synthesis of Organometallics by Electrolysis of Organoaluminum Complexes

Compound (Solvent) (Cathode)	Anode	$(CD\text{-}A/cm^2)$	Product(s)	(% Yield)	Ref.
$NaAl(C_2H_5)_4$	Al	(0.04)	$Al(C_2H_5)_3$	(99)	155
$NaAl(C_3H_7)_4$	Al	(0.04)	$Al(C_3H_7)_3$	(94)	155
$Al(C_2H_5)_3(Et_2O)(Pt)$	Pb	(0.70)	$Pb(C_2H_5)_4$	(99)	44
$KAl(C_2H_5)_4(Hg)$	Pb	(0.5)	$Pb(C_2H_5)_4$	(100)	156
$NaAl(C_2H_5)_4/NaAl(CH_3)_4(3/1)$ (Cu)	Pb	(0.25)	$Pb(C_2H_5)_4$	(89)	375
$NaF \cdot 2Al(C_2H_5)_3(Cu)$	Pb	(0.20)	$Pb(C_2H_5)_4$	(87)	42, 372
$K[Al(C_2H_5)_3(OC_4H_9)](Cu)$	Pb	(0.04)	$Pb(C_2H_5)_4$	(93)	383
$NaAl(CH_3)_4(THF)(Hg)$	Pb	(0.14)	$Pb(CH_3)_4$	(90)	23, 382
$NaF \cdot 2Al(C_6H_5)_3(C_6H_6)(Cu)$	Pb		$Pb(C_6H_5)_4$		372
$NaAl(C_2H_5)_4(Cu)$	Mg		$Mg(C_2H_5)_2$	(94)	155
$NaAl(CH_3)_4/NaAl(C_2H_5)_4(Cu)$	Zn	(0.35)	$Zn(C_2H_5)_2$	(80)	378
$Na[Al(C_2H_5)_3(OC_2H_5)](Cu)$	Pb	(0.04)	$Zn(C_2H_5)_2$	(100)	383
$NaAl(CH_3)_4/NaAl(C_2H_5)_4(Hg)$	Pb	(0.25)	$Sn(C_2H_5)_4$	(80)	383
$NaF \cdot 2Al(C_2H_5)_3(Cu)$	Sb		$Sb(C_2H_5)_3$	(80-90)	373
$NaF \cdot 2Al(C_2H_5)_3(Hg)$	Hg		$Hg(C_2H_5)_2$	(100)	432
$NaAl(C_2H_5)_4/Na[Al(C_2H_5)_3 (OC_4H_9)](Cu)$	Hg	(0.04)	$Hg(C_2H_5)_2$	(82)	383
$NaAl(C_2H_5)_4/Na[Al(C_2H_5)_3(OC_4H_9)](Cu)$	Cd	(0.04)	$Cd(C_2H_5)_2$	(81)	383
$NaAl(C_2H_5)_4/KCl(Cu)$	B_1		$B_1(C_2H_5)_3$		383
$NaF \cdot 2Al(C_2H_5)_3(Cu)$	In		$In(C_2H_5)_3$		42

Table 10.20. Electrochemical Synthesis of Organometallics by Electrolysis of Organoborons

Compound (Solvent) (Cathode)	Anode	(CD-A/cm²)	Product(s)	(% Yield)	Ref.
NaB(C₂H₅)₄/NaOH (water, ether) (Cu)	Pb	(0.01-1.0)	Pb(C₂H₅)₄	(88)	46
NaB(C₂H₅)F₈(THF) (Cu)	Pb		Pb(C₂H₅)₄	(High)	46, 377
(C₂H₅)₄N·B(C₂H₅)₄ (triethylamine) (Cu)	Pb		Pb(C₂H₅)₄	(100)	46, 377
NaB(C₂H₅)₄ (H₂O) (Hg)	Pb	(0.035-0.1)	Pb(C₂H₅)₄	(91)	434
NaB(C₂H₅)₄/ NaAl(CH₃)₄·NaAl(C₂H₅)₄ (H₂O) (Cu)	Pb	(0.3)	Pb(C₂H₅)₄	(90)	376
NaB(CH₃)₄ (H₂O) (Hg)	Pb	(0.035)	Pb(CH₃)₄	(78)	434
NaB(C₂H₅)₄ (H₂O) (Hg)	Hg	(0.13)	Hg(C₂H₅)₂	(81)	434
NaB(C₂H₅)₄ (H₂O) (Hg)	B₁	(0.035)	B₁(C₂H₅)₃	(98)	434
NaB(C₂H₅)₄ (H₂O) (Hg)	Mg	(0.035)	Mg(C₂H₅)₂	(73)	434
(CH₃)₄N·B(C₆H₅)₄(CH₃CN) (NaClO₄)	Pt		(C₆H₅)₂; diphenyl borinium perchlorate		157

Electrolysis of Organoboron Compounds

Tetraalkyllead compounds can also be formed upon electrolysis of organoboron compounds admixed with complexed organoaluminums. For example, Kobetz and Pinkerton [376] have reported that electrolysis of sodium tetraethylboron with a mixture of $NaAl(CH_3)_4$ and $NaAl(C_2H_5)_4$ gives tetraethyllead at 90% current efficiency:

$$4NaB(C_2H_5)_4 + 36[NaAl(CH_3)_4 - NaAl(C_2H_5)_4] + Pb \longrightarrow$$

$$Pb(C_2H_5)_4 + 4B(C_2H_5)_3 + 4Na + 36[NaAl(CH_3)_4 - NaAl(C_2H_5)_4]$$

Tetraethyllead is separated by distillation and sodium tetraethylboron is then regenerated by reaction of triethylboron with sodium, hydrogen, and ethylene.

$$B(C_2H_5)_3 + Na + \tfrac{1}{2}H_2 \qquad NaB(C_2H_5)_3H$$

$$NaB(C_2H_5)_3H + C_2H_4 \qquad NaB(C_2H_5)_4$$

Ether solutions of pure sodium tetraethylboron can also be electrolyzed to yield pure tetraethyllead. Representative examples are summarized in Table 10.20

4   ELECTROCHEMICAL REACTIONS AND PREPARATION OF ORGANO—TRANSITION METAL COMPOUNDS—ANODIC AND CATHODIC

The types of transition metal organics that have been dealt with by electrochemical techniques include mainly the transition element carbonyls, acetylacetonates, bipyridines, porphyrins, olefins, cyclopentadienlides, and some σ-bonded alkyls and halides.

In this area the greater number of electrochemical reports deal exclusively with polarography. Some of these polarographic reports deal with 2,2′-bipyridine complexes of iron [111, 348, 349], chromium [313] and manganese [314]; π-complexes of iron [11, 49, 189, 282], ruthenium [49], palladium [171], and osmium [49]; and other iron [289], cobalt [244, 245, 352], ruthenium [55], palladium [173], and manganese [68] complexes.

Within group III-b metals, polarographic studies of lanthanum [246] are reported. In 5-40% ethanol, only the organic ligands were reduced in the case of the lanthanum complexes I and II.

$$(\text{Ph}\overset{\displaystyle O}{\overset{\displaystyle \|}{C}}-CH-\overset{\displaystyle O}{\overset{\displaystyle \|}{C}}-R)_3 La \qquad\qquad (\text{Ph}\overset{\displaystyle O}{\overset{\displaystyle \|}{C}}CHC\overset{\displaystyle O}{\overset{\displaystyle \|}{}}CH_3)_4 La^- [H_2 N(Et)_2]^+$$

R = CH₃, Ph

I   **II**

Braithwaite [36] teaches the preparation of trialkyllanthanum compounds by anodic oxidation of Grignard reagents at a lanthanum anode but no specific examples are given:

$$6RMgX + 2La^{+3} \xrightarrow{\text{Electrolysis}} 2R_3 La + 3MgX_2 + 3Mg$$

Within group IV-b, titanium, zirconium, and hafnium, titanium has received the major share of attention with respect to the electrochemistry of its organometallic compounds. Among these, the most extensively studied have been π-cyclopentadienyltitanium halides [174, 175, 229, 240, 329, 387, 408].

In most cases, two successive polarographic waves are observed for π-dicyclopentadienyltitanium dihalides. The first reduction wave follows the scheme

$$(C_5H_5)_2 TiCl_2 + e \longrightarrow (C_5H_5)_2 TiCl + Cl^-$$

ESR techniques gave validity to the above trivalent titanium product formed in dimethylformamide [174, 329] or dimethoxyethane [84]. The second step was suggested as

$$(C_5H_5)_2 TiCl + e \longrightarrow (C_5H_5)_2 Ti + Cl^-$$

In aqueous systems [229, 387], instead of $(C_5H_5)_2 Ti$, the suggested product due to oxidation was $(C_5H_5)_2 TiOH^+$. Successive replacement of C_5H_5 groups by halogen did not alter the basic nature of the polarograms [174, 329].

Dessey et al. have reported the electrochemical reduction of two sulfur analogs of π-dicyclopentadienyltitanium dihalides, **III** and **IV**:

$$\pi(C_5H_5)_2 Ti(SC_6H_5)_2 \qquad\qquad \pi(C_5H_5)_2 Ti\overset{S}{\underset{S}{\diagdown}}\text{—}\underset{}{\diagup}\text{CH}_3$$

III   **IV**

Products were not isolated, but in the case of **IV** ring opening resulting in **V** was suggested [84]:

$$\pi(C_5H_5)_2Ti \underset{-S}{\overset{S}{\diagdown}} \begin{array}{c} CH_3 \end{array}$$

V

Polarographic studies [175] of π-complexes containing σ-bonded aliphatics suggest cleavage of the Ti-C σ-bond results when the complexes are reduced on mercury:

$$(C_5H_5)_2TiR_2 + e \longrightarrow [(C_5H_5)_2TiR] + R^-$$

$$(C_5H_5)_2TiR + e \longrightarrow [(C_5H_5)_2Ti] + R^-$$

Some of the interest in the oxidation reduction reactions of the π-cyclo-pentadienyltitanium complexes lies in their use in a number of homogeneous catalysis reactions.

Cyclooctatetraene complexes of titanium were recently synthesized by Lehmkuhl and Mehler [238]. Electrochemical reduction of Ti(IV) compounds in pyridine or THF in the presence of cycloctatetraene gave the complexes (Table 10.21). Aluminum cathodes and sacrificial aluminum anodes were involved.

No pertinent reports were located for organohafnium and only a few for organozirconium compounds [84, 408]. Zirconocene dichloride showed two reduction waves in dimethoxyethane, and at the second wave (-2.7 V versus Ag/Ag$^+$) a white solid precipitated at a mercury pool. The product was highly insoluble and the following reduction leading to polymer was suggested [84]:

$$\pi(C_5H_5)_2ZrCl_2 \longrightarrow \pi(C_5H_5)_2Zr \underset{Cl}{\overset{Cl}{\langle \rangle}} Zr(\pi C_5H_5)_2 \longrightarrow Polymer$$

Electrochemical synthesis of transition metal carbonyl compounds can be accomplished by electroreduction of acetylacetonates of first row transition elements (V, Cr, Mn, Fe, Co, Ni) from vanadium to nickel [112].

$$(CH_3-\underset{\underset{O}{|}}{C}-CH=\underset{\underset{O}{|}}{C}-CH_3)_3 \qquad (CH_3\underset{\underset{O}{|}}{C}-CH=\underset{\underset{O}{|}}{C}-CH_3)_2$$

$$\diagdown V \diagup \qquad\qquad \diagdown Ni \diagup$$

The reaction is preferably carried out in anhydrous pyridine under CO pressure at inert graphite or stainless steel cathodes. When an aluminum anode was used, the following electrochemical reactions were suggested [112]:

$$\text{(Anodic)}\qquad \text{Al} \longrightarrow \text{Al}^{+3} + 3e$$

$$\text{(Cathodic)}\qquad \text{M}^{n+} + x\text{CO} \begin{cases} + \text{ne} \longrightarrow \text{M(CO)}_x \\ + (n + 1)e \longrightarrow [\text{M(CO)}_x]^- \end{cases}$$

Because of incomplete replacement of coordinated pyridine, nonsimple carbonyl complexes can be formed such as $C_5H_5NCr(CO)_5$ or $C_5H_5NCr(CO)_4$.

Vanadium hexacarbonyl is formed at the anode by electrolysis of an alkali or alkaline earth metal-etherate salt, for example, VI, containing the hexacarbonyl vanadate anion:

$$[\text{Na(Et)}_2]^+ \text{ V(CO)}_6^-$$

VI

Suitable alkali metals are lithium, sodium, potassium, rubidium, and cesium, while alkaline earth metals are calcium, strontium, barium, and magnesium. Other appropriate metals are aluminum and zinc.

Giraitis et al. [162-164] have reported on the electrochemical synthesis of cyclopentadienyl manganese and cyclopentadienyl manganese tricarbonyl compounds. The latter are effective antiknock compounds for internal combustion engines. The syntheses employs a cyclopentadienide compound of a metal selected from groups I, II, III-A, IV-A, and VIII of the periodic table and utilizes manganese as anode material. The authors assert that cyclopentadienyl radicals are transferred from the cyclopentadienide compound of the electrolyte to the manganese anode. When the carbonyls are desired, CO is simply maintained in the electrolyte. The gross reaction for the formation of methylcyclopentadienyl manganese tricarbonyl is

Pertinent examples of these syntheses are summarized in Table 10.21. The above work also contains some interesting examples of the use of alternating electricity in electrolysis to minimize formation of insoluble products on the electrodes.

π-Cyclopentadienyltransition metal carbonyl hydrides of iron, manganese, and molybdenum were prepared in acidified tetrahydrofuran by electrochemical generation of a parent anion that was quickly protonated [167]:

$$[\pi(C_5H_5) \; Fe(CO)_2]_2 \; + \; e \longrightarrow [\pi(C_5H_5) \; Fe(CO)_2]^-$$

$$[\pi(C_5H_5) \; FeCO_2]_2^- \; + \; H^+ \longrightarrow \pi(C_5H_5) \; Fe(CO)_2H \; + \; ?$$

It was emphasized that careful adjustment of the potential was important to allow reduction of the parent molecule and prevent reduction of the product hydrides.

The same Co and Ni acetylacetonates as used above to obtain carbonyl compounds, when exhaustively reduced at mercury cathodes without CO in the system, formed Co and Ni metal [93]. An intermediate in the case of $Co(acac)_3$ was $Co(acac)_2$.

Reduction of iron acetylacetonates on an optically transparent thin-layer electrode in acetonitrile gave different product results depending on supporting electrolyte used [188, 279]. For example, when tetraethylammonium perchlorate was used in acetonitrile, the following was noted:

$$Fe(acac)_3 \; + \; e \longrightarrow Fe(acac)_3^-$$

When lithium perchlorate was present, a coordinative relaxation of the above product to the bis coordinated $Fe(acac)_2$ was observed:

$$Fe(acac)_3^- \; + \; 2Li^+ \longrightarrow Fe(acac)_2 \; + \; Li_2(acac)^+$$

Spectral measurements on the optically transparent electrode were used to identify products.

Lehmkuhl et al. [237] reported that complexed transition-metal organics could be obtained by electrochemical reduction of suitable transition metal salts (nickel acetylacetonates) in aprotic solvents in the presence of olefins or other ligands. For example, reduction of $Ni(acac)_2$ in pyridine in the presence of cyclooctadiene gave bis(cyclooctadiene-1,5)Ni(O) and a side product, $Al(acac)_3$. This method was applicable to preparation of other transition metal (Fe, Cr, Co) complexes.

Various types of σ-bonds associated with transition metal organic compounds are easily cleaved by electroreductive techniques. π-Cyclopentadienyltungstentricarbonyl compounds having various σ bonds such as metal-metal, VII, metal-halogen, VIII, and metal-carbon, IX, were reduced polarographically [75]. The reduction of VIII was considerably easier than either VII or IX. In the case of VIII the reduction was reported to yield symmetrical mercurials based on coincidence of half-wave value of known $[\pi(C_5H_5)W(CO)_3]_2Hg$ with the reduction product of the iodide.

Table 10.21. Electrochemical Reactions and Synthesis of Organotransition Element Compounds

Depolarizer	(Supporting Electrolyte, Solvent)	c, Cathode a, Anode (Potential, Other)	Products (% Yield)	Ref.
		Group IV-b — Organometallic Compounds		
$(C_5H_5)_2TiCl_2$	$(0.1\,N\,(Et)_4NClO_4,$ dimethylformamide)	c: Hg (-1.2 V vs SCE, redn. at Hg drop)	$(C_5H_5)_2TiCl$	329
Dicyclopentadienyl-titanium dichloride			Proposed on basis of EPR detection; not actually isolated	174
$(C_5H_5)_2\,TiCl_2$	$(LiCl, H_2O)$	c: Hg (-1.7 V vs SCE)	Coulometric measurement of $2e$/molecule of depolarizer suggested product was initially $(C_5H_5)_2Ti$ that oxidized to $(C_5H_5)_2TiOH^+$	229
$(C_5H_5)_2\,TiCl_2$	LiCl, dimethylformamide HCl or LiCl or HClO_4, H_2O	~-0.6 V vs SCE, separated cell c: Hg -0.6 to -0.8 V vs SCE, separated cell	Green solution of $(C_5H_5)_2Ti^+$	387
$TiCl_4(COT)$ COT = Cyclooctatetraene	$(Bu_4NBr, pyridine)$	c: Al $(0.3\ A/dm^2, 20-40°C)$	$(COT) TiCl·Pyr (77) + AlCl_3·Pyr$	238
$TiCl_4\,(COT)$	(Bu_4NBr, THF)	c: Al $(0.3\ A/dm^2, 20-40°C)$	$(COT) TiCl·THF (68) + AlCl_3·THF$	238

Compound	Conditions	Electrolysis	Products	Ref.
TiCl$_3$·3THF (COT)	(−, THF)	c: Al (0.6 A/dm^2, 40°C)	(COT)$_2$Ti (45) + AlCl$_3$·THF	238
Ti(OC$_4$H$_9$)$_4$ (COT)	(Bu$_4$NBr, THF)	c: Al (0.45 A/dm^2, 20-40°C)	(COT)$_3$Ti$_2$ (27) + Al(OR)$_3$	238
(C$_5$H$_5$)$_2$TiCl$_2$ (COT)	(Bu$_4$NBr, THF)	c: Al (0.3 A/dm^2, 0°C)	(C$_5$H$_5$)Ti(COT) (25) + AlCl$_3$·THF	238
(C$_5$H$_5$)TiCl$_3$ (COT)	(Bu$_4$NBr, THF)	c: Al (0.3 A/dm^2, 0°C)	(C$_5$H$_5$)Ti(COT) (4)	238
π(C$_5$H$_5$)$_2$ZrCl$_2$	(Bu$_4$NClO$_4$, dimethoxyethane)	c: Hg (-2.7 V vs Ag/Ag$^+$, −)	White, insoluble solid suggested polymer of $(\pi\text{-}C_5H_5)_2Zr \overset{Cl}{\underset{Cl}{\rightleftarrows}} Zr(\pi\text{-}C_5H_5)_2$	84
Vanadium				
V(acetylacetonate)$_3$	(Bu$_4$NBr, pyridine)	c: Stainless steel (1-10 mA/cm^2, CO pressure 85°C)	V(CO)$_6$ (30)	112
[Na(diglyme)$_2$]$^+$V(CO)$_6^-$	(Depolarizer, H$_2$O)	a: Pt (−, −)	V(CO)$_6$	404
Chromium				
(Ph)$_4$CrI	(Depolarizer, liquid NH$_3$)	c: (−, -40 to -50°C, diaphragm)	(Ph)$_4$Cr$^·$	182 184
(Ph)$_4$CrI	(Depolarizer, aqueous MeOH)	c: (−)	Final isolated (Ph)$_4$CrOH red soluble intermediate unidentified	181 186

Table 10.21. Electrochemical Reactions and Synthesis of Organotransition Element Compounds (cont.)

Depolarizer	(Supporting Electrolyte, Solvent)	c, Cathode a, Anode (Potential, Other)	Products (% Yield)	Ref.
$(Ph)_5CrOH$	(Depolarizer, liquid NH_3)	c: Pt (—)	$(Ph)Cr^{\cdot}$	184
$(Ph)_3CrI$	(0.1–0.2 g depolarizer salt per 30 ml liquid NH_3)	c: Pt (—, diaphragm -50 to -60°C)	$(Ph)_3Cr^{\cdot}$, less than 15 mg	184
$Cr(acetylacetonate)_3$	(Bu_4NBr, pyridine)	c: Stainless steel (1-10 mA/cm², CO pressure 81°C)	$Cr(CO)_6$ (60)	112
$Cr(acetylacetonate)_3$	(Bu_4NBr, pyridine)	c: Graphite (1-10 mA/cm², CO pressure 53°C)	$Cr(CO)_6$ (25)	112
$Cr(OAc)_3$	(Depolarizer, dimethyl-formamide)	c: Stainless steel (1-10 mA/cm², CO pressure 25°C)	$Cr(CO)_6$ (32)	112
Molybdenum				
$[\pi(C_5H_5)Mo(CO)_3]_2$	(Bu_4NClO_4, THF)	c: Hg (-0.9 V vs Ag/AgClO₄, CPE)	$\pi(C_5H_5)Mo(CO)_3H$	167
Tungsten				
$\pi(C_5H_5)W(CO)_3X$ X = Cl Br I	(Et_4NClO_4, CH_3CN)	c: Hg (V vs SCE $\begin{bmatrix} -0.82 \\ -0.73 \\ -0.50 \end{bmatrix}$, 25°C)	$[\pi(C_5H_5)W(CO)_3]_2Hg$ Not isolated but suggested on equivalence of $E_{1/2}$ with known mercurial	75

Manganese

Starting material	Conditions	Electrode conditions	Product (yield)	Ref.
Mn(acetylacetonate)$_2$	(Bu$_4$NBr, pyridine)	c: Stainless steel (1-10 mA/cm^2, CO pressure 84°C)	Mn(CO)$_5$ (45)	112
Mn(acetylacetonate)$_2$	(Bu$_4$NBr, pyridine)	c: Graphite (1-10 mA/cm^2, CO pressure 35°C)	Mn(CO)$_5$ (2)	112
NaCH$_3$ Cp Cp = cyclopentadienyl	(Depolarizer, THF-dimethoxyethane)	a: Mn (0.005-0.05 A/cm^2)	Mn(CH$_3$C$_5$H$_4$)$_2$ (25) current yield	164
NaCH$_3$ Cp	(As above but pure dimethoxyethane solvent used)	a: Mn (0.005-0.05 A/cm^2)	Mn(CH$_3$C$_5$H$_4$)$_2$ (3) current yield	164
NaCH$_3$ Cp	(Complex alkyl, dimethoxyethane-(Et)$_3$B)	a: Mn (0.005-0.05 A/cm^2)	Mn(CH$_3$C$_5$H$_4$)$_2$ "good yield"	164
NaCH$_3$ Cp	(Depolarizer, dimethoxyethane-pyridine)	a: Mn (0.005-0.05 A/cm^2, AC electricity used) c: Mn	Mn(CH$_3$C$_5$H$_4$)$_2$ "good yield"	164
NaCH$_3$ Cp	(Depolarizer, dimethoxyethane)	a: Mn (0.005-0.05 A/cm^2, AC current, CO pressure) c: Mn	CH$_3$C$_5$H$_4$Mn(CO)$_3$ "good yield"	164
Li Cp	(Depolarizer, dimethoxyethane)	a: Mn (0.005-0.05 A/cm^2, 180°C CO pressure)	(C$_5$H$_5$)Mn(CO)$_3$ "good yield"	164
KEt Cp	(Depolarizer, dimethoxyethane)	a: Mn (0.005-0.05 A/cm^2, 100°C CO pressure)	(EtC$_5$H$_4$)Mn(CO)$_3$ "good yield"	164

Table 10.21. Electrochemical Reactions and Synthesis of Organotransition Element Compounds (cont.)

Depolarizer	(Supporting Electrolyte, Solvent)	c, Cathode a, Anode (Potential, Other)	Products (% Yield)	Ref.
n-Decyl Cp Na n-Decyl Cp K	(Depolarizer, dibutoxyethane)	a: Mn (0.005-0.05 A/cm²)	(n-decyl C_5H_4)₂Mn	164
(Indenyl)₂Mg	(Depolarizer, dimethyl-ether)	a: Mn (0.005-0.05 A/cm²)	Mn₂	164
(Fluorenyl)₃Al	(Depolarizer, diethyl-ether)	a: Mn (0.005-0.05 A/cm²)	(Fluorenyl)₂Mn "good yield"	164
Cu(C₆H₅Cp)	(Depolarizer + Al(Et)₃ + Fe(CO)₅, benzene)	a: Mn (0.001-0.03 A/cm²)	(C₆H₅C₅H₄)Mn(CO)₃ "good yield"	164
(1,2-diethyl Cp)Tl	(Depolarizer, n-butrol-actone + nickel carbonyl)	a: Mn (0.005-0.05 A/cm²)	(1,2-diethyl-C₅H₃)Mn(CO)₃ "good yield"	164
(Butyl Cp)₂Fe	(Depolarizer, DMF)	a: Ferro manganese (0.005-0.05 A/cm², CO pressure)	(butyl C₅H₄)Mn(CO)₃ "good yield"	164
(CH₃ Cp)₂Mn CH₃ Cp	(MnCl₂ + depolarizer, DME)	a: Mn (0.005-0.05 A/cm², CO pressure)	(CH₃C₅H₄)Mn(CO)₃ "good yield"	164
[π(C₅H₅)Mn(CO)₃]₂	(Bu₄NClO₄, THF, 10⁻³ M/L HCl)	c: Hg (-1.3 V vs Ag/AgClO₄, CPE)	π(C₅H₅)Mn(CO)₃H	167
(MeCp)₂/Fe(CO)₅ Cp = cyclopentadienyl	(MnCl₂, dimethyl-formamide)	a: Mn (0.1 A/cm², CO pressure, 195°C)	MeCpMn(CO)₃ (7) current yield	132

MeCp/Ni(CO)$_4$	(MnBr$_2$, n-methyl-pyrrolidone)	a: Fe/Mn (—, CO pressure, 165°C)	MeCpMn (CO)$_3$ "good yield"	132
MeCp/Fe$_2$(CO)$_9$	(Na$_4$Mn(CN)$_6$, NaCl, diethyleneglycol-dimethylether)	a: Fe (—, CO pressure, 125°C)	MeCpMn(CO)$_3$	132
EtCp/Fe$_3$(CO)$_{12}$	(MnSO$_4$, dimethyl-formamide)	a: Graphite (—, CO pressure, 225°C)	EtCpMn(CO)$_3$	132
Indene/Co(CO)	(Mn(OAc)$_2$, LiBr, dicyclohexylamine)	a: Mn (0.1 A/cm^2, CO pressure, 200°C)	Indenyl Mn(CO)$_3$	132
Fluorene/Cr(CO)$_6$	(MnCl$_2$, hexamethyl-phosphoramide)	a: Mn (0.1 A/cm^2, CO pressure, 85°C)	Fluorenyl Mn(CO)$_3$	132
PhCp[Mn(CO)$_5$]$_2$	(MnI$_2$, dimethyl-formamide)	a: Mn (0.1 A/cm^2, CO pressure, 150°C)	PhCpMn(CO)$_3$	132

Iron

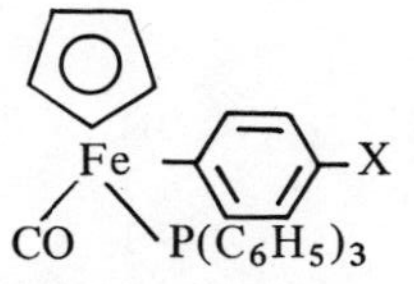

p-X = COOC$_2$H$_5$
F
H
OCH$_3$
CH$_3$

c: Hg (0.2 N Et$_4$NClO$_4$, CH$_3$CN)

$E_{1/2}$ V vs SCE, 25°C
-1.92
-1.99
-2.01
-2.03
-2.04

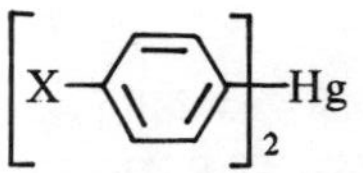

74

Not isolated but suggested on basis of equivalence of measured $E_{1/2}$ of products and corresponding symmetrical mercurial

519

Table 10.21. Electrochemical Reactions and Synthesis of Organotransition Element Compounds (cont.)

Depolarizer	(Supporting Electrolyte, Solvent)	c, Cathode a, Anode (Potential, Other)	Products (% Yield)	Ref.
$\pi(C_3H_5)Fe(CO)_3X$ X = Cl Br I NO₃	$(0.5\,N\ Et_4NClO_4,\ CH_3CN)$	c: Hg $E_{1/2}$ V vs SCE, 25°C $\begin{bmatrix} -0.44 \\ -0.39 \\ -0.37 \\ -0.48 \end{bmatrix}$	$\left[\pi(C_3H_5)Fe(CO)_3\right]^{\cdot}$ detected by ESR techniques	172
$1\text{-}C_6H_5\text{-}C_3H_4Fe(CO)_3Cl$ $2\text{-}C_6H_5\text{-}C_3H_4Fe(CO)_3$	$(0.5\,N\ Et_4NClO_4,\ CH_3CN)$	c: Hg $E_{1/2}$: V vs SCE $\begin{bmatrix} -0.43,\ 25°C \\ -0.35 \end{bmatrix}$	$[1\text{-}C_6H_5\text{-}C_3H_4\text{-}Fe(CO)_3]^{\cdot}$ $[2\text{-}C_6H_5\text{-}C_3H_4\text{-}Fe(CO)_3]^{\cdot}$ ESR techniques	172
$\pi(C_3H_5)Fe(CO)_2P(C_6H_5)_3I$	$(0.5\,N\ Et_4NClO_4,\ CH_3CN)$	c: Hg $(-0.47,\ 25°C)$	$[\pi(C_3H_5)Fe(CO)_2P(C_6H_5)_3]^{\cdot}$ ESR techniques	172
Fe(acetylacetonate)₃ acac = acetylacetonate	$(Bu_4NBr,\ pyridine)$	c: Stainless steel (1-10 mA/cm², CO pressure, 86°C)	$Fe(CO)_5$ (65)	112
Fe(acac)₃	$(Et_4NClO_4,\ CH_3CN)$	c: Gold (-1.6 V vs SCE, reduction electrochem. reversible)	$Fe(acac)_3^-$	188
Fe(acac)₃	$(Et_4NClO_4,\ CH_3CN)$ $(LiClO_4,\ CH_3CN)$	c: Gold (-0.8 V vs SCE, LiClO₄ forces different Rx than above)	$Fe(acac)_2$ Li_2acac	188
Fe(acac)₂⁺	$(Et_4NClO_4,\ CH_3CN)$	c: Gold (0.0 V vs SCE, —)	$Fe(acac)_3 + Fe^{+2}$	188
Fe(acac)₂⁺	$(Et_4NClO_4,\ CH_3CN)$	c: Gold (-0.8 V vs SCE, —)	$Fe(acac)_2$	188

Fe(acac)$^{2+}$	(Et$_4$NClO$_4$, CH$_3$CN)	c: Gold (0.0 V vs SCE, −)	Fe(acac)$_3$ + Fe^{+2}	188
[π(C$_5$H$_5$)Fe(CO)$_2$]$_2$	(Bu$_4$NClO$_4$, THF) HCl 10^{-3} M	c: Hg (-1.7 V vs Ag/AgClO$_4$,)	π(C$_5$H$_5$)Fe(CO)$_2$H without HCl present can obtain anion [π(C$_5$H$_5$)Fe(CO)$_2$]$_2^-$	167
π(C$_5$H$_5$)Fe$^+$(C$_6$H$_6$) BF$_4^-$ or PF$_6^-$	(Depolarizer + NaOH, 50% EtOH - H$_2$O)	c: Hg (-1.59 V vs SCE, CPE)	Benzene, cyclopentadiene, Fe	11
Tl(C$_5$H$_5$).	(−, dimethylformamide)	a: Fe (−, −) c: Fe	Fe(C$_5$H$_5$)$_2$ (90) Tl$^\circ$	385 386

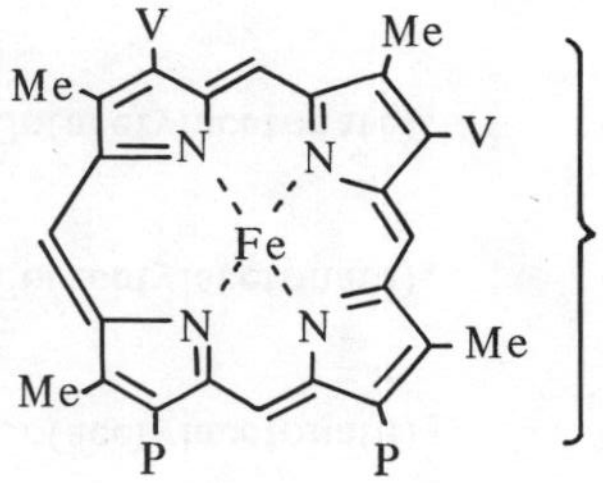

V = vinyl
P = propionic acid
= H

H(CN)$_2$ cyanide hemichrome or trans-dicyano-protoporphyrin-ferrate(III)	(1 M NaNO$_3$, 30% EtOH - H$_2$O)	c: Pt (-0.49 V vs SCE, −)	Cyanide hemochrome or trans-dicyanoprotopor-phyrin-ferrate(II)	67
H(pyridine)$_2$	(1 M NaNO$_3$, 30% EtOH - H$_2$O) pH 7.5-8.3	c: Pt (-0.128 V vs SCE)	Pyridine hemochrome	67
H(CN)(pyridine) cyanopyridine hemi-chrome	(1 M NaNO$_3$, 30% EtOH - H$_2$O)	c: Pt (-0.36) V vs SCE	Cyanopyridine hemochrome	67

521

Table 10.21. Electrochemical Reactions and Synthesis of Organotransition Element Compounds (cont.)

Depolarizer	(Supporting Electrolyte, Solvent)	c, Cathode a, Anode (Potential, Other)	Products (% Yield)	Ref.
Fe(C$_5$H$_5$)(C$_5$H$_4$NO$_2$) nitroferrocene	(pH 13.5, H$_2$O - EtOH)	Au/Hg (—, CPE)	Fe(C$_5$H$_5$)(C$_5$H$_4$NH$_2$) aminoferrocene	293
Iron tetraphenylporphyrin (cation) [Fe$^{(III)}$ TPP]$^+$	(n-Bu$_4$NClO$_4$, benzonitrile)	c: Pt (-0.32 V vs SCE), a: Pt (+1.18 V vs SCE), a: Pt (+1.50 V vs SCE)	[Fe(II) TPP], [Fe(III) TPP]·$^{2+}$ ligand, [Fe(III)TPP]$^{3+}$ oxidation	410
Cobalt				
Co(acetylacetonate)$_2$	(Bu$_4$NBr, pyridine)	c: Stainless steel (1-10 mA/cm^2, CO pressure, 85°C)	Co(CO)$_4$ (20)	112
Co(acetylacetonate)$_3$	(Bu$_4$NClO$_4$, dimethoxyethane)	c: Hg (-1.2 V vs Ag/Ag$^+$, CPE)	Co(acetylacetonate)$_2$	93
Co(acetylacetonate)$_2$	(Bu$_4$NClO$_4$, dimethoxyethane)	c: Hg (-2.6 V vs Ag/Ag$^+$)	Purple radical yields Co on decomposition	93
Co(OAc)$_2$	(100% Anhydrous acetic acid)	a: Pt (—, 65°C)	Co(OAc)$_3$	318
Cobalt tetraphenylporphyrin	(n-Pr$_4$NClO$_4$, DMSO)	c: Hg or Pt (-1.02 V vs SCE, CPE)	Cobalt tetraphenylporphyrin radical anion	126

Cobalt etioporphyrin

Cobalt etioporphyrin 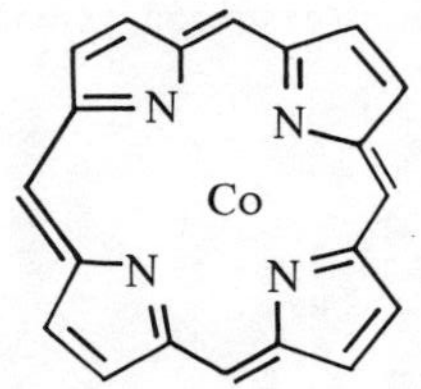	$(n\text{-}Pr_4NClO_4$, dimethylformamide$)$	c: Hg or Pt (-1.24 V vs SCE, CPE)	Cobalt etioporphyrin radical anion	126
Cobalt(II) octaethylporphyrin [$Co^{(II)}(OEP)$]	$(n\text{-}Pr_4NClO_4$, $CH_2Cl_2)$	a: Pt (2 oxidation steps, CPE)	[$Co^{(III)}(OEP)$]$^{+2}$	125
Sodium tetrasulfonated cobalt phthalocyanine [$Na_4Co^{(II)}(PTS)$]	$(Et_4NClO_4$, DMSO$)$	c: Hg $\quad -0.8$ V vs SCE $\quad\quad -1.5$ V vs SCE	[$Na_4Co^{(I)}\,PTS$]$^{+}$ [$Na_4Co^{(O)}\,PTS$]$^{+2}$	309
Cobalt tetraphenylporphyrin [$Co^{(II)}\,TPP$]	$(Bu_4NClO_4$, benzonitrile$)$	a: Pt $\quad +0.52$ V vs SCE $\quad\quad +1.19$ V vs SCE $\quad\quad +1.42$ V vs SCE	[$Co^{(III)}\,TPP$]$^{+}$ [$Co^{(III)}\,TPP$]$\cdot^{2+}$ ligand oxidation [$Co^{(III)}\,TPP$]$^{3+}$	410

Rhodium

[$(C_6H_5)_3P$]$_3$RhCl	$(Et_4NClO_4$, 85% CH_3CN $-$ 15% toluene$)$	Pt (-2.3 V vs Ag/Ag$^+$)	Rh[$(C_6H_5)_3P$]$_4$ (70)	287
[$(C_6H_5)_2PCH_3$]$_3$RhCl	$(Et_4NClO_4$, $CH_3CN)$	Pt (-2.2 V vs Ag/Ag$^+$)	Rh[$(C_6H_5)_2PCH_3$]$_4$ (74)	287

523

Table 10.21. Electrochemical Reactions and Synthesis of Organotransition Element Compounds (cont.)

Depolarizer	(Supporting Electrolyte, Solvent)	c, Cathode a, Anode (Potential, Other)	Products (% Yield)	Ref.
		Nickel		
Ni(acetylacetonate)$_2$	(Bu$_4$NBr, pyridine)	c: Stainless steel (1-10 mA/cm^2, CO pressure 25°C)	Ni(CO)$_4$ (40)	112
Ni(acetylacetonate)$_2$	(Bu$_4$NClO$_4$, dimethoxy-ethane)	c: Hg (-2.2 V vs Ag/Ag$^+$, —)	Deep blue radical yields Ni on decomposition	93
Ni(acetylacetonate)$_2$	(Bu$_4$NBr, pyridine + cyclooctadiene)	c: Al (1.5 mA/cm^2, 0°C) a: Al	Bis(cyclooctadiene-1,5)-Ni(O) (70) side product Al(acac)$_3$	237
Ni(acetylacetonate)$_2$	(Bu$_4$NBr, pyridine + cyclooctatetraene)	c: Al (1.5 mA/cm^2, 20°C) a: Al	Cyclooctatetraene Ni(O) (93)	237
Ni(acetylacetonate)$_2$	(Bu$_4$NBr, THF saturated with butadiene)	c: Al (1.5 mA/cm^2, 20°C) a: Al	Cyclododecatriene-1,5,9 Ni(O) *trans-trans-trans* (96) *trans-trans-cis* (3.4) *trans-cis-cis* (0.1)	237
Ni(acetylacetonate)$_2$	(Bu$_4$NBr, THF + (Ph)$_3$P)	c: Al (—, —) a: Al	[(Ph)$_3$P]$_4$Ni(O) (80) current yield	237

Ni(OAc)$_2$	(100% anhydrous acetic acid)	a: Pt (—, 65°C)	Ni(OAc)$_3$	318
Nickel tetraphenyl-porphyrin	(n-Pr$_4$NClO$_4$, benzene-dimethylformamide)	Hg or Pt (-1.38 V vs SCE, CPE)	Nickel tetraphenyl-porphyrin radical anion	126
Ni(*trans*-tetramine)(ClO$_4$)$_2$	(Depolarizer, CH$_3$CN)	c: Pt (-1.57 V vs Ag/Ag$^+$ in CH$_3$CN)	Ni(*trans*-tetramine)ClO$_4$	288
	(Et$_4$NClO$_4$, CH$_3$CN)	a: Pt (+1.7 V vs Ag/Ag$^+$ in CH$_3$CN)	Ni(*trans*-tetramine)(ClO$_4$)$_3$	288
Ni(*cis*-tetramine)(ClO$_4$)$_2$	(Depolarizer, CH$_3$CN)	c: Pt (-1.58 V vs Ag/Ag$^+$ in CH$_3$CN)	Ni(*cis*-tetramine)ClO$_4$	288
	(Et$_4$NClO$_4$, CH$_3$CN)	a: Pt (+1.7 V vs Ag/Ag$^+$ in CH$_3$CN)	Ni(*cis*-tetramine)(ClO$_4$)$_3$	288
Ni(*trans*-diene)(ClO$_4$)$_2$	(Depolarizer, CH$_3$CH)	c: Pt (-1.5 V vs Ag/Ag$^+$ in CH$_3$CN)	Ni(*trans*-diene)ClO$_4$	288
	(Et$_4$NClO$_4$, CH$_3$CN)	a: Pt (+1.7 V vs Ag/Ag$^+$ in CH$_3$CN)	Ni(*trans*-diene)(ClO$_4$)$_3$	288
Similarly for Ni(*cis*-diene)(ClO$_4$)$_2$				288
Sodium tetrasulfonated nickel phthalocyanine (Na$_4$Ni(II)PTS)	(Et$_4$NClO$_4$, DMSO)	c: Hg (-0.87 V vs SCE / ~-1.7 V vs SCE)	Na$_4$Ni(II)PTS$^+$ / Na$_4$Ni(II)PTS^{+2} } ligand reduction	309
Ni(II)tetraphenyl-porphyrin [Ni(II)TPP]	(Bu$_4$NClO$_4$, benzo-nitrile)	a: Pt (+1.0 V vs SCE / +1.1 V vs SCE / +1.4 V vs SCE)	[Ni(III)TPP]$^{+1}$ / [Ni(III)TPP]$^{.+2}$ / [Ni(III)TPP]$^{+3}$ } ligand oxidation	410

Nickel tetraphenyl-porphyrin

$$[\pi(C_5H_5)W(CO)_3]_2 \qquad \pi(C_5H_5)W(CO)_3-X \qquad \pi(C_5H_5)W(CO)_3R$$

$$X = Cl, Br, I$$

VII   VIII   IX

According to a polarographic study [76], electroreduction of pentacarbonyl derivatives of manganese and rehenium containing carbon-metal σ bonds always resulted in cleavage of the carbon-metal σ bond. Some polarographic detection of products supported the claim.

$$RMn(CO)_5 \qquad\qquad R-Re(CO)_5$$

$$R = \text{aryl or acyl}$$

In the case of cyclopentadienylironcarbonyls having an Fe—C σ bond, electrolytic or sodium amalgam reduction was claimed to preferentially cleave the Fe—C bond [74]. Benzene was isolated from the Na/Hg reduction of $\pi(C_5H_5)$-Fe(CO)$_2$ -σ - Ph, and this together with polarographic data were used to argue for the general electrolytic scheme.

$$R = CH_3, Ph, Pr, CN. \ldots$$

Mercury compounds were said [74] to result when one carbonyl group was replaced with either a triphenylphosphine or triphenylphosphite group. One-electron reductions were thought to proceed as follows:

A number of compounds for which $X = COOC_2H_5$, F, H, OCH_3, and CH_3 were said to give the corresponding mercurial which was not isolated, but identified by correspondence of its reduction wave to that of the proper mercurial.

Polarographic reduction of some π-allylirontricarbonyl halides [172] also resulted in cleavage of the Fe—X σ bond in a first reduction stage. The generated radical was detected by ESR techniques and was the same species as was generated by chemical reduction [278] of the Fe—X bond in the same compound using reducing agents such as Al_2O_3 or sodium salts of anions such as $[(C_3H_3)Fe(CO)_2]^-$. A second reduction stage of the π-allyl complexes was polarographically observed but in order to duplicate this stage with chemical reagents, more powerful reducing compounds were needed such as Na/Hg or naphthalene sodium.

$$(C_3H_5)Fe(CO)_3X + e \longrightarrow [(C_3H_5)Fe(CO)_3]^{\cdot} + X^-$$

The second stage was claimed to generate the 18 electron anion $[(C_3H_5)$-$Fe(CO)_3]^-$. The latter species was easily reoxidized to the corresponding radical by interaction with PhSnCl [172].

Valcher [385, 386] reported that the π-sandwich complex ferrocene, $Fe(C_5H_5)_2$, was formed when cyclopentadienylthallium was electrolyzed in dimethylformamide using iron electrodes. The equations given were

$$(\text{anode}) \quad Fe + 2TlCp \longrightarrow Fe(Cp)_2 + 2Tl^+ + 2e$$

$$(\text{cathode}) \quad Tl^+ + e \longrightarrow Tl^\circ$$

The same technique was also supposed to give nickelocene but in low yields due to the unstable nickelocinium ion [386].

Ferrocene, ruthenocene and osmocene were the subject of a chronopotentio-metric study [49].

The technique of controlled-potential reduction was found to be a convenient method for conversion of hemichromes to hemochromes [67]. Jordan et al. [206] showed by polarography at a dropping mercury electrode that iron(III), protoporphyrin chelate dimer undergoes a reversible two-electron reduction yielding two molecules of trans-diaquoprotoporphyrinferrate(II):

$$2 \left[H_2O\text{---}\left(\text{Fe} \right)\text{---}OH_2 \right]^{-2} + 2OH^-$$

A number of hemichromes such as dicyano, dipyridine and cyanopyridine, *trans*-protoporphyrins were also converted to the corresponding iron$^{(II)}$ hemochromes by electrolysis at a platinum cathode [67]. When hemichrome groups were π-bonded to polyvinylimidazole and polyhistidine chains, the electroreduction of the hemichrome groups was 15-28 orders of magnitude slower than ferriheme [177].

Electrochemical oxidation of metalloporphyrins of Fe, Co, Ni, Cu, and Zn showed that the first oxidation occurred at the metal atom in all but Cu and Zn where ligand oxidation was observed [410]. Stable products of either a one- or two-electron oxidation can be isolated [125], for example, cobalt octaethylporphyrin, $Co^{(II)}(OEP)$.

$$Co^{(II)}(OEP) \xrightarrow{-e} Co^{(III)}(OEP)^+ \longrightarrow Co^{(III)}(OEP)^{+2}$$

Controlled-potential electrolysis and voltammetric studies augmented by EPR and visible spectra studies [309] have shown Co, Ni, and Cu complexes of phthalocyanine, X, to be reducible in two steps to respective mono- and dications. In the case of Ni and Cu, reduction involved ligands but the cobalt complex was reduced at the metal atom.

X tetrasulfonated metal phthalocyanine

Hein has carried out extensive studies on the chemistry and preparation of polyphenylchromium compounds, including several studies on the electrolysis of these compounds [181, 182, 184, 186]. The materials are unstable salts of tri-, tetra-, and pentaphenylchromium.

$$(\phi)_n CrX \qquad \begin{aligned} &X = \text{anion} \\ &n = 3, 4, 5 \end{aligned}$$

Electrolysis of tetraphenylchromium iodide in methanol gave a red alcohol-soluble product that finally reverted to tetraphenylchromium hydroxide [181]. By employing low temperatures (-40 to $-60°C$), liquid ammonia solutions, and cell diaphragms, triphenylchromium iodide and tetraphenylchromium iodide were converted to free triphenylchromium and tetraphenylchromium, respectively [182, 184]:

$$(Ph)_n Cr^+ + e \longrightarrow (Ph)_n Cr^\cdot$$
$$n = 3, 4$$

The two materials have properties which are different than other analogous radicals such as the quaternary ammonium radicals discussed before. The polyphenylchromium radicals are insoluble in liquid ammonia (do not form bluish solutions), are nonconducting, and do not form amalgams.

Starting with nickel[III] tetraamine and diene, both *cis* and *trans*, stable nickel[I] and nickel[II] complexes were formed by reduction and oxidation, respectively, at platinum electrodes at acetonitrile [288]. For example, nickel[II] *trans*-tetraamine could be reduced or oxidized. Stability of the Ni[I] complexes to further reduction in acetonitrile was noted [288]. Products were isolated and identified in these cases.

Allyl palladium chloride when reduced polarographically showed two reduction waves [391]. The first wave was said to generate the radical $H_2C = CHCH_2Pd\cdot$, and the second wave the protonated material $H_2C = CHCH_3$ plus Pd metal.

A very large number of organotransition metal compounds have been examined by Dessy et al. [77, 78, 84, 88-93, 95-98]. The references cited contain extensive tabulation of results and are not repeated in the tables in this section. The survey of each compound involved all or some of the methods of (1) polarography, (2) multiple triangular-sweep voltammetry to establish chemical or electrochemical reversibility in the system, (3) exhaustive controlled-potential electrolysis and determination of the number of electrons involved in the polarographic step, (4) ESR to examine the resulting solutions, (5) re-oxidation or reduction of electrochemically generated species to a starting compound, and (6) polarography and spectroscopy studies of final solutions. Examples of electrochemically reversible and chemically reversible systems were observed along with some that were both electrochemically and chemically reversible. Species of reasonable lifetimes were generated by electron addition to, or removal from, depolarizing materials.

5 MISCELLANEOUS ELECTROCHEMICAL OXIDATIONS AND REDUCTIONS

A number of anodic oxidations exist in which the mercury pool anode is sacrificial and mercury atoms become a part of the electrolysis product. Two major routes for this can be envisaged: (1) anodic formation of mercury ions, which then enter into secondary reactions, and (2) the interaction of anodically formed radicals with the mercury pool in a manner analogous to the classical Paneth reactions of organic radicals with metals.

A study by Horner and Haufe [197] involved the anodic oxidation of a number of triorganophosphorous, arsenic, and antimony compounds. The half-wave oxidation potentials for a number of such compounds were measured in acetonitrile against a Ag/Ag$^+$ standard. For example, oxidation values for triphenylphosphine, triphenylarsine, and triphenylantimony were +0.12, +0.43, and +0.46 volts, respectively. Controlled-potential oxidations of phosphines and arsines in the same solvent at a stirred mercury pool yielded mercury complex salts (Table 10.22) of the formula I:

$$\text{I} \quad [R_3X]_2Hg(ClO_4)_2$$

R	X
C_6H_5	P
$p-CH_3-C_6H_4$	P
$p-CH_3OC_6H_4$	P
C_6H_5	As

Primary radicals were thought to form during electrolysis and react with the metallic mercury:

$$2[(C_6H_5)_3P^{\cdot+}] \;+\; Hg \longrightarrow \Big\{ [(C_6H_5)_3P]_2Hg \Big\}^{++}$$

When platinum or lead anodes were used [197] in a variety of supporting electrolytes, there was no oxidation of triphenylphosphine up to +2.0 V versus SCE.

It has been reported that cyclohexene reacts with anodically generated mercuric ions in acidified acetonitrile to form chloromercurials such as *trans*-2-chloromercuri-1-acetamidocyclohexane, **II** [402]:

II

The rate-determining step is thought to be the formation of mercurous ions at the mercury anode, followed by disproportionation to mercuric ions that react with the double bond of cyclohexene.

A large number of papers exist dealing with electron transfer reactions of metal-organic complexes with other complexes. The electrooxidation of a mercury[II] propene complex in 1 molar perchloric acid differs from the rest since actual cleavage of a mercury-carbon bond is considered and organic products are generated [135]. In addition to mercuric ion, the products were mainly acetic and formic acids. Based on isolated products, the suggested oxidation path is

$$Hg(CH_2{=}CHCH_3)^{+2} \xrightarrow{-2e} \begin{cases} \longrightarrow CH_3CH_2CHO \xrightarrow{-2e} CH_3CH_2COOH \\[2mm] \longrightarrow CH_3COCH_3 \xrightarrow{-6e} CH_3COOH + HCOOH \end{cases}$$

Propionaldehyde and acetone were thought to form via cleavage of a carbon-mercury σ bond in the complex, followed by a hydride shift. Thus for acetone,

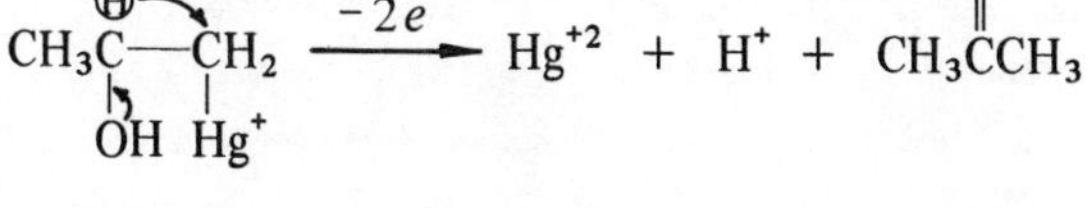

Table 10.22. Miscellaneous Anodic Reactions

Depolarizer or Starting Material	(Supporting Electrolyte, Solvent)	Anode (Potential, Other)	Products (% Yield)	Ref.
		Organomercury Compounds		
$(Ph)_3P$	(LiClO$_4$, acetonitrile)	Hg (+0.25 V vs SCE)	$[(Ph)_3P]_2Hg(ClO_4)_2$ (96)	197
$(CH_3{-}C_6H_4{-}P)_3$	(LiClO$_4$, acetonitrile)	Hg (+0.3 V vs SCE)	$[(CH_3{-}C_6H_4{-}P)_3]_2Hg(ClO_4)$ (68)	197
$(CH_3O{-}C_6H_4{-}P)_3$	(LiClO$_4$, acetonitrile)	Hg (+0.35 V vs SCE)	$[(CH_3O{-}C_6H_4{-}P)_3]_2Hg(ClO_4)_2$ (70)	197
$(Ph)_3As$	(LiClO$_4$, acetonitrile)	Hg (+0.5 V vs SCE)	$[(Ph)_3As]_2Hg(ClO_4)_2$ (85)	197
cyclohexene	(LiClO$_4$ (0.5 M), aceto-nitrile-acid system)	Hg (+1.2 V vs Ag/Ag$^+$)	cyclohexyl-HgCl, NHCOCH$_3$ (65% current efficiency); cyclohexyl-HgCl, R: R = OH (8% current efficiency) R = OCH$_3$ (60% current efficiency)	402
$PhSnCl_3$	(1 N NaOH, H$_2$O)	Hg (−0.25 V vs ?)	$(Ph)_2Hg$ + PhHgOH	103

532

		Major products (current yield)	
Mercury(II) Propene complex $(CH_3CH=CH_2)^{++}$ Hg	Pt (+2.7 V vs Ag/10^{-2} M Ag$^+$, potentials from +1.8 to +2.7 were studied	CH_3COCH_3 (2.3 current yield) HCOOH (55 current yield) CH_3COOH (59 current yield) CH_3CH_2COOH (16 current yield)	135
Mercury(II) Olefin complex with:	Pt (+2.4 V vs Ag/10^{-2} M Ag$^+$)	Major products (current yield)	134
Ethylene 1-Butene 2-Butene Isobutene 1-Pentene		CH_3COOH (33), $H_2C=O$ (10) n-Butyric acid (33), CH_3COOH (44) Isobutyric acid (18), CH_3COOH (73) CH_3COOH (95) n-Pentanoic aicd (18), n-butyric acid (20), HCOOH (20), propionic acid (14), CH_3COOH (>14)	
2-Pentene		α-Methylbutyric acid (17), propionic acid (38), CH_3COOH (>38)	
1-Hexene		n-Hexanoic acid (24), n-pentanoic acid (22), n-butyric acid (12), CH_3COOH (>12), HCOOH (22)	
2-Hexene		α-Methylpentanoic acid (17), propionic acid (28), n-butyric acid (24), CH_3COOH (>24)	
Cyclohexene		Cyclopentane carboxylic acid (10), adipic acid (<1), formylcyclopentane (18)	
1-Octene		n-Octanoic acid (1), n-heptanoic acid (3), HCOOH (3), n-hexanoic acid (2), CH_3COOH (>2)	134

	$(1 M$ HClO$_4$, H$_2$O)		
	$(1 M$ HClO$_4$, $1 M$ Hg(ClO$_4$)$_2$, H$_2$O)		

Table 10.22. Miscellaneous Anodic Reactions (cont.)

Depolarizer or Starting Material	(Supporting Electrolyte, Solvent)	Anode (Potential, Other)	Products (% Yield)	Ref.
		Arsenic Compounds		
$(CH_3)_2AsO_2H$	(Alkaline solution)	—	H_3AsO_4 (3) current yield / $CH_3AsO_3H_2$ (10.8) current yield	130
$CH_3AsO_3H_2$	(Alkaline solution)	—	H_3AsO_4 (5) current yield	130
		Tin Compounds		
BuBr, Br?	(—, BuOAc)	Sn (0.01 to 7A/dm²)	Bu_2SnBr_2 (100) current efficiency	7
		Silicon Compounds		
$(Et)_3Si$	$(CH_3)_4NI$, EtOH / $NaNO_3$ / $Ph(CH_3)_3NI$ / $(Et)_4NBr$ / $[C_{10}H_{21}]_4NBr,$ / NH_4NO_3	270 A/m², 78.5°C / 130 A/m², 78.5°C / 880–350 A/m², 78.5°C / 1200–530 A/m², 78.5°C / 860 A/m², 78.5°C / 670–400 A/m², 78.5°C	Ethoxytriethylsilane (30) (17) (80) (90) (95) (10)	212
$(Et)_3Si$	$([C_{10}H_{21}]_4NBr, n\text{-}C_3H_7OH)$	Pt (270–440 A/m², 97.1°C)	Propoxytriethylsilane (60)	212

534

Lead Compounds

$Pb(OAc)_2$	(Depolarizer, THF with any following alcohol—furfuryl alcohol, butanol, 1-pentanol, allyl alcohol, phenethyl alcohol)	Transparent tin oxide (12-V drop at 0.64 cm electrode spacing)	Amorphous anodic deposit of 82% total Pb	240
$Ca(CN)_2$	$(NaClO_3 + Ca(CN)_2, H_2O)$	Pb (1-3.5 A/dm^2)	$PbCN_2$ lead cyanamide (80-90% pure product)	269
$Pb(OAc)_2$	(2.1 N $Pb(OAc)_2$, 0.5 N KOAc, glacial CH_3COOH)	— (0.3 A/cm^2, 85°C)	$Pb(OAc)_4$ (98) current efficiency	132

Tin Compounds

$(CH_3)_4Sn$	(0.1 M NaOOCH, MeOH)	Hg (1.27 mA/cm^2, 15-hr run)	$(CH_3)_3SnOOCH$ [a]	291
$(CH_3)_4Sn$	(0.1 M $NaOOCCH_3$, MeOH)	Hg (1.27 mA/cm^2, 12-hr run)	$(CH_3)_3SnOOCCH_3$ [a]	291
$(CH_3)_4Sn$	(0.1 M $NaOOCCH_2Cl$, MeOH)	Hg (1.27 mA/cm^2, 16-hr run)	$(CH_3)_3SnOOCCH_2Cl$ [a]	291
$(CH_3)_4Sn$	(0.1 M $NaOOCC_2H_5$, MeOH)	Hg (1.27 mA/cm^2, 15-hr run)	$(CH_3)_3SnOOCC_2H_5$ [a]	291

[a] Dimethylmercury detected as a product in these preparations.

Since mercuric ion is generated, it may be recomplexed by bubbling propene through the electrolysis solution. The role of mercuric ion is mainly that of a catalyst. In a later report [134] 11 mercury[II] olefin complexes were similarly treated giving mercuric ion and carboxylic acids.

Anodic formation of a number of mercury 5,5-substituted barbiturates was studied potentiostatically on a stationary mercury drop electrode from solutions of various barbituric acids [8].

A Japanese patent [210] reported the anodic preparation of cadmium or zinc stearate by electrolysis of stearic acid at cadmium or zinc anodes respectively.

Ionization of a mercury pool anode to mercurous ions in an anolyte of methanol containing tetramethyltin and sodium carboxylates gave trimethyltin carboxylates [291]. Dimethylmercury was also detected as a product of the reaction. The following sequence of reactions was suggested [291]:

$$2Hg \longrightarrow Hg_2^{2+} + 2e$$

$$Hg_2^{2+} + 2Me_4Sn \longrightarrow 2Me_3Sn^+ + Hg + Me_2Hg$$

$$2Me_3Sn^+ + 2RCOO^- \longrightarrow 2Me_3SnOOCR$$

Trimethyltin formate, acetate, chloroacetate, and propionate were prepared in this manner [291].

Russian workers reported on the electrochemical alkoxylation of triethyl-silane with ethyl, propyl, and t-butyl alcohols [212]. These were anodic reactions carried out at the boiling point of the respective alcohol. Yields varied from 10-90%, depending on supporting electrolyte and alcohol.

Ziegler et al. have patented [424, 433] a process for the electrolytic deposition of aluminum onto other metals such as copper wires. The thin coating of aluminum may be oxidized to a good insulating coating that is resistant to chemical and thermal degradation. The process is carried out by electrolysis of an electrolyte which is a homogenous liquid phase consisting of (1) an organo-aluminum compound of the formula $AlR(R^1)_2$ (R is an alkyl radical and R^1 may also be an alkyl radical or hydrogen or halogen atom), and (2) a complex compound. The complex compound is formed from an alkali metal organic or halide and an aluminum compound as described above. The deposited aluminum is of very high purity.

Most of the studies on the electrolysis of organoberyllium compounds have been directed toward the synthesis of pure metallic beryllium via electro-deposition. Booth and Torrey [34], for example, found that electrolysis of beryllium basic acetate or beryllium acetylacetonate in nitrogen-containing solvents (e.g., ammonia, pyridine, or piperidine) yielded finely divided black colored deposits. The deposits were usually impure and the high resistance of

Table 10.23. Electrolysis of Ethyl Ether Solutions of $Be(Me)_2$ and $Be(Me)_2$ with $BeCl_2$

Bath Constitution	cd (A/dm^2)	Deposit	Beryllium Content (%)	Ref.
3 M $Be(Me)_2$	0.09	Brittle, black, treed	65	411
1.2 M $Be(Me)_2$	0.05	Brittle, black, treed	77	
3 M $Be(Me)_2$ + 0.3 M $BeCl_2$	0.1-0.3	Dark gray, metallic, brittle	80	
3 M $Be(Me)_2$ + 0.6 M $BeCl_2$	0.15	Black, brittle	–	
3 M $Be(Me)_2$ + 0.9 M $BeCl_2$	0.15	Black, brittle, thin	93	
3 M $Be(Me)_2$ + 2.3 M $BeCl_2$	0.1	Gray, metallic, brittle, thin ($BeCl_2 \cdot Et_2O$ crystallized out of solution, to leave about 1 M $BeCl_2$ in solution)	95	
2.8 M $BeCl_2$	0.1	Black powder	92	

Table 10.24. Miscellaneous Anodic and Cathodic Reactions

Depolarizer	(Supporting Electrolyte, Solvent)	C: Cathode A: Anode (Potential, Other)	Products (% Yield)	Ref.
			Organocopper Compounds	
Copper tetraphenylporphyrin	(n-Pr$_4$NClO$_4$, THF-dimethylformamide)	c: Hg or Pt (-1.40 V vs SCE, CPE)	Copper tetraphenylporphyrin radical anion	126
Copper etioporphyrin	(n-Pr$_4$NClO$_4$, dimethylformamide)	c: Hg or Pt (-1.66 V vs SCE, CPE)	Copper etioporphyrin radical anion	126
Sodium tetrasulfonated copper phthalocyanine [Na$_4$Cu(II) PTS]	(Et$_4$NClO$_4$, DMSO)	c: Hg -0.9 V vs SCE, CPE -1.7 V vs SCE, CPE	[Na$_4$Cu(II) PTS]$^{\cdot\,+}$ ligand [Na$_4$Cu(II) PTS]$^{2+}$ reduction	309
Copper tetraphenylporphyrin [Cu(II) TPP]	(Bu$_4$NClO$_4$, benzonitrile)	a: Pt +0.99 V vs SCE, CPE +1.33 V vs SCE, CPE	[Cu(II) TPP]$^{\cdot\,+}$ ligand [Cu(II) TPP]$^{2+}$ oxidation	410

Organozinc Compounds

Compound	Conditions	Electrode	Product	Ref.
Zinc tetraphenylporphyrin [Zn(II), Ph structure]	(n-Pr$_4$NClO$_4$, DMSO)	c: Pt or Hg (-1.51 V vs SCE, CPE)	Zinc tetraphenylporphyrin radical anion	126
Zinc etioporphyrin [Zn(II), Et/Me structure]	(n-Pr$_4$NClO$_4$, dimethylformamide)	c: Pt or Hg (-1.80 V vs SCE, CPE)	Zinc etioporphyrin radical anion	126
Zinc tetraphenylchlorin	(n-Pr$_4$NClO$_4$, DMSO)	c: Pt or Hg (-1.53 V vs SCE, CPE)	Zinc tetraphenylchlorin radical anion	126
Zinc tetraphenylporphyrin [Zn(II) TPP]	(Bu$_4$NClO$_4$, benzonitrile)	a: Pt +0.79 V vs SCE, CPE; +1.10 V vs SCE, CPE	[Zn(II) TPP]$\cdot^+$ ligand oxidation; [Zn(II) TPP]$^{2+}$	410
n-BuZn$^+$Br$^-$	(Depolarizer, diethylether)	a: Pt (—)	Octane (30)	271
n-amylZn$^+$Br$^-$	(Depolarizer, diethylether)	a: Pt (—)	Decane (32)	271

Similar dimerization and yields for RZn$^+$ where R is n-hexyl (37), n-heptyl (25), n-octyl (35), n-nonyl (32), n-decyl (25-30), n-dodecyl (28), tetradecyl (28), hexadecyl (55), octadecyl (71), α-naphthyl (75), cyclopentyl (28), cyclohexyl (30), farnesyl (25), fenchyl (20), geranyl (32), menthyl (25)

Table 10.24. Miscellaneous Anodic and Cathodic Reactions (cont.)

Depolarizer	(Supporting Electrolyte, Solvent)	c: cathode a: anode (Potential, Other)	Products (% Yield)	Ref.
(Et)$_2$Zn	(Depolarizer, Et$_2$O)	a: Pt (—)	Unidentified anode product but metallic zinc deposited on Pt cathode	306

Organocadmium Compounds

Depolarizer	(Supporting Electrolyte, Solvent)	c: cathode / a: anode (Potential, Other)	Products (% Yield)	Ref.
Cadmium tetraphenylporphyrin	(n-Pr$_4$NClO$_4$, DMSO)	c: Hg or Pt (-1.45 V vs SCE, CPE)	Cadmium tetraphenylporphyrin radical anion	126
n-BuCd$^+$Br$^-$	(Depolarizer, diethylether) a: Pt (—)		Octane (32)	271
n-amylCd$^+$Br$^-$	(Depolarizer, diethylether) a: Pt (—)		Decane (33)	271

Similar dimerization and yields for RCd$^+$ where R is n-hexyl (20-21), n-heptyl (20), n-octyl (30), n-nonyl (20), n-decyl (22-25), n-dodecyl (17), tetradecyl (66), hexadecyl (50-55), octadecyl (76), α-naphthyl (34), cyclopentyl (25-30), cyclohexyl (20), farnesyl (25), fenchyl (25), geranyl (10), menthyl (20)

Depolarizer	(Supporting Electrolyte, Solvent)	c: cathode / a: anode (Potential, Other)	Products (% Yield)	Ref.
Stearic acid	(HCl, H$_2$O, ethanolamine) a:	Cd (0.05 A/cm^2, 60-65°C) Zn	Cadmium stearate (95) current efficiency Zinc stearate	210

Aluminum Organometallics

$Al(Et)_3$ ($NaF \cdot 2Al(Et)_3$)

$$\overset{H}{AlH[CH_2-C(CH_3)_2]_2}$$
Aluminum diisobutylhydride

$$(KF + \overset{H}{AlH[CH_2-C(CH_3)_2]_2})$$

c: Cu (−, 100-120°C) Very pure Al metal deposited on cathode 424 433

c: Cu (−, 100-120°C) Very pure Al metal deposited on cathode 424 433

Similarly reduced are the systems $KF \cdot Al(i\text{-}C_4H_9)_3 \cdot Al(CH_3)_3$, $NaH \cdot HAl(i\text{-}C_4H_9)_2 \cdot Al(Et)_3$, $KCl + Al(Et)_2Cl$,
$(Et)_4NCl \cdot 2Al(Et)_3$, [pyridinium]$N\text{-}CH_3I \cdot 2Al(Et)_3$, $C_{12}H_{25}N(CH_3)_3Cl + Al(n\text{-}Bu)_3$ 424 433

the solutions made the process impractical. Booth and Torrey [34] suggested that the deposits contained metallic beryllium, but subsequent attempts by Wood and Brenner [411] to repeat this work were unsuccessful. Wood and Brenner [411], however, were successful in producing metallic beryllium by electrolyzing dimethylberyllium in ethyl ether solution. Pertinent data are summarized in Table 10.23. The beneficial effects of electrolyzing $Be(CH_3)_2$/ $BeCl_2$ mixtures are apparent. Hans similarly found that electrolysis of $Be(C_2H_5)_2$/KF gave metallic beryllium in good yield [179]. Attempts to produce metallic beryllium by the electrolysis of diphenylberyllium met with failure [411].

Strohmeir and Popp [336] found that electrolysis of diethylberyllium in pyridine at copper electrodes yielded pyridine stabilized ethylberyllium radicals, $C_6H_5N-BeC_2H_5^-$.

Electrolysis of sodium triphenylgermanide in ammonia yields triphenylgermane as well as dimeric hexaphenyldigermane [138].

$$2NaGe(C_6H_5)_3 \xrightarrow{-2e} [Ge(C_6H_5)_3]_2$$

$$6NaGe(C_6H_5)_3 + 2NH_3 \xrightarrow{-6e} 6Ge(C_6H_5)_3H + N_2$$

Depending on anode material, up to 35% hexaphenyldigermane is formed.

The electrolytic formation of the dication of magnesium octaethylporphyrin (MgOEP) and the radical cations of various porphyrins and ethyl chlorophyllide a has been reported by Felton and co-workers [125]. Electrolytic oxidation of MgOEP in methylene chloride solution yielded both the radical cation and dication. Subsequent reduction yielded starting material in 95% yield, illustrating the reversibility of the reaction. Magnesium tetraphenylporphyrin and ethyl chlorophyllide a was similarly found to undergo reversible one-electron electrolysis. Subsequent studies by Fuhrhop and Mauzerall [140] similarly illustrated that MgOEP underwent reversible one-electron electrolysis in methanol solvent. Elucidation of the nature of these redox systems is relevant to an understanding of their behavior in biological systems.

6 EXPERIMENTAL EXAMPLES

Cathodic Synthesis of Dibenzylmercury

The cathodic synthesis of dibenzylmercury can be conveniently performed in aqueous or aprotic solvents by the electroreduction of benzyldimethylsulfonium tosylate, I (BDMST) at a stirred mercury pool cathode [325a].

$$\langle\!\!\!\!\!\!\bigcirc\!\!\!\!\!\!\rangle\!-CH_2S^+\!\!\!\!\begin{array}{c}{}_{\nearrow CH_3}\\{}_{\searrow CH_3}\end{array}\qquad OTos^-$$

I

The expected products are toluene, dimethylsulfide, and dibenzylmercury. It will be possible to vary the yield of dibenzylmercury by varying the temperature and solvent medium employed.

The BDMST depolarizer may be synthesized by the displacement reaction of benzyl chloride and dimethylsulfide carried out in water-methanol solutions. Appropriate workup and ion exchange will yield the tosylate crystals. Use of the tosylate salt enables one to carry out the reaction in water or dimethylformamide. If only water is employed as solvent, benzyldmethylsulfonium chloride may be used, but yields of dibenzylmercury will be reduced a bit.

A glass electrolysis vessel that provides for separate anode and cathode compartments is employed. The separation can be provided by medium-porosity glass frits. Use of a central compartment makes for an even better separation of anolyte and catholyte:

catholyte ‖ central compartment ‖ anolyte

The cell should be fitted with ports to receive a reference electrode and stirring mechanism. Alternatively, the stirring can be provided by a magnetic stirring device. In either case, the mercury pool-solution interface should be the plane of agitation. A 200-ml cathode compartment is a convenient size, while smaller central and anolyte compartments can be employed. The electrolysis is performed potentiostatically.

A nitrogen purge of the catholyte may be employed, but is not necessary. It is suggested that the reaction be carried out under adequate hood ventilation. There are two reasons for this: the handling of mercury and the dimethylsulfide liberated in the reaction.

Three different experiments can be performed in a similar manner. In each experiment, the catholyte ($\sim$130 ml) contains about 10^{-2} moles of BDMST depolarizer while the entire cell contains 0.1 N potassium tosylate as supporting electrolyte.

Using water as the solvent, the electrolysis is carried out at $25°C$ and at $-1.30 V$ versus SCE. Again, with water as the solvent, the reduction is carried out at $90°C$ at a potential of -0.9 V versus SCE. The reductions are carried out until the electrolysis current drops to 1-3 mA in each case. At this point the depolarizer is almost entirely consumed and colorless crystalline dibenzylmercury is visible in the cathode compartment. The material may be filtered from the catholyte and is reasonably pure as formed. The yield of dibenzylmercury is 50-60% in water at 25°C and 85-95% in water at 90°C.

The same synthesis may be carried out in dimethylformamide using the same concentrations and materials. The yield of dibenzylmercury is about 90%.

Anodic Synthesis of Tetraethyllead

1700 cc of 1.27 M ethyl magnesium chloride in the dibutylether of diethylene glycol were charged to a stainless steel pressure cell having five stainless steel plate cathodes and six lead plate anodes spaced ¼ in. apart. The anode and cathode areas were each 310 cm^2 in area. The solution was magnetically stirred.

An electrolyzing current of 14 V at an average amperage of 0.9 A was passed through the solution for 30 hr, then 24 V starting at 1.5 A and dropping to 0.23 A in 60 hr. The temperature of the solution was 35-40°C. Ethyl chloride was added to the solution in a molar ratio of 0.9 mole of ethyl chloride per mole of ethyl magnesium chloride. At the end point when the conductance had dropped to about 0.3 A, analysis showed 0.69 mole of ethyl magnesium chloride remaining in the solution. Tetraethyllead formed in the solution which separated into two layers. The percentage conversion to tetraethyllead, based on consumption of ethyl magnesium chloride, was 68%, and the percentage yield based on consumption of ethyl magnesium chloride was 100%.

Acknowledgment

The authors wish to express their appreciation to the staff at The Dow Chemical Company Library in Midland, Michigan for much willing and able assistance with many chores in the preparation of this manuscript, and also to Mr. Phillip L. Fisher for many translations of foreign publications to English.

References

1. R. B. Allen, *Diss. Abstr.*, **20** (3), 897 (1959).
2. A. L. Allred and L. W. Bush, *J. Am. Chem. Soc.*, **90**, 3352 (1968).
3. C. P. Andrieux, L. Nadjo, and J. M. Savéant, *J. Electroanal. Chem. Interfacial Electrochem.*, **26**, 147 (1970).
4. C. P. Andrieux and J. M. Savéant, *Bull. Soc. Chim. France*, 4671 (1968).
5. T. Arai, *Bull. Chem. Soc. Japan*, **32**, 184 (1959).
6. T. Arai and T. Oguri, *Bull. Chem. Soc. Japan*, **33**, 1018 (1960).
7. L. V. Armenskaya, L. M. Monastyrskii, Z. S. Smolyan, and E. N. Lysenko, *Khim. Prom. (Moscow)*, **44**, 665 (1968) [*Chem. Abstr.*, **69**, 102461t (1968)].
8. R. D. Armstrong, M. Fleischmann, and J. W. Oldfield, *Trans. Faraday Soc.*, **65**, 3053 (1969).
9. T. Asahara, M. Senō, and H. Kaneko, *Bull. Chem. Soc. Japan*, **41**, 2985 (1968).
10. E. C. Ashby, *Quart. Rev.*, **21**, 259 (1967).

11. D. Astruc, R. Dabard, and E. Laviron, *Compt. Rend. Acad. Sci., Ser. C,* **269**, 608 (1969).

12. J. A. Azoo, F. G. Coll, and J. Grimshaw, *J. Chem. Soc. C,* **1969**, 2521.

13. E. H. Bachofner, *Diss. Abstr.,* **17**, 1897 (1957).

14. E. H. Bachofner, F. M. Beringer, and L. Meites, *J. Am. Chem. Soc.,* **80**, 4269 (1958).

15. E. H. Bachofner, F. M. Beringer, and L. Meites, *J. Am. Chem. Soc.,* **80**, 4274 (1958).

16. I. A. Bagotskaya and D. K. Durmanov, *Elektrokhimiya,* **4**, 115 (1968).

17. I. A. Bagotskaya and D. K. Durmanov, *Elektrokhimiya,* **4**, 1414 (1969).

18. M. M. Baizer, *Naturwiss.,* **56**, 405 (1969).

18.a.M. M. Baizer, *J. Electrochem. Soc.,* **111**, 215 (1964).

19. M. M. Baizer and J. P. Petrovich, *Prog. Phys. Org. Chem.,* **7**, 189 (1970).

20. H. J. Bar, *Z. Phys. Chem.,* **243**, 398 (1970).

21. I. P. Beletskaya, K. P. Butin, and O. A. Reutov, *Zh. Org. Khim.,* **3**, 231 (1967).

22. I. P. Beletskaya, K. P. Butin, and O. A. Reutov, *Zh. Org. Khim.,* **2**, 2094 (1966).

23. Belg. Patent 617,628 (1962).

24. R. Benesch and R. E. Benesch, *J. Am. Chem. Soc.,* **73**, 3391 (1951).

25. R. E. Benesch and R. Benesch, *J. Phys. Chem.,* **56**, 648 (1952).

26. R. A. Benkeser and E. M. Kaiser, *J. Am. Chem. Soc.,* **85**, 2858 (1963).

27. R. A. Benkeser, E. M. Kaiser, and R. F. Lambert, *J. Am. Chem. Soc.,* **86**, 5272 (1964).

28. R. A. Benkeser and C. A. Tincher, *J. Organometal Chem.,* **13**, 139 (1968).

29. F. M. Beringer, H. E. Bachofner, R. A. Falk, and M. Leff, *J. Am. Chem. Soc.,* **80**, 4279 (1958).

30. B. H. M. Billinge and B. G. Gowenlock, *J. Chem. Soc.,* **1962**, 1201.

31. A. J. Birch, *Nature,* **158**, 60 (1946).

32. W. A. Bonner, *J. Am. Chem. Soc.,* **73**, 464 (1951).

33. W. A. Bonner and J. E. Kahn, *J. Am. Chem. Soc.,* **73**, 2241 (1951).

34. H. S. Booth and G. G. Torrey, *J. Phys. Chem.,* **35**, 2465, 2492, 3111 (1931).

35. I. N. Brago, L.·V.·Kaabak, and A.·P. Tomilov, *Zh.· Vses. Khim. Obshch.,* **12**, 472 (1967) [*Chem. Abstr.,* **67**, 104513u (1967)].

36. D. G. Braithwaite, U.S. Patent 3,007,858 (1961) [*Chem. Abstr.,* **57**, 4471b (1962)].

37. C. Bratu and A. T. Balaban, *Rev. Roum. Chim.,* **13**, 625 (1968).

38. R. Brdicka, *Coll. Czech. Chem. Commun.,* **7**, 457 (1955).

39. R. Brettle and D. J. Seddon, *J. Chem. Soc. C,* **1970**, 1153.

40. R. Brettle, J. G. Parkin, and D. J. Seddon, *J. Chem. Soc. C,* **1970**, 1317.

41. B. Breyer, *Ber.,* **71B**, 163 (1938) [*Chem. Abstr.,* **32**, 28127 (1938)].

42. K. Ziegler, Br. Patent 814,609 (1959) [*Chem. Abstr.,* **53**, 17733c (1959)]

43. National Aluminate Corp., Br. Patent 839,172 (1960) [*Chem. Abstr.,* **54**, 24036i (1960)].

44. Ethyl Corp., Br. Patent 842,090 (1960) [*Chem. Abstr.,* **55**, 5199e (1961)].

45. Br. Patent 882,005 (1960); D. G. Braithwaite, U.S. Patent 3,007,858 [*Chem. Abstr.,* **56**, 4526i (1962)].

46. Br. Patent 895,457 (1963); R. C. Pinkerton, U.S. Patent 3,028,325 (1962) [*Chem. Abstr.*, **57**, 4471b (1962)].

47. O. R. Brown and J. A. Harrison, *Electroanal. Chem. Interfacial Electrochem.*, **21**, 387 (1969).

48. O. R. Brown and K. Lister, *Disc. Faraday Soc.*, 106 (1968).

49. D. E. Bublitz, G. Hoh, and T. Kuwana, *Chem. Ind. (London)*, 635 (1959).

50. K. P. Butin, I. P. Beletskaya, P. N. Belik, A. N. Ryabtsev, and O. A. Reutov, *J. Organometal Chem.*, **20**, 11 (1969).

51. K. P. Butin, I. P. Beletskaya, A. N. Kashin, and O. A. Reutov, *J. Organometal Chem.*, **10**, 197 (1967).

52. K. P. Butin, I. P. Beletskaya, and O. A. Reutov, *Elektrokhimiya*, **2**, 635 (1966).

53. K. P. Butin, I. P. Beletskaya, A. N. Ryabtsev, and O. A. Reutov, *Elektrokhimiya*, **3**, 1318 (1967).

54. G. H. F. Calingaert, U.S. Patent 1,539,297 (1925) [*Chem. Abstr.*, **19**, 2210 (1925)].

55. G. Ciantelli, F. Pantani, and P. Legittimo, *Ric. Sci.*, **38**, 947 (1968).

56. D. Chow, J. H. Robson, and G. F. Wright, *Can. J. Chem.*, **43**, 312 (1965).

57. E. L. Colichman, *Anal. Chem.*, **26**, 1204 (1954).

58. E. L. Colichman and H. P. Maffei, *J. Am. Chem. Soc.*, **74**, 2744 (1952).

59. E. L. Colichman and J. T. Matschiner, *J. Org. Chem.*, **18**, 1124 (1953).

60. G. Costa, *Ann. Chim. (Rome)*, **40**, 559 (1950).

61. G. Costa, *Ann. Chim. (Rome)*, **40**, 541 (1950).

62. G. Costa, *Gazz Chim. (Italy)*, **80**, 42 (1950).

63. G. Costa, *Ann. Chim. (Rome)*, **41**, 207 (1951).

64. B. Ćosović and M. Branica, *J. Electroanal. Chem. Interface Electrochem.*, **20**, 269 (1969).

65. W. R. T. Cottrell and R. A. N. Morris, *Chem. Commun.*, 409 (1968).

66. M. D. Curtis and A. L. Allred, *J. Am. Chem. Soc.*, **87**, 2554 (1965).

67. D. G. Davis and R. F. Martin, *J. Am. Chem. Soc.*, **88**, 1365 (1966).

68. D. G. Davis and J. G. Montalvo, Jr., *Anal. Lett.*, **1**, 641 (1968).

69. D. G. Davis and D. J. Orleron, *Anal. Chem.*, **38**, 179 (1966).

70. J. Decombe and C. Duval, *Compt. Rend.*, **206**, 1024 (1938).

71. C. Degrand and E. Laviron, *Bull. Soc. Chim. Fr.*, (5), 2233 (1968).

72. C. Degrand and E. Laviron, *Bull. Soc. Chim. Fr.*, (5), 2228 (1968) [*Chem. Abstr.*, **69**, 56514n (1968)].

73. W. M. Dehn, *Am. Chem. J.*, **40**, 88 (1908).

74. L. I. Denisovich, S. P. Gubin, and Y. A. Chapovskii, *Izv. Akad. Nauk SSSR, Ser. Khim.*, 2378 (1967).

75. L. I. Denisovich, S. P. Gubin, Y. A. Chapovskii, and N. A. Ustynok, *Izv. Akad. Nauk SSSR, Ser. Khim.*, 924 (1968).

76. L. I. Denisovich, A. A. Ioganson, S. P. Gubin, N. E. Kolobova, and K. N. Anisimov, *Izv. Akad. Nauk SSSR, Ser. Khim.*, 258 (1969).

77. R. E. Dessy, J. C. Charkoudian, T. P. Abeles, and A. L. Rheingold, *J. Am. Chem. Soc.*, **92**, 3947 (1970).

78. R. E. Dessy, J. C. Charkoudian, and A. L. Rheingold, *J. Am. Chem. Soc.,* **94**, 738 (1972).
79. R. E. Dessy, T. Chivers, and W. Kitching, *J. Am. Chem. Soc.,* **88**, 467 (1966).
80. R. E. Dessy, S. I. E. Green, and R. M. Salinger, *Tetrahedron Lett.,* **1964**, 1369.
81. R. E. Dessy and G. S. Handler, *J. Am. Chem. Soc.,* **80**, 5824 (1958).
82. R. E. Dessy, G. S. Handler, J. H. Wotig, and C. A. Hollingsworth, *J. Am. Chem. Soc.,* **79**, 3476 (1957).
83. R. E. Dessy and S. A. Kandil, *J. Org. Chem.,* **30**, 3857 (1965).
84. R. E. Dessy, R. B. King, and M. Waldrop, *J. Am. Chem. Soc.,* **88**, 5112 (1966).
85. R. E. Dessy, W. Kitching, and T. Chivers, *J. Am. Chem. Soc.,* **88**, 453 (1966).
86. R. E. Dessy, W. Kitching, T. Psarras, R. Salinger, A. Chen, and T. Chivers, *J. Am. Chem. Soc.,* **88**, 460 (1966).
87. R. E. Dessy M. Kleiner, and S. C. Cohen, *J. Am. Chem. Soc.,* **91**, 6800 (1969).
88. R. E. Dessy, R. Kornmann, C. Smith, and R. Haytor, *J. Am. Chem. Soc.,* **90**, 2001 (1968).
89. R. E. Dessy and R. L. Pohl, *J. Am. Chem. Soc.,* **90**, 1995 (1968).
90. R. E. Dessy and R. L. Pohl, *J. Am. Chem. Soc.,* **90**, 2005 (1968).
91. R. E. Dessy, R. L. Pohl, and R. B. King *J. Am. Chem. Soc.,* **88**, 5121 (1966).
92. R. E. Dessy, A. L. Rheingold, and G. D. Howard, *J. Am. Chem. Soc.,* **94**, 746 (1972).
93. R. E. Dessy, F. E. Stary, R. B. King, nad M. Waldrop, *J. Am. Chem. Soc.,* **88**, 471 (1966).
94. R. E. Dessy, P. M. Weissman, and R. L. Pohl, *J. Am. Chem. Soc.,* **88**, 5117 (1966).
95. R. E. Dessy and P. M. Weissman, *J. Am. Chem. Soc.,* **88**, 5124 (1966).
96. R. E. Dessy and P. M. Weissman, *J. Am. Chem. Soc.,* **88**, 5129 (1966).
97. R. E. Dessy, P. M. Weissman, and R. L. Pohl, *J. Am. Chem. Soc.,* **88**, 5117 (1966).
98. R. E. Dessy, and L. Wieczorek, *J. Am. Chem. Soc.,* **91**, 4963 (1969).
99. M. Devaud, *Compt. Rend. Ser. C,* **263**, 1269 (1966).
100. M. Devaud, *Compt. Rend. Ser. C,* **262**, 702 (1966).
101. M. Devaud, *J. Chim. Phys.,* **63**, 1335 (1966).
102. M. Devaud and E. Laviron, *Rev. Chim. Miner.,* **5**, 427 (1968).
103. M. Devaud and P. Souchay, *J. Chim. Phys.,* **64**, 1778 (1967).
104. M. Devaud, P. Souchay, and M. Person, *J. Chim. Phys.,* **64**, 646 (1967).
105. J. S. DiGregorio and M. D. Morris, *Anal. Chem.,* **40**, 1286 (1968).
106. L. Doretti and G. Tagliavini, *J. Organometal. Chem.,* **13**, 195 (1968).
107. L. Doretti and G. Tagliavini, *J. Organometal. Chem.,* **12**, 203 (1968).
108. E. I. du Pont de Nemours and Co., Netherlands Application 6,508,049,

Dec. 24, 1965; U.S. Appl. June 23, 1964, 46 pp. [*Chem. Abstr.*, **64**, 17,048 (1966)].

109. C. Duval, *Compt. Rend.*, **202**, 1184 (1936).

110. E. A. Efimov and I. G. Erusalimchik, *Russ. J. Phys. Chem.* (Eng. Transl.), **38**, 1560 (1964).

111. D. L. Ehman and D. T. Sawyer, *Inorg. Chem.*, **8**, 900 (1969).

112. R. Ercoli, M. Guainazzi, and G. Silvestri, *Chem. Commun.*, 927 (1967).

113. I. A. Esikova, O. N. Temkin, A. P. Tomilov, R. M. Flid, and N. N. Yakovleva, *Zh. Fiz. Khim.*, **44**, 264 (1970); *Russ. J. Phys.Chem,,* **44**, 147 (1970).

114. I. A. Esikova, O. N. Temkin, A. P. Tomilov, G. P. Pavlikova, and R. M. Flid, *Elektrokhimiya*, **6**, 743 (1970); *Sov. Electrochem.*, **6**, 727 (1970).

115. W. V. Evans and D. Braithwaite, *J. Am. Chem. Soc.*, **61**, 898 (1939).

116. W. V. Evans, D. Braithwaite, and E. Field, *J. Am. Chem. Soc.*, **62**, 534 (1940).

117. W. V. Evans and E. Field, *J. Am. Chem. Soc.*, **58**, 720 (1936).

118. W. V. Evans and E. Field, *J. Am. Chem. Soc.*, **58**, 2284 (1936).

119. W. V. Evans and F. Lee, *J. Am. Chem. Soc.*, **56**, 654 (1934).

120. W. V. Evans, F. Lee, and C. Lee, *J. Am. Chem. Soc.*, **57**, 489 (1935).

121. W. V. Evans and F. Lee, *Nanking J.*, **6**, 29 (1936).

122. W. V. Evans and R. Pearson, *J. Am. Chem. Soc.*, **64**, 2865 (1942).

123. W. V. Evans, R. Pearson, and D. Braithwaite, *J. Am. Chem. Soc.*, **63**, 2574 (1941).

124. S. Faleschini, G. Pilloni, and L. Doretti, *Electroanal. Chem. Interfacial Electrochem.*, **23**, 261 (1969).

125. R. H. Felton, D. Dolphin, D. C. Borg, and J. Fajer, *J. Am. Chem. Soc.*, **91**, 196 (1969).

126. R. H. Felton and H. Linschitz, *J. Am. Chem. Soc.*, **88**, 1113 (1966).

127. L. G. Feoktistov, A. P. Tomilov, Yu. D. Smirnov, and M. M. Goldin, *Elektrokhimiya*, **1**, 887 (1965).

128. L. G. Feoktistov and S. I. Zhdanov, *Electrochim. Acta*, **10**, 657 (1965).

129. F. Fichter, *Organische Elektrochemie*, Steinkopff, Dresden, 1942.

130. F. Fichter and E. Elkind, *Ber.*, **49**, 239 (1916) [*Chem. Abstr.*, **10**, 1038 (1916)].

131. C. G. Fink, and D. B. Summers, *Trans. Electrochem. Soc.*, **74**, 625 (1938).

132. M. Y. Fioshin and V. A. Guskov, *Dokl. Akad. Nauk SSSR*, **112**, 303 (1957) [*Chem. Abstr.*, **51**, 16,146e (1957)].

133. B. Fleet and R. D. Jee, *J. Electroanal. Chem. Interfacial Electrochem.*, **25**, 397 (1970).

134. M. Fleischmann, D. Pletcher, and G. M. Race, *J. Chem. Soc. (B)*, **1970**, 1746

135. M. Fleischmann, D. Pletcher, and G. M. Race, *J. Electro anal. Chem. Interfacial Electrochem.*, **23**, 369 (1969).

136. V. N. Flerov and Y. M. Tyurin, *Elektrokhimiya*, **6**, 1404 (1970).

137. V. N. Flerov and Y. M. Tyurin, *Zh. Obshch. Khim.*, **38**, 1669 (1968).

138. L. Foster and G. Hooper, *J. Am. Chem. Soc.*, **57**, 76 (1935).

139. A. N. Frumkin and A. B. Ershler, *Progress in Electrochemistry of Organic Compounds*, Plenum, New York, 1971, p. 241 (translation).

140. J. H. Fuhrhop and D. Mauzerall, *J. Am. Chem. Soc.*, **91**, 4174 (1969).

141. B. L. Funt, U.S. Patent 3,488,020 (June 3, (1969), Canadian Patent 839,692 (April 21, 1970).

142. B. L. Funt and S. N. Bhadani, *J. Polymer Sci. C*, No. 23(1), 1 (1968).

143. B. L. Funt, S. N. Bhadani, and D. Richardson, *J. Polymer Sci.*, *A-1*, **4**, 2871 (1966).

144. B. L. Funt and S. W. Laurent, *Can. J. Chem.*, **42**, 2728 (1964).

145. B. L. Funt, D. Richardson, and S. N. Bhadani, *Can. J. Chem.*, **44**, 711 (1966).

146. B. L. Funt and F. D. Williams, *J. Polymer Sci.*, *A*, **2**, 865 (1964).

147. L. W. Gaddum and H. E. French, *J. Am. Chem. Soc.*, **49**, 1295 (1927).

148. R. Galli, *Chim. Ind. (Milan)*, **50**, 977 (1968).

149. R. Galli, *J. Electroanal. Chem. Interfacial Electrochem.*, **22**, 75 (1969).

150. R. Galli and F. Olivani, *J. Electroanal Chem. Interfacial Electrochem.*, **25**, 331 (1970).

151. L. D. Gavrilova and S. I. Zhdanov, *Coll. Czech. Chem. Commun.*, **32**, 2215 (1967).

152. L. D. Gavrilova and S. I. Zhdanov, *Elektrokhim.*, **4**, 841 (1968).

153. R. Gelin, M. Breant, and D. Makula, *Compt. Rend.*, **260**, 5767 (1965).

154. K. Ziegler and H. Lehmkuhl, German Patent 1,153,371 (1963) [*Chem. Abstr.*, **60**, 546a (1964)].

155. K. Ziegler and H. Lehmkuhl, German Patent 1,150,078 (1963) [*Chem. Abstr.*, **59**, 4801a (1963)].

156. K. Ziegler and H. Lehmkuhl, German Patent 1,181,220 (1964) [*Chem. Abstr.*, **62**, 6156c (1965)].

157. D. H. Geske, *J. Phys. Chem.*, **66**, 1743 (1962).

158. A. Gillet, *Ind. Chim. Belge*, **16**, 74 (1951).

159. A. Gillet and F. Bloyaert, *Ind. Chim. Belge*, **16**, 78 (1951).

160. H. Gilman and R. G. Jones, *J. Am. Chem. Soc.*, **61**, 1513 (1939).

161. A. P. Giraitis, German Patent 1,046,617 (1958) [*Chem. Abstr.*, **55**, 383 (1961)].

162. A. P. Giraitis, T. H. Pearson, and R. C. Pinkerton, Br. Patent 845,074 (1960) [*Chem. Abstr.*, **55**, 4205g (1961)].

163. A. P. Giraitis, T. H. Pearson, and R. C. Pinkerton, U.S. Patent 2,915,440 (1959) [*Chem. Abstr.*, **54**, 17415g (1960)].

164. A. P. Giraitis, T. H. Pearson, and R. C. Pinkerton, U.S. Patent 2,960,450 (1960) [*Chem. Abstr.*, **55**, 8430h (1961)].

165. B. G. Gowenlock, P. P. Jones, and D. W. Ovenall, *J. Chem. Soc.*, **1958**, 535.

166. B. G. Gowenlock and J. Trotman, *J. Chem. Soc.*, **1957**, 2114.

167. D. Grěsová and A. A. Vlček, *Inorg. Chim. Acta.*, **1**, 482 (1967).

168. J. Grimshaw and J. S. Ramsey, *J. Chem. Soc. (B)*, **1968**, 60.

169. J. Grimshaw and J. S. Ramsey, *J. Chem. Soc. (B)*, **1968**, 63.

170. J. Grimshaw and E. J. F. Rea, *J. Chem. Soc. (C)*, **1967**, 2628.

171. S. P. Gubin and L. I. Denisovich, *Izv. Akad. Nauk SSSR, Ser. Khim.*, 149 (1966).
172. S. P. Gubin and L. I. Denisovich, *J. Organometal. Chem.*, **15**, 471 (1968).
173. S. P. Gubin, L. I. Denisovich, and A. Z. Rubezhov, *Dokl. Akad. Nauk SSSR*, **169**, 103 (1966).
174. S. P. Gubin and S. A. Smirnova, *J. Organometal. Chem.*, **20**, 229 (1969).
175. S. P. Gubin and S. A. Smirnova, *J. Organometal. Chem.*, **20**, 241 (1969).
176. V. Gutmann, P. Heilmayer, and G. Schoeber, *Monatsh.*, **92**, 240 (1961).
177. H. R. Gygax and J. Jordan, *Discussions Faraday Soc.*, 227 (1968).
178. C. J. Haggerty, *Trans. Am. Electrochem. Soc.*, **56**, 421 (1929).
179. G. Hans, German Patent 1,162,576 (1964) [*Chem. Abstr.*, **60**, 11624a (1964)].
180. F. Hein, *Z. Elektrochem.*, **28**, 469 (1922).
181. F. Hein, *Ber.*, **54**, 2708 (1921).
182. F. Hein and W. Eissner, *Ber.*, **59B**, 362 (1926).
183. F. Hein and A. Klein, *Chem. Ber.*, **71B**, 2381 (1938).
184. F. Hein and E. Markert, *Ber.*, **61B**, 2255 (1928).
185. F. Hein, K. Petzchner, K. Wagler, and F. A. Seigitz, *Z. Anorg. Allgem. Chem.*, **141**, 161 (1924).
186. F. Hein and O. Schwartzkopff, *Ber.*, **57B**, 8 (1924).
187. F. Hein and F. A. Segitz, *Z. Anorg. Allgem. Chem.*, **158**, 153 (1926).
188. W. R. Heineman, J. N. Burnett, and R. W. Murray, *Anal. Chem.*, **40**, 1970 (1968).
189. H. Hennig and O. Gurtler, *J. Organometal. Chem.*, **11**, 307 (1968).
190. L. Holleck, *J. Electroanal. Chem.*, **19**, 439 (1968).
191. L. Holleck and D. Marquarding, *Naturwiss.*, **49**, 468 (1962).
192. L. Horner, I. Ertel, H. D. Ruprecht, and O. Belovsky, *Chem. Ber.*, **103**, 1582 (1970).
193. L. Horner, and H. Fuchs, *Tetrahedron Lett.*, 203 (1962) [*Chem. Abstr.*, **57**, 5954d (1962)].
194. L. Horner and H. Fuchs, *Tetrahedron Lett.*, **1963**, 1573 [*Chem. Abstr.*, **59**, 15309d (1963).
195. L. Horner, H. Fuchs, H. Winkler, and A. Rapp, *Tetrahedron Lett.*, **1963**, 965 [*Chem. Abstr.*, **59**, 12837 (1963)].
196. L. Horner and J. Haufe, *Chem. Ber.*, **101**, 2903 (1968).
197. L. Horner and J. Haufe, *Chem. Ber.*, **101**, 2921 (1968).
198. L. Horner and A. Mentrup, *Ann. Chem.*, **646**, 65 (1961).
199. L. Horner, F. Rottger, and H. Fuchs, *Ber.*, **96**, 3141 (1963).
200. L. Horner H. Winkler, A. Rapp, A. Mentrup, H. Hoffmann, and P. Beck, *Tetrahedron Lett.*, **1961**, 161.
201. H. Hsiung and G. H. Brown, *J. Electrochem. Soc.*, **110**, 1085 (1963).
202. N. S. Hush and K. B. Oldham, *J. Electroanal. Chem.*, **6**, 34 (1963).
203. Imperial Chemical Inds., Ltd., Fr. Patent 1,565,463.
204. N. A. Izgaryshev and M. Ya. Fioshin, *Usp. Khim.*, **25**, 486 (1956).
205. R. G. Jones and H. Gilman, *Chem. Rev.*, **54**, 835 (1954).
206. J. Jordan and T. M. Bednarski, *J. Am. Chem. Soc.*, **86**, 5690 (1964).

207. L. V. Kaabak, A. P. Tomilov, and S. L. Varshavskii, *Zh. Vses. Khim. Obshch. D. I. Mendeleeva*, **9**, 700 (1964) [*Chem. Abstr.*, **62**, 12751h (1965)].

208. L. V. Kaabak and A. P. Tomilov, *Zh. Obshch. Khim.*, 33, 2808 (1963); *J. Gen. Chem. USSR* (Engl. Transl.), 33, 2734 (1963).

209. K. M. Kadish and J. Jordan, *Anal. Lett.*, 3, 113 (1970).

210. K. Kawasaki and T. Sato, Japan Patent 1116 (1962) [*Chem. Abstr.*, 59, 9602d (1963)].

211. V. M. Kazakova, I. G. Makarov, M. E. Kurek, and E. A. Chernyshev, *Zh. Strukt. Khim.*, 9, 525 (1968) [*Chem. Abstr.*, 69, 105703j (1968)].

212. N. P. Kharitonov, B. P. Nechaev, and G. T. Fedorova, *Zh. Obschch. Khim.* 34, 824 (1969).

213. V. G. Khomyakov and A. P. Tomilov, *Zh. Prikl. Khim.*, 36, 378 (1963).

214. V. G. Khomyakov, A. P. Tomilov, and B. G. Soldatov, *Elektrokhimiya*, 5, 850 (1969).

215. V. G. Khomyakov, A. P. Tomilov, and B. G. Soldatov, *Elektrokhimiya*, 5, 853 (1969).

216. A. Kirrmann and M. Kleine-Peter, *Bull. Soc. Chim. Erance*, 894 (1957) [*Chem. Abstr.*, 52, 5167d (1958)].

217. G. Klopman, *Helv. Chim. Acta*, 44, 1908 (1961).

218. I. L. Knunyants and N. S. Vyazankin, *Izv. Akad. Nauk SSSR Otd. Khim. Nauk*, 2, 238 (1957); English Trans., *Bull. Acad. Sci. USSR, Div. Chem. Sci.*, 253 (1957).

219. P. Kobetz and R. C. Pinkerton, U.S. Patent 3,028,319 (1962) [*Chem. Abstr.*, 57, 11231g (1962)].

220. D. A. Kochkin, T. L. Shkorbatova, L. D. Pegusova, and N. A. Voronkov, *Zh. Obshch. Khim.*, 39, 1777 (1969).

221. N. Kondyrev, *Zh. Russ. Fiz.-Khim. Obshch.*, 52, 17 (1920).

222. N. Kondyrev, *Ber.*, 58B, 459,464 (1925).

223. N. Kondyrev, *Ber.*, 61B, 208 (1928).

224. N. Kondyrev, and A. Ssusi, *Ber.*, 62B, 1856 (1929).

225. N. Kondyrev, *Zh. Russ. Fiz.-Khim. Obshch.*, 60, 545 (1928).

226. N. Kondyrev, *Zh. Obshch. Khim.*, 4, 203 (1934).

227. I. A. Korshunov and N. I. Malyugina, *Zh. Obshch. Khim.*, 31, 1062 (1961).

228. I. A. Korshunov and N. I. Malyugina, *Trud. Khim. Khim. Tekhnol.*, 4, 296 (1961).

229. I. A. Korshunov and N. I. Malyugina, *Zh. Obshch. Khim.*, 34, 734 (1964).

230. M. Ya. Kraft, O. I. Korzina, and A. S. Morozova, *Sbornik State Obshch. Khim.*, 2, 1356 (1953); *C. A.*, 49, 5347 (1955).

231. C. A. Kraus, *J. Am. Chem. Soc.*, 35, 1732 (1913).

232. Y. Kunihisa and S. Tsutsumi, *J. Org. Chem.*, 32, 468 (1967).

233. F. Laborie, *Double Liaisons*, 148, 1489 (1967).

234. D. Laurin and G. Parravano, *J. Polymer Sci. B*, 4, 797 (1966).

235. H. D. Law, *J. Chem. Soc.*, 101, 1016 (1912).

235.a. A. I. Lebedeva, *Zh. Obshch. Khim.*, 18, 1161 (1948) [*Chem. Abstr.*, 43, 994f (1949)].

236. M. LeBlanc, *Z. Phys. Chem.*, **5**, 467 (1890).
237. H. Lehmkuhl and W. Leuchte, *J. Organometal. Chem.*, **23**, C30 (1970).
238. H. Lehmkuhl and K. Mehler, *J. Organometal. Chem.*, **25**, C44 (1970).
239. H. Lehmkuhl and K. Ziegler, Belg. Patent 590,575 (1960).
240. C. W. Lewis and P. C. Edge, *Ind. Eng. Chem., Prod. Res. Dev.*, **8**, 399 (1969).
241. J. D. Littlehailes and B. J. Woodhall, *Chem. Commun.*, 665 (1967).
242. J. D. Littlehailes and B. J. Woodhall, *Discussions Faraday Soc.*, **45**, 187 (1968).
243. S. G. Mairanovskii, V. A. Ponomarenko, N. V. Barashkova, and M. A. Kadina, *Izv. Akad. Nauk SSSR, Ser. Khim.*, 1951 (1964) [*Chem. Abstr.*, **62**, 6378d (1965)].
244. N. Maki, Y. Ishiuchi, and S. Sakuraba, *Bull. Chem. Soc. Japan*, **42**, 3166 (1969).
245. N. Maki, K. Yamamoto, H. Sunahara, and S. Sakuraba, *Bull. Chem. Soc. Japan*, **42**, 3159 (1969).
246. N. I. Malyugina, I. A. Korshunov, and L. N. Vertyulina, *Tr. Khim. Khim. Tekhnol.*, 141 (1967) [*Chem. Abstr.*, **70**, 73556 (1969)].
247. C. K. Mann, J. L. Webb, and H. M. Walborsky, *Tetrahedron Lett.*, **1966**, 2249.
248. C. L. Mantell, *Electro-Organic Chemical Processing*, Noyes Development Corp., Park Ridge, N. J., 1968.
249. E. M. Marlett, *Ann. N. Y. Acad. Sci.*, **125**, 12 (1965).
250. L. W. Marple, L. E. I. Hummelstedt, and L. B. Rogers, *J. Electrochem. Soc.*, **107**, 437 (1960).
251. L. Martinot, *Bull. Soc. Chim. Belg.*, **75**, 711 (1966).
252. L. Martinot, *Bull. Soc. Chim. Belg.*, **76**, 617 (1967).
253. M. Maruyama and T. Furuya, *Bull. Chem. Soc. Japan*, **30**, 650 (1957).
254. M. Maruyama and T. Furuya, *Bull. Chem. Soc. Japan*, **30**, 657 (1957).
255. K. Matsumiya and H. Nakata, *Mem. Coll. Sci. Kyoto Imp. Univ.*, **A12**, 63 (1929) [*Chem. Abstr.*, **23**, 4939 (1929)].
256. K. Matsumiya and H. Nakata, *Mem. Coll. Sci. Kyoto*, **10**, 199 (1927) [*Chem. Abstr.*, **22**, 1337 (1928)].
257. H. Matsuo, *J. Sci. Hiroshima Univ.*, Ser. A, **22**, 51 (1958).
258. J. L. Maynard, *J. Am. Chem. Soc.*, **54**, 2108 (1932).
259. J. L. Maynard and H. C. Howard, *J. Chem. Soc.*, **1923**, 960.
260. H. N. McCoy and W. C. Moore, *J. Am. Chem. Soc.*, **33**, 273 (1911).
261. H. N. McCoy and F. L. West, *J. Phys. Chem.*, **16**, 261 (1912).
262. L. D. McKeever and R. W. Taft, *J. Am. Chem. Soc.*, **88**, 4544 (1966).
263. L. D. McKeever and R. Waack, *J. Organometal. Chem.*, **17**, 142 (1969).
264. P. S. McKinney and S. Rosenthal, *J. Electroanal. Chem.*, **16**, 261 (1968).
265. B. Mead, U.S. Patent 1,567,159 (1926) [*Chem. Abstr.*, **20**, 607 (1926)].
266. H. Mehner, H. Jehring, and H. Kriegsmann, *J. Organometal. Chem.*, **15**, 107 (1968).
267. N. N. Melnikov and M. S. Rokitskaya, *J. Gen. Chem. USSR*, **7**, 2596 (1937) [*Chem. Abstr.*, **32**, 2084^3 (1938)].

268. T. Jr., Midgley, C. A. Hochwalt, and G. Calingaert, *J. Am. Chem. Soc.*, **45**, 1821 (1923).

269. K. Miyazaki, *Bull. Chem. Soc. Japan*, **41**, 1730 (1968).

270. J. G. Montalvo and D. G. Davis, *J. Electroanal. Chem. Interfacial Electrochem.* **23**, 164 (1969).

271. J. L. Morgat and R. Pallaud, *Compt. Rend.*, **260**, 5579 (1965).

272. J. L. Morgat and R. Pallaud, *Compt. Rend.*, **260**, 574 (1965).

273. M. D. Morris, *J. Electroanal. Chem. Interfacial Electrochem.*, **20**, 263 (1969).

274. M. D. Morris, *J. Electroanal. Chem. Interfacial Electrochem.*, **16**, 569 (1968).

275. M. D. Morris, *Anal. Chem.*, **39**, 476 (1967).

276. M. D. Morris, P. S. McKinney, and E. C. Woodbury, *J. Electroanal. Chem.*, **10**, 85 (1965).

277. A. Mouneyrat, U.S. Patent 1,232,373 (1917) [*Chem. Abstr.*, **11**, 2600 (1917)].

278. N. D. Murdoch and E. A. Lucken, *Helv. Chim. Acta*, **47**, 162 (1967).

279. R. W. Murray and K. Hiller, *Anal. Chem.*, **39**, 1221 (1967).

280. J. Myatt and P. F. Todd, *Chem. Commun.*, 1033 (1967).

281. T. Nakabayashi, *J. Am. Chem. Soc.*, **82**, 3909 (1960).

282. L. N. Nekrasov, N. N. Nefedova, and A. D. Korsun, *Elektrokhimiya*, **5**, 889 (1969).

283. J. Neldon and W. V. Evans, *J. Am. Chem. Soc.*, **39**, 82 (1917).

284. M. L. O'Donnell and C. W. Kreke, *J. Am. Pharm. Assoc. Scient. Ed.*, **48**, 268 (1959).

285. M. L. O'Donnell, A. Schwarzkopf, and C. W. Kreke, *J. Pharm. Sci.*, **52**, 659 (1963).

286. K. Okamoto, *Nippon Kagaku Zasshi*, **81**, 125 (1960) [*Chem. Abstr.*, **56**, 490e (1962)].

287. D. C. Olson and W. Keim, *Inorg. Chem.*, **8**, 2028 (1969).

288. D. C. Olson and J. Vasilevskis, *Inorg. Chem.*, **8**, 1611 (1969).

289. F. Pantani and G. Ciantelli, *Ric. Sci.*, **38**, 721 (1968).

290. T. H. Pearson, U.S. Patent 2,915,440 (1959) [*Chem. Abstr.*, **54**, 17415g (1960)].

291. V. Peruzzo, G. Plazzogna, and G. Tagliavini, *J. Organometal. Chem.*, **18**, 89 (1969).

292. R. E. Plump and L. P. Hammett, *Trans. Electrochem. Soc.*, **73**, 523 (1938).

293. B. G. Podlibner and L. N. Nekrasov, *Elektrokhimiya*, **6**, 1155 (1970).

294. F. D. Popp and H. P. Schultz, *Chem. Rev.*, **62**, 19 (1962).

295. G. B. Porter, *J. Chem. Soc.*, **1954**, , 760.

296. J. H. Prescott, *Chem. Eng.*, **72**, 238, 249 (1965).

297. O. A. Ptitsyna, O. A. Reutov, and M. F. Turchinskii, *Nauch. Dokl. Vysshei Shkoly, Khim. Khim. Tekhnol.*, **138** (1959) [*Chem. Abstr.*, **53**, 17030i (1959)].

298. O. A. Ptitsyna, S. I. Orlov, and O. A. Reutov, *Vestnik Mosk. Univ. Ser. 11, Khim.*, **21**, 105 (1966) [*Chem. Abstr.*, **65**, 13,755 (1966)].

299.   O. A. Reutov and I. P. Beletskaya, *Reaction Mechanisms of Organometallic Compounds*, North Holland Amsterdam, 1968, p. 394.

300.   O. A. Reutov, O. A. Ptitsyna, and M. F. Turchinskii, *Dokl. Akad. Nauk SSSR*, **139**, 146 (1961).

301.   L. Riccoboni, *Gazz. Chim. Ital.*, **72**, 47 (1942) [*Chem. Abstr.*, **37**, 574⁵ (1943)].

302.   L. Riccoboni and P. Popoff, *Atti Veneto Sci.*, **107**, 123 (1949) [*Chem. Abstr.*, **44**, 6752ᵃ (1950)].

303.   F. O. Rice and B. L. Evering, *J. Am. Chem. Soc.*, **56**, 2105 (1934).

304.   A. Rius and H. Carrancio, *An. Real Soc. Espan. Fis. y Quim.*, *Ser. B.*,**47**, 767 (1951) [*Chem. Abstr.*, **46**, 10010 (1952)].

305.   G. Renger, *Ber.*, **44**, 337 (1911).

306.   W. H. Rodebush and J. M. Peterson, *J. Am. Chem. Soc.*, **51**, 638 (1929).

307.   B. E. Roetheli and I. B. Simpson, Br. Patent 797,093 (1958) [*Chem. Abstr.*, **53**, 930b (1959)].

308.   L. B. Rogers and A. J. Diefenderfer, *J. Electrochem. Soc.*, **114**, 942 (1967).

309.   L. D. Rollmann and R. T. Iwamoto, *J. Am. Chem. Soc.*, **90**, 1455 (1968).

310.   R. Routledge, *Chem. News*, **26**, 210 (1872).

311.   A. N. Ryabtsev, K. P. Butin, I. P. Beletskaya, and O. A. Reutov, *Zh. Org. Khim.*, **4**, 934 (1968).

312.   J. Sand and F. Singer., *Ber.*, **35**, 3170 (1902).

313.   Y. Sato and N. Tanaka, *Bull. Chem. Soc. Japan*, **42**, 1021 (1969).

314.   Y. Sato and N. Tanaka, *Bull. Chem. Soc. Japan*, **41**, 2064 (1968).

315.   H. Schäfer, *Chem. Ing. Tech.*, **42**, 164 (1970).

316.   H. Schäfer and H. Küntzel, *Tetrahedron Lett.*, **1970**, 3333.

317.   C. Schall and W. Kirst, *Z. Electrochem.*, **29**, 537 (1923).

318.   C. Schall and H. Markgraf, *Trans. Am. Electrochem. Soc.*, **45**, 161 (1924).

319.   W. Schepps, *Ber.*, **46**, 2564 (1913).

320.   W. Schlenk and W. Schlenk, *Ber.*, **62**, 920 (1929).

321.   H. Schmidt, *Z. Anorg. Allgem. Chem.*, **270**, 180 (1952).

322.   R. E. Seaver, NASA Tech. Note D-1089 (1961) [*Chem. Abstr.*, **55**, 26782 (1961)].

323.   M. Seebeck, *Ann. Chim. Phys.*, **66**, 191 (1808).

324.   T. Sekine, A. Yamura, and K. Sugino, *J. Electrochem. Soc.*, **112**, 439 (1965).

325.a.W. J. Settineri and R. A. Wessling, "Electroreduction of Sulfonium Compounds", Paper delivered at Eighth Annual E. C. Britton Symposium on Industrial Chemistry, Midland Michigan, April 30, 1970.

325.b.W. J. Settineri and R. A. Wessling, U.S. Patent 3,660,257 (1972).

326.   I. Shukoff, *Ber.*, **38**, 2691 (1905).

327.   E. F. Silversmith and W. J. Sloan, U.S. Patent 3,197,392 (1965).

328.   K. C. Smeltz, U.S. Patent 3,392,093 (1968).

329.   S. A. Smirnova and S. P. Gubin, *Izv. Akad. Nauk SSSR*, *Ser. Khim.*, 1890 (1969).

330.   B. C. Southworth and R. Osteryoung, *Anal. Chem.*, **33**, 208 (1961).

331.   H. W. Sternberg, R. Markby, and I. Wender, *J. Electrochem. Soc.*, **110**, 425 (1963).

332. H. Wl Sternberg, R. E. Markby, I. Wender, and D. M. Mohilner, *J. Am. Chem. Soc.*, **91**, 4191 (1969).
333. H. W. Sternberg, R. E. Markby, I. Wender, and D. M. Mohilner, *J. Am. Chem. Soc.*, **89**, 186 (1967).
334. H. W. Sternberg, R. E. Markby, I. Wender, and D. M. Mohilner, *J. Electrochem. Soc.*, **113**, 1060 (1966).
335. W. Strohmeier, *Z. Elektrochem.*, **60**, 396 (1956).
336. W. Strohmeier and G. Popp, *Z. Naturforsch.*, **B22**, 891 (1967).
337. W. Strohmeier and F. Seifert, *Z Elektrochem.*, **63**, 683 (1959).
338. K. Sugino and G. Yuki, *Kagaku Kyokai Shi*, **24**, 1170 (1966).
339. S. Swann, Jr., *Trans. Electrochem. Soc.*, **69**, 287 (1936); **77**, 459 (1940); **88**, 103 (1945).
340. S. Swann, Jr., *J. Electrochem. Soc.*, **99**, 219 (1952).
341. S. Swann, Jr., *Electrochem. Tech.*, **1**, 308 (1961); **4**, 550 (1966); **5**, 53, 101, 393, 467, 549 (1967); **6**, 59 (1968).
341.a.B. Szilard, *Z. Elektrochem.*, **12**, 393 (1906).
342. J. Tafel, *Ber.*, **39**, 3626 (1906).
343. J. Tafel, *Ber.*, **44**, 323 (1911).
344. J. Tafel, *Ber.*, **42**, 3146 (1909).
345. J. Tafel and K. Schmitz, *Z. Electrochem.*, **8**, 281 (1902).
346. G. Tagliavini and L. Doretti, *Chem. Commun.*, **562** (1966).
347. G. Tagliavini, S. Faleschini, and E. Genero, *Ric. Sci.*, **36**, 717 (1966).
348. N. Tanaka and Y. Sato, *Electrochim. Acta*, **13**, 335 (1968).
349. N. Tanaka and Y. Sato, *Bull. Chem. Soc. Japan*, **41**, 2059 (1968).
350. A. B. Thomas and E. G. Rochow, *J. Am. Chem. Soc.*, **79**, 1843 (1957).
351. A. B. Thomas and E. G. Rochow, *J. Inorg. Nucl. Chem.*, **4**, 205 (1957).
352. C. M. Tissier, *J. Chim. Phys. Physicochim. Biol*, **65**, 2037 (1968).
353. R. S. Tobias, *Organometal. Chem. Rev.*, **1**, 93 (1966).
354. A. P. Tomilov, *Zh. Obshch. Khim.*, **38**, 214 (1968).
355. A. P. Tomilov, *Izv. Khim.*, 481 (1968).
356. A. P. Tomilov and I. N. Brago, *Prog. Elektrokh. Org. Soed.*, **1**, 208 (1969).
357. A. P. Tomilov and M. Ya. Fioshin, *Russ. Chem. Rev.*, **32**, 30 (1963).
358. A. P. Tomilov and L. V. Kaabak, *Zh. Prik. Khim.*, **32**, 2600 (1959); English Transl., *J. Appl. Chem. USSR*, **32**, 2677 (1959).
359. A. P. Tomilov, L. V. Kaabak, and S. L. Varshavskii, *Zh. Obshch. Khim.*, **33**, 2811 (1963).
360. A. P. Tomilov, L. V. Kaabak, and S. L. Varshavskii, *Zh. Vses. Khim. Obshch. D. I. Mendeleeva*, **8**, 703 (1963) [*Chem. Abstr.*, **60**, 11618a (1964)].
361. A. P. Tomilov and B. L. Klyuev, *Elektrokhimiya*, **3**, 1168 (1967).
362. A. P. Tomilov and B. L. Klyuev, *Elektrokhimiya*, **2**, 1405 (1966).
363. A. P. Tomilov S. G. Mairanovskii, M. Ya. Fioshin, and V. A. Smirnov, "Electrochemistry of Organic Compounds". Chapter XI, p. 481, *Izd, Khimya*, (1968).
364. A. P. Tomilov and Yu. D. Smirnov, *Zh. Vses. Khim. Obshch. D. I. Mendeleeva*, **10**, 101 (1965) [*Chem. Abstr.*, **63**, 3897g (1965)].
365. A. P. Tomilov, Y. D. Smirnov, and S. L. Varshavskii, *Zh. Obshch. Khim.*,

35, 391 (1965).

366. A. P. Tomilov, O. N. Temkin, I. A. Esikova, R. M. Flid, S. M. Makaroch-kina, O. A. Kondakova, and V. A. Dolinchuk, *Elektrokhimiya*, 5, 722 (1969).

367. V. F. Toropova and M. K. Saikina, *Zh. Neorg. Khim.*, 10, 1166 (1965).

368. V. F. Toropova, M. K. Saikina, and M. G. Khakimov, *Zh. Obshch. Khim.*, 37, 47 (1967).

369. M. G. Townsend, *J. Chem. Soc.*, 1962, 51.

370. Y. M. Tyurin, V. N. Flerov, and Z. A. Nikitina, *Elektrokhimiya*, 5, 903 (1969).

371. H. E. Ulery, *J. Electrochem. Soc.*, 116, 1201 (1969).

372. A. P. Giraitis, U.S. Patent 2,944,948 (1960) [*Chem. Abstr.*, 54, 20591f (1960)].

373. K. Ziegler, Br. Patent 814,609 (1959) [*Chem. Abstr.*, 53, 17733c (1959)].

374. D. G. Braithwaite, U.S. Patent 3,007,957 (1961) [*Chem. Abstr.*, 56, 3280f (1962)].

375. P. Kobetz and R. C. Pinkerton, U.S. Patent 3,028,322 (1962) [*Chem. Abstr.*, 57, 11235i (1962)].

376. P. Kobetz and R. C. Pinkerton, U.S. Patent 3,028,323 (1962).

377. R. C. Pinkerton, U.S. Patent 3,028,325 (1962) [*Chem. Abstr.*, 57, 4471b (1962)].

378. P. Kobetz and R. C. Pinkerton, U.S. Patent 3,028,318 (1962) [*Chem. Abstr.*, 57, 11234d (1962)].

379. W. P. Hettinger, U.S. Patent 3,079,311 (1963) [*Chem. Abstr.*, 59, 2859f (1963)].

380. J. W. Ryznar and J. C. Premo, U.S. Patent 3,100,181 (1963) [*Chem. Abstr.*, 59, 9602b (1963)].

381. J. Linsk, U.S. Patent 3,118,825 (1964) [*Chem. Abstr.*, 61, 2736d (1964)].

K. Ziegler and H. Lehmkuhl, U. S. Patent 3,254,008 (1966).

383. K. Ziegler and H. Lehmkuhl, U.S. Patent 3,254,009, (1966).

384. D. G. Braithwaite, U.S. Patent 3,256,161 (1966) [*Chem. Abstr.*, 65, 6751d (1966)].

385. S. Valcher, *Corsi Semin. Chim.*, 37, (1968).

386. S. Valcher and E. Alunni, *Ric. Sci.*, 38, 527 (1968).

387. S. Valcher and M. Mastragostino, *J. Electroanal. Chem.*, 14, 219 (1967).

388. A. Vanachayangkul and M. D. Morris, *Anal. Lett.*, 1, 885 (1968).

389. S. V. Vasil'ev and G. D. Vovchenko, *Vestnik Moskov. Univ.*, 5, *Ser. Fiz.-Mat. i Estest. Nauk*, No. 2, 73 (1950) [*Chem. Abstr.*, 45, 6594 (1951)].

390. L. N. Vertyulina and I. A. Korshunov, *Khim. Nank. Prom.*, 4, 136 (1959).

391. L. N. Vertyulina and I. A. Korshunov, *Elektrokhim. Protesessy Uchastiem Org. Veshchestv*, 138 (1970) [*Chem. Abstr.*, 74, 18791w (1971)].

392. Voitkevich, Candidates dissertation to Perfumery Research Institute, Moscow (1949). Information from Ref. 354.

393. V. Vojir, *Coll. Czech. Chem. Commun.*, 16, 488 (1951).

394. M. E. Vol'pin, E. A. Tevdoradze, and K. P. Butin, *Zh. Obshch. Khim.*, 40, 315 (1970); Eng. Trans., *J. Gen. Chem. USSR*, 285 (1970).

395. J. H. Wagenknecht and M. M. Baizer, *J. Org. Chem.*, **31**, 3885 (1966).
396. J. H. Wagenknecht and M. M. Baizer, *J. Electrochem. Soc.*, **114**, 1095 (1967).
397. B. J. Wakefield, *Organometal. Chem. Rev.*, **1**, 131 (1966).
398. C. P. Wallis, *J. Electroanal. Chem.*, **1**, 307 (1960).
399. S. Wawzonek, *Anal Chem.*, **26**, 65 (1954).
400. S. Wawzonek, R. C. Duty, and J. H. Wagenknecht, *J. Electrochem. Soc.*, **111**, 74 (1964).
401. J. L. Webb, C. K. Mann, and H. M. Walborsky, *J. Am. Chem. Soc.*, **92**, 2042 (1970).
402. N. L. Weinberg, *Tetrahedron Lett.*, **1970**, 4823.
403. N. L. Weinberg and H. R. Weinberg, *Chem. Rev.*, **68**, 449 (1968).
404. R. P. M. Werner U.S. Patent 3,236,755 (1966). [*Chem. Abstr.*, **64**, 15395f (1966)].
405. R. A. Wessling and W. J. Settineri, U.S. Patent 3,480,525 (1969) [*Chem. Abstr.*, **72**, 32677d (1969)].
406. R. A. Wessling and W. J. Settineri, U.S. Patent 3,480,527 (1969) [*Chem. Abstr.*, **72**, 66593p (1970)].
407. J. Wiemann and M. L. Bouguerra, *Ann. Chim. (Paris)*, **3**, 215 (1968).
408. G. Wilkinson and J. M. Birmingham, *J. Am. Chem. Soc.*, **76**, 4281 (1954).
409. D. Wittenberg and H. Gilman, *J. Org. Chem.*, **23**, 1063 (1958).
410. A. Wolberg and J. Manassen, *J. Am. Chem. Soc.*, **92**, 2982 (1970).
411. G. B. Wood and A. Brenner, *J. Electrochem. Soc.*, **104**, 29 (1957).
412. B. J. Woodhall and J. D. Littlehailes, U.S. Patent 3,534,078.
413. W. L. Wuggatzer and J. M. Cross, *J. Am. Pharm. Assoc.*, **41**, 80 (1952).
414. N. Yamazaki, S. Nakahama, and S. Kambara, *Polymer Lett.*, **3**, 57 (1965).
415. N. Yamazaki, I. Tanaka, and S. Nakahama, *J. Macromol. Sci.*, A(2), **6**, 1121 (1968).
416. J. Y. Yang, W. E. McEwen, and J. Kleinberg, *J. Am. Chem. Soc.*, **79**, 5833 (1957).
417. K. Yang, J. D. Reddy, M. A. Johnson, and W. H. Harwood, Ger. Patent 1,955,211 [*Chem. Abstr.*, **74**, 31230f (1971)].
418. K. Yang, J. D. Reedy, M. A. Johnson, and W. H. Warwood, Ger. Patent 1,955,201 [*Chem. Abstr.*, **73**, 31079u (1970)].
419. K. Yasukouchi and H. Muto, *Denki Kagaku*, **35**, 420 (1967) [*Chem. Abstr.*, **68**, 45536z (1968)].
420. K. Yasukouchi and H. Muto, *Denki Kagaku*, **35**, 890 (1967) [*Chem. Abstr.*, **69**, 15327k (1968)].
421. K. Yasukouchi, H. Muto, A. Iwanaga, and M. Tetsuya, *Denki Kagaku*, **36**, 54 (1968) [*Chem. Abstr.*, **69**, 15328m (1968)].
422. K. Yosahida and S. Tsutsumi, *J. Org. Chem.*, **32**, 468 (1967).
423. E. V. Zappi and R. F. Mastrapaolo, *An. Asoc. Quim. Argentina*, **29**, 88 (1941).
424. K. Ziegler, Br. Patent 813,446 (1959) [*Chem. Abstr.*, **53**, 13845e (1959)].
425. K. Ziegler, Br. Patent 946,686 (1964).
426. K. Ziegler, *Brennstoff - Chem.*, **40**, 209 (1959).

427. K. Ziegler, *Angew. Chem.,* **71**, 628 (1959).
428. K. Ziegler and H. Lehmkuhl, Ger. Patent 1,212,085 (1966) [*Chem. Abstr.,* **64**, 19675a (1966)].
429. K. Ziegler and H. Lehmkuhl, U.S. Patent 2,985,568 (1961).
430. K. Ziegler and H. Lehmkuhl, Ger. Patent 1,161,562 (1964) [*Chem. Abstr.,* **60**, 11623g (1964)].
431. K. Ziegler and H. Lehmkuhl, *Z. Anorg. Ang. Chem.,* **283**, 414 (1956).
432. K. Ziegler adn H. Lehmkuhl, *Chem. Ing. Tech.,* **35**, 325 (1963).
433. K. Ziegler and K. Ruthardt, Br. Patent 816,574 (1959).
434. K. Ziegler and O. W. Steudel, *Ann. Chem.,* **652**, 1 (1962).
435. V. I. Zoltarev, Soviet Patent 53,766 (1938).

Chapter XI

ELECTROCHEMICAL SYNTHESIS OF POLYMERS

B. Lionel Funt and John Tanner

---

1 INTRODUCTION

Electrolysis is characterized by the occurrence of oxidation and reduction reactions at the anode and cathode of an electrolytic cell. For each electron transferred through the solution, a corresponding chemical reaction must occur at the electrode. Since a number of competing chemical processes may take place simultaneously, the determination of whether a particular oxidation or reduction reaction occurs is dependent upon the exact potential of the electrode in comparison to the discharge potential of the various species in solution. Inasmuch as the electrode potential can be varied over a broad range or maintained constant, a selectivity can be exercised over the possible chemical transformations that occur.

A superficially simple overall reaction may require a complex series of individual steps at the electrode. The voluminous literature on the Kolbe synthesis illustrates both the remarkable simplicity of electrode reaction and the complexity of intermediate processes which prevail [1]. Such factors as overpotential at an electrode, migration of ions to an electrode, adsorption on the electrode surface, interactions on the electrode material, and diffusion from the electrode all can play important roles in the determination of the particular products of an electrochemical transformation.

The tremendous intrinsic potentialities of electroorganic chemistry have been amply demonstrated. Difficult syntheses have been performed with remarkable simplicity and high yield. The application of electrochemical techniques to polymerization reactions has received much less attention. Indeed, in some of the earlier studies, the addition of a monomer to a reaction mixture was used as a convenient analytical technique to detect the presence of free radicals generated at the electrode, and the process of polymerization was quite incidental to the theme of the investigations [2].

It is apparent that a considerable spectrum of possibilities exists in the electrochemical formation of polymers. The passage of an electrolytic current in the presence of a monomer may cause the onset of polymerization in various ways. Ionic discharge at an electrode may form a free atom such as hydrogen, a radical such as methyl, or a radical anion or radical cation. Each of these in turn can add to monomer. If the potential for direct electron transfer to the monomer is lower than that required for transfer to a supporting electrolyte, a direct activation of the monomer to a radical ion may be achieved.

Other indirect effects may result from the electrolysis of the solution. The passage of current may form a stable substance that can act as an initiator of polymerization. This is equivalent to the electrochemical formation of the initiator in situ. Alternatively, the passage of current may take place in a reaction medium where the initiation of polymerization is prevented by the presence of an inhibitor. The electrolysis can be viewed as a scavenging technique to discharge or remove the inhibitor. This will then result in the onset of normal uninhibited thermal polymerization. Again, it is frequently observed that at the opposite electrode to that at which initiation takes place substances are generated that terminate or inhibit the electrochemical polymerization. The mixing of the contents of the electrode compartments, or the failure to prevent diffusion from cathode to anode, amy result in inhibition or termination of the chemical reaction.

Polymerization by electrochemical techniques may be studied from a variety of different standpoints. These can best be illustrated by some examples:

1. The rate of the reaction may be controlled by programming the current-time profile.

2. The molecular weight distributions can be controlled through variation of the impressed current.

3. The molecular architecture of the resultant polymers may be influenced by the conditions of electrolysis. Stereoregularity, due to absorption on the surface of the electrode, or due to complexing with the salts present in the electrolyte, may be induced and varied.

4. New syntheses of polymeric materials may be possible through the coupling

reactions generated by electrochemical means. This is particularly applicable to condensation reactions.

5. The properties of existing polymers may be varied or modified by cross-linking, degradation, grafting, or block formation reactions induced by the passage of electrolytic current.

6. Particular monomers may be initiated by selective voltage control.

Not all of these possibilities have as yet been surveyed experimentally. In a later section of this chapter, we will review the extant experimental work illustrative of these synthetic possibilities.

Inasmuch as electrochemical polymerization is relatively new, and certainly not thoroughly explored, the scope of the technique can best be illustrated by reference and analogy to photolytic polymerization. In each instance, the activation energy for the reaction is supplied by an external source. The number of photons absorbed in photolytic reactions is related directly to the number of molecules activated by Einstein's laws of photochemistry. In a similar manner a quantitative interpretation of the number of electrons transferred at the electrode is possible through the application of Faraday's laws of electrolysis. In photolytic reactions, not only the flux of photons, but their individual energy, can be regulated, measured, and controlled. An element of energy discrimination is thus introduced into the reaction systems, and by control of the activating wave length, specific chemical reactions can be induced to the exclusion of others. Similarly, control of the electrode potential can result in reactions taking place at a particular energy to the exclusion of reactions that require a higher activating potential.

It should be noted that, except for the initiating step, the electrochemical polymerization will proceed by the same mechanism as the thermal polymerization for the same monomer system in a similar environment. Once the initiating radical or anion is formed electrochemically, whether by direct electron transfer at the electrode or by interaction with transient species or stable substances formed at the electrode, the further course of the reaction is essentially unaffected by the passage of electrolytic current. Again, the analogy between photolysis and electrolysis in polymerization reactions is valid. The same general pattern can be applied to photolytic behavior, where the activation by light produces the initiating radical or other active substance. There is one particular feature of electrochemical polymerizations that differs from the photolytic ones. It is possible to reverse the current or to employ an alternating current so that the chemical environment at the electrode is changed as a function of time, the course of the reaction being changed accordingly. There is no photochemical process corresponding to the reversal of current or to the application of an alternating current, and there is a whole pattern of control over the course of the reaction and the resulting architecture of the final products that is possible in electrochemical work and that has no direct analog in photolytic studies.

Furthermore, in photolytic systems the specific absorption characteristics of the reacting species are utilized to determine their concentrations by absorption spectrophotometry. In electrochemical reactions the application of organic polarography, cyclic voltammetry, and other electroanalytical techniques provide analytical data that delineate whether a particular reaction will occur and that facilitate the analysis of products.

It is intrinsically far simpler to generate a large number of electrons by the passage of current than it is to generate the same number of photons. The maximum fluxes available are far greater in the electrochemical case than in the photolytic. The photon flux is itself generated electrically by a gas discharge, and this process is not particularly efficient. In any event, the highest intensity lamps do not provide photon fluxes greater than the electron fluxes easily generated in electrochemical work. The significance of this to a scale-up of operations is clear. By employing large electrode areas and high current densities, polymerization should be feasible on an industrial scale. The economics are favored by the fact that in some polymerizations a stoichiometric conversion can occur with one electron yielding one polymer molecule. In terms of pounds per kilowatt hour this is much more favorable than conventional electroorganic preparation because a thousand or more monomers units may combine to form one polymer molecule.

On the other hand, electrochemical techniques present their own experimental difficulties. One must seek systems that combine high conductivity, chemical compatibility, and adequate solubility characteristics. The choice of solvents and supporting electrolytes that fulfil these functions is limited, and as a result, many promising electrochemical systems cannot be utilized under ideal conditions.

Scope of the Work

We have limited our attention to those systems in which electrochemical transformations leading to polymerization take place in solution. We have thus not included the very major field of electrical discharge polymerization. This has fortunately been reviewed recently [3]. We have similarly not considered as electrochemical polymerization those situations in which the application of electric field has served to increase the rate of an electrochemical polymerization through orientation effects. The detailed studies of Ise and co-workers have been reviewed recently [4], but they are not considered directly within the scope of this work which is limited to electrochemical initiation of polymerization reactions.

Some of the most prolific and outstanding contributions to the understanding of electrochemical polymerization have stemmed from the work of Baizer and his co-workers [5-7]. However, inasmuch as the major part of Baizer's work was related to dimerization reactions and to the electrochemical production of polymerizable intermediates, this aspect of his work is not within the scope of the material covered in this chapter.

We are specifically excluding a consideration of those processes that are limited to dimerization and trimerization reactions and essentially limit our attention to those reactions that can be classed to form "high polymers" or generally substances with molecular weights in excess of 1000.

In later sections we review the extant literature and comment on the mechanism of the reaction.

2 EXPERIMENTAL

General Considerations

One can distinguish between those experimental features that are common to electroorganic synthesis and those that are specific to polymerization reactions. In general the apparatus and technique are identical with those employed in other electroorganic syntheses. As such, they are described elsewhere in this book. The major parameters to be controlled and scrutinized are the electrode potential, the current density, and the conductivity. These in turn control the chemical transformations occurring at the electrodes, the yield of product, and the applied voltage necessary to achieve a desired electrode potential and current density. Apparatus for research in electrochemical polymerization should preferably include a reference electrode as a standard comparison voltage for the control of electrode potential, an electronic feedback device for maintaining a prescribed potential, a reaction cell with suitable electrodes, and a solvent and supporting electrolytes with adequate conductance and compatible chemical reactivity to the reaction system.

A review of the experimental techniques for electrochemical polymerizations shows that relatively little sophisitication in the instrumentation has been employed. This has led to ambiguity in the determination of the primary electrode processes and has limited the interpretation which can be placed upon observed phenomena or postulated mechanisms.

Similarly there has been little use of electroanalytical techniques for following reactions in situ or in determining by separate analytical methods the likely course of the reaction.

Organic polarography has been employed as a screening device in some reactions, but there has been little definitive work performed in determining the chemical changes in solution during the course of a polymerization, utilizing electroanalytical measurements such as cyclic voltammetry or other electroanalytical tools. There is considerable scope for further work in this direction.

However, certain intrinsic limitations are present in electrochemical polymerization, and in retrospect one can justify the trade-off between simplicity of experimental technique versus sophistication and additional elucidation of mechanism that has in general thus far characterized this work.

Specifically, since most useful polymers are insoluble in aqueous systems, it is necessary to employ nonaqueous solvents for polymerization reactions. This places constraints on electrochemical techniques. It is a characteristic of polymer molecules that relatively low concentrations of polymer will produce very large increases in the viscosity of their solutions. Thus the viscosity changes during the course of a polymerization are far greater than those encountered in organic syntheses. Although the problem is not as severe in reactions performed in solution as contrasted to those conducted in bulk monomer, the very dramatic increase of viscosity and its resultant, and often ill-defined, effect upon the diffusion, migration, and other transport phenomena is a feature of electrochemical polymerization.

A given monomer may polymerize by a variety of mechanisms. For example, styrene can be induced to undergo free-radical, cationic, and anionic polymerizations. The experimental conditions necessary for the conduct of the polymerization after the electrochemical initiation step are peculiar and specific to each of these mechanisms of polymerization. For example, cationic polymerizations in general require very low temperatures in order to minimize the effects of chain transfer. This is the major process limiting the growth of individual chains in this type of polymerization. However, a reduction in temperature presents problems in terms of the solubility of the supporting electrolyte, the freezing point of the solvent employed, and the general decrease in conductivity of the solution. No simple recipe exists for an adequate choice of substances that can be used for any and all monomers, and a specific design of both the composition of the chemical system and the physical arrangement of the electrolysis cell must reflect the realities of experimental demands.

For some forms of anionic polymerization, particularly living anionic reactions, the demands in terms of chemical purity are extremely stringent and the solution must exclude any proton donating impurity at a concentration greater than about 10^{-4} molar. This limits the choice of solvent and supporting electrolyte.

The polarity of the solvent will affect the course of ionic reactions. The reactivity of ion pairs versus free ions may differ by a factor of 500 in the propagation step in an anionic polymerization. The concentration of supporting electrolyte such as sodium tetraphenylboride can be adjusted so that free ion formation is decreased and ion pair formation favored or vice versa. Similarly, the polarity of the solvent will have an important role in determining the degree of ionization of the propagating species and hence the overall rate of reaction.

The concentration of initiator that is formed electrochemically can be controlled by the current. However, it in turn will control the later stages of reaction. A high initiator concentration, regardless of its mechanism of formation, will result in a high rate of initiation in free-radical reactions, thus leading to a high rate of polymer formation but a corresponding decrease in the molecular weight of polymer. This is characteristic of free-radical polymerization, where, in general, molecular weight is inversely proportional to the rate of propagation. On the other hand, with anionic reactions, a high initial concentration of initiating anions will lead in the ideal case to each anion producing a growing chain. The limitation on molecular weight is determined by the concentration of impurities and by the total concentration of monomer and initiator present.

Rather than focus further on the specific features of various forms of polymerization, we will illustrate the design requirements by reference to some techniques employed and described in the literature on electrochemical work.

Solvents

The proper choice of solvent and of supporting electrolyte are crucial for successful electrochemical polymerization. Most polymers are insoluble in water. Thus, aqueous systems result in heterogeneous polymerization with deposition of the polymer on the electrode and consequent changes in operating parameters. Much of the change in outlook toward the practicality and importance of electrochemical polymerization has resulted from the adoption of solvents that facilitate polymerization in nonaqueous systems. However, the influence of the solvent upon the chemical aspects of the polymerization reaction must also be considered. For example, in living polymerizations the absence of proton donors is essential. Similarly, cationic and free-radical polymerizations cannot be conducted in solvents that lead to excessive chain transfer or termination.

An excellent review of the available nonaqueous solvents for electrochemical use has recently been published by C. K. Mann [9]. This author has given detailed information regarding the purification and use of the various solvents, suitable electrode materials, and reference electrodes. (See Table 11.1.)

A relatively small number of solvent and electrode systems have as yet been employed in electropolymerization. Some physical properties of the more useful solvents are presented in Table 11.2 [22]. Tetrahydrofuran, hexamethylphosphortriamide, and dimethoxyethane are noteworthy for their suitability for living anionic polymerizations. Dimethylformamide and dimethylsulfoxide have high dielectric constants and provide solutions of good conductance with many supporting electrolytes.

Reaction Cells

In the laboratory the design requirements of a reaction vessel for the conduct of an electrochemical polymerization are very modest. The desired results can and have been achieved in apparatus as simple as a U tube containing two

Table 11.1. Usable Range of Potentials of Solutions

Solvent	ϵ	Working Electrode	Supporting Electrolyte	Potential Range
Tetrahydrofuran	7.4	Pt	$LiClO_4$	+1.8 to -3.6
		Hg	$(C_4H_9)_4NI$	-1.2 to -3.6
Pyridine	12.3	Hg	$LiClO_4$	to -1.77
Dioxane (96%)	3.5	Hg	$(C_4H_9)_4NI$	to -2.32
Dimethyl sulfoxide	46	Hg	$(C_2H_5)_4NClO_4$	+.25 to -2.80
		Hg	NaOAc	-0.10 to -1.96
		Hg	LiCl	-.21 to 2.17
1,2-Dimethoxyethane	3.5		$(C_4H_9)_4NClO_4$	+0.65 to -2.95
Acetic anhydride	20.7	Pt	$LiClO_4$	+2.1 to -1.2
Hexamethylphosphoramide	30	Pt	$LiClO_4$	+0.75 to -3.6
Dimethylformamide	36.7	Pt	$(C_2H_5)_5NClO_4$	+1.6 to -2.1

Table 11.2. Physical Properties of Solvents

	Specific Conductance $(ohm^{-1}\ cm^{-1})$	Dielectric Constant	BP $(°C)$	FP $(°C)$	Viscosity (cP)
Acetic anhydride	$7.5 \times 10^{-7}{}_{20}$	20.7_{19}	1400	-73.0	0.783_{30}
Acetonitrile	$2 \times 10^{-8}{}_{25}$	37.45_{20}	81.6	-45.7	0.325_{30}
Acrylonitrile	—	38	78	-82	0.34_{25}
Dimethyl acetamide	2×10^{-7}	38.93	165	-20	0.92_{25}
Dimethylformamide	$3 \times 10^{-8}{}_{25}$	36.7	153	-61	0.796_{25}
Dimethyl sulfoxide	$3 \times 10^{-8}{}_{25}$	46.7	189	-18.55	2.003_{30}
1,2-Dimethoxyethane	—	3.49_{20}	85	-58	0.53_{10}
Methylene chloride	4.3×10^{-11}	8.93_{25}	39.95	-96.7	0.393_{30}
Propionitrile	8.5×10^{-8}	27.2_{20}	97.2	-91.9	0.389_{30}
Propylene carbonate	5×10^{-9}	36.67	241.7	-49.2	0.627_{25}
Pyridine	4.0×10^{-8}	12.3_{25}	115.6	-41.8	0.83_{30}

inserted electrodes. Most work has employed a simple H cell design with the cathode and anode compartments separated by a sintered glass disk or a porous membrane.

A good cell design must have provision for agitation, separation of anode and cathode compartments by a membrane, and the accommodation of a reference electrode.

At the present time there are no large-scale industrial processes in operation using electrochemical polymerization. Nevertheless, the problems of "scale-up" and the associated considerations of cell design are of interest. It is important for the economic success of electrolytic processes that the cell provide high current densities at low power consumption. The capital costs of such commercial electroorganic processes are generally not acceptable with current densities below 0.1 A/cm². A plate and frame cell for electroorganic preparations, such as shown in Fig. 11.1 [15], would no doubt be applicable for some electrochemical polymerizations. The new developments in fluidized bed cells, in which one or both electrodes consists of a fluidized bed of conducting particles, also merit careful study as potentially useful cell designs for polymerization.

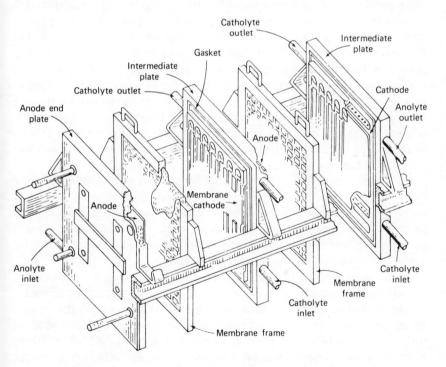

Fig. 11.1. Plate and frame cell for electroorganic processing. Reprinted by permission, from Ref. 15.

This type of cell provides a very high ratio of electrode area to volume, about 75 cm^2/cm^3, or a factor of 25 greater than that normally attainable in plate and frame cells [16, 17].

Three typical laboratory scale syntheses and the cell designs employed are described later in this section.

General Experimental Techniques

The course of the polymerization reaction can be followed by conventional techniques for measurement of polymer produced. Extraction of samples coupled with analysis by gas chromatography may indicate the consumption of monomer. Precipitation of polymer from aliquots removed from the solution can be used to determine polymer production as a function of time. It should be noted that the removal from a reaction cell of significant amounts of reactants can drastically alter the apparent kinetic behavior of the system. Nevertheless, it is usually possible to compensate for these changes. Again, sample removal, unless carried out with exceptional care, can introduce unwanted impurities to the reaction mixture. For these reasons, and for simplicity, in situ methods are to be preferred, and measurements of the density, the viscosity, or the spectral absorption of the solution can be performed. Each of these techniques has its limitations when applied to electrochemical polymerizations. The technique of dilatometry cannot be applied if a gas is produced in the solution, and furthermore it is difficult to perform such measurements in cells with frit-separated compartments. Viscosity measurements by falling-ball or capillary techniques may be employed, but they do not give unequivocal answers, since the viscosity increase is a function both of molecular weight of polymer and of concentration of polymer. Absorption measurements by spectrophotometry are often difficult in situ and cannot be made if the absorption bands of solvents or solute overlap the regions of interest.

Electroanalytical techniques are, in principle at least, very appealing for these types of investigations. However, organic polarography generally requires a sampling of the reaction mixture since the technique does not lend itself well to direct measurement within the reaction vessel. The techniques of chronopotentiometry and chronoamperometry shown in Table 11.3 appear suitable for continuous monitoring of the polymerization process.

Cyclic voltammetry is particularly interesting. It is a powerful technique for investigations conducted in conjunction with electrochemical polymerization reactions. During the reverse cycle, the products reduced during the forward sweep can be oxidized. For a reversible process, a symmetrical wave is obtained with a separation between positive and negative peaks of 0.1 RT/nF or 59 mV. However, when a rapid chemical reaction follows the reduction cycle, the corresponding reverse peak disappears. By varying the sweep rate, the kinetics of secondary electrochemical reactions can be studied directly in the reaction cell.

Table 11.3. Electroanalytical Techniques

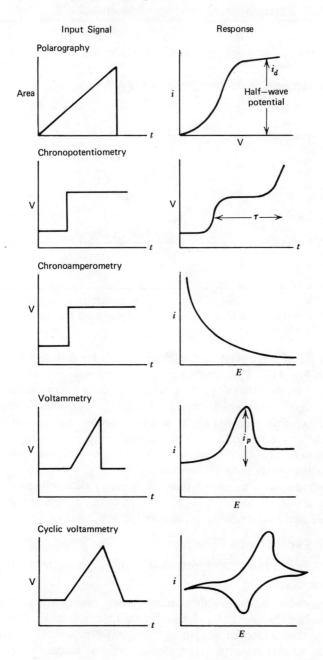

Table 11.3.   Electroanalytical Techniques[a] (continued)

Polarography

$$i = 607\, nD^{1/2} CM^{2/3} t^{1/6}$$

Chronopotentiometry

$$\tau^{1/2} = \frac{1}{2i}\, \pi^{1/2} D^{1/2} nFAC$$

Chronoamperometry

$$i_d = \frac{nFACD^{1/2}}{\pi^{1/2} t^{1/2}}$$

Voltammetry

$$i_p = \frac{0.447\, ACF^{3/2} n^{3/2} D^{1/2} V^{1/2}}{R^{1/2} T^{1/2}}$$

Cyclic Voltammetry

$$i_p = \frac{0.447\, ACF^{3/2} n^{3/2} D^{1/2} V^{1/2}}{R^{1/2} T^{1/2}}$$

---

[a]D = diffusion coefficient (Cm^2/sec); C = concentration, M = Hg flow rate (mg/sec); τ = transition time (sec); V = potential scan rate (volts/sec); A = area (Cm^2).

An example of a reversible cyclic voltammogram for diphenylpicrylhydrazyl is shown in Fig. 11.2. This substance is of particular interest as an inhibitor of free-radical polymerization [29]. A series of cyclic voltammograms on phenyl-substituted ethylenes is also instructive. The data [30] are illustrative of reversible one- and two-peak processes and show some reactions that are irreversible in the time scale of the experiment. In electrochemical polymerization, protonation and polymerization may be competitive and this technique may be used to delineate the optimum conditions for reaction.

The following is a sample of three fairly recent electrochemical polymerizations taken from the literature [152, 153] and from our own laboratories [177]. The experiments are described in considerable detail.

Specific Experimental Examples

Living Anionic Electropolymerization of 1,3 Butadiene [177]

INTRODUCTION

This procedure produces polybutadiene, whose molecular weight distribution is determined by the electrochemical conditions employed. The number of fractions, their molecular weight, and the weight percentage of each fraction can be controlled separately. The method employs successive equal pulses of current whose number determine the quantities of polymer produced, whereas

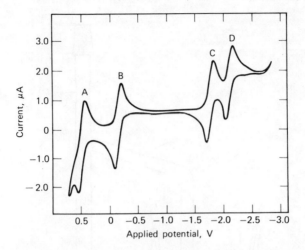

Fig. 11.2. Cyclic voltammogram of diphenylpicrylhydrazyl (1.0 mM) in 0.10M Bu$_4$NClO$_4$/ THF. Scan rate 0.07 V/sec. Reprinted by permission, from Ref. 29.

the molecular weights are determined by the time interval between initiating and terminating pulses:

$$CH_2{=}CH{-}CH{=}CH_2 \xrightarrow[\text{initiation}]{\text{Electrolysis}} \left[\begin{array}{c}CH{-}CH_2 \\ | \\ CH{=}CH_2\end{array}\right]_n^{\theta\theta} \xrightarrow[\text{termination}]{\text{Electrolysis}} \left[\begin{array}{c}CH{-}CH_2 \\ | \\ CH{=}CH_2\end{array}\right]_n$$

1.3 Butadiene   "Living" dianion   Polymer

APPARATUS

The apparatus consisted of an electrolytic cell, shown in Fig. 11.3, with anode and cathode compartments separated by a fine fritted disk. The cathode compartment contained 57 ml of solution and the anode compartment 28 ml. Square platinum electrodes, each 1 in.^{2} in area, were employed. A filling bulb (A) is joined to the cell as shown.

EXPERIMENTAL DESCRIPTION

First, 0.5 g of sodium tetraphenylboride (NaBPh$_4$), which was Fisher Certified grade and was not further purified, was placed into the filling bulb A and the system was evacuated for 24 hr with intermittent heating. Then, 80 ml of Fisher Certified tetrahydrofuran (THF) [purified according to the method of Fetters (27)] were distilled into the system followed by 5 ml of butadiene. Gaseous butadiene was passed through columns of molecular sieve and silica gel. After

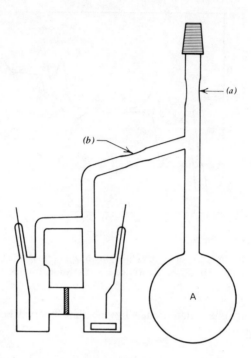

Fig. 11.3. Electrolysis cell [177]; for description see text.

thorough degassing of the system, the assembly was sealed off at *a* and the contents were transferred into the electrolytic cell, which, in turn, was isolated by sealing off at constriction *b*.

The electrolytic cell was then placed in a constant temperature bath at $-10°C$ and a constant current of 15 mA was passed through the solution until the yellow color formed at the cathode persisted. This indicated the complete removal of reactive impurities. Current reversal destroyed the living anions stoichiometrically, and thus zero concentration of the living anions was achieved readily by this process of electrochemical titration. The yellow color of polybutadiene anions was sufficiently intense to be detected visually against a white background even at low concentrations. However, the process could be followed spectrophotometrically if so desired.

For the polymerization, a current of 15 mA was passed for 33 sec and followed by 217 sec propagation time. A reverse current of the same intensity was then passed to destroy the living anions. To produce the required amount of polymer, this cycle was repeated 15 times. The propagation periods were thereafter increased for each subsequent cycle to compensate for the depletion of monomer. This is shown in Table 11.4.

Cycle No.	1	2	3	4	5	6	7	8	9	10	11	12	13	14	15
t_p (sec)	211	223	229	235	241	248	255	262	270	279	288	297	308	317	328

After the polymerization, the polymer was isolated by evaporation of THF solvent in a vacuum desiccator. The number average molecular weight was measured by gel permeation chromatography. It was determined to be 31,500 as compared with the value of 25,500 calculated from the yield and the concentration of living ends. The polydispersity ($M\overline{w}/M\overline{n}$) was determined to be 1.20. The factors determining the breadth of the molecular weight distribution are the ratio of the time of initiation to the time of propagation and also the rate of living-end attrition. The exact cause of the attrition of living ends is not known.

The Anionic and Cationic Electropolymerization of Acrolein

This work is taken directly from the papers of Schulz and Strobel [152, 153].

INTRODUCTION

Acrolein can polymerize in two ways,

to give either a vinyl-type polymerization (**II**) or a polymerization through the aldehyde function (**I**).

APPARATUS [152]

The electrolysis cell (see Fig. 11.4) is similar to the electrodialysis cell described by Thiele [28, 39]. The anode and cathode compartments consist of thick-walled galss cylinders, 3.3 cm in diameter with standard ground-glass joints and stopcock at the side. A glass frit (G4) served as a diaphragm separating the compartments. These cylinders were held between two chromium-plated brass plates that were thermostated and connected to the power source. A platinum sheet serving as electrode was placed on each plate. The magnetic stirrers on both electrodes provided for good agitation and cleaning of the surface. Stopcocks were provided for pressure equalization between both compartments, and thermometers (or self-sealing caps for sample withdrawal) were placed on the ground-glass joints. The volume of both the anode and cathode compartments was 25 ml. Each electrode surface was 8.5 cm^2 and the

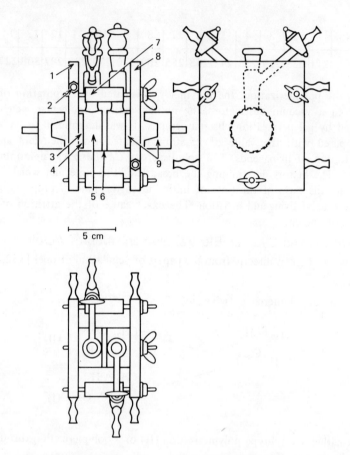

Fig. 11.4. Electrolysis cell [152,153]. 1. Chromium-plated brass plate. 2. Connection for the thermostating liquid. 3. Platinum sheet. 4. Gasket. 5. Glass cylinder. 6. Sintered glass frit. 7. PVC insulator. 8. Current terminal supply. 9. Stirring magnet.

distance between the electrodes was 5.6 cm. A galvanostat served as power source, being compensated for resistance changes in the cell so that the electrolysis proceeded at a constant current. Control was accomplished as follows: The electrolytic cell and a variable resistance were connected in series. The desired value was preselected on an amperometer that served as a two-way control. When the preselected value was exceeded, or when the current dropped below this value, a servo-motor changed the variable resistance so that the total resistance in the circuit remained constant. The voltage could be varied between 1 and 350 V, the current between 0.1 and 200 mA. The current remained constant within ±1.5%. Resistance changes in the cell of up to 10 kΩ were compensated for automatically. This range could be extended to 100 kΩ by manual control. To determine the percentage of reaction during the electrolysis, samples of 5µl were withdrawn from the cell with a syringe, and the amount of monomer was determined by gas chromatography. Low-molecular-weight by-products could not be detected under the electrolysis conditions described. After completion of the experiment, the contents of the cathode or anode compartment were dropped into a precipitating agent and the amount of precipitated polymer was determined gravimetrically. The error in percent conversion determined this way is approximately 10% within a given series of experiments. The diffusion of the monomer thorugh the diaphragm was determined by gas chromatography and can be neglected for the polymerization times used here.

EXPERIMENTAL DESCRIPTION

Tetrahydrofuran as Solvent with Sodium Tetraphenylboride as Supporting Electrolyte [152]. As a result of preliminary experiments with different solvents and conducting salts, a 0.5% solution of sodium tetraphenylboride in tetrahydrofuran was found to be a suitable electrolyte. The acrolein concentration was always approximately 12% and was determined at the beginning of the experiment by gas chromatography. At higher concentrations the polymerization proceeded too fast and the contents of the cathode compartment solidified to a gel. Figure 11.5 shows the time/conversion curves for the polymerization at − 20, 0, and +20°C. In all cases, an induction period of approximately 7 min is observed. After this time, a relatively fast polymerization reaction starts, but temperature influences the rate only slightly. The high current yield was remarkable for this polymerization. For instance, after completion of the induction period, at 0°C and under the conditions given in Fig. 11.5, approximately 1000 moles of acrolein were polymerized per Faraday, (approximately 56 kilograms of polymer formed per Faraday). This value decreased with increasing conversion and was equal to approximately 400 moles/F at 80%. The current yield was also lower at higher current intensities. For instance, 200 moles/F at 25 mA. Values between 10 and 80 moles/F are obtained from the data reported by Breitenbach [49] for the polymerization of AN by discharge of tetraalkylammonium ions. Conversion values of 3-36 moles/F were reported by Funt and Bhadani [96] for the polymerization of styrene in DMF. The influence of current intensity on the polymerization is shown in Fig. 11.6. Under otherwise identical conditions, the induction period decreases and the rate increases with increasing current intensity. However, there seemed to be no linear relationship between the total reaction rate and current intensity. Such a linear relationship was reported by Funt and co-workers [96, 78] for the polymerization of styrene and acrylonitrile in DMF. However, in accordance with the results reported by Funt, the monomer decrease was described by a first-order reaction, at current intensities between 1.3 and 25 mA, and at

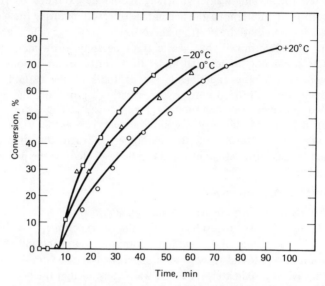

Fig. 11.5. Time-conversion curves for the electrochemical polymerization of acrolein in the catholyte at various temperatures. Electrolyte: 12% solution of Acrolein in THF with 0.5% sodium tetraphenylboride. Current strength: 1.3 mA.

conversions of up to 90% (see Fig. 11.7). If the current was switched off after approximately 50% conversion, the polymerization was not immediately stopped, but continued for some time. Under the conditions given in Fig. 11.5 at 0°C, the degree of conversion increased from 51 to 76% within 90 min after the current was switched off. On the other hand, no polymerization occurred without the diaphragm separating anolyte from catholyte.

In order to obtain information on the polymerization mechanism, inhibitors were added. The presence of methanol prevented the formation of polymers, only small amounts of oily products being formed. If m-dinitrobenzene was added to the electrolyte, the contents of the cathode compartment assumed a red-violet color after the current was switched off. This coloration was probably due to formation of a radical anion by one-electron transfer to m-dinitrobenzene:

NO₂

m-Dinitrobenzene   Red-violet
Radical anion

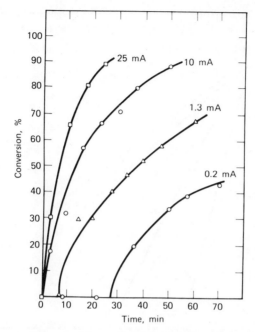

Fig. 11.6. Time-conversion curves for the electrochemical polymerization of acrolein in the catholyte at various current strengths. Electrolyte: 12% solution of acrolein in THF with 0.5% sodium tetraphenylboride. Temperature: 0°C.

It was concluded from these inhibition experiments that the polymerization proceeded by an anionic mechanism. The structure and properties of the polyacroleins also permitted conclusions with regard to the mode of their formation to be made. The polymers formed in the cathode compartment at 0°C and 1.3 mA were soluble in THF and dioxane. The molecular weights as determined by osmotic pressure methods were between 5,000 and 10,000. After storage, especially in the presence of air and humidity, the polymers become insoluble after some days. The soluble polymers soften at approximately 100°C. The IR spectra of the polymers formed at 0°C or at −20°C contain only weak carbonyl bands at 5.8 μ, but the typical absorption bands of the vinyl groups are observed at 3.25, 10.2, and 10.7μ. Such observations were in agreement with those made of polyacroleins formed in the presence of metallic sodium or sodium cyanide in THF at −50°C. This showed clearly that the polymers consist largely of unit I and that they are formed by an anionic mechanism. Electrolysis at higher temperature yielded polyacroleins containing larger portions of unit II. The IR spectra of these polymers showed a strong carbonyl band at 5.8 μ with the vinyl bands almost disappearing. Similar polymers were also formed in the polymerization initiated with cyanides above

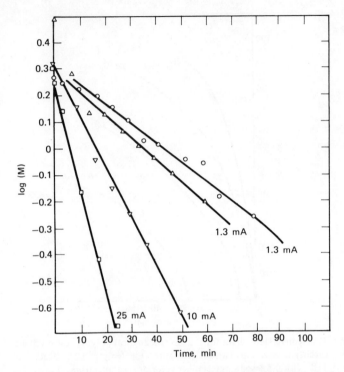

Fig. 11.7. Decrease in acrolein concentration in the catholyte during electrochemical polymerization. Electrolyte: 12% solution of Acrolein monomers in THF with 0.5% sodium tetraphenylboride. □ 25 mA, 0°C; ▽10 mA, 0°C; △ 1.3 mA, 0°C; ○ 1.3 mA, 20°C.

0°C. The current intensity produced no noticeable effect on the IR spectra, although at higher current intensities polymers were formed that became insoluble either during the electrolysis or during the work-up procedure. The inhibition experiments as well as the structure of the polymers showed that this is an anionic polymerization presumably initiated by metallic sodium formed at the cathode:

$$Na^+ + e \rightarrow Na$$

$$Na + CH_2 = \underset{\underset{H}{\overset{|}{\underset{|}{C}=O}}}{\overset{|}{C}H} \rightarrow polymeric\ products$$

Nitromethane as Solvent with Lithium Perchlorate as Supporting Electrolyte [153]. The initial monomer concentration was 2 mole/liter and the lithium perchlorate was 3 wt %. The estimation of percentage conversion, determined from the monomer consumption, was made by removing samples from the cell and analyzing gas chromatographically. At the end of the polymerization reaction the polymer was isolated by precipitation.

During the electrolysis, the reactants in the anolyte took on a red-brown coloration, and at about 60% conversion gelation occurred and the current ceased flowing. The catholyte remained colorless and lithium metal plated out on the cathode, little or no polymerization occurring.

Experiments were carried out at temperatures of between $-20°C$ and $+40°C$ at currents between 5 and 60 mA.

At low temperatures and small current densities, polymerization started after an induction period of about 20 min. The temperature had no effect on the rate of polymerization or the structure of the polymers (cf the anionic case). The rate of polymerization was independent of the monomer concentration for standard conditions, and proportional to the current density (see Fig. 11.8).

The current efficiency is 22-29 mole/F at current densities between 5 and 40 mA, and is considerably lower than in the previous (anionic) example.

Polyacroleins derived from these experiments were insoluble in organic solvents. From viscosity measurements taken in aqueous sulphuric acid solution, a degree of polymerization ($\overline{DP}$) of between 10 and 20 was found, lower than for the anionic polymers ($\overline{DP}$ approximately 100-200). From IR spectra it was inferred that polymerization had occurred at the olefinic bond to give predominantly structure II. The structure is similar to that found in polyacroleins initiated with Friedel-Crafts catalysts.

3 ELECTROINITIATED POLYMERIZATIONS

Perhaps the earliest mention of electroinitiated polymerization appeared in an article published at the turn of this century by Szarvasy [8]. No further results appeared until 1947, when a Ph.D. thesis by Rembold [31] mentioned the polymerization of methyl methacrylate by electrolytic initiation. In 1949 Wilson [2] made reference to the polymerization of acrylic acid while studying the electroreduction of various compounds.

In 1960, Fioshin and Tomilov published the first review of the subject [32], followed, in 1964, with surveys by Wilson [33], Ungureanu [34], and Chin-Yung Wu and Chang-Ya Hu [35]. Funt [36], Friedlander [37], Chao-Hua Chou, and Ching-Chen Li [38], Arnaud [40, 41], and Raff [42] all published their reviews in 1966. In 1967 Laborie [43], Asahara and Seno [44], and Ungureanu and Simionescu [45] surveyed the field, while a year later Tidswell [46] did likewise. Yamazaki [47], in an interesting article published in 1969, made mention of much of the current work being carried out in his laboratories and elsewhere. The review by Yoshizawa et al. [48] was the most recent available at the time of writing. An excellent survey by Breitenbach appeared

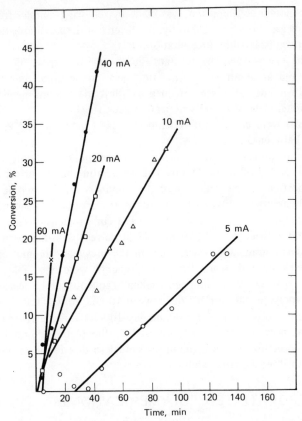

Fig. 11.8. Yield as a function of time in the electrochemically initiated cationic polymerization of acrolein at various currents. Acrolein $2M$ in nitromethane; 3 wt % $LiClO_4$. Reprinted with permission, from Ref. 153.

while this work was in the proof stage [203].

In various papers, there have been introductory summaries that, while not being full reviews, have nevertheless been very helpful in covering the earlier work. Among these are the papers of Breitenbach and Srna [49], Anderson [50], and some others. In 1969 Baizer [5] published a general review on trends in organic electrochemistry with general reference to electropolymerization reactions.

It is not intended here to cover the related fields of electric field polymerization and electric discharge polymerization. However, in 1969 Ise [4] published a survey on electric field polymerizations and, again in 1969, Hay published a review [3] on electric discharge polymerization. In this section it is hoped to cover all published work on electroinitiated polymerization to date. This will mean that a certain amount of the following section will duplicate previous reviews. However, it is thought that a complete review would be useful for comparative purposes and may serve to give a useful overall impression of the scope of electrically initiated polymerization reactions. It is not the intention of the authors to discuss deeply any work at this stage but rather to attempt to present a complete factual account of all published work. A more discursive treatment of various aspects of this subject will be contained in later sections.

Each monomer or group of monomers will be reviewed in turn.

Methyl Methacrylate (MMA)

In the first full paper published on electroinitiated polymerization Wilson and his co-workers in 1949 [51] reported the polymerization of MMA initiated by cathodically generated hydrogen. The reaction medium was aqueous sulphuric acid. In 1950 a note by Das and Palit [52] described the polymerization reaction in a solution of propylene glycol, with sodium acetate as supporting electrolyte. The polymerization occurred at the cathode, and gelation was reported. In 1951, Parravano [53] carried out the polymerization of MMA in aqueous sulfuric acid using a double cell. The two electrode compartments were separated by a sintered glass disk. Polymer yields were correlated with the hydrogen over-voltage at the various cathodes tested, and the polymerization was considered to occur through a radical mechanism initiated by cathodically generated hydrogen. Kern and Quast [54] confirmed Parravano's hydrogen overvoltage correlations and also noted that oxygen appeared to inhibit the polymerization or initiation reaction. In 1960 Smith and Gilde [55] utilized the Kolbe electrolysis of potassium acetate to produce free radicals in the anode compartment of a double cell:

a. Electrolytic step, Kolbe electrolysis

$$CH_3-\underset{\underset{O}{\parallel}}{C}-O^- \quad \xrightarrow[\text{Oxidation}]{-e} \quad CH_3-\underset{\underset{\underset{CH_3^{\cdot} + CO_2}{\downarrow}}{\underset{O}{\parallel}}}{C}-O^{\cdot}$$

These radicals were found to be capable of initiating MMA polymerization in aqueous solution.

b. Initiation of monomer by methyl radical

$$CH_3^{\cdot} + CH_2 = \underset{\underset{COOCH_3}{|}}{CH} \longrightarrow CH_3-CH_2-\underset{\underset{COOCH_3}{|}}{CH}^{\cdot}$$

Monomer radical
(initiating fragment)

c. Polymer propagation reaction

$$CH_3-CH_2-\underset{\underset{COOCH_3}{|}}{\overset{\cdot}{C}H} + n\,CH_2 = \underset{\underset{COOCH_3}{|}}{CH} \longrightarrow CH_3-[CH_2-\underset{\underset{COOCH_3}{|}}{C}H]_{\overline{n}}CH_2-\underset{\underset{COOCH_3}{|}}{\overset{\cdot}{C}H}$$

$$R-(M)^{\cdot}_{n+1}$$

Growing polymer radical

d. Polymer Termination

(1) $2R-(M)^{\cdot}_{n+1} \longrightarrow R-(M)_{\overline{2n+2}}R$ Dimerization

(2) $R-(M)^{\cdot}_{n+1} + SH \longrightarrow R-(M)_{\overline{n+1}}H + S$. Abstraction from SH

where SH could be polymer, monomer, solvent or some adventitious impurity which may or may not have been electrochemically generated.

(3) $2R-(M)^{\cdot}_{n+1} \longrightarrow R-(M)_{\overline{n+1}}H + R-(M)^{-}_{n}-CH = \underset{\underset{COOCH_3}{|}}{C}H$ Disproportionation.

They used ^{14}C labeled acetate ions to determine whether acetoxy or methyl radicals were responsible for the initiation of polymerization. However, their results were queried from an experimental point of view by Funt [36], who noted that the number of initiator fragments per polymer molecule was excessively large. Fedorova, Shelepin, and Moiseeva [56], using an experimental design similar to that of previous workers, reported a critical dependence of polymerization yield on oxygen concentration. When there was no oxygen in the anode area, or when the oxygen concentration was greater than 1.25×10^{-4} M, polymerization was completely inhibited. In work comparable to that of Parravano, Tsvetkov [57] investigated the polymerization of MMA at the cathode in an aqueous medium. The interesting feature of this work is the report that stirring decreases the rate of polymerization and also lowers the molecular weights. To explain these phenomena, it was suggested that an increase in the stirring rate would be likely to cause an increase in the rate of termination of the polymer chains. Tsvetkov also noted that increase in cathode current density, temperature, and time all tend to accelerate polymerization. The molecular weights of his polymers were found to increase with increasing current density. In 1962 Breitenbach and Srna [49] initiated the polymerization of MMA by free radicals generated anodically in an acetic acid-acetic anhydride solution saturated with lithium acetate. In the same year, Funt and Yu [58] published the first account of the homogeneous polymerization of MMA, in a medium of high dielectric constant. They discussed the earlier literature reports of low-molecular-weight polymers and heterogeneous reactions that tended to coat electrodes and make a study of the kinetics of polymerization difficult if not impossible. They employed a single cell in most experiments. This paper is also noteworthy for the first report of preelectrolysis for the removal of chemical impurities, and for the systematic attempts to correlate polymer yield with such factors as supporting electrolyte type, supporting electrolyte concentration, solvent, and electrode material. Among the salts tried, aluminum trichloride, zinc chloride, and zinc acetate proved the most useful. Dimethylsulfoxide (DMSO), dimethylformamide (DMF), and hydroxypropionitrile proved the most suitable solvents, while of the various electrode materials tried, under standard conditions, carbon and platinum were the most successful. Polymer molecular weights were high, frequently greater than 100,000. In a kinetic study of the MMA polymerization in DMSO with zinc acetate as supporting electrolyte and using graphite electrodes, a definite relationship was noted between the current impressed and the polymer yield. Equally important perhaps was the result that the initial monomer concentration was a considerable factor in the eventual polymer yield. The polymerization was reported as being free radical in nature. Fedorova, Kuo-Tung, and Shelepin [59] continued their studies on the effect of oxygen on the polymerization of MMA in saturated acid solution. They checked the hydrogen overvoltage in the presence and absence of monomer at a lead electrode, using polarographic techniques and also noted that there was no polymerization occurring at a lead cathode after current passage of from 6 to 20 hr in 1.5 M H_2SO_4 in the absence of oxygen. However, when oxygen was present under the same conditions, polymer was formed and the addition of hydrogen peroxide accelerated polymer formation. In 1964, Trifonov and Panayotov [60] attempted the polymerization of MMA in the presence of anthraquinone. Semiquinones inhibit free-radical polymerization, and thus anionic polymerization alone was expected to take place. When electrolysis was performed, the cathode solution turned red or purple and the presence of a free radical was indicated by its ESR (electron spin resonance) spectrum. When MMA was added to the colored solutions, polymers were obtained. However, copolymerization reactions indicated that the mechanism was not anionic but free radical in nature. This work was carried out in DMF solutions of either tetraethylammonium iodide salts or lithium chloride at a mercury cathode and a platinum anode. In the same year, Tsvetkov and Glotova [61] published further work on the cathodic hydrogen initiation of MMA in aqueous solution. They noted that the addition of sodium chloride to the solution increased the yield of polymer while an increased hydrogen chloride concentration decreased both the yield and the molecular weight of the polymer. They pointed out that the efficiency of polymerization was very much governed by the solubility of the monomer in the reaction medium. Epstein and Bar-Nun [62] reported the grafting of MMA to a cellophane sheet in aqueous sulphuric acid solution. They employed a mercury cathode and platinum anode although the cellophane sheet was not attached to either electrode. Shelepin and Fedorova [63] continued with their detailed studies of MMA polymerization at lead electrodes while Shelepin, Frumkin, Fedorova, and Vasina [64] discussed double layer structures in the electrochemical initiation of MMA polymerization. Tsvetkov and Koval'chuk [65] investigated the polymerization of MMA in aqueous solutions of phosphoric or sulphuric acids. The polymerization was initiated by the passage of alternating current, and the authors noted that the platinum electrodes gradually eroded and postulated that the (alternating) anode oxidized to PtO_2, which then reacted with the monomer to produce monomer peroxides and "catalytic" platinum, thus promoting the initiation of the polymerization reaction. At the cathode, hydrogen atoms were formed which would also initiate the monomer. Funt and Bhadani [66] carried out polymerization reactions in DMF with quaternary ammonium salts as supporting electrolytes. They reported polymer molecular weights of 50,000 and noted that the rate of polymer formation was directly proportional to the initial monomer concentration and to the square root of the impressed current. From inhibitor studies and copolymerization studies, they inferred that the mechanism was anionic, the first time this had properly been noted. Using a platinum anode and a mercury cathode, Tidswell [67] polymerized MMA in an acetic acid-acetic anhydride solution with sodium acetate as electrolyte. He reported a direct relationship between current and polymer yield, in agreement with the work of Funt and Yu [58]. He also noted that in experiments carried out with different concentrations of electrolyte, there was little effect on the polymerization rate. The polymerization occurred at the cathode. Tsvetkov and Koval'chuk [68, 69] studied the kinetics of the anodic polymerization of MMA in glycol-water solutions. The effect of water concentration on polymer yield was obtained. Both the reaction rate and the molecular weight passed through a maximum when the water content of the glycol solution was varied. Polymer yield and molecular weight also depended on the current density. Maznichenko and Fedorova [70] discussed the polymerization reaction at a dropping mercury electrode and the effect of various supporting electrolytes on the adsorption of polymer on the mercury drop. Lithium sulphate, caesium chloride, calcium chloride, and barium chloride prohibited adsorption, but magnesium sulphate gave magnesium hydroxide on reaction at the drop surface which promoted adsorption. Breitenbach, Olaj, and Sommer [71] carried out the anodic polymerization of MMA in DMF solution with alkali metal acetates. Yamazaki, Tanaka, and Nakahama [72] published the half-wave potentials for the reduction of MMA in dimethoxyethane (DME) solution.

Acrylonitrile (AN)

It is worth noting here that AN and MMA have been the two monomers most completely studied to date.

In 1951 Kolthoff and Ferstandig [73] published the first paper on the electro-initiated polymerization of AN. They employed a frit-separated cell, in one compartment of which was a standard calomel electrode (SCE). In aqueous sulphuric acid solution, they attempted to reduce various peroxides such as hydrogen peroxide, potassium persulfate, and cumene hydroperoxide. However, this appeared to fail and did not provide much AN polymer (PAN). Nevertheless, the use of a ferrous-ferric iron redox couple with hydrogen peroxide provided a good yield of polymer, according to the following mechanism:

Ferric-ferrous iron redox couple in the presence of peroxides (ROOR')

$$Fe^{3+} + e \xrightarrow{\text{Cathode}} Fe^{2+}$$

Ferrous ions then reduce peroxides

$$Fe^{2+} - ROOR' + \longrightarrow Fe^{3+} + RO^{\cdot} + \overset{\ominus}{O}R' \longrightarrow \text{may initiate polymerization}$$
$$\big\downarrow +e$$
$$RO^{\ominus} \text{ (and other reactions)}$$

They used a rotating carbon cathode that did not coat with polymer as did an ordinary (stationary) platinum electrode. In 1952 Goldschmidt and Stockl [74] attempted the anionic polymerization of AN using acetic acid and sodium acetate. However, the yields were very poor. Kern and Quast [54], in their 1953 publication, reported the initiation of polymerization by cathodically generated hydrogen. In that same year Friedlander, Schwan, and Marvel [75] failed to produce PAN during the electrolysis of potassium laurate. Breitenbach and Gabler [76] initiated the polymerization of acrylonitrile by the cathodic discharge of tetralkyl ions. In a 10^{-2} M solution of tetraethylammonium perchlorate in anhydrous AN, they found that, irrespective of the electrode material, PAN, which is insoluble in acrylonitrile, was formed at the cathode. They suggested that the ethyl radical initiated the polymerization, but in their next paper [77] disproved this thesis. Breitenbach, Srna, and Olaj [77] investigated the electroinitiated polymerization of AN, carrying out copolymerization studies that strongly indicated the existence of an anionic mechanism. Sisters Murphy, Carangelo, Ginaine, and Markham [21] attempted to correlate the polarographic reduction of AN with various concentrations of water present in solution. They also were able, after a fashion, to control the cathodic mercury potential of the reaction. In 1962 Breitenbach and Srna [49] produced PAN with a degree of polymerization of about 20 at the cathode, by electrolyzing a solution of tetralkylammonium salt in pure AN monomer. From molecular-weight studies they were able to prove that four polymer molecules were formed per primary act at the cathode. They suggested that this could probably be explained by the effect of chain transfer in the reaction. (A fuller account of the significance of chain transfer is found in Section 5 under the heading "Free-Radical Polymerizations" (p. 648). Another equally important feature of these experiments was that the number of electrochemical equivalents of salt converted was several orders of magnitude higher than the initial amount of salt present. There must, therefore, have been an efficient regeneration of salt ions in some way. With a mixture of equal volumes of AN and methanol, polymerization was completely suppressed. The kinetics of AN polymerization was studied in a DMF solution and showed that up to 330 moles of monomer were polymerized per initial mole of salt. However, it was stressed that the rate of polymerization was very much affected by the presence of small amounts of water and only had poor reproducibility. In the same paper, polymerization was effected by the anionic generation of free radicals in an acetic acid-acetic anhydride solution saturated with lithium acetate. High-molecular-weight polymers were formed in this reaction, in contrast to those formed in the previously discussed experiments. In 1964, Trifonov and Panyotov [60] prepared PAN at the cathode in DMF solutions with either tetraethylammonium iodide or lithium chloride as supporting electrolytes. In the solution anthraquinone was also present. This entity did not cause the monomer to polymerize ionically, as expected, but rather by a free-radical mechanism. In DMF solutions with sodium nitrate as supporting electrolyte, Funt and Williams [78] polymerized AN at the cathode. They suggested that the mechanism was anionic, with direct electron transfer to the double bond:

$$\underset{\underset{CN}{|}}{C}=CH \;+\; e \;\; \xrightarrow[\text{electrode}]{\text{at}} \;\; \overset{\cdot}{\underset{|}{C}}-\underset{\underset{CN}{|}}{C}{}^{\ominus}$$

Radical anion

followed either by:

(1) dimerization and pure anionic polymerization,

$$2(\cdot\overset{|}{\underset{|}{C}}-\overset{|}{\underset{|}{C}}{}^{\ominus}) \longrightarrow {}^{\ominus}\overset{|}{\underset{\underset{CN}{|}}{C}}-\overset{|}{\underset{|}{C}}-\overset{|}{\underset{\underset{CN}{|}}{C}}-\overset{|}{\underset{|}{C}}{}^{\ominus} \qquad \text{Dimeric dianion}$$

$$\Big\downarrow + \text{Monomer}$$

$$\underset{\underset{CN}{|}}{-(CH_2-CH)_n-} \quad \text{Polymer (PAN)}$$

or by

(2) both radical and anionic propagation,

$$\overset{\displaystyle .}{\underset{\underset{CN}{|}}{C}}-\overset{\ominus}{\underset{|}{C}} \quad + \quad \text{monomer} \quad \longrightarrow \quad \text{polymer} \quad \text{(PAN)}$$

The polymer yield was also compared with various salts used and molecular weights were of the order of 4000. Because of chain transfer, three polymer molecules were found formed per electron, which compared well with Breitenbach's data [49]. In 1967 Simionescu and Ungureanu [79] used a ferrous-ferric ion redox couple with potassium persulfate in 1 M sulfuric acid solution to graft AN on to a cellulose film. They noted that the grafting yield increased with current density and monomer concentration. This paper also reported the thermodegradation of the cellulose-AN film. Shapoval [80] used a 20% solution of AN in DMF with sodium acetate as supporting electrolyte to produce about 40% polymerization of the solution. Benzene, chlorobenzene, toluene, and hexane were added, and it was noted that the polymer yields decreased. Benzene caused the greatest decrease with a 25% reduction in polymer yield. The smallest decrease was produced by the addition of hexane, and these results were cited as evidence for a free-radical process. Yamazaki [72] reported the cathodically generated preparation of PAN from DME-tetrabutyl-ammonium perchlorate solutions of the monomer. A red viscous deposit was observed around the cathode; it was found to be PAN of molecular weight between 700 and 1800. Yamazaki suggested that the initiation reaction was a direct electron transfer to the double bond followed by anionic propagation, with chain transfer to monomer also occurring. The termination was due to the interaction of the growing carbonionic end with the tetrabutylammonium countercation. In a 1968 paper, Beck and Leitner [81] polymerized AN in the presence of quaternary ammonium salt and water at a pH of 9.5. They noted that increasing the concentration of water above 1% reduces the polymer yield. The polymer was colored and had a molecular weight of between 1,000 and 3,000. The use of quaternary phosphonium salts as electrolytes produced no polymer. Various cathode materials were used but appeared to have little effect upon the yield. This work was covered in a Beck and Leitner patent issued in 1968 [82]. Again, in 1968, Sommer, Breitenbach, and Olaj [83] investigated the cathodic polymerization of AN in quaternary ammonium salts and phosphonium salts. Chapiro and Herychowski [84] investigated the electroinitiated polymerization of AN using a free-radical mechanism. They used electrolysis solutions of AN in DMF in the presence of sodium nitrate. Ungureanu, Ungureanu, and Simionescu [85] polymerized AN anionically using alkali persulphates in DMSO. They considered that their results were similar to those reported in Funt and William's paper of 1964, and produced polymers with molecular weights of rather less than 5000. Olaj, Breitenbach, and Buchburger [86] reported that the cathodic initiation of AN was greatly affected by lowering the water concentration to less than 10^{-3} molar. They noted that the anionic mechanism was retained, but there were no living polymers. However, the polymer was white instead of yellow. Changing the quaternary ammonium salts or using phosphonium salts instead, caused large variations in yield and molecular weight. They studied the influence of the counterion on the rate of polymerization and noted that moderate water concentrations caused these effects to disappear. Mengoli, Farnia, and Vianello [87] in 1969 generated radical ions of nitrobenzene in DMF in the presence of AN monomer. The following mechanism was proposed:

$$Ph-NO_2 + e \longrightarrow Ph-NO_2^{-}$$

The nitrobenzene radical anion then abstracted a proton from the monomer (AN):

$$Ph-NO_2^{-} + CH_2=\underset{CN}{\overset{|}{CH}} \longrightarrow Ph-\dot{N}O_2H + CH_2=\underset{CN}{\overset{|}{C^-}}$$

$$\textbf{(I)} \qquad\qquad \textbf{(II)}$$

I then was said to decay to nitrobenzene and phenyl hydroxylamine: **II** propagated by an anionic process.

The possibility of direct electron transfer from the nitrobenzene radical anion to monomer followed by polymerization of the monomer was considered but was discarded in the basis of the experimental stoichiometric ratio of nitrobenzene to phenyl hydroxylamine, which was as predicted from a dismutation of the protonated species I. These radicals were not reactive towards MMA or styrene monomers but produced PAN of molecular weight 1200 in the presence of AN monomer.

The same year, Mengoli and Vidotto [10] produced benzophenone radical anions electrochemically, in DMF solutions with tetraethylammonium perchlorate as electrolyte, and studied their decay with the monomers AN and methacrylonitrile. High yields of PAN and polymethacrylonitrile were reported for radical anion concentrations of 2×10^{-2} mole liter. The kinetics of the radical anion decay were discussed in detail. Asahara, Seno, and Tsuchiya [88] discussed the effects of supporting electrolytes on the electrolytic polymerization of AN. In two interesting papers published in 1969, Tsvetkov and Koval-chuk [89] investigated the anion in polymerization of AN in sulphuric acid solution and also [90] reported the electrolysis of perchloric acid solutions in the presence of AN. Again, in 1969, Kikuchi [91] reported the polymerization of AN using both a direct current and a rectangular wave form current. His polymers had varying degrees of coloration, due in part to nitrile group intra-molecular interaction, depending on the magnitude of current used.

PAN

Intramolecular cyclic structure

The reaction was said to be "catalyzed" by an electron, the color being due to the C=N—H, iminonitrile, groups in the cyclic structure. In two patents published in 1966 and 1968, Baizer [92] polymerized AN to give low-molecular-weight oligomers by a reductive coupling mechanism. A Roumanian patent [93] issued in 1969 described a technique of polymerizing AN at relatively high current densities (300-500 mA/cm^2) in aqueous H_2SO_4 solution. The polymer was formed in the anode compartment in high yield and at a high molecular weight (400,000) and the initiating species was reported as being a hydroxyl radical. Kaneko [20] in 1970 reported the heterogeneous polymerization of AN in 1-10 vol % hexamethylphosphoramide solutions employing lithium chloride as supporting electrolyte. Initiation was considered to be via direct electron transfer.

Styrene and α-Methylstyrene

In 1951 Kolthoff and Ferstandig [73] reported their failure to polymerize styrene in aqueous solution by redox coupling techniques. Styrene was insoluble in the reaction medium, but even when solubilized in detergents to form an emulsion, it produced no polymer. Goldschmidt and Stöckl, a year later [74], produced very poor yields of styrene in acetic acid solution. This may be regarded as the first styrene polymer produced. Again, in 1953, Friedlander, Schwan, and Marvel [75] could produce no polymeric styrene during the electrolysis of potassium laurate. Yang, McEwen, and Kleinberg in 1957 [94] produced styrene at the cathode with molecular weights of between 2000 and 3000. They employed sodium iodide as electrolyte in anhydrous pyridine solution with magnesium electrodes. In a frit-separated cell, they produced no polymer in the anolyte but produced poor yields of polymer in the catholyte. They claimed that the polymerization was due to direct electron transfer since they were unable to find nitrogen in the polymers so formed. However, this would require that the reduction of sodium be at a greater negative potential than that of the olefinic bond for direct electron transfer to styrene. Breitenbach and Srna [49] produced polystyrene initiated by free radicals generated at the anode in acetic acid-acetic anhydride and lithium acetate solutions. The molecular weights of the polymers were low. In nitrobenzene solution, they were able to polymerize styrene cationically, though once again the molecular weights were very low. In similar experiments with α-methylstyrene, only very low-molecular-weight polymers were formed. In 1962, Smith and Manning [95] reported their inability to produce polymer using a dilute aqueous solution of potassium acetate. They suggested that appreciable water solubility was necessary to sustain polymerization. Funt and Bhadani [96] produced the first high-molecular-weight polymers of styrene (molecular weight 15,000-50,000). They employed DMF solutions of tetramethylammonium chloride and generated these polymers at the cathode by an anionic mechanism. They noted that methanol quenched the reaction. Polymer formation showed a first-order dependence on the impressed current and a first-order dependence on the initial monomer concentration. There was very much less chain transfer to monomer than in the AN case published in the same year by Funt and Williams [78]. Tsvetkov and Glotova [61] polymerized styrene in methanolic solution by cathodically generated hydrogen. Funt and Laurent [97] produced low-molecular-weight polystyrene (2000-5000 molecular weight) at relatively high electrical efficiencies using DMF solutions of alkali metal salts. They noted that the rate of polymerization, based on yield in a given time, increased with the impressed current and the initial monomer concentration. The type of supporting electrolyte employed also had a strong effect on the reaction.

KNO_3 proved the most advantageous in terms of the largest polymer yield for standard conditions. An anionic mechanism was proposed, but no reason was put forward for the preeminence of KNO_3 as supporting electrolyte, polarographic experiments being inconclusive on this point. In a very important paper published in 1965, Yamazaki, Nakahama, and Kambara [98] produced living poly(α-methylstyrene) in tetrahydrofuran (THF) solutions of sodium tetraethylaluminum salts and lithium aluminum hydride. They suggested that the initiation of the monomer was due to direct electron transfer to the double bond at the cathode:

$$CH_2=\underset{\underset{C_6H_5}{|}}{\overset{\overset{CH_3}{|}}{C}} + e \xrightarrow[\text{cathode}]{\text{at}} CH_2-\underset{\underset{C_6H_5}{|}}{\overset{\overset{CH_3}{|}}{C}} \overset{\ominus}{\cdot} \quad Na^{\oplus}$$

Styrene   Radical anion (I)

$$2 \text{ I} \longrightarrow \text{Na}^{\oplus} \ ^{\ominus}\underset{\underset{C_6H_5}{|}}{\overset{\overset{CH_3}{|}}{C}}-CH_2-CH_2-\underset{\underset{C_6H_5}{|}}{\overset{\overset{CH_3}{|}}{C}}^{\ominus} \ \text{Na}^{\oplus}$$

Dimeric dianion (II)

II + Monomer ⟶ "Living polymer" (III)

III ⟶ Termination by the addition of protic impurities such as methanol or water.

However, this idea was retracted in a later paper, the suggestion being that sodium cations were reduced at the cathode to sodium metal, which then initiated monomer polymerization in the normal way:

$$\text{Na}^{\oplus} + e \ \xrightarrow[\text{cathode}]{\text{at}} \ \text{Na Metal}$$

$$\text{Na metal} + CH_2{=}\underset{\underset{C_6H_5}{|}}{\overset{\overset{H}{|}}{C}} \longrightarrow \text{I, radical anion}$$

Yamazaki published a more comprehensive paper in 1967 on the same subject [99]. In 1966, Funt, Richardson, and Bhadani [100] initiated the polymerization of polystyrene in THF solutions of sodium tetraethylaluminum or sodium tetraphenylboride. They also added naphthalene to this solution and, without monomer, produced the green sodium naphthalene complex by electrolysis that itself initiated the polymerization of added monomer. When styrene monomer was added, living polystyrene anions were formed and were observed spectrophotometrically. In another paper [101, 102], Funt, Bhadani, and Richardson used THF solutions of sodium tetraphenylboride, potassium tetraphenylboride, and sodium tetraethylaluminum with α-methylstyrene. They determined that the number of electrons transferred at the cathode was strictly proportional to the absorbance of the living poly(α-methylstyrene) solutions and found that by reversing the polarity of the impressed current they were able to destroy quantitatively the living poly(α-methylstyrene) anions. This feature of electrolytically generated living polymer solutions is of great importance and will be discussed further. Anderson in 1958 [50] highlighted some of the outstanding questions about electroinitiated living anion polymers. He posed three questions. First, what solvents are satisfactory? Second, what electrolytes can be used? Finally, what is the mechanism of initiation? He noted first that he had been able to use hexamethylphosphoramide (HMPT) as a solvent but that the other solvents used by Yamazaki and Funt were quite satisfactory. The electrolyte question was also discussed and was shown to be of critical importance when considering the answer to the third and final question, that of the type of initiation mechanism. In a polarographic study, the reduction potentials of the sodium cation from sodium tetraphenylboride styrene were determined using a rotating platinum electrode in a DMF solution of 0.2 N tetrabutylammonium iodide as supporting electrolyte. The half-wave potential of the sodium ion was found to be -2.2 V, whereas that of styrene was -2.3 V (both with respect to the SCE). From these experiments, Anderson indicated that the expected method of initiation was via electron transfer from cathodically generated sodium. However, using tetramethylammonium ion, which reduces at a cathodic potential of -2.7 V under the polarographic conditions previously described, it seemed feasible that the initiation of polymerization should be due to direct electron transfer to the olefinic double bond of styrene. He supported these speculations with electroinitiated polymerization experiments on polystyrene. Yamazaki, Tanaka, and Nakahama [72, 103] published the results of polarographic studies on various monomer types. They also reported the polymerization of various of these monomers in tetrabutylammonium perchlorate solution in DME. Cathodically initiated polymerizations were performed using controlled-cathode potentials, at which the monomer alone could be reduced. In 1970, Funt and Richardson [104] produced living poly anions generated by the passage of an electrolytic current through solutions of THF with sodium tetraphenylboride as the supporting electrolyte. The concentration of living ends could be altered at will by varying the magnitude of the current, the time of passage of current, and the polarity of the current. Since the living-end concentration is a major factor in the final molecular weight of the polymer, so produced, it was found possible to produce predetermined molecular weight distributions by such techniques. Funt and Blain [105] performed detailed kinetic studies on the electroinitiation of styrene in methylene chloride as solvent with tetrabutylammonium perchlorate as supporting electrolyte. A green coloration in the solution supported the idea that perchlorate radicals formed at the anode oxidize the monomer to a radical cation:

$$ClO_4{}^{\ominus} \xrightarrow{\text{at anode}} ClO_4{}^{\cdot}$$

$$CH_2{=}C\underset{C_6H_5}{\overset{H}{\Big\langle}} + ClO_4{}^{\cdot} \rightarrow \overset{\cdot}{C}H{-}C\underset{C_6H_5}{\overset{H}{\overset{\oplus}{\Big\langle}}} \quad ClO_4{}^{\ominus}$$

Styrene   Green color, radical cation

Mengoli and Vidotto [190] recently reported that in nitrobenzene at $-10°C$ the cationic reaction exhibited a living character.

Tidswell and Doughty [191] electrolyzed a solution of styrene and sodium borofluoride in sulfolane. A cationic polymerization of styrene was initiated by BF_3 cocatalyzed by HF and/or H_2O. When the solvent was changed to dimethyl acetamide, the locus of reaction was changed to the cathode and the reaction proceeded by an anionic mechanism.

Mengoli and Vidotto have shown that radical cations of 9,10-diphenylanthracene can be generated electrochemically, have a long lifetime, and serve as efficient initiators of polymerization [193, 194].

Alkenes and Alkynes

Kolthoff and Ferstandig [73] in 1951 noted that they were unable to polymerize isoprene using a redox couple-peroxide system at the cathode in aqueous acid solution. In some of the earliest examples taken from the patent literature on electroinitiated polymierzation, Goerrig, Jonas, and Moschel [106, 107] in 1955 and 1956 published patents describing the polymerization of tetrafluorethylene and triphenylchloroethylene at atmospheric pressures by electrolytic techniques. High polymer yields were found. In 1957, Gehrke and Fechenheim [108] polymerized ethylene in aqueous benzene solution under high pressures employing K_2TiF_6 as supporting electrolyte. Carbon electrodes were used and a fairly high yield of polymer was reported with a degree of polymerization of approximately 50. A year later Loveland [109] reported the polymerization of ethylene in a dioxane solution of tetrabutylammonium bromide. He used platinum electrodes with stilbene present in the cathode compartment of a double cell. Using a single cell, he reported that it was possible to use an alternating current source to effect polymerization. In 1959, Lindsey and Peterson [110] and Smith and Gilde [111] attempted Kolbe-type free-radical polymerizations of butadiene and isoprene. They were able to report only dimerization of the monomers. Later in 1961, Smith and Gilde [207] discussed the stereochemistry of radical addition to the double bond of the monomers at the electrode surface. Smith and Manning [95], in a study of the polymerization of acrylic acid, reported their failure to polymerize butadiene in dilute aqueous solutions of potassium acetate. Funt and Bhadani [112] studied the polymerization of isoprene in THF solutions of sodium tetraphenylboride supporting electrolyte. Passage of current through this solution produced the yellow coloration of living anions. However, on cessation of current passage, even in solutions of the highest purity, a decrease in the absorption spectra of the isoprene living anion was noted with time. In fact, the anion concentration decreased to approximately half of its original value in 2 hr. A novel technique, involving the use of a feedback mechanism based on the spectrophotometric analysis of the reacting solution, was employed to maintain the living anion concentration constant. Yamazaki, Tanaka, and Nakahama [72] studied the polymerization of butadiene, isoprene, and chloroprene in dimethoxyethane solutions of tetrabutylammonium perchlorate. Butadiene and isoprene gave only low polymers in the catholyte solution, suggesting some form of termination reaction with the growing living anion and its counter cation:

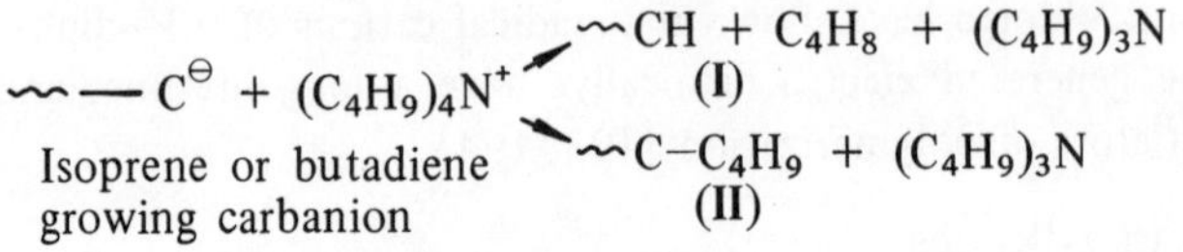

$$\sim\!\!\!-\!C^{\ominus} + (C_4H_9)_4N^+ \underset{\substack{\\ \sim\!C\!-\!C_4H_9\ +\ (C_4H_9)_3N \\ \textbf{(II)}}}{\overset{\substack{\sim\!CH\ +\ C_4H_8\ +\ (C_4H_9)_3N \\ \textbf{(I)}}}{\diagup}}$$

Isoprene or butadiene
growing carbanion

In the case of butadiene, NMR study showed the presence of **II**. However, with isoprene **II** was not formed and **I** was presumed. Similarly, with chloroprene (as well as vinylidine chloride) only low polymers were detected in the catholyte. However, in the latter case, it was thought that the propagation was via a free-radical mechanism rather than an anionic mechanism. In two patents published in 1965 and 1969, Korniker [113, 114] reported the polymerization of acetylene in acetonitrile solutions of nickel dibromide saturated with acetylene. In 1968 Yamazaki and Murai [115] attempted the polymerization of butadiene by the electrolysis of solutions of nickel (tetrakispyridine):

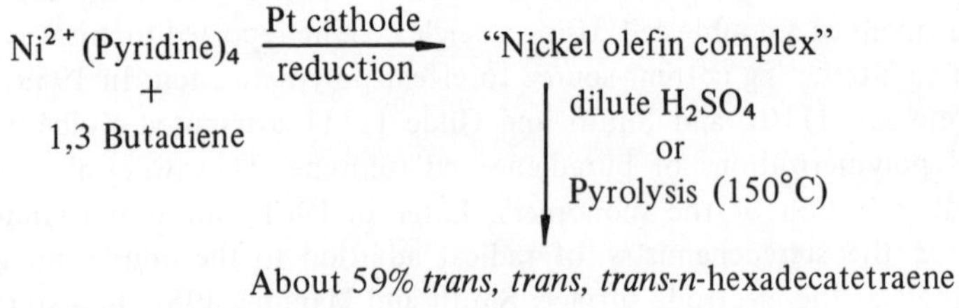

$$\begin{array}{c} Ni^{2+} (Pyridine)_4 \\ + \\ 1,3\ Butadiene \end{array} \xrightarrow[\text{reduction}]{\text{Pt cathode}} \text{"Nickel olefin complex"}$$

$$\Bigg\downarrow \begin{array}{l} \text{dilute } H_2SO_4 \\ \text{or} \\ \text{Pyrolysis } (150^{\circ}C) \end{array}$$

About 59% *trans, trans, trans-n*-hexadecatetraene

They were only able to isolate low-molecular-weight polymers (tetramers), which were partly hydrogenated.

Benzene Derivatives

In 1966, Gilch [116, 117], prepared poly(p-xylylenes) by the electrolytic reduction of α,α'-dihalo-p-xylenes at controlled cathodic potentials. The author considered that the reaction proceeded via xylylene intermediates formed at the cathode. Covitz [118] in 1967 studied the reaction more deeply and produced polarographic evidence for the formation of p-xylylene intermediates. He also noted a minor product of the reaction as being [2.2]-paracyclophane. The following reaction scheme was proposed:

$$Br-CH_2-\langle\!\!\!\bigcirc\!\!\!\rangle-CH_2Br \xrightarrow[\text{reduction}]{\text{Two-electron}} CH_2=\langle\!\!\!\bigcirc\!\!\!\rangle=CH_2 + 2Br^{\ominus}$$

α,α'-dibromo-p-xylene   p-xylylene

$$CH_2-\langle\!\!\!\bigcirc\!\!\!\rangle-CH_2$$
$$|\qquad\qquad|$$
$$CH_2-\langle\!\!\!\bigcirc\!\!\!\rangle-CH_2$$

Dimerized   Polymerized

$$-\!\!\lceil CH_2-\langle\!\!\!\bigcirc\!\!\!\rangle-CH_2\!\rceil\!-$$

5-10%   90-95%
Paracyclophane   Poly-p-xylylene

In similar studies, Ross and Kelley [119] and Ross, Peterson, and Finklestein [120, 121], in patents, were able to deposit adherent films of polyxylylenes on aluminum electrodes. They considered the reaction to be free-radical in nature.

Walker and Wisdom [122-124] in a series of interesting patents, prepared benzene polymers by electrolyzing benzene in a ternary complex solution of [RH·HX·2AIX$_3$], which consisted of an approximately 20-carbon-atom compound (RH) that was at least as basic as benzene, a hydrogen halide (HX), and aluminum trihalide (AlX$_3$). Platinum anodes were employed with aluminum or carbon cathodes and polymer was formed at the anode. [In an example with biphenyl as the monomer to be polymerized (instead of benzene) and with mesitylene as the C-20 compound in the ternary complex, a fair yield of sexiphenyl was formed.]

Sheppard and Dannels, in a paper in 1966 [125] and in a later patent [126], were able to polymerize benzene from aqueous hydrogen fluoride-benzene solutions. The polymers appeared to be formed from the interface of the benzene and aqueous hydrogen fluoride layers.

Poly Acids, α-ω Carboxylic Acids

In 1958 Smets, Poot, Mullier, and Bex [127] were able to graft vinyl monomers onto polymethacrylic acid by preelectrolyzing the polyacid solution to produce hydroperoxide side groups in a Kolbe-type reaction. These hydroperoxide-containing acid polymers were then isolated below 10°C and used for a second step polymerization with another monomer such as acrylamide. In 1964 Smets, Van Der Borght, and Van Haeren [128] and Smets and Bex [129], in a patent, electrolyzed methanolic solutions of polymethacrylic acid that had been partially neutralized. Carbon dioxide was evolved at the anode, and the degrees of decarboxylation, lactone formation, and unsaturated-group formation were studied. Crossed Kolbe-type reactions were also investigated by electrolyzing the polymeric acid solutions in the presence of various mono-carboxylic, low-molecular-weight acids. In 1969 Smets, Van Gorp, and Van Haeren [130] electrolyzed styrene in aqueous and alcoholic media with methacrylic acid-methylmethacryate and acrylic acid-methylmethacryate copolymers. Graft copolymers with as much as 15-20% by weight of styrene were thus prepared. It has long been hoped to prepare linear polymers from α-ω dicarboxylic acids in Kolbe-type electrolyses at the anode. Garrison [131] was able to report only moderate success in this respect. But a recent paper by Toy [132] reported the preparation of polyperfluoromethylene from the electrolysis of solutions of perfluoroglutaric acid with sodium in methanol. It was suggested that the mechanism involved diradical coupling at the electrode surface:

$$n-(CF_2)_3 \overset{\displaystyle COO^{\ominus}}{\underset{\displaystyle COO^{\ominus}}{<}} \xrightarrow[\text{platinum anode}]{\text{Oxidation}} n \cdot (CF_2)_3 \cdot \; + \; 2n\, CO_2$$

$$\downarrow \text{polymerizes}$$

$$-\!\!\left(CF_2\right)_{\!n}$$

Polydifluoromethylene

Caprolactam and Isocyanates

Gilch and Michael [133] in 1966 and Gilch in a 1966 patent [134] reported the anionic activated polymerization of caprolactam. A melt of caprolactam with an isocyanate as activator and an alkali salt was electrolyzed and a polyamide was formed as a coherent coating on the cathode.

In three quite recent papers, Shapoval, Skobets, and Markova [135] have reported the electroinitiation of polymerization of various isocyanates, principally phenylisocyanate, in DMF solutions of tetrabutylammonium salts. Polymerization was considered to be anionic in character with polymer formed only in the cathode compartment. A further feature of this work was the reported possibility of the stereospecific nature of the polymerization. This was shown from the x-ray analytical data that indicated the crystalline nature of the polymers.

Vinyl Amides

Shapoval and Shapoval in 1964 [135] polymerized acrylamide in aqueous solutions of potassium persulphate. They reported high yields at relatively low current densities (20-90 $\mu A/cm^2$) at platinum electrodes. Sobieski and Zerner [in a U.S. patent, 136] polymerized acrylamide and N,N'-methylenebisacrylamide in aqueous solutions of zinc chloride and copper chloride. A copolymer of the two monomers was formed on the cathode. The patent claims that this system is an excellent method of coating metal surfaces. Other zinc salts were also used, and polymerization occurred at the cathode. In the same year, Tsvetkov and Koval'chuk, and Fioshin, Koval'chuk, and Tsvetkov [137, 138] reported the polymerization of acrylamide and methacrylamide anodically in the presence of ethylene glycol and sodium acetate. In 1969, Tsvetkov and Koval'chuk reinvestigated a similar system [139]. In 1966 Ehrig and Kundell [140] electrolyzed an aqueous solution of diacetone acrylamide and recovered diacetone acrylamide polymer as a coating on the cathode.

Vinyl Ethers and Tetrahydrofuran

Breitenbach and Srna [49] in 1962 reported the cathodic polymerization of isobutyl vinylether in nitrobenzene solutions. This work has been extended in a recent paper by Funt and Blain [141], who initiated the polymerization of isobutyl vinylether in methylene chloride solutions of various tetraalkylammonium salts as supporting electrolytes. The course of the reaction was shown to be strongly dependent on the actual salt used and kinetic studies were reported in the case of tetrabutylammonium tetrafluoroborate and tetrabutylammonium perchlorate. Various reaction mechanisms were discussed. The polymerization of tetrahydrofuran has been reported by several authors, namely, Heins [142], Yamazaki [47], and Funt and Dyck [143]. In the case of both the vinyl ether and the tetrahydrofuran, the mechanism is considered to be cationic in nature. Mengoli and Vidotto have recently reported further studies of cationic polymerization [194-196].

Pyridine and Derivatives

Laurent and Parravano [144, 145] investigated the polymerization of 4-vinyl pyridine in liquid ammonia solutions of alkali metal salts. Rapid polymerization occurred at the cathode and polymers were formed exclusively there in the form of porous orange-red solid deposits. The authors considered the polymerization was initiated by a solvated electron formed around the cathode. In a later paper Bhadani and Parravano [14] reported the polymerization of the same monomer, this time in pyridine solutions of sodium tetraphenylboron. Once again, polymer was formed only at the cathode where the solution became red-orange after about 15 min of electrolysis. The results indicated that polymerization was an anionic one of the living type. Comparisons were made between the homogeneous polymerizations performed in pyridine solutions and heterogeneous polymerization carried out in ammonia solutions.

Fedorova and Vasina [146] produced a white solid product in the cathodic polymerization of an aqueous pyridine solution. The effect of pH on the rate of polymerization was studied. In a later paper Vasina, Astakhova, Karysheva, and Fedorova [147] again studied the polymerization of pyridine, this time in the presence of magnesium and beryllium ions. Once again, the polymerization was cathodic in nature.

Acrylic Acids

In 1949 Dineen, Schwan, and Wilson [51] were able to prepare poly(acrylic acid) from aqueous sulfuric acid solutions of the monomer. The polymerization was rapid at the cathode and was considered to have been initiated by hydrogen atoms electrolytically generated. Partial hydrogenation of the olefinic bonds was also reported. In 1962 Smith and Manning [95] reported a good yield of high-molecular-weight polymer from the Kolbe-type electrolysis of dilute aqueous solutions of acrylic acid with potassium acetate as supporting electrolyte. Polymer was formed at the anode. Porejko and Perec [148] in 1966 studied the polymerization of acrylic acid in aqueous solutions of sodium acetate. The polymerization occurred at the anode and was reported to be due to a free-radical mechanism initiated in a Kolbe reaction. In aqueous methanol solutions yields were higher than in aqueous solutions, a characteristic of the Kolbe reaction. In 1967 Izumi, Kungugi, and Nagura [149] again studied Kolbe's reaction in initiating acrylic acid polymerization. Lastly, Arnold and Swift [150] reported the polymerization of methacrylic acid using a bipyridylium salt as a source of free radicals:

$$\left[\underset{N}{\overset{\oplus}{\bigcirc}} \underset{CH_2-CH_2}{\overset{\oplus}{\bigcirc}} N \right] 2Br^{\ominus} \quad (DQ^{2+})$$

$$DQ^{2+} + e \xrightarrow{\text{Reduction}} DQ^{\overset{+}{\cdot}} + \xrightarrow{O_2} DQ^+-O_2\cdot$$

Radical-cation   Cationic peroxy radical
(initiation of polymerization)

Aldehydes and Vinylaldehydes

In 1967 Schlygin and Kondrikov [151] effected the polymerization of benzaldehyde in aqueous solutions at platinum anodes. A yellow polymer was formed with a melting point of approximately $300°C$. Schulz and Strobel in 1968 [152] reported the polymerization of THF solutions in acrolein with sodium tetraphenylboride as supporting electrolyte. The polymerization was considered to be anionic in nature with soluble polyacrolein being formed in the cathode compartment. Infrared spectra indicated the presence of both structures I and II, their relative proportions depending on the temperature of polymerization. Schulz and Strobel [153] in 1970 polymerized acrolein in nitromethane solutions of lithium perchlorate. Spectral studies indicated only vinyl-type units (II) and the mechanism was thought to be cationic:

n CH$_2$=CH
  |
  C=O
  |
  H

CH$_2$=CH
  |
$\sim\sim$ +C-O+$_n$ $\sim\sim$
  |
  H   (I)

$\sim\sim$ +CH$_2$-CH+$_n$ $\sim\sim$
            |
            CH=O   (II)

(These two acrolein polymerizations are reviewed in considerable detail in the experimental section, p. 575.)

Vinyl Acetate and Vinyl Chloride

A 1955 patent of Park and Bump [19] reported among other electroinitiated copolymerizations the emulsion electropolymerization of vinyl chloride. A mechanism was not suggested in this polymerization. In 1960, Smith and Gilde [55] produced poly(vinyl acetate) from a solution of potassium acetate and vinyl acetate. Polymer was formed at the anode and its hardness at first indicated some form of stereochemical control to give a crystalline polymer. However, experiments failed to confirm these ideas. Vinyl chloride under the same conditions gave only a very small yield of low-molecular-weight polymer. In 1962 Smith and Manning [95] again reported failure to produce poly(vinyl chloride) from aqueous potassium acetate solutions.

Acenaphthylene

Mitoguchi, Kikuchi, and Kimura [154] in 1969 investigated the electro-initiated polymerization of acenapthylene in acetic anhydride or acetonitrile solutions of various perchlorates. The average molecular weight of the polymers was between two and three thousand.

Maleic Anhydride

Cochran, in a patent issued in 1966 [155] polymerized maleic anhydride in DMF solutions of tetraethylammonium p-toluenesulphonate. High yields were reported of a polymer having molecular weight of about 2600. The polymer was mainly comprised of the polyanhydride along with some acid.

N-Vinyl Carbazole

In his 1962 paper, Breitenbach [49] noted the cationic polymerization of N-vinyl carbazole in nitrobenzene solution. Yields were quite good, but the molecular weights of the polymers were low. Tetraethylammonium perchlorate was used as the supporting electrolyte.

N-Vinyl Pyrollidone

Breitenbach and Srna [49] attempted to polymerize solutions of N-vinyl pyrollidone in which tetraethylammonium perchlorate was dissolved. However, on electrolysis of this solution, only an oily, possibly dimeric, product was formed at the anode. Funt and Williams in 1963 [156] polymerized vinyl pyrollidone in methanolic solutions of potassium acetate. The polymer formed insoluble sheets on the anode and prevented kinetic studies being performed. The absolute electrical efficiencies of this reaction were very low, and it was estimated that only 0.1% of the electrons transferred at the electrode ultimately resulted in the initiation of a polymer molecule.

Phenols

In 1969 a patent [157] was published describing the electrolytic polymerization of phenol with simultaneous deposition on the electrodes. An aqueous solution of phenol and caustic soda was placed in the anode compartment of an electrolytic cell. When eventually the current ceased to flow, the iron anode had a uniform reddish coating of polyphenyleneoxide. Borman in a patent [158] reported the polymerization of various substituted phenols using copper salts as catalysts with ammonia or amines as complexing agents in aqueous solution to give polyarylene oxides. Dijkstra and De Jonge [197] recently reported the formation of products of high molecular weight by the electrochemical oxidation of phenols in nonaqueous solvents.

Copolymer Studies

There have been very few studies of electroinitiated copolymerization per se, except those which have been used purely as a method of diagnosing the mechanism of propagation in electrolytic polymerization. In 1965, Kusnetsov and Bogoyavlenskaya [159] reported a polarographic study of the copolymerization kinetics of MMA with methylacrylic acid, in the presence of some univalent salts of methacrylic acid. The next year, Funt and Gray [160] reported the electrolytically initiated anionic copolymerization of styrene and MMA. They employed tetrabutylammonium salts in THF solutions. Fleischer and co-workers described cationic copolymerization of trioxane with 1,3-dioxolane [198].

Recently, there has been some investigation of the electroinitiation of monomer pair charge transfer complexes to form alternating 1:1 copolymers. Funt, McGregor, and Tanner [169, 170, 11] reported the copolymerization of MMA with styrene using zinc chloride as catalyst [169] and, similarly, the copolymerization of AN with styrene with zinc bromice as catalyst [170, 11]. Funt and Rybicky [12] copolymerized AN with butadiene using zinc chloride as catalyst. Phillips, Davies, and Smith [199] prepared an alternating copolymer of styrene with diethyl fumarate in the presence of zinc bromide. A further discussion of these systems appears in Section V, "Charge Transfer Complex Copolymerizations."

Miscellaneous

Two patents [161, 162] investigated the polymerization of various monomers using monomeric and polymeric quaternary ammonium salts. Forster [161] in 1964 reported the polymerization of AN and various substituted cyano-, and dicyanoethylenes in nonaqueous solvents such as DMF, DMSO, acetonitrile, and HMPT, with a simple quaternary ammonium salt. He noted the simultaneous occurrence of 1,2 and 1,4 polymerization (in the latter case nitrogen appears in the main chain). Bayer and Santiago [162] in 1970 noted the polymerization of widely differing monomers (mostly the common monomers previously discussed) in various aqueous and nonaqueous solvents. The novelty in this patent was the use of polymeric quaternary ammonium or polymeric onium salts. (References are made to patents involving the preparation of these salts [163, 164].) The authors claim, with AN as monomer, a superior (less colored) polymer with the polymeric electrolytes than with the monomeric salts used by Forster [161] and others.

Mengoli and Vidotto [13] reported the anionic polymerization of methyl vinylketone (MVK) initiated by electrochemically generated nitrobenzene radicals. The mechanism was similar to that reported in their work on the electro-initiated polymerization of AN [87]:

$$PhNO_2 + e \longrightarrow PhNO_2^{\ominus}$$

$$PhNO_2^{\ominus} + CH_3-\underset{\underset{O}{\|}}{C}-CH=CH_2$$

$$MVK$$

$$\longrightarrow PhNO_2H + \overset{\ominus}{C}H_2-\underset{\underset{O}{\|}}{C}-CH=CH_2$$

$$(I)$$

$$(I) + (MVK)_n \longrightarrow \text{Polymer, anionic propagation}$$

Mengoli, Vidotto, and Furlanetto [18] studied the reaction of electrolytically produced naphthalene radical anion $(N^{\ominus})$-DMF solutions, with the oxides of both ethylene and propylene. The supporting electrolyte was tetraethyl-ammonium perchlorate. The reaction was described as a second-order process, the decay of $N^{\ominus}$ obeying the following equation:

$$\frac{-d\,[N^{\ominus}]}{dt} = k\,[N^{\ominus}]\ \ [\text{epoxide}]$$

The reaction was thought to occur by direct electron transfer from $N^{\ominus}$ to the oxide:

$$N^{\ominus} + CH_2{-}CH_2 \longrightarrow \text{Naphthalene} + \cdot CH_2{-}CH_2{-}O^{\ominus}$$
$$\underset{O}{\diagdown \diagup} \qquad\qquad\qquad (N)$$

rather than by a method proposed by Szwarc [23] that consumes 50% of the reduced naphthalene. The authors noted no loss of naphthalene in their experiments.

Asahara, Seno, Tobayama, and Cheu [24] reported the polymerization of trioxane in dry THF to give poly(oxymethylene). The presence of water (0.6 ml/100 ml of electrolyte solution, 0.33 M) inhibited the polymerization. It was also noted that the melting point of the polymer was dependent on the polymerization temperature.

Recently, Mengoli and Vidotto [200] studied electroinitiated cationic polymerization of trioxane in acetonitrile and nitrobenzene. They reported that the living character of the polymerizations became more pronounced at elevated temperatures.

4 SUMMARY OF ELECTROINITIATED POLYMERIZATION (TABLES 11.5-11.7)

Section 4 is intended as a summary in tabular form of Section 3 and duplicates in simplified form the data contained therein. The following will explain the abbreviations used.

Cell type, electrode: "Single" denotes a cell where anode and cathode compartments are not separated in any way.
"Double" denotes a cell where the anolyte and catholyte are separated, usually by a glass frit.
(A) and (C) following a metal denote respectively the anode and cathode material.

Mixture homogeneity: "Heterog" and "homog" denote heterogeneous and homogeneous systems, respectively. In the heterogeneous case either the initial reaction mixture or the polymerizing reactants, during the course of the reaction, were inhomogeneous. Homogeneous denotes a single phase system.

Mode: I_c denotes a controlled current.

E_c a controlled potential, both in a direct current mode.

ac denotes alternating current.

606

Table 11.5. Summary of Section 3

Monomer	Cell Type/ Electrode	Mixture Homogeneity	Solvent (Medium)	Salt	Mode	Current	Electrode Surface Area	Current Density	Time	Temp. (°C)	Yield	Ref.
MMA	Single and double various (A),Hg(C)	Heterog.	Aq. MeOH	H_2SO_4	I_c	—	10 cm²	3.0 A/dm²	4 hr	50	54%	51
MMA	U-tube Pt (A, C)	Heterog.	P Propylene gglycol	NaOAc	I_c	23-32 mA	—	—	4 hr	30-35	To gelation	52
MMA	Double Pt (A), various (C)	Heterog.	H_2O	H_2SO_4	I_c	—	—	2.57 mA/cm²	Various	24	Various	53
MMA	Double Pt(A), various (C)	Heterog.	H_2O	HCl	I_c	Various	—	—	Various	25	Various	54
MMA	Single Pt (A,C)	Heterog.	H_2O	KOAc	I_c	1.5-2.5 A	2.25 in²	—	5-9 hr	25	15.5%	55
MMA	Double Pt/Pd(A), Hg(C)	Heterog.	H_2O and aq. alcohol	H_2SO_4	E_c	—	10 cm²	—	3 hr	20	—	56
MMA	Double Pt(A), Pb(C)	Heterog.	H_2O	HCl	I_c	150 mA	10 cm²	15 mA/cm²	0-6 hr	25	Less than 1% to 5%	57
MMA	Single Pt(A), Pt or Hg(C)	Heterog.	AcOH/ Ac_2O	LiOAc	I_c	7 mA	—	—	4 hr	—	—	49

MMA	Single graphite (A,C)	Homog.	Various (DMSO)	Various (An(OAc)$_2$)	I_c	1-110 mA	23.9 cm^2	Various	Various	R.T.	Various	58
MMA	Pb(C), -(A)	Heterog.	H$_2$O	H$_2$SO$_4$	E_c	—	2.0 cm^2	1 mA/cm^2	Various	20	—	59
MMA	Double Pt(A), Hg(C)	Homog.	DMF	LiCl or Et$_4$NI	—	—	—	—	—	—	—	60
MMA	Double Pt(A), Pb(C)	Heterog.	H$_2$O/added EtOH and MeOH	HCl or NaCl	I_c	100 mA	10.6 cm^2	—	6 hr	25	—	61
MMA	Double Pt(A), Hg(C)	Heterog.	H$_2$O	H$_2$SO$_4$	I_c	50-200 mA		—	6 hr	—	40-70% increase	62
MMA	Hg(C)	Heterog.	H$_2$O	H$_2$SO$_4$	E_c	—	—	—	Various	20	—	63
MMA	Dropping Hg(C)	Heterog.	H$_2$O	H$_2$SO$_4$	E_c	—	—	—	—	22	—	64
MMA	Single Pt(A,C)	Heterog.	H O	H SO	AC	—	—	—	—	—	—	65
MMA	Single Pt(A,C)	Homog.	DMF (and other)	Various quaternary ammonium salts	I_c	Various	—	Various	Various	25	Various	66
MMA	Single Pt(A), Hg(C)	Heterog.	AcOH/Ac$_2$O	NaOAc	I_c	Various	23 cm^2	Various	Various	Various	Various	67
MMA	—	Heterog.	Aq. glycol	NaOAc	I_c	—	—	125 mA/cm^2	Various	Various	Various	68

Table 11.5. (continued)

Monomer	Cell Type/ Electrode	Mixture Homogeneity	Solvent (Medium)	Salt	Mode	Current	Electrode Surface Area	Current Density	Time	Temp. (°C)	Yield	Ref.
MMA	—	Heterog.	H_2O, aq. acetone and various aq. alcohols	NaOAc	I_c	—	—	—	Various	25	Various	69
MMA	Double Pt(A), Hg(C)	Heterog.	H_2O	Various	E_c	—	—	—	—	—	—	70
MMA	Double Pt(A,C)	Homog.	DMF/AcOH	NaOAc and LiOAc	I_c	Various	—	—	Various	25	Various	71
Acylonitrile												
MMA	Double Various (A,C)	Heterog.	H_2O	H_2SO_4/Fe^{3+}	I_c	20 mA	94 cm²		Various 1-900 min	23-25	Various	73
MMA	Double Pt(A), Various(C)	Heterog.	H_2O	H_2SO_4, Various	I_c	Various	—	Various	—	25	—	54
MMA	Single Various (A,C)	Heterog.	AN	Et_4NClO_4	I_c	Various	—	Various	—	R.T.	—	76
MMA	Single Various (A,C)	Heterog.	AN	Et_4NClO_4	I_c	Various	—	Various	—	R.T.	—	77
MMA	Single Pt (A), Hg or Pt(C)	Heterog.	AN AcOH/Ac$_2$O	Et_4N salts various LiOAc	I_c	Various 0.3-7 mA	1 cm²	Various	Various	—	Various	49

Monomer	Electrode	Process	Solvent	Electrolyte		Current	Area	Current density	Time	Temp	Yield	Ref
MMA	Double Pt (A), Hg (C)	Homog.	DMF	LiCl or Et$_4$NI	I_c	—	—	—	—	—	—	60
MMA	Single and double Pt (A,C)	Homog.	DMF	NaNO$_3$	I_c	Various	—	Various	Various	R.T.	Various	78
MMA	Double Pt (A) Hg (C)	Heterog.	H$_2$O	K$_2$S$_2$O$_8$/Fe^{3+} H$_2$SO$_4$	I_c	50 mA	—	—	—	—	—	79
MMA	Single Ni (A,C)	Homog.	DMF	NaOAc	I_c	—	6.5 cm^2	0.5 mA/cm^2	6.5 hr	—	40%	80
MMA	Double Pt (A,C)	Homog.	Dimethoxyethane	Bu$_4$NClO$_4$	I_c	Various	—	Various	—	30	—	72
MMA	Double various Pt (A), Pb (C)	Heterog.	AN (0.1 → 1% H$_2$O)	Quaternary ammonium salts	I_c	Various	—	Various 4-150 mA/cm^2	—	5-60	30-80 g/An depending on salt	81
MMA	Double rotating Pt (A), Various (C)	Heterog.	AN (0.1 → 1% H$_2$O)	Quaternary ammonium salts	I_c	Various	—	1-200 mA/cm^2	—	0-80	—	82
MMA		Homog.	MeCN (or AN)	Bu$_4$NI or Et$_4$NClO$_4$	I_c	Various	—	—	30 min	25	—	83
MMA	Double graphite (A,C)	Heterog.	DMF	NaNO$_3$	I_c	—	—	—	—	0-40	—	84
MMA	U-tube, separated anode and cathode Pt (A,C)	Heterog.	DMSO	K, Na, NH$_4$—S$_2$O$_8$	I_c	Various 10-100 mA	1.2 cm^2	Various	Various	20	Various	85

Table 11.5. (continued)

Monomer	Cell Type/ Electrode	Mixture/ Homogeneity	Solvent (Medium)	Salt	Mode	Current	Electrode Surface Area	Current Density	Time	Temp. (°C)	Yield	Ref.
MMA	Double Pt(A,C)	Heterog.	DMF	Quaternary ammonium or phosphonium salts	I_c	2 mA	—	—	Various	Various	Various	86
MMA	Double Pt(A),Hg(C)	Homog	DMF	Bu$_4$NClO$_4$	E_c	Approx. 10 mA	—	—	2 min	—	<10%	87
MMA	Single Pt(A),Hg(C)	Homog.	DMF (MeCN H$_2$O	Various	I_c	Various	—	Various	Various	30	Various	88
MMA	Pt(A,C)	Heterog.	H$_2$O	H$_2$SO$_4$	I_c	Various 50-300 mA	50-300 mA	Various	Various	25	Various	89
MMA	—	Heterog.	H$_2$O	HClO$_4$	I_c	—	—	Various	Various	25	Various	90
MMA	—	Homog.	DMSO	NaNO$_3$, KClO$_4$, KBF$_4$	I_c	DC and rectangular wave current	—	20 mA/cm^2	—	30, 0, −78	DC yields greater than rectangular wave form yields	91
MMA	Double Pt(A),Hg(C)	Homog.	DMF	Et$_4$N-p-toluene-sulphonate	I_c	1.4 A	—	—	Approx. 1 hr	>40	59%	92
MMA	Double Ag(A),Hg(C)	Homog.	DMF	Bu$_4$NI	E_c	—	—	—	—	25	—	21
MMA	Double Ag(A),Hg(C)	Homog.	DMF	Et$_4$NClO$_4$	E_c	—	—	—	—	25	—	10

610

MMA	Double Pt(A,C)	Heterog.	H_2O	H_2SO_4	I_c	—	—	300-500 mA/cm^2	—	10-20	Approx. 85%	93
MMA	Al(C)	Heterog.	HMPT	LiCl	—	—	—	—	—	—	—	20

Styrenes

MMA	Single and double Pt(A,C)	Heterog.	AcOH	NaOAc	I_c	10 mA	—	1 mA/cm^2	180 hr	27	Approx. 0.5%	74
MMA	Double Mg(A,C)	Homog.	Pyridine	NaI	I_c	—	—	0.7-4 mA/cm^2	45 hr	30	—	94
MMA	Single Pt(A), Pt and Hg(C)	Heterog. Homog.	$AcOH/Ac_2O$ $Ph-NO_2$	NiOAc Et_4NBF_4	I_c	3 mA 0.35 mA	1 cm^2	—	15 hr 1 hr	R.T.	—	49
MMA	Single Pt(A,C)	Homog.	DMF	Me_4NCl	I_c	Various	Various	Various	Various	Various	Various	96
MMA	Double Pt(A),Pb(C)	Heterog.	Aq. alcohol (EtOH and MeOH)	HCl	I_c	100 mA	10.6 cm^2	—	6 hr	25	—	61
MMA	Single Pt(A,C)	Homog.	DMF	Various (particularly KNO_3)	I_c	Various 25-100 mA	1-in.^{2}	Various	Various	R.T.	Various, depends on salt	97
MMA	Double Pt(A,C)	Homog.	THF	$LiAlH_4$ or $NaAlEt_4$	I_c	Various	3.5 cm	Various	Various	Electrolysis at R.T. Polymerization at −78	Quantitative	98, 99

Table 11.5. (continued)

612

Monomer	Cell Type/ Electrode	Mixture Homogeneity	Solvent (Medium)	Salt	Mode	Current	Electrode Surface Area	Current Density	Time	Temp. (°C)	Yield	Ref.
MMA	Double Pt(A,C)	Homog.	THF	NaAlEt or NaPh$_4$B	I_c	Various	1-in.^{2}	Various	Various	R.T.	Quantitative	100
MMA	Double	Homog.	THF	NaPh$_4$B, (and potassium and lithium salts of same)	I_c	Various	1-in.^{2}	Various	Various	Electrolysis at R.T. Polymerization at -78	Quantitative for all but LiPh$_4$B	101, 102
MMA	Double Pt(A,C)	Homog.	Hexamethylphosphortriamide (HMPT)	Me$_4$NPh$_4$B	I_c	Various	–	Various	Various	–	Quantitative	50
MMA	Double Pt(A,C)	Homog.	Dimethoxyethane DME	Bu$_4$NClO$_4$	I_c	Various	–	Various	Various	25	Low	72, 103
MMA	Double Pt(A,C)	Homog.	THF	NaPh$_4$B	I_c	Various	–	Various	Various	–	Quantitative	104
MMA	Double Pt(A,C)	Homog.	Methylene chloride	Bu$_4$NClO$_4$	I_c	Various	1-cm diam. circular sheets	Various	25	Various	Various	105
MMA	Double Pt(A), Hg(C)	Homog.	Nitrobenzene	Bu$_4$NClO$_4$	I_c	Various	Circular mesh 3.5 cm diam., 1.5 cm height	Various	Various	-10	Various	190

MMA	Single and double Pt(A,C)	Homog.	Sulpholane	NaBF$_4$	I_c	Various	1 cm^2	Various	Various	Various	30	Various	191
MMA	Single and double Pt(A,C)	Homog.	N,N-di-methyl-acetamide	NaBF$_4$	I_c	Various	1 cm	Various	Various	Various	25	Various	192
MMA	–	Homog.	Nitro-benzene	Bu$_4$NClO$_4$	–	–	–	–	–	–	10	Various	193
Alkenes/Alkynes													
Ethylene	Double various (A,C)	Heterog.	Dioxane	Bu$_4$NBr (stilbene)	I_c	Various	Various	20–1000 mA/cm^2	–	–	–	–	109
Isoprene	Double various(A,C)	Heterog.	Dioxane	NaPh$_4$B or NaAlEt$_4$	I_c	Various	Various	20–1000 mA/cm^2	–	–	25	–	112
Butadiene isoprene	Double various(A,C)	Heterog.	DME	Bu$_4$NClO$_4$	I_c	Various	Various	20–1000 mA/cm^2	–	–	30	0.4–4.3% Butadiene 26–36% Isoprene	72
Acetylene	Single Ni(A), Pt(C)	Heterog.	MeCN	NiBr$_2$	I_c	200 mA	–	–	–	40 min	–	–	113, 114
Butadiene	Pt(C)	Homog.	EtOH or DME	NiCl$_2$ or Ni (pyridine)$_4$ Bu$_4$NClO$_4$	I_c	–	–	–	–	–	–	–	115

Table 11.5. (continued)

Monomer	Cell Type/ Electrode	Mixture Homogenity	Solvent (Medium)	Salt	Mode	Current	Electrode Surface Area	Current Density	Time	Temp. (°C)	Yield	Ref.
				Benzene Derivatives								
p-Xylylenes	Double carbon(A), Pb, Hg(C)	Heterog.	Aq. dioxane	HCl	I_c	Various	–	1-6 mA/cm²	–	-10—	Various	116, 117, 118
p-Xylylene	Single Pt(A), Al(C)	Heterog.	Various DMSO preferred	Monomer salt used	I_c	Various	15.0 cm²	Various	–	Various	Various	119, 120, 121
Benzene, etc.	U-tube Pt(A,C)	Heterog.		RH/HX/ 2AlX₃;e.g. mesitylene HCl/2AlCl₃	I_c	Various	–	10-500 mA/cm²	–	0-50	Various	122, 123, 124
Benzene	Single Ni(A,C)	Heterog.	H₂O	HF (aqueous and anhydrous)	I_c	–	81 cm²	–	–	–	Various	125, 126
				Polyacids-Carboxylic Acids								
Polymetha-crylic acid	Single Pt(A,C)	Homog.	H₂O	–	I_c	–	–	Various	Various	1-25	Various	127
Polymetha-crylic acid	Single Pt(A,C)	Homog.	Anh.MeOH	5-8% neutra-lized polyacid (with NaOMe)	I_c	100-1000 mA	–	Various	–	20	Various	129, 128
Copolymers of arylic acid	Single Pt(A,C)	Homog.	Aq. and anh. Me₂CO	–	I_c	–	–	–	–	–	Various	130

614

Monomer	Electrode		Solvent	Electrolyte				Current	Time	Temp		Ref.
Perfluoro-glutaric acid	Single Pt(A,C)	Heterog.	Anh. MeOH	NaOMe	I_c	—	—	40-100 mA/cm²	4 hr	25-55	—	132
Caprolactam-Isocyanate												
Caprolactam	Single and double steel(A), Fe(C)	Heterog.		Sodium benzoate	I_c	Various	—	Various	Various	130-185	Various	133, 134
Phenyliso-cyanate	Double	Heterog.		Bu$_4$NI	I_c	Various	—	Various	Various	Various	Various	204,205, 206
Vinyl Amides												
Acrylamide	Single and double Pt(A,C)	Heterog.	H$_2$O	K$_2$S$_2$O$_8$	I_c	Various	—	20-90 μA/cm²	3 hr	R.T.	50-100%	135
Acrylamide	Double various(A,C)	Heterog.	H$_2$O	ZnCl$_2$, trace CuCl$_2$	E_c	Various	—	Various	—	—	Surface coating	136
Acrylamide metha-crylamide	Single Pt (A,C)	Heterog.	Aq. glycol	NaOAc	Ic	Various	—	Various	Various	25	Various	137
Acrylamide metha-crylamide	Double –	Heterog.	Aq. glycol	NaOAc	I_c	100 mA	—	Various	Various	25	Various	138, 139
Diacetone acrylamide	Single or double various	Heterog.	H$_2$O	H$_2$SO$_4$	E_c	Various	—	Various	H	0-60	Various	140

Table 11.5. (continued)

Monomer	Cell Type/ Electrode	Mixture Homogeneity	Solvent (Medium)	Salt	Mode	Current	Electrode Surface Area	Current Density	Time	Temp. (°C)	Yield	Ref.
				Vinylethers-THF								
Isobutyl-vinyl ether	Single Pt(A), Pt or Hg(C)	Homog.	$PhNO_2$	Et_4NClO_4 $(AgClO_4)$	I_c	10 mA	1 cm²	—	3 min	R.T.	—	49
Isobutyl-vinyl ether	Double Pt(A,C)	Homog.	Methylene chloride	Bu_4N, BF_4 and others	I_c	Various	1-cm diameter	—	—	25	—	141
Isobutyl-vinyl ether	Double Pt(A), Hg(C)	Heterog.	CH_3CN or $C_6H_5NO_2$ and mixtures	$NaB(C_6H_5)$ Bu_4NClO_4	I_c	0.5-4 mA	Net electrode. Diam. 3.5 cm; height 1.5 cm	0.1 mA/ cm²	Various	-27 to 30	Various	A25
Ethylvinyl ether	Double Pt(A),	Homog.	CH_3CN or $C_6H_5NO_2$ and mixtures	$NaB(C_6H_5)_4$ Bu_4NClO_4		0.5-4 mA	Net electrode. Diam. 3.5 cm; height 1.5 cm	0.1 mA/ cm²	Various	-27 to 30	Various	A25
n-Butylvinyl ether	Double Pt(A),	Homog.	$C_6H_5NO_2$	$NaPh_4B$ plus Bu_4NClO_4	I_c	Various	Net electrode. Diam. 3.5 cm; height 1.5 cm	0.1 mA/ cm²	Various	10	15%	A26
THF	Double Pt(A,C)	Homog.	None	$LiClO_4$ or Et_4NClO_4	I_c	200 μA	5 cm²	—	5 days	—	34%	

Acrylic Acids

Acrylic acid	Single and double various (A), Hg(C)	Heterog.	Aq. MeOH	H_2SO_4	I_c	—	—	0.1 A/dm^2	4 hr	50	Various	51
Acrylic acid	Single Pt(A,C)	Homog.	H_2O	KOAc	I_c		1.2-in.^{2}		18 hr	—	Approx. 50%	95
Acrylic acid	Single and double	Homog.	H_2O	NaOAc	I_c	Various	—	Various	Various	-5 to 33	Various	148
Acrylic acid	Double Pt(A,C)	Homog.	H_2O	KOAc	E_c		9.8 cm^2	Various	Various	-5 to 33	Various	149
Methacrylic acid	Double Pt(A), Pt or Ta(C)	Heterog.	H_2O	N,N'-ethylene-2,2-bipyridylium dibromide (Diquat)	E_c	Various 1-5 mA	—	—	—	—	—	150

Pyridine and Derivatives

4-Vinyl pyridine	Double Pt or Carbon(A), Pt(C)	Heterog.	Liquid NH_3	NaN_3 or NaCl	I_c	Various	—	10-100 mA/cm^2	Various	-78 to	80-90%	144, 145
4-Vinyl pyridine	Double Pt(A,C)	Homog.	Pyridine	$NaPh_4B$	I_c	1-5 mA	1-in.^{2}	Various	21-65 min	25	Various	14
Pyridine	Hg(C)	Heterog.	H_2O	H_2SO_4	E_c	Various	—	Various	—	—	—	146
Pyridine	— Hg(C)	Heterog.	H_2O	Li_2SO_4 (+ Mg^{2+} Be^{2+})	E_c	Various	—	Various	—	—	—	147

Table 11.5. (continued)

Monomer	Cell Type/ Electrode	Mixture Homogeneity	Solvent (Medium)	Salt	Mode	Current	Electrode Surface Area	Current Density	Time	Temp. (°C)	Yield	Ref.
Aldehydes-Vinyl Aldehydes												
Benzalde-hyde	Double Pt(A,C)	–	H_2O	–	–	–	–	–	–	–	–	151
Acrolein	Double Pt(A,C)	Homog.	THF	$NaPh_4B$	I_c	Various 0.2-25 mA	8.5 cm²	Various	Various	-20, 0, +20	Various	152
Acrolein	Double Pt(A), Cu(C)	Homog.	Nitro-methane	$LiClO_4$	I_c	Various 5-60 mA	–	0.6-7 mA/cm²	Various	-20 to +40	Various	153
Vinyl Acetate												
	Single Pt(A,C)	Heterog.	H_2O	KOAc	–	1.5 A	1.5-in.²	–	6 hr	25	38%	55
	Double Pt(A,C)	Heterog. emulsion	H_2O	SO_2 and dioctyl succinate sodium sulphonate	–	–	–	–	8 hr	40	–	19
Acenapthylene												
	–	–	Ac_2O or MeCN	$LiClO_4$, $KClO_4$, KBF_4 or Bu_4NBr	–	–	–	–	–	-78 to	–	154

Maleic Anhydride

N-Vinyl Carbazole

N-Vinyl Pyrrolidone

Phenols

Copolymers

Substrate	Electrode	Type	Solvent	Electrolyte	Mode	Current	Area		Time	Temp	Yield	Ref
Maleic Anhydride												
	Double Pt(A), Hg(C)	Homog	DMF	Et$_4$N, paratoluene sulphonate	E_c	Various 0.4-1.8 A	—	—	3.5 hr	35 to 65	100%	155
N-Vinyl Carbazole												
	Single Pt(A), Pt or Hg(C)	Homog.	PhNO$_2$	Et$_4$NClO$_4$	I_c	2 mA	1 cm^2	—	5 min	25	—	49
N-Vinyl Pyrrolidone												
	Single Pt(A), Pt or Hg(C)	Homog.	Self-soluble	Et$_4$NClO$_4$	I_c	—	1 cm^2	—	—	25	—	49
	Single Pt(A,C)	Heterog.	MeOH	KOAc	I_c	2 mA	1-in.^{2}	—	—	—	1%	156
Phenols												
Phenols	—	Heterog. homog.	Various (nonaqueous)	—	—	—	—	—	—	R.T.	—	197
Phenol	Double Fe(A), Pt(C)	Heterog., homog.	H$_2$O	NaOH	I_c	40-100 mA	12.5 cm^2	Various	30 min	—	<1%	157
Phenol	Double	Heterog.	—	—	—	—	—	—	—	—	—	158
Copolymers												
MMA Methacrylic acid	—	—	—	Salts of methacrylic acid (0.2-3%)	—	—	—	—	—	50,60, 70	—	159

Table 11.5. (continued)

620

Monomer	Cell Type/ Electrode	Mixture Homogeneity	Solvent (Medium)	Salt	Mode	Current	Electrode Surface Area	Current Density	Time	Temp. (°C)	Yield	Ref.
Styrene/ MMA	Single Pt(A,C)	Homog.	THF, DMF or dichloro- ethane	Bu$_4$NBF$_4$, Bu$_4$NPh$_4$B		5-20 mA	1 in.^{2}	—	—	—	<10%	160
MMA/St	Single Al or Pt (A,C)	Homog.	Bulk	ZnCl$_2$	I_c	15 mA	1 in.^{2}	—	Various	0	<3%	169
St/AN	Single Pt(A,C)	Homog.	Bulk	ZnBr$_2$	I_c	15 mA	1 in.^{2}	—	Various	0	<4%	170
St/AN	Double Pt(A,C)	Homog.	Bulk	ZnBr$_2$	I_c	15 mA	1 in.^{2}	—	Various	0		11
But/AN	Single Pt(A,C)	Homog.	Bulk	ZnCl$_2$	I_c	8-45 mA	1 in.^{2}	—	Various	45		12
Diethyl fumarate styrene	Single Pt(A,C)	Heterog.	CH$_3$OH	ZnBr$_2$	I_c		10 cm^2			25		199
Trioxan-1,3-dioxolan	Double Cu(C), Pt(A)			LiClO$_4$		35-50 μA			Various	35	Various	A28
Miscellaneous												
Various monomers	Single Pt(A,C)	Both	Various	Monomeric or poly- meric quaternary ammonium compounds	I_c or E_c	Various	—	—	Various	Various	—	161, 163

Monomer	Electrodes		Solvent	Electrolyte							
Methyl vinyl ketone	Double Ag(A), Hg(C)	Homog.	DMF	Et_4NClO_4	E_c	–	–	–	25	>50%	13
Ethylene and propylene oxides	Double Ag(A), Hg(C)	Homog.	DMF	Et_4NClO_4	E_c	–	–	–	Various	–	18
Trioxane'	–	Homog.	THF (anhydrous)	Perchlorate	I_c	Various	–	Various	20-80	–	24
Trioxane	Double Pt(A), Hg(C)	Heterog.	CH_3CN, $C_6H_5NO_2$	Bu_4NClO_4	I_c	15-100	–	Various	0-75	Various	200

621

Table 11.6. Summary of Section 3

Molecular Weight	Character-ization Techniques	Proposed Mechanism	Means of Determining Mechanism	Kinetic Data	Remarks	Ref.
			METHYLMETHACRYLATE			
—	—	Cathodic, free radical	Inhibitors added	No	Paper is mainly concerned with acrylic acid and methacrylic acid.	51
—	—	Cathodic, initiated by sodium or H·	Not Kolbe free radical at anode as expected	No	Polymer insoluble (due to cross linking).	52
—	—	Cathodic, initiated by H· free radicals	Inferred	No	Low electrical efficiency. Mainly hydrogen gas evolved.	53
—	—	Cathodic, initiated by H· free radicals	Inferred	No	O_2 and air inhibit.	54
89 ml/g (MEK 25°C)	X-ray	Anodic, free radical	When double cell used, polymer only at anode. Tracer studies.	No	Hard, whie polymer formed.	55
Approx. 3×10^6	— —	Cathodic reduction of O_2 to give peroxides which initiate	Polarography, with and without O_2. Mechanism inferred.	No	O_2 inhibits at high and very low concentrations.	56
1.2-3.8 dl/g	—	Cathodic, hydrogen, free radical	Inferred	Yes	Stirring decreases the polymer yield.	57

					Low current efficiency.	49
Approx. 10^6	—	Anodic, free radical (not definite)	Inhibitors added; divided cell when polymer only at anode.	Yes	—	58
—	—	Peroxide formation at the cathode	Added H_2O_2 accelerates rate of polymerization	No	No polymer in absence of O_2.	59
—	—	Cathode, free radical	ESR, colored solutions, copolymer studies	No	—	60
4×10^5 - 28×10^5	—	Cathode, free radical	Inferred	Yes	No stirring. Increase in methanol concentration increased the yield.	61
—	IR of C=O bond	Possible chain transfer to cellulose	Speculative	No	—	62
3.2×10^4 - 7.2×10^4	—	Polymerization via autoxidation products	The rigorous exclusion of air or oxygen completely inhibits polymerization.	Yes	No polymer with rigorous exclusion of air.	63
—	—	Polymerization by hydrogen atoms	—	Yes	Adsorbtion of polymer on the mercury drop studied by differential capacitance measurements.	64
Various	—	Two separate reactions at each electrode, both free radical	Electrodes are consumed	No	Alternating anode and cathode. At anode PtO_2 formed which reacts with MMA to give (initiating) peroxides. At cathode hydrogen atoms initiate.	65

Table 11.6. (continued)

Molecular Weight	Characterization Techniques	Proposed Mechanism	Means of Determining Mechanism	Kinetic Data	Remarks	Ref.
Various, approx. 5×10^4	—	Cathode, anionic	Copolymerization; polymer at cathode in divided cell; inhibitors added	Yes	Chain dimerization (or association) occurs.	66
5.5×10^4 - 13.2×10^4 (acetone 25°C)	—	Cathode, nature of mechanism uncertain	Inhibitor added	Yes	Mechanism possibly a mixture of anionic and free-radical mechanisms.	67
Various, depending on CD	—	Anodic	—	Yes	Dependence of rate constant in the amount of water present shown.	68
—	—	Anodic, Kolbe	—	Yes	Rate of initiation highest for 25% HuO content.	69
—	—	—	—	—	Mainly concerned with differential capacitance studies of polymer absorbtion on a mercury drop.	70
150 ml/g, various	—	Free radical	Polymer only at anode	No	—	71
			ACRYLONITRILE			
—	—	Cathodic free radical	Inferred	No	—	73
—	—	—	—	No	—	54

DP = 40	—	Cathodic by ethyl radicals	See	No	A preliminary paper whose speculative mechanism was later changed.	76
—	—	Cathodic, anionic mechanism	Copolymer studies; addition of inhibitor	No	—	77
4-7 ml/g 100 ml/g	No details	Cathode, anionic Anode, free radical	Copolymer studies; addition of inhibitor	No	This is the full paper for the two preceding ones.	49
—	—	Free radical	Copolymer studies; ESR of colored solutions	No	No polymer details given. Mechanism not anionic as expected, but free radical.	60
0.05-0.15 dl/g (DMF at 30°C)	—	Anionic, bulk of polymer at cathode	Copolymer studies; polarography indicates direct electron transfer to the double bond	Yes	Yield versus monomer concentration maximizes at about 40% AN. Polymer is colored. High current efficiency; chain transfer operative.	78
—	—	Cathode, redox.	Radical grafting inferred	Yes	Polymerization in presence of air.	79
—	—	Anodic, free radical	Indicated by inhibiting effects of various hydrocarbon additives	No	—	80
700-1800	—	Direct electron transfer at cathode followed by anionic reaction	Reaction with counter ion terminates. Analysis (GLC) of products	No	Polymer colored. No chain transfer.	72

Table 11.6. (continued)

Molecular Weight	Character- ization Techniques	Proposed Mechanism	Means of Determining Mechanism	Kinetic Data	Remarks	Ref.
1000-3000	IR	Cathode, anionic	H_2O reduces yield; O_2 does not inhibit. Copolymer studies.	No	High current yield	81
1000-3000	IR	Cathode, anionic	—	No	A resin film used to separate two electrode compartments.	82
—	—	Cathode, anionic	Copolymerization studies	No	Mainly polarographic study.	83
Anode 9000-35000 Cathode 1000	Decreases with current Colored	Chief product at cathode anode free radical	Copolymerization; inhibitors; polymer colors	Yes	Cathode polymer contains solvent residue.	84
<5000	—	Cathode, anionic	—	Yes	Polymer colored. Two me- chanisms proposed. Anion radical from alkali metal or direct electron transfer to the double bond.	85
Various (DMF at 30°C)	Addition of H_2O lowers molecular weight con- siderably	Cathode, anionic (even when water present)	Copolymerization studies	Yes	No living polymers. Polymer is white and not yellow. Various salts cause wide differences.	86

1200	Yellow color	Anionic	Copolymer studies ESR	Yes	The nitrobenzene radical anion reacts with a proton on the monomer to form a monomer radical anion that itself polymerizes.	87
—	—	Depends on electrolyte	Polarography	Yes	Explains the maximization of yield with monomer concentration	88
Approx. 10^5	—	$SO_4^{\cdot-}$ initiates at anode		Yes	Oxidation of $SO_4^{=} \rightarrow SO_4^{\cdot-}$. Polymer yield increases with H_2SO_4 concentration up to $10\,M$ then drops off abruptly to zero.	89
—	—	Free radical at anode	Inhibited on O_2; voltage versus current measurements	—	Mechanism depends on strength of $HClO_4$ solution; $<5\,M$, O_2 produced which inhibits polymerization. $>5 \rightarrow$ radical initiation.	90
0.06–0.18 dl/g	IR	Cathodic, anionic (with radical polymerization occurring at 30 and 0°C)	—	No	KBF_4 best salt at all temperatures. Suggestion that positron annihilates anion to give radical. Polymer coloration studied.	91
714	Osmometry	Cathode, anionic	Termination by increasing the concentration of proton source	No	With increased concentration of proton source, dimerization occurs more and molecular weight lowers.	92

Table 11.6. (continued)

Molecular Weight	Characterization Techniques	Proposed Mechanism	Means of Determining Mechanism	Kinetic Data	Remarks	Ref.
—	IR	—	Comparison made with sodium initiated polymerization	No	Polarographic study also made.	21
1000–4000	—	Anionic	Initiated by benzophenone radical anions	Yes	Mainly a polarographic study of the decay of benzophenone radical anions with AN.	10
400,000	—	Hydroxyl radical initiation at anode	—	No	—	93
—	IR	Initiation by solvated electrons at cathode	Coloration around cathode	Yes	—	20
			STYRENES			
3200	Cryoscopic	Anode; Kolbe-type free radical	Inferred	No	Also dimer and low oligomers. Repeated with propionic acid for same result.	74
1800–2800	—	Cathode, anionic	No pyridine residue in polymer	No	Initiation via direct electron transfer.	94
27 ml/g	—	Kolbe free radical	Copolymer studies, inhibitors	No	The cationic mechanism is in addition to a free-radical process.	49
9 ml/g	—	Cationic	Copolymer studies	No	—	

Depends on temperature, 14,000-21,000	—	Cathodic, possibly anionic	Inhibitors, copolymer studies	Yes	Mechanism not certain.	96
—	—	—	—	—	Polystyrene produced because of its relative solubility in the (heterogeneous) medium	61
2000-5000	Osmometry	Cathode, anionic	Copolymer studies, inhibitors	Yes	High electrical efficiency. System not amenable to kinetic analysis.	97
Depends on current/time relationships	—	Cathode, living anionic	Color, molecular weight	No	—	98, 99
—	UV	Cathode, living anionic	Color (also napthalene added)	Yes	Single-cell experiments caused immediate living-end termination.	100
5×10^4-50×10^4	GPC, UV	Cathode, living anionic	With $LiPh_4B$, UV absorbance spectra slow self-terminating living end	Yes	With Bu_4NPh_4B no persistent color or polymer	101,
5×10^4	—	Cathode, living anionic Direct electron transfer	Polarography, no alkali ion, so direct transfer probable	No	With Me_4NPh_4B having living polymer in HMPT.	50

Table 11.6. (continued)

Molecular Weight	Character-ization Techniques	Proposed Mechanism	Means of Determining Mechanism	Kinetic Data	Remarks	Ref.
300–500	–	Direct transfer to double bond followed by anionic reaction	$-C^{\ominus} + Bu_4N^+ \rightarrow Bu_3N+$ "dead" products Bu_3N analyzed positively by GLC	No	Both polymers gave similar results.	72, 103
Various	GPC, UV	Cathode, living anionic	Color, UV	Yes	The low propagation constant of the monomer proved amenable to programming of molecular weights.	104
–	–	Cationic	Coloration, inferred	Yes	–	105
10 ml/g	Viscometry	Living cationic, anode	In nitrobenzene, polymer formed in anodic compartment	Yes	–	190
DP_n = 24–29	Osmometry	Cationic, anode	Copolymer data. Polymerization initiated by BF_3 in anodic compartment	Yes	Kinetics in a double cell vs that in a single cell.	191
Various	Osmometry	Anionic, cathode	Copolymer data. Polymerization initiated in cathodic compartment	Yes	Kinetics in a double cell vs that in a single cell.	192
–	–	Living cationic	Polymerization initiated by radical cations (produced electrolytically)	Yes	Radical-cation decay kinetics followed.	193

630

7000-10000	—	—	No	Alternating current works at high pressures of ethylene.	109
23,000-40,000	UV	Color	Yes	Anion decay due to interaction with THF Living-end concentration kept constant by feedback mechanism.	112

ALKENES/ALKYNES

Low polymer	—	Termination of carbanion by interaction with counter ion	No	—	113, 114
—	—	—	No	Patent	
Tetramer	—	—	No	An electrochemically produced nickel-butadiene intermediate is formed and is reacted further to give the tetramer.	115

BENZENE DERIVATIVES

—	Viscosities in α-chloronaphthalene at 170°C	Condensation reaction	Yes	High current efficiency. A condensation reaction of	116, 117, 118

Table 11.6. (continued)

Molecular Weight	Characterization Techniques	Proposed Mechanism	Means of Determining Mechanism	Kinetic Data	Remarks	Ref.
—	IR	Reduction at cathode followed by condensation	Product identification	No	Polymer formed as a coating on the aluminum cathode.	119, 120, 121
For benzene DP = 50	—	—	—	No		122, 123, 124
—	—	Cathode H$_2$ formed. Anodic polymer forms at benzene HF·H$_2$O interface	Oxidative polymerization of benzene or benzene/HG complex	No	Polymers contain fluorine residues.	125, 126
POLYACIDS-CARBOXYLIC ACIDS						
—	—	Kolbe-type free radical to give peroxides	-OOH groups determined iodometrically or polarographically	No	The hydroperoxide groups formed on the side chain and were then used in chemical grafting experiments with styrene.	127
—	—	Anodic decarboxylation	Addition of various low-molecular-weight acids caused crossed Kolbe reactions	No	Solely a side-chain reaction to form unsaturated groups, lactones, or chain degradation.	128, 129

130	—	No	Iodometric titration, styrene grafted	Hydroperoxide groups formed	—	—
132	The polymer coats the surface of the anode, perfluoromethylene formed.	No	Inferred	Kolbe-type anode surface link-up	From IR surface	1000
CAPROLACTAM-ISOCYANATE						
133, 134	—	Yes	—	Anionic-polymer coats cathode	—	Viscosity measured in m-cresol at 80°C
204,205, 206	Polymer was crystalline.	No	—	Cathode reaction	IR	—
VINYL AMIDES						
135	Yield and molecular weight depend on the current density.	Yes	—	—	—	1-4.5 dl/g
136	Rate and thickness of polymer coating are improved by adjusting pH to 5, increasing the monomer concentration, and by adding peroxides.	No	Copolymerized with , -methylenebisacrylamide	Cathode only	—	—

Table 11.6. (continued)

634

Molecular Weight	Characterization Techniques	Proposed Mechanism	Means of Determining Mechanism	Kinetic Data	Remarks	Ref.
—	—	Anode free radical	Addition of inhibitor decreases rate	Yes	Polymerization followed bromometrically.	137
—	—	Anode free radical	Inhibitor addition. Rate dependence on square root of current.	Yes	Polymerization followed bromometrically.	138, 139
—	—	—	—	No	—	140
			VINYL ETHERS-THF			
7.5 ml/g	—	Cationic	Copolymer study possible living polymer	No	High current efficiencies; rubberlike green colored polymer of	49
—	—	Cationic	Inferred, coloration	Yes	Various salts used. Onset of polymerization detected by Thermistor response.	141
0.59 inherent viscosity 0.1% solution in CHCl$_3$	IR	Cationic	Inferred—not studied	No	Red-brown viscous liquid. Also used Et$_4$NClO$_4$ as electrolyte. Et$_4$NI and Et$_4$NBr failed to support polymerization.	142

I-BUTYL VINYL ETHER						
Various	Osmometry	Cationic, anode	—	Yes	Chain transfer studies. Various solvents and temperatures.	195
ETHYL VINYL ETHER						
Various	Osmometry	Cationic, anode	—	Yes	Chain transfer studies. Various solvents and temperatures.	195
D-BUTYL VINYL ETHER						
Various	Viscometry	Cationic, anode	In nitrobenzene	Yes	Using $NaPh_4B$ the polymerization is perfectly controllable, without a violent exotherm.	196
PYRIDINES AND DERIVATIVES						
10,000–300,000	Softening points, IR discussed	Electron transfer from cathode to monomer. Anionic, living.	The color (red-orange) was sensitive to the addition of protic substances	Yes	Radical cations may or may not dimerize according to the charge density. Electrochemical yield depends on charge density.	144, 145

Table 11.6. (continued)

Molecular Weight	Characterization Techniques	Proposed Mechanism	Means of Determining Mechanism	Kinetic Data	Remarks	Ref.
190,000-570,000	IR	Cathode, living anionic	Direct electron transfer cyclic voltammetry; addition of protic substances	Yes	Electrochemical yield is 0.22 mole/F. Attributed to dimerization and impurities.	14
—	—	Radical at cathode	Polarography	Yes	Mainly a polarographic study.	146
—	—	$Mg(OH)_2$ initiates by polarization of C=N bond	Polarography; absorption studies by differential capacitance	Yes	—	147
ACRYLIC ACIDS						
—	—	Cathodic free radical	Inferred	—	Polymerization is compatible with hydrogen over voltage series.	51
—	—	Kolbe-type reaction at anode	Inferred	—	The addition of MeOH gave a better yield.	95
—	—	Kolbe-type reaction at anode	Inferred	Yes	Dimerization also a major process. Aq. MeOH gave a better yield.	148
—	—	At Anode: Kolbe At Cathode: hydrogen atoms	Inhibitors added; voltammetry	—	—	149

[η] / MW	Characterization	Mechanism	Observation		Remarks	Ref.
—	—	Cathode. O_2 + diquat radical cation initiates polymerization	Must have O_2 in small concentration for polymer. Coloration.	Yes	N,N'-ethylenebisacrylamide produces a gel. Green color due to radical cation.	150
ALDEHYDES/VINYL ALDEHYDES						
—		Free radical	Yellow color	No	Phenol also polymerizes under the same conditions. Low yield. Acetaldehyde does not polymerize.	151
1000-5000	IR suggests polymerization via $C=C$ bond and $C=O$	Cathode, anionic	Induction period, methanol prevents polymerization	Yes	Various other salts tried. Electrochemical yield higher than the cationic case [153].	152
560-1120	—	Anode, no polymer or cathode cationic	Coloration in anode while cathode remains colorless	Yes	Gelation occurred at 60% conversion. Induction period of about 20 min at low temperature and small currents. Other polymers prepared.	153
VINYL ACETATE						
80-99 ml/g (acetone, 25°C)	Polymer very hard. X-ray shows to be amorphous.	Anode: Kolbe free radical	With a double cell, polymer at anode only	No	Molecular weight depends on the current density. Under same conditions, a low yield of uncharacterized polyvinyl chloride.	55

Table 11.6. (continued)

Molecular Weight	Characterization Techniques	Proposed Mechanism	Means of Determining Mechanism	Kinetic Data	Remarks	Ref.
			VINYL CHLORIDE			
—	—	—	—	—	High voltage (1200 V) used.	19
			ACENAPTHYLENE			
2000–3000	IR and visible spectra	—	—	—	—	154
			MALEIC ANHYDRIDE			
2600	—	Cathode formation	—	No	—	155
			N-VINYL CARBAZOLE			
9.1 ml/g	—	Cationic, discharge of ClO_4^- at anode	Inferred, copolymer studies	No	"After-effect" noted.	49
			N-VINYL PYROLLIDONE			
		Cationic, discharge of ClO_4^- at anode	Inferred, copolymer studies	No	—	49

Polymer insoluble		Insoluble sheets of polymer formed on anode	The polymerization was not amenable to kinetic study		Low electrical efficiency.	
n = 3,000–12,000	Osmometry	Condensation, anode		Yes		156
			PHENOLS			
—	—	—		No	Variously substituted phenols.	197
—	—	—		No	—	157
$\overline{DP}$ 10–1500	—	Anodic		No	—	158
			COPOLYMERS			
—	—	—	MMA/methacrylic acid	Yes	Polarographic study	159
—	—	—	Styrene/MMA	Yes	Homopolystyrene also formed	160
Approx. 10^6	NMR, IR	Charge-transfer complex	Molecular weight, 1:1 alternating copolymer formed, anode	No		169
Approx. 10^6	NMR, IR	Charge-transfer complex	As before, cathode St/An	No		170
Approx. 10^6	NMR, IR	Charge-transfer complex	As before, cathode St/An	Yes	Mechanism of propagation proposed depending on a macrocyclic intermediate	11
Approx. 10^5	NMR, IR	Charge-transfer complex	As before, cathode Butadiene/AN	Yes		12

Table 11.6. (continued)

Molecular Weight	Characterization Techniques	Proposed Mechanism	Means of Determining Mechanism	Kinetic Data	Remarks	Ref.
4×10^5	NMR, IR, GPC	Charge-transfer complex	As before, cathode Diethylfumarate/St	Yes		199
—	NMR	Cationic	Dioxan/1,3-Dioxalan	Yes		198
			MISCELLANEOUS			
—	—	—	—	—	These are very general patents chiefly disclosing the salt types used.	161, 162
3000	IR	Anionic	Methyl vinyl ketone Inferred from kinetics and IR spectra	Yes	A polarographic study of the decay of nitrobenzene radical anions in methylvinyl ketone solutions.	13
—	—	Anionic	Ethylene and propylene oxides Inferred from kinetics	Yes	A polarographic study of the decay of naphthalene radical anions in olefin oxide solutions.	18
Viscosities given	—	Cationic	Trioxan Inferred, water inhibits polymerization	No	The melting point of the poly (oxymethylene) was dependent in the polymerization temperature.	24
$\leqslant 1840$	Viscometry	Cationic, anode	Trioxan	Yes	Some living characteristics	200

5 REACTION MECHANISMS

Effects of High Electric Fields

One should compare and distinguish the phenomena associated with electro-chemical polymerization from those existing in polymerizations conducted under the influence of an external electric field. The latter subject has been reviewed recently by Ise [4], who has been one of the prime investigators of polymerizations conducted in this manner. However, Ise has on a number of occasions expressed doubt about the very existence or validity of electrochemical polymerization [165, 4, 166, 167].

First, it should be noted that polymerizations conducted under a high electric field represent reactions in which an increase in rate, of an already existing polymerization, is noted. Ise has meticulously attempted to compensate for the Joule heating effects which accompany the passage of current through media of high resistance. This has now been done convincingly, and the enhancement in rate of polymerization is attributed to such factors as increase in degree of dissociation of ion pairs with field strength and to orientation effects in the electric field. In electrochemical polymerization, initiation of polymerization is produced and changes in *concentration* of the reacting species are found. This is best illustrated by reference to Fig. 11.9 where the application of an electric field changes the rate constant while apparently not changing the concentration of the living ends [166]. On the other hand, control of living-end concentrations as a function of the number of Faradays passed through the system has been illustrated by Funt, Bhadani, and Richardson in Fig. 11.10 [101], where the concentration of living ends measured spectrophotometrically agreed stoichio-metrically with the concentrations determined by Faraday's laws.

The magnitudes of fields necessary are completely different in the two types of phenomenon. In the work by Ise and co-workers, field strengths of 2000-10,000 V/cm were employed, and other workers have used even higher fields. In most electrochemical polymerizations, the electric fields are 100 times smaller while currents are several magnitudes larger than those employed by Ise.

Furthermore, the behavior of the monomer or of the growing polymer chain will be quite different in the presence or absence of a supporting electrolyte. The supporting electrolyte acts as the major ionic current carrier and prevents the migration and orientation effects that would exist in the system without salt. The voltage is not sufficient to cause second Wien effects in either salt or monomer.

Nevertheless, it is important in view of the results of Ise to monitor the voltages applied in electrochemical polymerizations in order to ensure that field effects are indeed negligible. In practice, however, it is difficult to obtain the currents necessary for electropolymerizations and the voltages necessary for electric field effects without great overheating of the reaction solution. For

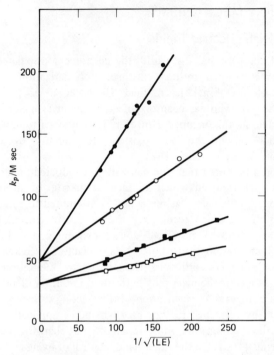

Fig. 11.9. Dependence of apparent propagation rate constant on living-end concentration; styrene-n-butyllithium-benzene-THF, 25°C. THF content (vol. %): circle, 60; square, 50. The filled marks are for 5 KV/cm and the blank ones for 0 KV/cm. The increased rate at a given concentration of living ends (LE) is attributed to the effect of electric field. Reprinted with permission, from Ref. 166.

example, the application of 1 kV to a solution at a current of 10 mA will produce 10 W and this would cause a considerable heating effect with the usual electrode geometry employed in polymerization cells.

A recent, and most novel, paper by Giusti and co-workers [189] has in fact put the whole interpretation of the "electric field" effect into some question. Their work deals with the attempted field enhancement of anethole and acenaphthylene monomers initiated by iodine or boron trifluoride etherate but has been tentatively shown to be more widely applicable. By periodic field reversal, the rate enhancement (noted in the same systems with unchanging field direction) was completely suppressed. The Ise interpretation of the effect (either second Wien effect or a desolvation process of the free ions) cannot explain these results, since these processes should not be affected by field reversal. Giusti et al. tentatively suggested that rate enhancement effects in an unidirectional field could be due to localized variations in the characteristics of the polymerizing solutions, particularly with respect to the concentration of the

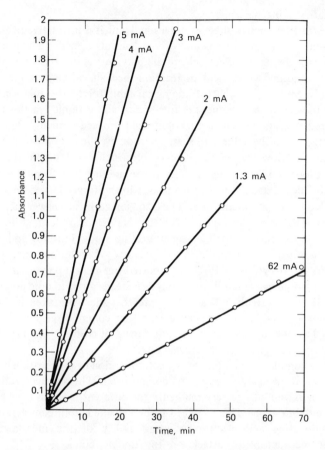

Fig. 11.10. Increase of absorbance due to increase of living end concentration with time of electrolysis. Solutions of α-methylstyrene in THF with NaPh₄B as supporting electrolyte. Reprinted with permission, from Ref. 107.

chemical species involved. These workers report that further studies are under-way in their laboratories to pursue this point.

The interrelationship between electric field effect and chemical initiation at the electrodes has been debated recently [201, 202].

Criteria for Determining Reaction Mechanisms

It is important to determine not only the nature of the electrode process by which an initiating species is generated, but also the mechanism by which the initiating species promotes the polymerization. Examples have now been reported in the literature of electrochemically initiated polymerization that proceed by free-radical, anionic, cationic, condensation, living anionic, and living cationic mechanisms.

Before viewing some of these examples in detail, it is instructive to examine the basis upon which the assignment of mechanism is made. There are specific conditions existing in electrochemical polymerization that complicate and thwart the normal assignment of reaction mechanism. One of the techniques commonly employed is that of adding an inhibitor to the solution. Thus, on the addition of benzoquinone, the free-radical and cationic reactions should be suppressed and anionic polymerization could proceed normally. This simplistic interpretation can be subject to some question even in conventional polymerization systems. However, in the case of electrochemical polymerization other complications are encountered. The benzoquinone may itself be subject to electrochemical attack. This in turn results in the formation of substances that are initiators of polymerization. Thus, an unequivocal answer cannot be obtained by reliance on this type of inhibitor study alone. Indeed, quite misleading results may be encountered. In an interesting study Trifonov and Panayotov [60] electrolyzed acrylonitrile in the presence of anthraquinone, with Et_4Ni or LiCl as electrolyte in DMF. They reported free-radical polymerization on the basis of copolymer compositional analysis instead of the anionic polymerization expected. Such a result indicates the care that must be taken in interpreting the results of "inhibited" experiments. Other common diagnostic techniques employed in polymer chemistry are not devoid of criticism when applied to the conditions existing in electrochemical systems.

Another common technique for the determination of propagation mechanisms is based upon the analysis copolymer compositions. The reactivity ratio of a pair of monomers will change very markedly for different reaction mechanisms and experimental systems. This is indeed to be expected from consideration of the electron-donating and electron-accepting ability of the monomer pair for different mechanisms of attack on the double bond.

This is shown quite clearly by the initial copolymer composition formed from a 1:1 styrene-MMA feed. If the polymerization takes place by a free-radical mechanism the percent of styrene in the initial polymer is found to be 51%. For carbonium ion propagation the styrene content in the initial polymer is greater than 99%; for carbanion polymerization the styrene content is less than 1%.

Again one must be wary of the results obtained by electrochemical techniques. The electrolyte frequently contains a metal salt as a supporting electrolyte. Preliminary results in the authors' laboratory indicate quite clearly that such electrolytes can drastically alter the mechanism of the propagation reaction, which may then proceed via a charge transfer complex [169, 170]. The nature of the electrode material is also of some importance in this respect.

Gaylord has shown quite clearly that with zinc chloride and other salts charge transfer complexes will be formed between donor and acceptor monomers. These salts facilitate the formation of such charge transfer complexes between suitable pairs of monomers. Various transition metal halides, Friedel Crafts catalysts and ethylaluminumhalides have been found to be particularly effective in initiating charge transfer complex polymerization [171-173].

A polymerization that occurs by this mechanism will in general lead to the formation of a 1:1 alternating copolymer. This copolymer composition will be obtained despite quite different reactivity ratios of the monomers, and will hold over a wide range of initial monomer feed concentrations and temperatures. This situation is further complicated by the occurrence of 1:2 charge transfer complexes in certain instances [174] and even by inadequate experimental techniques that fail to physically separate homopolymer and copolymer fractions (often difficult experimentally) from each other prior to compositional analysis.

It must be stressed here that the actual mechanism of propagation is similar to that of a conventional polymerization. In electroinitiated polymerization a reaction occurs at the electrode with either monomer or supporting electrolyte by which a chemical entity is formed which initiates a polymer chain. This chain propagates at a rate that is essentially not influenced by the electrode process, although the number of chains produced will be dependent on the current passage at the electrode. Some secondary effects may result which reflect the influence of the electrolyte on the reaction. For example, metal salts, such as zinc halides, can facilitate the formation of charge-transfer complexes between monomer pairs in copolymerization reactions. In living anionic reactions, the presence of sodium tetraphenyl boride salts, for example, can markedly effect the equilibrium between propagating free ions and ion pairs [25]:

$$\sim\!\!\sim\!C^{\ominus} Na^{\oplus} \rightleftharpoons \sim\!\!\sim\!C^{\ominus} + Na^{\oplus}$$

Ion pair   Free ions

This change in the equilibrium can have widely varying effects on the course of the reaction and the reaction rate [26]. Finally it should be noted that products formed at the counter electrode may serve to terminate the growing polymer chain.

Cationic Polymerizations

Surprisingly few investigations of cationic electropolymerizations have been reported. Despite the high yields and electrical efficiencies first reported by Breitenbach in his review [49], there are only a few fragmentary reports of additional investigations. In part this can be attributed to the difficulty of obtaining reproducible kinetic results in cationic polymerizations conducted in media containing several components.

Breitenbach reported the cationic polymerization of isobutylvinyl ether and of styrene by cationic techniques. In this pioneering report there were few details of kinetic data and kinetic analysis.

Recently, Heins [142] has reported on the polymerization of tetrahydrofuran. There is also mention of the polymerization of this solvent by Bhadani [175] and by Yamazaki [47]. A kinetic investigation of the polymerization have been conducted in the author's laboratory by Dyck [143].

The first detailed investigation of the kinetics of cationic polymerization has been published by Schulz and Strobel [153]. They have shown that acrolein may polymerize by a cationic mechanism, in contrast to an earlier paper [152] describing an anionic mechanism. The difference lies entirely in the reaction conditions employed in the solution. Cationic initiation and propagation is supported in nitromethane solutions with lithium perchlorate as electrolyte, while an anionic mechanism requires a THF solution of sodium tetraphenyl-boride. Some kinetic data obtained by these authors are shown in Fig. 11.8. Funt and Blain [141] have investigated the cationic polymerization of iso-butylvinyl ether. They have shown that successive pulses could be used to accelerate the reaction sequentially. A reversal of the current decreased the rate of polymerization in the cationic reaction. The same authors have investigated the kinetics of the polymerization of styrene in methylene chloride [105].

Anionic Polymerizations

It is possible to distinguish two types of anionic reactions initiated electro-chemically. In the first, the conditions of the polymerization are such that the lifetime of the growing polyanions is comparable to the time of growth of an individual chain. Propagation and termination are competitive processes, and indeed the length of chain or the degree of polymerization of the polymer will be dependent on the concentration of terminating substances.

In such systems the reaction will effectively cease when the current is interrupted. The data for the polymerization of the acrylonitrile are exemplary of such anionic reactions.

However, it is also possible to conduct anionic polymerizations under conditions in which the growing-chain ends preserve their activity, or ability

However, it is also possible to conduct anionic polymerizations under conditions in which the growing-chain ends preserve their activity, or ability to polymerize, indefinitely. Such polymerizations are referred to as "living" polymerizations and were first reported by Szwarc [176]. Yamazaki et al. intro-duced the use of tetrahydrofuran as a solvent in electropolymerization reactions and was able to obtain living polymers [98]. Funt, Bhadani and Richardson showed that the concentration of living ends could be increased or decreased by application of current and by the reversal of polarity [101]. They applied this technique to obtain polymers of controlled-molecular-weight distribution [104].

The ability to increase or to decrease the concentration of living ends by electrochemical means offers a fascinating spectrum of possibilities for the conduct of polymerization reactions. In principle it should be possible to control the rate profile of the reaction throughout the course of the polymerization. For example, Funt and Bhadani [112] compensated for the decrease in living-end concentration during the polymerization of isoprene by monitoring the living-end concentration spectrophotometrically and then feeding a compensating current through the solution to maintain the living-end concentration constant. It is in principle possible to maintain a constant rate of reaction by programming the current flow to compensate for the decrease in monomer concentration during the course of the reaction. Also for reactions where there is an abrupt increase in rate, the ability to decrease the rate of reaction electrochemically is a marked advantage.

On the other hand, there are obvious shortcomings to the technique. Conduct of living anionic polymerizations demands conditions of a very high purity and freedom from proton-donating substances. The electrochemical technique will provide a greater tolerance to the initial presence of impurities, inasmuch as it is possible to "preelectrolyze" the solution to minimize these effects. Nevertheless, the supporting electrolyte and solvent must be chosen to provide the desired electrochemical characteristics while avoiding porton-donating substances. Some quaternary ammonium salts cannot be employed because of proton abstraction, but cations such as tetramethylammonium, which do not contain β-hydrogen atoms [50], can support polymerization provided that the lifetime of the reaction is short, as in the case of styrene. Some hydrocarbon solvents cannot be utilized because of low conductance and the absence of suitable supporting electrolytes to furnish adequate conductance for electrolyses. In molecular weight control, the rate of growth of the polymerizing chain must be slow relative to the rate of initiation if monodisperse molecular weight distributions are to be obtained. This condition is not attainable with all monomers, and indirect techniques must therefore be used.

With suitable monomers the potentialities of the technique are quite remarkable. Hornof has recently shown that with polybutadiene the growth of the polyanion is easily controlled by the application of current, and that cycling techniques can produce either monodisperse or accurately defined polydisperse fractions [177].

The details of the initiation step have been investigated by Yamazaki [99]. He concluded from polarographic studies that direct electron transfer to the monomer occurs in solutions of dimethoxyethane [72]. The formation of living anions using tetramethylammonium cations has been shown by Gray in the author's laboratory [178] and by Anderson [50]. In such a system, the initiation by an alkali metal is ruled out completely. The addition of a quaternary ammonium salt to a solution of living anions will result in the destruction of the living anion activity by proton abstraction. With $(C_4H_9)_4N^+$ this takes place very quickly, but is much slower with $(CH_3)_4N^+$. With the latter the decrease in absorbance as a function of time can be measured spectrophotometrically over a period of minutes.

During the propagation step the composition of the electrolyte solution and the concentration of supporting electrolyte is important in determining the nature of the growing species. Free ions, ion pairs, or triple ions may be involved in the propagation step, and the rate constant will vary drastically from one type to another. Thus the reactivity of the free ion is some 500 times greater than that of the ion pair in styrene. These parameters have not been investigated separately in any electrochemical polymerization to date. Nevertheless, the detailed studies which have been performed on living anionic polymerizations indicate the need for considering of the effects of ionic environment upon the rate of the polymerization reaction.

Free-Radical Polymerizations

Electrochemical polymerizations proceeding by a free-radical mechanism were the first to be discovered and have been the most frequently investigated. In aqueous systems initiation by hydrogen atoms is postulated at the cathode, or by radicals such as methyl at the anode. In general the current efficiency is low. This is probably due to the short lifetime of radicals before recombination and their trapping by cage effects and by absorption in the vicinity of the electrode. Thus the surface concentration of radicals must exceed the bulk concentration even under conditions of rapid stirring. In aqueous systems there usually is a precipitation of polymer at the electrodes, and the kinetics of the reaction cannot be studied. In nonaqueous systems where both monomer and polymer are soluble in the solvent, detailed kinetic studies can be performed. These show a dependence of the rate of polymerization on the impressed current and an inverse dependence of the molecular weight on the rate of reaction.

In the polymerization of fluorinated hydrocarbons much higher current efficiencies have been reported [106-108].

A report that the rapid rate of stirring in a reaction vessel [57] caused a lowering in the rate of polymerization and a lowering in the polymer molecular weight is of some note. The author suggested that this could be due to an increase in the rate of termination caused by the increased stirring. Except for very viscous solutions such an effect seems unlikely.

It has been suggested that chain transfer could be responsible for low-molecular-weight polymers formed in PAN polymerization [49]. Chain transfer occurs when the growing polymer radical reacts with part of the substrate, X-H, and results in a "dead" polymer molecule:

$$\text{wwCH}_2-\underset{\underset{\text{CN}}{|}}{\text{CH}}\cdot \ + \ \text{X}-\text{H} \longrightarrow \text{wwCH}_2-\underset{\underset{\text{CN}}{|}}{\text{CH}}_2 \ + \ \text{X}\cdot$$

Growing polymer   Transfer   "dead"
   radical   agent   polymer   Further reaction

X-H is part of the substrate and called the transfer agent. It will usually contain an abstractable hydrogen or halogen atom. X-H may typically be

1. another monomer molecule,
2. a "living" or "dead" polymer chain,
3. some other, usually low-molecular-weight, compound, typically a halogenated hydrocarbon or a thiol.

Then X· may react further:

$$\text{X}\cdot \ + \ \text{CH}_2{=}\underset{\underset{\text{CN}}{|}}{\text{CH}} \longrightarrow \text{X}-\text{CH}_2-\underset{\underset{\text{CN}}{|}}{\text{CH}}\cdot$$

The probability of this reaction occurring will depend on the reactivity of X· towards the monomer.

The nature of X-H could well be a function of the type of electrode reaction products generated in solution.

Condensation Polymerizations

A molecule that possesses two functional groups, each of which is subject to attack by electrolysis, is potentially a substance that can lead to a stepwise condensation polymerization. Suppose a free radical is formed by the electrolysis of one of the functional groups. Then coupling will lead to two functional groups and the process would thus be repeatable. However, attempts to perform such condensation polymerizations with carboxylic acids are frustrated by the lack of Kolbe coupling in dicarboxylic acids.

Among specific successful examples of condensation polymerization is the recent work by Toy [132], who has shown that perfluoroglutaric acid will under such coupling to form polydifluoromethylene.

The formation of poly(p-xylylenes) by the electrolytic reduction of α,α-dihalo-p-xylenes at controlled cathodic potentials, has also been achieved [116]. The condensation of phenols was reported by Borman [158] and Dijkstra and De Jonge [197].

Charge-Transfer Complex Copolymerizations

Recently [169, 170, 11, 12, 199] there has been some investigation by the authors of the electroinitiation of "charge-transfer" copolymerization. This has been mentioned in another context (see p. 602) but merits a fuller description. In these systems suitable pairs of monomers react, with or without a catalyst, to form a donor-acceptor (or charge-transfer) complex. This complex then acts as a "monomer" and may polymerize to give a substantially 1:1 alternating copolymer. Consider, for example, styrene and acrylonitrile. Styrene can act as a donor of electrons to the π-electron system of acrylonitrile when the AN itself has been complexed and made more "electron-accepting" with a salt such as zinc bromide. A charge-transfer complex is thus formed:

$$CH_2=\underset{\underset{C\equiv N}{|}}{CH} \;+\; ZnBr_2 \;\rightleftharpoons\; CH_2=\underset{\underset{C\equiv N \,\cdots\, ZnBr_2}{|}}{CH}$$

Acrylonitrile   An isolatable complex

(I)

$$I \;+\; CH_2=\underset{\underset{C_6H_5}{|}}{CH} \;\rightleftharpoons\; \left[CH_2=\underset{\underset{C_6H_5}{|}}{CH} \;\cdots\; CH_2=\underset{\underset{CN \,\cdots\, ZnBr_2}{|}}{CH} \right]$$

Styrene   Donor-acceptor or charge-transfer
complex

(II)

II can then be thermally initiated to polymerize to give a regular 1:1 copolymer of alternating character:

$$n\,II \;\longrightarrow\; \text{polymerize to} \left[CH_2-\underset{\underset{C_6H_5}{|}}{CH}-CH_2-\underset{\underset{C\equiv N}{|}}{CH} \right]_n$$

1:1 alternating St/AN copolymer

However, thermal initiation of the system tends to disturb the various equilibria involved in the reaction. We have employed electrolytic reactions to initiate the polymerization which can then be performed at any temperature desired. So far four copolymerizing systems have been studied; butadiene-AN with zinc chloride catalyst [12], styrene-MMA with zinc chloride [169], styrene-AN with zinc bromide as catalyst [170, 11], and styrene-diethyl fumarate with zinc bromide [199]. Results show that the reactions are definitely initiated by passage of current. A mechanism of the initiating reaction has not been elucidated, and work is continuing on this point.

6 ELECTROCHEMICAL CONTROL OF MOLECULAR STRUCTURE

Molecular Weights

The controlled variation of current has been employed to porduce polymers of particular molecular structure. The control of molecular weight is exemplary of the physical modifications that can be induced in the growing polymer chains.

In free-radical reactions the rate of propagation, R_p, is

$$R_p = k_p [M] [M\cdot]$$

where $[M\cdot]$ is the concentration of growing radicals, $[M]$ the monomer concentration, and k_p the rate constant for propagation.

The kinetic chain length v, defined as the number of monomer units consumed per active center, is thus $v = R_p/R_i$, where R_i is the rate of initiation. With a conventional chemical initiator (Cat),

$$R_i = 2f k_d [Cat],$$

where f is the fraction of radicals that initiate chains, and the factor 2 arises from the assumption that each initiator molecule can provide two radicals, and k_d is the rate constant for decomposition of the initiator.

In the absence of chain-transfer reactions, the degree of polymerization $\overline{DP}$, and the molecular weight (which is merely $\overline{DP}$ times the molecular weight of the monomer unit) are directly related to v. For termination of the growing chains by combination, $\overline{DP} = 2v$, and for termination by disproportionation $\overline{DP} = v$. Thus for free-radical reactions initiated electrochemically, as the current is changed the molecular weight should vary accordingly. Funt and Yu [58] and Tidswell [67] have reported a linear rate dependence on the impressed current, while, in contrast, a radical initiator would normally have produced a dependence on $[Cat]^{1/2}$. The molecular weight was inversely proportional to the current, or to the rate of polymerization.

The electrochemical technique cannot produce polymers with sharply defined molecular weight distributions in free-radical polymerization. The process of radical termination by mutual combination is statistically controlled and inevitably leads to a broad molecular weight distribution. Secondary processes resulting from kinetic chain transfer, changes in diffusion due to increase in viscosity with reaction time, cage effects, and other factors all tend to broaden the molecular weight distribution even further. All this is inherent in the reaction mechanism and is found in polymerizations conducted with free-radical initiators generated by thermal decomposition or by photolysis. Thus, though the rate and molecular weight can be varied or controlled more readily in electrochemical polymerization, the sharpness of a particular molecular weight distribution cannot be increased by variation of the impressed current.

The method that will produce the most monodisperse molecular weight distributions is that based on anionic polymerizations. In the "living anionic" polymerizations pioneered by Szwarc [179], the growing chain ends referred to as "living ends" (LE) will not terminate with one another because of charge repulsion, and in a pure medium, free from proton-donating impurities and other destructive reactions, the living ends retain their reactivity indefinitely. The rate of reaction is given by

$$\text{Rate} = k_p \, [\text{LE}] \, [M]$$

The molecular weight in such polymerization is given simply by $\overline{DP} = n \, [M] \, / \, [\text{LE}]$, where n is the number of living ends per polymer molecule (usually one or two).

Funt, Bhadani, and Richardson [101] and Funt and Richardson [109] have shown that it is possible to apply electrochemical techniques to produce polymers with a controlled distribution of molecular weights. This is achieved by producing a specified concentration of growing ends and conducting the polymerization until a predetermined quantity of monomer has been polymerized. A reversal of the current then produces a stoichiometric destruction of living ends until the living-end concentration reaches a new and lower value. The polymerization is then allowed to proceed until a second portion of monomer or the remaining monomer is consumed.

A cycling technique has been applied to produce polymers of high degree of monodispersity by employing an anion whose rate of propagation is quite low. Polybutadiene with Mw/Mn ratios less than 1.10 have been produced, and accurately programmed molecular weight distributions have been formed by this technique as well [177].

Grafting Reactions

To date electrochemical grafting reactions may be categorized in two main types: (1) grafting reactions to cellulose and (2) grafting reactions to polyelectrolyte chains. These will be discussed separately.

Grafting to Cellulose

Epstein and Bar-Nun [62] in a paper published in 1964 reported the grafting of MMA on to a cellulose film. They placed a sheet of cellophane in a circular platinum screen in the cathode compartment of a double cell. The cathode itself was a pool of mercury and the anode was platinum. The reaction medium was sulphuric acid with a separate layer of MMA. At electrolysis currents of between 50 and 200 mA polymerization was allowed to continue for several hours, during which time the solution grew more turbid as the organic layer disappeared. The cellophane was washed with benzene several times and the graft yield was estimated by weight (ranging from 40% to 70%). The yield itself changed with the current and with the acid strength but both were varied contemporaneously, in the same sense, and the increase in graft yield could have been due to one or both of the parameters mentioned. The authors surmise that the mechanism could have been due either to chain transfer at cellulosic C-H bonds followed by radical initiation and grafting at the cellulose or by a polymer-cellulose termination reaction. In 1967 Simionescu and Ungureanu [79] grafted AN onto cellulose films using apparatus similar to that of Epstein and Bar-Nun. In this case, however, the cathode compartment contained a ferric sulfate-potassium persulfate redox couple in sulfuric acid solution with AN. The reaction was carried out in the presence of air at 25°C. The films were extracted with DMF to remove ungrafted PAN and the grafted film was subjected to elemental analysis to determine the amount of nitrogen present and thus the extent of grafting. It was shown that the amount of grafting was virtually independent of persulfate concentration but dependent on the ferric iron and monomer concentrations and also the current density and time of electrolysis. The mechanism was thought to be as follows,

$$Fe^{3+} + e \longrightarrow Fe^{2+}$$

$$Fe^{2+} + S_2O_8^= \longrightarrow Fe^{3+} + SO_4^= + SO_4^{\cdot-}$$

$$SO_4^{\cdot-} + H_2O \longrightarrow HSO_4^- + HO\cdot$$

followed by,

$$[Cellulose\ \bar{}H] + SO_4^{\cdot-} \longrightarrow [Cellulose]^{\cdot} + HSO_4^-$$

and/or

$$[Cellulose\ \bar{}H] + HO\cdot \longrightarrow [Cellulose]^{\cdot} + H_2O$$

The cellulose radical was then able to initiate AN polymerization in a grafting reaction.

It should perhaps be pointed out that the graft yields were in both cases determined in part by solvent extraction. If any crosslinking or other form of "insolubilization" reaction had occurred, then the polymer would not have dissolved away and would have tended to show higher graft yields than was the true case.

Grafting to Polyacids

This work may be entirely attributed to Smets and co-workers. Smets, Poot, Mullier, and Bex [127] electrolyzed partly neutralized polyacrylic or polymethacrylic acids in aqueous solution. They reported low electrical efficiencies for Kolbe-type decarboxylation reactions, the bulk of the current passed being used in the electrolysis of water. At 2-3°C, hydroperoxides were formed, and their analysis indicated a direct correspondence between the decarboxylation and hydroperoxidation reactions. At 25°C, no peroxides were detected. The peroxidized polymers were then used to initiate grafting reactions with such monomers as acrylamide, AN and vinyl pyrollidone:

$$\sim\text{CH}_2\text{–}\underset{\underset{\text{O}}{\overset{\|}{\underset{\|}{\text{C}}}}{\overset{\text{R}}{\text{C}}}\sim \xrightarrow[-e]{\text{Anode}} \sim\text{CH}_2\text{–}\overset{\text{R}}{\underset{\underset{\text{O}^{\cdot}}{\overset{\|}{\underset{\|}{\text{C}}}}}{\text{C}}}\sim$$

$$\sim\text{CH}_2\text{–}\overset{\text{R}}{\underset{|}{\text{C}}}\sim + \text{CO}_2$$

(P·)

$$\text{P·} + \text{O}_2 \xrightarrow{\text{H}_2\text{O}} \text{PO}_2\text{H} + \text{·OH}$$

Polymer hydroperoxide group

PO_2H may be isolated and later on heated to 25°C in the presence of a different monomer to initiate polymerization:

$$\text{PO}_2\text{H} \xrightarrow{\text{Heat}} \text{PO·} + \text{·OH}$$

$$\downarrow + \text{Monomer}$$

Graft copolymer

Smets, Van Der Borght, and Van Haeren [128] and Smets and Bex [129] electrolyzed partially neutralized polymethacrylic acid at 20°C in dry methanol solution. In a Kolbe-type reaction, CO_2 was evolved at the anode and the macroradicals so formed were found to react to give γ-lactones by intramolecular lactonization, unsaturated groups by degradation or bimolecular disproportionation, and finally methoxy groups by reaction with methoxy radicals. Macroradical coupling was suppressed by the use of dilute reaction conditions (less than 25% by weight of polyacid). Another feature of this work was the simultaneous electrolysis of the polyacid with some low-molecular-weight carboxylic acids (the subject of the patent [129]) such as acetic acid, N-acetylamino-caproic acid, and cyanoacetic acid. These reactions, termed by the authors "crossed-Kolbe" syntheses, gave grafts of methyl, alkylamino acid, and cyanomethylene groups, respectively:

$$NC-CH_2-COO^\ominus \xrightarrow{\;-e\;} NC-CH_2{}^\cdot + CO_2$$

Cyanoacetate anion

$$P^\cdot + NC-CH_2{}^\cdot \longrightarrow \quad \wedge\!\!\wedge CH_2 - \overset{\displaystyle R}{\underset{\displaystyle CH_2-CN}{\overset{|}{\underset{|}{C}}}}$$

Polymer   Cyanomethylene
radical   radical

Lastly, Smets, Van Gorp, and Van Haeren [130] electrolyzed aqueous acetone or dry methanol solutions of methacrylic acid-methylmethacrylate and acrylic acid-methylacrylate copolymers. Hydroperoxide groups were formed and graft terpolymers were prepared with styrene. The effect of copolymer acid content on the degree of decarbosylation and lactonization was studied as well as the electrolysis of dry methanol solutions of poly(methacrylic acid) in the presence of pyridine or transfer agents such as butylmercaptan. In this way it was found possible to graft foreign groups on to this polyacid chain.

Crosslinking

There has been no systematic study of crosslinking reactions brought about by electrolysis to date. However, crosslinking has sometimes been postulated by some authors to explain gelation and electrode coating. For example, Das and Palit [52] noted that the gelation of PMMA in their system was probably attributable to a crosslinking reaction at the electrode. Smets, Van Der Borght, and Van Haeren [128] were able to suppress crosslinking reactions in their Kolbe-type electrolyses of polyelectrolyte solutions by using dilute solutions, while in two patents [157, 158], poly(phenylene oxides) as three-dimensional networks were prepared by electrolysis of phenol or substituted phenol solutions. The copolymerization of acrylamide with N_1N'-methylenebisacryl-amide [136] gave a presumably crosslinked coating on the cathode while the copolymerization of N,N'-ethylenebisacrylamide with methacrylic acid gave a green-colored gel [150].

The possibility of coating an electrode (see next section) and inducing grafting (and crosslinking) reactions in a subsequent electrolysis has yet to be attempted:

$$CH_2=CH-CH=CH_2$$

1,3 Butadiene

Electrode surface →

Film of poly (1,2-butadiene)

Controlled potential Electrolysis + another monomer ↓ Copolymer film

The selectivity available in controlled electrode potential electrolysis may be of great value in this instance and in the discriminatory electroreaction of specific end or side groups.

Surface Coatings

In much of the early work on electroinitiated polymerization, the coating of an electrode surface by a polymeric material was an unwanted and unfortunate by-product of many heterogeneous reactive systems.

However, in quite recent years some patent work (which has formally been reviewed prior to this under monomer headings) has been published where polymer coatings on electrode surfaces have been the desired end products.

Ross, Peterson, and Finkelstein [121, 120] were able to produce poly(p-xylylene) films as coatings on aluminum cathodes. The electrodes were immersed in DMF solutions of mono or bis onium salts of p-xylylene, such as p-xylylenebis (trimethylammonium nitrate), for example, and electrolyzed at controlled currents of between 200 mA (initially) and 20 mA. A recent British patent, assigned to W. R. Grace & Co. [157] reports the polymerization of phenol producing a poly(arylene oxide) coating on an iron anode. The electrolyte/medium was aqueous sodium hydroxide and a film of approximately 2.5 microns thickness was deposited on the iron. Similarly, other metals (nickel, copper, titanium, tin, cadmium, and lead) were capable of taking coatings. The cathode was platinum. Ehrig and Kundel [140] formed a coating of poly-(diacetone acrylamide) on various cathode metals from aqueous solution, while Sobieski and Zerner [136] reported the copolymerization of a 10:1 (wt %) mixture of acrylamide and N,N'-methylenebisacrylamide in aqueous zinc chloride solutions containing a trace of copper chloride. The copolymer coating was formed as a 0.1-mm layer on the cathode, and the adhesion of the film was found to be strongly dependent on the pH of the solution. The addition of peroxides to the reactants increased the rate of copolymerization. In an article entitled, "Electropolymerization. A new technique for applying paint," Oliver

[180] reported the polymerization of monomeric diacetone acrylamide onto various electrodes from dilute sulphuric acid in the absence of oxygen. The films could be cured at 200°C to improve some of their physical properties.

However, the bulk of systematic research on this subject is attributed to the Tokyo University school of Asahara and his co-workers. Their first papers were general in nature and reported the surface coating of steel [181] and aluminum [182] by the electroinitiated polymerization of a variety of monomers (unsaturated carboxylic acids, aldehydes, trioxane, and vinyl compounds such as AN, MMA, acrylates, acetates, and styrenes). Various salts were used as electrolytes ($NaClO_4$, McKee's salt, KNO_3, and Et_4N-p-toluenesulphonate) with or without the use of certain solvents (DMF, MeOH, and methylethyl-ketone). The weight and thickness of the films generally depended on the current density and the time of electrolysis, while molecular weights were generally low (1000-2000). Asahara, Seno, and Tsuchia studied in some detail the formation of poly(AN) films on various metals [183, 184] from styrene or benzene solutions or from pure AN solution. Copper metal was the chosen electrode [185] for the deposition of films of poly(AN) from benzene solution. In this case, however, the film had poor adhering qualities (which could be improved by heat treatment at 150-100°C). Under certain circumstances, when a film of poly(AN) had already been formed on an aluminum electrode in the presence of benzene, styrene, toluene, or xylene [186], further polymerization caused "whiskers" of amorphous polymer to grow away from the electrode into the solution. These varied in length from 1 to 3 cm and in thickness from 1 to 3 mm. Their molecular weight was about 1000. Films of poly (MMA) were formed [187] at the cathode in $NaNO_3$/DMF solutions. Care had to be taken to ensure that the electrolyte was not codeposited with the polymer coating, since this tended to reduce the latter's adherent properties. Copolymers of AN and various vinylic monomers [188] were deposited on a steel electrode. It was noted that the electrical efficiency of the process depended solely on the monomers present and not on the metal used as the cathode. This suggested that the electrodes, mechanistically speaking, were a source only of electrons in the initiation process.

7 CONCLUSION

An examination of the number of papers published in the last five years, compared with the previous five-year period, indicates quite clearly the growing and continuing interest in the electroinitiation of polymerization. Of particular note is the increased number of patents and the increased industrial activity of which this is evidence. This growth and interest in the subject parallels the recent advances in the much broader field of electroorganic chemistry.

It is appropriate, but perhaps presumptuous, to try at this point to assess the contributions of the past and to attempt to predict developments for the future. It is clear that the experimental side of electropolymerization studies has lagged behind the conceptual aspects of the subject. For mechanistic determination, controlled-potential rather than controlled-current systems should increasingly be employed. There are now readily available quite sophisticated electronic control systems for analytical and preparative work in electropolymerization. As these become more generally employed, one can expect a much more reproducible and defensible data base from which to establish the mechanisms of the reactions and the control of the reaction systems.

Electropolymerization has not as yet made possible new syntheses that are unattainable by other techniques. Although syntheses of particular monomers have been proven feasible under conditions that are more favorable experimentally, better controlled, or more convenient, the substances thus far polymerized are capable of polymerization by conventional means. However, this is not to indicate that the advantages of electropolymerization for some monomers are to be considered trivial. For example, when it is possible to polymerize a substance such as ethylene or tetrafluoroethylene at normal pressures, in contrast to the high-pressure techniques employed in chemical polymerization, the advantages may be considerable. Nevertheless, on an industrial scale there has thus far not been any example of a polymerization that led to the choice of electropolymerization as the method to forming polymer.

The possibility still exists, however, that stepwise condensation reactions could fall into such a category. It is, in principle, possible to build up a polymer by a series of sequential steps in which the addition of each electron would result in the linking of further monomer units to the chain.

Another interesting possiblity of syntheses that is yet to be explored is that of electrosynthesis to produce particular polymeric configurations. For example, block polymer formation by electrochemical means could be developed. The authors have hypothesized that the termination of a growing polymer chain by an electroactive group could be employed as a technique for regenerating a growing polymer and adding to it a second monomer.

Similarly, the formation of star-type polymers may be postulated to proceed if an initial electroactive substance that can simultaneously start from one site a number of growing polymer chains is electrolyzed.

It is also possible to picture hybrid electropolymerization and photopolymerization in which a sequential addition takes place involving the activation of photosensitive and the electrosensitive groups. For example, if a photopolymerization were induced and the reaction terminated by electrochemical means to insert a new chromophoric group, the latter could again be initiated by photolysis. Similarly, if an electrochemical reaction were initiated and then terminated with a chromophoric group, a subsequent photolytic reaction could be initated. Quite interesting types of block polymers should be capable of synthesis by these techniques.

The control of electrochemical polymerizations has already been demonstrated, and there will no doubt be elaborations of these techniques in the future. The possibility of including very specific types of molecular weight distribution to produce polymers with desired properties due to plasticization has been alluded to. Molecular weight control is interesting for the horizons that it opens for the syntheses of polymers with unusual molecular distributions and therefore with intriguing properties.

An area of considerable industrial interest and fundamental concern is that of surface coatings. The ability to proceed from a monomer solution to an adherent and impermeable coating offers attractive possibilities. There are advantages in the ease and extent of coverage, the self-healing properties associated with electrochemical techniques, and the "throwing-power" to cover areas inaccessible by conventional coating techniques. It may also be possible to control the thickness of the coating more accurately than with mechanical application.

One can emphasize that fundamentally the economics of electrochemical polymerization appear quite favorable. One can reasonably expect 10^2-10^3 moles of monomer per Faraday at applied potentials up to 10 V. This is approximately 10-100 kg/kW hr or perhaps of the order of 10 kW hr/ton. Obviously, this would not be a limitation on the utilization of electrochemical methods for polymerization. The authors are well aware that the capital costs could be greater than for competing techniques, and that this is a very simplistic economic analysis. However, in comparison to any electroorganic technique, electropolymerization has an inherent advantage. In general, in an electroorganic syntheses, one mole of substance per Faraday represents 100% efficiency. In electropolymerization, one can exceed this by a factor of 100 or more, on the basis of monomer consumption.

In industrial terms, it is rather disappointing that the peculiar problems of scale-up for plant reaction have not to date been overcome. This is true, however, of the whole electrochemical field. With the exception of the reductive coupling work (Monsanto) and the preparation of tetraalkyl leads (Nalco) the increasing issues of new patents have not as yet produced new industrial products or processes. It will be interesting to note whether present developments in this fledgling field will herald renewed interest in the fundamental processes, creativity in investigation, and ingenuity in the application of electrochemical methods to polymerization.

References

1. B. C. Weedon, in R. A. Raphael, Ed., *Advances in Organic Chemistry* (R. A. Raphael, E. C. Taylor, and H. Wynberg, ed.; Intersci.), Vol. 1, 1960.
2. C. L. Wilson, *Rec. Chem. Prog.,* **10**, 25 (1949).
3. P. M. Hay, *Advan. Chem.,* **80**, 350 (1969).
4. N. Ise, *Advan. Polymer Sci.,* **6**, 347 (1969).
5. M. M. Baizer, *Naturwissenschaften,* **56**, 405 (1969).
6. M. M. Baizer, *Ann. N. Y. Acad. Sci.,* **147**, 614 (1969).
7. M. M. Baizer, Proc. 7th World Petrol. Congr. (1967), **5**, 311 (1968).
8. E. C. Szarvasy, *J. Chem. Soc.,* **77**, 207 (1900).
9. C. K. Mann, *Electroanalytical Chemistry* (A. J. Bard, ed.), Marcal Dekker, New York, 3, 1969.
10. G. Mangoli and G. Vidotto, *Makromol. Chem.,* **129**, 73 (1969).
11. B. L. Funt, I. McGregor, and J. Tanner, *Polymer Preprints,* **12**(1), 85 (1971).
12. B. L. Funt and J. Rybicky, *J. Polymer Sci, A-1,* **9**, 1441 (1971).
13. G. Mengoli and G. Vidotto, *Makromol. Chem.,* **133**, 279 (1970).
14. S. N. Bhadani and G. Parravano, *J. Polymer Sci.,* (A1), **8**, 225 (1970).
15. D. E. Danly, *Hydrocarbon Process.,* **48**, 159 (1969).
16. J. R. Backhurst, M. Fleischmann, F. Goodridge, and R. E. Plimley, Br. Patent Appl. 23070 (1966).
17. F. Goodridge, *Chem. Proc. Eng.,* **49**, 93 (1968).
18. G. Mengoli, G. Vidotto, and F. Furlanetto, *Makromol. Chem.,* **137**, 203 (1970).
19. H. F. Park and C. K. Bump, U. S. Patent 2,726,204 (1955).
20. H. Kaneko, *Kinzoku Hyomen Gijutsu,* **21**, 93 (1970).
21. M. Murphy, M. G. Carangelo, M. B. Ginaine, and M. C. Markham, *J. Polymer Sci.,* **54**, 107 (1961).
22. H. Strethlow, in J. J. Lagowski, Ed., *The Chemistry of Non-Aqueous Solvents,* Vol. 1, Academic, New York, 1966.
23. D. H. Richards and M. Szwarc, *Trans. Faraday Soc.,* **55**, 1644 (1959).
24. T. Asahara, M. Seno, M. Tobayama, and F. -Q. Cheu, *Seisan-Kenkyu,* **22**, 167 (1970).
25. F. S. Dainton, K. J. Ivin, and R. T. LaFlair, *Eur. Polymer J.,* **5**, 379 (1969).
26. V. Hornof, Ph.D. Thesis, Simon Fraser University (1971).
27. L. J. Fetters, *J. Res. Natl. Bur. Std. U.S.,* **70A**, 421 (1966).
28. H. Thiele and L. Langmaack, *Z. Phys. Chem.,* **207**, 118 (1957).
29. B. L. Funt and D. G. Gray, *Can. J. Chem.,* **46**, 1337 (1968).
30. B. L. Funt and D. G. Gray, *J. Electrochem. Soc.,* 1020 (1970).
31. E. A. Rembold, Ph.D. Thesis, Ohio State University, 1947.
32. M. Y. Fioshin and A. P. Tomilov, *Plasticheskie Massy,* **10**, 2 (1960).
33. C. L. Wilson, in *Encyclopedia of Electrochemistry* (C. A. Hampel, ed., Reinhold Pub. Co., New York), 963 (1964).
34. C. Ungureanu, *Mat. Plast.,* **1**, 193 (1964).
35. Chin-Yung Wu and Chang-Ya Hu, *Hua Hsuch Tung Pao,* 329 (1964).
36. B. L. Funt, *Macromol. Rev.,* **1**, 35 (1966).

37. H. Friedlander, *Encycl. Polymer Sci. Technol.*, **5**, 629 (1966).
38. Chao-Hua Chou and Ching-Chen Li, *Hua Hsueh Tung Pao*, **193** (1966).
39. H. Thiele and J. Lange, *Kolloid Z.*, **169**, 86 (1960).
40. Paul Arnaud, *Ind. Chim. Belge*, **31**, 896 (1966).
41. Paul Arnaud, *J. Four Elec.*, **71**, 189 (1966).
42. R. A. V. Raff, Wash. State Univ., College of Engineering, Circular No. 25 (1966).
43. F. Laborie, *Double Liaison*, **148**, 1489 (1967).
44. T. Asahara and M. Seno, *Yuki Gosei Kagaku Kyokai Shi*, **25**, 719 (1967).
45. C. Ungureanu and C. Simionescu, *Mat. Plast.*, **4**, 174 (1967).
46. B. M. Tidswell, *Rep. Prog. Appl. Chem.*, **53**, 516 (1968).
47. N. Yamazaki, *Adv. Polymer Sci.*, **6**, 377 (1969).
48. S. Yoshizawa, Z. Ogumi, and K. Fukuhara, *Denki Kagaku Oyobi Kogyo Butsuri Kagaku*, **37**, 740 (1969).
49. J. W. Breitenbach and C. Srna, *Pure Appl. Chem.*, **4**, 245 (1962).
50. J. D. Anderson, *J. Polymer Sci. (A1)*, **6**, 3185 (1968).
51. E. Dineen, T. C. Schwan, and C. L. Wilson, *Trans. Electrochem. Soc.*, **96**, 226 (1949).
52. M. N. Das and S. R. Palit, *Sci. and Culture*, **16**, 34 (1950).
53. G. Parravano, *J. Am. Chem. Soc.*, **73**, 628 (1951).
54. W. Kern and H. Quast, *Makromol. Chem.*, **10**, 202 (1953).
55. W. B. Smith and H. G. Gilde, *J. Am. Chem. Soc.*, **82**, 659 (1960).
56. A. I. Fedorova, I. V. Shelepin, and N. B. Moiseeva, *Dokl. Akad. Nauk SSSR*, **138**, 408 (1961).
57. M. S. Tsvetkov, *Vysokomol. Soyed.*, **3**, 549 (1961).
58. B. L. Funt and K. C. Yu, *J. Polymer Sci.*, **62**, 359 (1962).
59. A. I. Fedorova, L. Kuo-tung, and I. V. Shelepin, *Russ. J. Phys. Chem.*, **38**, 921 (1964).
60. A. Trifonov and I. M. Panayotov, *Makromol. Chem.*, **77**, 237 (1964).
61. N. S. Tsvetkov and Z. F. Glotova, *Vysokomol. Soyed.*, **5**, 997 (1963).
62. J. A. Epstein and A. Bar-Nun, *Polymer Let.*, **2**, 27 (1964).
63. I. V. Shelepin and A. I. Fedorova, *Zh. Fiz. Khim.*, **38**, 2676 (1964).
64. I. V. Shelepin, A. N. Frumkin, A. I. Fedorova, and S. Y. Vasina, *Dokl. Adad. Nauk SSR*, **154**, 203 (1964).
65. N. S. Tsvetkov and E. P. Koval'chuk, *Visn. L'viv. Derzh. Univ., Ser. Khim.*, **72** (1967).
66. B. L. Funt and S. N. Bhadani, *J. Polymer Sci. (A)*, **3**, 4191 (1965).
67. B. M. Tidswell, *Soc. Chem. Ind.*, Monograph No. 28, 130 (1966).
68. N. S. Tsvetkov and E. P. Koval'chuk, *Ser. Khim.*, Lvov, **72**, (1967).
69. N. S. Tsvetkov and E. P. Koval'chuk, *Visn. L'viv. Derzh. Univ., Ser. Khim.*, **24** (1965).
70. E. I. Maznichenko and A. I. Fedorova, *Elektrokhimiya*, **3**, 405 (1967).
71. J. W. Breitenbach, O. F. Olaj, and F. Sommer, *Monatsh. Chem.*, **99**, 203 (1968).
72. N. Yamazaki, I. Tanaka, and S. Nakahama, *J. Macromol. Sci. (Chem.)*, **6**, 1121 (1968).

73. I. M. Kolthoff and L. L. Ferstandig, *J. Polymer Sci.,* **5**, 563 (1951).
74. S. Goldschmidt and E. Stöckl, *Chem. Ber.,* **85**, 630 (1952).
75. H. Z. Friedlander, S. Swann, and C. S. Marvel, *J. Electrochem. Soc.,* **100**, 408 (1953).
76. J. W. Breitenbach and H. Gabler, *Monatsh. Chem.,* **91**, 202 (1960).
77. J. W. Breitenbach, C. Srna, and O. F. Olaj, *Makromol. Chem.,* **42**, 171 (1960).
78. B. L. Funt and F. D. Williams, *J. Polymer Sci. (A),* **2**, 865 (1964).
79. C. Simionescu and C. Ungureanu, *Cell., Chem. Technol. (Jassy),* 1, 33 (1967).
80. G. S. Shapoval, *Ukr. Khim. Zh.,* **33**, 946 (1967).
81. F. Beck and H. Leitner, *Angew. Makromol. Chem.,* **2**, 51 (1968).
82. Badische Anilin and Soda Fab., Ger. Patent 1,269,349 (1968).
83. F. Sommer, J. W. Breitenbach, and O. F. Olaj, *Monatsh. Chem.,* **99**, 2422 (1968).
84. A. Chapiro and E. Henrychowski, *J. Chim. Phys.,* **65**, 616 (1968).
85. C. Ungureanu, D. A. Ungureanu, and C. Simionescu, *Rev. Roum. Chim.,* **13**, 913 (1968).
86. O. F. Olaj, J. W. Breitenbach, and B. Buchberger, *Angew. Makromol. Chem.,* **2**, 160 (1968).
87. G. Mengoli, G. Farnia, and E. Vianello, *Eur. Polymer J.,* **5**, 61 (1969).
88. T. Asahara, M. Seno, and M. Tsuchiya, *Bull. Chem. Soc. Japan,* **42**, 2416 (1969).
89. M. S. Tsvetkov and E. P. Koval'chuk, *Elektrokhimiya,* **5**, 909 (1969).
90. M. S. Tsvetkov and E. P. Koval'chuk, *Elektrokhimiya,* **5**, 848 (1969).
91. Y. Kikuchi, *Chem. High Polymers, Tokyo,* **26**, 701 (1969).
92. M. M. Baizer, U. S. Patents 3,245,889 (1966) and 3,375,237 (1968).
93. R. Vasiliu and P. Andriescu, Roumainian Patent 51,659 (1969).
94. J. Y. Yang, W. E. McEwan, and J. Kleinberg, *J. Am. Chem. Soc.,* **79**, 5833 (1957).
95. W. B. Smith and D. T. Manning, *J. Polymer Sci.,* **59**, S45 (1962).
96. B. L. Funt and S. N. Bhadani, *Can. J. Chem.,* **42**, 2733 (1964).
97. B. L. Funt and S. W. Laurent, *Can. J. Chem.,* **42**, 2728 (1964).
98. N. Yamazaki, S. Nakahama, and S. Kambara, *Polymer Let.,* **3**, 57 (1965).
99. N. Yamazaki, *Kogyo Kagaku Zasshi,* **70**, 1978 (1967).
100. B. L. Funt, D. Richardson, and S. N. Bhadani, *Can. J. Chem.,* **44**, 711 (1966).
101. B. L. Funt, S. N. Bhadani, and D. Richardson, *J. Polymer Sci. (A-1),* **4**, 2871 (1966).
102. B. L. Funt, U. S. Patent 3,448,020 (1970).
103. N. Yamazaki, I. Tanaka, and S. Nakahama, *Asahi Gurasu Kogyo Gijutsu, Shorei-kai Kenkyu Hokuku,* **14**, 293 (1968).
104. B. L. Funt and D. Richardson, *J. Polymer Sic. (A-1),* **8**, 1055 (1970).
105. B. L. Funt and T. J. Blain, *J. Polymer Sci. (A-1),* **8**, 3339 (1970).
106. D. Goerrig, H. Jonas, and W. Moschel, Ger. Patent 935,867 (1955).
107. D. Goerrig and H. Jonas, Ger. Patent 937,919 (1956).
108. H. Gehrke and M. Fechenhein, Ger. Patent 1,014,744 (1957).
109. J. W. Loveland, Can. Patent 566,274 (1958).
110. R. V. Lindsey and M. R. Peterson, *J. Am. Chem. Soc.,* **81**, 2073 (1959).

111. W. B. Smith and H. G. Gilde, *J. Am. Chem. Soc.*, **81**, 5325 (1959).
112. B. L. Funt and S. N. Bhadani, *J. Polymer Sci. (C)*, **23**, 1 (1968).
113. W. A. Kornicker, Can. Patent 829,881 (1965).
114. W. A. Kornicker, U.S. Patent 3,474,012 (1969).
115. N. Yamazaki and S. Murai, *Chem. Commun.*, 147 (1968).
116. H. G. Gilch, *J. Polymer Sci. (A-1)*, **4**, 1351 (1966).
117. H. G. Gilch, U.S. Patent 3,399,124 (1968); Can. Patent 812,779 (1969).
118. F. H. Covitz, *J. Am. Chem. Soc.*, **89**, 5403 (1967).
119. S. D. Ross and D. J. Kelley, *J. Appl. Polymer Sci.*, **11**, *1209 (1967).*
120. S. D. Ross, R. C. Petersen, and M. Finkelstein, Can. Patent (1968).
121. S. D. Ross, R. C. Petersen, and M. Finkelstein, U.S. Patent 3,417,003 (1968).
122. N. E. Wisdom, U.S. Patent 3,437,570 (1967).
123. Esso, Br. Patent 1,136,699 (1968).
124. Esso, Fr. Patent 1,534,181 (1968).
125. A. F. Shepard and B. F. Dannels, *J. Polymer Sci. (A-1)*, **4**, 511 (1966).
126. A. F. Shepard and B. F. Dannels, U.S. Patent 3,386,899 (1968).
127. G. Smets, A. Poot, M. Mullier, and J. P. Bex, *J. Polymer Sci.*, **34**, 298 (1959).
128. G. Smets, X. VanDerBorght, and G. VanHaeren, *J. Polymer Sci. (A)*, **2**, 5187 (1964).
129. G. Smets and J. P. Bex, U.S. Patent 3,330,745 (1967).
130. G. Smets, R. VanGorp, and G. VanHaeren, *Eur. Polymer J. (Supplement)*, 215 (1969).
131. W. E. Garrison, *Diss. Abstr.*, **20**, 77 (1959).
132. S. M. Toy, *J. Electrochem. Soc.*, **114**, 1042 (1967).
133. H. Gilch and D. Michael, *Makromol. Chem.*, **99**, 103 (1966).
134. H. G. Gilch, U.S. Patent 3,419,482 (1969).
135. G. S. Shapoval and V. I. Shapoval, *Vysokomol. Soed.*, **6**, 121 (1964).
136. J. F. Sobieski and M. C. Zerner, U.S. Patent 3,464,960 (1969).
137. M. S. Tsvetkov and E. P. Koval'chuk, *Ukr. Khim. Zh.*, **35**, 1217 (1969).
138. M. Y. Fioshin, E. P. Koval'chuk, and M. S. Tsvetkov, *Elektrokhimiya*, **5**, 1188 (1969).
139. N. S. Tsvetkov and E. P. Koval'chuk, *Vysokomol. Soed. (B)*, **11**, 42 (1969).
140. R. J. Ehrig and F. A. Kundell, U.S. Patent 3,434,946 (1969).
141. B. L. Funt and T. J. Blain, *J. Polymer Sci.*, **9**, 115 (1971).
142. C. F. Heins, *J. Polymer Sci. (B)*, **7**, 625 (1969).
143. R. W. Dyck, M. Sc. Thesis, Simon Fraser University (1969).
144. D. Laurin and G. Parravano, *Polymer Let.*, **4**, 797 (1966).
145. D. Laurin and G. Parravano, *J. Polymer Sci. (C)*, **22**, 103 (1968).
146. A. I. Fedorova and S. Y. Vasina, *Elektrokhimiya*, **3**, 742 (1967).
147. S. Y. Vasina, G. F. Astakhova, R. F. Karysheva, and A. I. Fedorova, *Elektrokhimiya*, **4**, 468 (1968).
148. S. Porejko and L. Perec, *Polimery*, **11**, 68 (1966).
149. I. Izumi, A. Kungugi, and S. Nagura, *Osaka Shiritsu Daigaku Kogakula Memoirs*, **9**, 59 (1967).

150. R. Arnold and D. A. Swift, *Aust. J. Chem.*, **22**, 859 (1969).
151. A. I. Shlygin and N. B. Kondrikov, *Uch. Zap. Dal'nevost. Univ.*, **8**, 58 (1966).
152. R. C. Schulz and W. Strobel, *Monatsh. Chem.*, **99**, 1742 (1968).
153. W. Strobel and R. C. Schulz, *Makromol. Chem.*, **133**, 303 (1970).
154. H. Mitoguchi, Y. Kikuchi, and K. Kimura, *Yuki Gosei Kagaku Kyokai Shi*, **27**, 769 (1969).
155. C. C. Cochrane, U.S. Patent 3,427,233 (1969).
156. B. L. Funt and F. D. Williams, *Polymer Let.*, **1**, 181 (1963).
157. W. R. Grace and Co., Br. Patent 1,156,309 (1968).
158. W. F. H. Borman, Can. Patent 800,592 (1968).
159. E. V. Kusnetsov and L. A. Bogoyavlenskaya, *Vysokomol. Soed.*, **7**, 259 (1965).
160. B. L. Funt and D. G. Gray, *J. Macromol. Chem.*, **1**, 625 (1966).
161. E. O. Forster, U.S. Patent 3,140,276 (1964).
162. J. W. Bayer and E. Santiago, U.S. Patent 3,489,663 (1970).
163. D. M. Ritter, U.S. Patent 2,261,002 (1942).
164. N. E. Searle, U.S. Patent 2,271,378 (1942).
165. Y. Tanaka, N. Ise, and I. Sakurada, *Macromolecules*, **2**, 215 (1969).
166. N. Ise, H. Hirohara, T. Makino, and I. Sakurada, *J. Phys. Chem.*, **72**, 4543 (1968).
167. I. Sakurada, N. Ise, H. Hirohara, and T. Makino, *J. Phys. Chem.*, **71**, 3711 (1967).
168. I. Sakurada, Y. Tanaka, and N. Ise, *J. Polymer Sci. (A-1)*, **6**, 1463 (1968).
169. B. L. Funt, I. McGregor, and J. Tanner, *J. Polymer Sci.*, *B*, **8**, 695 (1970).
170. B. L. Funt, I. McGregor, and J. Tanner, *J. Polymer Sci.*, *B*, **8**, 699 (1970).
171. N. G. Gaylord and A. Takahaski, *Advan. Chem.*, **91**, 94 (1969).
172. H. Hirai, T. Ikegami, and S. Makishima, *J. Polymer Sci. (A-1)*, **7**, 2059(1969).
173. J. Furukawa, Y. Iseda, K. Hagai, and N. Kataoka, *J. Polymer Sci. (A-1)*, **8**, 1147 (1970).
174. T. Ikegami and H. Hirai, *J. Polymer Sci. (A-1)*, **8**, 195 (1970).
175. S. N. Bhadani, Ph.D. Thesis, University of Manitoba (1966).
176. M. Szwarc, *Nature*, **178**, 1168 (1956).
177. V. Hornof, Unpublished results.
178. D. G. Gray, Ph.D. Thesis, University of Manitoba (1968).
179. M. Szwarc, in *Carbanions, Living Polymers, and Electron Transfer Processes*, (Interscience, New York), 1968.
180. J. Oliver, *Prod. Finish (Cincinnati)*, **34**, 66 (1970).
181. T. Asahara, M. Seno, and M. Tsuchiya, *Kinzoku Hyomen Gijutsu*, **19**, 511 (1968).
182. T. Asahara, M. Seno, and M. Tsuchiya, *Kinzoku Hyomen Gijutsu*, **20**, 2 (1969).
183. T. Asahara, M. Seno, and M. Tsuchiya, *Kinzoku Hyomen Gijutsu*, **20**, 28 (1969).
184. T. Asahara, M. Seno, and M. Tsuchiya, *Kinzoku Hyomen Gijutsu*, **20**, 617 (1969).

185. T. Asahara, M. Seno, and M. Tsuchiya, *Kinzoku Hyomen Gijutsu,* 20, 414 (1969).

186. T. Asahara, M. Seno, and M. Tsuchiya, *Kinzoku Hyomen Gijutsu,* 20, 99 (1969).

187. T. Asahara, M. Seno, and M. Tsuchiya, *Kinzoku Hyomen Gijutsu,* 20, 411 (1969).

188. T. Asahara, M. Seno, and M. Tsuchiya, *Kinzoku Hyomen Gijutsu,* 20, 64 (1969).

189. P. Giusti, P. Cerrai, M. Tricoli, P. L. Maganini, and F. Andruzzi, *Makromol. Chem.,* 133, 299 (1970).

190. G. Mengoli and G. Vidotto, *Eur. Polymer J.,* 8, 661 (1972).

191. B. M. Tidswell and A. G. Doughty, *Polymer,* 12, 431 (1971).

192. B. M. Tidswell and A. G. Doughty, *Polymer,* 12, 760 (1971).

193. G. Mengoli and G. Vidotto, *Preprints,* 1, 285, IUPAC, Helsinki, 1972.

194. G. Mengoli and G. Vidotto, *Makromol. Chem.,* 150, 277 (1971).

195. G. Mengoli and G. Vidotto, *Makromol. Chem.,* 153, 57 (1972).

196. G. Mengoli and G. Vidotto, *Eur. Polymer J.,* 8, 671 (1972).

197. R. Kijkstra and J. De Jonge, *Extended Abstract,* IUPAC, Boston, 1971.

198. D. Fleischer, R. C. Schulz, and B. Turcsanyi, *Makromol. Chem.,* 152, 305 (1972).

199. D. C. Phillips, D. H. Davies, and J. D. Smith, *Makromol. Chem.,* 154, 32 (1972).

200. G. Mengoli and G. Vidotto, *Preprints,* 2, 639, IUPAC, Helsinki, 1972.

201. J. Tanner, V. Hornof, and B. L. Funt, *Makromol. Chem.,* 157, 227 (1972).

202. N. Ise, K. Takaya, and H. Hirohara, *Makromol. Chem.,* 157, 235 (1972).

203. J. W. Breitenbach, *Advan. Polymer Sci.,* 9, 47 (1972).

204. G. S. Shapoval, M. Skobets, and N. P. Markova, *Sintez Fiz. Khim. Polimerov,* 5, 76 (1968).

205. G. S. Shapoval, M. Skobets, and N. P. Markova, *Vysokomol. Soyed.,* 8, 1313 (1966).

206. G. S. Shapoval, E. M. Skobets, and N. P. Markova, *Dokl. Akad. Nauk., SSSR,* 173, 392 (1967).

207. W B. Smith and H. -G. Gilde, *J. Am. Chem. Soc.,* 83, 1355 (1961).

Appendix

OXIDATION AND REDUCTION HALF-WAVE POTENTIALS
OF ORGANIC COMPOUNDS

Howard Siegerman
Princeton Applied Research Corporation
Princeton, New Jersey 08540

INTRODUCTION

Many electroorganic syntheses require knowledge of voltammetric behavior of an organic substrate in a particular electrolyte. Electroanalysis at a microelectrode, that is, polarography, cyclic voltammetry, chronopotentiometry, etc., can yield basic informa-

tion such as optimum electrolyte, electrode
material, solution pH, current density, and cell
voltage for the macroscale electrolysis, as well as
sometimes offering a means by which starting
material and product can be assayed. Accordingly,
tabulations of half-wave potentials*--$E_{1/2}$'s--of
organic compounds have proved valuable for both the
synthetic and analytical areas of the electrochemical spectrum.

Several books[1-8] and reviews[9-11] contain $E_{1/2}$
tables of the polarographic reduction of organic
compounds. Similar tables for polarographic oxidation have appeared recently[7,11,12].

The polarographic half-wave potential has been
enthusiastically compared to the λ_{max} of absorption
spectroscopy in terms of analytical utility[13].
While the usefulness of an $E_{1/2}$ as an aid to organic
compound identification should not be denied,
neither should the niceties of specific adsorption,
of organics, junction potentials of interfaces,
influence of maxima and their suppressors, and
uncompensated IR effects be ignored in the shifting
of an $E_{1/2}$ from its "true" value.

The widespread availability of polarographs, ranging from the home-made variety to the highly sophisticated, multifunctional instrument, has brought the
technique into new fields and applications. In
their quest for new areas of endeavor, electroanalytical chemists have rotated and vibrated the dropping mercury electrode, streamed mercury through a
capillary, supported or hung a single mercury drop,
pasted carbon, impregnated graphite with wax, and
investigated a variety of electrode geometries such
as spheres, cylinders, wires, and planes. They have
perturbed their electrode assemblies with complex
waveforms including step functions (chronoamperometry),
triangles (cyclic voltammetry), voltage ramps
modulated with square waves (differential pulse
polarography) and sine waves (ac polarography).
Through it all, the half-wave potential of classical

* Loosely defined as the potential at which the
electrolysis current of the depolarizer is one-half
the diffusion current value.

dc polarography remains relatively unchallenged in terms of electroanalytical appeal and utility.*

In this appendix, 32 classes of compounds have been included in a compilation of oxidation and reduction half-wave potentials. The appendix does not claim 100% thoroughness for any given function- ality, but, rather, attempts to include a wide variety of compounds of general interest. Pesticides, pharmaceuticals, steroids, biochemicals, phenols, antioxidants, and others have been surveyed for pertinency to today's problems.

For those functionalities exhibiting dual behavior, oxidation half-wave potentials have been separated from, and generally precede, reduction half-wave potentials. Classification of substances with two or more electroactive groups is somewhat arbitrary. As an example, haloketones are classified for convenience under aliphatic ketones, since the aliphatic halide section is rather lengthy. This classification (and others throughout the Appendix) makes no assumptions as to which electroactive moiety is primarily involved in the electrode process. Such considerations are beyond the intended scope of these tables.

Wherever feasible, compounds in a given section have been arranged in order of increasing complexity, that is aliphatics generally increase in degree of substitution. Common names have sometimes been included for convenience.

The Solvent System defines the nature of the sup- porting electrolyte. Where no solvent has been specified, an aqueous solution was employed (e.g., $0.1\underline{M}$ Na_2SO_4).

---

* In addition, it is still possible to assemble a 2-electrode, manual, nonelectronic polarographic analyzer if Ohm's law is kept in mind.

Multiple half-wave potentials for a single compound are separated by commas; peak or half-peak
potentials are signified by E_p and $E_{p/2}$ respectively.*
The occasional chronopotentiometric quarter-wave
potentials that have been cited are indicated by
$E_{1/4}$.

Relations between half-wave potential and pH have
been quoted if given in the original reference
explicitly in the form $E_{1/2} = \underline{a} + \underline{b} \cdot pH$. No attempt
was made here to derive $E_{1/2}$-pH equations from tabu-
lated information. Similarly the half-wave potentials
quoted are those given explicitly by the author(s);
interpolation of $E_{1/2}$-pH or $E_{1/2}$-concentration plots
were not attempted.

A variety of reference electrodes have been
employed in the polarographic analysis of organic
compounds. Each supporting electrolyte has its own
requirements for a reference electrode. Some, like
the saturated calomel electrode and the silver-silver
ion couple, serve well in many media. An acceptable
reference electrode for a given medium is well
poised, drift free, and yields a reproducible
solution potential. Comparisons of reference electrode potentials in various solvents have been docu-
mented[7]. The electrode potentials (versus the
saturated calomel electrode which is itself + 0.2412 V
versus the normal hydrogen electrode at
25°C) of some frequently used reference systems are
listed in Table I.

Table I. Tabulation of Reference Electrode Potentials

Solvent	Reference System	E Volts vs sce	Reference
Water	Ag,AgCl($\underline{s}$),KCl(s)	-0.042	14
	Ag,AgCl($\underline{s}$)/KCl(0.1F)	+0.047	14
	Ag,AgCl($\underline{s}$),NaCl($\underline{s}$)	-0.047 $\pm$ 0.003	14

* For reversible electron transfers, E_p and $E_{p/2}$ are
related to $E_{1/2}$ via: $E_p = E_{1/2} - (0.029/n)V$ 25°C.
$$E_{p/2} = E_{1/2} + (0.028/n)V\ 25°C.$$

Water	$Hg,HgO(\underline{s})/NaOH(1\underline{F})$	-0.101	14
	$Hg,HgO(\underline{s})/NaOH(0.1\underline{F})$	-0.076	14
	$Hg,Hg_2SO_4(\underline{s})/H_2SO_4(0.5\underline{F})$	+0.441	14
	$Hg,Hg_2SO_4(\underline{s}),K_2SO_4(\underline{s})$	+0.41	14
Acetic	$Ag,AgCl(\underline{s}),KCl(\underline{s})$	+0.23	15
Acid	$Ag,AgNO_3(\underline{s})$	+0.87	15
	$Hg,Hg_2SO_4(\underline{s})KCl(\underline{s})$	+0.27	16
	$Hg,Hg_2SO_4(\underline{s}),K_2SO_4(\underline{s})$	+0.69	16
Acetoni-	$Ag/AgNO_3(0.1\underline{M})$	+0.337	17
trile	$Ag/AgNO_3(0.01\underline{M})$	+0.300	17
		+0.29	18
	$Ag/AgNO_3(0.1\underline{M}),NaClO_4$	+0.301	17
	$(0.1\underline{M})$		
	$Ag/AgNO_3(0.1\underline{M}),(Et)_4NClO_4$	+0.336	17
	$(0.1\underline{M})$		
N,N-Di-	$Hg/(\underline{n}-Bu)_4NI(0.1\underline{M})$	-0.55	19
methyl-	$Hg/(\underline{Et})_4NI(0.1\underline{M})$	-0.50	20
forma-	$Hg/LiCl(1\underline{M})$	-0.47	20
mide	$Cd(Hg)(\underline{s}),CdCl_2(\underline{s}),NaCl$	-0.7340 ± 0.0007	21
	$(\underline{s})/(E\underline{t})_4NClO_4$		
Dimethyl-	$Ag,AgCl(\underline{s})$	+0.30	22
sulfox-	$Zn(Hg)/\underline{Zn}(ClO_4)_2$	-1.08	23
ide			
Propylene	$Hg,Hg_2Cl_2(\underline{s}),(Et)_4NCl(\underline{s})$	$+0.0045\pm0.0010$	24
Carbo-			
nate			
Pyridine	$Ag/AgNO_3(1\underline{M})$	+0.09	25
	$Hg/LiCl(0.1\underline{M})$	-0.36	25
	$Hg/LiCl(0.3\underline{M})$	-0.36	25
	$Hg/LiNO_3(0.1\underline{M})$	0.00	25
	$Hg/(\underline{n}-Bu)_4NI(0.15\underline{M})$	-0.56	25

Some of the abbreviations used in the Appendix include:

Abbreviation	Full Name
$(Me)_4NClO_4$	Tetramethylammonium Perchlorate
$(Et)_4NClO_4$	Tetraethylammonium Perchlorate

$(n\text{-Propyl})_4NClO_4$	Tetra-n-propylammonium Perchlorate
$(n\text{-Bu})_4NClO_4$	Tetra-n-butylammonium Perchlorate
HOAc	Acetic Acid
OAc buffer	Acetate buffer
B-R buffer	Britton-Robinson buffer
C-L buffer	Clark-Lubs buffer
P-W buffer	Prideaux-Ward buffer
Py	Pyridine
DMF	N,N-Dimethylformamide
THF	Tetrahydrofuran
DMSO	Dimethylsulfoxide
PC	Propylene Carbonate
sce	Saturated Calomel Reference Electrode
NCE	Normal Calomel Reference Electrode

Buffer Constituents

McIlvaine buffer. $0.2\underline{M}$ disodium phosphate and $0.1\underline{M}$ citric acid mixed in various proportions to effect desired pH in range 2.2 to 8.

Britton-Robinson buffer. $0.04\underline{M}$ acetic acid, $0.04\underline{M}$ phosphoric acid, $0.04\underline{M}$ boric acid. Add $0.2\underline{M}$ sodium hydroxide to effect desired pH in range 1.8 to 12.

Space and simiplicity limitations preclude inclusion of detailed information concerning scan rates (where E_p or $E_{p/2}$'s are given), use of maximum suppressors, use of two- or three-electrode systems, electrode configuration, i.e., pyrolytic graphite versus wax-impregnated graphite, rotating platinum electrode (with rpm's specified) versus stationary platinum electrode, and electrode pretreatment rituals, if any. These same considerations prevented inclusion of data on electron transfer coefficients, and tabulations of electrons transferred per mole in the rate-controlling step in the overall electrode reaction, as well as other electrode kinetics information.

The original references, as always, are the best sources of information and perhaps the most useful aspect of the Appendix will be to guide the reader to the original material.

ACKNOWLEDGMENT

This compilation of half-wave potentials owes much

of its existence to Dr. Norman Weinberg, who suggeste
the preparation of these tables and made available
to the author many of his own personal files. Variou
individuals assisted greatly by supplying reprints
and preprints of papers, theses, and compendia of
references. These include: Professor R. N. Adams
(University of Kansas), Professor P. J. Elving
(University of Michigan), Professor H. Lund (Aarhus
University, Denmark), Professor R. F. Nelson
(University of Idaho, Moscow, Idaho), Dr. W. Settiner
(Dow Chemical Company, Midland, Michigan) and
Professor G. S. Wilson (University of Arizona).
Thanks are due Dr. G. Popp (Eastman Kodak, Rochester,
New York) for a careful reading of the manuscript
and several helpful suggestions.

References for Introduction

1. L. Meites, Polarographic Techniques, 2nd ed.,
    Wiley, New York, 1965.
2. J. Heyrovsky and J. Kuta, Principles of Polaro-
    graphy, Academic, New York, 1966.
3. M. Brezina and P. Zuman, Polarography in Medicine
    Biochemistry, and Pharmacy, Wiley, New York, 1958
4. K. Schwabe, Polarographic and chemische Konstitut
    organischer Verbindung, Akademie-Verlag, Berlin,
    1957.
5. J. Heyrovsky and P. Zuman, Einfuhrung in die
    praktische Polarographie, VEB Verlag Technik,
    Berlin, 1959.
6. I. M. Kolthoff and J. J. Lingane, Polarography,
    Vol. II, Wiley, New York, 1952.
7. C. K. Mann and K. K. Barnes, Electrochemical
    Reactions in Nonaqueous Systems, Marcel Dekker,
    New York, 1970.
8. J. Heyrovsky and P. Zuman, Practical Polarography
    Academic, New York, 1968.
9. P. Zuman, Collect. Czech. Chem. Commun., 15, 1107
    (1950).
10. G. Semerano and L. Griggio, Suppl. Riceria Sci.,
    27, 327 (1957).
11. N. L. Weinberg and H. R. Weinberg, Chem. Rev.,
    68, 449 (1968).
12. R. N. Adams, Electrochemistry at Solid Electrodes
    Marcel Dekker, New York, 1969.

13. P. Zuman, Chem. Eng. News, March 18, 94 (1968).
14. L. Meites, in Handbook of Analytical Chemistry, edited by L. Meites, McGraw-Hill, New York, 1963, pp. 5-13.
15. Tutundzic and Putanov, Glasnik Khem. Drushtva Beograd, 21, 19 (1956).
16. Tutundzic and Putanov, Glasnik Khem. Drushtva Beograd, 21, 257 (1956).
17. R. C. Larson, R. T. Iwamoto, and R. N. Adams, Anal. Chim. Acta, 25, 371 (1961).
18. A. I. Popov and D. H. Geske, J. Am. Chem. Soc., 79, 2074 (1957).
19. A. Streitweiser and I. Schwager, J. Phys. Chem., 66, 2316 (1962).
20. P. H. Given, M. E. Peover, and J. Schoen, J. Chem. Soc., 1958, 2674 (1958).
21. C. W. Manning and W. C. Purdy, Anal. Chim. Acta, 51, 124 (1970).
22. E. J. Johnson, K. H. Pool, and R. E. Hamm, Anal. Chem., 38, 183 (1966).
23. J. N. Butler, J. Electroanal. Chem., 14, 89 (1967)
24. I. Fried and H. Barak, J. Electroanal. Chem., 27, 167 (1970).
25. A. Cizak and P. J. Elving, J. Electrochem. Soc., 110, 160 (1963).

1. ALIPHATIC HYDROCARBONS

Oxidation

Compound	Solvent System	Working Electrode	$E_{1/2}$	Reference Electrode	Reference
2-Methylbutane	Fluorosulphonic Acid, 1.15M HOAc	Pt	+1.89	Pd/H_2	1
3-Methylbutane	Fluorosulphonic Acid, 1.15M HOAc	Pt	+1.73	Pd/H_2	1
2,2-Dimethylbutane	CH_3CN, $(Et)_4NBF_4$	Pt	+3.28	$Ag/0.01\underline{N}$ $Ag+$	2
n-Pentane	Fluorosulphonic Acid, 1.15M HOAc	Pt	+2.01	Pd/H_2	1
i-Pentane	CH_3CN, $(Et)_4NBF_4$	Pt	+3.00	$Ag/0.01\underline{N}$ $Ag+$	2
Cyclopentane	Fluorosulphonic Acid	Pt	+2.01	Pd/H_2	1
2-Methylpentane	CH_3CN, $(Et)_4NBF_4$	Pt	+3.01	$Ag/0.01\underline{N}$ $Ag+$	2
	Flūorosulphonic Acid, 1.15M HOAc	Pt	+1.68	Pd/H_2	1
3-Methylpentane	Fluorosulphonic Acid, 1.15M HOAc	Pt	+1.68	Pd/H_2	1
n-Hexane	Fluorosulphonic Acid, 1.15M HOAc	Pt	+1.86	Pd/H_2	1
Cyclohexane	CH_3CN, $(Et)_4NBF_4$	Pt	>+3.4	$Ag/0.01\underline{N}$ $Ag+$	2
	Fluorosulphonic Acid, 1.15M HOAc	Pt	+1.77	Pd/H_2	1
3-Methylhexane	Fluorosulphonic Acid, 1.15M HOAc	Pt	+1.60	Pd/H_2	1
n-Heptane	Fluorosulphonic Acid, 1.15M HOAc	Pt	+1.73	Pd/H_2	1

Compound	Electrolyte	Electrode	Potential	Reference	Ref.
n-Heptane	CH_3CN $(Et)_4NBF_4$	Pt	>+3.4	Ag/0.01N Ag+	2
Cycloheptane	Fluorosulphonic Acid, 1.15M HOAc	Pt	+1.67	Pd/H_2	1
n-Octane	Fluorosulphonic Acid, 1.15M HOAc	Pt	+1.64	Pd/H_2	1
Cyclooctane	CH_3CN, $(Et)_4NBF_4$	Pt	>+3.4	Ag/0.01N Ag+	2
	Fluorosulphonic Acid, 1.15M HOAc	Pt	+1.20	Pd/H_2	1
n-Decane	Fluorosulphonic Acid, 1.15M HOAc	Pt	+1.56	Pd/H_2	1
Phenylcyclopropane	CH_3CN, $LiClO_4$	Pt	+1.87	sce	3
trans-1-Methyl-2-phenylcyclopropane	CH_3CN, $LiClO_4$	Pt	+1.71	sce	3
trans-1-Methoxy-2-phenylcyclopropane	CH_3CN, $LiClO_4$	Pt	+1.46	sce	3
p-Methylphenylcyclopropane	CH_3CN, $LiClO_4$	Pt	+1.59	sce	3
p-Methoxyphenyl-cyclopropane	CH_3CN, $LiClO_4$	Pt	+1.59	sce	3
p-Methoxyphenyl-cyclopropane	CH_3CN, $LiClO_4$	Pt	+1.35	sce	3
p-Chlorophenylcyclo-propane	CH_3CN, $LiClO_4$	Pt	+1.97	sce	3
2,4,6-Trimethyl-cyclopropane	CH_3CN, $LiClO_4$	Pt	+1.67	sce	3
trans-1-Methyl-2-(2,4,6-trimethyl-phenyl)cyclopropane	CH_3CN, $LiClO_4$	Pt	+1.61	sce	3
trans-1-Methoxy-2-(2,4,6-trimethyl-phenyl)cyclopropane	CH_3CN, $LiClO_4$	Pt	+1.48	sce	3

(continued)

678 Appendix

References for Aliphatic Hydrocarbons

1. J. Bertram, M. Fleischmann, and D. Pletcher,
    Tetrahedron Letters, 349 (1971).
2. M. Fleischmann and D. Pletcher, Tetrahedron
    Letters, 6, 255 (1968).
3. T. Shono and Y. Matsumura, J. Org. Chem., 35,
    4157 (1970).

2. OLEFINS AND POLYOLEFINS

Oxidation

Compound	Solvent System	Working Electrode	$E_{1/2}$	Reference Electrode	Reference
Ethylene	CH_3CN, $(Et)_4NBF_4$	Pt	+2.90	Ag/0.01N Ag+	1
Styrene	CH_3CN, $(Et)_4NClO_4$	Pt	+1.99 to +2.05	Ag, AgCl	2
α-Methylstyrene	CH_3CN, $(Et)_4NClO_4$	Pt	+1.80 to +1.89	Ag, AgCl	2
4-Methylstyrene	CH_3CN, $(Et)_4NClO_4$	Pt	+1.80 to +1.89	Ag, AgCl	2
1,1-Diphenylethylene	HOAc, 0.5M NaOAc	Pt	+1.52	sce	3
trans-1,2-Diphenylethylene (trans-Stilbene)	HOAc, 0.5M NaOAc	Pt	+1.51	sce	3
Tetraphenylethylene	CH_3CN, Et_4NClO_4	Pt	+1.25, +1.40 E_p	sce	4
Tris-p-anisyl-ethylene	CH_3CN, Py, $LiClO_4$	Pt	+0.94, +1.27 E_p	sce	5
Tetra-p-anisyl-ethylene	CH_3CN, $LiClO_4$	Pt	+0.78 E_p	sce	6
Tetrakis(p-N,N-di-methylaminophenyl)ethylene	CH_3CN, $(n-Bu)_4NClO_4$	Pt	+0.895 E_p	sce	7
	CH_2Cl_2, $(n-Bu)_4NClO_4$	Pt	+0.99 E_p	sce	7
	CH_3CN, $(n-Bu)_4NClO_4$	Pt	+0.16 E_p	sce	7
1-Propene	CH_3CN, $(Et)_4NBF_4$	Pt	+2.84	Ag/0.01N Ag+	1
1-Butene	CH_3CN, $(Et)_4NBF_4$	Pt	+2.78	Ag/0.01N Ag+	1
2-Methyl-1-butene	CH_3CN, $NaClO_4$	Pt	+1.97	Ag/0.01N Ag+	8

(continued)

Compound	Solvent System	Working Electrode	$E_{1/2}$	Reference Electrode	Reference
2-Butene	CH_3CN, $(Et)_4NBF_4$	Pt	+2.26	Ag/0.01N Ag+	1
1-Pentene	CH_3CN, $(Et)_4NBF_4$	Pt	+2.74	Ag/0.01N Ag+	1
Cyclohexene	CH_3CN, $(Et)_4NBF_4$	Pt	+2.05	Ag/0.01N Ag+	1
	CH_3CN, $NaClO_4$	Pt	+1.89	Ag/0.01N Ag+	8
1-Octene	CH_3CN, $(Et)_4NBF_4$	Pt	+2.70	Ag/0.01N Ag+	1
2-Octene	CH_3CN, $(Et)_4NBF_4$	Pt	+2.23	Ag/0.01N Ag+	1
1,4-Cyclohexadiene	CH_3CN, $LiClO_4$	Pt	+1.60	Ag/0.1N Ag+	9
1,3-Butadiene	CH_3CN, $NaClO_4$	Pt	+2.03	Ag/0.1N Ag+	8
2-Methyl-1,3-butadiene	CH_3CN, $NaClO_4$	Pt	+1.84	Ag/0.1N Ag+	8
2,3-Dimethyl-1,3-butadiene	CH_3CN, $NaClO_4$	Pt	+1.83	Ag/0.1N Ag+	8
1,3,5-Cyclohepta-triene(Tropili-dene)	CH_3CN, $LiClO_4$	Pt	+1.13	Ag/0.1N Ag+	9
Bi-2,4,6-Cyclohep-tatrien-1-yl	CH_3CN, $LiClO_4$	Pt	+1.03	Ag/0.1N Ag+	9
Cyclooctatetraene	HOAc, 0.5M NaOAc	Pt	+1.42	sce	3

Reduction

Compound	Solvent System	Working Electrode	$E_{1/2}$	Reference Electrode	Reference
Ethylene	75% Dioxane, $(n-Bu)_4$-NBr	Hg	no wave	sce	10

Compound	Solvent, electrolyte	Electrode	E	Reference electrode	Ref.
Phenylethylene (Styrene)	75% Dioxane, (n-Bu)$_4$NI	Hg	-2.343	sce	11
	DMF, (n-Bu)$_4$NI	Hg	-1.96	Hg pool	12
	DMF, (n-Bu)$_4$NI	Hg	-2.451	sce	13
	THF, (n-Bu)$_4$NClO$_4$	Hg	-3.18	Ag	14
β-Methylstyrene	75% Dioxane, (n-Bu)$_4$NI	Hg	-2.537	sce	11
1,1-Diphenylethylene	75% Dioxane, (n-Bu)$_4$NI	Hg	-2.258	sce	11
	CH$_3$CN, (n-Bu)$_4$NBr	Hg	-1.92	Hg pool	12
	DMF, (n-Bu)$_4$NI	Hg	-2.302	sce	13
	THF, (n-Bu)$_4$NClO$_4$	Hg	-2.56	Ag	14
trans-1,2-Diphenyl-ethylene (trans-Stilbene)	75% Dioxane, (n-Bu)$_4$NI	Hg	-2.137	sce	11
	CH$_3$CN, (n-Bu)$_4$NBr	Hg	-1.73, -2.06	Hg pool	12
cis-1,2-Diphenyl-ethylene (cis-Stilbene)	THF, (n-Bu)$_4$NClO$_4$	Hg	-2.55, -2.90	Ag	14
	DMF, (n-Bu)$_4$NI	Hg	-1.61, -2.02	Hg pool	12
	DMF, (n-Bu)$_4$NI	Hg	-1.81, -2.11	sce	12
	DMF, (n-Bu)$_4$NI	Hg	-2.141, -2.498	sce	13
Tripehylethylene	DMF, (n-Bu)$_4$NI	Hg	-2.084	sce	13
	75% Dioxane, (n-Bu)$_4$NI	Hg	-2.115	sce	11
Triphenylmethyl-ethylene	CH$_3$CN, (n-Bu)$_4$NBr	Hg	-1.67	Hg pool	12
	THF, (n-Bu)$_4$NClO$_4$	Hg	-2.61	Ag	14
	75% Dioxane, (Et)$_4$NI	Hg	-2.24	sce	15

(continued)

681

Compound	Solvent System	Working Electrode	$E_{1/2}$	Reference Electrode	Reference
Tetraphenylethylene	75% Dioxane, $(n\text{-Bu})_4$NI	Hg	-2.046	sce	11
	DMF, $(n\text{-Bu})_4$NI	Hg	-2.020	sce	13
	CH_3CN, $(n\text{-Bu})_4$NBr	Hg	-1.62	Hg pool	12
	THF, $(n\text{-Bu})_4$NClO$_4$	Hg	-2.47	Ag	14
Allene	75% Dioxane, $(n\text{-Bu})_4$NBr	Hg	-2.29	sce	10
▷—C(CH$_3$)=C(CN)—CO$_2$C$_2$H$_5$	DMF, $(Heptyl)_4$NI	Hg	-1.72	sce	16
CH$_3$(CH$_2$)$_2$C(CH$_3$)=C-(CN)CO$_2$C$_2$H$_5$	DMF, $(Heptyl)_4$NI	Hg	-1.75	sce	16
CH$_3$CH$_2$C(CH$_3$)=C(CN)-CO$_2$C$_2$H$_5$	DMF, $(Heptyl)_4$NI	Hg	-1.74	sce	16

Diactivated Olefins: trans-X-CH=CH-Y

Compound	Solvent System	Working Electrode	$E_{1/2}$	Reference Electrode	Reference
X=C$_6$H$_5$CO Y=C$_6$H$_5$CO	DMF, $(Et)_4$NClO$_4$	Hg	-1.08, -1.73	sce	17
C$_6$H$_5$CO CN	DMF, $(Et)_4$NClO$_4$	Hg	-1.11	sce	17
C$_6$H$_5$CO CO$_2$C$_2$H$_5$	DMF, $(Et)_4$NClO$_4$	Hg	-1.23, -1.8	sce	17
C$_6$H$_5$CO 4-Py	DMF, $(Et)_4$NClO$_4$	Hg	-1.39, -2.20	sce	17
C$_6$H$_5$CO C$_6$H$_5$	DMF, $(Et)_4$NClO$_4$	Hg	-1.59, -2.21	sce	17
CN CO$_2$C$_2$H$_5$	DMF, $(Et)_4$NClO$_4$	Hg	-1.45, -2.4	sce	17
CN 4-Py	DMF, $(Et)_4$NClO$_4$	Hg	-1.64, -2.25	sce	17
CN C$_6$H$_5$	DMF, $(Et)_4$NClO$_4$	Hg	-1.99, -2.58	sce	17
CN CN	DMF, $(Et)_4$NClO$_4$	Hg	-1.36, -2.3	sce	17
CO$_2$C$_2$H$_5$ CO$_2$C$_2$H$_5$	DMF, $(Et)_4$NClO$_4$	Hg	-1.54, -2.1	sce	17
CO$_2$C$_2$H$_5$ 4-Py	DMF, $(Et)_4$NClO$_4$	Hg	-1.71, -2.36	sce	17
CO$_2$C$_2$H$_5$ C$_6$H$_5$	DMF, $(Et)_4$NClO$_4$	Hg	-2.00, -2.49	sce	17

X,Y,Z	Solvent, Electrolyte	Electrode	$E_{1/2}$	Ref. electrode	Ref.
X=CO$_2$C$_2$H$_5$ Y=CON(CH$_3$)$_2$	DMF, (Et)$_4$NClO$_4$	Hg	-1.76, -2.62	sce	17
CO$_2$C$_2$H$_5$ CF$_3$	DMF, (Et)$_4$NClO$_4$	Hg	-1.74	sce	17
C$_6$H$_5$ CON(CH$_3$)$_2$	DMF, (Et)$_4$NClO$_4$	Hg	-2.16, -2.58	sce	17
C$_6$H$_5$ C$_6$H$_5$	DMF, (Et)$_4$NClO$_4$	Hg	-2.36, -2.76	sce	17
CON(CH$_3$)$_2$ CON(CH$_3$)$_2$	DMF, (Et)$_4$NClO$_4$	Hg	-1.97, -2.28	sce	17

Miscellaneous Activated Olefins:

X,Y,Z	Solvent, Electrolyte	Electrode	$E_{1/2}$	Ref. electrode	Ref.
X=C$_6$H$_5$ Y=CN Z=CO$_2$C$_2$H$_5$	DMF, (Et)$_4$NClO$_4$	Hg	-1.32, -2.05	sce	17
CH$_3$CH=CH CN CO$_2$C$_2$H$_5$	DMF, (Et)$_4$NClO$_4$	Hg	-1.37	sce	17
CO$_2$C$_2$H$_5$ CH$_3$ CF$_3$	DMF, (Et)$_4$NClO$_4$	Hg	-1.91	sce	17
CO$_2$C$_2$H$_5$ CF$_3$ CF$_3$	DMF, (Et)$_4$NClO$_4$	Hg	-1.35, -2.57	sce	17
1,1-Dimethyl-3-bromo-1,2-propadiene	DMF, (Et)$_4$NBr	Hg	-1.06, -2.35	Ag, AgCl	18
Butadiene	75% Dioxane, (n-Bu)$_4$NBr	Hg	-2.59	sce	10
1,4-Diphenylbuta-diene-1,3	75% Dioxane, (n-Bu)$_4$NI	Hg	-1.980	sce	11
1,6-Diphenylhexa-triene	96% Dioxane, (n-Bu)$_4$NI	Hg	-1.76	sce	19
Cyclooctatetraene	Ethanol, (Et)$_4$NOH	Hg	-1.51	sce	20

(continued)

Compound	Solvent System	Working Electrode	$E_{1/2}$	Reference Electrode	Reference
cis,cis-1,4-Distyryl-benzene	Methanol, $(Et)_4NI$	Hg	-1.92, -2.14	sce	21
cis,trans-1,4-Distyrylbenzene	Methanol, $(Et)_4NI$	Hg	-1.88, -2.12	sce	21
trans,trans-1,4-Distyrylbenzene	Methanol, $(Et)_4NI$	Hg	-1.86, -2.12	sce	21
1,8-Diphenyloctatetraene	96% Dioxane, $(\underline{n}-Bu)_4NI$	Hg	-1.62, -2.02, -2.23, -2.59	sce	19
1,10-Diphenyldecapentaene	96% Dioxane, $(\underline{n}-Bu)_4NI$	Hg	-1.54, -2.02, -2.30, -2.55	sce	19
1,12-Diphenyldodecahexaene	96% Dioxane, $(\underline{n}-Bu)_4NI$	Hg	-1.45, -1.86, -2.05, -2.27	sce	19
11 Vinyl Derivatives of Aromatic compounds	DMF, $0.05\underline{M}$ $(Et)_4NI$				22
12 Divinylbenzene Derivatives	DMF, $0.05\underline{M}$ $(Et)_4NI$				23

See also Ethers (Section 17), Amines (Section 21), and Nitriles (Section 23) for substituted olefins.

References for Olefins and Polyolefins

1. M. Fleischman and D. Pletcher, *Tetrahedron Letters*, 60, 6255 (1968).
2. J. W. Breitenbach and F. Sommer, Abstracts CITCE Meeting, Prague, Czechoslovakia, 1970.
3. L. Eberson and K. Nyberg, *J. Am. Chem. Soc.*, 88, 1686 (1966).
4. J. D. Stuart and W. E. Ohnesorge, Abstract or Papers, Electrochemical Society Meeting, New York, May 1969, p. 305.
5. V. D. Parker and L. Eberson, *Chem. Commun.*, 451 (1969).
6. V. D. Parker, K. Nyberg, and L. Eberson, *J. Electroanal. Chem.*, 22, 150 (1969).
7. A. J. Bard and J. Phelps, *J. Electroanal. Chem.*, 25, Appendix 2 (1970).
8. W. C. Neikam, G. R. Dimeler, and M. M. Desmond, *J. Electrochem. Soc.*, 111, 1190 (1964).
9. D. H. Geske, *J. Am. Chem. Soc.*, 81, 4145 (1959).
10. M. von Stackelberg and W. Stracke, *Z. Elektrochem.*, 53, 118 (1949).
11. H. A. Laitinen and S. Wawzonek, *J. Am. Chem. Soc.*, 64, 1765 (1942).
12. S. Wawzonek, E. W. Blaha, R. Berkey, and M. E. Runner, *J. Electrochem. Soc.*, 102, 235 (1955).
13. P. G. Grodzka and P. J. Elving, *J. Electrochem. Soc.*, 110, 231 (1963).
14. B. L. Funt and D. G. Gray, *J. Electrochem. Soc.*, 117, 1020 (1970).
15. F. Goulden and F. L. Warren, *J. Biochem.*, 42 420 (1948).
16. M. M. Baizer, J. L. Chruma, and P. A. Berger, *J. Org. Chem.*, 35, 3569 (1970).
17. J. P. Petrovich, M. M. Baizer, and M. R. Ort, *J. Electrochem. Soc.*, 116, 743 (1969).
18. J. Simonet, H. Doupex, and D. Bretelle, *C. R. Acad. Sci. Paris*, t. 270, 59 (1970).
19. G. J. Hoijtink and J. v. Schooten, *Rec. trav. chim.*, 72, 691 (1953).
20. R. M. Elofson, *Anal. Chem.*, 21, 917 (1949).
21. V. N. Dmitrieva and V. D. Bezuglyi, *Zh. Anal. Khim.*, 22, 935 (1967).
22. V. D. Bezuglyi and Yu P. Ponomarev, *Zh. Anal. Khim.*, 24, 443 (1969).
23. N. P. Shimanskaya, V. D. Bezuglyi, L. Ya. Malkes, and L. V. Shubina, *Zh. Obsh. Khim.*, 37, 52 (1967).

3. ACETYLENES

Oxidation

Compound	Solvent System	Working Electrode	$E_{1/2}$	Reference Electrode	Reference
Phenylacetylene	CH_3CN, $(Et)_4NClO_4$	Pt	+2.12 to +2.30	Ag, AgCl	1

Reduction

Compound	Solvent System	Working Electrode	$E_{1/2}$	Reference Electrode	Reference
Acetylene	75% Dioxane, $(n\text{-}Bu)_4NBr$	Hg	no wave	sce	2
Diacetylene	75% Dioxane, $(n\text{-}Bu)_4NBr$	Hg	-2.27	sce	2
Vinylacetylene	75% Dioxane, $(n\text{-}Bu)_4NBr$	Hg	-2.40, -2.58	sce	2
Dimethylacetylene	Methanol, 0.1M $(n\text{-}Bu)_4NCl$	Hg	-1.948	Ag, AgCl	3
Divinylacetylene	75% Dioxane, $(n\text{-}Bu)_4NBr$	Hg	-2.07, -2.55	sce	2

Compound	Medium	Electrode	E	Reference electrode	Ref.
Diisopropylacety-lene	Methanol, 0.1M (n-Bu)4NCl	Hg	-1.954	Ag, AgCl	3
1,1-Dimethyl-1-bromopropyne	DMF, (Et)4NBr	Hg	-0.71, -2.35	Ag, AgCl	4
2-Butyne-1,4-diol	Methanol, 0.1M (n-Bu)4NCl	Hg	-1.886	Ag, AgCl	3
1,4-Dihydroxy-1,4-tetraphenyl 2-butyne	Methanol, 0.1M (n-Bu)4NCl	Hg	-2.024	Ag, AgCl	3
2-Hydroxy-3-hexyne	Methanol, 0.1M (n-Bu)4NCl	Hg	-1.808	Ag, AgCl	3
2,5-Dihydroxy-2,5-dimethyl-3-hexyne	Methanol, 0.1M (n-Bu)4NCl	Hg	-2.060	Ag, AgCl	3
Heptyne-1	75% Dioxane, (n-Bu)4NI	Hg	no wave	sce	5
Phenylacetylene	75% Dioxane, (n-Bu)4NI	Hg	-2.370	sce	2, 5
Diphenylacetylene	0.2M (n-Bu)4NI	Hg	-1.97	Hg pool	6
	75% Dioxane, (n-Bu)4NI	Hg	-2.195	sce	2, 5
o-Methyldipheny-lacetylene	0.2M (n-Bu)4NI	Hg	-1.69, -1.96	Hg pool	6
	0.03M (n-Bu)4NI	Hg	-2.11, -2.42	sce	7
	0.03M (n-Bu)4NI	Hg	-2.18, -2.49	sce	7
m-Methyldipheny-lacetylene	0.03M (n-Bu)4NI	Hg	-2.17, -2.48	sce	7
p-Methyldipheny-lacetylene	0.03M (n-Bu)4NI	Hg	-2.16	sce	7
p-Methoxydiphenyl-acetylene	0.03M (n-Bu)4NI	Hg	-2.24	sce	7

(continued)

Compound	Solvent System	Working Electrode	$E_{1/2}$	Reference Electrode	Reference
p-Nitrodiphenyl-acetylene	0.03M (n-Bu)$_4$NI	Hg	-0.89, -1.48, -2.24	sce	7
p,p'-Dimethyldi-phenylacetylene	0.03M (n-Bu)$_4$NI	Hg	-2.22	sce	7
	Methanol, 0.1M (n-Bu)$_4$NCl	Hg	-1.948	Ag, AgCl	3

References for Acetylenes

1. J. W. Breitenbach and F. Sommer, Abstracts CITCE Meeting, Prague, Czechoslovakia, 1970.
2. M. von Stackelberg and W. Stracke, *Z. Elektrochem.*, **53**, 118 (1949).
3. L. Horner and H. Roder, *Liebigs Ann. Chem.*, **723**, 11 (1969).
4. J. Simonet, H. Coupeux, and D. Bretelle, *C. R. Acad. Sci. Paris*, t. **270**, 59 (1970).
5. H. A. Laitinen and S. Wawzonek, *J. Am. Chem. Soc.*, **64**, 1765 (1942).
6. S. Wawzonek and D. Wearring, *J. Am. Chem. Soc.*, **81**, 2067 (1959).
7. R. E. Sioda, D. O. Cowan, and W. S. Koski, *J. Am. Chem. Soc.*, **89**, 230 (1967).

4. AROMATIC HYDROCARBONS

Oxidation

Compound	Solvent System	Working Electrode	$E_{1/2}$	Reference Electrode	Reference
Benzene	CH₃CN, NaClO₄	Pt	+2.08	Ag/0.1N Ag+	1,2
	CH₃CN, (Et)₄NClO₄	Pt	+2.38	sce	3
	CH₃CN, NaClO₄	Pt	+2.00	Ag/0.1N Ag+	4
	CH₃CN, (n-Bu)₄PF₆	Pt	+2.4 Ep	sce	5
	CH₃CN, NaClO₄	Pt	+2.30	sce	6
Toluene	CH₃CN, NaClO₄	Pt	+1.98	Ag/0.1N Ag+	1
	CH₃CN, NaClO₄	Pt	+1.93	Ag/0.1N Ag+	4
o-Xylene	CH₃CN, NaClO₄	Pt	+1.58, +2.04	Ag/0.1N Ag+	1
	CH₃CN, NaClO₄	Pt	+1.57	Ag/0.1N Ag+	2
	CH₃CN, NaClO₄	Pt	+1.89	sce	6
m-Xylene	CH₃CN, NaClO₄	Pt	+1.58	Ag/0.1N Ag+	1,2
	CH₃CN, NaClO₄	Pt	+1.91	sce	6
p-Xylene	CH₃CN, NaClO₄	Pt	+1.56	Ag/0.1N Ag+	1,2
	CH₃CN, NaClO₄	Pt	+1.77	sce	6
1,3,5-Trimethyl-benzene(Mesitylene)	CH₃CN, NaClO₄	Pt	+1.55	Ag/0.1N Ag+	2
	CH₃CN, NaClO₄	Pt	+1.80	sce	6
	HOAc, NaOAc	Pt	+1.90	sce	7
1,2,3-Trimethyl-benzene	CH₃CN, NaClO₄	Pt	+1.58	Ag/0.1N Ag+	2

1,2,4-Trimethyl-benzene	CH₃CN, NaClO₄	Pt	+1.41	Ag/0.1N Ag+	2
1,2,3,5-Tetramethyl-benzene	CH₃CN, NaClO₄	Pt	+1.50, +1.99	Ag/0.1N Ag+	1
1,2,4,5-Tetramethyl-benzene (Durene)	CH₃CN, NaClO₄	Pt	+1.43	Ag/0.1N Ag+	2
	CH₃CN, NaClO₄	Pt	+1.29	Ag/0.1N Ag+	2
Pentamethylbenzene	HOAc, NaOAc	Pt	+1.62	sce	7
	CH₃CN, NaClO₄	Pt	+1.28	Ag/0.1N Ag+	2
	HOAc, NaOAc	Pt	+1.62	sce	7
Hexamethylbenzene	CH₃CN, NaClO₄	Pt	+1.16	Ag/0.1N Ag+	2
	CH₃CN, NaClO₄	Pt	+1.124, +1.42	Ag/0.1N Ag+	8
	CH₃CN, LiClO₄	Pt	+1.58 Ep	sce	9
	CH₃CN, (Et)₄NClO₄	Pt	+1.58	sce	3
	HOAc, NaOAc	Pt	+1.52	sce	7
Ethylbenzene	CH₃CN, NaClO₄	Pt	+1.96	Ag/0.1N Ag+	10
Hexaethylbenzene	CH₃CN, LiClO₄	Pt	+1.49 Ep	sce	9
n-Propylbenzene	CH₃CN, NaClO₄	Pt	+1.97	Ag/0.1N Ag+	10
i-Propylbenzene	CH₃CN, NaClO₄	Pt	+1.87	Ag/0.1N Ag+	10
	CH₃CN, NaClO₄	Pt	+1.88	Ag/0.1N Ag+	1
t-Butylbenzene	CH₃CN, NaClO₄	Pt	+1.87	Ag/0.1N Ag+	10
Tropilidene	CH₃CN, LiClO₄	Pt	+1.13	Ag/AgNO₃	11
Biphenyl	CH₃CN, NaClO₄	Pt	+1.48	Ag/0.1N Ag+	4
	HOAc, NaOAc	Pt	+1.91	sce	7
	CH₃CN, (Et)₄NClO₄	Pt	+1.82	sce	3
	CH₃CN, (n-Bu)₄NClO₄	Pt	+1.85	sce	5

Polynuclear Hydrocarbons

Naphthalene	CH₃CN, NaClO₄	Pt	+1.34	Ag/0.1N Ag+	1,10

(continued)

691

Compound	Solvent System	Working Electrode	$E_{1/2}$	Reference Electrode	Reference
	HOAc, NaOAc	Pt	+1.72	sce	7
	CH$_3$CN, (Et)$_4$NClO$_4$	Pt	+1.65	sce	3
	CH$_3$CN, NaClO$_4$	Pt	+1.54	sce	6
	CH$_3$CN, NaClO$_4$	Pt	+1.31	Ag/0.1N̄ Ag+	4
1-Methylnaphthalene	CH$_3$CN, NaClO$_4$	Pt	+1.24	Ag/0.1N̄ Ag+	1,10
	HOAc, NaOAc	Pt	+1.53	sce	7
	CH$_3$CN, NaClO$_4$	Pt	+1.43	sce	6
2-Methylnaphthalene	CH$_3$CN, NaClO$_4$	Pt	+1.22	Ag/0.1N̄ Ag+	1,10
	HOAc, NaOAc	Pt	+1.55	sce	7
	CH$_3$CN, NaClO$_4$	Pt	+1.45	sce	6
2,3-Dimethylnaphthalene	CH$_3$CN, NaClO$_3$	Pt	+1.08, +1.34	Ag/0.1N̄ Ag+	1
	CH$_3$CN, NaClO$_4$	Pt	+1.08	Ag/0.1N̄ Ag+	2
	CH$_3$CN, NaClO$_4$	Pt	+1.35	sce	6
2,6-Dimethylnaphthalene	CH$_3$CN, NaClO$_4$	Pt	+1.08	Ag/0.1N̄ Ag+	2
	CH$_3$CN, NaClO$_4$	Pt	+1.36	sce	6
2,7-Dimethylnaphthalene	CH$_3$CN, NaClO$_4$	Pt	+1.12	Ag/0.1N̄ Ag+	2
1,2,3,4-Tetrahydronaphthalene	CH$_3$CN, NaClO$_4$	Pt	+1.57, +2.01	Ag/0.1N̄ Ag+	1
1,4,5,8-Tetraphenylnaphthalene	CH$_3$CN, (n-Propyl)$_4$-NClO$_4$	Pt	+1.39	sce	12
Indan	CH$_3$CN, NaClO$_4$	Pt	+1.59, +2.02	Ag/0.1N̄ Ag+	1
Indene	CH$_3$CN, NaClO$_4$	Pt	+1.23	Ag/0.1N̄ Ag+	4
	CH$_3$CN, (Et)$_4$NClO$_4$	Pt	+1.73 to +1.76	Ag, AgCl	13
Acenaphthene	CH$_3$CN, NaClO$_4$	Pt	+0.95	Ag/0.1N̄ Ag+	1
	CH$_3$CN, NaClO$_4$	Pt	+1.11	Ag/0.1N̄ Ag+	4

Compound	Electrolyte	Electrode	E	Reference electrode	Ref.
Acenaphthene	CH_3CN, $NaClO_4$	Pt	+1.21	sce	6
	HOAc, NaOAc	Pt	+1.36	sce	7
1,1,3-Triphenyl-indene	SO_2^a, $(n\text{-}Bu)_4NCl$	Pt	+1.35	Ag/0.1N Ag+, CH_3CN	14
1,2,3-Triphenyl,1-methoxyindene	SO_2^a, $(n\text{-}Bu)_4NCl$	Pt	+1.1	Ag/0.1N Ag+, CH_3CN	14
Anthracene	CH_3CN, $NaClO_4$	Pt	+0.84	Ag/0.1N Ag+	4,2
	CH_3CN, $NaClO_4$	Pt	+1.09	sce	6
	CH_3CN, $(Et)_4NClO_4$	Pt	+1.19	sce	3
	HOAc, NaOAc	Pt	+1.20	sce	7
	CH_3CN, $(Et)_4NClO_4$	Pt	+0.91	Ag/0.01N Ag+	15
	$(n\text{-}Bu)_4NNO_3$ melt^a	Pt	+0.96	Ag, AgCl	16
Dihydroanthracene	CH_3CN, $NaClO_4$	Pt	+1.53	Ag/0.1N Ag+	4
9-Methylanthracene	CH_3CN, $NaClO_4$	Pt	+0.96	sce	6
9-Phenylanthracene	CH_3CN, $LiClO_4$, 3,5-lutidine	Pt	+1.0, +1.4 E_p	sce	17
9,10-Dimethyl-anthracene	CH_3CN, $NaClO_4$	Pt	+0.65	sce	4,2
9,10-Diphenyl-anthracene	CH_3CN, $LiClO_4$	Pt	+1.0, +1.58	sce	18
	CH_3CN, $(Et)_4NClO_4-$	Pt	+0.92	Ag/0.01N Ag+	15
9,10-Bis(phenyl-ethynyl)anthracene	CH_3CN, $(n\text{-}Propyl)_4NClO_4$	Pt	+1.17	sce	12
Bianthryl	CH_3CN, $NaClO_4$	Pt	+0.87, +1.2	Ag/0.1N Ag+	4
Phenanthrene	CH_3CN, $NaClO_4$	Pt	+1.23	Ag/0.1N Ag+	4
	CH_3CN, $NaClO_4$	Pt	+1.50	sce	6
	CH_3CN, $(Et)_4NClO_4$	Pt	+1.28	Ag/0.01N Ag+	15
	HOAc, NaOAc	Pt	+1.68	sce	7
9,10-Diphenyl-phenanthrene	CH_3CN, $(Et)_4NClO_4$	Pt	+1.55, +1.75 E_p	sce	19

(continued)

Compound	Solvent System	Working Electrode	$E_{1/2}$	Reference Electrode	Reference
Fluorene	CH_3CN, $NaClO_4$	Pt	+1.25	Ag/0.1N Ag+	4
	HOAc, NaOAc	Pt	+1.65	sce	7
	CH_3CN, $(Et)_4NClO_4$	Pt	+1.50	sce	3
Retene	CH_3CN, $NaClO_4$	Pt	+1.18	Ag/0.1N Ag+	4
Biphenylene	HOAc, NaOAc	Pt	+1.30	sce	7
Chrysene	CH_3CN, $NaClO_4$	Pt	+1.13	Ag/0.1N Ag+	4
	CH_3CN, $NaClO_4$	Pt	+1.35	sce	6
	CH_3CN, $(Et)_4NClO_4$	Pt	+1.22	Ag/0.01N Ag+	15
	CH_3CN, $(Et)_4NClO_4$	Pt	+1.46	Ag/0.01N Ag+	15
Triphenylene	HOAc, NaOAc	Pt	+1.74	sce	7
	CH_3CN, $NaClO_4$	Pt	+1.55	sce	6
Fluoranthene	CH_3CN, $NaClO_4$	Pt	+1.18	Ag/0.1N Ag+	4
	CH_3CN, $NaClO_4$	Pt	+1.45	sce	6
	HOAc, NaOAc	Pt	+1.64	sce	7
Pyrene	CH_3CN, $NaClO_4$	Pt	+0.86, +1.12	Ag/0.1N Ag+	4
	CH_3CN, $NaClO_4$	Pt	+1.16	sce	6
	CH_3CN, $(Et)_4NClO_4$	Pt	+1.06	Ag/0.01N Ag+	15
	HOAc, NaOAc	Pt	+1.20	sce	7
	CH_3CN, $(Et)_4NClO_4$	Pt	+1.25	sce	3
Dihydropyrene	CH_3CN, $NaClO_4$	Pt	+0.87, +1.10	Ag/0.1N Ag+	4
Tetracene	CH_3CN, $NaClO_4$	Pt	+0.54, +1.20	Ag/0.1N Ag+	4
	CH_3CN, $NaClO_4$	Pt	+0.77	sce	6
5,12-Dihydrotetracene	$(n-Bu)_4NNO_3$ melt	Pt	+0.76	Ag, AgCl	16
	CH_3CN, $NaClO_4$	Pt	+1.16	Ag/0.1N Ag+	4
Methylcholanthrene	CH_3CN, $NaClO_4$	Pt	+0.69	Ag/0.1N Ag+	4
	CH_3CN, $NaClO_4$	Pt	+0.87	sce	6
Rubrene	CH_2Cl_2, $(n-Bu)_4NClO_4$	Pt	+0.82	sce	20

(9, 10, 11, 12-Tetra-

Compound	Solvent, electrolyte	Electrode	E	Reference	Ref.
1,2-Benzpyrene	CH_3CN, $(Et)_4NClO_4$	Pt	+1.34	sce	3
	CH_3CN, $NaClO_4$	Pt	+1.27	sce	6
	CH_3CN, $NaClO_4$	Pt	+0.76	Ag/0.1N Ag+	4
3,4-Benzpyrene	CH_3CN, $(Et)_4NClO_4$	Pt	+1.2 E_p	sce	21
	CH_3CN, $NaClO_4$	Pt	+0.94	sce	6
1,2,3,4-Dibenzpyrene	CH_3CN, $(Et)_4NClO_4$	Pt	+1.10	sce	3
1,2,4,5-Dibenzpyrene	CH_3CN, $NaClO_4$	Pt	+1.15	sce	6
3,4,9,10-Dibenzpy-rene	CH_3CN, $NaClO_4$	Pt	+1.01	sce	6
6-Acetoxybenz-o(α)pyrene	CH_3CN, $(Et)_4NClO_4$	Pt	+1.17, +1.98	sce	3
1,2-Benzanthracene	CH_3CN, $(Et)_4NClO_4$	Pt	+1.2 E_p	sce	21
	CH_3CN, $NaClO_4$	Pt	+0.92	Ag/0.1N Ag+	4
	CH_3CN, $NaClO_4$	Pt	+1.18	sce	6
	CH_3CN, $(Et)_4NClO_4$	Pt	+1.00	Ag/0.01N Ag+	15
	CH_3CN, $(Et)_4NClO_4$	Pt	+1.33, +1.83	sce	3
1-Methyl-1,2-benzanthracene	CH_3CN, $NaClO_4$	Pt	+1.14	sce	6
2-Methyl-1,2-benzanthracene	CH_3CN, $NaClO_4$	Pt	+1.14	sce	6
3-Methyl-1,2-benzanthracene	CH_3CN, $NaClO_4$	Pt	+1.14	sce	6
4-Methyl-1,2-benzanthracene	CH_3CN, $NaClO_4$	Pt	+1.14	sce	6
5-Methyl-1,2-benzanthracene	CH_3CN, $NaClO_4$	Pt	+1.15	sce	6
6-Methyl-1,2-benzanthracene	CH_3CN, $NaClO_4$	Pt	+1.15	sce	6

(continued)

Compound	Solvent System	Working Electrode	$E_{1/2}$	Reference Electrode	Reference
7-Methyl-1,2-benzanthracene	CH_3CN, $NaClO_4$	Pt	+1.08	sce	6
8-Methyl-1,2-benzanthracene	CH_3CN, $NaClO_4$	Pt	+1.13	sce	6
9-Methyl-1,2-benzanthracene	CH_3CN, $NaClO_4$	Pt	+1.15	sce	6
10-Methyl-1,2-benzanthracene	CH_3CN, $NaClO_4$	Pt	+1.14	sce	6
11-Methyl-1,2-benzanthracene	CH_3CN, $NaClO_4$	Pt	+1.14	sce	6
12-Methyl-1,2-benzanthracene	CH_3CN, $NaClO_4$	Pt	+1.07	sce	6
9,10-Dimethyl-1,2-benzanthracene	CH_3CN, $(Et)_4NClO_4$	Pt	+1.12	sce	3
4-Fluoro-10-methyl-1,2-benzanthracene	CH_3CN, $(Et)_4NClO_4$	Pt	+1.28, +1.54	sce	3
3-Fluoro-10-methyl-1,2-benzanthracene	CH_3CN, $(Et)_4NClO_4$	Pt	+1.27, +1.92	sce	3
1,2,3,4-Dibenzanthracene	CH_3CN, $NaClO_4$	Pt	+1.25	sce	6
	CH_3CN, $4(Et)$ $NClO_4$	Pt	+1.34, +1.71	sce	3
1,2,5,6-Dibenzanthracene	CH_3CN, $NaClO_4$	Pt	+1.19	sce	3
	CH_3CN, $(Et)_4NClO_4$	Pt	+1.40, +1.77	sce	6
	CH_3CN, $NaClO_4$	Pt	+1.00, +1.26	Ag/0.1$\underline{N}$ Ag+	4
1,2,7,8-Dibenzanthracene	CH_3CN, $NaClO_4$	Pt	+1.26	sce	3

Compound	Solvent System	Working Electrode	$E_{1/2}$	Reference Electrode	Reference
Perylene	CH_3CN, $NaClO_4$	Pt	+0.55	Ag/0.1N $\underline{Ag+}$	4
	CH_3CN, $NaClO_4$	Pt	+0.85	sce	6
	HOAc, NaOAc	Pt	+1.00	sce	7
	$(\underline{n}-Bu)_4NNO_3$ melt[a]	Pt	+0.94	Ag, AgCl	16
	CH_3CN, $(Et)_4NClO_4$	Pt	+1.04, +1.41	sce	3
1,2-Benzperylene	CH_3CN, $NaClO_4$	Pt	+1.01	sce	6
Picene	CH_3CN, $NaClO_4$	Pt	+1.33	sce	6
Pentacene	$(\underline{n}-Bu)_4NNO_3$ melt[a]	Pt	+0.74	Ag, AgCl	16
Coronene	CH_3CN, $NaClO_4$	Pt	+1.23	sce	6
Azulene	CH_3CN, $NaClO_4$	Pt	+0.71	sce	6
	HOAc, NaOAc	Pt	+0.91	sce	7

a, at 150°C

Reduction

Polynuclear Hydrocarbons:

Compound	Solvent System	Working Electrode	$E_{1/2}$	Reference Electrode	Reference
Naphthalene	CH_3CN, $(Et)_4NClO_4$	Hg	-2.63	sce	3
	75% Dioxane, $(\underline{n}-Bu)_4NI$	Hg	-2.50	sce	22
	THF $(\underline{n}-Bu)_4NI$	Hg	-3.32	Ag/$AgNO_3$	23
1-Acetylnaphthalene	McIlvaine buffer, KCl, pH 6.5	Hg	-1.44	Ag, AgCl	24

(continued)

Compound	Solvent System	Working Electrode	$E_{1/2}$	Reference Electrode	Reference
1,2-Dihydronaphthalene	75% Dioxane, $(n-Bu)_4NI$	Hg	-2.57	sce	22
α-Methylnaphthalene	75% Dioxane, $(n-Bu)_4NI$	Hg	-2.50	sce	22
β-Methylnaphthalene	75% Dioxane, $(n-Bu)_4NI$	Hg	-2.50	sce	22
Indene	75% Dioxane, $(n-Bu)_4NI$	Hg	-2.54	sce	22
Hydrindene	75% Dioxane, $(n-Bu)_4NI$	Hg	no wave	sce	22
3-Phenylindene	75% Dioxane, $(n-Bu)_4NI$	Hg	-2.33	sce	22
Anthracene	75% Dioxane, $(n-Bu)_4NI$	Hg	-1.94	sce	22
	THF, $(n-Bu)_4NI$	Hg	-1.77	sce	25
	CH_3CN, $(Et)_4NClO_4$	Hg	-2.75	Ag/AgNO₃	23
	$(n-Bu)_4NNO_3$ melt	Hg	-2.07, -2.52	sce	3
		Hg	-1.87	Ag, AgCl	16
1-Aminoanthracene	CH_3CN, $(Et)_4NClO_4$	Hg	-2.10, -2.51	sce	3
2-Aminoanthracene	CH_3CN, $(Et)_4NClO_4$	Hg	-2.17, -2.55	sce	3
1,2-Benzanthracene	CH_3CN, $(Et)_4NClO_4$	Hg	-2.11, -2.45	sce	3
	75% Dioxane, $(n-Bu)_4NI$	Hg	-2.03, -2.54	sce	22
9,10-Dimethyl-1,2-benzanthracene	75% Dioxane, $(n-Bu)_4NI$	Hg	-2.05, -2.53	sce	22
	CH_3CN, $(Et)_4NClO_4$	Hg	-2.15, -2.42	sce	3

Compound	Solvent, Electrolyte	Electrode	Potentials	Ref. electrode	Ref.
4-Fluoro-10-methyl-1,2-benzanthracene	CH_3CN, $(Et)_4NClO_4$	Hg	-2.04, -2.38	sce	3
3-Fluoro-10-methyl-1,2-benzanthracene	CH_3CN, $(Et)_4NClO_4$	Hg	-2.10, -2.40	sce	3
1,2,3,4-Dibenzanthracene	CH_3CN, $(Et)_4NClO_4$	Hg	-1.55, -1.93	sce	3
1,2,5,6-Dibenzanthracene	CH_3CN, $(Et)_4NClO_4$	Hg	-2.12, -2.42	sce	3
	75% Dioxane, $(n-Bu)_4N$, NI	Hg	-2.07, -2.53	sce	22
	75% Dioxane, $(n-Bu)_4N$, NI	Hg	-2.135, -2.505	sce	25
Acenaphthene	75% Dioxane, $(n-Bu)_4N$, NI	Hg	-2.58	sce	22
Fluorene	75% Dioxane, $(n-Bu)_4N$, NI	Hg	-2.65	sce	22
	CH_3CN, $(Et)_4NClO_4$	Hg	-2.77	sce	3
2-Acetylamino-fluorene	CH_3CN, $(Et)_4NClO_4$	Hg	-2.01, -2.73	sce	3
Phenanthrene	75% Dioxane, $(n-Bu)_4N$, NI	Hg	-2.46, -2.71	sce	22
9,10-Dihydrophenanthrene	75% Dioxane, $(n-Bu)_4N$, NI	Hg	-2.62	sce	22
Pyrene	75% Dioxane, $(n-Bu)_4N$, NI	Hg	-2.10, -2.46, -2.68	sce	22
	CH_3CN, $(Et)_4NClO_4$	Hg	-2.19, -2.64	sce	3
	CH_3CN, $(Et)_4NClO_4$	Hg	-2.22, -2.58	sce	3
1,2-Benzpyrene	75% Dioxane, $(n-Bu)_4N$, NI	Hg	-2.16, -2.36, -2.64	sce	25

(continued)

Compound	Solvent System	Working Electrode	E$_{1/2}$	Reference Electrode	Reference
3,4-Benzpyrene	75% Dioxane, (n-Bu)$_4$- NI	Hg	-1.88	sce	22
	CH$_3$CN, (Et)$_4$NClO$_4$	Hg	-1.95, -2.24	sce	3
3,4,9,10-Dibenz-pyrene	CH$_3$CN, (Et)$_4$NClO$_4$	Hg	-1.90, -2.10	sce	3
Tetracene	THF, (n-Bu)$_4$NI	Hg	-2.36	Ag/AgNO$_3$	23
	(n-Bu)$_4$NNO$_3$ melt	Hg	-1.42	Ag, AgCl	16
Fluoranthene	75% Dioxane, (n-Bu)$_4$- NI	Hg	-1.73, -2.11, -2.64	sce	22
Perylene	THF, (n-Bu)$_4$NI	Hg	-2.34	Ag/AgNO$_3$	23
	(n-Bu)$_4$NNO$_3$ melt b	Hg	-1.45	Ag, AgCl	16
	CH$_3$CN, (Et)$_4$NClO$_4$	Hg	-1.73, -2.21	sce	3
Pentacene	(n-Bu)$_4$NNO$_3$ melt	Hg	-1.20	Ag, AgCl	16
3-Methylcholanthrene	75% Dioxane, (n-Bu)$_4$- NI	Hg	-2.11, -2.51	sce	22
Biphenyl	75% Dioxane, (n-Bu)$_4$- NI	Hg	-2.70	sce	22
p-Triphenyl	THF, (n-Bu)$_4$NI	Hg	-3.38	Ag/AgNO$_3$	23
p-Tetraphenyl	THF, (n-Bu)$_4$NI	Hg	-3.06	Ag/AgNO$_3$	23
2,2'-Dinaphthyl	THF, (n-Bu)$_4$NI	Hg	-3.02	Ag/AgNO$_3$	23
	75% Dioxane, (n-Bu)$_4$- NI	Hg	-2.21, -2.49	sce	25
Tetralin	75% Dioxane, (n-Bu)$_4$- NI	Hg	no wave	sce	22

See also Aldehydes (Section 6), Ketones (Section 13), Carboxylic Acids (Section 8), Amines (Section 20), Halides (Section 22), and Nitriles (Section 23) for substituted aromatics.

References for Aromatic Hydrocarbons

1. J. W. Loveland and G. R. Dimeler, Anal. Chem., 33, 1196 (1961).
2. W. C. Neikam and M. M. Desmond, J. Am. Chem. Soc., 86, 4811 (1964).
3. T. A. Gough and M. E. Peover, Proceedings of the 3rd International Polarography Congress, Southhampton, 1965, Macmillan, London, 1966, p. 1017.
4. H. Lund, Acta Chem. Scand., 11, 1323 (1957).
5. T. Osa, A. Yildiz, and T. Kuwana, J. Am. Chem. Soc., 91, 3994 (1969).
6. E. S. Pysh and N. C. Yang, J. Am. Chem. Soc., 85, 2125 (1963).
7. L. Eberson and K. Nyberg, J. Am. Chem. Soc., 88, 1686 (1966).
8. A. E. Coleman, H. H. Richtol, and D. A. Aikens, J. Electroanal. Chem., 18, 165 (1968).
9. V. D. Parker, J. Electroanal. Chem., 21, App 1 (1969).
10. W. C. Neikam, G. R. Dimeler, and M. M. Desmond, J. Electrochem. Soc., 111, 1190 (1964).
11. D. H. Geske, J. Am. Chem. Soc., 81, 4145 (1959).
12. A. Zweig, A. H. Maurer, and B. G. Roberts, J. Org. Chem., 32, 1322 (1967).
13. J. W. Breitenbach and F. Sommer, Abstracts CITCE Meeting, Prague, Czechoslovakia, 1970.
14. L. L. Miller and E. A. Mayeda, J. Am. Chem. Soc., 92, 5818 (1970).
15. M. E. Peover and B. S. White, J. Electroanal. Chem., 13, 93 (1967).
16. B. J. Woodhall and G. R. Davies, Abstract of Papers, Society for Electrochemistry, Thornton Research Centre, England, 1969, p. 26.
17. V. D. Parker and L. Eberson, Tetrahedron Letters, 2843 (1969).
18. V. D. Parker, Chem. Commun., 848 (1969).
19. J. D. Stuart and W. E. Ohnesorge, Abstract of Papers, Electrochemical Society Meeting, New York, N.Y., May 1969, p. 305.
20. J. Phelps, K. S. V. Santhanam, and A. J. Bard, J. Am. Chem. Soc., 89, 1752 (1967).
21. L. Jeftić and R. N. Adams, J. Am. Chem. Soc., 92, 1332 (1970).
22. S. Wawzonek and H. A. Laitinen, J. Am. Chem. Soc., 64, 2365 (1942).
23. J. Perichon and R. Buvet, Electrochim. Acta, 9, 587 (1964).

24. J. Grimshaw and E. J. F. Rea, _J. Chem. Soc._, 1967, 2628 (1967).
25. G. J. Hoijtink and J. v. Schooten, _Rec. Trav. Chim._, 72, 691 (1953).

5. ALIPHATIC ALDEHYDES

Reduction

Compound	Solvent System	Working Electrode	$E_{1/2}$	Reference Electrode	Reference
Formaldehyde	1M Formaldehyde 0.05N KOH, 0.1N KCl	Hg	-1.50^a	---b	1
	0.1N HClO₄	Hg	-1.67	sce	2
	0.2N (Me)₄NOH	Hg	~-1.0	sce	3
Acetaldehyde	0.1N LiOH	Hg	-1.87	Hg pool	4
	0.2N (Me)₄NOH	Hg	-1.65	NHE	5
Propionaldehyde	0.1N LiOH	Hg	-1.92	Hg pool	4
	0.2N (Me)₄NOH	Hg	-1.68	NHE	5
n-Butyraldehyde	0.2N (Me)₄NOH	Hg	-1.90	Hg pool	4
Isobutyraldehyde	0.2N (Me)₄NOH	Hg	-1.91	Hg pool	4
n-Capronaldehyde	0.2N (Me)₄NOH	Hg	-1.98	Hg pool	4
n-Heptaldehyde	0.2N (Me)₄NOH	Hg	-1.90	Hg pool	4
α-Chloroisobutanal	DMF, (Et)₄NCl	Hg	~-1, -2.1	Hg pool	6
2-Chloro-2-methyl-pentanal	DMF, (Et)₄NCl	Hg	-1	Hg pool	6
α-Chlorobutanal	DMF, (Et)₄NCl	Hg	~-0.9, ~-1.5	Hg pool	6
α-Chloroheptanal	DMF, (Et)₄NCl	Hg	~-0.9, ~-1.5	Hg pool	6
α-Bromobutanal	DMF, (Et)₄NCl	Hg	~-0.9, ~-1.5	Hg pool	6
α-Fluorobutanal	DMF, (Et)₄NCl	Hg	~-1.6	Hg pool	6
α-Fluoroisobutanal	DMF, (Et)₄NCl	Hg	~-1.7	Hg pool	6

(continued)

Compound	Solvent System	Working Electrode	$E_{1/2}$	Reference Electrode	Reference
Acrolein	Aq LiCO₃, LiCl, pH 7.0	Hg	-1.04, -1.44	Hg pool	7
Crotonaldehyde	50% Ethanol, 0.2N (Et)₄NOH	Hg	-1.37, -1.80	Hg pool	4
	50% Ethanol, 0.1N NH₄Cl	Hg	-1.36, -1.64	Hg pool	4
	50% Dioxane, OAc buffer, pH 5	Hg	-1.13	sce	8
	50% Benzene, 50% Methanol, 0.3M LiCl	Hg	-1.55	sce	9
2,4-Hexandial	50% Benzene, 50% Methanol, 0.3M LiCl	Hg	-0.90	sce	9
2,4,6-Octantrienal	50% Benzene, 50% Methanol, 0.3M LiCl	Hg	-0.75	sce	9
2,4,6,8-Decatetrae-nal	50% Benzene, 50% Methanol, 0.3M LiCl	Hg	-0.64	sce	9
2,4,6,8,10-Dode-capentaneal	50% Benzene, 50% Methanol, 0.3M LiCl	Hg	-0.58, -1.32	sce	9
Citral	66% alcohol, LiCl	Hg	-1.55	sce	10
Citronellal	66% alcohol, LiCl	Hg	-1.9	sce	10
Glycolaldehyde	0.1N NaOH	Hg	-1.45	NHEC	5
Glyoxal	Buffer, pH 3.4	Hg	-1.41	sce	11
Methylglyoxal	Buffer, pH 4.5	Hg	-0.83	sce	12

Polyene Aldehydes:

$CH_3-(CH=CH)_xCHO$

		50% Aq buffer[d], 50% Dioxane		
x = 1	Hg	-0.853[e]	sce	13
2	Hg	-0.622[e]	sce	13
3	Hg	-0.480[e]	sce	13
4	Hg	-0.379[e]	sce	13
5	Hg	-0.315[e]	sce	13

a Molar reduction potential (Ref. 1).
b Not given.
c Normal hydrogen electrode.
d KCl, HOAc, H_3PO_4.
e Extrapolated at pH 0.

(continued)

705

References for Aliphatic Aldehydes

1. F. G. Jahoda, Collect. Czech. Chem. Commun., 7,
    415 (1935).
2. M. J. Boyd and K. Banibach, Ind. Eng. Chem.,
    Anal. Ed., 15, 314 (1943).
3. V. N. Epimakhov, Elektrokhim., 6, 322 (1970).
4. H. Adkins and F. W. Cox, J. Am. Chem. Soc., 60,
    1151 (1938).
5. R. Bieber and G. Trumpler, Helv. Chim. Acta, 30,
    2000 (1947).
6. N. S. Moe, Proceedings of the 3rd International
    Polarography Congress, Southampton 1965,
    Macmillan, London, 1966, p. 1077.
7. R. W. Moshier, Ind. Eng. Chem., Anal. Ed., 15,
    107 (1943).
8. M. Fields and E. R. Blout, J. Am. Chem. Soc., 70,
    930 (1948).
9. C. O. Willits, C. Ricciutti, H. B. Knight, and
    D. Swern, Anal. Chem., 24, 735 (1952).
10. M. I. Gerber, Z. B. Kuznetsova, and M. B. Neiman,
    Zh. Anal. Kim., 4, 103 (1949).
11. I. M. Kolthoff and J. J. Lingane, Polarography,
    Vol. II, Interscience, New York 1952, p. 676.
12. G. Mackinney and O. Teinmer, J. Am. Chem. Soc.,
    70, 3586 (1948).
13. D. M. Coulson, W. R. Crowell, and S. K. Tendrick,
    J. Am. Chem. Soc., 79, 1354 (1957).

6. AROMATIC ALDEHYDES

Oxidation

Compound	Solvent System	Working Electrode	$E_{1/2}$	Reference Electrode	Reference
Anisaldehyde	CH_3CN, $NaClO_4$	Pt	+1.63	sce	1

Reduction

Compound	Solvent System	Working Electrode	$E_{1/2}$	Reference Electrode	Reference
Benzaldehyde	Aq buffer, 50% Ethanol, pH 3.0	Hg	-1.06, -1.34	sce	2
	Aq buffer, 50% Dioxane	Hg	-0.869^b	sce	3
	50% Benzene, 50% Methanol, 0.3$\underline{M}$ LiCl	Hg	-1.58	sce	4
	66% Alcohol, 0.01$\underline{M}$ (Et)$_4$NBr	Hg	-1.506	sce	5
	Aq buffer, 60%^{c} Alcohol pH 1.73	Hg	-1.000	sce	5

(continued)

Compound	Solvent System	Working Electrode	$E_{1/2}$	Reference Electrode	Reference
	Aq buffer,[c] 60% Alcohol pH 3.16	Hg	-1.088	sce	5
Hexahydrobenzaldehyde[d]	40% Ethanol, 5% DMF, 0.2$\underline{M}$ (Me)4-NOH	Hg	-1.55, -2.10	NCE	6
1,2,3,6-Tetrahydrobenzaldehyde[d]	40% Ethanol, 5% DMF, 0.2$\underline{M}$ (Me)4-NOH	Hg	-2.025	NCE	6
2,-Phenyl-1,2,3,6-Tetrahydrobenzaldehyde[d]	40% Ethanol, 5% DMF, 0.2$\underline{M}$ (Me)4-NOH	Hg	-1.985	NCE	6
α-Thiophene aldehyde	40% Ethanol, 5% DMF, 0.2$\underline{M}$ (Me)4-NOH	Hg	-1.99	NCE	6
o-Methylbenzaldehyde	66% Alcohol, 0.01$\underline{M}$ (Et)4NBr	Hg	-1.45, -1.96	NCE	6
(o-Tolualdehyde)	Aq buffer,[c] 60% Alcohol, pH 1.73	Hg	-1.005	sce	5
	Aq buffer,[c] 60% Alcohol, pH 3.16	Hg	-0.358	sce	5
m-Methylbenzaldehyde	66% Alcohol, 0.01$\underline{M}$ (Et)4NBr	Hg	-0.525	sce	5
	Aq buffer,[c] 60% Alcohol, pH 1.73	Hg	-1.526	sce	5
	Alcohol, pH 1.73	Hg	-1.000	sce	5

Compound	Conditions	Electrode	E	Ref.	n
m-Methylbenzaldehyde	Aq buffer,c 60% Alcohol, pH 3.16	Hg	-1.092	sce	5
p-Methylbenzaldehyde	66% Alcohol, 0.01M (Et)₄NBr	Hg	-1.562	sce	5
	Aq buffer,c 60% Alcohol, pH 1.73	Hg	-1.020	sce	5
	Aq buffer,c 60% Alcohol, pH 3.16	Hg	-1.106	sce	5
	Aq buffer,a 50% Dioxane	Hg	-0.887[b]	sce	3
o-Methoxybenzal- dehyde	Aq buffer,a 50% Dioxane	Hg	-0.918[b]	sce	3
	Aq buffer, 50% Ethanol, pH 3.3	Hg	-1.01, -1.28	sce	2
	66% Alcohol, 0.01M (Et)₄NBr	Hg	-1.494	sce	5
	Aq buffer,c 60% Alcohol, pH 1.73	Hg	-0.960	sce	5
	Aq buffer,c 60% Alcohol, pH 2.59	Hg	-1.020	sce	5
o-Hydroxybenzal- dehyde (Salicylal- dehyde)	Aq buffer,a 50% Dioxane	Hg	-0.918[b]	sce	3
	Aq buffer, 50% Ethanol, pH 4.0	Hg	-1.16, -1.29	sce	2
	66% Alcohol, 0.01M (Et)₄NBr	Hg	-1.504	sce	5
	Aq buffer,c 60% Alcohol, pH 1.73	Hg	-1.050	sce	5

(continued)

709

Compound	Solvent System	Working Electrode	$E_{1/2}$	Reference Electrode	Reference
o-Hydroxybenzaldehyde	Aq buffer,[c] 60% Alcohol, pH 3.16	Hg	-1.134	sce	5
m-Hydroxybenzaldehyde	Aq buffer,[a] 50% Dioxane	Hg	-0.870[b]	sce	5
	66% Alcohol, 0.01M (Et)4NBr	Hg	-1.500	sce	5
	Aq buffer,[c] 60% Alcohol, pH 1.73	Hg	-1.008	sce	5
	Aq buffer,[c] 60% Alcohol, pH 3.16	Hg	-1.098	sce	5
p-Hydroxybenzaldehyde	Aq buffer,[a] 50% Dioxane	Hg	-0.991[b]	sce	3
	Aq buffer, 50% Ethanol, pH 5.5	Hg	-1.28, -1.54	sce	2
	66% Alcohol, 0.01M (Et)4NBr	Hg	-1.72	sce	5
	Aq buffer,[c] 60% Alcohol, pH 1.73	Hg	-1.116	sce	5
	Aq buffer,[c] 60% Alcohol, pH 3.16	Hg	-1.200	sce	5
m-Methoxybenzaldehyde	66% Alcohol, 0.01M (Et)4NBr	Hg	-1.622	sce	5
	Aq buffer,[c] 60% Alcohol, pH 1.73	Hg	-1.080	sce	5
	Aq buffer,[c] 60% Alcohol, pH 2.59	Hg	-1.122	sce	5
	Aq buffer,[a] 50% Dioxane	Hg	-0.845[b]	sce	3

Compound	Conditions	Electrode	E	Ref. electrode	Ref.
p-Methoxybenzal-dehyde (Anisaldehyde)	Aq buffer,[a] 50% Dioxane	Hg	-0.964^b	sce	3
o-Chlorobenzal-dehyde	Ethanol, pH 4.4	Hg	$-1.19, -1.38$	sce	2
	Aq buffer,[a] 50% Dioxane	Hg	-0.729^b	sce	3
	66% Alcohol, 0.01M (Et)$_4$NBr	Hg	-1.331	sce	5
	Aq buffer,[c] 60% Alcohol, pH 1.73	Hg	-0.868	sce	5
	Aq buffer,[c] 60% Alcohol, pH 3.16	Hg	-0.970	sce	5
m-Chlorobenzal-dehyde	Aq buffer,[a] 50% Dioxane	Hg	-0.798^b	sce	3
	66% Alcohol, 0.01M (Et)$_4$NBr	Hg	-1.384	sce	5
	Aq buffer,[c] 60% Alcohol, pH 1.73	Hg	-0.910	sce	5
	Aq buffer,[c] 60% Alcohol, pH 3.16	Hg	-1.030	sce	5
p-Chlorobenzal-dehyde	Aq buffer,[a] 50% Dioxane	Hg	-0.808^b	sce	3
	66% Alcohol, 0.01M (Et)$_4$NBr	Hg	-1.422	sce	5
	Aq buffer,[c] 60% Alcohol, pH 1.73	Hg	-0.952	sce	5
	Aq buffer,[c] 60% Alcohol, pH 3.16	Hg	-1.050	sce	5
o-Bromobenzaldehyde	Aq buffer,[c] 50% Dioxane	Hg	-0.733^b	sce	3

(continued)

711

Compound	Solvent System	Working Electrode	$E_{1/2}$	Reference Electrode	Reference
m-Bromobenzaldehyde	Aq buffer,[c] 50% Dioxane	Hg	-0.768[b]	sce	3
	66% Alcohol, 0.01M (Et)4NBr	Hg	-1.358	sce	5
	Aq buffer,[c] 60% Alcohol, pH 1.73	Hg	-0.904	sce	5
	Aq buffer,[c] 60% Alcohol, pH 3.16	Hg	-1.028	sce	5
p-Bromobenzaldehyde	Aq buffer,[a] 50% Dioxane	Hg	-0.812[b]	sce	3
	66% Alcohol, 0.01M (Et)4NBr	Hg	-1.410	sce	5
	Aq buffer,[c] 60% Alcohol, pH 1.73	Hg	-0.918	sce	5
	Aq buffer,[c] 60% Alcohol, pH 3.16	Hg	-1.032	sce	5
o-Iodobenzaldehyde	66% Alcohol, 0.01M (Et)4NBr	Hg	-1.248	sce	5
	Aq buffer,[c] 60% Alcohol, pH 1.73	Hg	-0.827	sce	5
	Aq buffer,[c] 60% Alcohol, pH 3.16	Hg	-0.914	sce	5
m-Iodobenzaldehyde	66% Alcohol, 0.01M (Et)4NBr	Hg	-1.390	sce	5
	Aq buffer,[c] 60% Alcohol, pH 1.73	Hg	-0.860	sce	5
	Aq buffer,[c] 60% Alcohol, pH 3.16	Hg	-0.952	sce	5

Compound	Solvent	Electrode	E	Ref. electrode	Ref.
p-Iodobenzaldehyde	66% Alcohol, 0.01M (Et)₄NBr	Hg	-1.400	sce	5
	Aq buffer, c 60% Alcohol, pH 1.73	Hg	-0.878	sce	5
	Aq buffer, c 60% Alcohol, pH 3.16	Hg	-0.980	sce	5
o-Aminobenzaldehyde	66% Alcohol, 0.01M (Et)₄NBr	Hg	-1.030	sce	5
	Aq buffer, c 60% Alcohol, pH 2.25	Hg	-1.52	sce	5
m-Aminobenzaldehyde	66% Alcohol, 0.01M (Et)₄NBr	Hg	-0.950	sce	5
	Aq buffer, c 60% Alcohol, pH 1.73	Hg	-1.030	sce	5
	Aq buffer, c 60% Alcohol, pH 3.16	Hg	-1.70	sce	5
p-Aminobenzaldehyde	66% Alcohol, 0.01M (Et)₄NBr	Hg	-1.064	sce	5
	Aq buffer, c 60% Alcohol, pH 1.73	Hg	-1.158	sce	5
	Aq buffer, c 60% Alcohol, pH 3.16	Hg		sce	5
p-Dimethylamino-benzaldehyde	50% Aq buffer, 50% Ethanol, pH 7.5	Hg	-1.35, -1.55	sce	2
1-Naphthaldehyde	50% Aq buffer, 50% Ethanol, pH 3.4	Hg	-0.96, -1.16	sce	2
2-Naphthaldehyde	50% Aq buffer, 50% Ethanol, pH 3.1	Hg	-0.99, -1.27	sce	2
9-Phenanthraldehyde	50% Aq buffer, 50% Ethanol	Hg	-0.91, -1.25	sce	2
Piperonal	Aq buffer, 50% Ethanol, pH 4.0	Hg	-1.13, -1.35	sce	2

(continued)

Compound	Solvent System	Working Electrode	$E_{1/2}$	Reference Electrode	Reference
Vanillin	Aq buffer, 50% Ethanol, pH 4.8	Hg	-1.23, -1.38	sce	2
Cinnamaldehyde	B-R buffer, 2% Ethanol, pH 4.7	Hg	-0.815, -1.19	sce	7
	DMF, (n-Bu)4NI	Hg	-0.97, -1.57	Hg pool	8
Jasmine aldehyde	66% Alcohol, LiCl	Hg	-1.55	sce	9

a KCl, HOAc, H3PO4.
b Extrapolated at pH 0.
c NaOAc, HCl, KCl.
d Included for comparison.

References for Aromatic Aldehydes

1. H. Lund, Acta Chem. Scand., 11, 491 (1957).
2. R. M. Powers and R. A. Day, Jr., J. Am. Chem. Soc., 80, 808 (1958).
3. D. M. Coulson, W. R. Crowell, and S. K. Tendick, J. Am. Chem. Soc., 79, 1354 (1957).
4. C. O. Willits, C. Ricciuti, H. B. Knight, and D. Swern, Anal. Chem., 24, 735 (1952).
5. E. Gergely and T. Iredale, J. Chem. Soc., 3226 (1953).
6. L. N. Nekrasov, D. N. Soshchin, and V. N. Gramenitskaya, Elektrokhim., 6, 1577 (1970).
7. D. Barnes and P. Zuman, Trans. Faraday Soc., 65, 1668 (1969).
8. S. Wawzonek and A. Gundersen, J. Electrochem. Soc., 111, 324 (1964).
9. M. I. Gerber, Z. B. Kuznetsova, and M. B. Neiman, Zh. Anal. Khim., 4, 103 (1949).

7. ALIPHATIC KETONES

Oxidation

Compound	Solvent System	Working Electrode	$E_{1/2}$	Reference Electrode	Reference
4-Methyl-2,6-hep-tanedione	CH_3CN, $LiClO_4$	Pt	+1.28	Ag/0.01$\underline{N}$ Ag+	1
4-Methyl-3,5-hep-tanediene-2-one	CH_3CN, $LiClO_4$	Pt	+0.64	Ag/0.01$\underline{N}$ Ag+	1
1,5-Diphenyl-1,5-pentanedione[a]	CH_3CN, $LiClO_4$	Pt	+2.10	Ag/0.01$\underline{N}$ Ag+	1
1,3,5-Triphenyl-1,5-pentanedione[a]	CH_3CN, $LiClO_4$	Pt	+1.80	Ag/0.01$\underline{N}$ Ag+	1

[a] Included for comparison.

716

Compound	Solvent System	Working Electrode	$E_{1/2}$	Reference Electrode	Reference
Acetone	0.025N (Me)$_4$NI 75% Dioxane,	Hg	-2.16	sce	2
	(Et)$_4$NI	Hg	-2.46	sce	3
	90% EtOH, (n-Bu)$_4$NI	Hg	-2.57	sce	4
Bromoacetone	OAc buffer, pH 4.6	Hg	-0.34	sce	4
Chloroacetone	OAc buffer, pH 4.6	Hg	-1.15	sce	4, 5
	CH$_3$CN, 1% H$_2$O, LiClO$_4$	Hg	-1.42	sce	5
1,1-Dichloroacetone	CH$_3$CN, 1% H$_2$O, LiClO$_4$	Hg	-1.09, -1.42	sce	5
	OAc buffer, pH 5.0	Hg	-0.57, -1.16	sce	5
1,3-Dichloroacetone	CH$_3$CN, 1.5% H$_2$O, LiClO$_4$	Hg	-1.09, -1.43	sce	5
	OAc buffer, pH 5.0	Hg	-0.83, -1.14	sce	5
Iodoacetone	90% EtOH, (n-Bu)$_4$NCl	Hg	-0.14	sce	4

(continued)

717

Compound	Solvent System	Working Electrode	$E_{1/2}$	Reference Electrode	Reference
Girard-T derivative of acetone	B-R buffer, pH 8.2	Hg	-1.52	sce	4
Girard-T derivative of cyclopentanone	B-R buffer, pH 8.2	Hg	-1.54	sce	6
Cyclohexanone	0.8M HCl, 0.4M KCl, pH 1.01	Hg	-1.25	sce	7
Cyclohex-2-en-1-one	0.8M HCl, 0.4M KCl, pH 1.06	Hg	-0.944, -1.026, -1.202		7
Cyclohexanone	75% Dioxane, 0.05M (Et)$_4$NI	Hg	-2.45	sce	6
Girard-T derivatives of 24 ketones	B-R buffer, KCl, 50% Ethanol			sce	8
Acetylacetone	0.1N LiCl	Hg	-1.07, -1.37	sce	9
Methyl ethyl ketone	0.025N (Me)$_4$NI	Hg	-2.25	NCE	2
Methyl vinyl ketone	0.1N KCl	Hg	-1.42	sce	10
Mesityl oxide	McIlvaine buffer, 50% Ethanol pH 4.9	Hg	-1.28	NCE	11
Biacetyl	0.1N NH$_4$Cl	Hg	-0.93, -1.68	Hg pool	12
	0.1N LiCl	Hg	-0.81, -1.93	sce	10
	OAc buffer, 50% Isopropanol, pH 5.6	Hg	-0.81	sce	13
	DMSO, (n-Proyl)$_4$-NClO$_4$	Hg	-1.27	sce	14

References for Aliphatic Ketones

1. C. Bratu and A. T. Balaban, Rev. Roumaine Chim.,
   10, 1001 (1965); Chem. Abstr., 64, 12489 (1966).
2. M. B. Neiman and Z. V. Markina, Zavod. Lab., 13,
   1177 (1947).
3. M. von Stackelberg and W. Stracke, Z. Elektrochem., 53, 118 (1949).
4. L. Meites, Polarographic Techniques, 2nd ed.,
   Interscience, New York, 1965, p. 675.
5. M. E. Hall and E. M. Harris, Anal. Chem., 41,
   1130 (1969).
6. Reference 4, p. 689.
7. E. J. Denney and B. Mooney, J. Chem. Soc. B,
   1968, 1410.
8. J. R. Young, J. Chem. Soc., 1955, 1516 (1955).
9. G. Semerano and A. Chisini, Gazz. Chim. Ital.,
   66, 504 (1936).
10. E. I. Fulmer, J. J. Kolfenbach, and L. A. Under-
    kofler, Ind. Eng. Chem., Anal. Ed., 16, 469 (1944).
11. R. Pasternak, Helv. Chim. Acta, 31, 753 (1948).
12. H. Adkins and F. W. Cox, J. Am. Chem. Soc., 60,
    1151 (1938).
13. N. J. Leonard, H. A. Laitinen, and E. H. Mottus,
    J. Am. Chem. Soc., 75, 3300 (1953).
14. G. A. Russell and S. A. Weiner, J. Am. Chem.
    Soc., 89, 6623 (1967).

8. AROMATIC KETONES

Oxidation

Compound	Solvent System	Working Electrode	$E_{1/2}$	Reference Electrode	Reference
2,3,4,5-Tetraphenyl-cyclopentadienone (tetracyclone)	CH_3CN, $(n-Bu)_4$ $NClO_4$	Pt	+1.52	sce	1
o-Hydroxyacetophen-one	OAc buffer, 50% Isopropanol, pH 5.6	Graphite	+0.801	sce	2
m-Hydroxyacetophen-one	OAc buffer, 50% Isopropanol, pH 5.6	Graphite	+0.754	sce	2
p-Hydroxyacetophen-one	OAc buffer, 50% Isopropanol, pH 5.6	Graphite	+0.791	sce	2
o-Aminoacetophen-one	OAc buffer, 50% Isopropanol, pH 5.6	Graphite	+0.847	sce	2
m-Aminoacetophen-one	OAc buffer, 50% Isopropanol, pH 5.6	Graphite	+0.758	sce	2
p-Aminoacetophenone	OAc buffer, 50% Isopropanol, pH 5.6	Graphite	+0.820	sce	2

Reduction

Compound	Solvent System	Working Electrode	$E_{1/2}$	Reference Electrode	Reference
Benzylideneacetone	DMF, (n-Bu)$_4$NI	Hg	-1.08, -2.06	sce	3
Propiophenone	OAc buffer, pH 6.0	Hg	-1.54	sce	4
	0.1M NH$_4$Cl+NH$_4$OH +0.9M KCl, pH 9	Hg	-1.52	sce	5
o-Hydroxypropio-phenone	OAc buffer, pH 6.0	Hg	-1.50	sce	4
	0.1M NH$_4$Cl+NH$_4$OH +0.9M KCl, pH 9	Hg	-1.56	sce	5
p-Hydroxypropio-phenone	OAc buffer, pH 6.0	Hg	-1.64	sce	4
	0.1M NH$_4$Cl+NH$_4$OH +0.9M KCl, 2%	Hg	-1.69	sce	5
1-Phenyl-1,3-butan-edione	B-R buffer, Ethanol, pH 4.99	Hg	-1.06, -1.23	sce	6
	DMSO, (n-Bu)$_4$NClO$_4$	Hg	-1.54, -2.41, -2.7	sce	7
1-Phenyl-1,2-propan-edione	DMSO, (n-Bu)$_4$NClO$_4$	Hg	-1.10, -2.16	sce	7
1,3-Bis(p-fluoro-phenyl)-1,3-propan-edione	DMSO, (n-Bu)$_4$NClO$_4$	Hg	-1.41, -1.73, -2.3	sce	7

(continued)

Compound	Solvent System	Working Electrode	$E_{1/2}$	Reference Electrode	Reference
1,3-Bis(p-methoxyphenyl)-1,3-propanedione	DMSO, $(n\text{-}Bu)_4NClO_4$	Hg	-1.56, -1.87, -2.5	sce	7
1,3-Diphenyl-1,3-propanedione	DMSO, $(n\text{-}Bu)_4NClO_4$	Hg	-1.38, -1.69, -2.25	sce	7
	Citrate buffer, NaCl, 50% Ethanol, pH 6.1	Hg	-1.22, -1.40	sce	8
1,4-Dibenzoyl-2,3-diphenyl-2,3-butanediol	DMSO, $(n\text{-}Bu)_4NClO_4$	Hg	-1.70, -1.95	sce	7
Benzil	McIlvaine buffer, pH 1.3	Hg	-0.31		9
	McIlvaine buffer, pH 11.3	Hg	-0.79	NCE	9
	DMSO, $(n\text{-}Propyl)_4NClO_4$	Hg	-1.16	sce	10
	DMSO, $(n\text{-}Bu)_4NClO_4$	Hg	-1.04, -1.76	sce	7
p,p'-Dimethoxybenzil	DMSO, $(n\text{-}Bu)_4NClO_4$	Hg	-1.20, -1.96	sce	7
Dibenzoylmethane	McIlvaine buffer, pH 1.3	Hg	-0.63	NCE	6
	McIlvaine buffer, pH 11.3	Hg	-1.34, -1.66	NCE	6
cis-Dibenzoylethylene	McIlvaine buffer, pH 1.3	Hg	-0.34	NCE	6
	McIlvaine buffer, pH 11.7	Hg	-0.66, -1.69	NCE	6

Compound	Medium	Electrode	E (V)	Reference electrode	Ref.
trans-Dibenzoyl-ethylene	McIlvaine buffer, pH 1.3	Hg	−1.6	NCE	6
	McIlvaine buffer, pH 11.3	Hg	−0.61, −1.57	NCE	6
13 Deriavtives of 2-Arylindandione	B-R buffer, Ethanol				11, 12
Benzophenone	McIlvaine buffer, 50% Ethanol, pH 1.3	Hg	−0.94	NCE	9
	McIlvaine buffer, 50% Ethanol, pH 11.3	Hg	−1.42	NCE	9
	OAc buffer, KCl, 40% Ethanol, pH 5.2	Hg	−1.140	3.5N calomel	13
	Citrate-Phosphate buffer, 25% Ethanol, pH 4.0	Hg	−0.99, −1.19	sce	14
	DMF, (Et)$_4$NI	Hg	−1.72, −1.99	sce	15
4-Aminobenzophenone	OAc buffer, KCl, 40% Ethanol pH 5.2	Hg	−1.161	3.5N calomel	13
4,4'-Dimethoxy-benzophenone	OAc buffer, KCl, 40% Ethanol, pH 5.2	Hg	−1.260	3.5N calomel	13
4,4'-Dimethyl-benzophenone	OAc buffer, KCl, 40% Ethanol, pH 5.2	Hg	−1.187	3.5N calomel	13
4-Methoxybenzo-phenone	OAc buffer, KCl, 40% Ethanol, pH 5.2	Hg	−1.194	3.5N calomel	13

(continued)

723

Compound	Solvent System	Working Electrode	$E_{1/2}$	Reference Electrode	Reference
4-t-Butylbenzophen-one	OAc buffer, KCl, 40% Ethanol, pH 5.2	Hg	-1.166	3.5$\underline{N}$ calomel	13
4-Methylbenzophen-one	OAc buffer, KCl, 40% Ethanol, pH 5.2	Hg	-1.165	3.5$\underline{N}$ calomel	13
4-Chloro-4'-methyl-benzophenone	OAc buffer, KCl, 40% Ethanol, pH 5.2	Hg	-1.102	3.5$\underline{N}$ calomel	13
4-Chlorobenzo-phenone	OAc buffer, KCl, 40% Ethanol, pH 5.2	Hg	-1.093	3.5$\underline{N}$ calomel	13
4-Bromobenzophenone	OAc buffer, KCl, 40% Ethanol, pH 5.2	Hg	-1.051	3.5$\underline{N}$ calomel	13
3-Bromobenzophenone	OAc buffer, KCl, 40% Ethanol, pH 5.2	Hg	-1.021	3.5$\underline{N}$ calomel	13
4,4'-Dichlorobenzo-phenone	OAc buffer, KCl, 40% Ethanol, pH 5.2	Hg	-1.002	3.5$\underline{N}$ calomel	13
3,3'-Dibromobenzo-phenone	OAc buffer, KCl, 40% Ethanol, pH 5.2	Hg	-0.913	3.5$\underline{N}$ calomel	13
2-Benzylbenzophenone	Aq NaOH, KCl, 60% Ethanol	Hg	-1.55	sce	16
4'-Chloro-2-benzyl-benzophenone	Aq NaOH, KCl, 60% Ethanol	Hg	-1.48	sce	16

Compound	Conditions				Ref.
3'-Trifluoromethyl-2-benzylbenzophenone	Aq NaOH, KCl, 60% Ethanol	Hg	-1.39	sce	16
2'-Fluoro-2-benzyl-benzophenone	Aq NaOH, KCl, 60% Ethanol	Hg	-1.48	sce	16
2'-Chloro-2-benzyl-benzophenone	Aq NaOH, KCl, 60% Ethanol	Hg	-1.53	sce	16
2'-Bromo-2-benzyl-benzophenone	Aq NaOH, KCl, 60% Ethanol	Hg	-1.41	sce	16
39 m- and p-substituted Benzophenones	Aq ethanol solution				17
Acetophenone	McIlvaine buffer, 50% Ethanol, pH 1.3	Hg	-1.12	NCE	9
	McIlvaine buffer, 50% Ethanol, pH 11.3	Hg	-1.64	NCE	9
	DMF, $(n-Bu)_4NI$, Et_4NI	Hg	-1.99, -2.46	sce	18
	Aq buffer[a], 50% Dioxane	Hg	-0.993[b]	sce	19
	DMSO, $(n-Bu)_4NClO_4$	Hg	-1.94	sce	7
o-Hydroxyaceto-phenone	0.1M KCl	Hg	-1.843[b]	Ag, AgCl	20
	Aq buffer[a], 50% Dioxane	Hg	-1.065[b]	sce	19
m-Hydroxyaceto-phenone	Aq buffer[a], 50% Dioxane	Hg	-1.000[b]	sce	19
p-Hydroxyaceto-phenone	Aq buffer[a], 50% Dioxane	Hg	-1.101[b]	sce	19

(continued)

725

Compound	Solvent System	Working Reference	$E_{1/2}$	Reference Electrode	Reference
o-Methylacetophenone	0.1M KCl	Hg	-1.818	Ag, AgCl	20
m-Methylacetophenone	0.1M KCl	Hg	-1.846	Ag, AgCl	20
p-Methylacetophenone	Aq buffer,[a] 50% Dioxane	Hg	-1.009[b]	sce	19
	Aq buffer,[a] 50% Dioxane	Hg	-1.019[b]	sce	19
o-Methoxyacetophenone	0.1M KCl	Hg	-1.940[b]	Ag, AgCl	20
	Aq buffer,[a] 50% Dioxane	Hg	-0.959[b]	sce	19
m-Methoxyacetophenone	0.1M KCl	Hg	-1.788	Ag, AgCl	20
	0.1M KCl	Hg	-1.802	Ag, AgCl	20
p-Methoxyacetophenone	Aq buffer,[a] 50% Dioxane	Hg	-0.974[b]	sce	19
	Aq buffer,[a] 50% Dioxane	Hg	-1.078[b]	sce	19
o-Chloroacetophenone	Aq buffer,[a] 50% Dioxane	Hg	-0.964[b]	sce	19
m-Chloroacetophenone	0.1M KCl	Hg	-1.845[b]	Ag, AgCl	20
	Aq buffer,[a] 50% Dioxane	Hg	-0.925[b]	sce	19
p-Chloroacetophenone	Aq buffer,[a] 50% Dioxane	Hg	-0.926[b]	sce	19
o-Bromoacetophenone	Aq buffer,[a] 50% Dioxane	Hg	-0.982[b]	sce	19
	0.1M KCl	Hg	-1.859	Ag, AgCl	20

Compound	Conditions	Electrode	E (V)	Ref. electrode	Ref.
m-Bromoacetophenone	0.1M KCl	Hg	-1.733[b]	Ag, AgCl	20
	Aq. buffer,[a] 50% Dioxane	Hg	-0.899[b]	sce	19
p-Bromoacetophenone	Aq. buffer,[a] 50% Dioxane	Hg	-0.928[b]	sce	19
o-Fluoroacetophenone	0.1M KCl	Hg	-1.786	Ag, AgCl	20
	0.1M KCl	Hg	-1.696	Ag, AgCl	20
m-Fluoroacetophenone	0.1M KCl	Hg	-1.752	Ag, AgCl	20
p-Fluoroacetophenone	0.1M KCl	Hg	-1.824	Ag, AgCl	20
C_6H_5COX					
X = CH_3 (Acetophenone)	0.1N LiOH, 50% Ethanol	Hg	-1.63	NCE	21
= CH_2CH_3	0.1N LiOH, 50% Ethanol	Hg	-1.65	NCE	21
= $(CH_2)_2CH_3$	0.1N LiOH, 50% Ethanol	Hg	-1.64	NCE	21
= $CH(CH_3)_2$	0.1N LiOH, 50% Ethanol	Hg	-1.70	NCE	21
= $(CH_2)_3CH_3$	0.1N LiOH, 50% Ethanol	Hg	-1.66	NCE	21
Benzylidene aceto-phenone (Chalcone)	B-R buffer, 20% Ethanol, pH 8.4	Hg	-0.96, -1.24, -1.4, -1.57	sce	22
	DMF, (n-Bu)$_4$NI	Hg	-0.86, -1.48	sce	3
Dihydrochalcone	B-R buffer, 20% Ethanol, pH 8.6	Hg	-1.42, -1.56	sce	22

(continued)

727

Compound	Solvent System	Working Electrode	$E_{1/2}$	Reference Electrode	Reference
α-Benzoylnaphthalene	Aq buffer, 50% Ethanol, pH 4.7	Hg	-1.17	sce	14
Benzanthrone	Aq buffer, 50% Ethanol, pH 1.6	Hg	-0.54	sce	14
Fluorenone	Citrate-phosphate buffer, 30% Ethanol, pH 4.1	Hg	-0.89, -1.02	sce	14
	Borate buffer, 30% Ethanol, pH 9.6	Hg	-0.97, -1.12, -1.46	sce	14
Xanthone	Citrate-phosphate buffer, 25% Ethanol, pH 4.1	Hg	-1.05	sce	23
1,4-Diphenyl-1,2,2-propanetrione	DMSO, $(\underline{n}\text{-Propyl})_4$NClO$_4$	Hg	-0.76, -2.25	sce	10
1,3-Diphenyl-1,2,3,4-butanetetraone	DMSO, $(\underline{n}\text{-Propyl})_4$NClO$_4$	Hg	-0.78, -2.04	sce	10
Alkyl-Substituted Deoxybenzoins: $C_6H_5COCHRC_6H_5$					
R = CH$_3$	Aq Borax buffer, pH 9.26	Hg	-1.63_5	sce	24_1
C$_2$H$_5$	Aq Borax buffer, pH 9.26	Hg	-1.64_6	sce	24
C$_3$H$_7$	Aq Borax buffer, pH 9.26	Hg	-1.65_6	sce	24

$i\text{-}C_3H_7$	Aq Borax buffer, pH 9.26	Hg	-1.66_2	sce	24
C_4H_9	Aq Borax buffer, pH 9.26	Hg	-1.66_0	sce	24
$i\text{-}C_4H_9$	Aq Borax buffer, pH 9.26	Hg	-1.65_8	sce	24
$s\text{-}C_4H_9$	Aq Borax buffer, pH 9.26	Hg	-1.66_3	sce	24
$t\text{-}C_4H_9$	Aq Borax buffer, pH 9.26	Hg	-1.69_3	sce	24
C_5H_{11}	Aq Borax buffer, pH 9.26	Hg	-1.65_9	sce	24

[a] KCl, HOAc, H_3PO_4.
[b] Extrapolated at pH 0.

(continued)

References for Aromatic Ketones

1. N. L. Weinberg, Can. J. Chem., 48, 1533 (1970).
2. J. C. Suatoni, R. E. Snyder, and R. O. Clark, Anal. Chem., 33, 1894 (1961).
3. S. Wawzonek and A. Gundersen, J. Electrochem. Soc., 111, 324 (1964).
4. N. v. Tkhan', I. A. Avrutskaya, and M. Ya Fioshin, Elektrokhim, 4, 1508 (1968).
5. I. A. Avrutskaya, S. F. Belevskii, M. Ya Fioshin, and N. Van Dang, Elektrokhim, 6, 683 (1970).
6. G. Nilsi, D. Barnes, and P. Zuman, J. Chem. Soc. B, 1970, 778 (1970).
7. R. C. Buchta and D. H. Evans, Anal. Chem., 40, 2181 (1968).
8. D. H. Evans and E. C. Woodbury, J. Org. Chem., 32, 2158 (1967).
9. R. Pasternak, Helv. Chim. Acta, 31, 753 (1948).
10. G. A. Russell and S. A. Weiner, J. Am. Chem. Soc., 89, 6623 (1967).
11. Ya. P. Stradyn and I. K. Tutane, Zh. Obsh. Khim., 37, 1956 (1967).
12. I. K. Tutane and Ya. P. Stradyn, Zh. Obsh. Khim., 37, 1962 (1967).
13. R. W. Brockman and D. E. Pearson, J. Am. Chem. Soc., 74, 4128 (1952).
14. R. A. Day Jr., S. R. Milliken, and W. D. Shults, J. Am. Chem. Soc., 74, 2741 (1952).
15. L. Meites, Polarographic Techniques, 2nd ed., Interscience, New York, 1965, p. 685.
16. M. O. L. Spangler, J. C. Wolford, G. E. Treadwell, and C. Crusenberry, J. Org. Chem., 34, 892 (1969).
17. P. Zuman, O. Exner, R. F. Rekker, and W. Th. Nauta, Collect. Czech. Chem. Commun., 33, 3213 (1968).
18. Reference 15, p. 675.
19. D. M. Coulson, W. R. Crowell, and S. K. Tendrick J. Am. Chem. Soc., 79, 1354 (1957).
20. J. R. Jones and J. A. Rowlinson, J. Electroanal. Chem., 19, 297 (1968).
21. W. P. Davies and D. P. Evans, J. Chem. Soc., 193 546 (1939).
22. A. Ryvolova-Kejharova and P. Zuman, J. Electroan Chem., 21, 197 (1969).

23. R. A. Day, Jr. and R. E. Biggers, J. Am. Chem.
     Soc., 75, 738 (1953).
24. P. Zuman, B. Tucsanyi, and A. K. Mills, Collect.
     Czech. Chem. Commun., 33, 3205 (1968).

9. ALCOHOLS, POLYOLS, AND CARBOHYDRATES

Oxidation

Alcohols

Compound	Solvent System	Working Electrode	$E_{1/2}$	Reference Electrode	Reference
Allyl alcohol	CH_3CN, $NaClO_4$	Pt	>+2.0	Ag/0.1N Ag+	1
Cyclohexanol	CH_3CN, $NaClO_4$	Pt	>+2.0	Ag/0.1N Ag+	1
o-Methoxybenzyl alcohol	CH_3CN, $NaClO_4$	Pt	+1.25	Ag/0.1N Ag+	1
m-Methoxybenzyl alcohol	CH_3CN, $NaClO_4$	Pt	+1.28	Ag/0.1N Ag+	1
p-Methoxybenzyl alcohol	CH_3CN, $NaClO_4$	Pt	+1.22, +1.64	Ag/0.1N Ag+	1
o-Chlorobenzyl alcohol	CH_3CN, $NaClO_4$	Pt	+1.84	Ag/0.1N Ag+	1
m-Chlorobenzyl alcohol	CH_3CN, $NaClO_4$	Pt	+1.85	Ag/0.1N Ag+	1
p-Chlorobenzyl alcohol	CH_3CN, $NaClO_4$	Pt	+1.79	Ag/0.1N Ag+	1
p-Bromobenzyl alcohol	CH_3CN, $NaClO_4$	Pt	+1.75	Ag/0.1N Ag+	1

Compound	Electrolyte	Electrode	Potential	Reference	Ref.
p-Iodobenzyl alcohol	CH$_3$CN, NaClO$_4$	Pt	+1.58, +1.91	Ag/0.1$\underline{N}$ Ag+	1
p-Methylbenzyl alcohol	CH$_3$CN, NaClO$_4$	Pt	+1.59	Ag/0.1$\underline{N}$ Ag+	1
Furfuryl alcohol	CH$_3$CN, NaClO$_4$	Pt	+1.33, +1.82	Ag/0.1$\underline{N}$ Ag+	1
Cinnamyl alcohol	CH$_3$CN, NaClO$_4$	Pt	+1.36, +1.77	Ag/0.1$\underline{N}$ Ag+	1
p-Nitrocinnamyl alcohol	CH$_3$CN, NaClO$_4$	Pt	+1.72	Ag/0.1$\underline{N}$ Ag+	1
Fluorenol	CH$_3$CN, NaClO$_4$	Pt	+1.31	Ag/0.1$\underline{N}$ Ag+	1
4-Methoxybenzyl-hydrol	CH$_3$CN, NaClO$_4$	Pt	+1.23	Ag/0.1$\underline{N}$ Ag+	1
4,4'-Dimethoxy-benzylhydrol	CH$_3$CN, NaClO$_4$	Pt	+1.22	Ag/0.1$\underline{N}$ Ag+	1
4,4'-Dichloro-benzylhydrol	CH$_3$CN, NaClO$_4$	Pt	+1.77	Ag/0.1$\underline{N}$ Ag+	1
Benzhydrol	CH$_3$CN, NaClO$_4$	Pt	> +2.0	Ag/0.1$\underline{N}$ Ag+	1
Benzyl alcohol	CH$_3$CN, NaClO$_4$	Pt	> +2.0	Ag/0.1$\underline{N}$ Ag+	1
p-Bromophenyl ethylene glycol	CH$_3$CN, NaClO$_4$	Pt	+1.62	Ag/0.1$\underline{N}$ Ag+	1
Benzopinacol	0.1$\underline{N}$ NaOH	Pt	-0.58	sce	2

(continued)

Reduction

t-Butyl alcohol, phenyl methyl carbinol, allyl alcohol, methallyl alcohol and methyl vinyl alcohol show no reduction waves in 1:1 benzene/methanol, 0.3$\underline{M}$ LiCl.

Compound	Solvent System	Working Electrode	$E_{1/2}$	Reference Electrode	Reference
Sugars					
d-Glucose (0.25$\underline{M}$)	Aq, pH 7.0	Hg	-1.59	NCE	3
d-Galactose (0.$\underline{25M}$)	Aq, pH 7.0	Hg	-1.60	NCE	3
d-Mannose (0.25$\underline{M}$)	Aq, pH 7.0	Hg	-1.57	NCE	3
l-Allose (0.25$\underline{M}$)	Aq, pH 7.0	Hg	-1.79	NCE	3
d-Lyxose (0.25$\underline{M}$)	Aq, pH 7.0	Hg	-1.60	NCE	3
d-Xylose (0.25$\underline{M}$)	Aq, pH 7.0	Hg	-1.54	NCE	3
l-Arabinose (0.25$\underline{M}$)	Aq, pH 7.0	Hg	-1.61	NCE	3
d-Ribose (0.25$\underline{M}$)	Aq, pH 7.0	Hg	-1.81	NCE	3
Fructose	0.1$\underline{N}$ LiCl	Hg	-1.80	NCE	4
Sorbose	0.1$\underline{N}$ LiCl	Hg	-1.80	NCE	4

Carbohydrate oximes are listed under "Oximes" (Section 25).

References for Alcohols, Polyols, and Carbohydrates

1. H. Lund, Acta Chem. Scand., 11, 491 (1957).
2. W. Kemula, Z. R. Grabowski, and M. K. Kalinowski,
    Collect. Czech. Chem. Commun., 25, 3306 (1960).
3. S. M. Cantor and D. P. Peniston, J. Am. Chem. Soc.,
    62, 2113 (1940).
4. J. Heyrovský and I. Smoler, Collect. Czech. Chem.
    Commun., 4, 521 (1932).

10. PHENOLS

Oxidation

Compound	Solvent System	Working Electrode	E 1/2	Reference Electrode	Reference
Phenol	OAc buffer, 50% Isopropanol, pH 5.6	Graphite	+0.633	sce	1
	OAc-Borate buffer, 50% Methanol	Pt	+0.92[a]	sce	2
	Phosphate buffer, 0.1N KNO_3, pH 9.0	Pt	+0.515	sce	3
	CH_3CN, $NaClO_4$	Pt	+1.04	Ag/0.1N $\underline{Ag^+}$	4
	CH_3CN, $(Et)_4NClO_4$[b] Universal buffer,	Pt	+1.47, +1.94	sce	5
	40% Isopropanol, pH 2.0	Graphite	(+1.06 −0.070 pH)	Cd/0.5M $\underline{Cd2c}$	6
	Py, 0.5M $LiClO_4$	Graphite	+0.75	Ag/0.1N $\underline{Ag^+}$, Pyd	7
o-Cresol	OAc buffer, 50% Isopropanol, pH 5.6	Graphite	+0.556	sce	1
	OAc-Borate buffer, 50% Methanol	Pt	+0.85[a]	sce	2
m-Cresol	OAc buffer, 50% Isopropanol, pH 5.6	Graphite	+0.607	sce	1

Compound	Solvent	Electrode	Potential	Reference	Ref.
	OAc-Borate buffer, 50% Methanol	Pt	+0.89[a]	sce	2
p-Cresol	OAc buffer, 50% Isopropanol, pH 5.6	Graphite	+0.543	sce	1
o-Ethylphenol	OAc-Borate buffer, 50% Methanol	Pt	+0.84[a]	sce	2
	OAc buffer, 50% Isopropanol, pH 5.6	Graphite	+0.551	sce	1
m-Ethylphenol	OAc-Borate buffer, 50% Methanol	Pt	+0.81[a]	sce	2
	OAc buffer, 50% Isopropanol, pH 5.6	Graphite	+0.616	sce	1
p-Ethylphenol	OAc-Borate buffer, 50% Methanol	Pt	+0.88[a]	sce	2
	OAc buffer, 50% Isopropanol, pH 5.6	Graphite	+0.567	sce	1
o-i-Propylphenol	OAc-Borate buffer, 50% Methanol	Pt	+0.82[a]	sce	2
p-i-Propylphenol	OAc-Borate buffer, 50% Methanol	Pt	+0.82[a]	sce	2
o-s-Butylphenol	OAc-Borate buffer, 50% Methanol	Pt	+0.82[a]	sce	2
o-s-Butylphenol	OAc-Borate buffer, 50% Methanol	Pt	+0.84[a]	sce	2
p-s-Butylphenol	OAc-Borate buffer, 50% Methanol	Pt	+0.86[a]	sce	2

(continued)

737

Compound	Solvent System	Working Electrode	$E_{1/2}$	Reference Electrode	Reference
o-t-Butylphenol	OAc buffer, 50% Isopropanol, pH 5.6	Graphite	+0.552	sce	1
	OAc-Borate buffer, 50% Methanol	Pt	+0.81[a]	sce	2
	Phosphate buffer, 0.1N KNO3, pH 9.0	Pt	+0.38	sce	3
m-t-Butylphenol	OAc-Borate buffer, 50% Methanol	Pt	+0.94[a]	sce	2
	Phosphate buffer, 0.1N KNO3, pH 9.0	Pt	+0.47	sce	3
p-t-Butylphenol	OAc buffer, 50% Isopropanol, pH 5.6	Graphite	+0.578	sce	1
	Phosphate buffer, 0.1N KNO3, pH 9.0	Pt	+0.41	sce	3
	OAc-Borate buffer, 50% Methanol	Pt	+0.84[a]	sce	2
p-1,1-Dimethyl-butylphenol	OAc-Borate buffer, 50% Methanol	Pt	+0.84[a]	sce	2
o-Phenylphenol	OAc buffer, 50% Isopropanol pH 5.6	Graphite	+0.563	sce	1
p-Phenylphenol	OAc buffer, 50% Isopropanol pH 5.6	Graphite	+0.534	sce	1
o-Methoxyphenol	OAc buffer, 50% Isopropanol, pH 5.6	Graphite	+0.456	sce	1

Compound	Electrolyte	Electrode	Potential	Reference	Ref.
m-Methoxyphenol	OAc buffer, 50% Isopropanol, pH 5.6	Graphite	+0.619	sce	1
p-Methoxyphenol	OAc buffer, 50% Isopropanol, pH 5.6	Graphite	+0.406	sce	1
o-Ethoxyphenol	2.04M H_2SO_4	C paste	+0.7 E_p	sce	8
	OAc buffer, 50% Isopropanol, pH 5.6	Graphite	+0.451	sce	1
m-Ethoxyphenol	OAc buffer, 50% Isopropanol, pH 5.6	Graphite	+0.620	sce	1
p-Ethoxyphenol	OAc buffer, 50% Isopropanol, pH 5.6	Graphite	+0.413	sce	1
o-Nitrophenol	OAc buffer, 50% Isopropanol, pH 5.6	Graphite	+0.846	sce	1
m-Nitrophenol	OAc buffer, 50% Isopropanol, pH 5.6	Graphite	+0.855	sce	1
p-Nitrophenol	OAc buffer, 50% Isopropanol, pH 5.6	Graphite	+0.924	sce	1
o-Chlorophenol	OAc buffer, 50% Isopropanol, pH 5.6	Graphite	+0.625	sce	1
m-Chlorophenol	OAc buffer, 50% Isopropanol, pH 5.6	Graphite	+0.734	sce	1

(continued)

Compound	Solvent System	Working Electrode	$E_{1/2}$	Reference Electrode	Reference
p-Chlorophenol	OAc buffer, 50% Isopropanol, pH 5.6	Graphite	+0.653	sce	1
p-Aminophenol	Diethylglycine, pH 11	Pt	+0.124	sce	9
p-Dimethylamino-phenol	2$\underline{M}$ H_2SO_4	C paste	+0.53, +1.2 E_p	sce	10
N-Phenyl-p-amino-phenol	H_2O, acetone, $HClO_4$	Pt	+0.45 $E_{p/2}$	sce	11
p-Azophenol	Borate buffer, KNO_3, pH 9.2	Hg	+0.17	sce	12
1-Hydroxynaphtha-lene (1-Naphthol)	CH_3CN, $(Et)_4NClO_4$	Pt	+1.14, +1.76	sce	5
	CH_3CN, $NaClO_4$	Pt	+0.74	Ag/0.1$\underline{N}$ Ag+	4
	B-R buffer, Na_2SO_4, pH 2.4	C paste	+0.63 $E_{p/2}$	sce	13
2-Hydroxynaphtha-lene)2-Naphthol)	CH_3CN, $(Et)_4NClO_4$	Pt	+1.27, +2.35	sce	5
1,5-Dihydroxy-naphthalene	CH_3CN, $NaClO_4$	Pt	+0.82	Ag/0.1$\underline{N}$ Ag+	4
1-Amino-7-naphthol	2$\underline{M}$ $HClO_4$	C paste	+0.54 E_p	sce	14
	B-R buffer, Na_2SO_4, pH 2.4	C paste	+0.45 $E_{p/2}$	sce	13
9-Hydroxyanthracene (9-Anthrol)	CH_3CN, $NaClO_4$	Pt	+0.44	Ag/0.1$\underline{N}$ Ag+	4
2-Hydroxydiphenyl	CH_3CN, $NaClO_4$	Pt	+0.97	Ag/0.1$\underline{N}$ Ag+	4
	OAc buffer, 50% Isopropanol, pH 5.6	Graphite	+0.563	sce	1

Compound	Electrolyte	Electrode	Potential	Reference	
4-Hydroxydiphenyl	CH$_3$CN, NaClO$_4$	Pt	+0.89	Ag/0.1N Ag+	4
	OAc buffer, 50% Isopropanol, pH 5.6	Graphite	+0.534	sce	1
2,3-Dimethylphenol	OAc-Borate buffer, 50% Methanol	Pt	+0.85[a]	sce	2
2,4-Dimethylphenol	OAc-Borate buffer, 50% Methanol	Pt	+0.76[a]	sce	2
	OAc buffer, 50% Isopropanol pH 5.6	Graphite	+0.459	sce	1
2,5-Dimethylphenol	OAc-Borate buffer, 50% Methanol	Pt	+0.82[a]	sce	2
2,6-Dimethylphenol	OAc-Borate buffer, 50% Methanol	Pt	+0.76[a]	sce	2
	OAc buffer, 50% Isopropanol pH 5.6	Graphite	+0.427	sce	1
3,4-Dimethylphenol	OAc buffer, 50% Isopropanol pH 5.6	Graphite	+0.513	sce	1
	OAc-Borate buffer, 50% Methanol	Pt	+0.81[a]	sce	2
3,5-Dimethylphenol	OAc-Borate buffer, 50% Methanol	Pt	+0.86[a]	sce	2
	OAc buffer, 50% Isopropanol, pH 5.6	Graphite	+0.587	sce	1
2,5-Diphenylphenol	OAc buffer, 50% Isopropanol, pH 5.6	Graphite	+0.471	sce	1

(continued)

741

Compound	Solvent System	Working Electrode	$E_{1/2}$	Reference Electrode	Reference
2,4-Di-t-butylphenol	OAc buffer, 50% Isopropanol pH 5.6	Graphite	+0.487	sce	1
	OAc-Borate buffer, 50% Methanol	Pt	+0.76[b]	sce	2
2,6-Di-t-butylphenol	OAc buffer, 50% Isopropanol, pH 5.6	Graphite	+0.378	sce	1
	OAc-Borate buffer, 50% Methanol	Pt	+0.68[b]	sce	2
2-Methyl-6-t-butyl-phenol	OAc buffer, 50% Isopropanol, pH 5.6	Graphite	+0.429	sce	1
2-t-Butyl-4-methyl-phenol	OAc buffer, 50% Isopropanol, pH 5.6	Graphite	+0.465	sce	1
2-Methyl-4-t-butyl-phenol	OAc buffer, 50% Isopropanol, pH 5.6	Graphite	+0.501	sce	1
2,6-Dimethoxyphenol	OAc buffer, 50% Isopropanol, pH 5.6	Graphite	+0.317	sce	1
	Universal buffer,[b] 40% Isopropanol, pH 2.0	Graphite	(+0.82 -0.069 pH)	Cd/0.5M $\underline{Cd^{2+}}$[c]	6
2-Methoxy-4-methyl-phenol	OAc buffer, 50% Isopropanol, pH 5.6	Graphite	+0.371	sce	1

Compound	Medium	Electrode			Reference electrode	Ref.
2,5-Dichlorophenol	OAc buffer, 50% Isopropanol, pH 5.6	Graphite	+0.695		sce	1
2,4-Dichlorophenol	Universal buffer,b 40% Isopropanol, pH 3.9	Graphite	(+0.93 pH)	−0.068	Cd/0.5M Cd^{2+}	6
4-Phenyl-2-chloro-phenol	0.14M LiCl	Pt	+0.66		sce	15
	0.14M LiCl	Pt	+0.56		sce	15
2-Chloro-4-bromo-phenol	0.14M LiCl	Pt	+0.67		sce	15
4-Carbomethoxy-2-chlorophenol	0.14M LiCl	Pt	+0.87		sce	15
4-Carboxy-2-chloro-phenol	0.14M LiCl	Pt	+0.92		sce	15
2,3,5-Trimethyl-phenol	OAc–Borate buffer, 50% Methanol	Pt	+0.78[a]		sce	2
2,4,5-Trimethyl-phenol	OAc–Borate buffer, 50% Methanol	Pt	+0.72[a]		sce	2
2,4,6-Trimethyl-phenol	OAc–Borate buffer, 50% Methanol	Pt	+0.67[a]		sce	2
	Universal buffer,b 40% Isopropanol, pH 2.0	Graphite	(+0.78 pH)	−0.062	Cd/0.5M Cd^{2+}	6
2,4,6-Tri-t-butylphenol	Universal buffer,b 40% Isopropanol, pH 2.0	Graphite	(+0.68 pH)	−0.062	Cd/0.5M Cd^{2+}	6

(Continued)

743

Compound	Solvent System	Working Electrode	$E_{1/2}$	Reference Electrode	Reference
	OAc-Borate buffer, 50% Methanol	Pt	+0.69	sce	2
4-Methyl-2,6-di-t-butylphenol	CH_3CN, (n-Bu)4-NOH, (Me)4NCl	Graphite	+0.07	Cd/0.5M Cd2+	16
	Universal buffer,b 40% Isopropanol, pH 2.0	Graphite	(+0.66 -0.065 pH)	sce	6
4-Phenyl-2,6-di-t-butylphenol	CH_3CN, (Et)4NClO4	Pt	+1.26, +1.63	sce	17
	CH_3CN, NaClO4	Pt	+0.93, +2.06	Ag/0.1N Ag+	4
	Universal buffer,b 40% Isopropanol, pH 2.0	Graphite	(+0.74 -0.066 pH)	sce	6
4-Methoxy-2,6-di-t-butylphenol	Universal buffer,b 40% Isopropanol, pH 2.0	Graphite	(+0.55 -0.059 pH)	sce	6
4-Carboxy-2,6-di-t-butylphenol	Universal buffer,b 40% Isopropanol, pH 2.0	Graphite	(+0.82 -0.071 pH)	sce	6
4-Hydroxymethyl-2,6-di-t-butylphenol	CH_3CN, (Et)4NOH	Pt	+1.68	sce	17
	Universal buffer,b 40% Isopropanol pH 2.0	Graphite	(+0.71 -0.071 pH)	sce	6
4-Hydroxymethyl-2,6-di-t-butylphenol	CH_3CN, (Et)4NOH	Pt	+1.24, +1.72	sce	17

744

Compound	Medium	Electrode	Potential	Reference	Ref.
4-Amino-2,6-di-t-butylphenol	CH$_3$CN, Perchlorate salt	Pt	+0.190, +1.800	Ag/0.01$\underline{N}$ Ag+	18
2,4,6-Trichloro-phenol	Universal buffer, 40% Isopropanol pH 3.9	Graphite	(+0.93 -0.068 pH)	Cd/0.5$\underline{M}$ Cd^{2+}	6
52 Aryl-substituted phenols	CH$_3$CN/H$_2$O and OAc buffer solutions				19
30 Sterically-hindered phenols	Borate, acetate, and phosphate buffers				6
Vanillate anion	CH$_3$CN, (Et)$_4$NClO$_4$	Pt	+0.22, +0.53	sce	17
2,4,6,7-Tetramethyl-5-hydroxycoumarin	0.1$\underline{M}$ Aniline buffer, 50% Methanol, pH 3.56	Hg	+0.219	sce	20
2,2,4,6,7-Penta-methyl-5-hydroxy-coumarin	0.1$\underline{M}$ Aniline buffer, 50% Methanol, pH 3.56	Hg	+0.219	sce	20
Adrenaline	1$\underline{M}$ H$_2$SO$_4$	C paste	+0.7 E_p	sce	21
Corypalline	Borate buffer, pH 9.0	Pt	+0.35 p	sce	22
Hydroquinone	OAc buffer, pH 4.5	Graphite	+0.234	Ag, AgCl	23
	CH$_3$CN, LiClO$_4$	Pt	+1.13 E_p	sce	24
	Py, 0.5$\underline{M}$ LiClO$_4$	Graphite	+0.31 $E_{p/2}$	Ag/1$\underline{M}$ Ag+ Py	7
Methylhydroquinone	Aq H$_2$SO$_4$	C paste	+0.537	sce	25
2,5-Dimethylhydro-quinone	Aq H$_2$SO$_4$	C paste	+0.537	sce	25
	CH$_3$CN, LiClO$_4$	Pt	+0.99 E_p	sce	24

(Continued)

745

Compound	Solvent System	Working Electrode	$E_{1/2}$	Reference Electrode	Reference
2,3,5,6-Tetramethyl-hydroquinone	CH_3CN, $LiClO_4$	Pt	$+0.83$ E_p	sce	24
Chlorohydroquinone	Aq H_2SO_4	C paste	$+0.558$	sce	25
2,3-Dichlorohydro-quinone	Aq H_2SO_4	C paste	$+0.498$	sce	25
2,5-Dichlorohydro-quinone	Aq H_2SO_4	C paste	$+0.502$	sce	25
2,6-Dichlorohydro-quinone	Aq H_2SO_4	C paste	$+0.482$, $+0.579$	sce	25
Bromohydroquinone	Aq H_2SO_4	C paste	No wave	sce	25
Resorcinol	Py, $0.5\underline{M}$ $LiClO4$	Graphite	$+0.61$, $+0.91$ $E_{p/2}$	Ag/1$\underline{M}$ Ag+, Pyd	7
	OAc buffer pH 4.5	Graphite	$+0.234$ $E_{p/2}$	Ag, AgCl	23
Catechol	OAc buffer pH 4.5	Graphite	$+0.613$	Ag, AgCl	23
	Py, $0.5\underline{M}$ $LiClO4$	Graphite	$+0.29$ $E_{p/2}$	Ag/1$\underline{M}$ Ag+, Pyd	7
4-Methylcatechol	1M $HClO4$	C paste	$+0.516$ E_p	sce	26
α-Tocopherol (Vitamin E)	CH_3CN, $LiClO_4$	Pt	$+0.76$ E_p	sce	27
	1N H_2SO_4, 75% Ethanol	Pt	$+0.515$	sce	29
β-Tocopherol	1N H_2SO_4, 75% Ethanol	Pt	$+0.600$	sce	29
γ-Tocopherol	1N H_2SO_4, 75% Ethanol	Pt	$+0.600$	sce	29
δ-Tocopherol	1N H_2SO_4, 75% Ethanol	Pt	$+0.685$	sce	29

a – Oxidation potential; extrapolated to pH 0.
b – Borate, Acetate, Phosphate.
c – Potential -0.678 V versus sce.
d – Potential +0.09 V versus sce.

Reduction

Compound	Solvent System	Working Electrode	$E_{1/2}$	Reference Electrode	Reference
Phenol	Py, $0.1\underline{M}$ (Et)$_4$NClO$_4$	Hg	-2.01	Ag/1$\underline{M}$ Ag+, Py$\underline{a}$	29
p-Chlorophenol	Py, $0.1\underline{M}$ (Et)$_4$NClO$_4$	Hg	-1.94	Ag/1$\underline{M}$ Ag+, Py$\underline{a}$	29
2,4-Dichlorophenol	Py, $0.1\underline{M}$ (Et)$_4$NClO$_4$	Hg	-1.76	Ag/1$\underline{M}$ Ag+, Py$\underline{a}$	29
3,4-Dimethylphenol	Py, $0.1\underline{M}$ (Et)$_4$NClO$_4$	Hg	-2.06	Ag/1$\underline{M}$ Ag+, Py$\underline{a}$	29
1-Hydroxynaphthalene (1-Naphthol)	CH$_3$CN, (Et)$_4$NClO$_4$	Hg	-2.47, -2.66	sce	5
2-Hydroxynaphthalene (2-Naphthol)	CH$_3$CN, (Et)$_4$NClO$_4$	Hg	-2.42, -2.63	sce	5

a – Potential +0.09 V versus sce.

(continued)

References for Phenols

1. J. C. Suatoni, R. E. Snyder, and R. O. Clark, Anal. Chem., 33, 1894 (1961).
2. G. E. Penketh, J. Appl. Chem., 7, 512 (1957).
3. J. F. Hedenburg and H. Freiser, Anal. Chem., 25, 1355 (1953).
4. C. Parkanyi and R. Zahradnik, Collect. Czech. Chem Commun., 30, 4288 (1965).
5. T. A. Gough and M. E. Peover, Proceedings of the 3rd International Polarography Congress, Southampton, 1965, Macmillan, London, 1966, p. 1017.
6. Yu. V. Vodzinskii, A. A. Vasil'eva, and I. A. Korshunov, Zh. Obshch. Khim., 39, 1196 (1969).
7. W. R. Turner and P. J. Elving, Anal. Chem., 37, 467 (1965).
8. D. Hawley and R. N. Adams, J. Electroanal. Chem., 8, 163 (1964).
9. D. B. Julian and W. R. Ruby, J. Am. Chem. Soc., 72, 4719 (1950).
10. M. F. Marcus and M. D. Hawley, J. Electroanal. Chem., 18, 175 (1968).
11. P. A. Malachesky, G. Petrie, D. W. Leedy, R. N. Adams, and R. L. Schowen, Preprints of Papers, Durham Electro-Organic Symposium, Durham, North Carolina, 1968, p. 235.
12. W. M. Lauer, H. P. Klug, and S. A. Harrison, J. Am. Chem. Soc., 61, 2775 (1939).
13. C. Olson and R. Adams, Anal. Chim. Acta., 22, 582 (1960).
14. L. Papouchado, G. Petrie, J. H. Sharp, and R. N. Adams, J. Am. Chem. Soc., 90, 5620 (1968).
15. H. N. Simpson, C. K. Hancock, and E. A. Meyers, J. Org. Chem., 30, 2678 (1965).
16. Yu. V. Vodzinskii, A. A. Vasil'eva, G. A. Abakumov, and I. A. Korshunov, Elektrokhim, 4, 1492 (1968).
17. F. J. Vermillion Jr., and I. A. Pearl, J. Electrochem. Soc., 111, 1392 (1964).
18. G. Cauquis, G. Fauvelot, and J. Rigaudy, Compt. Rend., 264, 1758 (1967).
19. F. W. Steuber and K. Dimroth, Chem. Ber., 99, 258 (1966).
20. L. I. Smith, I. M. Kolthoff, S. Wawzonek, and P. M. Ruoff, J. Am. Chem. Soc., 63, 1018 (1941).
21. M. D. Hawley, S. V. Tatawawadi, S. Piekarski, and R. N. Adams, J. Am. Chem. Soc., 89, 447 (196

22. G. F. Kirkbright, J. T. Stock, R. D. Pugliese, and J. M. Bobbitt, J. Electrochem. Soc., 116, 219 (1969).

23. P. J. Elving and A. F. Krivis, Anal. Chem., 30, 1645 (1958).

24. V. D. Parker, Chem. Commun., 1969, 716.

25. K. Sasaki and W. J. Newby, J. Electroanal. Chem., 20, 137 (1969).

26. R. N. Adams, M. D. Hawley, and S. W. Feldberg, J. Phys. Chem., 71, 851 (1967).

27. V. D. Parker, J. Am. Chem. Soc., 91, 5380 (1969).

28. G. Raspi, M. Cospito, L. Lucarini, Richercd. Sci., 39, 405 (1969).

29. K. Tsuji and P. J. Elving, Anal. Chem., 41, 286 (1969).

11. QUINONES

Oxidation

Compound	Solvent System	Working Electrode	$E_{1/2}$	Reference Electrode	Reference
Benzoquinones or Naphthoquinones	CH_3CN or CH_3NO_2 + Et_4NClO_4	Pt	No wave	sce	1
9,10-Anthraquinone	CH_3CN, Et_4NClO_4	Pt	+1.21	sce	1
	CH_3NO_2, Et_4NClO_4	Pt	+1.20	sce	1
1-Amino-9,10-anthraquinone	CH_3CN, Et_4NClO_4	Pt	+1.33	sce	1
	CH_3NO_2, Et_4NClO_4	Pt	+1.30	sce	1
1-Amino-4-hydroxy-9,10-anthraquinone	CH_3CN, Et_4NClO_4	Pt	+1.01	sce	1
	CH_3NO_2, Et_4NClO_4	Pt	+0.96	sce	1
1-Methylamino-9,10-anthraquinone	CH_3CN, Et_4NClO_4	Pt	+1.16	sce	1
	CH_3NO_2, Et_4NClO_4	Pt	+1.15	sce	1
1,4-Dihydroxy-9,10-anthraquinone	CH_3CN, Et_4NClO_4	Pt	+1.53	sce	1
	CH_3NO_2, Et_4NClO_4	Pt	+1.52	sce	1
1,2,4-Trihydroxy-9,10-anthraquinone	CH_3CN, Et_4NClO_4	Pt	+1.26	sce	1
	CH_3NO_2, Et_4NClO_4	Pt	+1.25	sce	1

Compound	Solvent System	Working Electrode	$E_{1/2}$	Reference Electrode	Reference
1,2,5,8-Tetra-hydroxy-9,10,-anthraquinone	CH_3CN, Et_4NClO_4	Pt	+1.21	sce	1
	CH_3NO_2, Et_4NClO_4	Pt	+1.22	sce	1
Phenanthraquinones	CH_3CN or CH_3NO_2 + Et_4NClO_4	Pt	No wave	sce	1

Reduction

Compound	Solvent System	Working Electrode	$E_{1/2}$	Reference Electrode	Reference
o-Benzoquinone	DMF, $(Et)_4NClO_4$	Hg	-0.07	Ag	2
	B-R buffer, pH 5.0	Hg	+0.29	sce	3
	B-R buffer, pH 7.0	Hg	+0.20	sce	4
	CH_3CN, $(Et)_4NClO_4$	Hg	-0.31, -0.90	sce	5
	0.1M Phosphate buffer, pH 7.0	Hg	+0.04	sce	3
p-Benzoquinone	Phosphate buffer, 20% Methanol, pH 7.0	Hg	-0.025	NCE	6
	DMSO, $(Et)_4NClO_4$	Hg	-0.40, -1.24	sce	3
	Py, $LiClO_4$	Graphite	+0.30 $E_p/2$	Ag/1M Ag+, Pya	7
	CH_3CN, $LiClO_4$	Pt	-0.34 E_p	sce	8

(Continued)

Compound	Solvent System	Working Electrode	$E_{1/2}$	Reference Electrode	Reference
	CH_3CN, $(Et)_4NClO_4$	Pt	-0.481, -1.030 Ep/2	sce	9
	CH_3CN, $(Et)_4NClO_4$	Pt	-0.48, -1.12 Ep	sce	10
	PC, $(Et)_4NClO_4$	Pt	-0.52, -1.04 Ep	sce	10
	DMF, $(Et)_4NClO_4$	Pt	-0.40, -1.25 Ep	sce	10
	$C_6H_5NO_2$, $(Et)_4NClO_4$	Pt	-0.51	sce	10
Chloro-1,4-benzoquinone	40% Alcohol, 10% HOAc, 50% H_2O	Hg	$+0.265$	NCE	11
2,5-Dichloro-1,4-benzoquinone	CH_3CN, $(Et)_4NClO_4$	Hg	-0.34, -0.92	sce	5
2,6-Dichloro-1,4-benzoquinone	CH_3CN, $(Et)_4NClO_4$	Hg	-0.18, -0.81	sce	5
Trichloro-1,4-benzoquinone	CH_3CN, $(Et)_4NClO_4$	Hg	-0.18, -0.81	sce	5
Tetrachloro-1,4-benzoquinone (Chloranil)	CH_3CN, $(Et)_4NClO_4$	Hg	-0.08, -0.78	sce	5
	CH_3CN, $(Et)_4NClO_4$	Hg	$+0.01$, -0.71	sce	5
	Aq buffer, pH 4.8, 50% Dioxane	Hg	-0.20	Hg, Hg_2SO_4	12
Tetrabromo-1,4-benzoquinone	CH_3CN, $(Et)_4NClO_4$	Hg	0.00, -0.72	sce	5
Tetrafluoro-1,4-benzoquinone	CH_3CN, $(Et)_4NClO_4$	Hg	-0.04, -0.82	sce	5

Compound	Supporting electrolyte	Electrode	$E_{1/2}$	Ref. electrode	Ref.
2-Methyl-1,4-benzoquinone (Toluquinone)	OAc buffer, 50% Methanol, pH 5.40	Hg	+0.090	sce	13
	Phosphate buffer, 20% Methanol, pH 7	Hg	-0.025	NCE	6
2,3-Dimethyl-1,4-benzoquinone (o-Xyloquinone)	Phosphate buffer, 20% Methanol, pH 7	Hg	-0.135	NCE	6
2,5-Dimethyl-1,4-benzoquinone (m-Xyloquinone)	Phosphate buffer, 20% Methanol, pH 7	Hg	-0.137	NCE	6
2,6-Dimethyl-1,4-benzoquinone (p-Xyloquinone)	OAc buffer, 50% Methanol, pH 5.40	Hg	+0.036	sce	13
	OAc buffer, 50% Methanol, pH 5.40	Hg	+0.031	sce	13
	Phosphate buffer, 20% Methanol, pH 7	Hg	-0.138	NCE	6
2,3,5-Trimethyl-1,4-benzoquinone	40% Alcohol, 10% HOAc, 50% H2O	Hg	+0.140	NCE	11
	Phsophate buffer, 20% Methanol, pH 7	Hg	-0.215	NCE	6
2,3,5,6-Tetramethyl-1,4-benzoquinone (Duroquinone)	Phosphate buffer, 20% Methanol, pH 6	Hg	-0.260	NCE	6
2,6-Diethyl-1,4-benzoquinone	OAc buffer, 50% Methanol, pH 5.40	Hg	-0.093	sce	13
	40% Alcohol, 10% HOAc, 50% H2O	Hg	+0.140	NCE	11

(Continued)

753

Compound	Solvent System	Working Electrode	$E_{1/2}$	Reference Electrode	Reference
Phenyl-1,4-benzoquinone	Phosphate buffer, 20% Methanol, pH 7	Hg	-0.045	NCE	6
2,5-Diphenyl-1,4-benzoquinone	Phosphate buffer, 20% Methanol, pH 7	Hg	-0.032	NCE	6
2-Methyl-5-isopropyl-1,4-benzoquinone	Phosphate buffer, 20% Methanol, pH 7	Hg	-0.140	NCE	6
Hydroxy-1,4-benzoquinone	Phosphate buffer, 20% Methanol, pH 7	Hg	-0.205	NCE	6
Methoxy-1,4-benzoquinone	Phosphate buffer, 20% Methanol, pH 7	Hg	-0.133	NCE	6
2,3,5-Trimethoxy-1,4-benzoquinone	Phosphate buffer, 20% Methanol, pH 7	Hg	-0.203	NCE	6
2,5-Di-t-butyl-1,4-benzoquinone	Phosphate buffer, 20% Methanol, pH 7	Hg	-0.168	NCE	6
2,6-Polymethylene-1,4-benzoquinones					

$(H_2C)_{n-3}$

	Solvent	Electrode	E	Ref. electrode	Ref.
n = 9	40% Alcohol, 10% HOAc, 50% H_2O	Hg	-0.029	NCE	11
" = 10	40% Alcohol, 10% HOAc, 50% H_2O	Hg	+0.026	NCE	11
" = 11	40% Alcohol, 10% HOAc, 50% H_2O	Hg	+0.086	NCE	11
" = 12	40% Alcohol, 10% HOAc, 50% H_2O	Hg	+0.117	NCE	11
" = 13	40% Alcohol, 10% HOAc, 50% H_2O	Hg	+0.139	NCE	11
" = 14	40% Alcohol, 10% HOAc, 50% H_2O	Hg	+0.143	NCE	11
" = 15	40% Alcohol, 10% HOAc, 50% H_2O	Hg	+0.160	NCE	11
" = 16	40% Alcohol, 10% HOAc, 50% H_2O	Hg	+0.151	NCE	11
" = 17	40% Alcohol, 10% HOAc, 50% H_2O	Hg	+0.158	NCE	11
" = 18	40% Alcohol, 10% HOAc, 50% H_2O	Hg	+0.144	NCE	11
" = 19	40% Alcohol, 10% HOAc, 50% H_2O	Hg	+0.144	NCE	11
1,2-Naphthoquinone	B-R buffer, pH 4.5	Hg	-0.34	sce	14
1,4-Naphthoquinone	CH_3CN, $(Et)_4NClO_4$	Hg	-0.56, -1.02	sce	5
	Phosphate buffer, pH 3	Hg	+0.12	sce	15
	Phosphate buffer, pH 7	Hg	-0.13	sce	15
2-Methyl-1,4-naphthoquinone	B-R buffer, pH 2.3	Hg	+0.02	sce	15

(Continued)

Compound	Solvent System	Working Electrode	$E_{1/2}$	Reference Electrode	Reference
(Vitamin K₃)	OAc buffer, pH 4.98	Hg	(-0.085, -0.068 pH)	Ag, AgCl	16
	B-R buffer, pH 6.3	Hg	-0.22	sce	15
	B-R buffer, pH 12.0	Hg	-0.47	sce	15
2,3-Dimethyl-1,4-naphthoquinone	OAc buffer, 50% Methanol, pH 5.40	Hg	-0.21	sce	15
1,4-Anthraquinone	DMF, Et₄NClO₄	Hg	-0.37	Ag, AgCl	2
	CH₃CN, (Et)₄NClO₄	Hg	-0.75, -1.25	sce	5
9,10-Anthraquinone	OAc buffer, 40% Dioxane, pH 7.4	Hg	-0.54	sce	17
	DMF, (Et)₄NI	Hg	-0.83, -1.46	sce	17
	DMF, (Et)₄NClO₄	Hg	-0.93, -1.63	sce	10
	PC, (Et)₄NClO₄	Hg	-0.97, -1.43	sce	10
	CH₃CN, (Et)₄NClO₄	Hg	-0.98, -1.50	sce	10
	0.1N NaOH, 50% Ethanol	Hg	-0.766	sce	18
9,10-Anthraquinone	Universal buffer,[b] 70% Methanol, pH 4.5	Hg	-0.460	sce	19
1-Fluoro-9,10-Anthraquinone	Universal buffer,[b] 70% Methanol, pH 4.5	Hg	-0.414	sce	19
	0.1N NaOH, 50% Ethanol	Hg	-0.739	sce	18
2-Fluoro-9,10-Anthraquinone	Universal buffer,[b] 70% Methanol, pH 4.5	Hg	-0.391	sce	19

Compound	Conditions				Ref
1-Chloro-9,10-Anthraquinone	0.1N NaOH, 50% Ethanol	Hg	-0.736	sce	18
	Universal buffer,[b] 70% Methanol, pH 4.5	Hg	-0.423	sce	19
2-Chloro-9,10-Anthraquinone	0.1N NaOH, 50% Ethanol	Hg	-0.770	sce	18
	Universal buffer,[b] 70% Methanol, pH 4.5	Hg	-0.398	sce	19
1-Bromo-9,10-Anthraquinone	0.1N NaOH, 50% Ethanol	Hg	-0.730	sce	18
	Universal buffer,[b] 70% Methanol, pH 4.5	Hg	-0.431	sce	19
2-Bromo-9,10-Anthraquinone	0.1N NaOH, 50% Ethanol	Hg	-0.769	sce	18
	Universal buffer,[b] 70% Methanol, pH 4.5	Hg	-0.441	sce	19
1-Iodo-9,10-Anthraquinone	0.1N NaOH, 50% Ethanol	Hg	-0.717	sce	18
	Universal buffer,[b] 70% Methanol, pH 4.5	Hg	-0.445	sce	19
2-Iodo-9,10-Anthraquinone	0.1N NaOH, 50% Ethanol	Hg	-0.751	sce	18
	Universal buffer,[b] 70% Methanol, pH 4.5	Hg	-0.417	sce	19

(continued)

757

Compound	Solvent System	Working Electrode	$E_{1/2}$	Reference Electrode	Reference
1-Sulfato-9,10-Anthraquinone	0.1N NaOH, 50% Ethanol	Hg	-0.734	sce	18
2-Sulfato-9,10-Anthraquinone	0.1N NaOH, 50% Ethanol	Hg	-0.712	sce	18
2-Sulfato-9,10-Anthraquinone	0.1N NaOH, 50% Ethanol	Hg	-0.690	sce	18
2-Chloro-6-Sulfato-9,10-Anthraquinone	0.1N NaOH, 50% Ethanol	Hg	-0.616	sce	18
2-Bromo-6-Sulfato-9,10-Anthraquinone	0.1N NaOH, 50% Ethanol	Hg	-0.652	sce	18
6,12-Benzo(a)pyrenequinone	CH_3CN, $(Et)_4NClO_4$	Hg	-0.58, -0.90	sce	10
	PC, $(Et)_4NClO_4$	Hg	-0.59, -0.92	sce	10
	DMF, $(Et)_4NClO_4$	Hg	-0.54, -1.07	sce	10
	$C_6H_5-NO_2$, $(Et)_4NClO_4$	Hg	-0.65	sce	10
1,6-Benzo(a)pyrenequinone	CH_3CN, $(Et)_4NClO_4$	Hg	-0.57, -0.77	sce	10
	CH_3CN, $(Et)_4NClO_4$	Hg	-0.60, -0.86	sce	10
	CH_3CN, $(Et)_4NClO_4$	Hg	-0.685	sce	20

(quinone structure)	CH_3CN, $(Et)_4NClO_4$	Hg	-0.695	sce	20
(quinone structure)	CH_3CN, $(Et)_4NClO_4$	Hg	-0.748	sce	20
(quinone structure)	CH_3CN, $(Et)_4NClO_4$	Hg	-0.854	sce	20
(quinone structure, CH_3, CH_3)	CH_3CH, $(Et)_4NClO_4$	Hg	-0.846	sce	20
30 Quinones	CH_3CN, $(Et)_4NClO_4$				5
8 Quinones of the Triptycene Series	DMF, 0.1M $\underline{KNO_3}$				21

a Potential +0.09 V versus SCE.
b HCl, citric acid, veronal, phenol, LiOH.

(continued)

References for Quinones

1. M. Ahsraf and J. B. Headridge, _Talanta_, 16, 1439 (1969).
2. T. G. Edwards and R. Grinter, _Trans. Faraday Soc._, 69, 1070 (1968).
3. L. Meites, Polarographic Techniques, 2nd ed., Interscience, New York, 1965, p.685.
4. J. Heyrovský and J. Kůta, Principles of Polarography, Academic, New York, 1966, p.548.
5. M. E. Peover, _J. Chem. Soc._, 1962, 4540.
6. W. Flaig, H. Beutelspacher, H. Riemer, and E. Kälke, _Lieb. Ann. Chem._, 719, 96 (1968).
7. W. R. Turner and P. J. Elving, _J. Electrochem. Soc._, 112, 1215 (1965).
8. V. D. Parker, _Chem. Commun._, 1969, 716.
9. B. R. Eggins, _Chem. Commun._, 1969, 1267.
10. L. Jeftič and G. Manning, _J. Electroanal. Chem._, 26, 195 (1970).
11. V. Prelog, O. Haflinger, and K. Wiesner, _Helv. Chim. Acta_, 31, 877 (1948).
12. D. K. Gullstrom and H. P. Burchfield, _Anal. Chem._, 20, 1174 (1948).
13. L. I. Smith, I. M. Kolthoff, S. Wawzonek, and P. M. Ruoff, _J. Am. Chem. Soc._, 63, 1018 (1941).
14. Reference 3, p.699.
15. Reference 3, p.700.
16. G. J. Patriarche and J. J. Lingane, _Anal. Chim. Acta_, 49, 241 (1970).
17. Reference 3, p.678.
18. B. Mooney and H. I. Stonehill, _J. Chem. Soc. A_, 1967, 1.
19. V. D. Bezuglyi, L. Ya. Kheifets, and N. A. Sobina Zh. Obshch. Khim., 37, 1433 (1967).
20. R. D. Rieke, W. E. Rich, and T. H. Ridgway, _Tetrahedron Letters_, 50, 4381 (1967).
21. E. I. Klabunovskii, R. Yu. Mamedzade-Alieva, and A. A. Balandin, _Russ. J. Phys. Chem._, 42, 615 (1968).

12. ALIPHATIC CARBOXYLIC ACIDS

Reduction

Compound	Solvent System	Working Electrode	$E_{1/2}$	Reference Electrode	Reference
Formic Acid	Aq LiCl, Li$_2$SO$_4$, (Me)$_4$NI	Hg	-1.74 to -1.85[a]	sce	1
	Py, 0.1M (Et)$_4$NClO$_4$	Hg	-1.57	Ag/1M Ag+, Py [b]	2
Acetic Acid	Aq LiCl, Li$_2$SO$_4$, (Me)$_4$NI	Hg	-1.76 to -1.86[a]	sce	1
	Py, 0.1M (Et)$_4$NClO$_4$	Hg	-1.71	Ag/1M Ag+, Py [b]	2
Chloroacetic Acid	Py, 0.1M (Et)$_4$NClO$_4$	Hg	-1.49	Ag/1M Ag+, Py [b]	2
Dichloroacetic Acid	Py, 0.1M (Et)$_4$NClO$_4$	Hg	-1.38	Ag/1M Ag+, Py [b]	2
Trifluoroacetic Acid	Py, 0.1M (Et)$_4$NClO$_4$	Hg	-1.35	Ag/1M Ag+, Py [b]	2
Isobutyric Acid	Aq LiCl, Li$_2$SO$_4$, (Me)$_4$NI	Hg	-1.81 to -1.87[a]	sce	1
Isovaleric Acid	Aq LiCl, Li$_2$SO$_4$, (Me)$_4$NI	Hg	-1.75 to -1.82[a]	sce	1
Oxalic Acid	Aq LiCl, Li$_2$SO$_4$, (Me)$_4$NI	Hg	-1.66 to -1.80[a]	sce	1

(continued)

Compound	Solvent System	Working Electrode	$E_{1/2}$	Reference Electrode	Reference
Malonic Acid	Py, 0.1M (Et)4 NClO4	Hg	-1.37, -1.78	Ag/1M Ag+, Py[b]	2
	Aq LiCl, Li2SO4, (Me)4NI	Hg	-1.69 to -1.74[a]	sce	1
Tartaric Acid	Aq LiCl, Li2SO4, (Me)4NI	Hg	-1.64 to -1.77[a]	sce	1
	Py, 0.1M (Et)4 NClO4	Hg	-1.42, -1.65	Ag/1M Ag+, Py[b]	2
Citric Acid	Aq LiCl, Li2SO4, (Me)4NI	Hg	-1.64 to -1.77[a]	sce	1
Malic Acid	Aq LiCl, Li2SO4, (Me)4NI	Hg	-1.66 to -1.74[a]	sce	1
Succinic Acid	Aq LiCl, Li2SO4, (Me)4NI	Hg	-1.80[a]	sce	1
Adipic Acid	Py, 0.1M (Et)4 NClO4	Hg	-1.50, -1.93	Ag/1M Ag+, Py[b]	2
	Py, 0.1M (Et)4 NClO4	Hg	-1.65, -1.76	Ag/1M Ag+, Py[b]	2
	Aq LiCl, Li2SO4, (Me)4NI	Hg	-1.76 to -1.80[a]	sce	1
Fumaric Acid	0.9N KCl+HCl, pH 2.56	Hg	-0.79	sce	3
	0.9N K2HPO4, pH 8.78	Hg	-1.71	sce	3
Maleic Acid	Py, 0.1M (Et)4 NClO4	Hg	-1.46, -1.78	Ag/1M Ag+, Py[b]	2
	0.9N KCl+HCl, pH 2.43	Hg	-0.73	sce	3

	0.9N NH$_3$+NH$_4$Cl, pH 9.63	Hg	-1.42	sce	3
	Py, 0.1M (Et)$_4$NClO$_4$	Hg	-2.01	Ag/1M Ag+, Py[b]	2
Pyruvic Acid	B-R buffer, pH 6.8	Hg	-1.22, -1.53	sce	4
α-Ketoglutaric Acid	Ammoniacal buffer	Hg	-1.30	sce	4
Glycolic, lactic, and glyceric acids show only hydrogen waves	0.1N (Et)$_4$NCl				5
Muconic Acid	OAc buffer, pH 4.5	Hg	-0.97	NCE	6
Citraconic Acid	1M HCl	Hg	-0.58[c]	NCE	7
Mesaconic Acid	1M HCl	Hg	-0.57[c]	NCE	7
Itaconic Acid	1M HCl	Hg	No wave	NCE	7
Aconitic Acid	1M HCl	Hg	-0.55[c]	NCE	7
	50% Benzene, 50% Methanol, 0.3M LiCl	Hg	-1.24	sce	8
Acetylenedicarboxylic Acid	1M HCl	Hg	-0.45[c]	NCE	7
Crotonic Acid	0.1N LiOH	Hg	-2.09	NCE	9
Sorbic Acid	0.1N LiOH	Hg	-2.01	NCE	9
Piperic Acid	0.1N LiOH	Hg	-1.85	NCE	9
β-Ethylacrylic Acid	Buffer	Hg	No wave	NCE	10

[a] Discharge of hydrogen ions.
[b] Potential +0.09 V versus sce.
[c] Tangent potential.

(continued)

References for Aliphatic Carboxylic Acids

1. I. A. Korshunov, E. B. Kuznetsova, and M. K. Shchennikova, Zh. Fiz. Khim., 23, 1292 (1949).
2. K. Tsuji and P. J. Elving, Anal. Chem., 41, 286 (1969).
3. P. J. Elving and C. Teitelbaum, J. Am. Chem. Soc., 71, 3916 (1949).
4. J. Heyrovský and J. Kůta, Principles of Polarography, Academic, New York, 1966, p. 554.
5. I. M. Kolthoff and J. J. Lingane, Polarography, 2nd ed., Interscience, New York, 1952, p. 720.
6. Z. Fencl, Chem. Listy, 46, 76, 300 (1952).
7. L. Schwaer, Collect. Czech. Chem. Commun., 7, 326 (1935).
8. C. Ricciuti, C. O. Willits, H. B. Knight, and D. Swern, Anal. Chem., 25, 933 (1953).
9. F. Santavy and O. Hanc, Časopis českého lékárnic 59, 40 (1946).
10. V. Zambotti, Arch. Sci. Biol. (Bologna), 26, 80 (1940).

13. AROMATIC CARBOXYLIC ACIDS

Oxidation

Compound	Solvent System	Working Electrode	$E_{1/2}$	Reference Electrode	Reference
o-Hydroxybenzoic Acid (Salicyclic Acid)	OAc buffer, 50% Isopropanol, pH 5.6	Graphite	+0.845	sce	1
p-Hydroxybenzoic Acid	OAc buffer, 50% Isopropanol, pH 5.6	Graphite	+0.716	sce	1
o-Aminobenzoic Acid (Anthranilic Acid)	OAc buffer, 50% Isopropanol, pH 5.6	Graphite	+0.676	sce	1
m-Aminobenzoic Acid	OAc buffer, 50% Isopropanol, pH 5.6	Graphite	+0.668	sce	1
p-Aminobenzoic Acid	B-R buffer, pH = pKa amine	C paste	+0.84	sce	2
2-Biphenylcarboxylic Acid	HOAc, NaOAc	Pt	+1.71	sce	3

(continued)

Reduction

Compound	Solvent System	Working Electrode	$E_{1/2}$	Reference Electrode	Reference
Benzoic Acid	Aq LiCl, Li$_2$SO$_4$, (Me)$_4$NI	Hg	-1.56 to -1.72[a]	sce	4
	50% Dioxane, (n-Bu)$_4$NI	Hg	-1.87[a]	sce	5
	75% Dioxane, (Et)$_4$NI	Hg	-1.94 to -2.46[a]	sce	4
	Py, 0.1M (Et)$_4$- NClO$_4$	Hg	-1.59	Ag/1M Ag+, Py[b]	6
Mandelic Acid	Aq LiCl, Li$_2$SO$_4$, (Me)$_4$NI	Hg	-1.70 to -1.78[a]	sce	4
Gallic Acid	Aq LiCl, Li$_2$SO$_4$, (Me)$_4$NI	Hg	-1.71 to -1.73[a]	sce	4
Salicylic Acid	Ag LiCl, Li$_2$SO$_4$, (Me)$_4$NI	Hg	-1.66 to -1.83[a]	sce	4
	Py, 0.1M (Et)$_4$- NClO$_4$	Hg	-1.40	Ag/1M Ag+, Py[b]	6
Acetylsalicyclic Acid	Py, 0.1M (Et)$_4$- NClO$_4$	Hg	-1.52 to -1.6[a]	sce	4
o-Aminobenzoic Acid (Anthranilic Acid)	Py, 0.1M (Et)$_4$- NClO$_4$	Hg	-1.60[a]	sce	4
Sulfanilic Acid	Py, 0.1M (Et)$_4$- NClO$_4$	Hg	-1.54[a]	sce	4

Naphthionic Acid	Py, 0.1M (Et)$_4$ NClO$_4$	Hg	-1.42 to -1.52[a]	sce	4
Phenylglyoxylic Acid	B-R buffer, pH 7.2	Hg	-0.98, -1.25	sce	7
Phthalic Acid	Py, 0.1M (Et)$_4$ NClO$_4$	Hg	-1.35	Ag/1M Ag+, Py[b]	6
Terephthalic Acid	Py, 0.1M (Et)$_4$ NClO$_4$	Hg	-1.53, -1.68	Ag/1M Ag+, Py[b]	6
Cinnamic Acid	1M HCl	Hg	-1.04[c]	NCE	8
	DMF, (Et)$_4$NI	Hg	-1.85, -2.60	sce	9
β-Benzoylacrylic Acid	Phosphate buffer, 50% Dioxane	Hg	-1.04	NCE	10
Retinic Acid	DMF, (Et)$_4$NI	Hg	-1.62, -2.13, -2.75	sce	9

a Discharge of hydrogen ion.
b Potential +0.09 V versus sce.
c Tangent potential.

(continued)

References for Aromatic Carboxylic Acids

1. J. C. Suatoni, R. E. Snyder, and R. O. Clark, Anal. Chem., 33, 1894 (1961).
2. J. Bacon and R. N. Adams, J. Am. Chem. Soc., 90, 6596 (1968).
3. L. Eberson and K. Nyberg, J. Am. Chem. Soc., 88, 1686 (1966).
4. I. A. Korshunov, E. B. Kuznetsova, and M. K. Shchennikova, Zh. Fiz, Khim., 23, 1292 (1949).
5. M. von Stackelberg and W. Stracke, Z. Electrochem, 53, 118 (1949).
6. K. Tsuji and P. J. Elving, Anal. Chem., 41, 286 (1969).
7. J. Heyrovský and J. Kůta, Principles of Polarography, Academic, New York, 1966, p. 554.
8. L. Schwaer, Collect. Czech, Chem. Commun., 7, 326 (1935).
9. V. G. Mairanovskii, I. E. Valashek, and G. I. Samokhvalov, Elektrokhim., 3, 611 (1967).
10. S. Wawzonek and J. H. Fossum, Sbornik mezinárod. polarog. sjezdu, praze, 1st Congr., 1951, Pt. I. Proc., p. 548.

14. ANHYDRIDES

Reduction

Compound	Solvent System	Working Electrode	$E_{1/2}$	Reference Electrode	Reference
Maleic Anhydride	50% Benzene, 50% Methanol, 0.3M LiCl	Hg	-0.72	sce	1
	DMF, (Et)$_4$NClO$_4$	Hg	-0.88	sce	2
	DMF, (Et)$_4$NClO$_4$	Hg	-0.85	sce	3
	CH$_3$CN, (Et)$_4$NClO$_4$	Hg	-0.84	sce	3
Butyl ester of aconitic acid anhydride	50% Benzene, 50% Methanol, 0.3M LiCl	Hg	-0.77	sce	1
Citraconic Anhydride	50% Benzene, 50% Methanol, 0.3M LiCl	Hg	-0.84	sce	1
Crotonic Anhydride	50% Benzene, 50% Methanol, 0.3M LiCl	Hg	-1.62	sce	1
n-Butyric Anhydride	50% Benzene, 50% Methanol, 0.3M LiCl	Hg	No wave	sce	1
Caproic Anhydride	50% Benzene, 50%	Hg	No wave	sce	1

(continued)

769

Compound	Solvent System	Working Electrode	$E_{1/2}$	Reference Electrode	Reference
Succinic Anhydride	Methanol, 0.3$\underline{M}$ LiCl	Hg	No wave	sce	1
Benzoic Anhydride	50% Benzene, 50% Methanol, 0.3$\underline{M}$ LiCl	Hg	-2.53	sce	3
	CH_3CN, $(n\text{-}Bu)_4NI$	Hg	-1.62	sce	1
Phthalic Anhydride	50% Benzene, 50% Methanol, 0.3$\underline{M}$ LiCl	Hg	-1.12	sce	1
Phthalic Anhydride	DMF, $(Et)_4NClO_4$	Hg	-1.36	sce	2
	DMF, $(Et)_4NClO_4$	Hg	-1.27	sce	3
	DMF, $(n\text{-}Bu)_4NI$	Hg	-1.19	sce	4
	CH_3CN, $(Et)_4NClO_4$	Hg	-1.31	sce	3
Tetrachlorophthalic Anhydride	CH_3CN, $(Et)_4NClO_4$	Hg	-0.86	sce	3
2-Benzoylbenzoic Anhydride	DMF, $(Et)_4NClO_4$	Hg	-0.83	sce	3
	50% Dioxane, 0.1$\underline{M}$ $(n\text{-}Bu)_4NI$	Hg	-1.28, -1.58, -1.90	sce	5
Acetic 2-benzoyl-benzoic Anhydride	50% Dioxane, 0.1$\underline{M}$ $(n\text{-}Bu)_4NI$	Hg	-1.48, -1.92	sce	5
Pyromellitic Dianhydride	CH_3CN or DMF, $(Et)_4NClO_4$	Hg	-0.60	sce	2
	CH_3CN, $(Et)_4NClO_4$	Hg	-0.55	sce	3
	DMF, $(Et)_4NClO_4$	Hg	-0.52	sce	3

Dibromopyromellitic Anhydride	CH_3CN, $(Et)_4NClO_4$	Hg	-0.32	sce	3
Naphthalic Anhydride	DMF, $(Et)_4NClO_4$	Hg	-0.29	sce	3
	DMF, $(\underline{n}-Bu)_4NI$	Hg	-1.12	sce	4

(continued)

References for Anhydrides

1. Ricciuti, C. O. Willits, H. B. Knight, and D. Swern, _Anal. Chem._, 25, 933 (1953).
2. M. E. Peover, _Nature_, 191, 702 (1961).
3. M. E. Peover, _Trans. Faraday Soc._, 58, 2370 (1962).
4. R. E. Sioda and W. S. Koski, _J. Am. Chem. Soc._, 89, 475 (1967).
5. S. Wawzonek, H. A. Laitinen, and S. J. Kwiatkowski, _J. Am. Chem. Soc._, 66, 827 (1944).

15. ESTERS

Reduction

Compound	Solvent System	Working Electrode	$E_{1/2}$	Reference Electrode	Reference
Phenyl acetate	0.5M NaOAc	Hg	-1.30	sce	1
p-Bromoethylphenyl acetate	Methanol, 0.25M LiOMe	Hg	-0.916	Ag, AgCl	2
	Methanol, 0.25M LiCl	Hg	-1.045	Ag, AgCl	2
Ethyl 3-bromo-propionate	DMF, (Et)$_4$NBr	Hg	-2.039	sce	3
4-Bromoethylbutyrate	DMF, (Et)$_4$NBr	Hg	-2.177	sce	3
Ethyl α-nitro-i-butyrate	CH$_3$CN, (Et)$_4$NBr	Hg	-1.32	sce	4
Ethyl α-nitroso-i-butyrate	CH$_3$CN, (Et)$_4$NBr	Hg	-1.70	sce	4
Ethyl 5-bromo-valerate	DMF, (Et)$_4$NBr	Hg	-2.180	sce	3
Ethyl oxalate	DMSO, (n-Propyl)$_4$-NClO$_4$	Hg	-1.85	sce	5
Ethyl mesoxalate	DMSO, (n-Propyl)$_4$-NClO$_4$	Hg	-0.84	sce	5
Methyl pyruvate	B-R buffer, pH 7.0	Hg	-0.87	sce	6

(continued)

Compound	Solvent System	Working Electrode	$E_{1/2}$	Reference Electrode	Reference
Ethyl pyruvate	DMSO, $(\underline{n}\text{-Propyl})_4$NClO$_4$	Hg	-1.51	sce	5
Diethyl fumarate	0.9N KCl + HCl, pH 2.16	Hg	-0.75	sce	7
	0.9N NH$_3$+NH$_4$Cl, pH 9.63	Hg	-1.15, -1.37	sce	7
	50% Ethanol, 0.1$\underline{M}$ LiClO$_4$	Hg	-1.18	sce	8
Diethyl chloro-fumarate	50% Ethanol, 0.1$\underline{M}$ LiClO$_4$	Hg	-1.05, -1.17	sce	8
	25% Ethanol, 0.1$\underline{M}$ LiClO$_4$, 0.5$\underline{M}$ HClO$_4$	Hg	-0.71	sce	8
Diethyl dichloro-fumarate	50% Ethanol, 0.1$\underline{M}$ LiClO$_4$	Hg	-0.86, -1.07, -1.23	sce	8
	25% Ethanol, 0.1$\underline{M}$ LiClO$_4$ 0.5$\underline{M}$ HClO$_4$	Hg	-0.67	sce	8
Diethyl dibromo-fumarate	50% Ethanol, 0.1$\underline{M}$ LiClO$_4$	Hg	-0.38, -0.82, -1.24	sce	8, 9
	50% Dioxane, 0.1$\underline{M}$ LiClO$_4$	Hg	-0.55, -1.02, -1.31	sce	9
	McIlvaine buffer, 4% Ethanol, pH 2	Hg	-0.25, -0.55, -0.73	sce	9
Diethyl diiodo-fumarate	50% Ethanol, 0.1$\underline{M}$ LiClO$_4$	Hg	-0.17, -0.85, -1.24	sce	9
	50% Dioxane, 0.1$\underline{M}$ LiClO$_4$	Hg	-0.19, -1.01, -1.32	sce	9
Diethyl chloro-iodofumarate	50% Ethanol, 0.1$\underline{M}$ LiClO$_4$	Hg	-0.25, -0.84, -1.25	sce	9

Compound	Electrolyte	Electrode	$E_{1/2}$ (V)	Reference electrode	Ref.
	50% Dioxane, 0.1M LiClO$_4$	Hg	-0.28, -0.98, -1.33	sce	9
Diethyl maleate	0.9N KCl+HCl, pH 2.20	Hg	-0.87	sce	7
	0.9N NH$_3$+NH$_4$Cl, pH 8.68	Hg	-1.05, -1.46	sce	7
	50% Ethanol, 0.1M LiClO$_4$	Hg	-1.18	sce	8
Diethyl dichloro-maleate	50% Ethanol, 0.1M LiClO$_4$	Hg	-0.96, -1.11, -1.24	sce	8
	25% Ethanol, 0.1M LiClO$_4$, 0.5M HClO$_4$	Hg	-0.79	sce	8
Diethyl dibromo-maleate	50% Ethanol, 0.1M LiClO$_4$	Hg	-0.51, -0.90, -1.24	sce	9
	50% Dioxane, 0.1M LiClO$_4$	Hg	-0.63, -1.04, -1.27	sce	9
Diethyl diiodo-maleate	McIlvaine buffer, 4% Ethanol, pH 2	Hg	-0.30, -0.56, -0.71	sce	9
	50% Ethanol, 0.1M LiClO$_4$	Hg	-0.18, -0.82, -1.19	sce	9
	50% Dioxane, 0.1M LiClO$_4$	Hg	-0.20, -0.97, -1.30	sce	9
Butyl acrylate	20% Ethanol, 0.1N LiCl	Hg	-1.99	NCE	10
Methyl methacrylate	30% Ethanol, 0.1N LiCl	Hg	-2.00	NCE	10
Ethyl methacrylate	25% Ethanol, 0.1N LiCl	Hg	-1.92	NCE	11
	25% Ethanol, 0.1N LiCl, 0.1N (Me)$_4$NI	Hg	-1.70, -2.60	NCE	12

(continued)

775

Compound	Solvent System	Working Electrode	$E_{1/2}$	Reference Electrode	Reference
Butyl methacrylate	30% Ethanol, 0.01$\underline{N}$ LiCl	Hg	-2.11	NCE	10
Methyl sorbate	DMF, (Et)$_4$NI	Hg	-2.00, -2.66	sce	13
Methyl benzoate	DMF, (Et)$_4$NI	Hg	-2.32, -2.83	sce	13
Ethyl benzoate	75% Dioxane, 0.05$\underline{M}$ (Et)$_4$NI	Hg	-2.14	sce	14
	50% Dioxane, 0.1$\underline{M}$ (n-Bu)$_4$NI	Hg	-2.14	sce	14
	75% Dioxane, 0.175$\underline{M}$ (n-Bu)$_4$NI	Hg	-2.13	sce	16
Tetraethylammonium benzoate[a]	75% Dioxane, 0.05$\underline{M}$ (Et)$_4$NI	Hg	-1.98, -2.38	sce	14
Ethyl benzoylformate	DMSO, (n-Propyl)$_4$NClO$_4$	Hg	-1.48	sce	5
Ethyl β-benzoyl-acrylate	50% Dioxane, Phosphate buffer	Hg	-0.92, -1.65	NCE	17
Methyl cinnamate	DMF, (Et)$_4$NI	Hg	-1.90, 2.33	sce	13
Methyl ester of phthalaldehydic acid	50% Dioxane, 0.1$\underline{M}$ (n-Bu)$_4$NI	Hg	-1.35, -2.07	sce	15
Ethyl ester of phthalaldehydic acid	50% Dioxane, 0.1$\underline{M}$ (n-Bu)$_4$NI	Hg	-1.37, -2.05	sce	15
Monomethyl phthalate	50% Benzene, 50% Methanol, 0.3$\underline{M}$ LiCl	Hg	-1.48	sce	18
Monocyclohexyl phthalate	50% Benzene, 50% Methanol, 0.3$\underline{M}$ LiCl	Hg	-1.56	sce	18

Compound	Solvent / electrolyte	Electrode	E (V)	Reference	Ref.
Di-2-ethylhexyl phthalate	50% Benzene, 50% Methanol, 0.3M LiCl	Hg	-1.30	sce	18
Dimethyl phthalate	75% Ethanol, 0.1M (Me)$_4$NCl	Hg	-1.83, -2.17	sce	19
Diethyl phthalate	75% Ethanol, 0.1M (Me)$_4$NCl	Hg	-1.87, -2.17	sce	19
Dibutyl phthalate	75% Ethanol, 0.1M (Me)$_4$NCl	Hg	-1.89, -2.17	sce	19
Diphenyl phthalate	75% Ethanol, 0.1M (Me)$_4$NCl	Hg	-1.65, -2.08	sce	19
Dioctyl phthalate	75% Ethanol, 0.1M (Me)$_4$NCl	Hg	-1.93, -2.18	sce	19
Methylphthalyl ethyl glycolate	75% Ethanol, 0.1M (Me)$_4$NCl	Hg	-1.79, -2.15	sce	19
Ethyl 1-naphthoate	75% Dioxane, 0.175M (n-Bu)$_4$NI	Hg	-1.86, -2.09	sce	16
Ethyl 2-naphthoate	75% Dioxane, 0.175M (n-Bu)$_4$NI	Hg	-1.95, -2.10	sce	16
Ethyl 1-anthroate	75% Dioxane, 0.175M (n-Bu)$_4$NI	Hg	-1.61, -1.83	sce	16
Ethyl 2-anthroate	75% Dioxane, 0.175M (n-Bu)$_4$NI	Hg	-1.63, -1.87, -2.20	sce	16
Ethyl chrysene-2-carboxylate	75% Dioxane, 0.175M (n-Bu)$_4$NI	Hg	-1.75, -1.89	sce	16
Ethyl 2-phen-anthroate	75% Dioxane, 0.175M (n-Bu)$_4$NI	Hg	-1.93, -2.1	sce	16
Ethyl 9-phen-anthroate	75% Dioxane, 0.175M (n-Bu)$_4$NI	Hg	-1.82, -1.99	sce	16
Ethyl pyrene-3-carboxylate	75% Dioxane, 0.175M (n-Bu)$_4$NI	Hg	-1.63, -2.02	sce	16

(continued)

778

Compound	Solvent System	Working Electrode	$E_{1/2}$	Reference Electrode	Reference
Ethyl pyrene-4-carboxylate	75% Dioxane, 0.175M $(n\text{-}Bu)_4NI$	Hg	-1.96, -2.10	sce	16
Ethyl fluoranthene-1-carboxylate	75% Dioxane, 0.175M $(n\text{-}Bu)_4NI$	Hg	-1.44, -1.74	sce	16
Ethyl fluoranthene-3-carboxylate	75% Dioxane, 0.175M $(n\text{-}Bu)_4NI$	Hg	-1.39, -1.73	sce	16
Ethyl fluoranthene-8-carboxylate	75% Dioxane, 0.175M $(n\text{-}Bu)_4NI$	Hg	-1.54, -1.78	sce	16
Tosylate Esters of:					
Isopropyl alcohol	Ethanol, $(Me)_4NCl$	Hg	-1.96	sce	20
2-Butanol	Ethanol, $(Me)_4NCl$	Hg	-1.98	sce	20
Methylbenzylcarbinol	Ethanol, $(Me)_4NCl$	Hg	-1.96	sce	20
l-Menthol	Ethanol, $(Me)_4NCl$	Hg	-1.95	sce	20
l-Borneol	Ethanol, $(Me)_4NCl$	Hg	-1.99	sce	20
Cholesterol	Ethanol, $(Me)_4NCl$	Hg	-1.95	sce	20
Methanol	CH_3CN, $(n\text{-}Propyl)_4NClO_4$	Hg	-2.61	Ag/0.01N Ag+	21
Ethanol	CH_3CN, $(n\text{-}Propyl)_4NClO_4$	Hg	-2.52	Ag/0.01N Ag+	21
n-Butanol	CH_3CN, $(n\text{-}Propyl)_4NClO_4$	Hg	-2.61	Ag.0.01N Ag+	21
Neopentyl alcohol	CH_3CN, $(n\text{-}Propyl)_4NClO_4$	Hg	-2.52	Ag/0.01N Ag+	21
Cyclohexyl alcohol	CH_3CN, $(n\text{-}Propyl)_4NClO_4$	Hg	-2.57	Ag/0.01N Ag+	21

a Included for comparison.

References for Esters

1. L. Eberson and K. Nyberg, J. Am. Chem. Soc., 88, 1686 (1966).
2. G. Klopman, Helv. Chim. Acta, 44, 1908 (1961).
3. F. L. Lambert, J. Org. Chem., 31, 4184 (1966).
4. H. Sayo, Y. Tsukitani, and M. Masui, Tetrahedron, 24, 1717 (1968).
5. G. A. Russell and S. A. Weiner, J. Am. Chem. Soc., 89, 6623 (1967).
6. J. Heyrovský and J. Kůta, Principles of Polarography, Academic, New York, 1966, p. 554.
7. P. J. Elving and C. Teitelbaum, J. Am. Chem. Soc., 71, 3916 (1949).
8. L. G. Feoktistov and I. G. Markova, Zh. Obshch. Khim., 39, 512 (1969).
9. I. G. Markova and L. G. Feoktistov, Zh. Obshch. Khim., 38, 970 (1968).
10. A. W. Rjabow and G. D. Panowa, Ber. Akad. Wiss. UdSSR (N.S.), 99, 547 (1954).
11. A. C. Cope and E. M. Hardy, J. Am. Chem. Soc., 62, 3319 (1940).
12. M. B. Neiman and M. A. Shubenko, Betriebs-Lab., 14, 394 (1948).
13. V. G. Mairanovskii, I. E. Valashek, and G. I. Samokhvalov, Elektrokhim., 3, 611 (1967).
14. M. von Stackelberg and W. Stracke, Z. Elektrochem., 53, 118 (1949).
15. S. Wawzonek and J. H. Fossum, J. Electrochem. Soc., 96, 234 (1949).
16. G. Klopman and J. Nasielski, Bull. Soc. Chim. Belges, 70, 490 (1961).
17. S. Wawzonek and J. H. Fossum, Sborník mezinárod. polarog. sjezdu, praze, 1st Cong., 1951, Pt#1. Proc., p. 548.
18. C. Rucciutti, C. O. Willits, H. B. Knight, and D. Swern, Anal. Chem., 25, 933 (1953).
19. Cr. C. Whitnack and E. St. C. Gantz, Anal. Chem., 25, 553 (1953).
20. L. Horner and R. Singer, Chem. Ber., 101, 3329 (1968).
21. P. Yousefzadeh and C. K. Mann, J. Org. Chem., 33, 2716 (1968).

16. PEROXIDES, PERACIDS, AND PERESTERS

Oxidation

Compound	Solvent System	Working Electrode	$E_{1/2}$	Reference Electrode	Reference
Hydrogen Peroxide	0.1M Li$_2$SO$_4$, 0.01M LiOH	Hg	-0.11	sce	1

Reduction

Peroxides

Compound	Solvent System	Working Electrode	$E_{1/2}$	Reference Electrode	Reference
Hydrogen Peroxide	0.1M Li$_2$SO$_4$	Hg	-0.88	sce	1
	50% Benzene, 50% Methanol, 0.3M LiCl	Hg	-1.16	Hg pool	2
	DMF, (n-Bu)$_4$NClO$_4$	Hg	-2.15	sce	3
Methyl hydroper-oxide	0.1M Li$_2$SO$_4$	Hg	-0.64	sce	1

Compound	Conditions	Electrode	Potential	Ref. electrode	Ref.
Ethyl hydroperoxide	$0.1M$ Li_2SO_4	Hg	-0.42	sce	1
n-Butyl hydroperoxide	$0.1F$ H_2SO_4, 5% Ethanol	Hg	-0.26	sce	4
s-Butyl hydroperoxide	$0.1F$ H_2SO_4, 5% Ethanol	Hg	-0.28	sce	4
t-Butyl hydroperoxide	$0.1F$ H_2SO_4, 5% Ethanol	Hg	-0.34	sce	4
t-Butyl hydroperoxide	DMF, $(n-Bu)_4-NClO_4$	Hg	-2.33	sce	3
	$0.1M$ Li_2SO_4	Hg	-0.28	sce	1
	50% Benzene, 50% Methanol, $0.3M$ LiCl	Hg	-1.15	Hg pool	2
	50% Benzene, 50% Ethanol, $0.3M$ LiCl	Hg	-1.22	sce	5
	50% Benzene, 50% Methanol, $0.3M$ LiCl	Hg	-0.96	sce	6
n-Pentyl hydroperoxide	$0.1F$ H_2SO_4, 5% Ethanol	Hg	-0.20	sce	4
	$0.1N$ NaOH	Hg	-0.330	sce	7
s-Pentyl hydroperoxide	$0.1F$ H_2SO_4, 50% Ethanol	Hg	-0.24	sce	4
t-Pentyl hydroperoxide	$0.1F$ H_2SO_4, 50% Ethanol	Hg	-0.22	sce	4
	50% Benzene, 50% Methanol, $0.3M$ LiCl	Hg	-1.19	sce	5
3-Methyl-1-butyl hydroperoxide	5% Ethanol, $0.1F$ H_2SO_4	Hg	-0.23	sce	4

(continued)

Compound	Solvent System	Working Electrode	$E_{1/2}$	Reference Electrode	Reference
Cyclopentyl hydro-peroxide	5% Ethanol, 0.1F H_2SO_4	Hg	-0.25	sce	4
n-Hexyl hydroper-oxide	5% Ethanol, 0.1F H_2SO_4	Hg	-0.12	sce	4
s-Hexyl hydroper-oxide	5% Ethanol, 0.1F H_2SO_4	Hg	-0.16	sce	4
t-Hexyl hydroper-oxide	5% Ethanol, 0.1F H_2SO_4	Hg	-0.16	sce	4
Cyclohexyl hydro-peroxide	5% Ethanol, 0.1F H_2SO_4	Hg	-0.14	sce	4
n-Heptyl hydroper-oxide	5% Ethanol, 0.1F H_2SO_4	Hg	-0.03	sce	4
s-Heptyl hydroper-oxide	5% Ethanol, 0.1F H_2SO_4	Hg	-0.12	sce	4
n-Octyl hydroper-oxide	5% Ethanol, 0.1F H_2SO_4	Hg	-0.02	sce	4
s-Octyl hydroper-oxide	5% Ethanol, 0.1F H_2SO_4	Hg	-0.80	sce	4
n-Nonyl hydroper-oxide	5% Ethanol, 0.1F H_2SO_4	Hg	-0.01	sce	4
Formyl hydroper-oxide	0.1M Li_2SO_4, 0.004N H_2SO_4	Hg	$+0.2$	sce	1
Acetyl hydroper-oxide	0.1M Li_2SO_4, 0.004N H_2SO_4	Hg	$+0.2$	sce	1
Propionyl hydro-peroxide	0.1N $HClO_4$	Hg	$+0.095$	sce	7
	0.1M Li_2SO_4, 0.004N H_2SO_4	Hg	$+0.2$	sce	1
Cumyl hydroperoxide	DMF, (n-Bu)$_4NClO_4$	Hg	-2.10	sce	3

Compound	Solvent/Electrolyte	Electrode	Potential	Reference electrode	Ref.
Cumene hydroperoxide	50% Benzene, 50% Ethanol, 0.3M LiCl	Hg	-0.97	sce	5
	50% Benzene, 50% Methanol, 0.3M LiCl	Hg	-0.68	sce	6
	50% Benzene, 50% Methanol, 0.3M LiCl	Hg	-0.879	sce	8
	50% Benzene, 50% Methanol, 0.3M LiCl	Hg	-1.08	Hg pool	2
Pinyl hydroperoxide	50% Benzene, 50% Ethanol, 0.3M LiCl	Hg	-0.13	sce	5
Pinane hydroperoxide	50% Benzene, 50% Methanol, 0.3M LiCl	Hg	-0.80	sce	6
	50% Benzene, 50% Methanol, 0.3M LiCl	Hg	-1.08	Hg pool	2
α-Pinene hydroperoxide	50% Benzene, 50% Methanol, 0.3M LiCl	Hg	-0.82	sce	6
Tetralyl hydroperoxide	50% Benzene, 50% Ethanol, 0.3M LiCl	Hg	-0.88	sce	5
Tetralin hydroperoxide	50% Benzene, 50% Methanol, 0.3M LiCl	Hg	-0.73	sce	6

(continued)

Compound	Solvent System	Working Electrode	$E_{1/2}$	Reference Electrode	Reference
1,1-Diphenylethyl hydroperoxide	50% Benzene, 50% Ethanol, 0.3$\underline{M}$ LiCl	Hg	-0.82	sce	5
p-Menthane hydro-peroxide	50% Benzene, 50% Methanol 0.3$\underline{M}$ LiCl	Hg	-1.06	Hg pool	2
Cyclohexene hydro-peroxide	50% Benzene, 50% Methanol, 0.3$\underline{M}$ LiCl	Hg	-0.77	sce	6
t-Butylisopropyl-phenyl hydroper-oxide	50% Benzene, 50% Methanol, 0.3$\underline{M}$ LiCl	Hg	-1.06	Hg pool	2
Diisopropylphenyl hydroperoxide	50% Benzene, 50% Methanol, 0.3$\underline{M}$ LiCl	Hg	-1.10	Hg pool	2
Phenycyclohexane hydroperoxide	50% Benzene, 50% Methanol, 0.3$\underline{M}$ LiCl	Hg	-0.66, -1.08	Hg pool	2
Methyl oleate hydroperoxide	50% Benzene, 50% Methanol, 0.3$\underline{M}$ LiCl	Hg	-0.61	sce	6
Heptanoyl peroxide	50% Benzene, 50% Ethanol, 0.3$\underline{M}$ LiCl	Hg	-0.20	sce	9
Methyl ethyl ketone peroxide	50% Benzene, 50% Methanol, 0.3$\underline{M}$ LiCl	Hg	-0.60, -1.26	Hg pool	2
	50% Benzene, 50% Methanol, 0.3$\underline{M}$	Hg	-0.949	sce	8

Compound	Solvent/Electrolyte	Electrode	E	Ref.	Ref. No.
Diethyl peroxide	0.1M Li_2SO_4	Hg	-0.65	sce	1
Acetyl peroxide	50% Benzene, 50% Methanol, 0.3M LiCl	Hg	-0.28	Hg pool	2
Di-t-butyl peroxide	50% Benzene, 50% Methanol, 0.3M LiCl	Hg	No wave	sce	6
	50% Benzene, 50% Methanol, 0.3M LiCl	Hg	No wave	Hg pool	2
	50% Benzene, 5% Water, 37% Ethanol, NH_4OAc	Hg	-0.96	sce	10
Bis(1-hydroxyheptyl) peroxide	50% Benzene, 50% Methanol, 0.3M LiCl	Hg	0.00, -1.20	Hg pool	2
Succinic acid peroxide	50% Benzene, 50% Methanol, 0.3M LiCl	Hg	0.00	Hg pool	2
	50% Benzene, 50% Methanol, 0.3M LiCl	Hg	-0.19	sce	6
	50% Benzene, 50% Methanol, 0.3M LiCl	Hg	-0.15	Hg pool	2
Lauroyl peroxide	50% Benzene, 50% Ethanol, 0.3M LiCl	Hg	-0.22	sce	9
Stearoyl peroxide	50% Benzene, 50% Ethanol, 0.3M LiCl	Hg	-0.22	sce	9

(continued)

785

Compound	Solvent System	Working Electrode	$E_{1/2}$	Reference Electrode	Reference
Benzoyl peroxide	50% Benzene, 50% Methanol, 0.3M LiCl	Hg	0.00	Hg pool	2
	50% Benzene, 50% Methanol, 0.3M LiCl	Hg	0.00	sce	6
	50% Benzene, 50% Ethanol, 0.3M LiCl	Hg	-0.16	sce	9
o-Chlorobenzoyl peroxide	50% Benzene, 50% Ethanol, 0.3M LiCl	Hg	-0.124	sce	9
m-Chlorobenzoyl peroxide	50% Benzene, 50% Ethanol, 0.3M LiCl	Hg	-0.129	sce	9
p-Chlorobenzoyl peroxide	50% Benzene, 50% Ethanol, 0.3M LiCl	Hg	-0.134	sce	9
2,4-Dichlorobenzoyl peroxide	50% Benzene, 50% Ethanol, 0.3M LiCl	Hg	-0.198	sce	9
Bis(2,4-dichloro-benzoyl) peroxide	50% Benzene, 50% Methanol, 0.3M LiCl	Hg	0.00	Hg pool	2
Ascaridole	50% Benzene, 50% Methanol, 0.3M LiCl	Hg	-1.22	Hg pool	2
1-Phenylmethyl-t-butyl peroxide	50% Benzene, 50% Methanol, 0.3M	Hg	No wave	Hg pool	2

Peracids

Peroxyacetic acid	50% Benzene, 50% Methanol, 0.3M LiCl	Hg	0.00	Hg pool	2
Peroxybenzoic acid	50% Benzene, 50% Methanol, 0.3M LiCl	Hg	0.00	Hg pool	2
Percaproic acid	50% Benzene, 50% Methanol, 0.3M LiCl	Hg	0.00 to -0.06	sce	11
Percaprylic acid	50% Benzene, 50% Methanol, 0.3M LiCl	Hg	0.00 to -0.06	sce	11
Perpelargonic acid	50% Benzene, 50% Methanol, 0.3M LiCl	Hg	0.00 to -0.06	sce	11
Percapric acid	50% Benzene, 50% Methanol, 0.3M LiCl	Hg	0.00 to -0.06	sce	11
Perhendecanoic acid	50% Benzene, 50% Methanol, 0.3M LiCl	Hg	0.00 to -0.06	sce	11
Perlauric acid	50% Benzene, 50% Methanol, 0.3M LiCl	Hg	0.00 to -0.06	sce	11
Pertridecanoic acid	50% Benzene, 50% Methanol, 0.3M LiCl	Hg	0.00 to -0.06	sce	11
Permyristic acid	50% Benzene, 50% Methanol, 0.3M LiCl	Hg	0.00 to -0.06	sce	11

(continued)

Compound	Solvent System	Working Electrode	$E_{1/2}$	Reference Electrode	Reference
Perpalmitic acid	50% Benzene, 50% Methanol, 0.3M LiCl	Hg	0.00 to -0.06	sce	11
Perstearic acid	50% Benzene, 50% Methanol, 0.3M LiCl	Hg	0.00 to -0.06	sce	11
Peresters					
t-Butyl peracetate	50% Benzene, 50% Methanol, 0.3M LiCl	Hg	-1.02	Hg pool	2
Di-t-butyl perphthalate	50% Benzene, 50% Methanol, 0.3M LiCl	Hg	-0.70, -1.05	Hg pool	2
t-Butyl perbenzoate	50% Benzene, 50% Methanol, 0.3M LiCl	Hg	-0.95	Hg pool	2
	50% Benzene, 50% Ethanol, 0.3M LiCl	Hg	-0.89	sce	12
	B-R buffer, 20% Ethanol, pH 4.6	Hg	-0.215	sce	12
t-Butyl-o-chloro-perbenzoate	50% Benzene, 50% Ethanol, 0.3M LiCl	Hg	-0.74	sce	12
t-Butyl-m-chloro-perbenzoate	50% Benzene, 50% Ethanol, 0.3M LiCl	Hg	-0.78	sce	12

Compound	Solvent	Electrode	E	Ref. electrode	Ref.
t-Butyl-p-chloro-perbenzoate	50% Benzene, 50% Ethanol, 0.3M LiCl	Hg	-0.85	sce	12
t-Butyl-2,4-dichloroperbenzoate	50% Benzene, 50% Ethanol, 0.3M LiCl	Hg	-0.74	sce	12
t-Butyl-p-nitro-perbenzoate	50% Benzene, 50% Ethanol, 0.3M LiCl	Hg	-0.17, -0.262, -0.77, -1.07	sce	12
t-Butyl-o-pertoluylate	50% Benzene, 50% Ethanol, 0.3M LiCl	Hg	-0.97	sce	12
t-Butyl-m-pertoluylate	50% Benzene, 50% Ethanol, 0.3M LiCl	Hg	-0.91	sce	12
t-Butyl-p-pertoluylate	50% Benzene, 50% Ethanol, 0.3M LiCl	Hg	-0.96	sce	12
t-Butyl-peracrylate	50% Benzene, 50% Ethanol, 0.3M LiCl	Hg	-1.02	sce	13
t-Butyl-percrotonate	50% Benzene, 50% Ethanol, 0.3M LiCl	Hg	-0.97	sce	13
t-Butyl-percapronate	50% Benzene, 50% Ethanol, 0.3M LiCl	Hg	-1.08	sce	13
Catalytically-cracked gasoline[a]	50% Benzene, 50% Methanol, 0.3M LiCl	Hg	-0.881	sce	8

(continued)

789

Compound	Solvent System	Working Electrode	$E_{1/2}$	Reference Electrode	Reference
Thermally-cracked gasoline[a]	50% Benzene, 50% Methanol, 0.3$\underline{M}$ LiCl	Hg	-0.887	sce	8
Platformate gasoline[a]	50% Benzene, 50% Methanol, 0.3$\underline{M}$ LiCl	Hg	-0.974	sce	8

[a] Included for comparison.

References for Peroxides, Peracids and Peresters

1. H. Brüschweiler and G. J. Minkoff, Anal. Chim. Acta, 12, 186 (1955).
2. E. J. Kuta and F. W. Quackenbush, Anal. Chem., 32, 1069 (1960).
3. A. V. Yamshchikov and E. S. Levin, Elektrokhim., 6, 588 (1970).
4. D. A. Skoog and A. B. H. Lauwzecha, Anal. Chem., 28, 825 (1956).
5. V. L. Antonovskii, Z. S. Frolova, E. V. Skorobog-atova, and M. M. Buzlanova, Zh. Obshch. Khim., 39, 518 (1969).
6. C. O. Willits, C. Ricciuti, H. B. Knight, and D. Swern, Anal. Chem., 24, 735 (1952).
7. E. S. Levin and A. V. Yamshchikov, Elektrokhim., 4, 54 (1968).
8. M. L. Whisman and B. H. Eccleston, Anal. Chem., 30, 1638 (1958).
9. V. L. Antonovskii, Z. S. Frolova, T. T. Shleina, and M. M. Buglanova, Zh. Obshch. Khim., 39, 368 (1969).
10. J. Heyrovský and J. Kuta, Principles of Polaro-graphy, Academic, New York 1966, p. 562.
11. W. E. Parker, C. Ricciuti, C. L. Ogg, and D. Swern, J. Am. Chem. Soc., 77, 4037 (1955).
12. Z. S. Frolova, M. M. Buzlanova, O. N. Romantsova, and V. L. Antonovskii, Zh. Obshch. Khim., 38, 1944 (1968).
13. V. L. Antonovskii, Z. S. Frolova, E. V. Skoroboga-tova, O. N. Romantsova, and M. M. Buzlanova, Zh. Obshch. Khim., 38, 1948 (1968).

17. ETHERS

Oxidation

Methyl Ethers

Compound	Solvent System	Working Electrode	$E_{1/2}$	Reference Electrode	Reference
Methoxybenzene (Anisole)	CH_3CN, $(n\text{-Propyl})_4$-$NClO_4$	Pt	+1.76	sce	1
	CH_3CN, $NaClO_4$	Pt	+1.35	sce	2
	CH_3CN, $(Et)_4NClO_4$	Pt	+1.7 E_p	sce	3
	$HOAc$, $NaOAc$	Pt	+1.67	sce	4
1,2-Dimethoxybenzene	CH_3CN, $(n\text{-Propyl})_4$-$NClO_4$	Pt	+1.45	sce	1
1,3-Dimethoxybenzene	CH_3CN, $(Et)_4NCN$	Pt	+1.38	sce	3
1,4-Dimethoxybenzene	CH_3CN, $(n\text{-Propyl})_4$-$NClO_4$	Pt	+1.34	sce	1
	CH_3CN, $(Et)_4NClO_4$	Pt	+1.15; +1.75 E_p	sce	3
1,2,3-Trimethoxy-benzene	CH_3CN, $(n\text{-Propyl})_4$-$NClO_4$	Pt	+1.42	sce	1
1,2,4-Trimethoxy-benzene	CH_3CN, $(n\text{-Propyl})_4$-$NClO_4$	Pt	+1.12	sce	1
1,3,5-Trimethoxy-benzene	CH_3CN, $(n\text{-Propyl})_4$-$NClO_4$	Pt	+1.49	sce	1

Compound	Solvent, Electrolyte	Electrode	E	Ref. electrode	Ref.
1,2,3,4-Tetramethoxybenzene	CH_3CN, (n-Propyl)$_4$NClO$_4$	Pt	+1.25	sce	1
1,2,4,5-Tetramethoxybenzene	CH_3CN, (n-Propyl)$_4$NClO$_4$	Pt	+0.81	sce	1
Pentamethoxybenzene	CH_3CN, (n-Propyl)$_4$NClO$_4$	Pt	+1.07	sce	1
Hexamethoxybenzene	CH_3CN, (n-Propyl)$_4$NClO$_4$	Pt	+1.24	sce	1
4-Methoxystyrene	CH_3CN, (Et)$_4$NClO$_4$	Pt	+1.45, +1.60 to +1.80	Ag,AgCl	5
3,4-Dimethoxypropenylbenzene	CH_3CN, (Et)$_4$NClO$_4$	Pt	+1.37, +1.62	sce	6
	CH_3CN, NaClO$_4$	Pt	+0.98, +1.2, +1.4	sce	7
4,4'-Dimethoxystilbene	CH_3CN, NaClO$_4$[a]	Pt	+1.09	sce	7
	CH_3CN, LiClO$_4$	Pt	+0.90, +1.15	sce	8
Methyl ether of hydroquinone	0.1N LiCl, 0.05M Potassium Acid Phthalate, 80% Alcohol	C	+0.74 E_p	NCE	9
Durohydroquinone dimethyl ether	CH_3CN, LiClO$_4$	Pt	+1.28 E_p	sce	10
1-Methoxynaphthalene	CH_3CN, (n-Propyl)$_4$NClO$_4$	Pt	+1.38	sce	11
2-Methoxynaphthalene	CH_3CN, (n-Propyl)$_4$NClO$_4$	Pt	+1.52	sce	11
1,3-Dimethoxynaphthalene	CH_3CN, (n-Propyl)$_4$NClO$_4$	Pt	+1.27	sce	11
1,4-Dimethoxynaphthalene	CH_3CN, (n-Propyl)$_4$NClO$_4$	Pt	+1.10	sce	11

(continued)

793

Compound	Solvent System	Working Electrode	$E_{1/2}$	Reference Electrode	Reference
1,5-Dimethoxy-naphthalene	CH_3CN, $(\underline{n}$-Propyl$)_4$-$NClO_4$	Pt	+1.28	sce	11
1,6-Dimethoxy-naphthalene	CH_3CN, $(\underline{n}$-Propyl$)_4$-$NClO_4$	Pt	+1.28	sce	11
1,7-Dimethoxy-naphthalene	CH_3CN, $(\underline{n}$-Propyl$)_4$-$NClO_4$	Pt	+1.28	sce	11
1,8-Dimethoxy-naphthalene	CH_3CN, $(\underline{n}$-Propyl$)_4$-$NClO_4$	Pt	+1.17	sce	11
2,3-Dimethoxy-naphthalene	CH_3CN, $(\underline{n}$-Propyl$)_4$-$NClO_4$	Pt	+1.39	sce	11
2,6-Dimethoxy-naphthalene	CH_3CN, $(\underline{n}$-Propyl$)_4$-$NClO_4$	Pt	+1.33	sce	11
2,7-Dimethoxy-naphthalene	CH_3CN, $(\underline{n}$-Propyl$)_4$-$NClO_4$	Pt	+1.47	sce	11
1,4,5,8-Tetra-methoxynaphthalene	CH_3CN, $(\underline{n}$-Propyl$)_4$-$NClO_4$	Pt	+0.70	sce	11
1,5-Dimethoxy-4,8-diphenoxynaph-thalene	CH_3CN, $(\underline{n}$-Propyl$)_4$-$NClO_4$	Pt	+0.98	sce	11
9-Methoxyanthracene	CH_3CN, $(\underline{n}$-Propyl$)_4$-$NClO_4$	Pt	+1.05	sce	11
9,10-Dimethoxyan-thracene	CH_3CN, $(\underline{n}$-Propyl$)_4$-$NClO_4$	Pt	+0.98	sce	11
9,10-Bis(2,6-dimethoxyphenyl)-anthracene	CH_3CN, $(\underline{n}$-Propyl$)_4$-$NClO_4$	Pt	+1.18	sce	11
4-Methoxybiphenyl	CH_3CN, $(\underline{n}$-Propyl$)_4$-$NClO_4$	Pt	+1.53	sce	11

Compound	Solvent, Electrolyte	Electrode	Potential	Reference	Ref.
4,4'-Dimethoxy-biphenyl	CH_3CN, $NClO_4$	Pt	+1.30	sce	11
3,3'-Dimethoxy-biphenyl	CH_3CN, $NClO_4$	Pt	+1.60	sce	11
2,2'-Dimethoxy-biphenyl	CH_3CN, $NClO_4$	Pt	+1.51	sce	11
10,10'-Dimethoxy-9,9'-bianthracenyl	CH_3CN, $NClO_4$	Pt	+1.10	sce	11
1,6-Dimethoxypyrene	CH_3CN, $NClO_4$	Pt	+0.82	sce	11

Vinyl Ethers

Compound	Solvent, Electrolyte	Electrode	Potential	Reference	Ref.
Vinyl isobutyl ether	CH_3CN, $(Et)_4NClO_4$	Pt	+2.07 to 2.11	Ag, AgCl	5
Vinyl acetate	CH_3CN, $(Et)_4NClO_4$	Pt	No wave	Ag, AgCl	5

Phenyl Ethers

Compound	Solvent, Electrolyte	Electrode	Potential	Reference	Ref.
9,10-Diphenoxy-anthracene	CH_3CN, $NClO_4$	Pt	+1.20	sce	11

Aryl Acetates

Compound	Solvent, Electrolyte	Electrode	Potential	Reference	Ref.
Phenyl acetate	HOAc, NaOAc	Pt	+1.30	sce	4
o-Acetoxyanisole	HOAc, NaOAc	Pt	+1.74	sce	4
m-Acetoxyanisole	HOAc, NaOAc	Pt	+1.25	sce	4
p-Acetoxyanisole	HOAc, NaOAc	Pt	+1.12	sce	4
1,2-Diacetoxybenzene	HOAc, NaOAc	Pt	+2.5	sce	4
1,3-Diacetoxybenzene	HOAc, NaOAc	Pt	+1.46	sce	4
1,4-Diacetoxybenzene	HOAc, NaOAc	Pt	+2.5	sce	4

(continued)

795

Compound	Solvent System	Working Electrode	$E_{1/2}$	Reference Electrode	Reference
1-Acetoxy-naphthalene	HOAc, NaOAc	Pt	+1.67	sce	4
2-Acetoxy-naphthalene	HOAc, NaOAc	Pt	+1.86	sce	4
2-Acetoxybiphenyl	HOAc, NaOAc	Pt	+2.04	sce	4
4-Acetoxybiphenyl	HOAc, NaOAc	Pt	+1.90	sce	4
Durohydroquinone diacetate	CH_3CN, $LiClO_4$	Pt	+1.90 E_p	sce	10

a 1.0$\underline{M}$ Py.

Reduction

Compound	Solvent System	Working Electrode	$E_{1/2}$	Reference Electrode	Reference
Vinylpropyl, Vinylbutyl, Vinylisoamyl, Vinylphenyl ethers show no reduction waves in 75% Ethanol, (Me)4NI					12

9,10-Epoxyoctadecanol, 9,10-Epoxystearic acid, Methyl 9,10-Epoxystearate, 1,2-Epoxydecane, 3,4-Epoxy-1-butene and 1-Phenyl-1,2-epoxyethane show no reduction waves in 1:1 Benzene-Methanol, 0.3$\underline{M}$ LiCl

Thioethers are tabulated
Sulfur Functions (Section 27)

References for Ethers

1. A. Zweig, W. G. Hodgson and W. H. Jura, J. Am. Chem. Soc., 86, 4124 (1964).
2. H. Lund, Acta Chem. Scand., 11, 1323 (1957).
3. S. Andreades and E. W. Zahnow, J. Am. Chem. Soc., 91, 4181 (1969).
4. L. Eberson and K. Nyberg, J. Am. Chem. Soc., 88, 1686 (1966).
5. J. W. Breitenbach and F. Sommer, Abstracts CITCE Meeting, Prague, Czechoslovakia, 1970.
6. U. Kuenkel, R. Landsberg, and S. Mueller, Z. Chem., 10, 303 (1970).
7. J. J. O'Connor and I. A. Pearl, J. Electrochem. Soc., 111, 335 (1964).
8. V. D. Parker and L. Eberson, Chem. Commun., 1969, 340.
9. J. R. Covington and R. J. Lacoste, Anal. Chem., 37, 420 (1965).
10. V. D. Parker, Chem. Commun., 1969, 610.
11. A. Zweig, A. H. Maurer, and B. G. Roberts, J. Org. Chem., 32, 1322 (1967).
12. M. I. Bobrova, and A. N. Matveeva, Zh. Obshch. Khim., 24, 1713 (1954).

18. CARBOHYDRATE LACTONES

Reduction

Compound	Solvent System	Working Electrode	$E_{1/2}$	Reference Electrode	Reference
l-Threono-γ-lactone	0.1M (Et)₄NCl	Hg	-2.54	Hg pool	1
d-Xylono-γ-lactone	0.1M (Et)₄NCl	Hg	-2.48 to -2.56	Hg pool	1
d-Glucono-γ-lactone	0.1M (Et)₄NCl	Hg	-2.15 to -2.30	Hg pool	1
d-β-Glucohepto-γ-lactone	0.1M (Et)₄NCl	Hg	-2.53	Hg pool	1
l-Arabono-γ-lactone	0.1M (Et)₄NCl	Hg	-2.60	Hg pool	1
d-Galactono-γ-lactone	0.1M (Et)₄NCl	Hg	-2.62	Hg pool	1
d-α-Mannoheptono-γ-lactone	0.1M (Et)₄NCl	Hg	-2.65	Hg pool	1
d-β-Guloheptono-γ-lactone	0.1M (Et)₄NCl	Hg	-2.6	Hg pool	1
l-Ribono-γ-lactone	0.1M (Et)₄NCl	Hg	-2.50	Hg pool	1
l-Allono-γ-lactone	0.1M (Et)₄NCl	Hg	-2.52	Hg pool	1
d-α-Guloheptono-γ-lactone	0.1M (Et)₄NCl	Hg	-2.55	Hg pool	1
d-β-Mannoheptone-γ-lactone	0.1M (Et)₄NCl	Hg	-2.50 to -2.56	Hg pool	1

Compound					
d-Erythrono-γ-lactone	0.1M (Et)$_4$NCl	Hg	-2.65	Hg pool	1
d-Lyxono-γ-lactone	0.1M (Et)$_4$NCl	Hg	-2.6	Hg pool	1
d-Mannono-γ-lactone	0.1M (Et)$_4$NCl	Hg	-2.65	Hg pool	1
L-Rhamnono-γ-lactone	0.1M (Et)$_4$NCl	Hg	-2.68	Hg pool	1
d-Gulono-γ-lactone	0.1M (Et)$_4$NCl	Hg	-2.60	Hg pool	1
d-Galaheltono-γ-lactone	0.1M (Et)$_4$NCl	Hg	-2.65	Hg pool	1
d-α-Glucoheptono-γ-lactone	0.1M (Et)$_4$NCl	Hg	-2.56 to -2.64	Hg pool	1

(continued)

References for Carbohydrate Lactones

1.   H. Matheson, H. S. Isbell, and E. R. Smith, <u>J.</u>
     <u>Res. Natl. Bur. Standards</u>, <u>28</u>, 95 (1942).

19. ALIPHATIC AMINES

Oxidation

Compound	Solvent System	Working Electrode	$E_{1/2}$	Reference Electrode	Reference
Hydroxylamine	Borate buffer, pH 9.18	Hg	-0.024	sce	1
n-Propylamine	CH_3CN, $NaClO_4$	Pt	+1.63 E_p	NHE	2
n-Butylamine	CH_3CN, $NaClO_4$	Pt	+1.63 E_p	NHE	2
i-Butylamine	CH_3CN, $NaClO_4$	Pt	+1.62 E_p	NHE	2
t-Butylamine	CH_3CN, $NaClO_4$	Pt	+1.64 E_p	NHE	2
n-Pentylamine	CH_3CN, $NaClO_4$	Pt	+1.69 E_p	NHE	2
n-Nonylamine	CH_3CN, $NaClO_4$	Pt	+1.72 E_p	NHE	2
Dimethylamine	B-R buffer, pH 11.9	Glassy C	+1.03 E_p	sce	3
N,N-Dimethyl-ethylamine	B-R buffer, pH 11.9	Glassy C	+0.74, +1.00 E_p	sce	3
N,N-Dimethyl-n-propylamine	B-R buffer, pH 11.9	Glassy C	+0.75, +0.98 E_p	sce	3
N,N-Dimethyl-i-propylamine	B-R buffer, pH 11.9	Glassy C	+0.72, +0.99 E_p	sce	3
N,N-Dimethyl-t-butylamine	B-R buffer, pH 11.9	Glassy C	+0.70 E_p	sce	3
N,N-Dimethylglycine	B-R buffer, pH 11.9	Glassy C	+0.73, +0.97 E_p	sce	3

(continued)

Compound	Solvent System	Working Electrode	$E_{1/2}$	Reference Electrode	Reference
Dimethylaminoaceto-nitrile	B-R buffer, pH 11.9	Glassy C	+0.96, E_p / +1.01, E_p	sce	3
N,N-Dimethylglycin-amide	CH₃CN, NaClO₄	Pt	+1.59 E_p	NHE	2
	B-R buffer, pH 11.9	Glassy C	+0.93 E_p	sce	3
N,N-Dimethylethanol-amine	B-R buffer, pH 11.9	Glassy C	+0.76, +0.99 E_p	sce	3
N,N-Dimethylbenzyl-amine	B-R buffer, pH 11.9	Glassy C	+0.74 E_p	sce	3
	CH₃CN, LiClO₄	Pt	+0.92	sce	4
	DMF, (n-Bu)₄-NClO₄	Pt	+1.08	sce	5
N,N-Dimethyl-ethylenediamine	B-R buffer, pH 11.9	Glassy C	+0.77, +1.01 E_p	sce	3
N-Methyl-di-n-propylamine	B-R buffer, pH 11.9	Glassy C	+0.65, +0.91 E_p	sce	3
Diethylamine	B-R buffer, pH 11.9	Glassy C	+1.00 E_p	sce	3
Di-n-propylamine	CH₃CN, NaClO₄	Pt	+1.89	Ag/0.1N Ag⁺	6
	B-R buffer, pH 11.9	Glassy C	+0.90 E_p	sce	3
Di-n-butylamine	CH₃CN, NaClO₄	Pt	+1.26 E_p	NHE	2
Di-s-butylamine	CH₃CN, NaClO₄	Pt	+1.31 E_p	NHE	2
Di-n-pentylamine	CH₃CN, NaClO₄	Pt	+1.40 E_p	NHE	2
Dibenzylamine	CH₃CN, NaClO₄	Pt	+1.35 E_p	NHE	2
Trimethylamine	CH₃CN, NaClO₄	Pt	+1.49 E_p	NHE	2
	CH₃CN, NaClO₄	Pt	+1.29 E_p	NHE	2
	B-R buffer, pH 11.9	Glassy C	+0.76, +0.98 E	sce	3

Compound	Solvent, electrolyte	Electrode	E	Reference	Ref.
Triethylamine	B-R buffer, pH 11.9	Glassy C	+0.69, +1.00 E_p	sce	3
	CH_3CN, $NaClO_4$	Pt	+1.19 E_p	NHE	2
	CH_3CN, $NaClO_4$	Pt	+0.85, +1.88	Ag/0.1N Ag+	6
	$DMSO$, $Pb(NO_3)_2$	Pt	+1.470a $E_{1/4}$	Pb/Pb2+	7
Tripropylamine	CH_3CN, $NaClO_4$	Pt	+1.02 E_p	NHE	2
Tri-n-butylamine	CH_3CN, $NaClO_4$	Pt	+1.02 E_p	NHE	2
Tribenzylamine	CH_3CN, $NaClO_4$	Pt	+1.27 E_p	NHE	2
Tripentylamine	CH_3CN, $NaClO_4$	Pt	+1.13 E_p	NHE	2
Triethylenediamine	CH_3CN	Pt	+0.68	sce	8

Vinylamines (Enamines)b

Compound	Solvent, electrolyte	Electrode	E	Reference	Ref.
1,1,4,4-Tetrakis-(dimethylamine)-2,3-dimethylbuta-diene	CH_3CN, $(Et)_4NClO_4$	Hg	-0.901	sce	9
Tetrakis(dimethyl-amino)ethylene	CH_3CN, $(Et)_4NClO_4$	Pt	>-0.78, <-1.02	sce	9
1,1,4,4-Tetrakis-(dimethylamino)-butadiene	CH_3CN, $(Et)_4NClO_4$	Pt	-0.77, -0.65	sce	9
	CH_3CN, $(Et)_4NClO_4$	Hg	-0.615	sce	9
1,4-Bis(dimethyl-amino)-1,4-diphenylbutadiene	CH_3CN, $(Et)_4NClO_4$	Pt	-0.58, -0.66	sce	9
	CH_3CN, $(Et)_4NClO_4$	Pt	-0.40, -0.17	sce	9

(continued)

803

Compound	Solvent System	Working Electrode	$E_{1/2}$	Reference Electrode	Reference
2,5-Bis(dimethyl-amino)-2,4-hexadiene	CH_3CN, $(Et)_4NClO_4$	Pt	-0.36, -0.18	sce	9
1,4-Bis(dimethyl-amino)-1,4-diphenylbutadiene	CH_3CN, $(Et)_4NClO_4$	Pt	-0.20, +0.02	sce	9
1,3-Bis[bis(di-methylamino)-methylene] cyclobutane	CH_3CN, $(Et)_4NClO_4$	Pt	0.00	sce	9
1,1,5,5-Tetrakis-(dimethylamino)-1,4-pentadiene	CH_3CN, $(Et)_4NClO_4$	Pt	0.00	sce	9
2-Methylpropyl-idenebis(dimethyl-amine)	CH_3CN, $(Et)_4NClO_4$	Pt	+0.05	sce	9
Propylidenebis-(dimethylamine)	CH_3CN, $(Et)_4NClO_4$	Pt	+0.23	sce	9
9-[Bis(dimethyl-amino)methylene]-fluorene	CH_3CN, $(Et)_4NClO_4$	Pt	+0.24	sce	9
9-(1-Dimethylamino)-ethylidenefluorene	CH_3CN, $(Et)_4NClO_4$	Pt	+0.32	sce	9
Tetrakis(dimethyl-amino)ethylene	CH_3CN, $(Et)_4NClO_4$	Hg	-0.75, -0.61	sce	10
1,1,4,4-Tetrakis-(dimethylamino)-butadiene	DMF, $(Et)_4NClO_4$	Pt	-0.58 $E°$	sce	11

Compound	Conditions	Electrode	$E_{1/2}$		Ref.
Tetrakis-(p-di-methylaminophenyl)-ethylene	CH_3CN, $(n\text{-}Bu)_4$-$NClO_4$	Hg	+0.25	sce	12
2,6-Dimethylcyclo-hexenyl-1-dimethyl-amine	CH_3CN, $(Et)_4$-$NClO_4$	Pt	+0.38	sce	9
1-Cyclohexenyldi-methylamine	CH_3CN, $(Et)_4$-$NClO_4$	Pt	+0.42	sce	9
Vinylidene bis(di-methylamine)	CH_3CN, $(Et)_4$-$NClO_4$	Pt	+0.48	sce	9
9-(Dimethylamino)-methylenefluorene	CH_3CN, $(Et)_4$-$NClO_4$	Pt	+0.48	sce	9
2,2-Dichlorovinyl-idenebis(dimethyl-amine)	CH_3CN, $(Et)_4$-$NClO_4$	Pt	+0.52	sce	9
2,3-Bis(dimethyl-amino)butadiene	CH_3CN, $(Et)_4$-$NClO_4$	Pt	+0.58, +1.10	sce	9
Tetrakis(dimethyl-amino)methane	CH_3CN, $(Et)_4$-$NClO_4$	Pt	+0.65	sce	9
2,2-Difluorovinyl-idenebis(dimethyl-amine)	CH_3CN, $(Et)_4$-$NClO_4$	Pt	+0.7	sce	9
1-Dimethylamino-styrene	CH_3CN, $(Et)_4$-$NClO_4$	Pt	+0.70	sce	9
1-Methyl-3-dimethyl-amino-2-butenyl-idenedimethyl-ammonium fluorobor-ate	CH_3CN, $(Et)_4$-$NClO_4$	Pt	+1.30	sce	9

a Current density 0.842 ma/cm^2.
b Some of these compounds are oxidized extremely easily -- yielding negative $E_{1/2}$'s.

(continued)

References for Aliphatic Amines

1. G. R. Rao and L. Meites, _J. Phys. Chem._, _70_, 3620 (1966).
2. C. K. Mann, _Anal. Chem._, 36, 2424 (1964).
3. M. Masui, H. Sayo, and Y. Tsuda, _J. Chem. Soc._ _B_, 1968, 973.
4. N. L. Weinberg, _J. Org. Chem._, _33_, 4326 (1968).
5. N. L. Weinberg, unpublished results.
6. J. W. Loveland and G. R. Dimeler, _Anal. Chem._, _33_, 1196 (1961).
7. R. F. Dapo and C. K. Mann, _Anal. Chem._, _35_, 677 (1963).
8. T. M. McKinney and D. H. Geske, _J. Am. Chem. Soc._, _87_, 3013 (1965).
9. J. M. Fritsch, H. Weingarten, and J. D. Wilson, _J. Am. Chem. Soc._, _92_, 4038 (1970).
10. K. Kuwata and D. H. Geske, _J. Am. Chem. Soc._, _86_, 2101 (1964).
11. J. M. Fritsch and H. Weingarten, _J. Am. Chem. Soc._, _90_, 793 (1968).
12. R. M. Elofson and K. F. Schulz, _Can. J. Chem._, _47_, 4447 (1969).

20. AROMATIC AMINES

Oxidation

Compound	Solvent System	Working Electrode	$E_{1/2}$	Reference Electrode	Reference
Aniline	CH_3CN, $(Et)_4$NClO$_4$	Pt	+0.98	sce	1
	CH_3CN, NaClO$_4$a	Pt	+0.70	Ag/0.1N $\underline{Ag^+}$	2
	CH_3CN, NaClO$_4$	Pt	+0.54	Ag/0.1N $\underline{Ag^+}$	2
	B-R buffer, pH=pKa amine	C paste	+0.72 $E_{p/2}$	sce	3
	C-L buffer, pH 1.0	Graphite	+0.864	sce	4
	50% OAc buffer, 50% Methanol	Pt	+0.982b	sce	5
Mono-Substituted Anilines					
o-Methylaniline (o-Toluidine)	CH_3CN, $(Et)_4$NClO$_4$	Pt	+0.85	sce	1
	C-L buffer, pH 1.0	Graphite	+0.804	sce	4
	50% OAc buffer, 50% Isopropanol, pH 5.6	Graphite	+0.573	sce	6

(continued)

Compound	Solvent System	Working Electrode	$E_{1/2}$	Reference Electrode	Reference
	B-R buffer, pH=pKa amine	C paste	+0.67 $E_{p/2}$	sce	3
	50% OAc buffer, 50% Methanol	Pt	+0.982[b]	sce	5
m-Methylaniline (m-Toluidine)	CH3CN, (Et)4-NClO4	Pt	+0.84	sce	1
	C-L buffer, pH 1.0	Graphite	+0.829	sce	4
	50% OAc buffer, 50% Isopropanol, pH 5.6	Graphite	+0.606	sce	6
p-Methylaniline (p-Toluidine)	CH3CN, (Et)4-NClO4	Pt	+0.78, +1.07	sce	1
	C-L buffer, pH 1.0	Graphite	+0.780	sce	4
	50% OAc buffer, 50% Isopropanol, pH 5.6	Graphite	+0.537	sce	6
N-Methylaniline	B-R buffer, Na2-SO4, pH 2.4	C paste	+0.7 $E_{p/2}$	sce	7
	CH3CN, (Et)4-NClO4	Pt	+0.77	sce	1
	50% OAc buffer, 50% Methanol	Pt	+0.822[b]	sce	5
o-Ethylaniline	50% OAc buffer, 50% Methanol	Pt	+0.942[b]	sce	5
o-Isopropylaniline	50% OAc buffer, 50% Methanol	Pt	+0.960[b]	sce	5

Compound	Solvent/Electrolyte	Electrode	Potential	Reference	Ref.
p-t-Butylaniline	50% OAc buffer, 50% Methanol	Pt	+0.940^b	sce	5
o-Nitroaniline	CH$_3$CN, NaClO$_4$	Pt	+1.07	Ag/0.1N Ag+	8
	50% OAc buffer, 50% Isopropanol, pH 5.6	Graphite	+0.989	sce	6
m-Nitroaniline	CH$_3$CN, NaClO$_4$	Pt	+0.90	Ag/0.1N Ag+	8
	50% OAc buffer, 50% Isopropanol, pH 5.6	Graphite	+0.854	sce	6
p-Nitroaniline	CH$_3$CN, NaClO$_4$	Pt	+0.97, +1.14	Ag/0.1N Ag+	8
	50% OAc buffer, 50% Isopropanol, pH 5.6	Graphite	+0.935	sce	6
	B-R buffer, pH=pKa amine	C paste	+1.07 $E_{p/2}$	sce	3
m-Bromoaniline	CH$_3$CN, NaClO$_4$	Pt	+0.70	Ag/0.1N Ag+	8
	CH$_3$CN, (Et)$_4$-NClO$_4$	Pt	+1.08	sce	1
p-Bromoaniline	CH$_3$CN, NaClO$_4$	Pt	+0.61	Ag/0.1N Ag+	8
	CH$_3$CN, (Et)$_4$-NClO$_4$	Pt	+0.97	sce	1
o-Chloroaniline	50% OAc buffer, 50% Isopropanol, pH 5.6	Graphite	+0.742	sce	6
m-Chloroaniline	50% OAc buffer, 50% Isopropanol, pH 5.6	Graphite	+0.774	sce	6
	CH$_3$CN, (Et)$_4$-NClO$_4$	Pt	+1.05	sce	1

(continued)

Compound	Solvent System	Working Electrode	$E_{1/2}$	Reference Electrode	Reference
p-Chloroaniline	B-R buffer, pH=pKa amine	C paste	+0.73	sce	3
	CH$_3$CN, NaClO4	Pt	+0.60	Ag/0.1N Ag+	8
	CH$_3$CN, (Et)4-NClO4	Pt	+0.94	sce	1
	OAc buffer, 50% Isopropanol, pH 5.6	Graphite	+0.675	sce	6
o-Methoxyaniline (o-Anisidine)	CH$_3$CN, NaClO4	Pt	+0.34	Ag/0.1N Ag+	8
	OAc buffer, 50% Isopropanol, pH 5.6	Graphite	+0.498	sce	6
m-Methoxyaniline (m-Anisidine)	OAc buffer, 50% Isopropanol, pH 5.6	Pt	+0.615	sce	6
p-Methoxyaniline (p-Anisidine)	CH$_3$CN, NaClO4	Pt	+0.26	Ag/0.1N Ag+	8
	OAc buffer, 50% Isopropanol, pH 5.6	Graphite	+0.393	sce	6
p-Ethoxyaniline (p-Phenetidine)	B-R buffer, pH=pKa amine	C paste	+0.44 $E_{p/2}$	sce	3
	B-R buffer, pH=pKa amine	C paste	+0.46 $E_{p/2}$	sce	3
p-Cyanoaniline	B-R buffer, pH=pKa amine	C paste	+1.03 $E_{p/2}$	sce	3

Di-Substituted Anilines

2,4-Dimethylaniline	CH$_3$CN, (Et)$_4$-NClO$_4$	Pt	+0.70, +1.00	sce	1
2,5-Dimethylaniline	CH$_3$CN, (Et)$_4$-NClO$_4$	Pt	+0.79	sce	1
3,5-Dimethylaniline	CH$_3$CN, (Et)$_4$-NClO$_4$	Pt	+0.81	sce	1
2,6-Diethylaniline	50% OAc buffer, 50% Methanol	Pt	+0.882[b]	sce	5
2-Ethyl-p-toluidine	50% OAc buffer, 50% Methanol	Pt	+0.860[b]	sce	5
2-Isopropyl-p-toluidine	50% OAc buffer, 50% Methanol	Pt	+0.854[b]	sce	5
2-t-Butyl-p-toluidine	50% OAc buffer, 50% Methanol	Pt	+0.862[b]	sce	5
6-Ethyl-o-toluidine	50% OAc buffer, 50% Methanol	Pt	+0.891[b]	sce	5
6-Isopropyl-o-toluidine	50% OAc buffer, 50% Methanol	Pt	+0.884[b]	sce	5
4-t-Butyl-o-toluidine	50% OAc buffer, 50% Methanol	Pt	+0.897[b]	sce	5
N,N-Dimethylaniline	B-R buffer, pH 2.4	C paste	+0.680 E$_{p/2}$	sce	7
	CH$_3$CN, NaClO$_4$	Pt	+0.68	sce	9
	CH$_3$CN, (Et)$_4$-NClO$_4$	Pt	+0.73	sce	1
	CH$_3$CN, (Et)$_4$-NClO$_4$		+0.71 E$_{p/2}$	sce	10

(continued)

Compound	Solvent System	Working Electrode	$E_{1/2}$	Reference Electrode	Reference
	CH_3CN, $(Et)_4$-$NClO_4$	Pt	+0.72 $E_{p/2}$	sce	11
	50% OAc buffer, 50% Methanol	Pt	+0.864[b]	sce	5
N,N-Dimethylaniline	DMF, $(n\text{-}Bu)_4$-$NClO_4$	Pt	+0.88	sce	12
N,N-Diethylaniline	CH_3CN, $NaClO_4$	Pt	+0.34	Ag/0.1N Ag+	8
	CH_3CN, $(Et)_4$-$NClO_4$	Pt	+0.70 $E_{p/2}$	sce	11
N,N-Di-n-butyl-aniline	CH_3CN, $(Et)_4$-$NClO_4$	Pt	+0.69 $E_{p/2}$	sce	11
N,N-Di-s-butyl-aniline	CH_3CN, $(Et)_4$-$NClO_4$	Pt	+0.70 $E_{p/2}$	sce	11
N,N-Di-n-decyl-aniline	CH_3CN, $(Et)_4$-$NClO_4$	Pt	+0.69 $E_{p/2}$	sce	11
N,N-Dimethyl-p-anisidine	CH_3CN, $(Et)_4$-$NClO_4$	Pt	+0.50 $E_{p/2}$	sce	11
o-Ethyl-N-methyl-aniline	50% OAc buffer, 50% Methanol	Pt	+0.859[b]	sce	5
o-Isopropyl-N-methylaniline	50% OAc buffer, 50% Methanol	Pt	+0.877[b]	sce	5
2,6-Diisopropyl-aniline	50% OAc buffer, 50% Methanol	Pt	+0.880[b]	sce	5
2,4-Dinitroaniline	CH_3CN, $NaClO_4$	Pt	+1.48	Ag/0.1N Ag+	8
2,4-Dichloroaniline	CH_3CN, $NaClO_4$	Pt	+0.78	Ag/0.1N Ag+	8
p-Amino-N,N-dialkylanilines- -60 compounds	Aq. solution, pH 11	Pt			12A

Compound	Solvent / Electrolyte	Electrode	E	Reference electrode	Ref.
o-Dianisidine	1M H₂SO₄	B₄C	+0.594	sce	13
	1M H₂SO₄	Pt	+0.592	sce	13
	B–R buffer, Na₂SO₄, pH 2.4	C paste	+0.46	sce	14

Tri-Substituted Aniline

Compound	Solvent / Electrolyte	Electrode	E	Reference electrode	Ref.
2,6-Diethyl-N-methylaniline	50% OAc buffer, 50% Methanol	Pt	+0.828[b]	sce	5
2,6-Diethyl-p-toluidine	50% OAc buffer, 50% Methanol	Pt	+0.880[b]	sce	5
2,6-Diisopropyl-p-toluidine	50% OAc buffer, 50% Methanol	Pt	+0.812[b]	sce	5
2,6-Di-t-butyl-p-toluidine	50% OAc buffer, 50% Methanol	Pt	+0.816[b]	sce	5
2,4,6-Tri-t-butyl-aniline	50% OAc buffer, 50% Methanol	Pt	+0.762[b]	sce	5
2,4,6-Tri-t-butyl-aniline	50% OAc buffer, 50% Methanol	Pt	+0.774[b]	sce	5
	CH₃CN, LiClO₄	Pt	+0.530	Ag/0.1N Ag+	15, 16
	Methanol, LiClO₄, MgO[c]	Pt	+0.230	Ag/0.1N Ag+	15
	CH₃CN, LiClO₄, Py	Pt	+0.515	Ag/o.1N Ag+	15
	CH₃NO₂, LiClO₄	Pt	+1.10	Ag/0.1N Ag+	16
N,N-Dimethyl-p-chloroaniline	CH₃CN, (Et)₄-NClO₄	Pt	+0.84 $E_{p/2}$	sce	10
N,N-Dimethyl-p-nitroaniline	CH₃CN, (Et)₄-NClO₄	Pt	+1.19 $E_{p/2}$	sce	10
N,N-Dimethyl-o-anisidine	CH₃CN, (n-Propyl)₄-NClO₄	Pt	+0.48	Ag/0.1N Ag+	17

(continued)

Compound	Solvent System	Working Electrode	$E_{1/2}$	Reference Electrode	Reference
N,N-Dimethyl-m-anisidine	CH₃CN, (n-Propyl)₄-NClO₄	Pt	+0.49	Ag/0.1$\underline{N}$ Ag+	17
N,N-Dimethyl-p-anisidine	CH₃CN, (n-Propyl)₄-NClO₄	Pt	+0.33	Ag/0.1$\underline{N}$ Ag+	17
	CH₃CN, (Et)₄-NClO₄	Pt	+0.50 $E_{p/2}$	sce	11
	CH₃CN, (Et)₄-NClO₄	Pt	+0.49 $E_{p/2}$	sce	10
	B-R buffer, pH=pKa amine	C paste	+0.40 $E_{p/2}$	sce	3
N,N-Dimethyl-p-toluidine	CH₃CN, (Et)₄-NClO₄	Pt	+0.65 $E_{p/2}$	sce	10
	CH₃CN, (Et)₄-NClO₄	Pt	+0.7	sce	18
N,N-Diethyl-p-chloroaniline	CH₃CN, NaClO₄	Pt	+0.47	Ag/0.1$\underline{N}$ Ag+	8
1,3,5-Trichloro-aniline	CH₃CN, NaClO₄	Pt	+0.95	Ag/0.1$\underline{N}$ Ag+	8
2,4,6-Tribromo-aniline	CH₃CN, LiClO₄	Pt	+1.08, +1.76	Ag/0.1$\underline{N}$ Ag+	19
p-Dimethylamino-benzaldehyde	DMF, (n-Bu)₄-NClO₄	Pt	+1.13	sce	12
2,4,6-Tri-t-butyl-N-methylaniline	50% OAc buffer, 50% Methanol	Pt	+0.862[b]	sce	5
3,4-Dimethoxy-N,N-dimethylaniline	CH₃CN, (n-Propyl)₄-NClO₄	Pt	+0.20	Ag/0.1$\underline{N}$ Ag+	17
3,5-Dimethoxy-N,N-dimethylaniline	CH₃CN, (n-Propyl)₄-NClO₄	Pt	+0.50	Ag/0.1$\underline{N}$ Ag+	17

2,4-Dimethoxy-N,N-dimethylaniline	CH_3CN, $(n\text{-Propyl})_4NClO_4$	Pt	+0.27	$Ag/0.1N\ Ag^+$	17

Diamines

o-Phenylenediamine	OAc buffer, pH 4.5	Graphite	+0.354	Ag, AgCl	20
	C-L buffer, pH 1	Graphite	+0.495	sce	4
	CH_3CN, $NaClO_4$	Pt	+0.06, +0.75	$Ag/0.1N\ Ag^+$	21
	CH_3CN, $NaClO_4$[b]	Pt	+0.25	$Ag/0.1N\ Ag^+$	21
m-Phenylenediamine	OAc buffer, pH 4.5	Graphite	+0.723	Ag, AgCl	20
	C-L buffer, pH 1	Graphite	+0.811	sce	4
	1M HCl	Pt	+0.95, --[d]	sce	22
p-Phenylenediamine	OAc buffer, pH 4.5	Graphite	+0.197, +0.462	Ag, AgCl	20
	C-L buffer, pH 1	Graphite	+0.495	sce	4
	CH_3CN, $NaClO_4$[f]	Pt	-0.16, +0.36	$Ag/0.1N\ Ag^+$	21
	CH_3CN, $NaClO_4$	Pt	-0.15, +0.13	$Ag/0.1N\ Ag^+$	21
	1M HCl	Pt	+0.57, +0.72	sce	22
	Universal buffer[f]	Pt	(+0.72 -0.059 pH[g])	sce	23
Benzidine	50% Phosphate buffer, 50% Ethanol, pH 0.7	Graphite	+0.78	sce	24
	CH_3CN, $NaClO_4$	Pt	+0.475, +0.635	sce	25

(continued)

Compound	Solvent System	Working Electrode	$E_{1/2}$	Reference Electrode	Reference
2,2'-Diaminoazobenzene	CH_3CN, $(Et)_4$-$NClO_4$	Pt	+0.52, +0.73	sce	1
	CH_3CN, $NaClO_4$	Pt	+0.41	Ag/0.1N $\underline{Ag^+}$	21
4,4'-Diaminoazobenzene	CH_3CN, $NaClO_4$	Pt	+0.23	Ag/0.1N $\underline{Ag^+}$	21
2,4-Diaminotoluene	B-R buffer, Na_2SO_4, pH 2.4	C paste	+0.71	sce	14
3,4-Diaminotoluene	B-R buffer, Na_2SO_4, pH 4.4	C paste	+0.28	sce	14
Diphenylbenzidine	CH_3CN, $NaClO_4$	Pt	+0.570, +0.700	sce	19
4-Amino-N,N-diethylaniline	Diethylglycine buffer, pH 11	Pt	+0.25	sce	26
4-Amino-N,N-diethyl-$\underline{m}$-toluidine	Diethylglycine buffer, pH 10	Pt	+0.17	sce	26
N,N',N-Tetramethylbenzidine	Diethylglycine buffer, pH 10	Pt	+0.315, +0.490	sce	25
N,N',N-Tetramethylbenzidine	CH_3CN, ($\underline{n}$-Propyl)$_4$-$NClO_4$	Pt	+0.43	sce	27
	Liquid SO_2 [h]	Pt	+0.49, +0.75, +1.31	Ag, AgBr	28
1,4-Phenylenediamine	Diethylglycine buffer, pH 10	Pt	+0.185, +0.730, +1.040	sce	25
N,N,N',N-Tetramethyl-1,4-phenylenediamine	Diethylglycine buffer, pH 10	Pt	+0.010, +0.595	sce	25

Compound	Supporting electrolyte	Electrode	E (V)	Reference	Ref.
N,N'-Diphenyl-1,4-phenylenediamine	Diethylglycine buffer, pH 10	Pt	+0.345, +0.690, +0.950	sce	25
N,N'-Bis(4-dimethyl-aminophenyl)-1,4-phenylenediamine	Diethylglycine buffer, pH 10	Pt	+0.015, +0.225, +0.590	sce	25
4-Dimethylamino-N-methyldiphenylamine	Diethylglycine buffer, pH 10	Pt	+0.175, +0.685	sce	25
4-Dimethylamino-diphenylamine	Diethylglycine buffer, pH 10	Pt	+0.185, +0.520, +0.910	sce	25
4-Aminodiphenyl-amine	Diethylglycine buffer, pH 10	Pt	+0.275, +0.715, +0.915	sce	25
4,4'-Diamino-diphenylamine	Diethylglycine buffer, pH 10	Pt	+0.120	sce	25
Di-Aryl Amines					
Diphenylamine	CH_3CN, (Et)$_4$-NClO$_4$	Pt	+0.86, +1.03	sce	1
Di-4-tolylamine	CH_3CN, NaClO$_4$	Pt	+0.835	sce	9
N-Methyldiphenyl-amine	CH_3CN, NaClO$_4$	Pt	+0.705, 1.54	sce	9
	CH_3CN, (Et)$_4$-NClO$_4$	Pt	+0.84 $E_{p/2}$	sce	10
N-Methyl-di-p-tolylamine	CH_3CN, (Et)$_4$-NClO$_4$	Pt	+0.60 $E_{p/2}$	sce	10
N-Methyl-di-p-anisylamine	CH_3CN, (Et)$_4$-NClO$_4$	Pt	+0.65 $E_{p/2}$	sce	10

(continued)

817

Compound	Solvent System	Working Electrode	$E_{1/2}$	Reference Electrode	Reference
N-Methyl-N-phenyl-p-anisylamine	CH_3CN, $(Et)_4$-$NClO_4$	Pt	+0.77 $E_{p/2}$	sce	10
N,N,N',N'-Tetramethyl-p-phenyl-enediamine	CH_3CN, $(\underline{n}$-Propyl$)_4$-$NClO_4$	Pt	-0.10	Ag/0.01N Ag+	17
N,N,N',N'-Tetramethyl-m-phenyl-enediamine	CH_3CN, $(\underline{n}$-Propyl$)_4$-$NClO_4$	Pt	+0.32	Ag/0.01N Ag+	17
N,N,N',N'-Tetramethyl-o-phenyl-enediamine	CH_3CN, $(\underline{n}$-Propyl$)_4$-$NClO_4$	Pt	+0.28	Ag/0.01N Ag+	17

	Solvent System	Working Electrode	$E_{1/2}$	Reference Electrode	Reference
R = H	CH_3CN, $NaClO_4$	Pt	+1.28	Ag/0.01$\underline{N}$ Ag+	29
= $\underline{p}$-OCH_3	CH_3CN, $NaClO_4$	Pt	+0.97	Ag/0.01$\underline{N}$ Ag+	29
= $\underline{p}$-$N(CH_3)_2$	CH_3CN, $NaClO_4$	Pt	+0.40	Ag/0.01$\underline{N}$ Ag+	29
= $\underline{p}$-$N(C_6H_5)_2$	CH_3CN, $NaClO_4$	Pt	+0.66	Ag/0.01$\underline{N}$ Ag+	29
= $3,4(OCH_3)_2$	CH_3CN, $NaClO_4$	Pt	+0.95	Ag/0.01$\underline{N}$ Ag+	29

R = H	CH_3CN, $NaClO_4$	Pt	+1.18	Ag/0.01N Ag+	29
= p-OCH₃	CH_3CN, $NaClO_4$	Pt	+0.90	Ag/0.01N Ag+	29
= p-N(CH₃)₂	CH_3CN, $NaClO_4$	Pt	+0.28	Ag/0.01N Ag+	29
= p-N(C₆H₅)₂	CH_3CN, $NaClO_4$	Pt	+0.60	Ag/0.01N Ag+	29
= 3,4(OCH₃)₂	CH_3CN, $NaClO_4$	Pt	+0.84	Ag/0.01N Ag+	29

R = H; = p-OCH₃; = p-N(CH₃)₂; = p-N(C₆H₅)₂; = 3,4(OCH₃)₂

(structure: aryl ring with R substituent, CH–N, attached ring bearing NO₂ and NO₂ groups)

R = H	CH_3CN, $NaClO_4$	Pt	+1.24	Ag/0.01N Ag+	29
= p-OCH₃	CH_3CN, $NaClO_4$	Pt	+0.90	Ag/0.01N Ag+	29
= p-N(CH₃)₂	CH_3CN, $NaClO_4$	Pt	+0.40	Ag/0.01N Ag+	29
= p-N(C₆H₅)₂	CH_3CN, $NaClO_4$	Pt	+0.66	Ag/0.01N Ag+	29
= 3,4(OCH₃)₂	CH_3CN, $NaClO_4$	Pt	+0.90	Ag/0.01N Ag+	29

Tri-Aryl Amines

(structure: triarylamine N with rings bearing R_1, R_2, R_3)

R_1	R_2	R_3					
H	H	H	CH_3CN, (Et)₄-NClO₄	Pt	+0.92 $E_{p/2}$	sce	30
H	H	H	CH_3CN, $NaClO_4$	Pt	+0.860	sce	9

(continued)

Compound R₁	R₂	R₃	Solvent System	Working Electrode	$E_{1/2}$	Reference Electrode	Reference
p-OCH₃	p-OCH₃	p-OCH₃	CH₃CN, (Et)₄NClO₄	Pt	+0.52 $E_{p/2}$	sce	30
p-CH₃	p-CH₃	p-CH₃	CH₃CN, (Et)₄NClO₄	Pt	+0.75 $E_{p/2}$	sce	30
p-F	p-F	p-F	CH₃CN, (Et)₄NClO₄	Pt	+0.95 $E_{p/2}$	sce	30
p-Cl	p-Cl	p-Cl	CH₃CN, (Et)₄NClO₄	Pt	+1.04 $E_{p/2}$	sce	30
p-Br	p-Br	p-Br	CH₃CN, (Et)₄NClO₄	Pt	+1.05 $E_{p/2}$	sce	30
p-CO₂-CH₃	p-CO₂-CH₃	p-CO₂-CH₃	CH₃CN, (Et)₄NClO₄	Pt	+1.26 $E_{p/2}$	sce	30
p-OCH₃	p-OCH₃	p-NO₂	CH₃CN, (Et)₄NClO₄	Pt	+0.86 $E_{p/2}$	sce	30
p-CH₃	p-CH₃	p-NO₂	CH₃CN, (Et)₄NClO₄	Pt	1.03 $E_{p/2}$	sce	30
p-OCH₃	p-OCH₃	H	CH₃CN, (Et)₄NClO₄	Pt	+0.63 $E_{p/2}$	sce	30
p-CH₃	p-CH₃	H	CH₃CN, (Et)₄NClO₄	Pt	+0.82 $E_{p/2}$	sce	30
p-Cl	p-Cl	H	CH₃CN, (Et)₄NClO₄	Pt	1.01 $E_{p/2}$	sce	30
p-NO₂	p-NO₂	H	CH₃CN, (Et)₄NClO₄	Pt	1.36 $E_{p/2}$	sce	30
p-OCH₃	p-NO₂	H	CH₃CN, (Et)₄NClO₄	Pt	+0.98 $E_{p/2}$	sce	30

p-OCH$_3$	H	H	CH$_3$CN, (Et)$_4$-NClO$_4$	Pt	+0.76 E$_{p/2}$	sce	30
p-C$_6$H$_5$	H	H	CH$_3$CN, (Et)$_4$-NClO$_4$	Pt	+0.89 E$_{p/2}$	sce	30
p-CH$_3$	H	H	CH$_3$CN, (Et)$_4$-NClO$_4$	Pt	+0.88 E$_{p/2}$	sce	30
p-Cl	H	H	CH$_3$CN, (Et)$_4$-NClO$_4$	Pt	+0.99 E$_{p/2}$	sce	30
p-CN	H	H	CH$_3$CN, (Et)$_4$-NClO$_4$	Pt	+1.14 E$_{p/2}$	sce	30
p-NO$_2$	H	H	CH$_3$CN, (Et)$_4$-NClO$_4$	Pt	+1.17 E$_{p/2}$	sce	30
o-OCH$_3$	H	H	CH$_3$CN, (Et)$_4$-NClO$_4$	Pt	+0.96 E$_{p/2}$	sce	30
o-CH$_3$	H	H	CH$_3$CN, (Et)$_4$-NClO$_4$	Pt	+1.02 E$_{p/2}$	sce	30
o-Cl	H	H	CH$_3$CN, (Et)$_4$-NClO$_4$	Pt	+1.08 E$_{p/2}$	sce	30
o-OCH$_3$	o-OCH$_3$	o-OCH$_3$	CH$_3$CN, (Et)$_4$-NClO$_4$	Pt	+0.80 E$_{p/2}$	sce	30
o-CH$_3$	o-CH$_3$	o-CH$_3$	CH$_3$CN, (Et)$_4$-NClO$_4$	Pt	+1.01 E$_{p/2}$	sce	30
o-Cl	o-Cl	o-Cl	CH$_3$CN, (Et)$_4$-NClO$_4$	Pt	+1.44 E$_{p/2}$	sce	30
m-OCH$_3$	H	H	CH$_3$CN, (Et)$_4$-NClO$_4$	Pt	+0.91 E$_{p/2}$	sce	30
m-OCH$_3$	m-OCH$_3$	m-OCH$_3$	CH$_3$CN, (Et)$_4$-NClO$_4$	Pt	+0.92 E$_{p/2}$	sce	30
m-NO$_2$	m-NO$_2$	m-NO$_2$	CH$_3$CN, (Et)$_4$-NClO$_4$	Pt	+1.44 E$_{p/2}$	sce	30

(continued)

821

Amine-Substituted Polynuclear Hydrocarbons

Compound	Solvent System	Working Electrode	$E_{1/2}$	Reference Electrode	Reference
1-Naphthylamine	B-R buffer, pH 5.1	C paste	+0.53	sce	14
	CH_3CN, $NaClO_4$	Pt	+0.44	Ag/0.1N Ag+	2
	CH_3CN, $NaClO_4$	Pt	+0.54	sce	31
	CH_3CN, $(Et)_4-NClO_4$	Pt	+0.68, +1.25	sce	1
2-Naphthylamine	CH_3CN, $NaClO_4$	Pt	+0.54	Ag/0.1N Ag+	2
	CH_3CN, $NaClO_4$	Pt	+0.64	sce	31
	CH_3CN, $(Et)_4-NClO_4$	Pt	+0.75, +0.94	sce	1
1-Dimethylamino-naphthalene	CH_3CN, $(\underline{n}-Propyl)_4-NClO_4$	Pt	+0.75	sce	27
2-Dimethylamino-naphthalene	CH_3CN, $(\underline{n}-Propyl)_4-NClO_4$	Pt	+0.67	sce	27
1,5-Bis(dimethyl-amino)naphthalene	CH_3CN, $(\underline{n}-Propyl)_4-NClO_4$	Pt	+0.59	sce	27
2,6-Bis(dimethyl-amino)naphthalene	CH_3CN, $(\underline{n}-Propyl)_4-NClO_4$	Pt	+0.26	sce	27
2,7-Bis(dimethyl-amino)naphthalene	CH_3CN, $(\underline{n}-Propyl)_4-NClO_4$	Pt	+0.57	sce	27
1-Aminoanthracene	CH_3CN, $NaClO_4$	Pt	+0.31	Ag/0.1N Ag+	2
	CH_3CN, $(Et)_4NClO_4$	Pt	+0.55, +1.27	sce	1
2-Aminoanthracene	CH_3CN, $NaClO_4$	Pt	+0.33	Ag/0.1N Ag+	2
	CH_3CN, $NaClO_4$	Pt	+0.44	sce	31
	CH_3CN, $(Et)_4NClO_4$	Pt	+0.56, +1.30	sce	1
9-Amino-10-phenyl-	CH_3CN, $NaClO_4$	Pt	+0.170, +1.030	Ag/0.1N Ag+	2

Compound	Solvent, salt	Electrode	E	Reference electrode	Ref.
9-Aminoanthracene	CH₃CN, NaClO₄	Pt	+0.15	Ag/0.1N Ag+	2
2-Aminophenanthrene	CH₃CN, NaClO₄	Pt	+0.59	Ag/0.1N Ag+	2
9-Aminophenanthrene	CH₃CN, NaClO₄	Pt	+0.46	Ag/0.1N Ag+	2
2-Aminofluorene	CH₃CN, NaClO₄	Pt	+0.53	sce	31
	CH₃CN, NaClO₄	Pt	+0.46	Ag/0.1N Ag+	2
2-Acetylaminofluorene	CH₃CN, (Et)₄NClO₄	Pt	+0.64, +1.35	sce	1
	CH₃CN, (Et)₄NClO₄	Pt	+1.14, +2.40	sce	1
1-Aminopyrene	CH₃CN, NaClO₄	Pt	+0.32	Ag/0.1N Ag+	2
2-Aminopyrene	CH₃CN, NaClO₄]	Pt	+0.57	Ag/0.1N Ag+	2
6-Aminochrysene	CH₃CN, NaClO₄	Pt	+0.38	Ag/0.1N Ag+	2
2-Aminobiphenyl	CH₃CN, NaClO₄	Pt	+0.65	Ag/0.1N Ag+	2
4-Aminobiphenyl	CH₃CN, (Et)₄NClO₄	Pt	+0.93, +1.35	sce	1
	CH₃CN, (Et)₄NClO₄	Pt	+0.76, +2.51	sce	1
	CH₃CN, NaClO₄	Pt	+0.55	Ag/0.1N Ag+	2
9-Phenylamino-anthracene	CH₃CN, Perchlorate salt	Pt	+0.420	Ag/0.01N Ag+	32
9-p-Tolylamino-anthracene	CH₃CN, Perchlorate salt	Pt	+0.390	Ag/0.01N Ag+	32
9-p-Anisylamino-anthracene	CH₃CN, Perchlorate salt	Pt	+0.330	Ag/0.01N Ag+	32
9-p-Dimethylamino-phenylaminoanthra-cene	CH₃CN, Perchlorate salt	Pt	-0.074	Ag/0.01N Ag+	32
9-p-Carbomethoxy-phenylaminoanthra-cene	CH₃CN, Perchlorate salt	Pt	+0.510	Ag/0.01N Ag+	32
9-p-Nitrophenylamino-anthracene	CH₃CN, Perchlorate salt	Pt	+0.615	Ag/0.01N Ag+	32
9-Phenylamino-10-phenylanthracene	CH₃CN, Perchlorate salt	Pt	+0.460, +0.780	Ag/0.01N Ag+	32

(continued)

Compound	Solvent System	Working Electrode	$E_{1/2}$	Reference Electrode	Reference
9-p-Tolylamino-10-phenylanthracene	CH₃CN, Perchlorate salt	Pt	+0.420, +0.760	Ag/0.01N Ag+	32
9-p-Anisylamino-10-phenylanthracene	CH₃CN, Perchlorate salt	Pt	+0.370, +0.745	Ag/0.01N Ag+	32
9-m-Anisylamino-10-phenylanthracene	CH₃CN, Perchlorate salt	Pt	+0.465, +0.755	Ag/0.01N Ag+	32
9-p-Dimethylamino-10-phenylanthracene	CH₃CN, Perchlorate salt	Pt	-0.075, +0.600	Ag/0.01N Ag+	32

See also Aromatic Ketones (Section 7) for Amino Ketones

a Pyridine Added.
b Extrapolated value at pH 0.
c MgO in suspension.
d Py added.
e Second $E_{1/2}$ not given.
f Acetate, citrate, phosphate.
g pH 1 to 4.5.
h At -20°C.

References for Aromatic Amines

1. T. A. Gough, and M. E. Peover, Proc. 3rd International Polarography Congress, Southampton, 1965, MacMillan, London, 1966, p. 1017.
2. C. Párkányi and R. Zahrádnik, Collect. Czech. Chem. Commun., 30, 4288 (1965).
3. J. Bacon and R. N. Adams, J. Am. Chem. Soc., 90, 6596 (1968).
4. S. S. Lord, Jr. and L. B. Rogers, Anal. Chem., 26, 284 (1954).
5. M. Fedtke and S. Stöck, Z. Chem., 9, 196 (1969).
6. J. C. Suantoni, R. E. Snyder, and R. O. Clark, Anal. Chem., 33, 1894 (1961).
7. Z. Galus and R. N. Adams, J. Phys. Chem., 67, 862 (1963).
8. S. Wawzonek and T. W. McIntyre, J. Electrochem. Soc., 114, 1025 (1967).
9. V. Dvořák, I. Nemec, and J. Zyka, Microchem. J., 12, 99 (1967).
10. E. T. Seo, R. F. Nelson, J. M. Fritsch, L. S. Marcoux, D. W. Leedy, and R. N. Adams, J. Am. Chem. Soc., 88, 3498 (1966).
11. R. H. Hand and R. F. Nelson, J. Electrochem. Soc., 117, 1353 (1970).
12. N. L. Weinberg, unpublished results.
2A. R. L. Bent, J. C. Dessloch, F. C. Duennebier, D. W. Fassett, D. B. Glass, T. H. James, D. B. Julian, W. R. Ruby, J. N. Snell, J. H. Sterner, J. R. Thirtle, P. W. Vittum, and A. Weissberger, J. Am. Chem. Soc., 73, 3100 (1951).
13. T. R. Mueller and R. N. Adams, Anal. Chim. Acta, 25, 482 (1961).
14. C. Olson and R. N. Adams, Anal. Chim. Acta, 22, 582 (1960).
15. G. Cauquis, J. L. Cros, and M. Genies, Preprints of Papers, Symposium on Electro-Organic Chemistry, Durham, North Carolina, October 14-16, 1968, p. 249.
16. G. Cauquis and M. Genies, Compt. Rend., 265, 1340 (1967).
17. A. Zweig, J. E. Lancaster, M. T. Neglia, and W. H. Jura, J. Am. Chem. Soc., 86, 4130 (1964).
18. M. Melicharek and R. F. Nelson, J. Electroanal. Chem., 26, 201 (1970).
19. G. Cauquis, J. Coquand and J. Rigaudy, Compt. Rend., 268, 2265 (1969).
20. P. J. Elving and A. F. Krivis, Anal. Chem., 30, 1645 (1958).

826

21. S. Wawzonek, T. H. Plaisance, L. M. Smith, Jr., and E. B. Buchanan, Jr., Preprints of Papers, Durham Sumposium on Electro-Organic Chemistry, Durham, North Carolina, 1968, p. 247.
22. R. E. Parker and R. N. Adams, Anal. Chem., 28, 828 (1956).
23. L. S. Reishakhrit, T. B. Argova, and L. V. Vesheva, Elektrokhim., 4, 653 (1968).
24. M. Rifi, Tetrahedron Letters, 58, 5089 (1969).
25. V. Dvorak, I. Nemec, J. Zyka, Microchem. J., 12, 324 (1967).
26. D. B. Julian and W. R. Ruby, J. Am. Chem. Soc., 72, 4719 (1950).
27. A. Zweig, A. H. Maurer, and B. G. Roberts, J. Or Chem., 32, 1322 (1967).
28. D. A. Hall, M. Sakuma, and P. J. Elving, Electrochim. Acta, 11, 337 (1966).
29. A. T. Balaban, P. T. Frangopol, M. Frangopol, an N. Negoita, Tetrahedron, 23, 4661 (1967).
30. R. F. Nelson and R. N. Adams, J. Am. Chem. Soc., 90, 3925 (1968).
31. E. S. Pysh and N. C. Yang, J. Am. Chem. Soc., 85 2125 (1963).
32. G. Cauquis, J. P. Billon, J. Raison, and Y. Thibaud, Compt. Rend., 257, 2128 (1963).

21. ALIPHATIC HALIDES

Oxidation

Compound	Solvent System	Working Electrode	$E_{1/2}$	Reference Electrode	Reference
Methyl iodide	CH_3CN, $LiClO_4$	Pt	+2.12	$Ag/0.01\underline{N}$ $Ag+$	1
i-Propyl iodide	CH_3CN, $LiClO_4$	Pt	+2.04	$Ag/0.01\underline{N}$ $Ag+$	1
t-Butyl iodide	CH_3CN, $LiClO_4$	Pt	+1.87	$Ag/0.01\underline{N}$ $Ag+$	1
Neopentyl iodide	CH_3CN, $LiClO_4$	Pt	+2.14	$Ag/0.01\underline{N}$ $Ag+$	1

Reduction

Compound	Solvent System	Working Electrode	$E_{1/2}$	Reference Electrode	Reference
Methyl chloride	75% Dioxane, $0.05M$ $(Et)_4NBr$	Hg	-2.23	sce	2
Triphenylchloro-methane	80% Ethanol, $0.01M$ $(Me)_4NBr$	Hg	-1.25	NCE	3
Methyl bromide	75% Dioxane, $0.05M$ $(Et)_4NBr$	Hg	-1.63	sce	2
	DMF, $(Et)_4NBr$	Hg	-1.964	sce	4

(continued)

Compound	Solvent System	Working Electrode	$E_{1/2}$	Reference Electrode	Reference
Methyl iodide	CH$_3$CN, (n-Bu)$_4$NBr	Pb	-1.68 E$_{p/2}$	sce	5
	CH$_3$CN, (n-Bu)$_4$NBr	Hg	-1.77	sce	5
	CH$_3$CN, (n-Bu)$_4$NBr	Sn	-2.45	sce	5
	75% Dioxane, 0.05M (Et)$_4$NBr	Hg	-1.63	sce	2
	0.5N KCl	Ga	-1.2	sce	6
	CH$_3$CN, (n-Bu)$_4$NI	Pb	-1.17 E$_{p/2}$	sce	5
	50% Ethanol, 0.05M (Et)$_4$NI	Hg	-1.39	sce	7
Methylene chloride- (Dichloromethane)	75% Dioxane, 0.05M (Et)$_4$NBr	Hg	-2.33	sce	2
Methylene bromide- (Dibromomethane)	DMF, (n-Bu)$_4$NBr	Hg	-2.14	sce	8
	75% Dioxane, 0.05M (Et)$_4$NBr	Hg	-1.48	sce	2
Methylene iodide- (Diiodomethane)	0.5N KCl	Ga	-1.2	sce	6
	75% Dioxane, 0.05M (Et)$_4$NBr	Hg	-1.12, -1.53	sce	2
	50% Ethanol, 0.05M (Et)$_4$NBr	Hg	-0.92, -1.40	sce 7	7
Chloroform	75% Dioxane, 0.05M (Et)$_4$NBr	Hg	-1.67	sce	2
	DMF, (n-Bu)$_4$NBr	Hg	-1.45, -2.14	sce	9
	67% Methanol, 0.1M (Me)$_4$NBr	Hg	-1.71	sce	10
Bromoform	0.5N KCl	Ga	-1.2, -1.5	sce	6
	75% Dioxane, 0.05M (Et)$_4$NBr	Hg	-0.64, -1.51	sce	2
Iodoform	75% Dioxane, 0.05M (Et)$_4$NBr	Hg	-0.49, -1.09, -1.50	sce	2
	0.05M (Et)$_4$NBr				

Compound	Electrolyte	Electrode	Potential	Ref. electrode	Ref.
Carbon tetrachloride	75% Dioxane, 0.05M (Et)$_4$NBr	Hg	-0.78, -1.71	sce	2
	DMF, (n-Bu)$_4$NBr	Hg	-0.25, -1.49, -2.17	sce	9
	67% Methanol, 0.05M (Me)$_4$NBr	Hg	-0.75, -1.70	sce	10
	20% Methanol, 0.05M KSCN	Hg	-0.44	sce	9
Carbon tetrabromide	75% Dioxane, 0.05M (Et)$_4$NBr	Hg	-0.3, -0.75, -1.49	sce	2
Ethyl chloride	75% Dioxane, 0.05M (Et)$_4$NBr	Hg	No wave	sce	2
Ethyl bromide	75% Dioxane, 0.05M (Et)$_4$NBr	Hg	-2.08	sce	2
	DMF, (Et)$_4$NBr	Hg	-2.13	sce	11
	DMF, (Et)$_4$NBr	Hg	-1.70	Ag, AgBr	12
	0.5N KCl	Ga	-1.7a	sce	6
	CH$_3$CN, (n-Bu)$_4$NBr	Pb	-2.04 E$_p$/2	sce	5
Ethyl iodide	75% Dioxane, 0.05M (Et)$_4$NBr	Hg	-1.67	sce	2
	50% Ethanol, 0.05M (Et)$_4$NI	Hg	-1.57	sce	7
	0.5N KCl	Ga	-1.2	sce	6
1,2-Dichloroethane	CH$_3$CN, (n-Bu)$_4$NI	Pb	-1.22 E$_p$/2	sce	5
1,1-Dibromoethane	0.5N KCl	Ga	-1.85$_p$/2	sce	6
1,2-Dibromoethane	75% Dioxane, 0.05M (Et)$_4$NBr	Hg	-1.62	sce	2
	0.5N KCl	Ga	-1.2	sce	6
	DMF, (n-Bu)$_4$NClO$_4$	Hg	-1.38	sce	13
	75% Dioxane, 0.05M (Et)$_4$NBr	Hg	-1.52	sce	2

(continued)

829

Compound	Solvent System	Working Electrode	$E_{1/2}$	Reference Electrode	Reference
Tribromoethane	20% Methanol, 1% Na$_2$SO$_3$	Hg	-1.42	NCE	14
	75% Dioxane, 0.05$\underline{M}$ (Et)$_4$NBr	Hg	-0.70	sce	2
1,1,2,2-Tetrabromo-ethane	80% HOAc, 3$\underline{M}$ NaOAc	Hg	-0.30	sce	15
Pentabromoethane	75% Dioxane, 0.05$\underline{M}$ (Et)$_4$NBr	Hg	-0.90, -1.16	sce	2
Hexachloroethane	75% Dioxane, 0.05$\underline{M}$ (Et)$_4$NBr	Hg	-0.62, -1.73, -1.96	sce	2
Hexabromoethane	75% Dioxane, 0.05$\underline{M}$ (Et)$_4$NBr	Hg	-0.68, -1.18	sce	2
1,2-Dibromo-1-chloroethane	80% HOAc, 3$\underline{M}$ NaOAc	Hg	-0.85	sce	15
1,2-Dibromo-2-chloroethane	50% Methanol, 1% Na$_2$SO$_3$	Hg	-0.46	NCE	14
1,2-Dibromo-1,2-dichloroethane	80% HOAc, 3$\underline{M}$ NaOAc	Hg	-0.55	sce	15
	50% Methanol, 0.1$\underline{N}$ LiCl	Hg	-0.31	NCE	14
1,2-Dibromo-1,1,2-trichloroethane	80% HOAc, 3$\underline{M}$ NaOAc	Hg	-0.05	sce	15
Diphenyldichloro-ethane	80% Ethanol, 0.01$\underline{M}$ (Me)$_4$NBr	Hg	-0.93	NCE	3
1,1,1-Trichloro-2,2-di-4-chloropenyl-ethane (DDT)	80% Ethanol, 0.01$\underline{M}$ (Me)$_4$NBr	Hg	-0.88	sce	15
1,1,1-Trichloro-2,2-di-4-bromophenyl...	80% Ethanol, 0.01$\underline{M}$ (Me)$_4$NBr	Hg	-0.86	sce	15

Compound	Solvent / Electrolyte		$E_{1/2}$	Ref. electrode	Ref.
1,1,1-Trichloro-2,2-diphenylethane	80% Ethanol, 0.01M (Me)₄NBr	Hg	-0.97	sce	15
1,1,1-Trichloro-2,2-di-4-tolylethane	80% Ethanol, 0.01M (Me)₄NBr	Hg	-0.95	sce	15
p,p-Dichlorodiphenyl-tetrachloroethane	80% Ethanol, 0.01M (Me)₄NBr	Hg	-2.02	NCE	3
Vinyl chloride	75% Dioxane, 0.05M (Et)₄NBr	Hg	No wave	sce	2
Vinyl bromide	75% Dioxane, 0.05M (Et)₄NBr	Hg	-2.47	sce	2
1,1-Dichloroethylene	DMF, (n-Bu)₄NBr	Hg	-2.46	sce	16
	75% Dioxane, 0.05M (Et)₄NBr	Hg	~ -2.4	sce	2
1,2-Dichloroethylene	75% Dioxane, 0.05M (Et)₄NBr	Hg	No wave	sce	2
Trichloroethylene	75% Dioxane, 0.05M (Et)₄NBr	Hg	-2.14	sce	2
Tetrachloroethylene	75% Dioxane, 0.05M (Et)₄NBr	Hg	-1.88	sce	2
Triphenylbromoethylene	DMF, (n-Bu)₄NI	Hg	-1.60, -1.99	sce	16
1-Anisyl-2,2-diphenyl-bromoethylene	DMF, (n-Bu)₄NI	Hg	-1.66, -2.09	sce	16
2-Anisyl-1,2-diphenyl-bromoethylene	DMF, (n-Bu)₄NI	Hg	-1.63, -2.04	sce	16
1,1-Diphenyl-2-bromoethylene	DMF, (n-Bu)₄NI	Hg	-1.80, -2.08	sce	16
1,1-Di-(p-2-chloro-phenyl)-2,2-dichloro-ethylene	80% Ethanol, 0.01M (Me)₄NBr	Hg	-1.98	NCE	3
Diiodoacetylene	75% Dioxane, 0.05M (Et)₄NBr	Hg	-0.25, -0.61	sce	2

(continued)

Compound	Solvent System	Working Electrode	$E_{1/2}$	Reference Electrode	Reference
Acetyl chloride	Acetone, LiCl	Hg	-1.25	asce[b]	17
Acetyl bromide	Acetone, LiCl	Hg	-1.18	asce[b]	17
Iodoacetic acid	50% Ethanol, 0.05M (Et)$_4$NI, pH 3-8	Hg	-0.50	sce	7
Ethyl chloroacetate	CH$_3$CN, (n-Bu)$_4$NBr	Hg	-0.39, -1.74	Hg pool	18
Ethyl bromoacetate	CH$_3$CN, (n-Bu)$_4$NBr	Hg	-0.43	Hg pool	18
n-Propyl bromide	DMF, (Et)$_4$NBr	Hg	-2.20	sce	11
i-Propyl bromide	DMF, (Et)$_4$NBr	Hg	-2.26	sce	11
	DMF, (n-Bu)$_4$NBr	Pb	-2.30 $E_{p/2}$	sce	5
Cyclopropyl bromide	DMF, (Et)$_4$NBr	Hg	-2.36	sce	11
n-Propyl iodide	50% Ethanol, 0.05M (Et)$_4$NI	Hg	-1.63	sce	7
i-Propyl iodide	DMF, (n-Bu)$_4$NI	Pb	-1.25 $E_{p/2}$	sce	5
Vinyl bromide	DMF, (n-Bu)$_4$NI	Hg	-2.46	sce	16
1,3-Dibromopropane	DMF, (n-Bu)$_4$NClO$_4$	Hg	-1.91	sce	13
2-Methyl-2,3-di-bromopropane	30% Ethanol, 0.1N LiCl	Hg	-1.04	NCE	14
1-Chloro-2-bromo-cyclopropane	30% Ethanol, 0.1M (Me)$_4$NI	Hg	-1.68	NCE	19
1-Bromo-2,2-diphenyl-cyclopropane carboxylic acid	95% Ethanol, 0.1M (Et)$_4$NBr, 0.01M NH$_3$	Hg	-1.72	sce	20
Methyl-1-bromo-2,2-diphenylcyclopro-pane carboxylate	95% Ethanol, 0.1M (Et)$_4$NBr, 0.01M NH$_3$	Hg	-1.10	sce	20
1-Bromo-1-methyl-2,2-diphenylcyclopropane	95% Ethanol, 0.1M (Et)$_4$NBr	Hg	-2.3	sce	20

Compound	Electrolyte/Solvent	Electrode	E	Reference electrode	Ref.
2-Bromopropionic acid	0.5M KCl + HCl, pH 2.0	Hg	-0.39	sce	21
	0.5M OAc buffer, pH 5.2	Hg	-0.89	sce	21
	0.5M NH$_4$Cl+NH$_3$, pH 8.2	Hg	-1.14	sce	21
Propionyl chloride	Acetone, LiCl	Hg	-1.24	asce[b]	17
Propionyl bromide	Acetone, LiCl	Hg	-1.24	asce[b]	17
Allyl chloride	75% Dioxane, 0.05M (Et)$_4$NBr	Hg	-1.91	sce	2
Allyl bromide	75% Dioxane, 0.05M (Et)$_4$NBr	Hg	-1.29	sce	2
Allyl iodide	75% Dioxane, 0.05M (Et)$_4$NBr	Hg	-0.23, -1.16	sce	2
1-Bromo-2-methyl-propene	DMF, (n-Bu)$_4$NI	Hg	-2.6	sce	16
Trimethylene dibromide	95% DMF, (Me)$_4$NBr	Hg	-1.65	sce	22
n-Butyl bromide	DMF, (Et)$_4$NBr	Hg	-2.23	sce	11
	DMF, (Me)$_4$NClO$_4$	Hg	-1.985	sce	23
	75% Dioxane, 0.05M (Et)$_4$NBr	Hg	-2.27	sce	2
i-Butyl bromide	DMF, (Et)$_4$NBr	Hg	-2.32	sce	11
t-Butyl bromide	DMF, (Et)$_4$NBr	Hg	-2.19	sce	11
t-Butyl iodide	CH$_3$CN, (n-Bu)$_4$NBr	Pb	-1.84	sce	5
Cyclobutylbromide	CH$_3$CN, (n-Bu)$_4$NBr	Pb	-1.35	sce	5
1-Chloro-3-bromo-cyclobutane	DMF, (Et)$_4$NBr	Hg	-2.36	sce	11
	DMF, (Et)$_4$NBr	Hg	-1.80	sce	24
1-Bromo-3-bromo-methylcyclobutane	DMF, (n-Bu)$_4$NClO$_4$	Hg	-2.08	sce	13

(continued)

Compound	Solvent System	Working Electrode	E$_{1/2}$	Reference Electrode	Reference
1,3-Dibromo-1,3-dimethylcyclobutane-(cis and trans)	DMF, (n-Bu)$_4$NClO$_4$	Hg	-2.02	sce	13
1,2-Dibromobutane	50% Methanol, 1% Na$_2$SO$_3$	Hg	-1.45	NCE	14
1,4-Dibromobutane	DMF, (Me)$_4$NClO$_4$	Hg	-1.754, -2.027	sce	23
Butyryl chloride	DMF, (n-Bu)$_4$NClO$_4$	Hg	-1.99	sce	13
	Acetone, LiCl	Hg	-1.24	asce[b]	17
2-Bromo-n-butyric acid	0.5M KCl + HCl, pH 2.0	Hg	-0.37	sce	21
	0.5M OAc buffer, pH 5.2	Hg	-0.81	sce	21
	0.5M NH$_4$Cl+NH$_3$, pH 8.2	Hg	-1.15	sce	21
n-Pentyl bromide	DMF, (Et)$_4$NBr	Hg	-2.26	sce	11
i-Pentyl bromide	DMF, (Et)$_4$NBr	Hg	-2.273	sce	4
Neopentyl bromide	DMF, (Et)$_4$NBr	Hg	-2.317	sce	4
Cyclopentyl bromide	DMF, (Et)$_4$NBr	Hg	-2.451	sce	4
	DMF, (Et)$_4$NBr	Hg	-2.19	sce	11
1,4-Dibromopentane	DMF, (Me)$_4$NClO$_4$	Hg	-1.794, -2.030	sce	23
1,5-Dibromopentane	DMF, (N-Bu)$_4$NClO$_4$	Hg	-2.14	sce	13
trans-1,2-Dibromo-cyclopentane	DMF, (Me)$_4$NBr	Hg	-0.94	sce	22
Isovaleryl chloride	Acetone, LiCl	Hg	-1.24	asce[b]	17
2-Bromo-n-valeric acid	0.5M OAc buffer, pH 5.1	Hg	-0.64	sce	21
n-Hexyl bromide	DMF, (Et)$_4$NBr	Hg	-2.288	sce	4

Compound	Solvent, electrolyte	Electrode	E	Ref. electrode	Ref.
n-Hexyl bromide	DMF, (Et)$_4$NBr	Hg	-2.288	sce	4
i-Hexyl bromide	DMF, (Et)$_4$NBr	Hg	-1.81	Ag, AgBr	12
Cyclohexyl bromide	DMF, (Et)$_4$NBr	Hg	-2.294	sce	4
Cyclohexyl iodide	DMF, (Et)$_4$NBr	Hg	-2.29	sce	11
	90% Alcohol, OAc buffer, 0.06N LiCl	Hg	-1.47	sce	25
Cyclohexylmethyl bromide	DMF, (Et)$_4$NBr	Hg	-2.339	sce	4
trans-4-Bromo-t-butylcyclohexane	DMF, (Et)$_4$NBr	Hg	-2.45	sce	26
cis-4-Bromo-t-butylcyclohexane	DMF, (Et)$_4$NBr	Hg	-2.32	sce	26
1,6-Dibromohexane	DMF, (n-Bu)$_4$NClO$_4$	Hg	-2.13	sce	13
2,5-Dibromohexane	DMF, (Me)$_4$NClO$_4$	Hg	-1.849, -2.067	sce	23
cis-1,2-Dibromocyclohexane	DMF, (Me)$_4$NBr	Hg	-1.64	sce	22
trans-1,2-Dibromocyclohexane	DMF, (Me)$_4$NBr	Hg	-1.04	sce	22
α-Hexachlorocyclohexane	80% Ethanol, 0.1N (Et)$_4$NI	Hg	-2.10	NCE	27
β-Hexachlorocyclohexane	80% Ethanol, 0.1N (Et)$_4$NI	Hg	-2.23	NCE	27
γ-Hexachlorocyclohexane	80% Ethanol, 0.1N (Et)$_4$NI	Hg	-1.69, -2.66	NCE	27
	80% Ethanol, 0.1N (Et)$_4$NBr	Hg	-1.85	NCE	3
	50% Ethanol, 1% KI	Hg	-1.39	NCE	28

(continued)

Compound	Solvent System	Working Electrode	$E_{1/2}$	Reference Electrode	Reference
δ-Hexachlorocyclo-hexane	Phosphate buffer, pH 11.5	Hg	-1.27	NCE	29
	B-R buffer, pH 5.1	Hg	-1.33	NCE	30
α-Heptachlorocyclo-hexane	80% Ethanol, 0.1N (Et)4NI	Hg	-2.12	NCE	27
Perchloro-1,5-hexadiene	80% Ethanol, 0.1N (Et)4NI	Hg	-0.99	NCE	31
	80% Ethanol, 0.1N (Et)4NI	Hg	+0.1		32
Perchloro-1,3,5-hexatriene	3M LiCl, 70% EtOH	Hg	-1.5		32
	Me4NI, 70% Ethanol	Hg	-1.13[c]		32
	Me4NI, 70% Ethanol	Hg	-1.16[d]		32
Octafluorocyclohexa-1,4-diene	60% Ethanol, 0.4M (Me)4NCl	Hg	-1.49	sce	33
1H-Heptafluorocyclo-hexa-1,4-diene	60% Ethanol, 0.4M (Me)4NCl	Hg	-1.54	sce	33
1H,2H-Hexafluoro-cyclohexa-1,4-diene	60% Ethanol, 0.4M (Me)4NCl	Hg	-1.51	sce	33
1H,5H-Hexafluoro-cyclohexa-1,4-diene	60% Ethanol, 0.4M (Me)4NCl	Hg	-1.63	sce	33
1H,4H-Hexafluoro-cyclohexa-1,4-diene	60% Ethanol, 0.4M (Me)4NCl	Hg	-1.64	sce	33
1H,2H,4H-Pentafluoro-cyclohexa-1,4-diene	60% Ethanol, 0.4M (Me)4NCl	Hg	-1.61	sce	33
Heptafluoro-1-methyl-cyclohexa-1,4-diene	60% Ethanol, 0.4M (Me)4NCl	Hg	-1.73	sce	33
Octafluorocyclohexa-1,3-diene	60% Ethanol, 0.4M (Me)4NCl	Hg	-1.19	sce	33

Compound	Solvent/Electrolyte	Electrode	Potential	Ref. Electrode		Ref.
1H-Heptafluorocyclo-hexa-1,3-diene	60% Ethanol, 0.4M (Me)4NCl	Hg	-1.24	sce		33
2H-Heptafluorocyclo-hexa-1,3-diene	60% Ethanol, 0.4M (Me)4NCl	Hg	-1.22	sce		33
1H,2H-Hexafluoro-cyclohexa-1,3-diene	60% Ethanol, 0.4M (Me)4NCl	Hg	-1.22	sce		33
2H,3H-Hexafluoro-cyclohexa-1,3-diene	60% Ethanol, 0.4M (Me)4NCl	Hg	-1.21	sce		33
1H,4H-Hexafluoro-cyclohexa-1,3-diene	60% Ethanol, 0.4M (Me)4NCl	Hg	-1.25	sce		33
Heptafluoro-2-methyl-cyclohexa-1,3-diene	60% Ethanol, 0.4M (Me)4NCl	Hg	-1.37	sce		33
Heptafluoro-1-methyl-cyclohexa-1,3-diene	60% Ethanol, 0.4M (Me)4NCl	Hg	-1.42	sce		33
2-Bromo-n-hexanoic acid	0.5M OAc buffer, pH 5.4	Hg	-0.54	sce		21
Cycloheptyl bromide	DMF, (Et)4NBr	Hg	-2.27	sce		11
1-Bromobicyclo[2.2.1]heptane	DMF, (Et)4NBr	Hg	-2.17	Ag	Ag, AgBr	12
endo-Norbornyl bromide	DMF, (Et)4NBr	Hg	-2.43	sce		26
exo-Norbornyl bromide	DMF, (Et)4NBr	Hg	-2.34	sce		26
7-Bromo-7-chloro-bicyclo[4.1.0]heptane	95% Ethanol, 0.1M (Et)4NBr	Hg	-1.39	sce		12
7,7-Dibromobicyclo-[4.1.0] heptane	95% Ethanol, 0.1M (Et)4NBr	Hg	-1.26	sce		12
trans-1,2-Dibromo-cycloheptane	DMF, (Et)4NBr	Hg	-1.00	sce		22
exo-cis-2,3-Dibromo-bicyclo[2.2.1]heptane	DMF, (Et)4NBr	Hg	-1.53	sce		22

(continued)

Compound	Solvent System	Working Electrode	$E_{1/2}$	Reference Electrode	Reference
endo,cis-2,3-Dibromo-bicyclo [2.2.1] heptane	DMF, (Et)$_4$NBr	Hg	-1.21	sce	22
trans-2,3-Dibromo-bicyclo [2.2.1] heptane	DMF, (Et)$_4$NBr	Hg	-1.56	sce	22
2-Bromo-n-heptanoic acid	0.5M OAc buffer, pH 5.4	Hg	-0.44	sce	21
n-Octyl bromide	DMF, (Et)$_4$NBr	Hg	-2.291	sce	4
	75% Dioxane, 0.05M (Et)$_4$NBr	Hg	-2.38	sce	2
1-Bromobicyclo [2.2.2] octane	DMF, (Et)$_4$NBr	Hg	-2.48	sce	26
trans-1,2-Dibromo-cyclooctane	DMF, (Et)$_4$NBr	Hg	-1.79	sce	12
	DMF, (Me)$_4$NBr	Hg	-1.05	sce	22
cis-2,3-Dibromo-bicyclo [2.2.2] octane	DMF, (Me)$_4$NBr	Hg	-1.28	sce	22
trans-2,3-Dibromo-bicyclo [2.2.2] octane	DMF, (Me)$_4$NBr	Hg	-1.335	sce	22
2-Bromo-n-octanoic acid	0.5M OAc buffer, pH 5.4	Hg	-.38	sce	22
trans-1,2-Dibromo-cyclodecane	DMF, (Me)$_4$NBr	Hg	-1.44	sce	22
cis-1,6-Dibromocyclo-decane	DMF, (Me)$_4$NBr	Hg	-1.78	sce	22

trans-1,6-Dibromo-cyclodecane	DMF, $(Me)_4NBr$	Hg	-1.73	sce	22
erythro-5,6-Dibromo-decane	DMF, $(Me)_4NBr$	Hg	-1.14	sce	22
threo-5,6-Dibromo-decane	DMF, $(Me)_4NBr$	Hg	-1.24	sce	22
5,5,8,8-Tetramethyl-trans-1,2-dibromo-cyclodecane	DMF, $(Me)_4NBr$	Hg	-1.15	sce	22
cis-1,2-Dibromo-cyclotetradecane	DMF, $(Me)_4NBr$	Hg	-1.33	sce	22
1-Bromohexadecane	DMF, $(Et)_4NBr$	Hg	-1.79	Ag, AgBr	12
cis-1,2-Dibromo-cyclohexadecane	DMF, $(Me)_4NBr$	Hg	-1.21	sce	22

a Foot of the wave.
b Acetone - saturated calomel electrode.
c Liquid isomer.
d Solid isomer.

(continued)

References for Aliphatic Halides

1. L. L. Miller and A. K. Hoffmann, J. Am. Chem. Soc., 89, 593 (1967).
2. M. von Stackelberg and W. Stracke, Z. Elektrochem 53, 118 (1949).
3. H. Keller, M. Hochweber, and H. v. Halban, Helv. Chim. Acta, 29, 761 (1946).
4. F. L. Lambert, J. Org. Chem., 31, 4184 (1966).
5. H. E. Ulery, J. Electrochem. Soc., 116, 1201 (196
6. I. A. Bagotskaya and D. K. Burmanov, Elektrokhim. 4, 115 (1968).
7. R. A. Caldwell and S. Hacobian, Aust. J. Chem., 21, 1 (1968).
8. L. Meites, Polarographic Techniques, 2nd ed., Interscience, New York, 1965, p. 697.
9. Reference 8, p. 687.
10. I. M. Kolthoff, T. S. Lee, D. Stocesova, and E. P. Parry, Anal. Chem., 22, 521 (1950).
11. F. L. Lambert and K. Kobayashi, J. Am. Chem. Soc. 82, 5342 (1960).
12. J. W. Sease, P. Chang, and J. L. Groth, J. Am. Chem. Soc., 86, 3154 (1964).
13. M. R. Rifi, Tetrahedron Letters, 13, 1043 (1969).
14. A. W. Rjabow, G. D. Panova, and A. N. Frumkin, Ber. Akad. Wiss. UdSSR. (N.S.), 99, 547 (1954).
15. Reference 8, p. 691.
16. L. L. Miller and E. Riekena, J. Org. Chem., 34, 3359 (1969).
17. P. Arthur and H. Lyons, Anal. Chem., 24, 1422 (1952).
18. S. Wawzonek, R. C. Duty, and J. H. Wagenknecht, J. Electrochem. Soc., 111, 74 (1964).
19. M. B. Neiman, A. W. Rjabow, and J. M. Schejanow, Ber. Akad. Wiss. UdSSR (N.S.), 68, 1065 (1949).
20. R. Annino, R. E. Erickson, J. Michalovic, and B. McKay, J. Am. Chem. Soc., 88, 4424 (1966).
21. I. Rosenthal, C. H. Albright, and P. J. Elving, J. Electrochem. Soc., 99, 227 (1952).
22. J. Zavada, J. Krupicka, and J. Sicher, Collect. Czech. Chem. Commun., 28, 1664 (1965).
23. J. A. Dougherty and A. J. Diefenderfer, J. Electroanal. Chem., 21, 531 (1969).
24. M. R. Rifi, J. Am. Chem. Soc., 89, 4442 (1967).
25. E. L. Colichman and S. K. Liu, J. Am. Chem. Soc. 76, 913 (1954).
26. F. Lambert, A. H. Albert, and J. P. Hardy, J. Am Chem. Soc., 86, 3155 (1964).

27. K. Schwabe, Z. Naturforsch., 3, 217 (1948).
28. G. B. Ingram and H. K. Southern, Nature (London), 161, 437 (1948).
29. D. Monnier, L. Roesgen, and Y. Monnier, Anal. Chim. Acta, 4, 309 (1950).
30. K. Schwabe and H. Frind, Z. Phys. Chem., 196, 242 (1951).
31. M. Nakazima, S. Kioka, and Y. Katamura, Botyu-Kagaku, 13, 14 (1949).
32. V. D. Simonov, V. A. Semenov, and M. V. Fominykh, Zh. Ohshch. Khim., 40, 9870 (1970).
33. A. M. Doyle, A. E. Pedler, and J. C. Tatlow, J. Chem. Soc. C, 1968, 2740.

22. AROMATIC HALIDES

Oxidation

Compound	Solvent System	Working Electrode	$E_{1/2}$	Reference Electrode	Refer- ence
Chlorobenzene	CH_3CN, $NaClO_4$	Pt	+2.07	Ag/0.1N Ag+	1
Bromobenzene	CH_3CN, $NaClO_4$	Pt	+1.98	Ag/0.1N Ag+	1
	CH_3CN, $NaClO_4$	Pt	>+2.20	sce	2
Iodobenzene	CH_3CN, $NaClO_4$	Pt	+1.77	Ag/0.1N Ag+	1
p-Chlorotoluene	CH_3CN, $NaClO_4$	Pt	+1.76	Ag/0.1N Ag+	1
p-Bromotoluene	CH_3CN, $NaClO_4$	Pt	+1.72	Ag/0.1N Ag+	1
4-Bromobiphenyl	CH_3CN, $NaClO_4$	Pt	+1.95	sce	2
1-Bromonaphthalene	CH_3CN, $NaClO_4$	Pt	+1.85	sce	2
2-Bromonaphthalene	CH_3CN, $NaClO_4$	Pt	+1.90	sce	2
9-Bromoanthracene	CH_3CN, $NaClO_4$	Pt	+1.33	sce	2
9,10-Dibromoanthracene	CH_3CN, $NaClO_4$	Pt	+0.99, +1.48	Ag/0.1N Ag+	3
	CH_3CN, $NaClO_4$	Pt	+1.15, +1.47	Ag/0.1N Ag+	3
	CH_3CN, $LiClO_4$	Pt	+1.42 Ep	sce	4
1,5-Dichloroanthracene	CH_3CN, $NaClO_4$	Pt	+1.040	Ag/0.1N Ag+	5
9-Bromophenanthrene	CH_3CN, $NaClO_4$	Pt	+1.79	sce	2
6-Bromochrysene	CH_3CN, $NaClO_4$	Pt	+1.60	sce	2

Reduction

Compound	Solvent System	Working Electrode	$E_{1/2}$	Reference Electrode	Reference
Chlorobenzene	75% Dioxane, 0.05M (Et)$_4$NBr	Hg	No wave	sce	6
	DMF, (Et)$_4$NBr	Hg	-2.58	sce	7
	DMF, (Et)$_4$NClO$_4$	Hg	-2.60	sce	7
	DMF, (n-Bu)$_4$NClO$_4$	Hg	-2.00	Hg pool	8
	DMF, (Et)$_4$NBr	Hg	-2.13	Ag, AgBr	9
	75% Dioxane, 0.05M (Et)$_4$NBr	Hg	-2.32	sce	6
Bromobenzene	DMF, (n-Bu)$_4$NI	Hg	-2.38	sce	10
	0.1M (Me)$_4$NCl	Hg	-2.07	sce	11
	0.5N KCl	Ga	-1.7a	sce	12
	DMF, (Et)$_4$NBr	Hg	-1.81	Ag, AgBr	9
	75% Dioxane, 0.05M (Et)$_4$NBr	Hg	-1.62	sce	6
Iodobenzene	1M LiClO$_4$	Hg	-1.655	sce	11
	0.1M (Me)$_4$NCl	Hg	-1.467	sce	11
	50% Ethanol, 0.05M (Et)$_4$NI	Hg	-1.540	sce	13
	90% Alcohol, 0.06M LiCl, OAc buffer, pH 7.00	Hg	-1.68	sce	14
	66% Ethanol, 0.01M (Et)$_4$NI	Hg	-1.616	sce	15

(continued)

Compound	Solvent System	Working Electrode	$E_{1/2}$	Reference Electrode	Reference
Iodobenzene	0.5N KCl	Ga	~ -1.7	sce	16
	DMF, (Et)$_4$NBr	Hg	-1.21	Ag, AgBr	17
o-Dichlorobenzene	75% Dioxane, 0.05M (Et)$_4$NBr	Hg	-2.51	sce	6
m-Dichlorobenzene	75% Dioxane, 0.05M (Et)$_4$NBr	Hg	-2.48	sce	6
p-Dichlorobenzene	75% Dioxane, 0.05M (Et)$_4$NBr	Hg	-2.49	sce	6
	DMF, (Et)$_4$NBr	Hg	-1.85	Ag, AgBr	17
	CH$_3$CN, (Et)$_4$NBr	Hg	-2.05	Ag, AgBr	17
o-Dibromobenzene	DMF, (n-Bu)$_4$NBr	Hg	-1.28	Hg pool	8
	CH$_3$CN, (n-Bu)$_4$NBr	Hg	-1.41	Hg pool	8
	75% Dioxane, 0.2M (n-Bu)$_4$NBr	Hg	-1.40	Hg pool	8
	DMF, (n-Bu)$_4$NI	Hg	-1.83	sce	10
m-Dibromobenzene	DMF, (n-Bu)$_4$NI	Hg	-1.94, -2.42	sce	10
	DMF, (n-Bu)$_4$NBr	Hg	-1.40, -1.88	Hg pool	8
	CH$_3$CN, (n-Bu)$_4$NBr	Hg	-1.55, -2.05	Hg pool	8
	75% Dioxane, 0.2M (n-Bu)$_4$NBr	Hg	-1.56	Hg pool	8
	DMF, (Et)$_4$NBr	Hg	-1.45	Ag, AgBr	17
p-Dibromobenzene	DMF, (Et)$_4$NBr	Hg	-1.54	Ag, AgBr	17
	CH$_3$CN, (Et)$_4$NBr	Hg	-1.62	Ag, AgBr	17
	75% Dioxane, 0.05M (Et)$_4$NBr	Hg	-2.10	sce	6
	DMF, (n-Bu)$_4$NBr	Hg	-1.53, -1.82	Hg pool	18
	CH$_3$CN, (n-Bu)$_4$NBr	Hg	-1.70, -2.02	Hg pool	18
	0.5N KCl	Ga	-1.8a	sce	16

Compound	Solvent / Electrolyte	Electrode	E (V)	Reference electrode	Ref.
o-Diiodobenzene	67% Ethanol, 0.01M (Et)$_4$NBr	Hg	-1.23	sce	19
m-Diiodobenzene	67% Ethanol, 0.01M (Et)$_4$NBr	Hg	-1.38	sce	19
p-Diiodobenzene	DMF, (Et)$_4$NBr	Hg	-0.92	sce	17
	CH$_3$CN, (Et)$_4$NBr	Hg	-0.96	sce	17
	67% Ethanol, 0.01M (Et)$_4$NBr	Hg	-1.50	NCE	20
	50% Ethanol, 0.05M (Et)$_4$NI	Hg	-1.480	sce	13
o-Bromochlorobenzene	DMF, (Et)$_4$NBr	Hg	-1.01	Ag, AgBr	17
	0.1M (Me)$_4$NCl	Hg	-1.69	sce	11
	DMF, (n-Bu)$_4$NBr	Hg	-1.38	Hg pool	8
	CH$_3$CN, (n-Bu)$_4$NBr	Hg	-1.54	Hg pool	8
m-Bromochlorobenzene	DMF, (n-Bu)$_4$NI	Hg	-1.94, -2.55	sce	10
	0.1M (Me)$_4$NCl	Hg	-1.87	sce	11
	DMF, (Et)$_4$NBr	Hg	-1.53	Ag, AgBr	17
	CH$_3$CN, (Et)$_4$NBr	Hg	-1.59	Ag, AgBr	17
	DMF, (n-Bu)$_4$NI	Hg	-2.02, -2.57	sce	10
p-Bromochlorobenzene	0.1M (Me)$_4$NCl	Hg	-1.96	sce	11
	DMF, (Et)$_4$NBr	Hg	-1.61	Ag, AgBr	17
	CH$_3$CN, (Et)$_4$NBr	Hg	-1.72	Ag, AgBr	17
	DMF, (n-Bu)$_4$NBr	Hg	-1.60, -2.02	Hg pool	8
	CH$_3$CN, (n-Bu)$_4$NBr	Hg	-1.81	Hg pool	8
	DMF, (n-Bu)$_4$NI	Hg	-2.14, -2.55	sce	10
o-Chloroiodobenzene	90% Alcohol, 0.06M LiCl, OAc buffer, pH 7.00	Hg	-1.48	sce	14
	1M LiClO$_4$	Hg	-1.375	sce	11
	0.1M (Me)$_4$NCl	Hg	-1.203	sce	11

(continued)

Compound	Solvent System	Working Electrode	$E_{1/2}$	Reference Electrode	Reference
m-Chloroiodobenzene	90% Alcohol, 0.06M LiCl, OAc buffer, pH 7.00	Hg	-1.56	sce	14
	1M LiClO4	Hg	-1.509	sce	11
	0.1M (Me)4NCl	Hg	-1.337	sce	11
	DMF, (Et)4NBr	Hg	-0.98	Ag, AgBr	17
	67% Ethanol, 0.01M (Et)4NBr	Hg	-1.45	sce	19
p-Chloroiodobenzene	90% Alcohol, 0.06M LiCl, OAc buffer, pH 7.00	Hg	-1.61	sce	14
	1M LiClO4	Hg	-1.557	sce	11
	0.1M (Me)4NCl	Hg	-1.388	sce	11
	DMF, (Et)4NBr	Hg	-1.06	Ag, AgBr	17
	50% Ethanol, 0.05M (Et)4NI	Hg	-1.475	sce	13
o-Bromofluorobenzene	0.1M (Me)4NCl	Hg	-1.69	sce	11
p-Bromofluorobenzene	0.1M (Me)4NCl	Hg	-2.00	sce	11
o-Bromoiodobenzene	1M LiClO4	Hg	-1.322	sce	11
	0.1M (Me)4NCl	Hg	-1.153	sce	11
	90% Alcohol, 0.06M LiCl, OAc buffer, pH 7.00	Hg	-1.50	sce	14
m-Bromoiodobenzene	1M LiClO4	Hg	-1.482	sce	11
	0.1M (Me)4NCl	Hg	-1.318	sce	11
	DMF, (Et)4NBr	Hg	-0.96	Ag, AgBr	17
	67% Ethanol, 0.01M (Et)4NBr	Hg	-1.13	sce	19

Compound	Conditions	Electrode	E	Reference electrode	Ref.
p-Bromoiodobenzene	90% Alcohol, 0.06M LiCl, OAc buffer, pH 7.00	Hg	−1.61	sce	14
	1M $LiClO_4$	Hg	−1.550	sce	11
	0.1M $(Me)_4NCl$	Hg	−1.383	sce	11
	DMF, $(Et)_4NBr$	Hg	−1.08	Ag, AgBr	17
o-Fluoroiodobenzene	1M $LiClO_4$	Hg	−1.374	sce	11
	0.1M $(Me)_4NCl$	Hg	−1.205	sce	11
p-Fluoroiodobenzene	1M $LiClO_4$	Hg	−1.593	sce	11
	0.1M $(Me)_4NCl$	Hg	−1.420	sce	11
	90% Alcohol, 0.06M LiCl, OAc buffer, pH 7.0	Hg	−0.72, −1.66	sce	14
3,4-Dichlorobromobenzene	0.1M $(Me)_4NCl$	Hg	−1.77	sce	11
1,2,4,5-Tetrabromobenzene	75% Dioxane, 0.05M $(Et)_4NBr$	Hg	−1.45, −2.05	sce	6
Hexachlorobenzene	75% Dioxane, 0.05M $(Et)_4NBr$	Hg	−1.44, −1.69, −1.90	sce	6
Hexabromobenzene	75% Dioxane, 0.05M $(Et)_4NBr$	Hg	−0.75, −1.45	sce	6
m-Chlorotoluene	DMF, $(Et)_4NBr$	Hg	−2.16	Ag, AgBr	17
p-Chlorotoluene	DMF, $(Et)_4NBr$	Hg	−2.16	Ag, AgBr	17
o-Bromotoluene	0.1M $(Me)_4NCl$	Hg	−2.12	sce	11
m-Bromotoluene	0.1M $(Me)_4NCl$	Hg	−2.09	sce	11
	DMF, $(Et)_4NBr$	Hg	−1.85	Ag, AgBr	17
	CH_3CN, $(Et)_4NBr$	Hg	−1.95	Ag, AgBr	17
p-Bromotoluene	0.1M $(Me)_4NCl$	Hg	−2.11	sce	11
	DMF, $(Et)_4NBr$	Hg	−1.84	Ag, AgBr	17
	CH_3CN, $(Et)_4NBr$	Hg	−1.96	Ag, AgBr	17

(continued)

Compound	Solvent System	Working Electrode	$E_{1/2}$	Reference Electrode	Reference
o-Iodotoluene	1M LiClO₄	Hg	-1.688	sce	11
	0.1M (Me)₄NCl	Hg	-1.495	sce	11
	OAc buffer, 0.06M LiCl, 90% alcohol	Hg	-1.70	sce	14
	66% Ethanol, 0.01M (Et)₄NBr	Hg	-1.656	sce	15
m-Iodotoluene	1M LiClO₄	Hg	-1.673	sce	11
	0.1M (Me)₄NCl	Hg	-1.48	sce	11
	OAc buffer, 0.06M HCl, 90% Alcohol	Hg	-1.71	sce	14
	66% Ethanol, 0.01M (Et)₄NBr	Hg	-1.613	sce	15
p-Iodotoluene	DMF, (Et)₄NBr	Hg	-1.22	Ag, AgBr	17
	1M LiClO₄	Hg	-1.687	sce	11
	0.1M (Me)₄NCl	Hg	-1.497	sce	11
	OAc buffer, 0.06M LiCl, 90% Alcohol	Hg	-1.74	sce	14
	0.05M (Et)₄NI, 50% Alcohol	Hg	-1.580	sce	13
	DMF, (Et)₄NBr	Hg	-1.23	Ag, AgBr	17
	66% Ethanol, 0.01M (Et)₄NBr	Hg	-1.660	sce	15
o-Iodoethylbenzene	1M LiClO₄	Hg	-1.685	sce	11
	0.1M (Me)₄NCl	Hg	-1.478	sce	11
2,4-Dimethyliodo-	1M LiClO₄	Hg	-1.718	sce	11

2,4-Dimethyliodo-benzene	0.1M (Me)$_4$NCl	Hg	-1.527	sce	11
2,6-Dimethyliodo-benzene	1M LiClO$_4$	Hg	-1.718	sce	11
	0.1M (Me)$_4$NCl	Hg	-1.516	sce	11

Trifluoromethylbenzenes

X = H	DMF, (n-Bu)$_4$NI, (Et)$_4$NI	Hg	-1.98	Hg pool	21
= o-NH$_2$	DMF, (n-Bu)$_4$NI, (Et)$_4$NI	Hg	-2.14	Hg pool	21
= m-NH$_2$	DMF, (n-Bu)$_4$NI, (Et)$_4$NI	Hg	-2.09	Hg pool	21
= o-SO$_2$NH$_2$	DMF, (n-Bu)$_4$NI, (Et)$_4$NI	Hg	-1.35, -1.98	Hg pool	21
= m-SO$_2$NH$_2$	DMF, (n-Bu)$_4$NI, (Et)$_4$NI	Hg	-1.51, -1.97	Hg pool	21
= p-SO$_2$NH$_2$	DMF, (n-Bu)$_4$NI, (Et)$_4$NI	Hg	-1.40, -1.91	Hg pool	21
= o-CO$_2$H	DMF, (n-Bu)$_4$NI, (Et)$_4$NI	Hg	-1.32, -2.10	Hg pool	21
= m-CO$_2$H	DMF, (n-Bu)$_4$NI, (Et)$_4$NI	Hg	-1.33, -2.03	Hg pool	21

(continued)

Compound	Solvent System	Working Electrode	$E_{1/2}$	Reference Electrode	Reference
= p-CO₂H	DMF, (n-Bu)₄NI, (Et)₄NI	Hg	-1.26, -1.90	Hg pool	21
o-Trifluoromethyl-bromobenzene	0.1M (Me)₄NCl	Hg	-1.67	sce	11
m-Trifluoromethyl-bromobenzene	0.1M (Me)₄NCl	Hg	-1.89	sce	11
p-Trifluoromethyl-bromobenzene	DMF, (Et)₄NBr	Hg	-1.52	Ag, AgBr	17
	CH₃CN, (Et)₄NBr	Hg	-1.59	Ag, AgBr	17
	0.1M (Me)₄NCl	Hg	-1.90	sce	11
	DMF, (Et)₄NBr	Hg	-1.53	AgBr	17
o-Trifluoromethyl-iodobenzene	CH₃CN, (Et)₄NBr	Hg	-1.61	AgBr	17
	1M LiClO₄	Hg	-1.343	sce	11
o-Trifluoromethyl-bromobenzene	0.1M (Me)₄NCl	Hg	-1.190	sce	11
	0.1M (Me)₄NCl	Hg	-1.67	sce	11
m-Trifluoromethyl-iodobenzene	1M LiClO₄	Hg	-1.517	sce	11
p-Trifluoromethyl-iodobenzene	0.1M (Me)₄NCl	Hg	-1.343	sce	11
	DMF, (Et)₄NBr	Hg	-1.00	Ag, AgBr	17
	1M LiClO₄	Hg	-1.515	sce	11
m-Bromobiphenyl	0.1M (Me)₄NCl	Hg	-1.351	sce	11
	DMF, (Et)₄NBr	Hg	-1.01	Ag, AgBr	17
	DMF, (Et)₄NBr	Hg	-1.58	Ag, AgBr	17
	CH₃CN, (Et)₄NBr	Hg	-1.73	Ag, AgBr	17
p-Bromobiphenyl	0.1M (Me)₄NCl	Hg	-1.99	sce	11

Compound	Conditions	Electrode	E	Reference electrode	Ref.
o-Iodobiphenyl	DMF, (Et)₄NBr	Hg	-1.56	Ag, AgBr	17
	CH₃CN, (Et)₄NBr	Hg	-1.70	Ag, AgBr	17
	1M LiClO₄	Hg	-1.570	sce	11
	0.1M (Me)₄NCl	Hg	-1.327	sce	11
	OAc Buffer, 0.06M LiCl, 90% Ethanol, pH 7.0	Hg	-1.61	sce	14
m-Iodobiphenyl	DMF, (Et)₄NBr	Hg	-1.15	Ag, AgBr	17
	1M LiClO₄	Hg	-1.622	sce	11
	0.1M (Me)₄NCl	Hg	-1.413	sce	11
p-Iodobiphenyl	DMF, (Et)₄NBr	Hg	-1.16	Ag, AgBr	17
	CH₃CN, (Et)₄NBr	Hg	-1.20	Ag, AgBr	17
	0.05M (Et)₄NI, 50% Ethanol	Hg	-1.506	sce	13
	OAc Buffer, 0.06M LiCl, 90% Ethanol, pH 7.0	Hg	-1.68	sce	14
o-Bromophenol	0.1M (Me)₄NCl	Hg	-1.97	sce	11
m-Bromophenol	0.1M (Me)₄NCl	Hg	-2.06	sce	11
p-Bromophenol	0.1M (Me)₄NCl	Hg	-2.14	sce	11
o-Iodophenol	1M LiClO₄	Hg	-1.535	sce	11
	0.1M (Me)₄NCl	Hg	-1.391	sce	11
	0.05M (Et)₄NI, 50% Ethanol, pH 4.0	Hg	-1.41	sce	13
	0.05M (Et)₄NI, 50% Ethanol, pH 10.0	Hg	-1.560	sce	13

(continued)

851

Compound	Solvent System	Working Electrode	$E_{1/2}$	Reference Electrode	Reference
	OAc Buffer, 0.06M LiCl, 90% Alcohol, pH 7.0	Hg	-1.55	sce	14
m-Iodophenol	1M LiClO4	Hg	-1.637	sce	11
	0.1M (Me)4NCl	Hg	-1.463	sce	11
p-Iodophenol	1M LiClO4	Hg	-1.700	sce	11
	0.1M (Me)4NCl	Hg	-1.520	sce	11
	0.05M (Et)4N1, 50% Ethanol, pH 4.0	Hg	-1.542	sce	13
	0.05M (Et)4N1, 50% Ethanol, pH 11.0	Hg	-1.66	sce	13
o-Iodobenzyl alcohol	1M LiClO4	Hg	-1.555	sce	11
	0.1M (Me)4NCl	Hg	-1.377	sce	11
p-Iodobenzyl alcohol	1M LiClO4	Hg	-1.636	sce	11
	0.1M (Me)4NCl	Hg	-1.456	sce	11
α-Bromostyrene	DMF, (n-Bu)4NI	Hg	-1.86, -2.31	sce	25
trans-β-Bromostyrene	DMF, (n-Bu)4NI	Hg	-1.98, -2.33	sce	25
α,α'-Dibromo-o-xylene	DMF, (Et)4NBr	Hg	-0.61, -1.58	sce	23
α,α'-Dibromo-m-xylene	DMF, (Et)4NBr	Hg	-1.32	sce	23
α,α'-Dibromo-p-xylene	DMF, (Et)4NBr	Hg	-0.80, -1.72	sce	23
N,N,N-Triethyl-N-(4-bromomethylbenzyl)-ammonium bromide	DMF, (Et)4NBr	Hg	-0.79, -1.72	sce	23
α,α'-Dibromo-2-chloro-p-xylene	DMF, (Et)4NBr	Hg	-0.61, -1.59	sce	23

Compound	Solvent, electrolyte	Electrode	E (V)	Ref. electrode	Ref.
α,α'-Dibromo-2,4-dichloro-p-xylene	DMF, (Et)$_4$NBr	Hg	-0.45, -1.36	sce	23
4,4'-Bis(bromomethyl)-bibenzyl	DMF, (Et)$_4$NBr	Hg	-1.13	sce	23
[2.2]Paracyclophane b	DMF, (Et)$_4$NBr	Hg	No wave	sce	23
α-Iodonaphthalene	66% Ethanol, 0.01M (Et)$_4$NBr	Hg	-1.506	sce	15
	OAc Buffer, 0.06M LiCl, 90% Alcohol, pH 7.0	Hg	-1.62	sce	14
β-Iodonaphthalene	66% Ethanol, 0.01M (Et)$_4$NBr	Hg	-1.560	sce	15
4-Iodoacenaphthene	66% Ethanol, 0.01M (Et)$_4$NBr	Hg	-1.620	sce	15
Benzoyl chloride	Acetone, LiCl	Hg	-1.07	asce c	26
β-Phenylpropionyl chloride	Acetone, LiCl	Hg	-1.24	asce c	26
3-Phenylpropyl bromide	DMF, (Et)$_4$NBr	Hg	-2.178	sce	27
2-Phenylethyl bromide	DMF, (Et)$_4$NBr	Hg	-2.149	sce	27
4-Phenoxybutyl bromide	DMF, (Et)$_4$NBr	Hg	-2.134	sce	27
3-Phenoxypropyl bromide	DMF, (Et)$_4$NBr	Hg	-2.119	sce	27
2-Phenoxyethyl bromide	DMF, (Et)$_4$NBr	Hg	-2.008	sce	27
o-Iodobenzoic acid	0.05M (Et)$_4$N1, 50% Ethanol, pH 3.9	Hg	-1.17	sce	13

(continued)

Compound	Solvent System	Working Electrode	$E_{1/2}$	Reference Electrode	Reference
	0.05M $(Et)_4N1$, 50% Ethanol, pH 8.2	Hg	-1.504	sce	13
	OAc Buffer, 0.06M LiCl, 90% Alcohol, pH 7.0	Hg	-1.47	sce	14
m-Iodobenzoic acid	0.05M $(Et)_4N1$, 50% Ethanol, pH 3.9	Hg	-1.443	sce	13
	0.05M $(Et)_4N1$, 50% Ethanol, pH 8.2	Hg	-1.507	sce	13
	OAc Buffer, 0.06M LiCl, 90% Alcohol, pH 7.0	Hg	-1.57	sce	14
p-Iodobenzoic acid	0.05M $(Et)_4N1$, 50% Ethanol, pH 3.9	Hg	-1.437	sce	13
	0.05M $(Et)_4N1$, 50% Ethanol, pH 8.5	Hg	-1.500	sce	13
	1M $LiClO_4$	Hg	-1.505	sce	11
	0.1M $(Me)_4NCl$	Hg	-1.377	sce	11
	OAc Buffer, 0.06M LiCl, 90% Alcohol,	Hg	-1.57	sce	14

Compound	Solution	Electrode	E	Reference electrode	Ref.
Methyl o-iodobenzoate	0.05M (Et)$_4$N1, 50% Ethanol	Hg	-1.268	sce	13
Methyl-m-iodobenzoate	0.05M (Et)$_4$N1, 50% Ethanol	Hg	-1.433	sce	13
Methyl-p-iodobenzoate	0.05M (Et)$_4$N1, 50% Ethanol	Hg	-1.427	sce	13
o-Bromoaniline	0.1M (Me)$_4$NCl	Hg	-2.01	sce	11
m-Bromoaniline	0.1M (Me)$_4$NCl	Hg	-2.07	sce	11
p-Bromoaniline	0.1M (Me)$_4$NCl	Hg	-2.15	sce	11
	DMF, (Et)$_4$NBr	Hg	-1.96	Ag, AgBr	17
	CH$_3$CN, (Et)$_4$NBr	Hg	-2.07	Ag, AgBr	17
o-Iodoaniline	1M LiClO$_4$	Hg	-1.565	sce	11
	0.1M (Me)$_4$NCl	Hg	-1.416	sce	11
m-Iodoaniline	1M LiClO$_4$	Hg	-1.650	sce	11
	0.1M (Me)$_4$NCl	Hg	-1.463	sce	11
p-Iodoaniline	1M LiClO$_4$	Hg	-1.710	sce	11
	0.1M (Me)$_4$NCl	Hg	-1.529	sce	11
	OAc Buffer, 0.06M LiCl, 90% Alcohol, pH 7.0	Hg	-1.76	sce	14
	50% Ethanol, 0.05M (Et)$_4$N1, pH 4.0	Hg	-1.42	sce	13
	50% Ethanol, 0.05M (Et)$_4$N1, pH 8.0	Hg	-1.595	sce	13
m-Chloro-N,N-dimethylaniline	DMF, (Et)$_4$NBr	Hg	-2.23	Ag, AgBr	17

(continued)

Compound	Solvent System	Working Electrode	$E_{1/2}$	Reference Electrode	Reference
p-Bromo-N,N-dimethylaniline	DMF, (Et)4NBr	Hg	-1.97	Ag, AgBr	17
p-Iodo-N,N-dimethyl-aniline	CH3CN, (Et)4NBr	Hg	-2.04	Ag, AgBr	17
	DMF, (Et)4NBr	Hg	-1.35	Ag, AgBr	17
p-Chloroanisole (p-Methoxychloro-benzene)	CH3CN, (Et)4NBr	Hg	-1.36	Ag, AgBr	17
	50% Ethanol, 0.05M (Et)4N1, pH 8.5	Hg	-1.615	sce	13
	DMF, (Et)4NBr	Hg	-2.15	Ag, AgBr	17
o-Bromoanisole	0.1M (Me)4NCl	Hg	-1.93	sce	11
m-Bromoanisole	DMF, (Et)4NBr	Hg	-1.76	Ag, AgBr	17
	CH3CN, (Et)4NBr	Hg	-1.84	Ag, AgBr	17
p-Bromoanisole	DMF, (Et)4NBr	Hg	-1.84	Ag, AgBr	17
	CH3CN, (Et)4NBr	Hg	-1.95	Ag, AgBr	17
	0.1M (Me)4NCl	Hg	-2.10	sce	11
o-Iodoanisole	1M LiClO4	Hg	-1.567	sce	11
	0.1M (Me)4NCl	Hg	-1.393	sce	11
	OAc Buffer, 0.06M LiCl, 90% Alcohol, pH 7.0	Hg	-1.60	sce	14
m-Iodoanisole	1M LiClO4	Hg	-1.603	sce	11
	0.1M (Me)4NCl	Hg	-1.421	sce	11
	DMF, (Et)4NBr	Hg	-1.19	Ag, AgBr	17
	CH3CN, (Et)4NBr	Hg	-1.23	Ag, AgBr	17

Compound	Supporting electrolyte	Electrode	E	Ref. electrode	Ref.
	OAc Buffer, 0.06M LiCl, 90% Alcohol, pH 7.0	Hg	-1.64	sce	14
p-Iodoanisole	1M $LiClO_4$	Hg	-1.672	sce	11
	0.1M $(Me)_4NCl$	Hg	-1.488	sce	11
	DMF, $(Et)_4NBr$	Hg	-1.25	Ag, AgBr	17
	CH_3CN, $(Et)_4NBr$	Hg	-1.26	Ag, AgBr	17
	OAc Buffer, 0.06M LiCl, 90% Alcohol, pH 7.0	Hg	-1.75	sce	14
o-Bromophenetole (o-Ethoxybromobenzene)	50% Ethanol, 0.05M $(Et)_4Nl$	Hg	-1.562	sce	13
p-Bromophenetole	0.1M $(Me)_4NCl$	Hg	-1.99	sce	11
o-Iodophenetole	0.1M $(Me)_4NCl$	Hg	-2.10	sce	11
	DMF, $(Et)_4NBr$	Hg	-1.82	Ag, AgBr	17
	CH_3CN, $(Et)_4NBr$	Hg	-1.93	Ag, AgBr	17
	1M $LiClO_4$	Hg	-1.587	sce	11
	0.1M $(Me)_4NCl$	Hg	-1.405	sce	11
	OAc Buffer, 0.06M LiCl, 90% Alcohol, pH 7.0	Hg	-1.64	sce	13
p-Iodophenetole	OAc Buffer, 0.06M LiCl, 90% Alcohol, pH 7.0	Hg	-1.75	sce	13

(continued)

Compound	Solvent System	Working Electrode	$E_{1/2}$	Reference Electrode	Reference
p-Phenoxybromobenzene	1M LiClO4	Hg	-1.672	sce	11
	0.1M (Me)4NCl	Hg	-1.488	sce	11
p-Chlorobenzaldehyde	DMF, (Et)4NBr	Hg	-1.73	Ag, AgBr	17
	CH3CN, (Et)4NBr	Hg	-1.81	Ag, AgBr	17
p-Iodobenzaldehyde	DMF, (Et)4NBr	Hg	-1.20	Ag, AgBr	17
	CH3CN, (Et)4NBr	Hg	-1.26	Ag, AgBr	17
p-Chlorobenzonitrile	DMF, (Et)4NBr	Hg	-0.96	Ag, AgBr	17
	CH3CN, (Et)4NBr	Hg	-1.01	Ag, AgBr	17
m-Bromobenzonitrile	DMF, (Et)4NBr	Hg	-1.36	Ag, AgBr	17
	CH3CN, (Et)4NBr	Hg	-1.48	Ag, AgBr	17
p-Bromobenzonitrile	DMF, (Et)4NBr	Hg	-1.29	Ag, AgBr	17
	CH3CN, (Et)4NBr	Hg	-1.41	Ag, AgBr	17
p-Iodobenzonitrile	DMF, (Et)4NBr	Hg	-1.26	Ag, AgBr	17
	CH3CN, (Et)4NBr	Hg	-1.37	Ag, AgBr	17
	1M LiClO4	Hg	-1.417	sce	11
	0.1M (Me)4NCl	Hg	-1.287	sce	11
m-Bromoacetophenone	DMF, (Et)4NBr	Hg	-1.19	Ag, AgBr	17
	CH3CN, (Et)4NBr	Hg	-1.29	Ag, AgBr	17
p-Bromoacetophenone	DMF, (Et)4NBr	Hg	-1.15	Ag, AgBr	17
	CH3CN, (Et)4NBr	Hg	-1.24	Ag, AgBr	17
p-Iodoacetophenone	DMF, (Et)4NBr	Hg	-1.04	Ag, AgBr	17
	CH3CN, (Et)4NBr	Hg	-1.12	Ag, AgBr	17
p-Bromobenzophenone	DMF, (Et)4NBr	Hg	-1.06	Ag, AgBr	17
	CH3CN, (Et)4NBr	Hg	-1.13	Ag, AgBr	17
p-Iodobenzophenone	DMF, (Et)4NBr	Hg	-0.96	Ag, AgBr	17
	CH3CN, (Et)4NBr	Hg	-1.01	Ag, AgBr	17
p-Chloroacetanilide	DMF, (Et)4NBr	Hg	-2.09	Ag, AgBr	17
	CH3CN, (Et)4NBr	Hg	-2.16	Ag, AgBr	17

Compound	Conditions	Electrode	E	Ref. electrode	Ref.
p-Bromoacetanilide	DMF, (Et)₄NBr	Hg	-1.88	Ag, AgBr	17
p-Iodoacetanilide	CH₃CN, (Et)₄NBr	Hg	-1.83	Ag, AgBr	17
	50% Ethanol, 0.05M (Et)₄Nl	Hg	-1.510	sce	13
	OAc Buffer, 0.06M LiCl, 90% Alcohol, pH 7.0	Hg	-1.66	sce	14
Benzyl chloride	0.5N KCl	Ga	~-1.3	sce	16
	DMF, (n-Bu)₄NBr	Hg	-0.84, -1.82	Hg pool	22
	CH₃CN, (n-Bu)₄NBr	Hg	-0.83, -1.88	Hg pool	22
	CH₃CN, (n-Bu)₄Nl	Hg	-0.12, -1.89	Hg pool	22
	75% Dioxane, (Et)₄NBr	Hg	-1.94	sce	6
Benzyl bromide	DMF, (n-Bu)₄NBr	Hg	-0.83	Hg pool	22
	CH₃CN, (n-Bu)₄NBr	Hg	-0.84	Hg pool	22
	CH₃CN, (n-Bu)₄NBr	Hg	-0.19	Hg pool	22
	DMF, (n-Bu)₄NBr	Hg	-1.22	sce	23
		Hg	-1.330	Ag, AgCl	24
Benzyl bromide	Methanol, 0.25M LiOMe	Hg	-1.324	Ag, AgCl	24
	Methanol, 0.25M LiCl				
	DMF, (Et)₄NBr	Hg	-0.82	Ag, AgBr	17
	DMF, (Et)₄NBr	Hg	-1.22	sce	23
Benzyl iodide	CH₃CN, (n-Bu)₄NBr	Hg	-0.24	Hg pool	22
	CH₃CN, (n-Bu)₄Nl	Hg	-0.27	Hg pool	22

Substituted Benzyl Bromides

Substituent:

| p-CN | CH₃CN, Et₄NBr | Hg | -0.42 | Ag, AgBr | 17 |

(continued)

859

Compound	Solvent System	Working Electrode	E 1/2	Reference Electrode	Reference
m-Br	CH₃CN, Et4NBr	Hg	-0.62	Ag, AgBr	17
m-Cl	CH₃CN, Et4NBr	Hg	-0.71	Ag, AgBr	17
m-F	CH₃CN, Et4NBr	Hg	-0.71	Ag, AgBr	17
p-Br	CH₃CN, Et4NBr	Hg	-0.72	Ag, AgBr	17
p-Cl	CH₃CN, Et4NBr	Hg	-0.73	Ag, AgBr	17
m-OMe	CH₃CN, Et4NBr	Hg	-0.80	Ag, AgBr	17
p-CMe	CH₃CN, Et4NBr	Hg	-0.87	Ag, AgBr	17
p-F	CH₃CN, Et4NBr	Hg	-0.84	Ag, AgBr	17
m-Me	CH₃CN, Et4NBr	Hg	-0.78	Ag, AgBr	17
p-Me	CH₃CN, Et4NBr	Hg	-0.80	Ag, AgBr	17

Nitrobenzyl Halides

Compound	Solvent System	Working Electrode	E 1/2	Reference Electrode	Reference
o-Nitrobenzyl chloride	CH₃CN, (Et)4-NClO4	Pt	-1.00, -1.26, -1.36 Ep	sce	28
m-Nitrobenzyl chloride	CH₃CN, (Et)4-NClO4	Pt	-1.12 Ep	sce	28
p-Nitrobenzyl chloride	CH₃CN, (Et)4-NClO4	Pt	-0.97, -1.22 Ep	sce	28
o-Nitrobenzyl bromide	CH₃CN, (Et)4-NClO4	Pt	-0.91, --, d --d Ep	sce	28
m-Nitrobenzyl bromide	CH₃CN, (Et)4-NClO4	Pt	-1.12 Ep	sce	28
p-Nitrobenzyl bromide	CH₃CN, (Et)4-NClO4	Pt	-0.86, --, d --d Ep	sce	28

Twenty-seven sets of
half-wave potentials
of ortho-substituted
benzene derivatives.

including halogen
and nonhalogen
derivatives

a Foot of the wave.
b Included for comparison.
c Acetone-saturated calomel electrode.
d Numerical data not given.

(continued)

References for Aromatic Halides

1. W. C. Neikam, G. R. Dimeler, and M. M. Desmond, J. Electrochem. Soc., 111, 1190 (1964).
2. C. Parkanyi and R. Zahradnik, Collect. Czech. Chem. Commun., 30, 4288 (1965).
3. H. Lund, Acta Chem. Scand., 11, 1323 (1957).
4. V. D. Parker and L. Eberson, Chem. Commun., 1969 973.
5. A. E. Coleman, H. H. Richtol, and D. A. Aikens, J. Electroanal. Chem., 18, 165 (1968).
6. M. von Stackelberg and W. Stracke, Z. Elektrochem., 53, 118 (1945).
7. F. L. Lambert and K. Kobayashi, J. Org. Chem., 23, 773 (1958).
8. S. Wawzonek and J. H. Wagenknecht, J. Electrochem. Soc., 110, 420 (1963).
9. J. W. Sease, F. G. Burton, and S. L. Nickol, J. Am. Chem. Soc., 90, 2595 (1968).
10. L. Meites, Polarographic Techniques, 2nd ed., Interscience, New York, 1965, p. 681.
11. W. W. Hussey and A. J. Diefenderfer, J. Am. Chem. Soc., 89, 5359 (1967).
12. I. Bagotskaya and D. K. Durmanov, Elektrokhim, 4, 115 (1968).
13. R. A. Caldwell and S. Hacobian, Aust. J. Chem., 21, 1 (1968).
14. E. L. Colichman and S. K. Liu, J. Am. Chem. Soc. 76, 913 (1954).
15. E. Gergely and T. Iredale, J. Chem. Soc., 1951, 13.
16. I. Bagotskaya and D. K. Durmanov, Elektrokhim, 4, 115 (1968).
17. J. W. Sease, F. G. Burton, and S. K. Nickol, J. Am. Chem. Soc., 90, 2595 (1968).
18. H. Wittig, Angew. Chem., 69, 245 (1957).
19. Reference 10, p. 682.
20. E. Gergely and T. Iredale, J. Chem. Soc., 1953, 3226.
21. A. I. Cohen, B. T. Keeler, N. H. Coy, and H. L. Yale, Anal. Chem., 34, 216 (1962).
22. S. Wawzonek, R. C. Duty, and J. H. Wagenknecht, J. Electrochem. Soc., 111, 74 (1964).
23. F. H. Covitz, J. Am. Chem. Soc., 89, 5403 (1967)
24. G. Klopman, Helv. Chim. Acta, 44, 1908 (1961).
25. L. L. Miller and E. Riekena, J. Org. Chem., 34, 3359 (1969).
26. P. Arthur and H. Lyons, Anal. Chem., 24, 1422 (1952).

27. F. L. Lambert, J. Org. Chem., 31, 4184 (1966).
28. J. G. Lawless, D. E. Bartak, and M. D. Hawley, J. Am. Chem. Soc., 91, 7121 (1969).
29. M. Charton and B. I. Charton, J. Org. Chem., 36, 260 (1971).

23. CARBON-NITROGEN FUNCTION

Oxidation

Schiff Bases

$R_1 = H$, $R_2 = H$, with structure CH=N

Compound	Solvent System	Working Electrode	$E_{1/2}$	Reference Electrode	Reference
$R_1 = H$, $R_2 = H$	CH_3CN, 0.01$\underline{M}$-LiClO$_4$	Pt	+1.29, +1.60	Ag/0.01$\underline{M}$ Ag+	1
= m-CH$_3$	CH_3CN, 0.01$\underline{M}$-LiClO$_4$	Pt	+1.32, +1.60	Ag/0.01$\underline{M}$ Ag+	1
= p-CH$_3$	CH_3CN, 0.01$\underline{M}$-LiClO$_4$	Pt	+1.25, +1.77	Ag/0.01$\underline{M}$ Ag+	1
= m-OCH$_3$	CH_3CN, 0.01$\underline{M}$-LiClO$_4$	Pt	+1.33, +1.86	Ag/0.01$\underline{M}$ Ag+	1
= p-OCH$_3$	CH_3CN, 0.01$\underline{M}$-LiClO$_4$	Pt	+1.18, +1.78	Ag/0.01$\underline{M}$ Ag+	1
= m-OH	CH_3CN, 0.01$\underline{M}$-LiClO$_4$	Pt	+1.30	Ag/0.01$\underline{M}$ Ag+	1
= p-OH	CH_3CN, 0.01$\underline{M}$-LiClO$_4$	Pt	+0.86, +1.20	Ag/0.01$\underline{M}$ Ag+	1

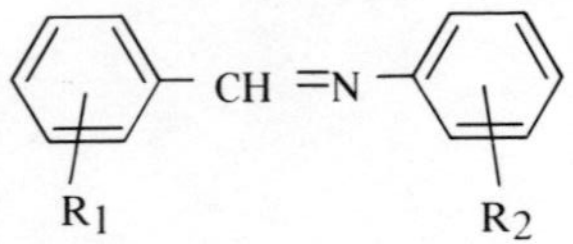

		Solvent, Electrolyte	Electrode	E	Reference Electrode	Ref.
= m-Cl	= H	CH_3CN, $LiClO_4$ 0.01$\underline{M}$	Pt	+1.47	Ag/0.01$\underline{M}$ Ag+	1
= p-Cl	= H	CH_3CN, $LiClO_4$ 0.01$\underline{M}$	Pt	+1.36	Ag/0.01$\underline{M}$ Ag+	1
= p-N(CH_3)_2	= H	CH_3CN, $LiClO_4$ 0.01$\underline{M}$	Pt	+0.55, +1.32	Ag/0.01$\underline{M}$ Ag+	1
= H	= p-CH_3	CH_3CN, $LiClO_4$ 0.01$\underline{M}$	Pt	+1.21, +1.67	Ag/0.01$\underline{M}$ Ag+	1
= H	= p-OCH_3	CH_3CN, $LiClO_4$ 0.01$\underline{M}$	Pt	+0.98, +1.61	Ag/0.01$\underline{M}$ Ag+	1
= H	= p-OH	CH_3CN, $LiClO_4$ 0.01$\underline{M}$	Pt	+0.80, +1.57	Ag/0.01$\underline{M}$ Ag+	1
= H	= p-NH_2	CH_3CN, $LiClO_4$ 0.01$\underline{M}$	Pt	+0.53, +1.14	Ag/0.01$\underline{M}$ Ag+	1

Aliphatic Amides

	Solvent, Electrolyte	Electrode	E	Reference Electrode	Ref.
N,N-Dimethylformamide	CH_3CN, $NaClO_4$	Pt	+1.21 E_p	Ag/0.1$\underline{N}$ Ag+	2
	CH_3CN, $NaClO_4$	Pt	+1.90	sce	3
Acetamide	CH_3CN, $NaClO_4$	Pt	Near +2.0 E_p	sce	4
N-Methylacetamide	CH_3CN, $NaClO_4$	Pt	+1.51	Ag/0.1$\underline{N}$ Ag+	2
	CH_3CN, $NaClO_4$	Pt	+1.81 E_p	sce	4
N,N-Dimethylacetamide	CH_3CN, $NaClO_4$	Pt	+1.02 E_p	Ag/0.1$\underline{N}$ Ag+	2
	CH_3CN, $NaClO_4$	Pt	+1.32 E_p	sce	4
N,N-Diethylacetamide	CH_3CN, $NaClO_4$	Pt	+0.97 E_p	Ag/0.1$\underline{N}$ Ag+	2
N-Propylacetamide	CH_3CN, $NaClO_4$	Pt	+1.46 E_p	Ag/0.1$\underline{N}$ Ag+	2
N,N-Dipropylacetamide	CH_3CN, $NaClO_4$	Pt	+0.97 E_p	Ag/0.1$\underline{N}$ Ag+	2
N,N-Diamylacetamide	CH_3CN, $NaClO_4$	Pt	+0.99 E_p	Ag/0.1$\underline{N}$ Ag+	2
N,N-Dimethylpropion-amide	CH_3CN, $NaClO_4$	Pt	+0.94 E_p	Ag/0.1$\underline{N}$ Ag+	2

(continued)

Compound	Solvent System	Working Electrode	$E_{1/2}$	Reference Electrode	Reference
N,N-Dipropylpropion-amide	CH_3CN, $NaClO_4$	Pt	+0.96 E_p	Ag/0.1$\underline{N}$ Ag+	2
N,N-Dimethylbutyramide	CH_3CN, $NaClO_4$	Pt	+1.26 E_p	sce	4
N,N-Diethylisobutyr-amide	CH_3CN, $NaClO_4$	Pt	+0.89 E_p	Ag/0.1$\underline{N}$ Ag+	2
	CH_3CN, $NaClO_4$	Pt	+1.02 E_p	Ag/0.1$\underline{N}$ Ag+	2
N,N-Dimethylstear-amide	CH_3CN, $NaClO_4$	Pt	+0.92 E_p	Ag/0.1$\underline{N}$ Ag+	2

Reduction

Aliphatic Amides

Compound	Solvent System	Working Electrode	$E_{1/2}$	Reference Electrode	Reference
Acrylamide	Methanol, 0.01$\underline{M}$ (Et)$_4$N1	Hg	-1.72	NHE	5

Aromatic Amides

Compound	Solvent System	Working Electrode	$E_{1/2}$	Reference Electrode	Reference
Benzamide	Ethanol, 0.1$\underline{M}$ (Me)$_4$NCl	Hg	-2.15	Ag, AgCl	6
p-Tolylamide	Ethanol, 0.1$\underline{M}$ (Me)$_4$NCl	Hg	-2.32	Ag, AgCl	6

p-Anisylamide

NHCOCH$_n$Cl$_m$ (substituted benzene with R)

R	n	m	Solvent	Electrode	E	Ref. electrode	Ref.
			Ethanol, 0.1$\underline{M}$ (Me)$_4$NCl	Hg	-2.33	Ag, AgCl	6
H	2	1	0.5$\underline{M}$ (Me)$_4$NI, 50% Alcohol, B-W buffer, pH 7.75	Hg	-1.42	sce	7
H	1	2	0.5$\underline{M}$ (Me)$_4$NI, 50% Alcohol, B-W buffer, pH 7.75	Hg	-1.05, -1.40	sce	7
H	0	3	0.5$\underline{M}$ (Me)$_4$NI, 50% Alcohol, B-W buffer, pH 7.75	Hg	-0.09, -1.06, -1.41	sce	7
$\underline{m}$-CH$_3$	1	2	0.5$\underline{M}$ (Me)$_4$NI, 50% Alcohol, B-W buffer, pH 7.75	Hg	-0.98, -1.31	sce	7
$\underline{p}$-CH$_3$	1	2	0.5$\underline{M}$ (Me)$_4$NI, 50% Alcohol, B-W buffer, pH 7.75	Hg	-0.98, -1.31	sce	7

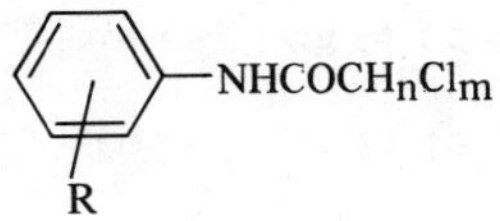

(continued)

Compound		Solvent System	Working Electrode	$E_{1/2}$	Reference Electrode	Reference
m-CH$_3$	0 3	0.5M (Me)$_4$NI, 50% Alcohol, B-W buffer^a, pH 7.75	Hg	-0.09, -0.93, -1.36	sce	7
p-CH$_3$	0 3	0.5M (Me)$_4$NI, 50% Alcohol, B-W buffer^a, pH 7.75	Hg	-0.12, -0.95, -1.36	sce	7
2,3-(CH$_3$)$_2$	2 1	0.5M (Me)$_4$NI, 50% Alcohol, B-W buffer^a, pH 7.75	Hg	-1.34	sce	7
3,4-(CH$_3$)$_2$	2 1	0.5M (Me)$_4$NI, 50% Alcohol, B-W buffer^a, pH 7.75	Hg	-1.30	sce	7

Anilides

Compound	Solvent System	Working Electrode	$E_{1/2}$	Reference Electrode	Reference
p-Chloroacetanilide	DMF, 0.1M (Et)$_4$NI	Hg	-2.06	Hg pool	8
Benzanilide	Ethanol, 0.1M (Me)$_4$NCl	Hg	-2.02	Ag, AgCl	6
p-Tolylanilide	Ethanol, 0.1M (Me)$_4$NCl	Hg	-2.08	Ag, AgCl	6
p-Anisylanilide	Ethanol, 0.1M (Me)$_4$NCl	Hg	-2.11	Ag, AgCl	6
p-Hydroxybenzanilide	Ethanol, 0.1M (Me)$_4$NCl	Hg	-2.20	Ag, AgCl	6

p-Chlorobenzanilide	Ethanol, 0.1$\underline{M}$ (Me)$_4$NCl	Hg	-1.90	Ag, AgCl	6
p-Fluorobenzanilide	Ethanol, 0.1$\underline{M}$ (Me)$_4$NCl	Hg	-2.04	Ag, AgCl	6
m-Trifluoromethyl-benzanilide	Ethanol, 0.1$\underline{M}$ (Me)$_4$NCl	Hg	-1.86	Ag, AgCl	6
N-Benzoyl-$\underline{p}$-toluidide	Ethanol, 0.1$\underline{M}$ (Me)$_4$NCl	Hg	-2.01	Ag, AgCl	6
N-Benzoyl-$\underline{p}$-chloro-anilide	Ethanol, 0.1$\underline{M}$ (Me)$_4$NCl	Hg	-1.95	Ag, AgCl	6
N-Benzoyl-2,4-di-chloroanilide	Ethanol, 0.1$\underline{M}$ (Me)$_4$NCl	Hg	-1.91	Ag, AgCl	6
N-Benzoyl-N-methyl-anilide	Ethanol, 0.1$\underline{M}$ (Me)$_4$NCl	Hg	-1.93	Ag, AgCl	6
N-$\underline{p}$-Tolyl-N-methyl-anilide	Ethanol, 0.1$\underline{M}$ (Me)$_4$N	Hg	-2.06	Ag, AgCl	6
N-$\underline{p}$-Anisyl-N-methyl-anilide	Ethanol, 0.1$\underline{M}$ (Me)$_4$NCl	Hg	-2.08	Ag, AgCl	6
N-Benzoyldiphenyl-amide	Ethanol, 0.1$\underline{M}$ (Me)$_4$NCl	Hg	-1.83	Ag, AgCl	6
N-$\underline{p}$-Tolyldiphenyl-amide	Ethanol, 0.1$\underline{M}$ (Me)$_4$NCl	Hg	-1.90	Ag, AgCl	6
N-$\underline{p}$-Anisyldiphenyl-amide	Ethanol, 0.1$\underline{M}$ (Me)$_4$NCl	Hg	-1.93	Ag, AgCl	6
o-Phthalylanilide	Ethanol, 0.1$\underline{M}$ (Me)$_4$NCl	Hg	-1.90	Ag, AgCl	6

Isocyanates

| 1,2,4-Trichloroben-zene-3,5-diisocyanate | DMF, ($\underline{n}$-Bu)$_4$NI | Hg | -1.99 | Hg pool | 9 |

(continued)

Compound	Solvent System	Working Electrode	$E_{1/2}$	Reference Electrode	Reference
1,5-Dichlorobenzene-2,4-diisocyanate	DMF, (n-Bu)$_4$NI	Hg	-2.07	Hg pool	9
5-Chloro-2,4-toluidine diisocyanate	DMF, (n-Bu)$_4$NI	Hg	-2.09	Hg pool	9
Metaphenylene diisocyanate	DMF, (n-Bu)$_4$NI	Hg	-2.10	Hg pool	9
2,4-Toluidine diisocyanate	DMF, (n-Bu)$_4$NI	Hg	-2.15	Hg pool	9
1,4-Methoxybenzene diisocyanate	DMF, (n-Bu)$_4$NI	Hg	-2.17	Hg pool	9
1-Methoxybenzene diisocyanate	DMF, (n-Bu)$_4$NI	Hg	-2.22	Hg pool	9
2,6-Toluidine diisocyanate	DMF, (n-Bu)$_4$NI	Hg	-2.25	Hg pool	9
1,4-Cyclohexane diisocyanate	DMF, (n-Bu)$_4$NI	Hg	-2.25	Hg pool	9
Tolylene diisocyanate	75% Dioxane, 25% DMF, (n-Bu)$_4$NI, (Et)$_4$NI	Hg	-1.74	Hg pool	10
Phenyl isocyanate	75% Dioxane, 25% DMF, (n-Bu)$_4$NI, (Et)$_4$NI	Hg	-1.55	Hg pool	10

Schiff Bases

X =		Solvent / Electrolyte		Metal	E		Ref. electrode	Ref.
X = H		DMF,	0.1M (Et)$_4$NBr	Hg	-2.15,	-2.35	sce	11
= m-CH$_3$		DMF,	0.1M (Et)$_4$NBr	Hg	-2.08,	-2.30	sce	11
= m-OCH$_3$		DMF,	0.1M (Et)$_4$NBr	Hg	-2.02,	-2.25	sce	11
= p-OCH$_3$		DMF,	0.1M (Et)$_4$NBr	Hg	-2.15,	-2.36	sce	11

X =	Y =	Solvent / Electrolyte		Metal	E		Ref. electrode	Ref.
X = H	Y = o-OH	DMF,	0.05N (Et)$_4$NI	Hg	-1.451, -1.636,	-2.016	sce	12
= p-Cl	= o-OH	DMF,	0.05N (Et)$_4$NI	Hg	-1.346, -1.533,	-1.849	sce	12
= p-CH$_3$	= o-OH	DMF,	0.05N (Et)$_4$NI	Hg	-1.505, -1.780,	-2.140	sce	12
= p-OCH$_3$	= o-OH	DMF,	0.05N (Et)$_4$NI	Hg	-1.560, -1.838,	-2.210	sce	12
= p-N(CH$_3$)$_2$	= o-OH	DMF,	0.05N (Et)$_4$NI	Hg	-1.600, -1.850,	-2.225	sce	12
= m-Cl	= o-OH	DMF,	0.05N (Et)$_4$NI	Hg	-1.341, -1.521,	-1.901	sce	12
= m-F	= o-OH	DMF,	0.05N (Et)$_4$NI	Hg	-1.327, -1.502,	-1.927	sce	12
= o-Cl	= o-OH	DMF,	0.05N (Et)$_4$NI	Hg	-1.304, -1.481,	-1.866	sce	12
= o-F	= o-OH	DMF,	0.05N (Et)$_4$NI	Hg	-1.378, -1.652,	-2.085	sce	12
= o-CH$_3$	= o-OH	DMF,	0.05N (Et)$_4$NI	Hg	-1.420, -1.687,	-2.055	sce	12

(continued)

Compound		Solvent System	Working Electrode	$E_{1/2}$	Reference Electrode	Reference
= o-OCH₃	= o-OH	DMF, 0.05N (Et)₄NI	Hg	-1.418, -1.638, -2.023	sce	12
= o-OC₄H₉	= o-OH	DMF, 0.05N (Et)₄NI	Hg	-1.430, -1.688, -2.012	sce	12
= 2,4-Cl₂	= o-OH	DMF, 0.05N (Et)₄NI	Hg	-1.210, -1.412, -1.822	sce	12
= 2,6-Cl₂	= o-OH	DMF, 0.05N (Et)₄NI	Hg	-1.185, -1.417, -1.840	sce	12
= H	= p-OH	DMF, 0.05N (Et)₄NI	Hg	-1.768, -2.070	sce	12
= p-Cl	= p-OH	DMF, 0.05N (Et)₄NI	Hg	-1.688, -1.947	sce	12
= p-CH₃	= p-OH	DMF, 0.05N (Et)₄NI	Hg	-1.831, -2.146	sce	12
= p-OCH₃	= p-OH	DMF, 0.05N (Et)₄NI	Hg	-1.840, -2.215	sce	12
= p-N(CH₃)₂	= p-OH	DMF, 0.05N (Et)₄NI	Hg	-1.890, -2.280	sce	
= m-Cl	= p-OH	DMF, 0.05N (Et)₄NI	Hg	-1.647, -1.967	sce	12
= m-F	= p-OH	DMF, 0.05N (Et)₄NI	Hg	-1.700, -1.950	sce	12
= o-Cl	= p-OH	DMF, 0.05N (Et)₄NI	Hg	-1.594, -1.926	sce	12

= o-F = p-OH	DMF, 0.05N-(Et)4N1	Hg	-1.654, -1.950	sce	12
= o-CH3 = p-OH	DMF, 0.05N-(Et)4N1	Hg	-1.738, -2.035	sce	12
= o-OCH3 = p-OH	DMF, 0.05N-(Et)4N1	Hg	-1.776, -2.122	sce	12
= 2,4-Cl2 = p-OH	DMF, 0.05N-(Et)4N1	Hg	-1.515, -1.818	sce	12
= o-Cl = p-OCH3	DMF, 0.05N-(Et)4N1	Hg	-1.631, -1.911	sce	12
= o-Cl = o-OCH3	DMF, 0.05N-(Et)4N1	Hg	-1.636, -1.886	sce	12
Benzylidene Aniline	DMF, 0.05N-(Et)4N1	Hg	-1.740	sce	12
p-Chlorobenzylidene aniline	DMF, 0.05N-(Et)4N1	Hg	-1.660	sce	12
p-Methylbenzylidene aniline	DMF, 0.05N-(Et)4N1	Hg	-1.753	sce	12
p-Methoxybenzylidene aniline	DMF, 0.05N-(Et)4N1	Hg	-1.818	sce	12
p-Dimethylaminobenzylidene aniline	DMF, 0.05N-(Et)4N1	Hg	-1.863	sce	12

Schiff Bases Derived from:

Amine	Aldehyde					
Aniline	Benzaldehyde	DMF, (n-Propyl)4-NClO4	Hg	-1.83	sce	13
Aniline	Cinnamaldehyde	DMF, (n-Propyl)4-NClO4	Hg	-1.61	sce	13

(continued)

873

Compound	Solvent System	Working Electrode	$E_{1/2}$	Reference Electrode	Reference
Aniline	DMF, $(\underline{n}\text{-Propyl})_4\text{-}$ $NClO_4$	Hg	-1.69	sce	13
Aniline	DMF, $(\underline{n}\text{-Propyl})_4\text{-}$ $NClO_4$	Hg	-1.75	sce	13
Aniline	DMF, $(\underline{n}\text{-Propyl})_4\text{-}$ $NClO_4$	Hg	-1.65	sce	13
Aniline	DMF, $(\underline{n}\text{-Propyl})_4\text{-}$ $NClO_4$	Hg	-1.36	sce	13
Aniline	DMF, $(\underline{n}\text{-Propyl})_4\text{-}$ $NClO_4$	Hg	-1.77	sce	13
4-Amino-biphenyl	DMF, $(\underline{n}\text{-Propyl})_4\text{-}$ $NClO_4$	Hg	-1.56	sce	13
4-Amino-biphenyl	DMF, $(\underline{n}\text{-Propyl})_4\text{-}$ $NClO_4$	Hg	-1.65	sce	13
4-Amino-biphenyl	DMF, $(\underline{n}\text{-Propyl})_4\text{-}$ $NClO_4$	Hg	-1.68	sce	13
4-Amino-biphenyl	DMF, $(\underline{n}\text{-Propyl})_4\text{-}$ $NClO_4$	Hg	-1.61	sce	13
4-Amino-biphenyl	DMF, $(\underline{n}\text{-Propyl})_4\text{-}$ $NClO_4$	Hg	-1.32	sce	13
2-Amino-naphthalene	DMF, $(\underline{n}\text{-Propyl})_4\text{-}$ $NClO_4$	Hg	-1.75	sce	13
2-Amino-naphthalene	DMF, $(\underline{n}\text{-Propyl})_4\text{-}$ $NClO_4$	Hg	-1.55	sce	13
2-Amino-naphthalene	DMF, $(\underline{n}\text{-Propyl})_4\text{-}$ $NClO_4$	Hg	-1.62	sce	13
2-Amino-naphthalene	DMF, $(\underline{n}\text{-Propyl})_4\text{-}$ $NClO_4$	Hg	-1.67	sce	13

Compound descriptors (second line per group):
1-Naphthal-dehyde, 2-Naphthal-dehyde, 9-Phenanthral-dehyde, 9-Anthracenal-dehyde, Benzaldehyde, Cinnamal-dehyde, 1-Naphthal-dehyde, 2-Naphthal-dehyde, 9-Phenanthral-dehyde, 9-Anthracenal-dehyde, Benzaldehyde, Cinnamal-dehyde, 1-Naphthal-dehyde, 2-Naphthal-dehyde

Amine	Aldehyde	Solvent/electrolyte		E (V)	Ref. electrode	Ref.
2-Amino-naphthalene	9-Phenanthral-dehyde	DMF, $(n\text{-Propyl})_4\text{NClO}_4$	Hg	-1.59	sce	13
2-Amino-naphthalene	9-Anthracenal-dehyde	DMF, $(n\text{-Propyl})_4\text{NClO}_4$	Hg	-1.32	sce	13
6-Amino-chrysene	Benzaldehyde	DMF, $(n\text{-Propyl})_4\text{NClO}_4$	Hg	-1.66	sce	13
6-Amino-chrysene	Cinnamal-dehyde	DMF, $(n\text{-Propyl})_4\text{NClO}_4$	Hg	-1.47	sce	13
6-Amino-chrysene	1-Naphthal-dehyde	DMF, $(n\text{-Propyl})_4\text{NClO}_4$	Hg	-1.56	sce	13
6-Amino-chrysene	2-Naphthal-dehyde	DMF, $(n\text{-Propyl})_4\text{NClO}_4$	Hg	-1.58	sce	13
6-Amino-chrysene	9-Phenanthral-dehyde	DMF, $(n\text{-Propyl})_4\text{NClO}_4$	Hg	-1.53	sce	13
6-Amino-chrysene	9-Anthracenal-dehyde	DMF, $(n\text{-Propyl})_4\text{NClO}_4$	Hg	-1.27	sce	13

Azomethines

$$\text{RHC}=N-\text{(C}_6\text{H}_4)-N=\text{CHR}$$

			E (V)		
R = Phenyl	DMF, $0.05N$-$(Et)_4NI$	Hg	-1.63, -1.91	sce	14
= o-Hydroxyphenyl	DMF, $0.05N$-$(Et)_4NI$	Hg	-1.40, -1.66, -2.10	sce	14
= 1-Naphthyl	DMF, $0.05N$-$(Et)_4NI$	Hg	-1.51, -1.71, -2.45	sce	14

(continued)

875

Compound	Solvent System	Working Electrode	$E_{1/2}$	Reference Electrode	Reference
$N=CHR$ (attached to benzene ring), $RHC=N$ = 2-Hydroxy-1-naphthyl	DMF, $0.05N$-$(Et)_4NI$	Hg	-1.34, -1.60, -2.20	sce	14
R = o-Hydroxyphenyl	DMF, $0.05N$-$(Et)_4NI$	Hg	-1.46, -1.95	sce	14
= 2-Hydroxy-1-naphthyl	DMF, $0.05N$-$(Et)_4NI$	Hg	-1.43, -2.06	sce	14
$RHC=N$ (biphenyl) $N=CHR$ R = Phenyl	DMF, $0.05N$-$(Et)_4NI$	Hg	-1.64	sce	14
= o-Hydroxyphenyl	DMF, $0.05N$-$(Et)_4NI$	Hg	-1.40, -2.21	sce	14
= 1-Naphthyl	DMF, $0.05N$-$(Et)_4NI$	Hg	-1.52	sce	14
= 2-Hydroxy-1-naphthyl	DMF, $0.05N$-$(Et)_4NI$	Hg	-1.34, -2.05	sce	14
$RHC=N$—CH_2—$N=CHR$ (diphenylmethane) R = Phenyl	DMF, $0.05N$-$(Et)_4NI$	Hg	-1.65, -1.77	sce	14

	Conditions	Electrode	Potentials	Ref.
= o-Hydroxyphenyl	DMF, 0.05N (Et)₄NI	Hg	-1.47, -1.83, -2.03 sce	14
= 1-Naphthyl	DMF, 0.05N (Et)₄NI	Hg	-1.56, -2.37, sce	14
= 2-Hydroxyl-1-naphthyl	DMF, 0.05N (Et)₄NI	Hg	-1.37, -1.74, -2.09 sce	14

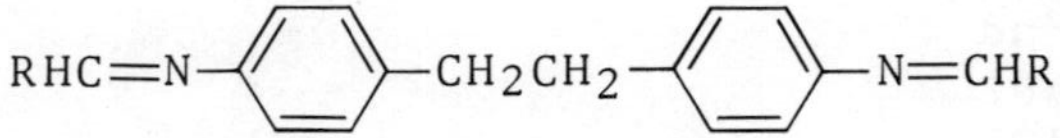

RHC=N—⟨benzene⟩—CH₂CH₂—⟨benzene⟩—N=CHR

	Conditions	Electrode	Potentials	Ref.
R = Phenyl	DMF, 0.05N (Et)₄NI	Hg	-1.77 sce	14
= o-Hydroxyphenyl	DMF, 0.05N (Et)₄NI	Hg	-1.55, -1.80, sce	14
= 1-Naphthyl	DMF, 0.05N (Et)₄NI	Hg	-1.63, -2.40, sce	14
= 2-Hydroxyl-1-naphthyl	DMF, 0.05N (Et)₄NI	Hg	-1.43, -1.69, -2.05 sce	14

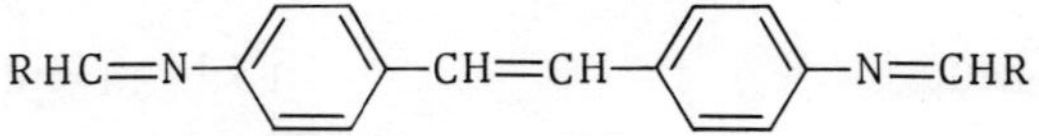

RHC=N—⟨benzene⟩—CH=CH—⟨benzene⟩—N=CHR

	Conditions	Electrode	Potentials	Ref.
R = Phenyl	DMF, 0.05N (Et)₄NI	Hg	-1.65, -2.53, sce	14
= o-Hydroxyphenyl	DMF, 0.05N (Et)₄NI	Hg	-1.47, -1.95, -1.73, sce	14
= 2-Hydroxyl-1-naphthyl	DMF, 0.05N (Et)₄NI	Hg	-1.37, -1.65, -2.03 sce	14

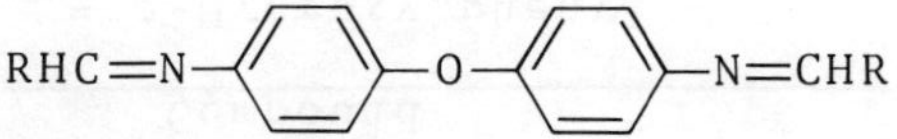

RHC=N—⟨benzene⟩—O—⟨benzene⟩—N=CHR

	Conditions	Electrode	Potentials	Ref.
R = Phenyl	DMF, 0.05N (Et)₄	Hg	-1.77 sce	14

(continued)

877

Compound	Solvent System	Working Electrode	$E_{1/2}$	Reference Electrode	Reference
= o-Hydroxylphenyl	DMF, 0.05N (Et)$_4$NI	Hg	-1.50, -1.78, -2.03	sce	14
= 2-Hydroxyl-1-naphthyl	DMF, 0.05N (Et)$_4$NI	Hg	-1.43, -2.12	sce	14

RHC=N=CHR

Compound	Solvent System	Working Electrode	$E_{1/2}$	Reference Electrode	Reference
R = Phenyl	DMF, 0.05N (Et)$_4$NI	Hg	-1.65	sce	14
= o-Hydroxylphenyl	DMF, 0.05N (Et)$_4$NI	Hg	-1.40, -1.91, -2.19	sce	14
= 2-Hydroxy-1-naphthyl	DMF, 0.05N (Et)$_4$NI	Hg	-1.34, -2.00	sce	14

Phenylhydrazone derivatives of:

Compound	Solvent System	Working Electrode	$E_{1/2}$	Reference Electrode	Reference
Benzaldehyde	Phosphate-Borate buffer, 20% Methanol, pH 6.0	Hg	-1.065	NCE	15
	Phosphate-Citrate buffer, 1M KCl, 40% Alcohol, pH 7.40	Hg	-1.24	sce	16
p-Isopropylbenzaldehyde	Phosphate-Borate buffer, 20% Methanol, pH 6.0	Hg	-1.068	NCE	15

Compound	Conditions		Potential	Electrode	Ref
p-Dimethylaminobenzaldehyde	Phosphate-Borate buffer, 20% Methanol, pH 6.0	Hg	-1.070	NCE	15
	Phosphate-Citrate buffer, 1M KCl, 40% - Alcohol, pH 7.40	Hg	-1.27	NCE	16
m-Nitrobenzaldehyde	Phosphate-Borate buffer, 20% Methanol, pH 6.0	Hg	-1.068	NCE	15
p-Nitrobenzaldehyde	Phosphate-Borate buffer, 20% Methanol, pH 6.0	Hg	-1.070	NCE	15
Cinnamaldehyde	Phosphate-Citrate buffer, 1M KCl, 40% - Alcohol, pH 7.40	Hg	-1.21	sce	16
Acetone	0.5N H2SO4	Hg	-1.211	NCE	17
Acetophenone	Phosphate-Borate buffer, 20% Methanol, pH 6.0	Hg	-1.170	NCE	15
p-Chloroacetophenone	Phosphate-Borate buffer, 20% Methanol, pH 6.0	Hg	-1.154	NCE	15

(continued)

Compound	Solvent System	Working Electrode	$E_{1/2}$	Reference Electrode	Reference
p-Aminoacetophenone	Phosphate-Citrate buffer, 1M KCl, 40% - Alcohol, pH 7.40	Hg	-1.38	sce	16
Semicarbazone Derivatives of:					
Formaldehyde	OAc buffer, 50% - Ethanol, pH 3.85	Hg	-1.081	sce	18
Acetaldehyde	OAc buffer, 50% - Ethanol, pH 3.85	Hg	-1.141	sce	18
Hexaldehyde	OAc buffer, 50% - Ethanol, pH 3.85	Hg	-1.120	sce	18
Heptaldehyde	OAc buffer, 50% - Ethanol, pH 3.85	Hg	-1.170	sce	18
Benzaldehyde	OAc buffer, 50% - Ethanol, pH 3.85	Hg	-1.058	sce	18
	Phosphate-Citrate buffer, 1M KCl, 40% - Alcohol, pH 5.20	Hg	-1.20	sce	16
	Phosphate-Borate buffer, 20% Methanol, pH 6.0	Hg	-1.065	NCE	15
	0.5M H_2SO_4, 50% Ethanol	Hg	-0.729, pH -0.061 pH	sce	19
2-Methylbenzaldehyde	0.5M H_2SO_4, 50% Ethanol	Hg	-0.705, pH -0.060 pH	sce	19

Compound	Conditions	Electrode	Potential	Reference electrode	Ref.
3-Methylbenzaldehyde	0.5M H_2SO_4, 50% Ethanol	Hg	-0.711, -0.061 pH	sce	19
4-Methylbenzaldehyde	0.5M H_2SO_4, 50% Ethanol	Hg	-0.733, -0.062 pH	sce	19
	OAc buffer, 50% Ethanol, pH 3.85	Hg	-1.105	sce	18
2,4,6-Trimethylbenzaldehyde	OAc buffer, 50% Ethanol, pH 3.85	Hg	-1.112	sce	18
2-Ethylbenzaldehyde	0.5M H_2SO_4, 50% Ethanol	Hg	-0.700, -0.063 pH	sce	19
3-Ethylbenzaldehyde	0.5M H_2SO_4, 50% Ethanol	Hg	-0.712, -0.061 pH	sce	19
4-Ethylbenzaldehyde	0.5M H_2SO_4, 50% Ethanol	Hg	-0.719, -0.061 pH	sce	19
2-i-Propylbenzaldehyde	0.5M H_2SO_4, 50% Ethanol	Hg	-0.712, -0.062 pH	sce	19
3-i-Propylbenzaldehyde	0.5M H_2SO_4, 50% Ethanol	Hg	-0.717, -0.062 pH	sce	19
4-i-Propylbenzaldehyde	0.5M H_2SO_4, 50% Ethanol	Hg	-0.726, -0.061 pH	sce	19
2-t-Butylbenzaldehyde	0.5M H_2SO_4, 50% Ethanol	Hg	-0.741, -0.065 pH	sce	19
3-t-Butylbenzaldehyde	0.5M H_2SO_4, 50% Ethanol	Hg	-0.716, -0.059 pH	sce	19
4-t-Butylbenzaldehyde	0.5M H_2SO_4, 50% Ethanol	Hg	-0.727, -0.059 pH	sce	19
3-Methoxybenzaldehyde	OAc buffer, 50% Ethanol, pH 3.85	Hg	-1.100	sce	18

(continued)

Compound	Solvent System	Working Electrode	$E_{1/2}$	Reference Electrode	Reference
4-Methoxybenzaldehyde	OAc buffer, 50% Ethanol, pH 3.85	Hg	-1.252	sce	18
	0.005M H_2SO_4, 50% Ethanol	Hg	-0.876	sce	19
3,4-Dimethoxybenzaldehyde	0.005M H_2SO_4, 50% Ethanol	Hg	-0.851	sce	19
3-Hydroxybenzaldehyde	0.5M H_2SO_4, 50% Ethanol	Hg	-0.751, -0.085 pH	sce	19
3,4-Dihydroxybenzaldehyde	OAc buffer, 50% Ethanol, pH 3.85	Hg	-0.979	sce	18
3-Chlorobenzaldehyde	0.5M H_2SO_4, 50% Ethanol	Hg	-0.689, -0.065 pH	sce	19
4-Chlorobenzaldehyde	0.5M H_2SO_4, 50% Ethanol	Hg	-0.700, -0.065 pH	sce	19
3,4-Dichlorobenzaldehyde	0.5M H_2SO_4, 50% Ethanol	Hg	-0.688, -0.061 pH	sce	19
4-Bromobenzaldehyde	OAc buffer, 50% Ethanol, pH 3.85	Hg	-0.948	sce	18
4-Cyanobenzaldehyde	0.5M H_2SO_4, 50% Ethanol	Hg	-0.657, -0.082 pH	sce	19
3-Trifluoromethylbenzalde-hyde	o.5M H_2SO_4, 50% Ethanol	Hg	-0.711, -0.090 pH	sce	19
4-Nitrilobenzaldehyde	OAc buffer, 50% Ethanol, pH 3.85	Hg	-0.902	sce	18

Compound	Conditions				Ref.
p-Isopropylbenzaldehyde	Phosphate-Borate buffer, 20% Methanol, pH 6.0	Hg	-1.138	sce	15
p-Dimethylaminobenzaldehyde	Phosphate-Borate buffer, 20% Methanol, pH 6.0	Hg	-1.190	sce	15
m-Nitrobenzaldehyde	Phosphate-Borate buffer, 20% Methanol, pH 6.0	Hg	-1.130	NCE	15
p-Nitrobenzaldehyde	Phosphate-Borate buffer, 20% Methanol, pH 6.0	Hg	-1.174	NCE	15
Cinnamaldehyde	Phosphate-Citrate buffer, 1M KCl, 40% Alcohol, pH 7.40	Hg	-1.22, -1.59	sce	16
Acetone	OAc buffer, 50% Ethanol, pH 3.85	Hg	-1.249	sce	18
Benzalacetone	0.5N H2SO4	Hg	-1.270	sce	17
	Phosphate-Citrate buffer, 1M KCl, 40% Alcohol, pH 7.40	Hg	-1.21	sce	16

(continued)

Compound	Solvent System	Working Electrode	$E_{1/2}$	Reference Electrode	Reference
Methylisopropyl ketone	OAc buffer, 50% Ethanol, pH 3.85	Hg	-1.200	sce	18
Methyl isobutyl ketone	OAc buffer, 50% Ethanol, pH 3.85	Hg	-1.268	sce	18
Acetophenone	OAc buffer, 50% Ethanol, pH 3.85	Hg	-1.119	sce	18
	Phosphate-Borate buffer, 20% Methanol, pH 6.0	Hg	-1.276	NCE	15
p-Chloroacetophenone	Phosphate-Borate buffer, 20% Methanol, pH 6.0	Hg	-1.220	NCE	15
p-Aminoacetophenone	Phosphate-Citrate buffer, 1M KCl, 40% Alcohol, pH 7.40	Hg	-1.39	sce	16
Benzophenone	OAc buffer, 50% Ethanol, pH 3.85	Hg	-1.125	sce	18
Cyclohexanone	OAc buffer, 50% Ethanol, pH 3.85	Hg	-1.144	sce	18

Thiosemicarbazone Derivatives of:

Benzaldehyde	Phosphate-Borate buffer, 20% Methanol, pH 6.0	Hg	-1.110	NCE	15
p-Isopropylbenzaldehyde	Phosphate-Borate buffer, 20% Methanol, pH 6.0	Hg	-1.123	NCE	15
p-Dimethylaminobenzaldehyde	Phosphate-Borate buffer, 20% Methanol, pH 6.0	Hg	-1.150	NCE	15
m-Nitrobenzaldehyde	Phosphate-Borate buffer, 20% Methanol, pH 6.0	Hg	-1.115	NCE	15
p-Nitrobenzaldehyde	Phosphate-Borate buffer, 20% Methanol, pH 6.0	Hg	-1.144	NCE	15
Acetophenone	Phosphate-Borate buffer, 20% Methanol, pH 6.0	Hg	-1.210	NCE	15
p-Chloroacetophenone	Phosphate-Borate buffer, 20% Methanol, pH 6.0	Hg	-1.162	NCE	15

(continued)

885

Girard-T Derivatives of:

Compound	Solvent System	Working Electrode	$E_{1/2}$	Reference Electrode	Reference
Dimethyl ketone	Universal buffer,[b] 0.2M KCl, 50% Ethanol, pH 8	Hg	-1.50	NCE	20
Ethyl methyl ketone	Universal buffer,[b] 0.2M KCl, 50% Ethanol, pH 8	Hg	-1.51	NCE	20
n-Propyl methyl ketone	Universal buffer,[b] 0.2M KCl, 50% Ethanol, pH 8	Hg	-1.51	NCE	20
i-Propyl methyl ketone	Universal buffer,[b] 0.2M KCl, 50% Ethanol, pH 8	Hg	-1.54	NCE	20
n-Butyl methyl ketone	Universal buffer,[b] 0.2M KCl, 50% Ethanol, pH 8	Hg	-1.51	NCE	20
i-Butyl methyl ketone	Universal buffer,[b] 0.2M KCl, 50% Ethanol, pH 8	Hg	-1.57	NCE	20
t-Butyl methyl ketone	Universal buffer,[b] 0.2M KCl, 50% Ethanol, pH 8	Hg	-1.55	NCE	20
n-Pentyl methyl ketone	Universal buffer,[b] 0.2M KCl, 50% Ethanol, pH 8	Hg	-1.51	NCE	20
n-Hexyl methyl ketone	Universal buffer,[b]	Hg	-1.50	NCE	20

Compound	Solution				
Cyclohexyl methyl ketone	Universal buffer, 0.2M KCl, 50% Ethanol, pH 8	Hg	−1.54	NCE	20
Diethyl ketone	Universal buffer, 0.2M KCl, 50% Ethanol, pH 8	Hg	−1.52	NCE	20
n-Propyl ethyl ketone	Universal buffer, 0.2M KCl, 50% Ethanol, pH 8	Hg	−1.53	NCE	20
i-Propyl ethyl ketone	Universal buffer, 0.2M KCl, 50% Ethanol, pH 8	Hg	−1.55	NCE	20
t-Butyl ethyl ketone	Universal buffer, 0.2M KCl, 50% Ethanol, pH 8	Hg	−1.60	NCE	20
Cyclohexyl ethyl ketone	Universal buffer, 0.2M KCl, 50% Ethanol, pH 8	Hg	−1.54	NCE	20
Di-n-propyl ketone	Universal buffer, 0.2M KCl, 50% Ethanol, pH 8	Hg	−1.55	NCE	20
i-Propyl-n-propyl ketone	Universal buffer, 0.2M KCl, 50% Ethanol, pH 8	Hg	−1.57	NCE	20
t-Butyl-n-propyl ketone	Universal buffer, 0.2M KCl, 50% Ethanol, pH 7	Hg	−1.31, −1.55	NCE	20
Cyclohexyl-n-propyl ketone	Universal buffer, 0.2M KCl, 50% Ethanol, pH 8	Hg	−1.56	NCE	20

(continued)

887

Compound	Solvent System	Working Electrode	$E_{1/2}$	Reference Electrode	Reference
Di-i-propyl ketone	Universal buffer,b 0.2M KCl, 50% Ethanol, pH 8	Hg	-1.55	NCE	20
t-Butyl-i-propyl ketone	Universal buffer,b 0.2M KCl, 50% Ethanol, pH 8	Hg	No wave	NCE	20
Cyclohexyl-i-propyl ketone	Universal buffer,b 0.2M KCl, 50% Ethanol, pH 8	Hg	-1.56	NCE	20
Di-i-butyl ketone	Universal buffer,b 0.2M KCl, 50% Ethanol, pH 8	Hg	-1.31	NCE	20
Cyclohexanone	Universal buffer,b 0.2M KCl, 50% Ethanol, pH 7	Hg	-1.47	NCE	20
Norcamphor Anil	DMF, (Et)4NBr	Hg	-1.98	Ag, AgBr	21
Camphor Anil	DMF, (Et)4NBr	Hg	-2.14	Ag, AgBr	21
α-Methyl benzylidene-α-methylbenzylamine	DMF, (Et)4NBr	Hg	-1.75	Ag, AgBr	21
21 Immonium salts	CH3CN, (Et)4NClO4				22, 23
5 Azomethine derivatives of Phenyl- and 2-Furylglyoxal	20% alcohol-B-R buffers, pH 2-12				24
19 Azomethines derived from aniline and benzidine and chloro derivatives of these amines					25

21 Azomethines with benzal, salicylal, and 2-hydroxy-1-naphthal functions   DMF, $(Et)_4N1$

Bisazomethines

$RCH=N$ ... $N=CHR$ (with X bridge)

X =	R =					
–	C_6H_5	DMF, $0.05\underline{N}$ $(Et)_4NI$	Hg	-1.708	sce	27
–	$\underline{o}-C_6H_4Cl$	DMF, $0.05\underline{N}$ $(Et)_4NI$	Hg	-1.410, -1.877	sce	27
–	$\underline{p}-C_6H_4Cl$	DMF, $0.05\underline{N}$ $(Et)_4NI$	Hg	-1.687	sce	27
CH_2	C_6H_5	DMF, $0.05\underline{N}$ $(Et)_4NI$	Hg	-1.721	sce	27
CH_2	$\underline{o}-C_6H_4Cl$	DMF, $0.05\underline{N}$ $(Et)_4NI$	Hg	-1.542	sce	27
CH_2	$\underline{p}-C_6H_4Cl$	DMF, $0.05\underline{N}$ $(Et)_4NI$	Hg	-1.620	sce	27
CH_2	$\underline{o}-C_6H_4OCH_3$	DMF, $0.05\underline{N}$ $(Et)_4NI$	Hg	-1.746	sce	27
CH_2	$\underline{p}-C_6H_4OCH_3$	DMF, $0.05\underline{N}$ $(Et)_4NI$	Hg	-1.742	sce	27
CH_2	$\underline{o}-C_6H_4OH$	DMF, $0.05\underline{N}$ $(Et)_4NI$	Hg	-1.448, -1.888	sce	27
CH_2	$\underline{m}-C_6H_4OH$	DMF, $0.05\underline{N}$ $(Et)_4NI$	Hg	-1.567, -1.937	sce	27
CH_2	$\underline{p}-C_6H_4N-(CH_3)_2$	DMF, $0.05\underline{N}$ $(Et)_4NI$	Hg	-1.820	sce	27
CH_2	$\alpha-C_{10}H_7$	DMF, $0.05\underline{N}$ $(Et)_4NI$	Hg	-1.539, -1.784	sce	27
CH_2	$\beta-C_{10}H_7$	DMF, $0.05\underline{N}$ $(Et)_4NI$	Hg	-1.598	sce	27

(continued)

Compound	Solvent System	Working Electrode	$E_{1/2}$	Reference Electrode	Reference
= CH_2 = $2\text{-}C_{10}H_6OH$	DMF, 0.05N (Et)4NI	Hg	-1.408, -2.170	sce	27
= NH = $o\text{-}C_6H_4OH$	DMF, 0.05N (Et)4NI	Hg	-1.610, -1.830	sce	27
= CO = $o\text{-}C_6H_4OH$	DMF, 0.05N (Et)4NI	Hg	-1.004, -1.476, -1.837	sce	27
= $CONH$ = $o\text{-}C_6H_4OH$	DMF, 0.05N (Et)4NI	Hg	-1.360, -1.752	sce	27
= SO_2 = $o\text{-}C_6H_4OH$	DMF, 0.05N (Et)4NI	Hg	-1.196, -1.374, -1.633	sce	27
= $N=N$ = $o\text{-}C_6H_4OH$	DMF, 0.05N (Et)4NI	Hg	-1.253, -1.825	sce	27

Nitriles (Saturated, Olefinic, and Aromatic)

Compound	Solvent System	Working Electrode	$E_{1/2}$	Reference Electrode	Reference
Acetonitrile	No wave in aqueous-alcoholic solutions containing any of: (Et)4NCl, (Et)4NI, (n-Bu)4NI, LiCl, $C_3H_4(OH)(COOLi)_3$				28
Acrylonitrile	0.2M (Et)4NI	Hg	-1.84	Hg pool	29
	DMF, (Et)4NI	Hg	-1.67	Hg pool	29
	DMF, 5% H_2O, (n-Bu)4NI	Hg	-1.63	Hg pool	30
	0.1M LiCl	Hg	-1.84	Hg pool	29
	50% Ethanol, LiCl	Hg	-2.34	sce	28

Compound	Electrolyte/Solvent	Electrode	E (V)	Ref. electrode	Ref.
Methacrylonitrile	0.02M LiCl	Hg	-1.91	Hg pool	29
	0.2M (Et)4NI	Hg	-2.05	Hg pool	29
	DMF, (Et)4NI	Hg	-1.79	Hg pool	29
	0.1M LiCl	Hg	-1.86	Hg pool	29
	0.02M LiCl	Hg	-2.05	Hg pool	29
Crotonitrile	DMF, (Et)4NI	Hg	-1.91	Hg pool	29
	0.1M LiCl	Hg	-1.37	Hg pool	29
	0.02M LiCl	Hg	-1.47	Hg pool	29
Vinylacrylonitrile	0.2M (Et)4NI	Hg	-1.47	Hg pool	29
	DMF, (Et)4NI	Hg	-1.31, -1.91	Hg pool	29
Cyanobut-1-ene	DMF, (Et)4NI	Hg	-1.91	Hg pool	29
Dicyanobut-1-ene	0.2M (Et)4NI	Hg	-1.87	Hg pool	29
	DMF, (Et)4NI	Hg	-1.73	Hg pool	29
	0.1M LiCl	Hg	-1.74	Hg pool	29
	0.02M LiCl	Hg	-1.87	Hg pool	29
1,4-Dicyano-2-butene	0.2M (Et)4NI	Hg	No wave	Hg pool	29
	DMF, (Et)4NI	Hg	No wave	Hg pool	29
Fumaronitrile	50% Ethanol, 0.14M LiCl	Hg	-1.27, -2.10	sce	31
	0.2M (Et)4NI	Hg	-1.50	sce	28
cis-Fumaronitrile	0.2M LiCl	Hg	-1.275, -1.785	sce	32
	40% Ethanol, (Et)4NI	Hg	-0.82, -1.07	sce	32
trans-Fumaronitrile	0.2M LiCl	Hg	-1.29, -1.99	sce	32
	40% Ethanol, (Et)4NI	Hg	-0.99, -1.56	sce	32
	50% Ethanol, (n-Bu)4NI	Hg	-1.32	sce	32
2,2'-Azobisisobutyronitrile	0.14M LiCl	Hg	-1.71	sce	31

(continued)

Compound	Solvent System	Working Electrode	$E_{1/2}$	Reference Electrode	Reference
Tetracycanoquinodi-methane	CH$_3$CN, (n-Propyl)$_4$NClO$_4$	Hg	-0.19, -0.75	Ag/0.1N Ag+	33
Tetracyanoethylene	CH$_3$CN, (n-Propyl)$_4$NClO$_4$	Hg	-0.17, -1.17	Ag/0.1N Ag+	33
	CH$_3$CN, (Et)$_4$NClO$_4$	Hg	-0.24	sce	34
	DMF, (Et)$_4$NClO$_4$	Hg	-0.16	sce	34
Tetramethylammonium-1,1,2,3,3-pentacyano-propenide	DMF, (n-Propyl)$_4$NClO$_4$	Hg	-1.76, -2.30c	Ag/0.1N Ag+	33
Sodium 1,1,3,3-tetra-cyano-2-dimethylamino-propenide	DMF, (n-Propyl)$_4$NClO$_4$	Hg	-2.75	Ag/0.1N Ag+	33
Sodium 1,1,3,3-tetra-cyano-2-ethoxypropenide	DMF, (n-Propyl)$_4$NClO$_4$	Hg	-2.25, -2.40	Ag/0.1N Ag+	33
Cinnamonitrile	75% Ethanol, 0.14M LiCl	Hg	-1.90	sce	31
	0.1M LiCl	Hg	-1.36	Hg pool	29
	0.02M LiCl	Hg	-1.50	Hg pool	29
	0.2M (Et)$_4$NI	Hg	-1.5	Hg pool	29
	DMF, (Et)$_4$NI	Hg	-1.32, -1.81	Hg pool	29
α-Phenylcinnamonitrile	75% Ethanol, 0.14M LiCl	Hg	-1.83	sce	31
Benzonitrile	DMF, (n-Propyl)$_4$NClO$_4$	Hg	-2.74	Ag	33
Benzonitrile	0.2M (Et)$_4$NI	Hg	No wave	Hg pool	29
	DMF, (Et)$_4$NI	Hg	-1.82	Hg pool	29
3-Cyanobenzoic acid	5% Ethanol, Borate-buffer, pH 9.3	Hg	-1.85	sce	34

Compound	Conditions	Electrode	E	Ref. electrode	Ref.
4-Cyanobenzoic acid	5% Ethanol, Borate-buffer, pH 9.3	Hg	-1.776	sce	34
4-Cyanobenzoic acid	DMF, (n-Propyl)$_4$-NClO$_4$	Hg	-1.91, -2.53	Ag/0.1N Ag+	33
Methyl-m-cyanobenzoate	Borate buffer, pH 9.3, 5% Ethanol	Hg	-1.761	sce	34
Methyl-p-cyanobenzoate	Borate buffer, pH 9.3, 5% Ethanol	Hg	-1.527	sce	34
Phenyl-m-cyanobenzoate	Borate buffer, pH 9.3, 5% Ethanol	Hg	-1.624	sce	34
Phenyl-p-cyanobenzoate	Borate buffer, pH 9.3, 5% Ethanol	Hg	-1.416	sce	34
Phthalonitrile(o-dicyanobenzene)	DMF, (n-Propyl)$_4$-NClO$_4$	Hg	-2.12, -2.76	Ag/0.1N Ag+	33
Isophthalonitrile(m-Dicyanobenzene)	DMF, (n-Propyl)$_4$-NClO$_4$	Hg	-2.17	Ag/0.1N Ag+	33
	5% Ethanol, Borate-buffer, pH 9.3	Hg	-1.811	sce	34
	0.15N LiCl, 50%-Alcohol	Hg	-1.92 to -1.98	sce	35
	0.2M Li3Cit, 50% Alcohol	Hg	-1.92 to -1.96	sce	35
	0.1M (Et)$_4$NI, 50% Alcohol	Hg	-1.80 to -1.90	sce	35
	DMF, 5% H$_2$O, 0.1M (Et)$_4$NOH	Hg	-1.82 to -1.86	sce	35
	DMF, 5% H$_2$O, 0.1M (Et)$_4$NI	Hg	-1.80 to -1.90	sce	35
Terephthalonitrile (p-Dicyanobenzene)	5% Ethanol, Borate-buffer, pH 9.3	Hg	-1.612	sce	34

(continued)

893

Compound	Solvent System	Working Electrode	$E_{1/2}$	Reference Electrode	Reference
	DMF, (n-Propyl)4-NClO4	Hg	-1.97, -2.64	Ag/0.1N Ag+	33
	0.15N LiCl, 50% Alcohol	Hg	-1.72 to -1.73, -1.98 to -2.05	sce	35
	0.2M Li3Cit, 50% Alcohol	Hg	-1.72 to -1.81, -1.97 to -1.98	sce	35
	DMF, 5% H_2O, 0.1M (Et)4NOH	Hg	-1.61 to -1.65, -1.98 to -2.07	sce	35
	DMF, 5% H_2O, 0.1M (Et)4N1	Hg	-1.60 to -1.70, -2.40 to -2.60	sce	35
4-Chlorobenzonitrile	DMF, (n-Propyl)4-NClO4	Hg	-2.4	Ag/0.1N Ag+	33
4-Fluorobenzonitrile	DMF, (n-Propyl)4-NClO4	Hg	-2.69	Ag.0.1N Ag+	33
4-Cyanoacetophenone	0.3N H_2SO_4	Hg	-0.674	sce	36
4-Cyanobenzophenone	50% Ethanol, 0.1N HCl	Hg	-0.723, -0.788, -0.943	sce	36
	50% Ethanol, borate-buffer, pH 9.3	Hg	-1.210	sce	36

Compound	Electrolyte	Electrode	E	Reference electrode	Ref.
4-Cyanobenzophenone	50% Ethanol, 0.1N NaOH	Hg	-1.208	sce	36
4-Anisonitrile	DMF, (n-Propyl)4-NClO4	Hg	-2.95	Ag/0.1N Ag+	33
Pyromellitonitrile (1,2,4,5-tetracyano-benzene)	DMF, (n-Propyl)4-NClO4	Hg	-1.02, -2.07	Ag.0.1N Ag+	33
4-Tolunitrile	DMF, (n-Propyl)4-NClO4	Hg	-2.75	Ag/0.1N Ag+	33
4-Aminobenzonitrile	DMF, (n-Propyl)4-NClO4	Hg	-3.12	Ag.0.1N Ag+	33
4-Nitrobenzonitrile	DMF, (n-Propyl)4-NClO4	Hg	-1.25, -1.9	Ag/0.1N Ag+	33
3,5-Dinitrobenzonitrile	DMF, (n-Propyl)4-NClO4	Hg	-0.96, -1.5[d]	Ag/0.1N Ag+	33
Benzoylacetonitrile	DMF, (n-Propyl)4-NClO4	Hg	-2.09	Ag.0.1N Ag+	33
Benzoyl cyanide	DMF, (n-Propyl)4-NClO4	Hg	-1.45	Ag.0.1N Ag+	33
Dinitrile of 3-cyano-benzylamine	DMF, 5% H2O, 0.1M-(Et)4Nl	Hg	-2.40 to -2.60	sce	35
Dinitrile of 4-cyano-benzylamine	DMF, 5% H2O, 0.1M-(Et)4Nl	Hg	-2.40 to -2.60	sce	35

a Britton and Welford.
b Borate, phosphate, acetate.
c Third 1-electron wave at -2.75 V.
d 3-Electron wave at -2.1 V.

(continued)

References for Carbon-Nitrogen Function

1. P. Martinet, J. Simonet, and J. Tendil, Compt. Rend., 268, 2329 (1969).
2. J. F. O'Donnell, Ph.D. Thesis, Florida State University, 1966.
3. L. Eberson and K. Nyberg, J. Am. Chem. Soc., 88, 1686 (1966).
4. J. F. O'Donnell and C. K. Mann, J. Electroanal. Chem., 13, 157 (1967).
5. A. S. Gorokhovskaya and L. F. Markova, Zh. Obshch Khim., 38, 967 (1968).
6. L. Horner and R. Singer, Liebigs Ann. Chem., 723, 1 (1969).
7. Yu. V. Svetkin and L. N. Andreeva, Zh. Obshch. Khim., 37, 1948 (1967).
8. R. Jones and B. C. Page, Anal. Chem., 36, 35 (196
9. G. S. Shapoval, L. S. Sheinina, V. A. Zhabenko, M. A. Morozov, and M. A. Lagutin, Elektrokhim, 6, 872 (1970).
10. G. S. Shapoval, E. M. Skobets, and N. P. Markova, Dokl. Akad. Nauk SSSR, 173, 392 (1967).
11. P. Martinet, J. Simonet, and J. Tendil, Compt. Rend., 268, 303 (1969).
12. N. A. Rozanel'skaya, V. N. Dmitrieva, B. I. Stepa and V. D. Bezuglyi, Zh. Obshch. Khim., 38, (11), 2421 (1968).
13. J. M. W. Scott and W. H. Jura, Can. J. Chem., 45, 2375 (1967).
14. V. N. Dmitrieva, N. I. Mal'tseva, V. D. Bezuglyi, and B. M. Krasovitskii, Zh. Obshch. Khim., 37 (2), 372 (1967).
15. Yu. P. Kitaev, G. K. Budnikov, T. V. Troepol'ska and A. E. Arbuzov, Dokl. Akad. Nauk SSSR, 137 (4), 862 (1961).
16. H. Lund, Acta Chem. Scand., 13, 249 (1959).
17. D. K. Banerjee, G. C. Riechmann, and C. C. Budke Anal. Chem., 36, 2220 (1964).
18. B. Fleet, Anal. Chim. Acta, 36, 304 (1966).
19. B. Fleet and P. Zuman, Collect. Czech. Chem. Com 32, 2066 (1967).
20. J. R. Young, J. Chem. Soc., 1955, 1516.
21. A. J. Fry and R. G. Reed, J. Am. Chem. Soc., 91, 6448 (1969).
22. C. P. Andrieux and J. M. Saveant, J. Electroanal Chem., 26, 223 (1970).
23. C. P. Andrieux and J. M. Saveant, Bull. Soc. Chi Fr., 4671 (1968).

24. Ya. P. Stradyn', I. Ya. Kravis, and N. O. Saldabol, Zh. Obshch. Khim., 37 (5), 977 (1967).

25. V. N. Dmitrieva, V. B. Smelyakova, B. M. Krasovitskii, and V. D. Bezughji, Zh. Obshch. Khim., 36 (3), 405 (1966).

26. N. F. Levchenko, L. Sh. Afanasiadi, and V. D. Bezughji, Zh. Obshch. Khim., 37 (3), 666 (1967).

27. V. N. Dmitrieva, A. I. Nazarenko, B. M. Krasovitskii, and V. D. Bezuglyi, Zh. Obshch. Khim., 37 (9), 1967 (1967).

28. M. I. Bobrova and A. N. Matveeva, Zh. Obshch. Khim., 27, 1137 (1957).

29. I. G. Sevast'yanova and A. P. Tomilov, Zh. Obshch. Khim., 33, 2815 (1963).

30. G. C. Claver and M. E. Murphy, Anal. Chem., 31, 1682 (1959).

31. M. I. Bobrova and A. N. Matveeva-Kudasheva, Zh. Obshch. Khim., 28, 2929 (1958).

32. S. K. Smirnov, I. G. Sevast'yanova, A. P. Tomilov, L. A. Fedorova, and O. G. Strukov, Zh. Org. Khim., 5, 1392 (1969).

33. P. H. Rieger, I. Bernal, W. H. Reinmuth, and B. K. Fraenkel, J. Am. Chem. Soc., 85, 683 (1963).

34. M. E. Peover, Trans. Faraday Soc., 58, 2370 (1962).

35. L. S. Reishakhrit, T. B. Argova, L. V. Vesheva, and T. V. Forsova, Elektrokhim., 5, 1017 (1969).

36. P. Zuman and O. Manousek, Collect. Czech. Chem. Commun., 34, 1580 (1969).

24. NITROGEN-NITROGEN FUNCTION

Oxidation

Azo Compounds

Compounds	Solvent System	Working Electrode	$E_{1/2}$	Reference Electrode	Reference
Azobenzene	CH_3CN, $NaClO_4$	Pt	+1.33	Ag/0.1N Ag+	1
4-Aminoazobenzene	CH_3CN, $(Et)_4NClO_4$	Pt	+1.89	sce	2
4-Dimethylaminoazobenzene	CH_3CN, $(Et)_4NClO_4$	Pt	+0.92, +1.48	sce	2
4-Dimethylaminoazobenzene	CH_3CN, $(Et)_4NClO_4$	Pt	+0.86, +1.40	sce	2
4-Diethylaminoazobenzene	CH_3CN, $(Et)_4NClO_4$	Pt	+0.84, +1.30	sce	2
4-Dimethyl-2'-methyl-aminoazobenzene	CH_3CN, $(Et)_4NClO_4$	Pt	+0.90, +1.33	sce	2
4-Dimethyl-3'-methyl-aminoazobenzene	CH_3CN, $(Et)_4NClO_4$	Pt	+0.91, +1.32	sce	2
4-Dimethyl-4'-methyl-aminoazobenzene	CH_3CN, $(Et)_4NClO_4$	Pt	+0.90, +1.34	sce	2
4-Dimethyl-3'-fluoro-aminoazobenzene	CH_3CN, $(Et)_4NClO_4$	Pt	+0.90, +1.27	sce	2
2-Methyl-4-dimethyl-aminoazobenzene	CH_3CN, $(Et)_4NClO_4$	Pt	+0.80, +1.22	sce	2
4,4'-Dichloroazobenzene	CH_3CN, $NaClO_4$	Pt	+1.44	Ag/0.1N Ag+	1

Compound	Electrolyte	Electrode	Potential	Reference	Ref
4,4'-Dimethoxyazobenzene	CH_3CN, $NaClO_4$	Pt	+0.98	Ag/0.1$\underline{N}$ Ag+	1
	CH_3CN, $NaClO_4$[a]	Pt	+0.98, +1.25	Ag/0.1$\underline{N}$ Ag+	1
4,4'-Dinitroazobenzene	CH_3CN, $NaClO_4$[a]	Pt	No wave	Ag/0.1$\underline{N}$ Ag+	1
2,2',4,4'-Tetranitro-azobenzene	CH_3CN, $NaClO_4$[a]	Pt	No wave	Ag/0.1$\underline{N}$ Ag+	1
2,2',4,4'-Tetrachloro-azobenzene	CH_3CN, $NaClO_4$[a]	Pt	+1.59	Ag/0.1$\underline{N}$ Ag+	1
sym-Hexachloroazoben-zene	CH_3CN, $NaClO_4$	Pt	+1.63	Ag/0.1$\underline{N}$ Ag+	1

Hydrazo Compounds

Compound	Electrolyte	Electrode	Potential	Reference	Ref
Hydrazobenzene	OAc buffer, 30% Methanol, pH 4.0	Hg	-0.16	sce	3
	Ammonia buffer, 30% Methanol, pH 9.2	Hg	-0.33	sce	3
	McIlvaine buffer, 50% Ethanol, pH 3.3	Graphite	-0.11 Ep/2	sce	4
	Ammonia buffer, 50% Ethanol, pH 6.6	Graphite	-0.18 Ep/2	sce	4
	KCl-KOH buffer, 50% Ethanol, pH 12.5	Graphite	-0.39 Ep/2	sce	4
4,4'-Dichlorohydrazo-benzene	CH_3CN, $NaClO_4$	Pt	+0.18, +1.35	Ag/0.1$\underline{N}$ Ag+	1
	CH_3CN, $NaClO_4$[a]	Pt	+0.05	Ag/0.1$\underline{N}$ Ag+	1
	CH_3CN, $NaClO_4$	Pt	+0.26, +1.44	Ag/0.1$\underline{N}$ Ag+	1

(continued)

Compound	Solvent System	Working Electrode	$E_{1/2}$	Reference Electrode	Reference
9-Hydrazoacridine	CH_3CN, $NaClO_4$[a]	Pt	+0.16	Ag/0.1N Ag+	1
	CH_3CN, 0.01M Diphenyl-guanidine, $0.1\underline{M}$ $LiClO_4$	Pt	-0.350	Ag.0.01N̄ Ag+	5
P,p'-Hydrazotoluene	CH_3CN, $(Et)_4NClO_4$	Pt	+0.34	sce	2
Hydrazine Compounds					
Hydrazine	$0.1\underline{N}$ NaOH	Hg	-0.548	Hg, Hg_2SO_4	6
Methyl hydrazine	$0.1\underline{N}$ NaOH	Hg	-0.634	Hg, Hg_2SO_4	6
n-Propyl hydrazine	$0.1\underline{N}$ NaOH	Hg	-0.689	Hg, Hg_2SO_4	6
$\underline{n}$-Hexyl hydrazine	$0.1\underline{N}$ NaOH	Hg	-0.791	Hg, Hg_2SO_4	6
$\underline{1}$,1-Dimethylhydrazine	$0.1\underline{N}$ NaOH	Hg	-0.38	sce	7
	$DMS\underline{O}$, $(Et)_4NClO_4$	Hg	+0.03 E1/4	sce	8
1,2-Dimethylhydrazine	Aq KOH, **p**H 13	Hg	-0.39	sce	7
	$0.1\underline{N}$ NaOH	Hg	-0.698	Hg, Hg_2SO_4	6
	$DMS\underline{O}$, $(Et)_4NClO_4$	Hg	+0.02 E1/4	sce	8
1,1-Diallylhydrazine	Aq KOH, **p**H 13	Hg	-0.40	sce	7
1,1-Diethylhydrazine	Aq KOH, pH 13	Hg	-0.44	sce	7
1,1-Di-n-butylhydrazine	Aq KOH, pH 13	Hg	-0.48	sce	7
1,2-Di-$\underline{n}$-butylhydrazine	Aq KOH, pH 13	Hg	-0.39	sce	7
1,1-Diisobutylhydrazine	Aq KOH, pH 13	Hg	-0.39	sce	7
1,2-Diisobutylhydrazine	Aq KOH, pH 13	Hg	-0.49	sce	7
Phenylhydrazine	$0.1\underline{N}$ NaOH	Hg	-0.757	Hg, Hg_2SO_4	6
Phenylhydrazine	$0.1\underline{N}$ HaOH	Hg	-0.752	Hg, Hg_2SO_4	6
1,1-Diphenylhydrazine	$CH_3\underline{C}N$, $LiClO_4$	Pt	+0.035	Ag/0.01N̄ Ag+	9
	CH_3CN, $LiClO_4$	Pt	+0.175, +1.10	Ag/0.01N̄ Ag+	10
	CH_3CN, $LiClO_4$, $HClO_4$	Pt	+1.050	Ag/0.01N̄ Ag+	10
	CH_3CN, $LiClO_4$, Py	Pt	+0.10	Ag/0.01N̄ Ag+	10

Compound	Medium	Electrode	E		Ref.
1-Methyl-1-phenyl-hydrazine	Aq KOH, pH 13	Hg	-0.38	sce	7
1-Methyl-1-benzyl-hydrazine	Aq KOH, pH 13	Hg	-0.41	sce	7
2-Carboxyphenyl-hydrazine	Aq KOH, pH 13	Hg	-0.39	sce	7
4-Carboxyphenyl-hydrazine	Aq KOH, pH 13	Hg	-0.47	sce	7
N-Aminopyrrolidine	Aq KOH, pH 13	Hg	-0.46	sce	7
N-Aminopiperidine	Aq KOH, pH 13	Hg	-0.41	sce	7
N-Aminomorpholine	Aq KOH, pH 13	Hg	-0.33	sce	7
Hydrazides					
Acetyhydrazine	Aq KOH, pH 13	Hg	-0.28	sce	7
Propionylhydrazine	Aq KOH, pH 13	Hg	-0.29	sce	7
n-Butyrylhydrazine	Aq KOH, pH 13	Hg	-0.30	sce	7
Chloromethanesulfonyl-hydrazine	Aq KOH, pH 13	Hg	-0.38	sce	7
Benzenesulfonylhydra-zine	Aq KOH, pH 13	Hg	-0.45	sce	7
Isonicotinoylhydrazine	Aq KOH, pH 13	Hg	-0.32	sce	7
1-Isonicotinoyl-2-phenylhydrazine	Aq Solution, pH 13	Pt	-0.24	sce	11
Isonicotinic hydrazide	Aq Solution, pH 13	Pt	-0.28	sce	11
2-Thiazolecarboxy-hydrazide	Aq KOH, pH 13	Hg	-0.33	sce	7
1-Methyl-2-acetyl-hydrazine	Aq KOH, pH 13	Hg	-0.39	sce	7
1-Phenyl-2-acetyl-hydrazine	Aq KOH, pH 13	Hg	-0.36	sce	7

(continued)

Compound	Solvent System	Working Electrode	$E_{1/2}$	Reference Electrode	Reference
Semicarbazide	Aq KOH, pH 13	Hg	-0.33	sce	7
Thiosemicarbazide	Aq KOH, pH 13	Hg	-0.51	sce	7
$(C_6H_5)_2NNH$ (2,6-dinitrophenyl)	CH_3CN, $NaClO_4$	Pt	+0.69	Ag/0.01N Ag+	12
$(C_6H_5)_2NNH$ (2,4-dinitrophenyl)	CH_3CN, $NaClO_4$	Pt	+0.81	Ag/0.01N Ag+	12
$(C_6H_5)_2NNH$ (2,4,6-trinitrophenyl)	CH_3CN, $NaClO_4$	Pt	+1.28	Ag/0.01N Ag+	12
Diphenylpicrylhydrazyl	CH_3CN, $NaClO_4$	Pt	+0.70	sce	13
	CH_3CN, $NaClO_4$	Pt	+0.82	Ag/0.01N Ag+	12
	CH_3CN, $NaClO_4$	Pt	+0.693	sce	13
	CH_3CN, $NaClO_4$	Pt	+0.70	sce	14
	CH_3CN, $(Et)_4NClO_4$	Pt	+0.43	Ag/0.1N Ag+	14
	Methanol, $NaClO_4$	Pt	+0.727	sce	13
	Ethanol, $NaClO_4$	Pt	+0.731	sce	13
	Acetone, $NaClO_4$	Pt	+0.772	sce	13
	DMSO, $NaClO_4$	Pt	+0.798	sce	13
	Py. $(Et)_4NClO_4$	Graphite	+0.31, +0.86	Ag/1N Ag+	14

Compound	Electrolyte	Electrode	E	Reference	Ref.
Py		Graphite	+0.32, +0.79, +1.02	Ag/1N Ag+	14
SO₂ᵇ		Graphite	+0.92, +1.28, +2.29	Ag, AgBr/SO₂	14
Diphenyldiazomethane	CH₃CN, LiClO₄	Pt	+0.95, +1.73	Ag/0.1N Ag+	15
Tetraphenyltetrazene	CH₃CN, LiClO₄	Pt	+0.520	Ag/0.01N Ag+	10
Phthalhydrazide	0.5M NaOH	Pt	+0.42 Ep	sce	16
3-Methylphthalhydrazide	0.1M NaOH	Pt	+0.32 Ep	sce	16
Luminol	0.1M NaOH	Pt	+0.33 Ep	sce	16

Diazomethanes

Compound	Electrolyte	Electrode	E	Reference	Ref.
1-Diazo-2,3,4,5-tetraphenylcyclopentadiene	CH₃CN, LiClO₄	Pt	+1.01	sce	17
p-Phenyldiphenyldiazomethane	CH₃CN, LiClO₄	Pt	+0.89	sce	17
Diphenyldiazomethane	CH₃CN, LiClO₄	Pt	+0.95	sce	17
p-Nitrodiphenyldiazomethane	CH₃CN, LiClO₄	Pt	+1.14	sce	17
p-Nitrophenylmethyldiazomethane	CH₃CN, LiClO₄	Pt	+1.16	sce	17
p-Nitrophenyldiazomethane	CH₃CN, LiClO₄	Pt	+1.33	sce	17
9-Diazofluorene	CH₃CN, LiClO₄	Pt	+1.22	sce	17
Diazomethane	CH₃CN, LiClO₄	Pt	+1.70	sce	17
Ethyl diazoacetate	CH₃CN, LiClO₄	Pt	+2.10	sce	17

Substituted Diphenyl Diazomethanes

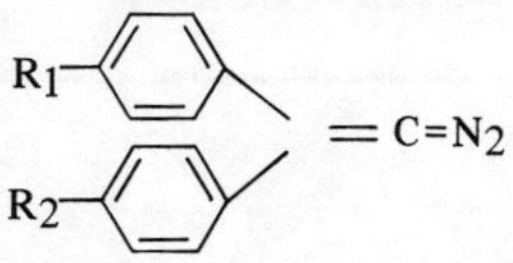

$=C=N_2$

R_1

R_2

(continued)

Compound		Solvent System	Working Electrode	$E_{1/2}$	Reference Electrode	Reference
R_1 = OCH$_3$	R_2 = H	CH$_3$CN, LiClO$_4$	Pt	+0.79	sce	18
= CH$_3$	= H	CH$_3$CN, LiClO$_4$	Pt	+0.89	sce	18
= H	= H	CH$_3$CN, LiClO$_4$	Pt	+0.95	sce	18
= Cl	= H	CH$_3$CN, LiClO$_4$	Pt	+0.99	sce	18
= Br	= H	CH$_3$CN, LiClO$_4$	Pt	+0.99	sce	18
= I	= H	CH$_3$CN, LiClO$_4$	Pt	+1.00	sce	18
= Br	= Br	CH$_3$CN, LiClO$_4$	Pt	+1.04	sce	18
= CN	= H	CH$_3$CN, LiClO$_4$	Pt	+1.11	sce	18
= NO$_2$	= H	CH$_3$CN, LiClO$_4$	Pt	+1.14	sce	18

$$R_1 - \text{C}_6\text{H}_4 - \overset{}{\underset{N_2}{C}} - \overset{O}{\underset{}{C}} - R_2$$

Compound		Solvent System	Working Electrode	$E_{1/2}$	Reference Electrode	Reference
R_1 = CH$_3$	R_2 = C$_6$H$_5$	CH$_3$CN, LiClO$_4$	Pt	+1.36	sce	19
= F	= C$_6$H$_5$	CH$_3$CN, LiClO$_4$	Pt	+1.46	sce	19
= H	= C$_6$H$_5$	CH$_3$CN, LiClO$_4$	Pt	+1.47	sce	19
= Cl	= C$_6$H$_5$	CH$_3$CN, LiClO$_4$	Pt	+1.53	sce	19
= Br	= C$_6$H$_5$	CH$_3$CN, LiClO$_4$	Pt	+1.54	sce	19
= NO$_2$	= C$_6$H$_5$	CH$_3$CN, LiClO$_4$	Pt	+1.75	sce	19
= H	= p-CH$_3$O-C$_6$H$_4$	CH$_3$CN, LiClO$_4$	Pt	+1.43	sce	19
= H	= p-C$_2$H$_5$-C$_6$H$_4$	CH$_3$CN, LiClO$_4$	Pt	+1.47	sce	19
= H	= p-Cl-C$_6$H$_4$	CH$_3$CN, LiClO$_4$	Pt	+1.50	sce	19
= CH$_3$	= OC$_2$H$_5$	CH$_3$CN, LiClO$_4$	Pt	+1.32	sce	19
= H	= OC$_2$H$_5$	CH$_3$CN, LiClO$_4$	Pt	+1.43	sce	19
= NO$_2$	= OC$_2$H$_5$	CH$_3$CN, LiClO$_4$	PT	+1.71	sce	19

= NO_2 = OCH_3 CH_3CN, $LiClO_4$ Pt   +1.75   sce   19
= NO_2 = CH_3 CH_3CN, $LiClO_4$ Pt   +1.76   sce   19

a Pyridine added.
b At -20°C.

Reduction

Azo Compounds

Compound	Solvent System	Working Electrode	$E_{1/2}$	Reference Electrode	Reference
Azobenzene	DMF, 0.1$\underline{M}$ (Et)$_4$-NClO$_4$	Hg	-1.81, -2.29	Ag/0.1$\underline{N}$ Ag+	20
	DMF, 0.1$\underline{M}$ (n-Bu)$_4$-NClO$_4$	Hg	-1.36, -2.03	sce	21
	PC, 0.1$\underline{M}$ (n-Bu)$_4$-NClO$_4$	Hg	-1.40	sce	22
	CH_3CN, (Et)$_4$NClO$_4$	Hg	-1.45, -1.74	sce	2
	KCl-HCl buffer, 50% Ethanol, pH 1.6	Graphite	-0.03 E$_{p/2}$	sce	4
	McIlvaine buffer, 50% Ethanol, pH 5.4	Graphite	0.32 E$_{p/2}$	sce	4
	KCl-KOH buffer, 50% Ethanol, pH 12.5	Graphite	0.76 E$_{p/2}$	sce	4
cis-Azobenzene	OAc buffer, 30%-Methanol, pH 5.1	Hg	-0.22, -0.97	sce	3

(continued)

Compound	Solvent System	Working Electrode	$E_{1/2}$	Reference Electrode	Reference
trans-Azobenzene	Ammoniacal buffer, 30% Methanol, pH 9.9	Hg	-0.52	sce	3
	OAc buffer, 30%-Methanol, pH 5.1	Hg	-0.24	sce	3
	Ammoniacal buffer, 30% Methanol, pH 9.9	Hg	-0.52	sce	3
cis- or trans-Azobenzene	Aq buffers, 10% Ethanol, pH 2.85 to 12.50	Hg	0.060, -0.062 pH	sce	23
o-Methylazobenzene	Aq buffer, 50% Alcohol, pH 4.6	Graphite	-0.35	NCE	24
o,o'-Dimethylazobenzene	Aq buffer, 50% Alcohol, pH 4.6	Graphite	-0.38	NCE	24
Phenylazomesitylene	Aq buffer, 50% Alcohol, pH 4.6	Graphite	-0.43	NCE	24
o-Tolylazomesitylene	Aq buffer, 50% Alcohol, pH 4.6	Graphite	-0.58	NCE	24
4-Aminoazobenzene	CH_3CN, $(Et)_4NClO_4$	Hg	-1.54, -1.74	sce	2
4-Dimethylaminoazobenzene	CH_3CN, $(Et)_4NClO_4$	Hg	-1.58, -1.84	sce	2
4-Dimethyl-2'-methyl-aminoazobenzene	CH_3CN, $(Et)_4NClO_4$	Hg	-1.59, -1.84	sce	2
4-Dimethyl-3'-methyl-aminoazobenzene	CH_3CN, $(Et)_4NClO_4$	Hg	-1.57, -1.80	sce	2
4-Dimethyl-4'-methyl-aminoazobenzene	CH_3CN, $(Et)_4NClO_4$	Hg	-1.58, -1.86	sce	2
4-Dimethyl-3'-fluoro-aminoazobenzene	CH_3CN, $(Et)_4NClO_4$	Hg	-1.50, -1.66	sce	2

Compound	Medium	Electrode	E	Ref. electrode	Ref.
2-Methyl-4-dimethyl-aminoazobenzene	CH_3CN, $(Et)_4NClO_4$	Hg	-1.60, -1.88	sce	2
4-Diethylaminoazobenzene	CH_3CN, $(Et)_4NClO_4$	Hg	-1.62, -1.81	sce	2
3,3'-Azotoluene	DMF, $0.1\underline{M}$ $(\underline{n}\text{-Bu})_4\text{-}NClO_4$	Hg	-1.38, -2.01	sce	21
1,1'-Azonaphthalene	DMF, $0.1\underline{M}$ $(\underline{n}\text{-Bu})_4\text{-}NClO_4$	Hg	-1.13, -1.69	sce	21
2,2'-Azonaphthalene	OAc buffer, 75% Ethanol	Hg	-0.335	NCE	25
	DMF, $0.1\underline{M}$ $(\underline{n}\text{-Bu})_4\text{-}NClO_4$	Hg	-1.22, -1.75	sce	21
	OAc buffer, 75% Ethanol	Hg	-0.345	NCE	25
4,4'-Azobiphenyl	DMF, $0.1\underline{M}$ $(\underline{n}\text{-Bu})_4\text{-}NClO_4$	Hg	-1.22, -1.77	sce	21
4,4'-Azopyridine	DMF, $0.1\underline{M}$ $(\underline{n}\text{-Bu})_4\text{-}NClO_4$	Hg	-0.80, -1.53	sce	21
o-Bisazobenzene	Aq buffers, pH 4-12, 80% Methanol	Hg	+0.085, -0.059 pH, -0.125- -0.059 pH	sce	26
m-Bisazobenzene	Aq buffers, pH 1-12, 80% Methanol	Hg	0.00, -0.059 pH	sce	26
p-Bisazobenzene	Aq buffers, pH 4.3-12.7, 80% Methanol	Hg	+0.165, -0.059 pH	sce	27
	Methanol, $0.15\underline{M}$ LiCl	Hg	-0.49, -0.79	sce	27

Diazonium Salts

Compound	Medium	Electrode	E	Ref. electrode	Ref.
Benzenediazonium bi-sulphate	Walpole, OAc buffer, KCl, pH 4.12	Hg	-0.188	sce	28

(continued)

Compound	Solvent System	Working Electrode	$E_{1/2}$	Reference Electrode	Reference
Benzenediazonium chloride	0.1N HCl	Hg	-0.18, -0.67	sce	29
	C-L buffer, pH 4.02	Hg	+0.029, -0.057, -0.690	sce	29
	0.1M (Me)4NCl 0.1M Phosphate-buffer, pH 4.0	Hg	-0.13, -0.97	sce	30
	Walpole, OAc buffer, KCl, pH 4.12	Hg	-0.190	sce	
	C-L buffer, pH 6.0	Hg	+0.05, -0.05, -0.92	sce	31
Benzenediazonium tetrafluoroborate	Sulfolane, 0.1M-(n-Bu)4NClO4	Hg	+0.295	sce	32
o-Methylbenzenediazonium tetrafluoroborate	Sulfolane, 0.1M-(n-Bu)4NClO4	Hg	+0.228	sce	32
m-Methylbenzenediazonium tetrafluoroborate	Sulfolane, 0.1M-(n-Bu)4NClO4	Hg	+0.285	sce	32
p-Methylbenzenediazonium tetrafluoroborate	Sulfolane, 0.1M-(n-Bu)4NClO4	Hg	+0.250	sce	32
	Walpole, OAc buffer, KCl, pH 4.12	Hg	-0.153	sce	28
o-Methoxybenzenediazonium tetrafluoroborate	Sulfolane, 0.1M-(n-Bu)4NClO4	Hg	+0.153	sce	32
p-Methoxybenzenediazonium tetrafluoroborate	Sulfolane, 0.1M-(n-Bu)4NClO4	Hg	+0.140	sce	32
	Walpole, OAc buffer,	Hg	-0.116	sce	28

Substance	Medium	Electrode	Potential (V)	Reference	Ref.
Diazotized o-amino-benzoic acid	C-L buffer, pH 4.2	Hg	-0.056, -0.575	sce	29
Diazotized m-amino-benzoic acid	0.1M K_2SO_4, 0.1M-Phosphate buffer, pH 3.6	Hg	-0.11, -0.81	sce	30
Diazotized p-amino-benzoic acid	0.1M K_2SO_4, 0.1M-Phosphate buffer, pH 3.5	Hg	-0.09, -0.70	sce	30
	0.1M K_2SO_4, 0.1M-Phosphate buffer, pH 3.6	Hg	-0.05, -0.80	sce	30
m-Nitrobenzenediazon-ium bisulfate	Sulfolane, 0.1M-$(n\text{-}Bu)_4NClO_4$	Hg	+0.328	sce	32
p-Nitrobenzenediazon-ium bisulfate	Walpole, OAc buffer, KCl, pH 4.12	Hg	-0.152	sce	28
p-Nitrobenzenediazon-ium tetrafluoroborate	Walpole, OAc buffer, KCl, pH 4.12	Hg	-0.366	sce	28
o-Chlorobenzenediazon-ium tetrafluoroborate	Walpole, OAc buffer, KCl, pH 4.12	Hg	-0.300	sce	28
p-Nitrobenzenediazon-ium tetrafluoroborate	Sulfolane, 0.1M-$(n\text{-}Bu)_4NClO_4$	Hg	+0.450	sce	32
p-Cyanobenzenediazon-ium tetrafluoroborate	Sulfolane, 0.1M-$(n\text{-}Bu)_4NClO_4$	Hg	+0.433	sce	32
o-Chlorobenzenediazon-ium tetrafluoroborate	Sulfolane, 0.1M-$(n\text{-}Bu)_4NClO_4$	Hg	+0.410	sce	32
m-Chlorobenzenediazon-ium tetrafluoroborate	Sulfolane, 0.1M-$(n\text{-}Bu)_4NClO_4$	Hg	+0.410	sce	32
p-Chlorobenzenediazon-ium tetrafluoroborate	Sulfolane, 0.1M-$(n\text{-}Bu)_4NClO_4$	Hg	+0.350	sce	32
p-Chlorobenzenediazon-ium bisulfate	Walpole, OAc buffer, KCl, pH 4.12	Hg	-0.179	sce	28

(continued)

Compound	Solvent System	Working Electrode	$E_{1/2}$	Reference Electrode	Reference
Diazotized dichloro-aniline	C–L buffer, pH 3.92	Hg	-0.022, -0.525, -0.875	sce	29
2,5-Dichlorobenzene-diazonium tetrafluoro-borate	Sulfolane, 0.1M (n-Bu)$_4$NClO$_4$	Hg	$+0.297$	sce	32
m-Bromobenzenediazon-ium chloride	Walpole, OAc buffer, KCl, pH 4.12	Hg	-0.162	sce	28
m-Bromobenzenediazon-ium tetrafluoroborate	Sulfolane, 0.1M (n-Bu)$_4$NClO$_4$	Hg	$+0.429$	sce	32
p-Bromobenzenediazon-ium tetrafluoroborate	Sulfolane, 0.1M (n-Bu)$_4$NClO$_4$	Hg	$+0.383$	sce	32
p-Iodobenzenediazonium tetrafluoroborate	Sulfolane, 0.1M (n-Bu)$_4$NClO$_4$	Hg	$+0.383$	sce	32
Diazotized sulfanilic acid	Sulfolane, 0.1M (n-Bu)$_4$NClO$_4$	Hg	$+0.295$	sce	32
	0.1N HCl	Hg	-0.326, -0.555, -0.831	sce	29
	C–L buffer, pH 4.02	Hg	-0.095, -0.260, -0.626, -0.970	sce	29
	Walpole, OAc buffer, KCl, pH 4.12	Hg	-0.270	sce	29
3-Diazocamphor	B–R buffer, pH 9.40	Hg	-0.99, -1.23	sce	33

Azines

Compound	Conditions	Electrode	E (V)	Ref. electrode	Ref.
Dibenzalazine	0.02N (Et)$_4$NI, 92% Methanol	Hg	-1.42, -2.12	sce	34
p-Methoxybenzalazine	0.02N (Et)$_4$NI, 92% Methanol	Hg	-1.44, -2.08	sce	34
Bis(4-methoxybenzal)-azine	0.02N (Et)$_4$NI, 92% Methanol	Hg	-1.48, -2.08	sce	34
Bis(4-dimethylamino-benzal)azine	0.02N (Et)$_4$NI, 92% Methanol	Hg	-1.60, -2.20	sce	34
Bis(4-nitrobenzal)azine	0.02N (Et)$_4$NI, 92% Methanol	Hg	-0.76, -0.88, -1.63, -2.26	sce	34
Bis(1-naphthal)azine	0.02N (Et)$_4$NI, 92% Methanol	Hg	-1.24, -1.74	sce	34
Bis(2-naphthal)azine	0.02N (Et)$_4$NI, 92% Methanol	Hg	-1.28, -1.74	sce	34
Bis(4-phenylbenzal)-azine	0.02N (Et)$_4$NI, 92% Methanol	Hg	-1.29, -1.79	sce	34

Azides

Compound	Conditions	Electrode	E (V)	Ref. electrode	Ref.
X = H	DMF, 0.05N (Et)$_4$NI	Hg	-1.460	a	35
= m-NO$_2$	DMF, 0.05N (Et)$_4$NI	Hg	-1.201	a	35
= p-NO$_2$	DMF, 0.05N (Et)$_4$NI	Hg	-1.206	a	35
= m-CH$_3$	DMF, 0.05N (Et)$_4$NI	Hg	-1.409	a	35
= p-CH$_3$	DMF, 0.05N (Et)$_4$NI	Hg	-1.519	a	35

(continued)

Compound	Solvent System	Working Electrode	$E_{1/2}$	Reference Electrode	Reference
= p-OCH₃	DMF, 0.05N (Et)₄NI	Hg	-1.544	--[a]	35
Y = =CH=Y=CH=	DMF, 0.05N (Et)₄NI	Hg	-1.000	--[a]	35
"	DMF, 0.05N (Et)₄NI	Hg	-0.935	--[a]	35
"	DMF, 0.05N (Et)₄NI	Hg	-1.010	--[a]	35
"	DMF, 0.05N (Et)₄NI	Hg	-1.098	--[a]	35
" = =N-N=	DMF, 0.05N (Et)₄NI	Hg	-1.163	--[a]	35
Picrylhydrazyls					
Diphenylpicrylhydrazyl	CH₃CN, NaClO₄	Pt	+0.178	sce	13
	Methanol, NaClO₄	Pt	+0.205	sce	13
	Ethanol, NaClO₄	Pt	+0.212	sce	13
	Acetone, NaClO₄	Pt	+0.223	sce	13

DMSO, NaClO$_4$	Pt	+0.348	sce	13
CH$_3$CN	Pt	+0.15	sce	14
CH$_3$CN, (Et)$_4$NClO$_4$	Pt	-0.10	Ag/0.1$\underline{N}$ Ag+	14
Py	Graphite	-0.62, -1.22, -1.50, -1.80, -2.20	Ag/1$\underline{N}$ Ag+	14
Py, (Et)$_4$NClO$_4$	Graphite	-0.54, -1.24, -1.47	Ag/1$\underline{N}$ Ag+	14
SO$_2$[b]	Graphite	+0.48, +0.06, -0.64	Ag, AgBr/SO$_2$	14
Ethanol, 0.1$\underline{N}$ LiCl	Hg	-0.69, -0.84	Ag·0.01$\underline{N}$ Ag+	12
Ethanol, 0.1$\underline{N}$ LiCl	Hg	-0.67, -0.90	Ag·0.01$\underline{N}$ Ag+	12
Ethanol, 0.1$\underline{N}$ LiCl	Hg	-0.57, -0.75, -0.95	Ag/0.01$\underline{N}$ Ag+	12

[a] Not specified.
[b] At -20°C.

(continued)

References for Nitrogen-Nitrogen Functions

1. S. Wawzonek and T. McIntyre, J. Electrochem. Soc. 114, 1025 (1967).
2. T. A. Gough and M. E. Peover, Proc. 3rd International Polarography Congress, Southampton, 1965 Macmillan, London, 1966, p. 1017.
3. S. Wawzonek and J. D. Fredrickson, J. Amer. Chem. Soc., 77, 3985 (1955).
4. L. Chang, I. Fried, and P. J. Elving, Anal. Chem. 37, 1528 (1965).
5. G. Cauquis and G. Fauvelot, Bull. Soc. Chim. Fr., 2014 (1964).
6. A. F. Krivis and G. R. Supp, Abstracts American Chemical Society Meeting, Division of Fuels Chemistry, Miami Beach, Fla., 1967, p. 137.
7. P. E. Iversen and H. Lund, Anal. Chem., 41, 1322 (1969).
8. M. Michlmayr and D. T. Sawyer, J. Electroanal. Chem., 23, 375 (1969).
9. G. Cauquis and M. Genies, Tetrahedron Letters, 3537 (1968).
10. G. Cauquis, J. Cros, and M. Genies, Preprints of Papers, Durham Symposium on Electro-Organic Chemistry, Durham, North Carolina, 1968, p. 249.
11. H. Lund, Acta Chem. Scand., 17, 1077 (1963).
12. A. T. Balaban, P. T. Frangopol, M. Frangopol, an N. Negoita, Tetrahedron, 23, 4661 (1967).
13. E. Solon and A. J. Barn, J. Amer. Chem. Soc., 86 1926 (1964).
14. D. A. Hall and P. J. Elving, Electrochim. Acta, 12, 1363 (1967).
15. W. Jugelt and F. Pragst, Angew. Chem. Internat. Edit., 7, 290 (1968).
16. B. Epstein and T. Kuwana, J. Electroanal. Chem., 15, 389 (1967).
17. F. Pragst and W. Jugelt, Electrochim. Acta, 15, 1543 (1970).
18. W. Jugelt, and F. Pragst, Tetrahedron, 24, 5123 (1968).
19. F. Pragst, W. Hubner, and W. Jugelt, J. Prakt. Chem., 312, 105 (1970).
20. G. H. Aylward, J. L. Garnett, and J. H. Sharp, Anal. Chem., 39, 457 (1967).
21. J. L. Sadler and A. J. Bard, J. Amer. Chem. Soc. 90, 1979 (1968).
22. J. E. McClure and D. L. Maricle, Anal. Chem., 39, 236 (1967).

23.   C. R. Castor and J. H. Saylor, *J. Amer. Chem. Soc.*, **75**, 1427 (1953).
24.   C. Prevost, P. Souchay, and C. Malen, *Bull. Soc. Chim. Fr.*, **78**, (1953).
25.   I. F. Wladimirzew and I. J. Posstowski, *Ber. Akad. Wiss. UdSSR*, **83**, 855 (1952).
26.   T. M. H. Saber, M. L. Ali, and N. M. Abed, *J. Electroanal. Chem.*, **28**, 177 (1970).
27.   A. M. Shams-El-Din, T. M. H. Saber, and N. M. Abed, *J. Electroanal. Chem.*, **21**, 377 (1969).
28.   J. K. Kochi, *J. Amer. Chem. Soc.*, **77**, 3208 (1955).
29.   R. M. Elofson, R. L. Edsberg, and P. A. Mecherly, *J. Electrochem. Soc.*, **97**, 166 (1950).
30.   E. R. Atkinson, H. L. Warren, P. I. Abell, and R. E. Wing, *J. Am. Chem. Soc.*, **72**, 915 (1950).
31.   L. Meites, Polarographic Techniques, 2nd ed., Interscience, New York, 1965, p. 683.
32.   R. M. Elofson and F. F. Gadallah, *J. Org. Chem.*, **34**, 854 (1969).
33.   M. E. Cardinali, I. Carelli, and A. Trazza, *J. Electroanal. Chem.*, **23**, 399 (1969).
34.   V. D. Bezuglyi and N. P. Shimanskaya, *Zh. Obshch. Khim.*, **35**, 17 (1965).
35.   L. V. Kononenko, T. A. Yurre, V. N. Dmitrieva, L. S. Efros, and V. D. Bezuglyi, *Zh. Obshch. Khim.*, **40** (6), 1359 (1970).

25. NITROGEN-OXYGEN FUNCTION

Oxidation

Aliphatic Nitro Compounds

Compound	Solvent System	Working Electrode	$E_{1/2}$	Reference Electrode	Reference
Nitromethane	Aq buffer, pH 11.2	Pt	+0.91	sce	1
Nitroethane	Aq buffer, pH 11.2	Pt	+0.91	sce	1
1-Nitropropane	Aq buffer, pH 11.2	Pt	+0.90	sce	1
2-Nitropropane	Aq buffer, pH 11.2	Pt	+0.95	sce	1
2,5-Dinitrohexane	Aq buffer, pH 11.2	Pt	+0.94	sce	1
3,5-Dinitroheptane	Aq buffer, pH 11.2	Pt	+1.01	sce	1
Sodium nitritea	Aq buffer, pH 11.2	Pt	+1.05	sce	1
Potassium ethanenitron- ate	CH_3CN, 0.5M $NaClO_4$	Hg	+1.06	Hg pool	1
	DMF, 0.5M $NaClO_4$	Hg	+1.16	Hg pool	1
	DMF, 0.1M $NaNO_3$	Hg	+0.93	Hg pool	1
	DMSO, 0.1M KNO_3	Hg	+1.00	Hg pool	1
	DMSO, 0.2M $NaNO_3$	Hg	+0.95	Hg pool	1
Sodium ethanenitronate	CH_3CN, 0.5M $NaClO_4$	Hg	+0.80	Hg pool	1
	DMSO, 0.3M KNO_3	Hg	+0.76	Ag, AgCl	1
Potassium 1-propane- nitronate	Methanol, 0.1M $NaClO_4$	Hg	+0.76	sce	1
	Methanol, 0.1M $KOCH_3$	Hg	+0.68	sce	1

Compound	Solvent, Electrolyte	Electrode	E	Reference	Ref.
Potassium 2-propane-nitronate	CH_3CN, $0.5\underline{M}$ $NaClO_4$	Hg	+1.04	Ag, AgCl	1
Sodium 2-propanenitron-ate	Methanol, $0.1\underline{M}$ $KOCH_3$	Hg	+0.65	sce	1
	Methanol, $0.1\underline{M}$ $LiNO_3$	Hg	+0.70	sce	1
	CH_3CN, $0.5\underline{M}$ $NaClO_4$	Hg	+0.92	Ag, AgCl	1
Lithium 2-propane-nitronate	DMF, $0.2\underline{M}$ $NaNO_3$	Hg	+0.95	Hg pool	1
	DMSO, $0.3\underline{M}$ KNO_3	Hg	+0.56	Ag, AgCl	1
	DMF, $0.1\underline{M}$ $LiNO_3$	Hg	+1.21	Hg pool	1
	DMSO, $0.1\underline{M}$ $LiNO_3$	Hg	+0.91	Hg pool	1

Aromatic Nitro Compounds

Compound	Solvent, Electrolyte	Electrode	E	Reference	Ref.
2-Nitroterphenyl	CH_3CN, $0.1\underline{M}$ $(\underline{n}-Bu)_4$ $NClO_4$	Pt	+1.88	sce	2
3-Nitroterphenyl	CH_3CN, $0.1\underline{M}$ $(\underline{n}-Bu)_4$ $NClO_4$	Pt	+1.78	sce	2
4-Nitroterphenyl	CH_3CN, $0.1\underline{M}$ $(\underline{n}-Bu)_4$ $NClO_4$	Pt	+1.76, +1.95	sce	2
2,4-Dinitroterphenyl	CH_3CN, $0.1\underline{M}$ $(\underline{n}-Bu)_4$ $NClO_4$	Pt	+1.86	sce	2
2,6-Dinitroterphenyl	CH_3CN, $0.1\underline{M}$ $(\underline{n}-Bu)_4$ $NClO_4$	Pt	+1.76	sce	2
4,4"-Dinitroterphenyl	CH_3CN, $0.1\underline{M}$ $(\underline{n}-Bu)_4$ $NClO_4$	Pt	+2.06	sce	2
2,4,6-Trinitroterphenyl	CH_3CN, $0.1\underline{M}$ $(\underline{n}-Bu)_4$ $NClO_4$	Pt	+2.00	Ag/$0.1\underline{N}$ Ag+	2
1-Nitronaphthalene	CH_3CN, $0.5\underline{M}$ $NaClO_4$	Pt	+1.62	Ag/$0.1\underline{N}$ Ag+	3
9-Nitroanthracene	CH_3CN, $0.5\underline{M}$ $NaClO_4$	Pt	+1.25	Ag/$0.1\underline{N}$ Ag+	3

(continued)

Hydroxylamines

Compound	Solvent System	Working Electrode	E$_{1/2}$	Reference Electrode	Reference
Hydroxylamine	B-R buffer, pH 4.6	Hg	-1.46	sce	4
	B-R buffer, pH 6.8	Hg	-1.52	sce	4
	B-R buffer, pH 10.4	Hg	-1.75	sce	4
	H$_2$O, pH 13	Hg	-0.35	sce	5
N-Methylhydroxylamine	H$_2$O, pH 13	Hg	-0.48	sce	5
N-Ethylhydroxylamine	H$_2$O, pH 13	Hg	-0.49	sce	5
N-Propylhydroxylamine	H$_2$O, pH 13	Hg	-0.49	sce	5
N-Isopropylhydroxyla-mine	H$_2$O, pH 13	Hg	-0.48	sce	5
N-t-Butylhydroxylamine	H$_2$O, pH 13	Hg	-0.47	sce	5
N-t-Octylhydroxylamine	H$_2$O, pH 13	Hg	-0.49	sce	5
N-Cyclohexylhydroxyla-mine	H$_2$O, pH 13	Hg	-0.47	sce	5
N-Phenylhydroxylamine	KCl+HCl buffer, pH 1.6	Graphite	+0.261 E$_{p/2}$	sce	6
	50% Ethanol, pH 5.4	Graphite	-0.010 E$_{p/2}$	sce	6
	KCl+HCl buffer, pH 8.6	Graphite	-0.188, -0.757 E$_{p/2}$	sce	6
	50% Ethanol, pH 12.5	Graphite	-0.416, -0.785, -0.921 E$_{p/2}$	sce	6
N-Benzylhydroxylamine	Borate buffer, 1M KCl, 40% Alcohol, pH 9.8	Hg	-0.11	sce	7
N-Benzylhydroxylamine	H$_2$O, pH 13	Hg	-0.52	sce	5
2-Phenyl-2-hydroxyl-	H$_2$O, pH 13	Hg	-0.47	sce	5

Compound	Solvent System	Working Electrode	$E_{1/2}$	Reference Electrode	Reference
3-Methyl-3-hydroxyl-amino-2-butanol	H_2O, pH 13	Hg	-0.51	sce	5
2-Methyl-2-hydroxyl-amino-1-propanol	H_2O, pH 13	Hg	-0.49	sce	5
2-Methyl-2-hydroxyl-amino-1,3-propanediol	H_2O, pH 13	Hg	-0.49	sce	5
2-Hydroxymethyl-2-hydroxyl-1,3-propane-diol	H_2O, pH 13	Hg	-0.49	sce	5
N,N-Dibenzylhydroxyl-amine	Citrate-phosphate buffer, 1M KCl, 40% Alcohol, pH 3.55	Hg	-1.29	sce	7
	Citrate phosphate buffer, 1M KCl, 40% Alcohol, pH 5.20	Hg	-1.38	sce	7
	H_2O, pH 13	Hg	-0.38	sce	5

a Included for comparison.

Reduction

Aliphatic Nitro Compounds

Compound	Solvent System	Working Electrode	$E_{1/2}$	Reference Electrode	Reference
Nitromethane	HOAc, 1M NH_4OAc	Hg	-1.125	Hg pool	8

(continued)

Compound	Solvent System	Working Electrode	$E_{1/2}$	Reference Electrode	Reference
	0.05M H₂SO₄	Hg	-1.137	Hg, Hg₂SO₄	9
	0.05$\underline{M}$ Na₂SO₄	Hg	-1.35	Hg, Hg₂SO₄	9
	McIlvaine buffer, pH 2.1	Hg	-0.60	sce	10
	McIlvaine buffer, pH 7.0	Hg	-0.88	sce	10
	Sorensen buffer[a], pH 11.9	Hg	-0.96	sce	10
Tribromonitromethane	3% NaCl	Graphite	+0.06	sce	11
Dinitromethane	B-R buffer, 0.1$\underline{M}$ KCl, pH 2.0	Hg	-0.280, -0.665, -1.075	sce	12
	B-R buffer, 0.1$\underline{M}$ KCl, pH 6.0	Hg	-0.632	sce	12
	B-R buffer, 0.1$\underline{M}$ KCl, pH 12.0	Hg	-0.720, -1.405	sce	12
Trinitromethane	Aq buffers, pH 2.0	Hg	-0.16	sce	13
	Aq buffers, pH 6.0	Hg	-0.37	sce	13
Trinitrochloromethane	3% NaCl	Graphite	+0.24	sce	11
Trinitroiodomethane	3% NaCl	Graphite	+0.36	sce	11
Tetranitromethane	3% NaCl	Graphite	+0.42	sce	11
	Aq buffers, 10% Ethanol, pH 2.0	Hg	+0.06	sce	13
Tetranitromethane	Aq buffers, 10% Ethanol, pH 8.0	Hg	+0.06	sce	13
	Aq buffers, 10% Ethanol	Hg	-0.16	sce	13

Compound	Medium	Electrode	E	Ref. electrode	Ref.
Tris(hydroxymethyl)-nitromethane	Methanol	Hg	+0.45, +0.29	sce	14
	CH$_3$CN, 0.1M-(Et)$_4$NBr	Hg	−1.45	sce	15
gem-Chloronitrodimethyl-methane	Aprotic medium, 0.1M (Et)$_4$NClO$_4$	Hg	−0.91	sce/DMF	16
	Protic medium, 0.1M (Et)$_4$NClO$_4$	Hg	−0.75	sce/DMF	16
gem-Bromonitrodimethyl-methane	Aprotic medium, 0.1M (Et)$_4$NClO$_4$	Hg	−0.29	sce/DMF	16
	Protic medium, 0.1M (Et)$_4$NClO$_4$	Hg	−0.23	sce/DMF	16
Dimethylnitromethane	Aprotic medium, 0.1M (Et)$_4$NClO$_4$	Hg	−1.70	sce/DMF	16
	Protic medium, 0.1M (Et)$_4$NClO$_4$	Hg	−1.59	sce/DMF	16
Nitroethane	HOAc, 1M NH$_4$OAc	Hg	−1.100	Hg pool	8
	0.05M H$_2$SO$_4$	Hg	−1.115	Hg, Hg$_2$SO$_4$	9
	0.05M Na$_2$SO$_4$	Hg	−1.36	Hg, Hg$_2$SO$_4$	9
	McIlvaine buffer, pH 2.1	Hg	−0.63	sce	10
	McIlvaine buffer, pH 7.0	Hg	−0.90	sce	10
	Sorensen buffer[a], pH 11.9	Hg	−0.95	sce	10
	50% Methanol, 50% Benzene, 0.3M LiCl	Hg	−1.16	sce	17
Nitroethane	Methanol, 0.5M LiCl	Hg	−1.12	sce	17
	Isobutanol, 0.2M-LiCl	Hg	−1.15	sce	17

(Continued)

Compound	Solvent System	Working Electrode	$E_{1/2}$	Reference Electrode	Reference
1,1-Dinitroethane	Ethylene glycol, 0.3M LiCl	Hg	-1.03	sce	17
	Glycerol, 0.3M LiCl	Hg	-0.93	sce	17
	B-R buffer, 0.1M KCl, pH 2.0	Hg	-0.331, -0.806, -1.151	sce	12
	B-R buffer, pH 6.0 0.1M KCl	Hg	-0.579	sce	12
	B-R buffer, pH 12.0 0.1M KCl	Hg	-0.765, -1.490	sce	12
1-Nitropropane	0.05M H2SO4	Hg	-1.069	Hg, Hg2SO4	9
	0.05M Na2SO4	Hg	-1.35	Hg, Hg2SO4	9
	0.1N HCl	Hg	0.605	sce	18
	0.1N H2SO4	Hg	-0.620	sce	18
	0.1N H2SO4, 0.1N KCl	Hg	-0.618	sce	18
	0.1N H2SO4, 0.1N Br	Hg	-0.634	sce	18
	0.1N H2SO4, 0.1N KI	Hg	-0.660	sce	18
	McIlvaine buffer, pH 2.1	Hg	-0.56	sce	10
	McIlvaine buffer, pH 2.2	Hg	-0.668	sce	18
	McIlvaine buffer, pH 5.1	Hg	-0.800	sce	18
	McIlvaine buffer, pH 7.0	Hg	-0.86	sce	10
	McIlvaine buffer, pH 8.1	Hg	-0.890	sce	18
	Sorensen buffer[a], pH 11.9	Hg	-0.96	sce	10

Compound	Solvent / electrolyte	Electrode	Potential	Reference	Ref.
	50% Methanol, 50% Benzene, 0.3M LiCl	Hg	-1.16	sce	17
	Methanol, 0.5M LiCl	Hg	-1.09	sce	17
	Isobutanol, 0.2M LiCl	Hg	-1.11	sce	17
	Ethylene Glycol, 0.3M LiCl	Hg	-1.02	sce	17
2-Nitropropane	Glycerol, 0.3M LiCl	Hg	-0.98	sce	17
	HOAc, 1M NH4OAc	Hg	-1.10	Hg pool	8
	HOAc, 1M NH4OAc	Hg	-1.140	Hg pool	8
	0.05M H$_2$SO$_4$	Hg	-1.129	Hg, Hg$_2$SO$_4$	9
	0.05M Na$_2$SO$_4$	Hg	-1.40	Hg, Hg$_2$SO$_4$	9
	McIlvaine buffer, pH 2.1	Hg	-0.49	sce	10
	McIlvaine buffer, pH 7.0	Hg	-0.90	sce	10
	Sorensen buffer[a], pH 11.9	Hg	-0.98	sce	10
	50% Methanol, 50% Benzene, 0.3M LiCl	Hg	-1.31	sce	17
	Methanol, 0.5M LiCl	Hg	-1.20	sce	17
	Isobutanol, 0.2M LiCl	Hg	-1.16	sce	17
	Ethylene glycol, 0.3M LiCl	Hg	-1.08	sce	17
t-Nitropropane (sic)	Glycerol, 0.3M LiCl	Hg	-1.00	sce	17
	DMF, 0.1M LiCl	Hg	-1.100	Hg pool	19
1,3-Dinitropropane	DMF, 0.1M (n-Bu)4N1	Hg	-1.64	sce	20
	50% Methanol, 50%- Benzene, 0.3M LiCl	Hg	-1.16	sce	17

(continued)

Compound	Solvent System	Working Electrode	$E_{1/2}$	Reference Electrode	Reference
	Methanol, 5M LiCl	Hg	-1.14	sce	17
	Isobutanol, 0.2M LiCl	Hg	-1.00	sce	17
	Ethylene glycol, 0.3M LiCl	Hg	-0.97	sce	17
2,2-Dinitropropane	50% Methanol, 50% Benzene, 0.3M LiCl	Hg	-0.86	sce	17
	Methanol, 0.5M LiCl	Hg	-0.70	sce	17
	Isobutanol, 0.2M LiCl	Hg	-0.70	sce	17
	Ethylene glycol, 0.3M LiCl	Hg	-0.61	sce	17
	B-R buffer, 0.1M KCl, 10% Ethanol, pH 2.0	Hg	-0.348	sce	12
	B-R buffer, 0.1M KCl, 10% Ethanol, pH 6.0	Hg	-0.488	sce	12
	B-R buffer, 0.1M KCl, 10% Ethanol, pH 12.0	Hg	-0.499	sce	12
2,2-Dimethyl-1,3-dinitropropane	50% Methanol, 50%-Benzene, 0.3M LiCl	Hg	-1.22	sce	17
	Methanol, 0.5M LiCl	Hg	-1.10	sce	17
	Isobutanol, 0.2M LiCl	Hg	-1.05	sce	17
	Ethylene glycol, 0.3M LiCl	Hg	-1.01	sce	17
1-Bromo-1,1-dinitropropane	3% NaCl	Graphite	+0.25	sce	11
3-Nitropropanoic Acid	HCl-NaCl buffer, pH 1.1	Hg	-0.57	sce	21
	OAc buffer, pH 4.7	Hg	-0.78	sce	21
	Borate buffer, pH 9.3	Hg	-0.99	sce	21

Compound	Supporting electrolyte / solvent	Electrode	E	Reference electrode	Ref.
2-Methyl-2-nitro-1-propanol	CH_3CN, 0.1M $(Et)_4NBr$	Hg	-1.52	sce	14
1-Nitrobutane	0.05M H_2SO_4	Hg	-1.017	Hg, Hg_2SO_4	9
	0.05M Na_2SO_4	Hg	-1.39	Hg, Hg_2SO_4	9
	50% Methanol, 50%-Benzene, 0.3M LiCl	Hg	-1.22	sce	17
	Methanol, 0.5M LiCl	Hg	-1.11	sce	17
	Isobutanol, 0.2M LiCl	Hg	-1.17	sce	17
	Ethylene glycol, 0.3M LiCl	Hg	-0.93	sce	17
	McIlvaine buffer, 0.5M KCl, pH 5.0	Hg	-0.73	sce	22
	McIlvaine buffer, 0.5M KCl, pH 7.0	Hg	-0.83	sce	22
2-Nitrobutane	50% Methanol, 50%-Benzene, 0.3M LiCl	Hg	-1.31	sce	22
	Methanol, 0.5M LiCl	Hg	-1.21	sce	22
	Isobutanol, 0.2M LiCl	Hg	-1.28	sce	22
	Ethylene glycol, 0.3M LiCl	Hg	-1.11	sce	17
	McIlvaine buffer, 0.5M KCl, pH 5.0	Hg	-0.79	sce	22
	McIlvaine buffer, 0.5M KCl, pH 7.0	Hg	-0.85	sce	22
t-Nitrobutane	DMF, 0.1M $(n-Bu)_4NI$	Hg	-1.64	sce	20
	Dimethoxyethane, 0.1M $(n-Bu)_4NClO_4$	Hg	-1.77	sce	23
	CH_3CN, 0.1M $(Et)_4NBr$	Hg	-1.62, -2.40	sce	14
	CH_3CN, 10% H_2O, 0.1M $(Et)_4NBr$	Hg	-1.37	sce	14

(continued)

Compound	Solvent System	Working Electrode	$E_{1/2}$	Reference Electrode	Reference
Ethyl-α-nitroisobutyrate	CH₃CN, 0.1M (Et)₄NBr	Hg	-1.32	sce	14
	CH₃CN, 10% H₂O, 0.1M (Et)₄NBr	Hg	-1.18, -1.97	sce	14
α-Nitroisobutyronitrile	CH₃CN, 0.1M (Et)₄NBr	Hg	-1.08	sce	14
	CH₃CN, 10% H₂O, 0.1M (Et)₄NBr	Hg	-0.92, -1.69	sce	14
α-Nitroisobutyramide	CH₃CN, 0.1M (Et)₄NBr	Hg	-1.31, -1.71	sce	14
	CH₃CN, 10% H₂O, 0.1M (Et)₄NBr	Hg	-1.11, -1.61	sce	14
2,3-Dinitro-2,3-dimethylbutane	Dimethoxyethane, 0.1M (n-Bu)₄NClO₄	Hg	-1.38	sce	23
	Aprotic medium, 0.1M (Et)₄NClO₄	Hg	-1.42	sce/DMF	16
1-Nitro-2-butanol	McIlvaine buffer, 0.5M KCl, pH 5.0	Hg	-0.69	sce	22
	McIlvaine buffer, 0.5M KCl, pH 7.0	Hg	-0.82	sce	22
2-Nitro-1-butanol	McIlvaine buffer, 0.5M KCl, pH 5.0	Hg	-0.73	sce	22
	McIlvaine buffer, 0.5M KCl, pH 7.0	Hg	-0.81	sce	22
2-Nitro-1-butyl formate	McIlvaine buffer, 0.5M KCl, pH 5.0	Hg	-0.69	sce	22
	McIlvaine buffer, 0.5M KCl, pH 7.0	Hg	-0.82	sce	22
2-Nitro-1-butyl acetate	McIlvaine buffer, 0.5M KCl, pH 5.0	Hg	-0.66	sce	22
	McIlvaine buffer, 0.5M KCl, pH 7.0	Hg	-0.75	sce	22

Compound	Conditions	Electrode	E	Ref.	
1-Methoxy-2-nitrobutane	McIlvaine buffer, 0.5M KCl, pH 5.0	Hg	-0.71	sce	22
	McIlvaine buffer, 0.5M KCl, pH 7.0	Hg	-0.81	sce	22
1-Ethoxy-2-nitrobutane	McIlvaine buffer, 0.5M KCl, pH 5.0	Hg	-0.69	sce	22
	McIlvaine buffer, 0.5M KCl, pH 7.0	Hg	-0.80	sce	22
1-Propoxy-2-nitrobutane	McIlvaine buffer, 0.5M KCl, pH 5.0	Hg	-0.65	sce	22
	McIlvaine buffer, 0.5M KCl, pH 7.0	Hg	-0.74	sce	22
1,5-Dinitropentane	50% Methanol, 50%-Benzene, 0.3M LiCl	Hg	-1.17	sce	17
	Methanol, 0.5M LiCl	Hg	-1.06	sce	17
	Isobutanol, 0.2M LiCl	Hg	-1.05	sce	17
	Ethylene glycol, 0.3M LiCl	Hg	-0.99	sce	17
2-Nitro-2,4,4-trimethylpentane	Dimethoxyethane, 0.1M (n-Bu)4NClO4	Hg	-1.67	sce	23
2,5-Dinitrohexane	Aq buffer, pH 8	Hg	-0.82	sce	1
3,5-Dinitroheptane	Aq buffer, pH 8	Hg	-0.84, -1.41	sce	1
Nitroguanidine	B-R buffer, pH 2.4	Hg	-0.73, -1.13	sce	24
	B-R buffer, pH 6.9	Hg	-1.24	sce	24
	B-R buffer, pH 11.8	Hg	-1.15	sce	24
	0.1N KCl, pH 3	Hg	-0.56	sce	25
Methyl nitrolic acid	0.2M LiOH	Hg	-0.6, -1.0	sce	26

(continued)

Aromatic Nitro Compounds

Compound	Solvent System	Working Electrode	$E_{1/2}$	Reference Electrode	Reference
Nitrobenzene	HOAc, 1M NH$_4$OAc	Hg	-0.675	Hg pool	8
	CH$_3$CN, 0.1M-(n-Propyl)$_4$NClO$_4$	Hg	-1.151, -1.9	sce	27
	CH$_3$CN, 0.1M-(n-Propyl)$_4$NClO$_4$	Hg	-1.147, -1.9	sce	28
	DMF, 0.1M (n-Propyl)$_4$-NClO$_4$	Hg	-1.082	sce	29
	DMF, 0.1M (Et)$_4$NClO$_4$	Hg	-1.18, -1.98 $E_{p/2}$	sce	30
	DMF, 0.1M (n-Bu)$_4$-NClO$_4$	Hg	-1.12	sce	2
	DMF, 0.2M NaNO$_3$	Hg	-1.01, -1.44	sce	31
	0.01M (Et)$_4$NBr, 66% Alcohol	Hg	-0.935	sce	32
	0.5M HCl, LiCl, 10% Ethanol	Hg	-1.50, -0.665	sce	33
	0.5M HCl, LiCl, 70% Ethanol	Hg	-2.05, -0.695	sce	33
	0.1M HCl, LiCl, pH 1	Hg	-0.185, -0.680	sce	33
	0.1M HCl, LiCl, 40% Ethanol, pH 1	Hg	-0.240, -0.740	sce	33
	0.1M HCl, LiCl, 90% Ethanol, pH 1	Hg	-0.255, -0.800	sce	33
	0.01M HCl, LiCl, 40% Ethanol, pH 2	Hg	-0.310, -0.775	sce	33

Nitrobenzene

0.01M HCl, LiCl, 80% Ethanol, pH 2	Hg	-0.385, -0.810	sce	33
0.22N H₂SO₄, 55% Alcohol	Hg	-0.270	sce	34
Aq buffer^b, 60%-Alcohol, pH 1.73	Hg	-0.326	sce	32
Aq buffer,^{b} 60%-Alcohol, pH 3.16	Hg	-0.430	sce	32
Citrate buffer, LiCl, 10% Ethanol, pH 2.75	Hg	-0.292, -0.798	sce	33
Citrate buffer, LiCl, 50% Ethanol, pH 2.75	Hg	-0.396, -0.859	sce	33
Citrate buffer, LiCl, 10% Ethanol, pH 3.9	Hg	-0.355, -0.905	sce	33
Citrate buffer, LiCl, 50% Ethanol, pH 3.9	Hg	-0.470, -0.952	sce	33
Citrate buffer, LiCl, 10% Ethanol, pH 5.3	Hg	-0.425	sce	33
Borate buffer, LiCl, 10% Ethanol, pH 7.18	Hg	-0.570	sce	33
Borate buffer, LiCl, 50% Ethanol, pH 7.18	Hg	--0.694	sce	33
Borate buffer, LiCl, 10% Ethanol, pH 10.0	Hg	-0.657	sce	33
Borate buffer, LiCl, 50% Ethanol, pH 10.0	Hg	-0.775	sce	33
0.01M NaOH, LiCl	Hg	-0.695	sce	33
0.01M NaOH, LiCl, 40% Ethanol	Hg	-0.785	sce	33
0.01M NaOH, LiCl, 90% Ethanol	Hg	-0.885	sce	33

(continued)

Compound	Solvent System	Working Electrode	$E_{1/2}$	Reference Electrode	Reference
	0.5$\underline{M}$ NaOH, LiCl, 10% Ethanol	Hg	-0.720	sce	33
	0.5$\underline{M}$ NaOH, LiCl, 50% Ethanol	Hg	-0.794	sce	33
	0.5$\underline{M}$ NaOH, LiCl, 70% Ethanol	Hg	-0.801	sce	33
	0.1$\underline{N}$ NaCl	Hg	-0.83	sce	35
	0.1$\underline{N}$ NaCl[c]	Hg	-0.79	sce	35
2-Nitrodiphenylamine	0.25$\underline{M}$ (Me)$_4$NCl, 75% Ethanol, 10% Acetone	Hg	-0.835	Hg pool	36

Substituted p-Nitrotriphenylamines $(R_1R_2R_3N)$

Compound	Solvent System	Working Electrode	$E_{1/2}$	Reference Electrode	Reference
$R_1=C_6H_4NO_2$ $R_2=C_6H_5$ $R_3=C_6H_5$	CH_3CN, 0.1$\underline{M}$ (Et)$_4$-NClO$_4$	Pt	-1.18 $E_{p/2}$	sce	37
$R_1=C_6H_4NO_2$ $R_2=C_6H_4OMe$ $R_3=C_6H_5$	CH_3CN, 0.1$\underline{M}$ (Et)$_4$-NClO$_4$	Pt	-1.21 $E_{p/2}$	sce	37
$R_1=C_6H_4NO_2$ $R_2=C_6H_4OMe$ $R_3=C_6H_4OMe$	CH_3CN, 0.1$\underline{M}$ (Et)$_4$-NClO$_4$	Pt	-1.24 $E_{p/2}$	sce	37
$R_1=C_6H_4NO_2$ $R_2=C_6H_4Me$ $R_3=C_6H_4Me$	CH_3CN, 0.1$\underline{M}$ (Et)$_4$-NClO$_4$	Pt	-1.22 $E_{p/2}$	sce	37
$R_1=C_6H_4NO_2$ $R_2=C_6H_4NO_2$ $R_3=C_6H_5$	CH_3CN, 0.1$\underline{M}$ (Et)$_4$-NClO$_4$	Pt	-1.02 $E_{p/2}$	sce	37
$R_1=C_6H_4NO_2OMe\underline{d}$ $R_2=C_6H_4NO_2$ $R_3=C_6H_5$	CH_3CN, 0.1$\underline{M}$ (Et)$_4$-NClO$_4$	Pt	-1.05 $E_{p/2}$	sce	37
1-Nitronaphthalene	HOAc, 1$\underline{M}$ NH$_4$OAc	Hg	-0.450	Hg pool	8

	Electrolyte / Solvent	Electrode	E (V)	Ref. electrode	Ref.
	KCl-HCl buffer, 80% Ethanol, pH 2.1	Hg	-0.30, -0.65	sce	38
2-Nitronaphthalene	DMF, 0.2M NaNO₃	Hg	-0.97, -1.41	sce	39
	KCl-HCl buffer, 80% Ethanol, pH 2.1	Hg	-0.30, -0.72	sce	38
γ-Nitrotropolone	DMF, 0.2M NaNO₃	Hg	-0.98, -1.42	sce	39
5-Nitrotetralin	HOAc, 1M NH₄OAc	Hg	-0.450	Hg pool	8
	KCl-HCl buffer, 80% Ethanol, pH 2.1	Hg	-0.36, -0.56	sce	40
6-Nitrotetralin	KCl-HCl buffer, 80% Ethanol, pH 2.1	Hg	-0.35, -0.55	sce	40
2-Nitrofluorenone	HCl buffer, 50%-Acetone, pH 1.4	Hg	-0.20, -0.68, -0.98	sce	41
	Citrate-Phosphate-buffer, 50% Acetone, pH 6.5	Hg	-0.54, -1.03	sce	41
	Phosphate buffer, pH 11.0, 50% Acetone	Hg	-0.85, -1.21	sce	41

Mono-Substituted Nitrobenzene

	Electrolyte / Solvent	Electrode	E (V)	Ref. electrode	Ref.
o-Nitrotoluene	HOAc, 1M NH₄OAc	Hg	-0.715	Hg pool	8
	CH₃CN, 0.1M (n-Propyl)₄NClO₄	Hg	-1.263, -2.1	sce	42
	CH₃CN, 0.1M (Et)₄-NClO₄	Hg	-1.29 E_p	sce	43
	DMF, 0.1M (n-Bu)₄N	Hg	-1.28	sce	20
	0.01M (Et)₄NBr, 66% Alcohol	Hg	-1.005	sce	32
	0.05M (Me)₄N1, 50% Methanol	Hg	-0.80	Hg pool	44

(continued)

931

Compound	Solvent System	Working Electrode	$E_{1/2}$	Reference Electrode	Reference
	Aq buffer[b], 66%-Alcohol, pH 1.73	Hg	-0.358	sce	32
	Aq buffer[b], 66%-Alcohol, pH 3.16	Hg	-0.525	sce	32
	$0.22N$ H_2SO_4, 55% Alcohol	Hg	-0.300	sce	34
m-Nitrotoluene	HOAc, 1M NH_4OAc	Hg	-0.635	Hg pool	8
	CH_3CN, $\underline{0.1M}$-(n-Propyl)$_4NClO_4$	Hg	-1.180, -1.9	sce	27
	CH_3CN, $\underline{0.1M}$ (Et)$_4$-$NClO_4$	Hg	-1.20 E_p	sce	43
	$\underline{0.01M}$ (Et)$_4NBr$, 66% Alcohol	Hg	-0.936	sce	32
	$0.22N$ H_2SO_4, 55% Alcohol	Hg	-0.245	sce	34
	Aq buffer[b], 60%-Alcohol, pH 1.73	Hg	-0.305	sce	32
	Aq buffer[b], 60%-Alcohol, pH 3.16	Hg	-0.416	sce	32
	$\underline{0.05M}$ (Me)$_4NI$, 50% Methanol	Hg	-0.67	Hg pool	44
p-Nitrotoluene	HOAc, 1M NH_4OAc	Hg	-0.665	Hg pool	8
	CH_3CN, $\underline{0.1M}$-(n-Propyl)$_4NClO_4$	Hg	-1.180, -1.9	sce	42
	CH_3CN, $\underline{0.1M}$ (Et)$_4$-$NClO_4$	Hg	-1.23 E_p	sce	43
	DMF, 0.1M (n-Bu)$_4NI$	Hg	-1.44	sce	20
	$\underline{0.01M}$ (Et)$_4\underline{N}Br$, 66% Alcohol	Hg	-1.035	sce	32

Compound	Conditions	Electrode	E	Reference	Ref.
	0.22N H_2SO_4, 55%-Alcohol	Hg	-0.270	sce	34
	Aq buffer^b, 60%-Alcohol, pH 1.73	Hg	-0.336	sce	32
	Aq buffer^b, 60%-Alcohol, pH 3.16	Hg	-0.442	sce	32
	0.05M (Me)$_4$NI, 50% Methanol	Hg	-0.82	Hg pool	44
o-t-Butylnitrobenzene	DMF, 0.1M (n-Bu)$_4$NI	Hg	-1.30	sce	20
2-t-Butylnitrobenzene	CH_3CN, 0.1M-(n-Propyl)$_4$NClO$_4$	Hg	-1.355, -1.98	sce	42
o-Nitroanisole	0.22N H_2SO_4, 55%-Alcohol	Hg	-0.285	sce	45
	NaCl, HCl buffer, 10% Ethanol, pH 1.0	Hg	-0.210	sce	46
	McIlvaine buffer, 10% Ethanol, pH 4.0	Hg	-0.460	sce	46
	McIlvaine buffer, 10% Ethanol, pH 8.0	Hg	-0.675	sce	46
	Phosphate buffer, 10% Ethanol, pH 12.0	Hg	-0.795	sce	46
m-Nitroanisole	0.22N H_2SO_4, 55% Alcohol	Hg	-0.225	sce	45
	NaCl, HCl buffer, 10% Ethanol, pH 1.0	Hg	-1.90, -0.590	sce	46
	McIlvaine buffer, 10% Ethanol, pH 4.0	Hg	-0.405, -0.925	sce	46
	McIlvaine buffer, 10% Ethanol, pH 8.0	Hg	-0.630	sce	46
	Phosphate buffer, 10% Ethanol, pH 12.0	Hg	-0.755	sce	46

(continued)

Compound	Solvent System	Working Electrode	$E_{1/2}$	Reference Electrode	Reference
	CH₃CN, 0.1M-(n-Propyl)₄NClO₄	Hg	-1.147, -1.8	sce	27
	DMF, 0.1M-(n-Propyl)₄NClO₄	Hg	-1.150	sce	29
p-Nitroanisole	NaCl-HCl buffer, 10% Ethanol, pH 1.0	Hg	-0.260, -0.600	sce	46
	McIlvaine buffer, 10% Ethanol, pH 4.0	Hg	-0.460, -0.840	sce	46
	McIlvaine buffer, 10% Ethanol, pH 8.0	Hg	-0.715	sce	46
	Phosphate buffer, 10% Ethanol, pH 12.0	Hg	-0.855	sce	46
	CH₃CN, 0.1M (Et)₄-NClO₄, 0.01M o-Phthalic Acid	Pt	-0.82 $E_{p/2}$	sce	47
	CH₃CN, 0.1M-(n-Propyl)₄NClO₄	Hg	-1.250, -2.1	sce	27
o-Fluoronitrobenzene	0.22N H₂SO₄, 55% Alcohol	Hg	-0.320	sce	45
m-Fluoronitrobenzene	0.22N H₂SO₄, 55% Alcohol	Hg	-0.235	sce	34
	0.22N H₂SO₄, 55% Alcohol	Hg	-0.215	sce	34
p-Fluoronitrobenzene	HOAc, 1M NH₄OAc	Hg	-0.635	Hg pool	8
	0.22N H₂SO₄, 55% Alcohol	Hg	-0.275	sce	34
o-Chloronitrobenzene	HOAc, 1M NH₄OAc	Hg	-0.705	Hg pool	8
	0.22N H₂SO₄, 55% Alcohol	Hg	-0.200	sce	34

Compound	Electrolyte	Electrode	E	Reference electrode	Ref.
	HOAc, 1M NH$_4$OAc	Hg	−0.645	Hg pool	8
	DMF, 0.1$\underline{M}$ (Et)$_4$-NClO$_4$	Hg	−1.14, −1.69 $E_{p/2}$	sce	30
	0.01M (Et)$_4$NBr, 66% Alcohol	Hg	−0.866	sce	31
	0.22N H$_2$SO$_4$, 55% Alcohol	Hg	−0.175	sce	34
m-Chloronitrobenzene	HOAc, 1M NH$_4$OAc	Hg	−0.580	Hg pool	8
	DMF, 0.1$\underline{M}$ (Et)$_4$NClO$_4$	Hg	−1.08, −1.95 $E_{p/2}$	sce	30
	0.01M (Et)$_4$NBr, 66% Alcohol	Hg	−0.830	sce	32
	0.22N H$_2$SO$_4$, 55% Alcohol	Hg	−0.205	sce	34
p-Chloronitrobenzene	HOAc, 1M NH$_4$OAc	Hg	−0.615	Hg pool	8
	CH$_3$CN, 0.1$\underline{M}$-(n-Propyl)$_4$NClO$_4$	Hg	−1.063, −1.8	sce	27
	DMF, 0.1$\underline{M}$ (Et)$_4$NClO$_4$	Hg	−1.08, −1.72 $E_{p/2}$	sce	30
	DMF, 0.1$\underline{M}$ (Et)$_4$NClO$_4$, 0.01$\underline{M}$ o-Phthalic acid	Hg	−0.88 $E_{1/4}$	sce	47
	PC, 0.1$\underline{M}$ (Et)$_4$NClO$_4$, 0.0024 $\underline{M}$ o-Phthalic acid	Hg	−0.515 $E_{p/2}$	sce	47
	CH$_3$CN, 0.1$\underline{M}$ (Et)$_4$-NClO$_4$, 0.01$\underline{M}$ o-Phthalic acid	Pt	−0.69 $E_{p/2}$	sce	47
	CH$_3$CN, 0.1$\underline{M}$ (Et)$_4$-NBr, 0.008$\underline{M}$ o-Phthalic acid	Hg	−0.76 $E_{p/2}$	sce	47

(continued)

Compound	Solvent System	Working Electrode	$E_{1/2}$	Reference Electrode	Reference
o-Bromonitrobenzene	DMSO, 0.1M (Et)4NClO4, 0.01M o-Phthalic acid	Hg	-0.81 $E_{p/2}$	sce	47
	0.01M (Et)4NBr, 66% Alcohol	Hg	-0.930	sce	32
	0.22N H2SO4, 55%- Alcohol	Hg	-0.180	sce	34
	HOAc, 1M NH4OAc	Hg	-0.620	Hg pool	8
	DMF, 0.1M (Et)4NClO4	Hg	-1.15, -1.77 $E_{p/2}$	sce	30
m-Bromonitrobenzene	0.01M (Et)4NBr, 66% Alcohol	Hg	-0.860 $E_{p/2}$	sce	32
	Aq bufferb, 60%-Alcohol, pH 1.73	Hg	-0.263	sce	32
	Aq buffer, 60%- Alcohol, pH 3.16	Hg	-0.396	sce	32
	0.22N H2SO4, 55% Alcohol	Hg	-0.160	sce	34
	HOAc, 1M NH4OAc	Hg	-0.550	Hg pool	8
	DMF, 0.1M (Et)4NClO4	Hg	-1.15, -1.73 $E_{p/2}$	sce	30
p-Bromonitrobenzene	0.01M (Et)4NBr, 66% Alcohol	Hg	-0.816	sce	32
	Aq bufferb, 60%- Alcohol, pH 1.73	Hg	-0.208	sce	32
	Aq bufferb, 60%- Alcohol, pH 3.16	Hg	-0.308	sce	32
	0.22N H2SO4, 55%- Alcohol	Hg	-0.170	sce	34

Compound	Supporting electrolyte/solvent	Electrode	E	Reference electrode	Ref.
	HOAc, 1M NH4OAc	Hg	-0.575	Hg pool	8
	CH3CN, 0.1M (n-Propyl)4NClO4	Hg	-1.050, -1.49	sce	27
	DMF, 0.1M (Et)4-NClO4	Hg	-1.15, -1.75 Ep/2	sce	30
	DMF, 0.1M (Et)4-NClO4	Hg	-1.00, -1.94, -2.16 Ep	sce	48
p-Bromonitrobenzene	0.01M (Et)4NBr, 66% Alcohol	Hg	-0.854	sce	32
	Aq bufferb, 60%-Alcohol, pH 1.73	Hg	-0.230	sce	32
	Aq bufferb, 60%-Alcohol, pH 3.16	Hg	-0.342	sce	32
o-Iodonitrobenzene	0.22N H2SO4, 55% Alcohol	Hg	-0.160	sce	34
	0.01M (Et)4NBr, 66% Alcohol	Hg	-0.816	sce	32
	Aq bufferb, 60%-Alcohol, pH 1.73	Hg	-0.204	sce	32
	Aq bufferb, 60%-Alcohol, pH 3.16	Hg	-0.306	sce	32
m-Iodonitrobenzene	0.22N H2SO4, 55% Alcohol	Hg	-0.120	sce	34
	HOAc, 1M NH4OAc	Hg	-0.515	Hg pool	8
	DMF, 0.1M (Et)4NClO4	Hg	-1.10, -1.25, -1.75 Ep/2	sce	30
	0.01M (Et)4NBr, 66% Alcohol	Hg	-0.757	sce	32
	Aq bufferb, 60%-Alcohol, pH 1.73	Hg	-0.176	sce	32

(continued)

937

Compound	Solvent System	Working Electrode	$E_{1/2}$	Reference Electrode	Reference
p-Iodonitrobenzene	Aq buffer[b], 60%-Alcohol, pH 3.16	Hg	-0.276	sce	32
	0.22N H_2SO_4, 55% Alcohol	Hg	-0.130	sce	34
	HOAc, 1M NH_4OAc	Hg	-0.535	Hg pool	8
	CH_3CN, 0.1M (n-Propyl)$_4$NClO$_4$	Hg	-1.050, -1.8	sce	27
	DMF, 0.1M (Et)$_4$NClO$_4$	Hg	-1.09, -1.23, -1.77 Ep/2	sce	30
	0.01M (Et)$_4$NBr, 66% Alcohol	Hg	-0.780	sce	32
	Aq buffer[b], 60%-Alcohol, pH 1.73	Hg	-0.194	sce	32
	Aq buffer, 60%-Alcohol, pH 3.16	Hg	-0.294	sce	32
o-Nitrobenzonitrile	0.22N H_2SO_4, 55% Alcohol	Hg	-0.115	sce	34
m-Nitrobenzonitrile	0.22N H_2SO_4, 55% Alcohol	Hg	-0.155	sce	34
	CH_3CN, 0.1M (n-Propyl)$_4$NClO$_4$	Hg	-0.938, -1.5	sce	27
p-Nitrobenzonitrile	0.22N H_2SO_4, 55% Alcohol	Hg	-0.100	sce	34
	CH_3CN, 0.1M (n-Propyl)$_4$NClO$_4$	Hg	-0.875, -1.48	sce	27
	0.22N H_2SO_4, 55% Alcohol	Hg	-0.200	sce	45
o-Nitrophenol	HOAc, 1M NH_4OAc	Hg	-0.600	Hg pool	8
	DMF, 0.1M (Et)$_4$NClO$_4$	Hg	-0.85,	sce	49

Compound	Medium	Electrode	$E_{1/2}$	Ref.	Ref. No.
o-Nitrophenol	0.01M (Et)4NBr, 66% Alcohol	Hg	-0.990	sce	32
	Sorensen buffer, 10% Ethanol, pH 1.0	Hg	-0.185, -0.635	sce	46
	Aq buffer[b], 60%- Alcohol, pH 1.73	Hg	-0.246	sce	32
	Aq buffer[b], 60%- Alcohol, pH 3.16	Hg	-0.362	sce	32
	McIlvaine buffer, 10% Ethanol, pH 4.0	Hg	-0.355, -0.940	sce	46
	McIlvaine buffer, 10% Ethanol, pH 8.0	Hg	-0.640	sce	46
	Phosphate buffer, 10% Ethanol, pH 12.0	Hg	-0.900	sce	46
	0.22N H2SO4, 55% Alcohol	Hg	-0.250	sce	45
	HOAc, 1M NH4OAc	Hg	-0.605	Hg pool	8
m-Nitrophenol	0.01M (Et)4NBr, 66% Alcohol	Hg	-0.960	sce	32
	Aq buffer[b], 60%- Alcohol, pH 1.73	Hg	-0.310	sce	32
	Aq buffer[b], 60%- Alcohol, pH 3.16	Hg	-0.418	sce	32
	Sorensen buffer, 10% Ethanol, pH 1.0	Hg	-0.190, -0.690	sce	46
	McIlvaine buffer, 10% Ethanol, pH 4.0	Hg	-0.365	sce	46
	McIlvaine buffer, 10% Ethanol, pH 8.0	Hg	-0.630	sce	46
	Phosphate buffer, 10% Ethanol, pH 12.0	Hg	-0.940, -1.145	sce	46

(continued)

Compound	Solvent System	Working Electrode	$E_{1/2}$	Reference Electrode	Refer-ence
p-Nitrophenol	0.22N H_2SO_4, 55% Alcohol	Hg	-0.355	sce	45
	HOAc, 1M NH4OAc	Hg	-0.670	Hg pool	8
	0.01M (Et)4NBr, 66% Alcohol	Hg	-1.350	sce	32
	Aq buffer^b, 60%- Alcohol, pH 1.73	Hg	-0.408	sce	32
	Aq buffer^b, 60%- Alcohol, pH 3.16	Hg	-0.566	sce	32
	Sorensen buffer, 10% Ethanol, pH 1.0	Hg	-0.260, -0.755	sce	46
	McIlvaine buffer, 10% Ethanol, pH 4.0	Hg	-0.470, -0.975	sce	46
	McIlvaine buffer, 10% Ethanol, pH 8.0	Hg	-0.810, -1.260	sce	46
	Phosphate buffer, 10% Ethanol, pH 12.0	Hg	-0.975, -1.565	sce	46
o-Nitroaniline	0.22N H_2SO_4, 55% Alcohol	Hg	-0.335	sce	34
	HOAc, 1M NH4OAc	Hg	-0.735	Hg pool	8
	0.01M (Et)4NBr, 66% Alcohol	Hg	-1.030	sce	32
	Aq buffer^b, 60%- Alcohol, pH 1.73	Hg	-0.372	sce	32
	Aq buffer^b, 60%- Alcohol, pH 3.16	Hg	-0.488	sce	32
m-Nitroaniline	0.22N H_2SO_4, 55% Alcohol	Hg	-0.270	sce	34
	HOAc, 1M NH4OAc	Hg	-0.630	Hg pool	8

Compound	Electrolyte / Solvent	Electrode	E	Reference	Ref.
p-Nitroaniline	CH₃CN, 0.1M (n-Propyl)₄NClO₄	Hg	-1.208, -1.90	sce	27
	DMF, 0.1M (n-Propyl)₄NClO₄	Hg	-1.150	sce	29
	0.01M (Et)₄NBr, 66% Alcohol	Hg	-0.935	sce	30
	Aq buffer,b 60%-Alcohol, pH 1.73	Hg	-0.320	sce	30
	Aq buffer,b 60%-Alcohol, pH 3.16	Hg	-0.422	sce	30
	0.22N H₂SO₄, 55% Alcohol	Hg	-0.360	sce	45
	HOAc, 1M NH₄OAc	Hg	-0.740	Hg pool	8
	CH₃CN, 0.1M (n-Propyl)₄NClO₄	Hg	-1.358, -2.0	sce	27
	CH₃CN, 0.1M (Et)₄NClO₄	Hg	-0.76 $E_{p/2}$	sce	47
	DMF, 0.1M (n-Propyl)₄NClO₄	Hg	-1.335	sce	29
	DMF, 0.1M (n-Bu)₄NI	Hg	-1.37	sce	20
	0.01M (Et)₄NBr, 66% Alcohol	Hg	-1.128	sce	32
	Aq buffer,b 60%-Alcohol, pH 1.73	Hg	-0.438	sce	32
	Aq buffer,b 60%-Alcohol, pH 3.16	Hg	-0.554	sce	32
o-Nitrobenzaldehyde	OAc buffer, pH 1.31	Hg	-0.14, -0.90	sce	50
	OAc buffer, pH 4.01	Hg	-0.34, -1.26	sce	50
	Borate-Phosphate-buffer, pH 8.26	Hg	-0.42, -1.40	sce	50

(continued)

Compound	Solvent System	Working Electrode	$E_{1/2}$	Reference Electrode	Reference
	Borate-Phosphate buffer, pH 11.95	Hg	-0.49, -1.61	sce	50
	0.01M (Et)4NBr, 66% Alcohol	Hg	-0.690	sce	32
	Aq buffer, b 60%- Alcohol, pH 1.73	Hg	-0.228	sce	32
	Aq buffer, b 60%- Alcohol, pH 3.16	Hg	-0.346	sce	32
	P-W buffer, pH 9.11	Hg	-e, -1.35, -1.50	sce	51
	P-W buffer, pH 11.94	Hg	-e, -1.42, -1.57	sce	51
m-Nitrobenzaldehyde	CH_3CN, 0.1M (n-Propyl)4NClO4	Hg	-1.016, -1.6, -1.95	sce	27
	OAc buffer, pH 1.31	Hg	-0.21, -1.01	sce	50
	OAc buffer, pH 4.01	Hg	-0.50, -1.41	sce	50
	Borate-Phosphate buffer, pH 8.26	Hg	-0.74, -1.53	sce	50
	Borate-Phosphate buffer, pH 11.95	Hg	-0.78, -1.56	sce	50
	0.01M (Et)4NBr, 66% Alcohol	Hg	-0.822	sce	32
	Aq buffer, b 60%- Alcohol, pH 1.73	Hg	-0.257	sce	32
	Aq buffer, b 60%- Alcohol, pH 3.16	Hg	-0.380	sce	32
	P-W buffer, pH 6.84	Hg	-1.32	sce	51
	P-W buffer, pH 9.11	Hg	-1.41	sce	51
	P-W buffer, pH 11.94	H	-1.48	sce	51

p-Nitrobenzaldehyde

Medium	Electrode	E	Ref. electrode	Ref.
CH_3CN, 0.1M (n-Propyl)$_4$NClO$_4$	Hg	-0.863, -1.34	sce	27
CH_3CN, 0.1M (Et)$_4$-NClO$_4$	Hg	-0.67 $E_{p/2}$	sce	47
OAc buffer, pH 1.62	Hg	-0.19, -1.06	sce	50
OAc buffer, pH 4.68	Hg	-0.44, -1.35	sce	50
Borate-Phosphate buffer, pH 8.67	Hg	-0.47, -1.29, -1.63	sce	50
Borate-Phosphate buffer, pH 11.74	Hg	-0.56, -1.37, -1.68	sce	50
0.01M (Et)$_4$NBr, 66% Alcohol	Hg	-0.632	sce	32
Aq buffer,b 60%-Alcohol, pH 1.73	Hg	-0.174	sce	32
Aq buffer,b 60%-Alcohol, pH 3.16	Hg	-0.322	sce	32
P-W buffer, pH 6.84	Hg	-e, -1.27, -1.40	sce	51
P-W buffer, pH 9.11	Hg	-e, -1.31, -1.60	sce	51
P-W buffer, pH 11.94	Hg	-e, -1.40, -1.68	sce	51

o-Nitrobenzoic Acid

Medium	Electrode	E	Ref. electrode	Ref.
0.22N H$_2$SO$_4$, 55% Alcohol	Hg	-0.230	sce	52
0.01M (Et)$_4$NBr, 66% Alcohol	Hg	-0.526, -0.968	sce	32
Aq buffer,b 60%-Alcohol, pH 1.73	Hg	-0.304	sce	32
Aq buffer,b 60%-Alcohol, pH 3.16	Hg	-0.420	sce	32
P-W buffer, pH 2.0	Hg	-0.26	sce	51

(continued)

943

Compound	Solvent System	Working Electrode	$E_{1/2}$	Reference Electrode	Reference
o-Nitrobenzoic Acid	P-W buffer, pH 3.0	Hg	-0.35	sce	51
	P-W buffer, pH 4.0	Hg	-0.41	sce	51
	P-W buffer, pH 10.0	Hg	-0.82	sce	51
	NaCl-HCl buffer, pH 2.0	Hg	-0.225, -0.725	sce	46
	McIlvaine buffer, pH 6.0	Hg	-0.530	sce	46
	Glycine buffer, pH 10.0	Hg	-0.855, -1.045	sce	46
	Phosphate buffer, pH 12.0	Hg	-0.845, -1.075	sce	46
m-Nitrobenzoic Acid	0.22N H_2SO_4, 55% Alcohol	Hg	-0.190	sce	34
	HOAc, 1M NH_4OAc	Hg	-0.575	Hg pool	8
	0.01M (Et)4NBr, 66% Alcohol	Hg	-0.440, -1.026	sce	32
	Aq buffer,b 60%-Alcohol, pH 1.73	Hg	-0.252	sce	32
	Aq buffer,b 60%-Alcohol, pH 3.16	Hg	-0.362	sce	32
	P-W buffer, pH 2.0	Hg	-0.22	sce	51
	P-W buffer, pH 3.0	Hg	-0.30	sce	51
	P-W buffer, pH 4.0	Hg	-0.35	sce	51
	P-W buffer, pH 10.0	Hg	-0.72	sce	51
	NaCl-HCl buffer, pH 2.0	Hg	-2.00, -0.695	sce	46
	McIlvaine buffer, pH 6.0	Hg	-0.515	sce	46
	pH 6.0	Hg	0.770	sce	46

Compound	Medium	Electrode	E	Reference	Ref.
m-Nitrobenzoic Acid	Phosphate buffer, pH 12.0	Hg	-0.760, -1.070	sce	46
p-Nitrobenzoic Acid	0.22N H$_2$SO$_4$, 55% Alcohol	Hg	-0.100	sce	34
	HOAc, 1M NH$_4$OAc	Hg	-0.490	Hg pool	8
	0.01M (Et)$_4$NBr, 66% Alcohol	Hg	-0.390, -0.972	sce	32
	Aq buffer,b 60%- Alcohol, pH 1.73	Hg	-0.186	sce	32
	Aq buffer,b 60%- Alcohol, pH 3.16	Hg	-0.304	sce	32
	P-W buffer, pH 2.0	Hg	-0.16	sce	51
	P-W buffer, pH 3.0	Hg	-0.23	sce	51
	P-W buffer, pH 4.0	Hg	-0.30	sce	51
	P-W buffer, pH 10.0	Hg	-0.65	sce	51
	NaCl-HCl buffer, pH 2.0	Hg	-0.170, -0.740	sce	46
	McIlvaine buffer, pH 6.0	Hg	-0.470	sce	46
	Glycine buffer, pH 10.0	Hg	-0.705, -0.990	sce	46
	Phosphate buffer, pH 12.0	Hg	-0.705, -1.010	sce	46
Methyl o-nitrobenzoate	0.22N H$_2$SO$_4$, 55% Alcohol	Hg	-0.210	sce	52
	NaCl-HCl buffer, 10% Ethanol, pH 1.0	Hg	-0.180, -0.530	sce	46
	McIlvaine buffer, 10% Ethanol, pH 4.0	Hg	-0.355	sce	46
	McIlvaine buffer, 10% Ethanol, pH 8.0	Hg	-0.590	sce	46

(continued)

Compound	Solvent System	Working Electrode	$E_{1/2}$	Reference Electrode	Reference
	Phosphate buffer, 10% Ethanol, pH 12.0	Hg	-0.675	sce	46
Methyl m-nitrobenzoate	0.22N H_2SO_4, 55% Alcohol	Hg	-0.180	sce	52
	NaCl-HCl buffer, 10% Ethanol, pH 1.0	Hg	-0.170, -0.510	sce	46
	McIlvaine buffer, 10% Ethanol, pH 4.0	Hg	-0.340	sce	46
	McIlvaine buffer, 10% Ethanol, pH 8.0	Hg	-0.570	sce	46
	Phosphate buffer, 10% Ethanol, pH 12.0	Hg	-0.730, -1.070	sce	46
	CH_3CN, 0.1M (n-Propyl)$_4$NClO$_4$	Hg	-1.044, -1.7, -2.3	sce	27
Methyl p-nitrobenzoate	CH_3CN, 0.1M (n-Propyl)$_4$NClO$_4$	Hg	-0.947, -1.4	sce	27
	NaCl-HCl buffer, 10% Ethanol, pH 1.0	Hg	-0.125, -0.555	sce	46
	McIlvaine buffer, 10% Ethanol, pH 4.0	Hg	-0.300	sce	46
	McIlvaine buffer, 10% Ethanol, pH 8.0	Hg	-0.510	sce	46
	Phosphate buffer, 10% Ethanol, pH 12.0	Hg	-0.700, -1.020	sce	46
Ethyl o-nitrobenziate	0.01M (Et)$_4$NBr, 66% Alcohol	Hg	-0.826	sce	32
	Aq buffer,b 60%-Alcohol, pH 1.73	Hg	-0.282	sce	32

Compound	Conditions		Potential (V)		Ref.
	Aq buffer, b 60%-Alcohol,	Hg	-0.400	sce	32
	0.22N H2SO4, 55% Alcohol	Hg	-0.230	sce	51
Ethyl m-nitrobenzoate	0.01M (Et)4NBr, 66% Alcohol	Hg	-0.850	sce	32
	Aq buffer, b 60%-Alcohol, pH 1.73	Hg	-0.214	sce	32
	Aq buffer, b 60%-Alcohol, pH 3.16	Hg	-0.330	sce	32
	0.22N H2SO4, 55% Alcohol	Hg	-0.160	sce	52
Ethyl p-nitrobenziate	0.01M (Et)4NBr, 66% Alcohol	Hg	-0.770	sce	32
	Aq buffer, b 60%-Alcohol, pH 1.73	Hg	-0.166	sce	32
	Aq buffer, b 60%-Alcohol, pH 3.16	Hg	-0.282	sce	32
	0.22N H2SO4, 55% Alcohol	Hg	-0.115	sce	52
o-Nitrobenzenethiocyan-ate	B-R buffer, 50%-Alcohol, pH 3.10	Hg	-0.19, -1.07	sce	53
	B-R buffer, 50%-Alcohol, pH 4.85	Hg	-0.34, -1.27	sce	53
	B-R buffer, 50%-Alcohol, pH 8.70	Hg	-0.57, -0.89, -1.26	sce	53
m-Nitrobenzenethiocyan-ate	B-R buffer, 50%-Alcohol, pH 3.10	Hg	-0.24, -0.83, -1.19	sce	53
	B-R buffer, 50%-Alcohol, pH 4.85	Hg	-0.42, -1.44	sce	53

(continued)

947

Compound	Solvent System	Working Electrode	E$_{1/2}$	Reference Electrode	Reference
p-Nitrobenzenethiocyanate	B-R buffer, 50%-Alcohol, pH 8.70	Hg	-0.62, -0.80, -1.39	sce	53
	B-R buffer, 50%-Alcohol, pH 3.10	Hg	-0.17, -0.79, -1.20	sce	53
	B-R buffer, 50%-Alcohol, pH 4.85	Hg	-0.39, -1.39	sce	53
	B-R buffer, 50%-Alcohol, pH 8.70	Hg	-0.60, -0.82, -1.39	sce	53
	CH$_3$CN, 0.1M (n-Propyl)4NClO4	Hg	-0.96, -1.3	sce	27
o-Trifluoromethylnitro-benzene	DMF, 0.1M (n-Propyl)4NClO4	Hg	-0.72, -1.42	sce	54
m-Trifluoromethylnitro-benzene	DMF, 0.1M (n-Propyl)4NClO4	Hg	-0.70, -1.50	sce	54
o-Nitrobiphenyl	0.01M (Et)4NBr, 66% Alcohol	Hg	-0.928	sce	32
	Aq buffer,b 60%-Alcohol, pH 1.73	Hg	-0.290	sce	32
	Aq buffer,b 60%-Alcohol, pH 3.16	Hg	-0.412	sce	32
	DMF, 0.1M (n-Bu)4-NClO4	Hg	-1.16	sce	2
	DMF, 0.1M (n-Propyl)4NClO4	Hg	-1.14, -1.78	sce	55
m-Nitrobiphenyl	0.01M (Et)4NBr, 66% Alcohol	Hg	-0.844	sce	32
	Aq buffer,b 60%-Alcohol, pH 1.73	Hg	-0.228	sce	32

Compound	Conditions	Electrode	E	Ref. electrode	Ref.
p-Nitrobiphenyl	Aq buffer,[b] 60%-Alcohol, pH 3.16 0.01M (Et)4NBr, 66% Alcohol	Hg	-0.348	sce	32
	Aq buffer,[b] 60%-Alcohol, pH 1.73	Hg	-0.854	sce	32
	Aq buffer,[b] 60%-Alcohol, pH 3.16	Hg	-0.228	sce	32
	CH3CN, 0.1M (n-Propyl)4NClO4	Hg	-0.342	sce	32
	DMF, 0.1M (n-Bu)4NClO4	Hg	-1.103, -1.5	sce	27
	Aq buffer, 50%-Acetone, pH 1.4	Hg	-1.03	sce	2
	Aq buffer, 50%-Acetone, pH 6.8	Hg	-0.20, -0.74	sce	41
	Aq buffer, 50%-Acetone, pH 11.2	Hg	-0.63	sce	41
	DMF, 0.1M (n-Bu)4NClO4	Hg	-0.79, -1.00	sce	41
2-Nitroterphenyl	DMF, 0.1M (n-Bu)4NClO4	Hg	-1.30	sce	2
3-Nitroterphenyl	DMF, 0.1M (n-Bu)4NClO4	Hg	-0.98	sce	2
4-Nitroterphenyl	DMF, 0.1M (n-Bu)4NClO4	Hg	-1.14	sce	2
o-Nitroacetanilide	Ethanol, HClO4, LiCl, pK=1	Hg	-0.12	Hg, Hg2Cl2/- sat'd LiCl, Ethanol[f]	56
	Ethanol, Chloro- acetic acid, LiCl, pK=6	Hg	-0.53	Hg, Hg2Cl2/- sat'd LiCl, Ethanol[f]	56

(continued)

Compound	Solvent System	Working Electrode	$E_{1/2}$	Reference Electrode	Reference
	Ethanol, LiOEt, LiCl, pK=17	Hg	-1.05	Hg, Hg2Cl2/- sat'd LiCl, Ethanol[f]	56
m-Nitroacetanilide	Ethanol, HClO4, LiCl, pK=1	Hg	-0.15	Hg, Hg2Cl2/- sat'd LiCl, Ethanol[f]	56
	Ethanol, Chloro-acetic acid, LiCl, pK=6	Hg	-0.62	Hg, Hg2Cl2/- sat'd LiCl, Ethanol[f]	56
	Ethanol, LiOEt, LiCl, pK=17	Hg	-0.91	Hg, Hg2Cl2/- sat'd LiCl, Ethanol[f]	56
p-Nitroacetanilide	HOAc, 1M NH4OAc	Hg	-0.580	Hg pool	8
	Ethanol, HClO4, LiCl, pK=1	Hg	-0.18	Hg, Hg2Cl2/- sat'd LiCl, Ethanol[f]	56
	Ethanol, Chloro-acetic acid, LiCl, pK=6	Hg	-0.66	Hg, Hg2Cl2/- sat'd LiCl, Ethanol[f]	56
	Ethanol, LiOEt, LiCl, pK=17	Hg	-1.01	Hg, Hg2Cl2/- sat'd LiCl, Ethanol[f]	56
Phthalic acid[9]	HOAc, 1M NH4OAc	Hg	-0.625	Hg pool	8
3-Nitrophthalic acid	P-W buffer, pH 2.0	Hg	-1.23	sce	56
4-Nitrophthalic acid	P-W buffer, pH 2.0	Hg	-0.27, -1.19	sce	56
4-Nitrophthalic acid	P-W buffer, pH 2.0	Hg	-0.14, -1.18	sce	56
o-Nitrobenzyl chloride	CH3CN, 0.1M (Et)4-NClO4	Hg	-1.00, -1.26, -1.36 E_p	sce	43

m-Nitrobenzyl chloride	CH$_3$CN, 0.1M (Et)$_4$NClO$_4$	Hg	-1.13, --,^{h}- E_p	sce	43
p-Nitrobenzyl chloride	CH$_3$CN, 0.1M (Et)$_4$NClO$_4$	Hg	-0.97, -1.22 E_{ph}	sce	43
o-Nitrobenzyl bromide	CH$_3$CN, 0.1M (Et)$_4$NClO$_4$	Hg	-0.91, -^{h}, E_p	sce	43
m-Nitrobenzyl bromide	CH$_3$CN, 0.1M (Et)$_4$NClO$_4$	Hg	-1.08, --,^{h}- E_p	sce	43
p-Nitrobenzyl bromide	CH$_3$CN, 0.1M (Et)$_4$NClO$_4$	Hg	-0.86, --,^{h}- E_p	sce	43
o-Nitrobenzyl alcohol	P-W buffer, pH 2.0	Hg	-0.28	sce	51
m-Nitrobenzyl alcohol	P-W buffer, pH 2.0	Hg	-0.33	sce	51
m-Nitroacetophenone	P-W buffer, pH 9.11	Hg	-1.54	sce	51
p-Nitroacetophenone	P-W buffer, pH 9.11	Hg	-1.43, -1.73	sce	51
m-Dimethylaminonitro-benzene	HOAc, 1M NH$_4$OAc	Hg	-0.605	Hg pool	8
p-Dimethylaminonitro-benzene	HOAc, 1M NH$_4$OAc	Hg	-0.720	Hg pool	8
m-Nitrobenzene sul-phonic acid	HOAc, 1M NH$_4$OAc	Hg	-0.595	Hg pool	8
p-Nitrobenzene sul-phonic acid	HOAc, 1M NH$_4$OAc	Hg	-0.570	Hg pool	8

Di-Substituted Nitrobenzene

(NO$_2$-benzene ring with substituents R$_1$, R$_2$)

R$_1$=2-CH$_3$ R$_2$=2-CH$_3$	0.22N H$_2$SO$_4$, 55% Alcohol	Hg	-0.295	sce	34

(continued)

Compound	Solvent System	Working Electrode	$E_{1/2}$	Reference Electrode	Reference
$R_1=2\text{-}CH_3$, $R_2=5\text{-}CH_3$	CH_3CN, $0.1M$ $(n\text{-}Propyl)_4NClO_4$	Hg	-1.318, -2.1	sce	42
	$0.22N$ H_2SO_4; 55% Alcohol	Hg	-0.295	sce	34
$R_1=2\text{-}CH_3$, $R_2=6\text{-}CH_3$	CH_3CN, $0.1M$ $(n\text{-}Propyl)_4NClO_4$	Hg	-1.280	sce	42
	$0.22N$ H_2SO_4; 55% Alcohol	Hg	-0.500	sce	34
$R_1=3\text{-}CH_3$, $R_2=4\text{-}CH_3$	CH_3CN, $0.1M$ $(n\text{-}Propyl)_4NClO_4$	Hg	-1.402, -2.3	sce	42
$R_1=3\text{-}CH_3$, $R_2=5\text{-}CH_3$	CH_3CN, $0.1M$ $(n\text{-}Propyl)_4NClO_4$	Hg	-1.233, -2.1	sce	42
$R_1=2\text{-}t\text{-}C_4H_9$, $R_2=5\text{-}C_4H_9$	CH_3CN, $0.1M$ $(n\text{-}Propyl)_4NClO_4$	Hg	-1.190, -2.0	sce	42
$R_1=2\text{-}Cl$, $R_2=6\text{-}Cl$	CH_3CN, $0.1M$ $(n\text{-}Propyl)_4NClO_4$	Hg	-1.360	sce	42
	DMF, $0.1M$ $(Et)_4NClO_4$	Hg	-1.13 E_p	sce	48
$R_1=3\text{-}Cl$, $R_2=5\text{-}Cl$	$HOAc$, $1M$ NH_4OAc	Hg	-0.490	Hg pool	8
$R_1=2\text{-}CO_2H$, $R_2=3\text{-}CO_2H$	$0.22N$ H_2SO_4; 55% Alcohol	Hg	-0.185	sce	52
$R_1=2\text{-}CO_2H$, $R_2=4\text{-}CO_2H$	$0.22N$ H_2SO_4; 55% Alcohol	Hg	-0.075	sce	52

Compound / Substituents	Conditions	Electrode	E	Ref. Electrode	Ref.
$R_1=2\text{-}CO_2H$, $R_2=5\text{-}CO_2H$	0.22N H_2SO_4, 55% Alcohol	Hg	-0.135	sce	52
$R_1=3\text{-}CO_2H$, $R_2=4\text{-}CO_2H$	0.22N H_2SO_4, 55% Alcohol	Hg	-0.100	sce	34
$R_1=3\text{-}CO_2H$, $R_2=5\text{-}CO_2H$	0.22N H_2SO_4, 55% Alcohol	Hg	-0.090	sce	34
$R_1=2\text{-}OCH_3$, $R_2=4\text{-}OCH_3$	0.22N H_2SO_4, 55% Alcohol	Hg	-0.300	sce	45
$R_1=3\text{-}OCH_3$, $R_2=4\text{-}OCH_3$	0.22N H_2SO_4, 55% Alcohol	Hg	-0.285	sce	45
$R_1=2\text{-}CH_3$, $R_2=3\text{-}Cl$	0.22N H_2SO_4, 55% Alcohol	Hg	-0.215	sce	34
$R_1=2\text{-}CH_3$, $R_2=5\text{-}Cl$	0.22N H_2SO_4, 55% Alcohol	Hg	-0.200	sce	34
$R_1=3\text{-}CH_3$, $R_2=4\text{-}Cl$	0.22N H_2SO_4, 55% Alcohol	Hg	-0.200	sce	34
$R_1=2\text{-}Cl$, $R_2=5\text{-}CH_3$	0.22N H_2SO_4, 55% Alcohol	Hg	-0.190	sce	34
$R_1=3\text{-}Cl$, $R_2=4\text{-}CH_3$	0.22N H_2SO_4, 55% Alcohol	Hg	-0.185	sce	34

Poly-Substituted Nitrobenzene

Compound	Conditions	Electrode	E	Ref. Electrode	Ref.
4-Nitrophthalide	P-W buffer, pH 0	Hg	-0.07	sce	51
4-Nitrophthalide	P-W buffer, pH 6.84	Hg	-0.39	sce	51
5-Nitrophthalide	P-W buffer, pH 0	Hg	-0.09	sce	51
5-Nitrophthalide	P-W buffer, pH 6.84	Hg	-0.37	sce	51
6-Nitrophthalide	P-W buffer, pH 0	Hg	-0.07	sce	51
6-Nitrophthalide	P-W buffer, pH 6.84	Hg	-0.40	sce	51
2,4,6-Trimethylnitrobenzene (Nitromesitylene)	DMF, 0.1M (n-Bu)$_4$NI	Hg	-1.38	sce	20

(continued)

953

Compound	Solvent System	Working Electrode	$E_{1/2}$	Reference Electrode	Reference
	DMF, 0.1$\underline{M}$ (n-Propyl)$_4$NClO$_4$	Hg	-1.380	sce	29
	DMF, 0.2$\underline{M}$ NaNO$_3$	Hg	-1.26	sce	39
	DMF, 0.1$\underline{M}$ (n-Bu)$_4$-NClO$_4$	Hg	-1.32, -1.62	Ag/AgClO$_4$	57
	CH$_3$CN, 0.1$\underline{M}$ (n-Propyl)$_4$NClO$_4$	Hg	-1.442	sce	42
2,3,5,6-Tetramethyl-nitrobenzene (Nitrodurene)	DMF, 0.1$\underline{M}$ (n-Bu)$_4$-NClO$_4$	Hg	-1.39, -1.72	Ag/AgClO$_4$	49
Aminonitrodurene	DMF, 0.1$\underline{M}$ (n-Bu)$_4$NI	Hg	-1.42	sce	20
	CH$_3$CN, 0.1$\underline{M}$ (n-Propyl)$_4$NClO$_4$	Hg	-1.442, -2.6	sce	42
	DMF, 0.1$\underline{M}$ (n-Bu)$_4$NI	Hg	-1.52	sce	20
Pentamethylnitrobenzene	DMF, 0.1$\underline{M}$ (n-Propyl)$_4$NClO$_4$	Hg	-1.428	sce	29

Dinitro Aromatic Compounds

Compound	Solvent System	Working Electrode	$E_{1/2}$	Reference Electrode	Reference
o-Dinitrobenzene	CH$_3$CN, 0.1$\underline{M}$ (n-Propyl)$_4$NClO$_4$	Hg	-0.81,	sce	23
	DMF, 0.1$\underline{M}$ (n-Propyl)$_4$NClO$_4$	Hg	-1.06, -2.54		
		Hg	-0.72, -1.08,	sce	49
			-1.66,		
			-2.24 $E_p/2$		
	0.01$\underline{M}$ (Et)$_4$NBr, 66% Alcohol	Hg	-0.570, -1.030	sce	32
	Phthalate buffer, 8% Ethanol, pH 3.8	Hg	-0.22, -0.45	sce	59
	Phosphate buffer	Hg	-0.418,	sce	58

m-Dinitrobenzene

HOAc, 1M NH4OAc	Hg	-0.560	Hg pool	8
CH3CN, 0.1M (n-Propyl)4NClO4	Hg	-0.90, -1.25, -2.01	sce	23
DMF, 0.1M (n-Propyl)4NClO4	Hg	-0.806, -1.245	sce	29
DMF, 0.1M (n-Bu)4-NClO4	Hg	-0.83, -1.27	sce	60
DMF, 0.1M (n-Bu)4NI	Hg	-0.80	sce	20
0.01M (Et)4NBr, 66% Alcohol	Hg	-0.684, -0.935	sce	32
Phthalate buffer, pH 3.8 8% Ethanol	Hg	-0.23, -0.38	sce	59
Ammonia, 1 NH4NO3	Pt	-0.08, -0.25	Pb/Pb(NO3)2, LiNO3, NH3	61
Phosphate buffer, 6.5% Ethanol, pH 8.95	Hg	-0.542, -0.713	sce	58

p-Dinitrobenzene

HOAc, 1M NH4OAc	Hg	-0.430, -0.730	Hg pool	8
DMF, 0.1M (n-Propyl)4NClO4	Hg	-0.535, -0.84, -2.24 Ep/2	sce	49
CH3CN, 0.1M (n-Propyl)4NClO4	Hg	-0.69, -0.89, -2.5	sce	29
DMF, 0.1M (n-Propyl)4NClO4	Hg	-0.543, -0.872	sce	29
0.01M (Et)4NBr, 66% Alcohol	Hg	-0.436, -1.128	sce	32
Phthalate buffer, pH 3.8 8% Ethanol	Hg	-0.19, -0.45	sce	59

(continued)

Compound	Solvent System	Working Electrode	$E_{1/2}$	Reference Electrode	Reference
2,4,6-Trimethyl-1,3-dinitrobenzene	Phosphate buffer, pH 8.95 6.5% Ethanol	Hg	-0.336, -0.855	sce	58
	CH₃CN, 0.1M (n-Propyl)₄NClO₄	Hg	-1.22, -1.47	sce	42
Tetramethyl-1,2-dinitrobenzene	CH₃CN, 0.1M (n-Propyl)₄NClO₄	Hg	-1.13, -1.28	sce	42
Tetramethyl-1,4-dinitrobenzene	CH₃CN, 0.1M (n-Propyl)₄NClO₄	Hg	-1.216	sce	42
2,2'-Dinitrobiphenyl	DMF, 0.1M (n-Propyl)₄NClO₄	Hg	-0.99, -1.32, -2.0	sce	55
2,4'-Dinitrobiphenyl	Aq buffer, 50%-Acetone, pH 1.3	Hg	-0.24, -0.82	sce	41
	Aq buffer, 50%-Acetone, pH 4.0	Hg	-0.40, -0.62	sce	41
	Aq buffer, 50%-Acetone, pH 7.9	Hg	-0.67, -0.82	sce	41
	Aq buffer, 50%-Acetone, pH 11.7	Hg	-0.65, -1.09	sce	41
4,4'-Dinitrobiphenyl	Aq buffer, 50%-Acetone, pH 1.4	Hg	-0.15, -0.78	sce	41
	Aq buffer, 50%-Acetone, pH 7.0	Hg	-0.54, -0.68	sce	41
	Aq buffer, 50%-Acetone, pH 11.7	Hg	-0.70, -1.18	sce	41
	CH₃CN, (n-Propyl)₄NClO₄	Hg	-1.004	sce	62
	DMSO, (n-Propyl)₄NClO₄	Hg	-0.891	sce	62
	DMF, 0.1M (n-Bu)...	Hg	-0.81	sce	2

2,6-Dinitroterphenyl	DMF, 0.1M (n-Bu)$_4$-NClO$_4$	Hg	-1.04	sce	2
4,4"-Dinitroterphenyl	DMF, 0.1M (n-Bu)$_4$-NClO$_4$	Hg	-1.02	sce	2
2,2'-Dinitrobibenzyl	CH$_3$CN, 0.1M (Et)$_4$-NClO$_4$	Hg	-1.23, -1.33 Ep	sce	43
4,4'-Dinitrobibenzyl	CH$_3$CN, 0.1M (Et)$_4$-NClO$_4$	Hg	-1.23 Ep	sce	43
1,2-Dinitronaphthalene	KCl-HCl buffer, 80% Ethanol, pH 2.1	Hg	-0.12, -0.38	sce	38
1,3-Dinitronaphthalene	KCl-HCl buffer, 80% Ethanol, pH 2.1	Hg	-0.16, -0.27, -0.93	sce	38
1,4-Dinitronaphthalene	KCl-HCl buffer, 80% Ethanol, pH 2.1	Hg	-0.06, -0.36	sce	38
1,5-Dinitronaphthalene	KCl-HCl buffer, 80% Ethanol, pH 2.1	Hg	-0.22, -0.78	sce	38
1,6-Dinitronaphthalene	KCl-HCl buffer, 80% Ethanol, pH 2.1	Hg	-0.22, -0.82	sce	38
1,7-Dinitronaphthalene	KCl-HCl buffer, 80% Ethanol, pH 2.1	Hg	-0.22, -0.86	sce	38
1,8-Dinitronaphthalene	KCl-HCl buffer, 80% Ethanol, pH 2.1	Hg	-0.26, -0.69	sce	38
2,3-Dinitronaphthalene	KCl-HCl buffer, 80% Ethanol, pH 2.1	Hg	-0.20, -0.67	sce	38
2,6-Dinitronaphthalene	KCl-HCl buffer, 80% Ethanol, pH 2.1	Hg	-0.14, -0.28, -0.90	sce	38
2,7-Dinitronaphthalene	KCl-HCl buffer, 80% Ethanol, pH 2.1	Hg	-0.23, -0.81	sce	38
5,6-Dinitroacenaphthene	KCl-HCl buffer, 80% Ethanol, pH 2.1	Hg	-0.26, -0.52	sce	63

(continued)

Compound	Solvent System	Working Electrode	$E_{1/2}$	Reference Electrode	Reference
3,4-Dinitroacenaphthe-nequinone	KCl-HCl buffer, 80% Ethanol, pH 2.1	Hg	-0.21	sce	63
4,5-Dinitronaphthalic anhydride	KCl-HCl buffer, 80% Ethanol, pH 2.1	Hg	-0.17	sce	63
5,6-Dinitrotetralin	KCl-HCl buffer, 80% Ethanol, pH 2.1	Hg	-0.21, -0.38	sce	40
6,7-Dinitrotetralin	KCl-HCl buffer, 80% Ethanol, pH 2.1	Hg	-0.18, -0.34	sce	40
5,7-Dinitrotetralin	HCl-KCl buffer, 80% Ethanol, pH 2.1	Hg	-0.20, -0.34, -0.72	sce	40
5,8-Dinitrotetralin	HCl-KCl buffer, 80% Ethanol, pH 2.1	Hg	-0.19, -0.40	sce	40
2,5-Dinitrofluorene	Aq buffer, 50%-Acetone, pH 1.4	Hg	-0.20	sce	41
	Aq buffer, 50%-Acetone, pH 8.5	Hg	-0.71	sce	41
	Aq buffer, 50%-Acetone, pH 11.6	Hg	-0.72, -1.09	sce	41
2,6-Dinitrofluorenone	Aq buffer, 50%-Acetone, pH 1.5	Hg	-0.20, -0.61, -1.01	sce	41
	Aq buffer, 50%-Acetone, pH 6.3	Hg	-0.42, -0.56, -1.04	sce	41
	Aq buffer, 50%-Acetone, pH 11.5	Hg	-0.72, -1.16	sce	41
2,7-Dinitrofluorenone	Aq buffer, 50%-Acetone, pH 1.4	Hg	-0.25, -0.68, -1.00	sce	41
	Aq buffer, 50%-Acetone, pH 6.6	Hg	-0.42, -0.56, -1.07	sce	41

Trinitro Aromatic Compounds

Compound	Medium		Potential	Reference	Ref.
sym-Trinitrobenzene	HOAc, 1M NH4OAc	Hg	-0.450	Hg pool	8
	DMF, 0.1M (n-Bu)4-NClO4	Hg	-0.46	sce	2
	Phthalate buffer, 8% Ethanol, pH 3.8	Hg	-0.16, -0.24, -0.34	sce	59
	Phosphate buffer, 6.5% Ethanol, pH 8.95	Hg	-0.336, -0.855	sce	58
Trinitromesitylene	DMF, 0.1M (n-Propyl)4NClO4	Hg	-0.928, -1.4, -1.7	sce	29
	DMF, 0.1M (n-Bu)4-NClO4	Hg	-1.00, -1.15, -1.60	Ag/AgClO4	57
2,4,6-Trinitroterphenyl	DMF, 0.1M (n-Bu)4-NClO4	Hg	-0.67	sce	2
4 Substituted Nitro-phenylhydrazines	Aqueous buffer-20% Methanol				64
19 Substituted Nitro-phenylhydrazones	Aqueous buffer-20% Methanol				65

N-Nitrosoamines

Compound	Medium		Potential	Reference	Ref.
N-Nitroso dimethylamine	4N HCl, 50% Ethanol	Hg	-0.93	sce	66
	Aq buffer, 1M KCl, 20% Ethanol, pH 3.60	Hg	-1.21	sce	67
	Aq buffer, 1M KCl, 20% Ethanol, pH 12.5	Hg	-1.55	sce	67
N-Nitroso diethylamine	4N HCl, 50% Ethanol	Hg	-0.92	sce	66
	Aq buffer, 1M KCl, 20% Ethanol, pH 3.60	Hg	-1.07	sce	67

(continued)

959

Compound	Solvent System	Working Electrode	$E_{1/2}$	Reference Electrode	Reference
N-Nitroso diethylamine	Aq buffer, 1M KCl, 20% Ethanol, pH 12.5	Hg	-1.35	sce	67
N-Nitroso di-n-propyl-amine	B-R buffer, 20%- Ethanol, pH 2.0	Hg	-0.858	sce	68
	B-R buffer, 20%- Ethanol, pH 6.93	Hg	-1.289	sce	68
	B-R buffer, 20%- Ethanol, pH 7.47	Hg	-1.312	sce	68
N-Nitroso diisopropyl-amine	4N HCl, 50% Ethanol	Hg	-0.88	sce	66
N-Nitroso dibutylamine	4N HCl, 50% Ethanol	Hg	-0.77	sce	66
N-Nitroso diisobutyl-amine	4N HCl, 50% Ethanol	Hg	-0.82	sce	66
N-Nitroso methylcyclo-hexylamine	4N HCl, 50% Ethanol	Hg	-0.79	sce	66
N-Nitroso dicylohexyl-amine	4N HCl, 50% Ethanol	Hg	-0.74	sce	66
N-Nitroso diallylamine	4N HCl, 50% Ethanol	Hg	-0.79	sce	66
N-Nitroso methylbenzyl-amine	4N HCl, 50% Ethanol	Hg	-0.81	sce	66
N-Nitroso dibenzylamine	4N HCl, 50% Ethanol	Hg	-0.64	sce	66
N-Nitroso methylaniline	4N HCl, 50% Ethanol	Hg	-0.68	sce	66
N-Nitroso N-methyl-	Aq buffer, 1M KCl, pH 3.60	Hg	-0.92	sce	67
aniline	Aq buffer, 1M KCl, 20% Ethanol, pH 12.5	Hg	-1.31	sce	67
N-Nitroso ethylaniline	4N HCl, 50% Ethanol	Hg	-0.62	sce	66
N-Nitroso diphenylamine	50% Acetone, 50%- Methanol, 0.1M (n-	Hg	-1.38	sce	69

Compound	Supporting electrolyte	Electrode	E	Reference electrode	Ref.
N-Nitroso phenylglycine	Aq buffer, 1M KCl, 20% Ethanol, pH 3.60	Hg	-0.765	sce	67
	Aq buffer, 1M KCl, 20% Ethanol, pH 12.5	Hg	-1.03	sce	67
N-Nitroso pyrrolidone	4N HCl, 50% Ethanol	Hg	-0.64	sce	66
N-Nitroso piperidine	4N HCl, 50% Ethanol	Hg	-0.81	sce	66
N-Nitroso hexamethyl-enimine	4N HCl, 50% Ethanol	Hg	-0.78	sce	66
	4N HCl, 50% Ethanol	Hg	-0.87	sce	66
N-Nitroso morpholine	4N HCl, 50% Ethanol	Hg	-0.80	sce	66
	Aq buffer, 1M KCl, pH 3.60	Hg	-1.09	sce	67
	Aq buffer, 1M KCl, 20% Ethanol, pH 12.5	Hg	-1.40	sce	67
N-Nitroso tetrahydro-quinoline	4N HCl, 50% Ethanol	Hg	-0.55	sce	66
N-Nitroso tetrahydro-isoquinoline	4N HCl, 50% Ethanol	Hg	-0.62	sce	66
N,N'-Dinitroso pipera-zine	Aq buffer, 1M KCl, 20% Ethanol, pH 3.60	Hg	-1.07	sce	67
	Aq buffer, 1M KCl, 20% Ethanol, pH 12.5	Hg	-1.35	sce	67

Nitroso Compounds

Compound	Supporting electrolyte	Electrode	E	Reference electrode	Ref.
Nitrosobenzene	HOAc, 1M NH$_4$OAc	Hg	-0.13, -0.89, -1.20	Hg pool	8
	KCl + HCl buffer, pH 1.6, 50% Ethanol	Graphite	+0.240 $E_{p/2}$	sce	6
	McIlvaine buffer, pH 5.3, 50% Ethanol	Graphite	-0.004 $E_{p/2}$	sce	6

(continued)

Compound	Solvent System	Working Electrode	$E_{1/2}$	Reference Electrode	Reference
	KCl + KOH buffer, pH 12.5	Graphite	-0.412 $E_{p/2}$	sce	6
Nitrosoguanidine	50% Ethanol, pH 12.5				
	B-R buffer, pH 2.4	Hg	$-0.73,$ -1.15	sce	24
	B-R buffer, pH 6.1	Hg	$-0.91,$ -1.18	sce	24
	B-R buffer, pH 11.7	Hg	-1.12	sce	24
p-Nitrosodiphenylamine	50% Acetone, 50%- Methanol, 0.1M (n-Bu)$_4$NOH	Hg	-0.68	sce	69
p-Nitrosophenol	Aq buffer, 10%- Methanol, pH 7.0	Hg	-0.632	Hg,Hg$_2$SO$_4$	70
p-Nitrosodimethyl-aniline	Aq buffer, 10%- Methanol, pH 7.0	Hg	-0.615	Hg,Hg$_2$SO$_4$	70
p-Nitrosotoluene	Aq buffer, 10%- Methanol, pH 7.0	Hg	-0.525	Hg,Hg$_2$SO$_4$	70
p-Nitrosobenzene	Aq buffer, 10%- Methanol, pH 7.0	Hg	-0.479	Hg,Hg$_2$SO$_4$	70
p-Chloronitrosobenzene	Aq buffer, 10%- Methanol, pH 7.0	Hg	-0.469	Hg,Hg$_2$SO$_4$	70
p-Bromonitrosobenzene	Aq buffer, 10%- Methanol, pH 7.0	Hg	-0.471	Hg,Hg$_2$SO$_4$	70
p-Iodonitrosobenzene	Aq buffer, 10%- Methanol, pH 7.0	Hg	-0.485	Hg,Hg$_2$SO$_4$	70
p-Nitrosonitrobenzene	Aq buffer, 10%- Methanol, pH 7.0	Hg	-0.430	Hg,Hg$_2$SO$_4$	70
p-Nitrosobenzaldehyde	Aq buffer, 10%- Methanol, pH 7.0	Hg	-0.417	Hg,Hg$_2$SO$_4$	70
1-Nitroso-2-naphthol	Aq buffer, 50%- Ethanol, pH 2.90	Hg	$+0.11,$ -0.08	sce	70
	Aq buffer, 50%	Hg	-0.13	sce	70

Compound	Conditions	Electrode	E	Ref. electrode	Ref.
2-Nitroso-1-naphthol	Aq buffer, 50%-Ethanol, pH 12.40	Hg	-0.12, -0.35	sce	70
	Aq buffer, 50%-Ethanol, pH 2.90	Hg	+0.07	sce	70
	Aq buffer, 50%-Ethanol, pH 7.35	Hg	-0.17	sce	70
	Aq buffer, 50%-Ethanol, pH 12.40	Hg	-0.13, -0.44	sce	70

Azoxy Compounds

Compound	Conditions	Electrode	E	Ref. electrode	Ref.
Azoxybenzene	HOAc, 1M NH4OAc	Hg	-0.735	Hg pool	8
	Sorensen buffer, 50% Isopropanol, pH 1.6	Hg	-0.29	sce	71
	Formate buffer, 30% Methanol, pH 3.0	Hg	-0.29, -1.02	sce	71
	OAc buffer, 30%-Methanol, pH 4.0	Hg	-0.42, -1.04	sce	71
	OAc buffer, 50%-Isopropanol, pH 4.5	Hg	-0.63	sce	71
	Ammoniacal buffer, 30% Methanol, pH 9.2	Hg	-0.70	sce	71

Oximes

Compound	Conditions	Electrode	E	Ref. electrode	Ref.
i-Butyraldoxime	Citrate-Phosphate buffer, 1M KCl, 40% Alcohol, pH 5.20	Hg	-1.37	sce	7
Phenylacetaldoxime	Citrate-Phosphate buffer, 1M KCl, 40% Alcohol, pH 5.20	Hg	-1.28	sce	7

(continued)

Compound	Solvent System	Working Electrode	$E_{1/2}$	Reference Electrode	Reference
Benzylacetoneoxime	Citrate-Phosphate buffer, 1M KCl, 40% Alcohol, pH 3.55	Hg	-1.34	sce	7
Crotonaldoxime	Citrate-Phosphate buffer, 1M KCl, 40% Alcohol, pH 5.20	Hg	-1.17	sce	7
Mesityloxide oxime	Citrate-Phosphate buffer, 1M KCl, 40% Alcohol, pH 5.20	Hg	-1.25	sce	7
Carvone oxime	Citrate-Phosphate buffer, 1M KCl, 40% Alcohol, pH 5.20	Hg	-1.16	sce	7
Testosterone propion- ate oxime	Citrate-Phosphate buffer, 1M KCl, 40% Alcohol, pH 5.20	Hg	-1.22	sce	7
anti-Cinnamaldoxime	Citrate-Phosphate buffer, 1M KCl, 40% Alcohol, pH 5.20	Hg	-1.00	sce	7
syn-Cinnamaldoxime	Citrate-Phosphate buffer, 1M KCl, 40% Alcohol, pH 5.20	Hg	-1.40	sce	7
Benzalacetone oxime	Citrate-Phosphate buffer, 1M KCl, 40% Alcohol, pH 5.20	Hg	-1.32	sce	7
anti-Benzaldoxime	Citrate-Phosphate buffer, 1M KCl, 40% Alcohol, pH 7.40	Hg	-1.63	sce	7
syn-Benzaldoxime	Citrate-Phosphate buffer, 1M KCl,	Hg	-1.60	sce	7

Compound	Conditions				Ref.
p-Aminoacetophenone oxime	Citrate-Phosphate buffer, 1M KCl, 40% Alcohol, pH 5.20	Hg	-1.24	sce	7
p-Dimethylaminobenzald-oxime	Citrate-Phosphate buffer, 1M KCl, 40% Alcohol, pH 5.20	Hg	-1.07	sce	7
2,4-Dihydroxybenzophen-one oxime	Citrate buffer, pH 3.05	Hg	-0.72, -0.80	sce	72
N-Benzylbenzaldoxime	Citrate-Phosphate buffer, 1M KCl, 40% Alcohol, pH 5.20	Hg	-0.97	sce	7
Mesoxalic acid ester oxime	Citrate-Phosphate buffer, 1M KCl, 40% Alcohol, pH 5.20	Hg	-1.05	sce	7
Benzamideoxime	Citrate-Phosphate buffer, 1M KCl, 40% Alcohol, pH 5.20	Hg	-1.36	sce	7
Ribose oxime	McIlvaine buffer, pH 2.5	Hg	-1.17	sce	73
Lyxose oxime	McIlvaine buffer, pH 2.5	Hg	-1.17	sce	73
Arabinose oxime	McIlvaine buffer, pH 2.5	Hg	-1.17	sce	73
Xylose oxime	McIlvaine buffer, pH 2.5	Hg	-1.16	sce	73
Mannose oxime	McIlvaine buffer, pH 2.5	Hg	-1.17	sce	73
Galactose oxime	McIlvaine buffer, pH 2.5	Hg	-1.16	sce	73
Glucose oxime	McIlvaine buffer, pH 2.5	Hg	-1.17	sce	73

(continued)

Nitrate Esters

Compound	Solvent System	Working Electrode	$E_{1/2}$	Reference Electrode	Reference
Ethyl nitrate	0.5M LiCl, 11.25%- Ethanol	Hg	-0.82	sce	75
n-Butyl nitrate	50% Methanol, 50%- Benzene, 0.3M LiCl	Hg	-1.32	sce	17
	Methanol, 0.5M LiCl	Hg	-1.10	sce	17
	Isobutanol, 0.2M- LiCl	Hg	-1.05	sce	17
	Ethylene glycol, 0.3M LiCl	Hg	-1.01	sce	17
n-Hexyl nitrate	0.5M LiCl, 11.25%- Ethanol	Hg	-0.54	sce	75
Cyclohexyl nitrate	0.5M LiCl, 11.25%- Ethanol	Hg	-0.63	sce	75
Glycerol nitrate- (nitroglycerine)	75% Ethanol, 0.5M- (Et)4NCl	Hg	-0.70	sce	74
	70% Ethanol, 10%- Acetone, 0.25M- (Me)4NCl	Hg	-0.735	Hg pool	36
	50% Methanol, 0.05M (Me)4NI	Hg	-0.25, -0.45, -0.75	Hg pool	44
Pentaerythritol trini-trate	70% Ethanol 10%- Acetone, 0.25M- (Me)4NCl	Hg	-0.772	Hg pool	36
Dinitroglycol	50% Methanol, 0.05M (Me)4NI	Hg	-0.45, -0.75	Hg pool	44

Heterocyclic Amine N-Oxides

Compound					
Pyridine N-oxide	DMF, 0.1M (n-Propyl)$_4$NClO$_4$	Hg	-2.297	sce	76
4-Methylpyridine N-oxide	DMF, 0.1M (Et)$_4$NBr	Hg	-1.79, -2.22	Hg pool	77
	DMF, 0.1M (n-Propyl)$_4$NClO$_4$	Hg	-2.375	sce	76
4-Ethylpyridine N-oxide	DMF, 0.1M (n-Propyl)$_4$NClO$_4$	Hg	-2.370	sce	76
4-Chloropyridine N-oxide	DMF, 0.1M (n-Propyl)$_4$NClO$_4$	Hg	-1.889	sce	76
4-Methoxypyridine N-oxide	DMF, 0.1M (n-Propyl)$_4$NClO$_4$	Hg	-2.402	sce	76
4-Ethoxypyridine N-oxide	DMF, 0.1M (n-Propyl)$_4$NClO$_4$	Hg	-2.445	sce	76
4-Carbethoxypyridine N-oxide	DMF, 0.1M (n-Propyl)$_4$NClO$_4$	Hg	-1.606	sce	76
4-Cyanopyridine N-oxide	DMF, 0.1M (n-Propyl)$_4$NClO$_4$	Hg	-1.557	sce	76
4-Nitropyridine N-oxide	DMF, 0.1M (n-Propyl)$_4$NClO$_4$	Hg	-0.768	sce	76
3-Methoxypyridine N-oxide	DMF, 0.1M (n-Propyl)$_4$NClO$_4$	Hg	-2.315	sce	76
3-Carbethoxypyridine N-oxide	DMF, 0.1M (n-Propyl)$_4$NClO$_4$	Hg	-1.670	sce	76
3-Cyanopyridine N-oxide	DMF, 0.1M (n-Propyl)$_4$NClO$_4$	Hg	-1.667	sce	76
Pyrazine mono-N-oxide	DMF, 0.1M (n-Propyl)$_4$NClO$_4$	Hg	-1.809	sce	76
Pyrazine di-N-oxide	DMF, 0.1M (n-Propyl)$_4$NClO$_4$	Hg	-1.616	sce	76

(continued)

Compound	Solvent System	Working Electrode	$E_{1/2}$	Reference Electrode	Reference
Pyridazine mono-N-oxide	DMF, 0.1$\underline{M}$ (n-Propyl)$_4$NClO$_4$	Hg	-1.898	sce	76
Pyrimidine mono-N-oxide	DMF, 0.1$\underline{M}$ (n-Propyl)$_4$NClO$_4$	Hg	-1.949	sce	76
3,5-Lutidine N-oxide	DMF, 0.1$\underline{M}$ (n-Propyl)$_4$NClO$_4$	Hg	-2.365	sce	76
Quinoline N-oxide	DMF, 0.1$\underline{M}$ (n-Propyl)$_4$NClO$_4$	Hg	-1.80	sce	76
Quinoline N^{15}-oxide	DMF, 0.1$\underline{M}$ (Et)$_4$NBr; DMF, 0.1$\underline{M}$ (n-Propyl)$_4$NClO$_4$	Hg; Hg	-1.30, -2.05; -1.805	Hg pool; sce	77; 76
Isoquinoline N-oxide	DMF, 0.1$\underline{M}$ (n-Propyl)$_4$NClO$_4$	Hg	-1.946	sce	76
Phthalazine mono-N-oxide	DMF, 0.1$\underline{M}$ (Et)$_4$NBr; DMF, 0.1$\underline{M}$ (n-Propyl)$_4$NClO$_4$	Hg; Hg	-1.43, -2.18; -1.716	Hg pool; sce	77; 76
Quinoxaline mono-N-oxide	DMF, 0.1$\underline{M}$ (n-Propyl)$_4$NClO$_4$	Hg	-1.419	sce	76
Quinoxaline di-N-oxide	DMF, 0.1$\underline{M}$ (n-Propyl)$_4$NClO$_4$	Hg	-1.241	sce	76
Acridine N-oxide	DMF, 0.1$\underline{M}$ (n-Propyl)$_4$NClO$_4$	Hg	-1.30	sce	76
Phenanthridine N-oxide	DMF, 0.1$\underline{M}$ (Et)$_4$NBr; DMF, 0.1$\underline{M}$ (n-Propyl)$_4$NClO$_4$	Hg; Hg	-0.81, -1.51, -1.78; -1.774	Hg pool; sce	77; 76
Phenazine mono-N-oxide	DMF, 0.1$\underline{M}$ (n-Propyl)$_4$NClO$_4$	Hg	-0.972	sce	76

Compound	Electrolyte	Electrode	$E_{1/2}$	Reference	Ref.
4-Nitropyridine-nitro-N15N-oxide	DMF, 0.1M (n-Propyl)4NClO4	Hg	-0.792	sce	76
4-Cyanopyridine 2,6-d2-N-oxide	DMF, 0.1M (n-Propyl)4NClO4	Hg	-1.595	sce	76
4-Nitropyridine 2,6-d2-N-oxide	DMF, 0.1M (n-Propyl)4NClO4	Hg	-0.793	sce	76
4,4'-Azobispyridine-1,1'-dioxide	DMF, 0.1M (n-Bu)4-NClO4	Hg	-0.73, -1.35	sce	78
1-Azaphenanthrene N-oxide	DMF, 0.1M (Et)4NBr	Hg	-1.38, -1.93	Hg pool	77
4-Azaphenanthrene N-oxide	DMF, 0.1M (Et)4NBr	Hg	-1.42, -1.75, -2.06	Hg pool	77
9-Azaphenanthrene N-oxide	DMF, 0.1M (Et)4NBr	Hg	-1.26, -1.88, -2.12	Hg pool	77
Adenine 1-N-oxide	KCl-HCl buffer, pH 1.43	Hg	-0.88, -1.14, -1.25	sce	79
	McIlvaine buffer, pH 3.72	Hg	-1.11, -1.32	sce	79
	McIlvaine buffer, pH 5.49	Hg	-1.46	sce	79
Dimethyldodecylamine oxide	McIlvaine buffer, pH 2.2	Hg	-0.85	sce	80
	McIlvaine buffer, pH 5.3	Hg	-0.97	sce	80
	McIlvaine buffer, pH 6.9	Hg	-1.12	sce	80

See also Heterocycles (Section 30)

a Glycine, NaCl, NaOH.
b NaOc, HCl, 0.2N KCl.
c In presence of charcoal.
d The methoxy group is in the 2 position.
e First $E_{1/2}$ not given.
f Potential +0.05V versus sce.
g Included for comparison.
h $E_{1/2}$ not given.
i At 0°C.

(continued)

Reference for Nitrogen-Oxygen Functions

1. T. Su, Ph.D. Thesis, University of Iowa, 1967.
2. R. L. Hansen, P. E. Toren, and R. H. Young,
    J. Phys. Chem., 70, 1653 (1966).
3. H. Lund, Acta Chem. Scand., 11, 1323 (1950).
4. F. Petrů, Collect. Czech. Chem. Commun., 12,
    620 (1947).
5. P. E. Iversen and H. Lund, Anal. Chem., 41,
    1322 (1969).
6. L. Chang, I. Fried, and P. J. Elving, Anal.
    Chem., 36, 2427 (1964).
7. H. Lund, Acta Chem. Scand., 13, 249 (1959).
8. I. Bergman, and J. C. James, Trans. Faraday
    Soc., 48, 956 (1952).
9. T. DeVries and R. W. Ivett, Anal. Chem., 13,
    339 (1941).
10. E. W. Miller, A. P. Arnold, and M. J. Astle,
    J. Amer. Chem. Soc., 70, 3971 (1949).
11. V. G. Korsakov, V. B. Aleskovskii, and I. A.
    Kendrinskii, Electrokhim., 4(1), 77 (1968).
12. M. Masui and H. Sayo, J. Chem. Soc., 1961, 5325.
13. L. Meites, Polarographic Techniques, 2nd ed.,
    Interscience, New York, 1965, p.698.
14. V. A. Petrosyan, V. I. Slovetskii, S. G. Mairano
    and A. A. Fainzil'berg, Electrokhim., 6, (10),
    1595 (1970).
15. H. Sayo, Y. Tsukitani, and M. Masui, Tetrahedron
    24, 1717 (1968).
16. J. Armand, J. Pinson, and J. Simonet, Anal.
    Letters, 4, 219 (1971).
17. N. Radin and T. DeVries, Anal. Chem., 24, 971 (1
18. G. Grandi, R. Andreoli, and G. B. Gavioli, J.
    Electroanal. Chem., 27, 177 (1970).
19. R. A. Wasserman and W. C. Purdy, J. Electroanal.
    Chem., 9, 51 (1965).
20. M. E. Peover and J. S. Powell, J. Electroanal.
    Chem., 20, 427 (1969).
21. M. M. Frodyma, L. H. Muramoto, D. J. Williams,
    and H. Matsumoto, Anal. Chem., 35, 1403 (1963).
22. W. J. Seagers and P. J. Elving, J. Am. Chem. Soc
    72, 5183 (1950).
23. A. K. Hoffmann, W. G. Hodgson, D. L. Maricle, an
    W. H. Jura, J. Am. Chem. Soc., 86, 631 (1964).
24. G. C. Whitnack and E. St. C. Gantz, J. Electroch
    Soc., 106, 422 (1959).
25. M. Yamashita and K. Sugino, J. Electrochem. Soc.
    104, 100 (1957).

26. A. P. Ballod, S. I. Molchanova, I. V. Patsevich, A. V. Topchiev, and V. Ya. Shtern, *Zh. Anal. Khim.*, **14**, 201 (1959).

27. R. A. Wasserman and W. C. Purdy, *J. Electroanal. Chem.*, **9**, 51 (1965).

28. A. H. Maki and D. H. Geske, *J. Am. Chem. Soc.*, **83**, 1852 (1961).

29. R. D. Allendoerfer and P. H. Rieger, *J. Amer. Chem. Soc.*, **88**, 3711 (1966).

30. T. Kitagawa, T. P. Layloff, and R. N. Adams, *Anal. Chem.*, **35**, 1086 (1963).

31. W. Kemula and R. Sioda, *Bull. Acad. Polon. Sci., Ser. Sci. Chim.*, **10**, 107 (1962).

32. E. Gergely and T. Iredale, *J. Chem. Soc.*, **1953**, 3226.

33. S. K. Vijayalakshamma and R. S. Subrahmanya, *J. Electroanal. Chem.*, **23**, 99 (1969).

34. M. LaGuyader, *Bull. Soc. Chim. Fr.*, 1848 (1966).

35. V. Yu. Glushchenko, *Zh. Obshch. Khim.*, **38**, 940 (1967).

36. W. M. Ayres and G. W. Leonard, *Anal. Chem.*, **31**, 1485 (1959).

37. R. F. Nelson and R. N. Adams, *J. Phys. Chem.*, **72**, 4336 (1968).

38. R. N. Boyd and A. A. Redlinger, *J. Electrochem. Soc.*, **107**, 611 (1960).

39. W. Kemula and R. Sioda, *J. Electroanal. Chem.*, **7**, 223 (1964).

40. R. N. Boyd, A. A. Redlinger, and M. J. Sher, *J. Electrochem. Soc.*, **107**, 302 (1960).

41. J. T. Gary and R. A. Day, Jr., *J. Electrochem. Soc.*, **107**, 616 (1960).

42. D. H. Geske, J. L. Ragle, M. A. Bambenek, and A. L. Balch, *J. Am. Chem. Soc.*, **86**, 987 (1964).

43. J. G. Lawless, D. E. Bartak, and M. D. Hawley, *J. Am. Chem. Soc.*, **91**, 7121 (1969).

44. A. F. Williams and D. Kenyon, *Talanta*, **3**, 160 (1959).

45. M. LeGuyader, *Bull. Soc. Chim. Fr.*, 1858 (1966).

46. J. E. Page, J. W. Smith, and J. G. Waller, *J. Phys. Chem.*, **53**, 545 (1949).

47. S. H. Cadle, P. R. Tice, and J. Q. Chambers, *J. Phys. Chem.*, **71**, 3517 (1967).

48. J. G. Lawless and M. D. Hawley, *J. Electroanal. Chem.*, App. 1-5 (1969).

49. J. Q. Chambers and R. N. Adams, *J. Electroanal. Chem.*, **9**, 400 (1965).

50. I. A. Korshunov, and L. N. Sazanova, Zh. Fiz. Khim., 23, 1299 (1949).
51. J. Tirouflet, Bull. Soc. Chim. Fr., 274 (1956).
52. M. LeGuyader, Bull. Soc. Chim. Fr., 1867 (1966).
53. V. Bellavita, N. Fedi, and N. Cagnoli, Ric. Sci. 25, 504 (1955).
54. W. N. Grieg and J. W. Rogers, J. Am. Chem. Soc., 91, 5495 (1969).
55. J. W. Rogers and W. H. Watson, Anal. Chim. Acta, 54, 41 (1971).
56. M. E. Runner and E. C. Wagner, J. Am. Chem. Soc. 74, 2529 (1952).
57. I. Bernal and G. K. Fraenkel, J. Am. Chem. Soc., 86, 1671 (1964).
58. T. L. Marple and L. B. Rogers, Anal. Chem., 25, 1351 (1953).
59. J. Pearson, Trans. Faraday Soc., 44, 683 (1948).
60. W. Kemula and R. Sioda, Naturwissenschaften, 50, 708 (1963).
61. W. H. Tiedemann and D. N. Bennion, J. Electroche Soc., 117, 203 (1970).
62. P. T. Cottrell and P. H. Rieger, Mol. Phys., 12, 149 (1967).
63. G. M. LoPresti, S. Huang, and A. A. Reidlinger, J. Electrochem. Soc., 115, 1135 (1968).
64. Yu. P. Kitaev and I. M. Skrebkova, Zh. Obshch. Khim., 37 (6), 1198 (1967).
65. Yu. P. Kitaev and I. M. Skrebkova, Zh. Obshch. Khim., 37 (6), 1142 (1967).
66. P. E. Iversen, Acta. Chem. Scand., 25, 2337 (19
67. H. Lund, Acta. Chem. Scand., 11, 990 (1957).
68. F. Pulidori, G. Borghesani, C. Bighi, and R. Pe J. Electroanal. Chem., 27, 385 (1970).
69. H. Siegerman, unpublished results.
70. L. Holleck and R. Schindler, Z. Elektrochem., 6 1138 (1956).
71. S. Wawzonek and J. D. Fredrickson, J. Am. Chem. Soc., 77, 3988 (1955).
72. H. Lund, Acta. Chem. Scand., 18, 563 (1964).
73. J. W. Haas, Jr., J. D. Storney, and C. C. Lynch Anal. Chem., 34, 145 (1962).
74. G. C. Whitnack, M. M. Mayfield, and E. St. C. G Anal. Chem., 27, 899 (1955).
75. F. Kaufman, H. J. Cook, and S. M. Davis, J. Am. Chem. Soc., 74, 4997 (1952).
76. T. Kubota, K. Nishikida, H. Miyazaki, K. Iwatan and Y Oishi, J. Am. Chem. Soc., 90, 5080 (1968)
77. G. Anthoine, J. Nasielski, E. V. Donckt, and

N. Vanlautem, _Bull. Soc. Chim. Belges._, 76, 230 (1967).

78. J. L. Sadler and A. J. Bard, _J. Electrochem. Soc._, 115, 343 (1968).

79. C. R. Warner and P. J. Elving, _Collect. Czech. Chem. Commun._, 30, 4210 (1965).

80. L. M. Chambers, _Anal. Chem._, 36, 2431 (1964).

26. CARBON-PHOSPHORUS FUNCTION

Oxidation

Compound	Solvent System	Working Electrode	$E_{1/2}$	Reference Electrode	Reference
2-Methyl-1,4-naptho-quinol-1-phosphate	$1\underline{M}$ H_2SO_4	C paste	+0.6	sce	1

Reduction

Compound	Solvent System	Working Electrode	$E_{1/2}$	Reference Electrode	Reference
Phosphorus Trichloride[a]	CH_3CN	Hg	-0.93	Hg pool	2
Phosphoryl Chloride[a]	CH_3CN	Hg	-0.89, -2.74	Hg pool	2
Bis-p-nitrophenyl phosphate	DMF, $0.1\underline{M}$ $(\underline{n}-Bu)_4NI$	Hg	-0.94, -1.23, -2.6	sce	3
Triphenylphosphine	DMF, $(\underline{n}-Bu)_4NI$	Hg	-2.70	sce	4
Triphenylphosphine	DMF, $(\underline{n}-Bu)_4NI$	Hg	-2.08	Hg pool	5
Triphenylphosphine oxide	DMF, $(\underline{n}-Bu)_4NI$	Hg	-2.51, -2.84	sce	4
	DMF, $(\underline{n}-Bu)_4NI$	Hg	-1.91	Hg pool	5

Compound	Solvent, Electrolyte	Electrode	E (V)	Reference electrode	Ref.
Triphenylarsine[a]	DMF, (n-Bu)₄NI	Hg	-2.19	Hg pool	5
Triphenylarsine oxide[a]	DMF, (n-Bu)₄NI	Hg	-1.74	Hg pool	5
Tetraphenylarsonium oxide[a]	DMF, (n-Bu)₄NI	Hg	-1.01, -2.20	Hg pool	5
Triphenylstibine[a]	DMF, (n-Bu)₄NI	Hg	-2.03	Hg pool	5
Triphenylbismuthine[a]	DMF, (n-Bu)₄NI	Hg	-1.90	Hg pool	5

Phosphonium Salts

Compound	Solvent, Electrolyte	Electrode	E (V)	Reference electrode	Ref.
Tetraphenylphosphonium chloride	DMF, (n-Bu)₄NI	Hg	-1.20, -1.87, -2.08	Hg pool	5
Triphenyl-n-butyl-phosphonium chloride	DMF, (n-Bu)₄NI	Hg	-1.24, -1.60, -2.02	Hg pool	5
Benzyltriphenylphosphonium chloride	Methanol, (Me)₄NCl	Hg	-1.826	Ag, AgCl	6
p-Methoxybenzyltriphenylphosphonium chloride	Methanol, (Me)₄NCl	Hg	-1.845	Ag, AgCl	6
p-Butylbenzyltriphenylphosphonium chloride	Methanol, (Me)₄NCl	Hg	-1.842	Ag, AgCl	6
p-Chlorobenzyltriphenylphosphonium chloride	Methanol, (Me)₄NCl	Hg	-1.717	Ag, AgCl	6
m-Methoxybenzyltriphenylphosphonium chloride	Methanol, (Me)₄NCl	Hg	-1.758	Ag, AgCl	6
m-Bromobenzyltriphenylphosphonium chloride	Methanol, (Me)₄NCl	Hg	-1.712	Ag, AgCl	6
Phenyltrimethylphosphonium iodide	95% Alcohol, (Et)₄NBr	Hg	-2.19, -2.30	sce	7

(continued)

975

Compound	Solvent System	Working Electrode	$E_{1/2}$	Reference Electrode	Reference
Tetra-o-tolylphosphon-ium iodide	95% Alcohol, $(Et)_4NBr$	Hg	-2.39	sce	7
Dialkyl Aroyl Phosphonates					
$R-C_6H_4 - CO-PO(OC_2H_5)$					
R=H	CH_3CN, $(n-Bu)_4NClO_4$	Hg	-1.27	ace b	8
R=p-t-C_4H_9	CH_3CN, $(n-Bu)_4NClO_4$	Hg	-1.29	ace b	8
R=o-Cl	CH_3CN, $(n-Bu)_4NClO_4$	Hg	-1.06	ace b	8
R=p-Cl	CH_3CN, $(n-Bu)_4NClO_4$	Hg	-1.14	ace b	8
R=o-CH_3O	CH_3CN, $(n-Bu)_4NClO_4$	Hg	-1.07	ace b	8
R=p-CH_3	CH_3CN, $(n-Bu)_4NClO_4$	Hg	-1.29	ace b	8
66 Phosphonium and Arsonium Salts					10

a Included for comparison.
b Acetone calomel electrode (9).

References for Carbon-Phosphorus Function

1. C. A. Chambers and J. Q. Chambers, *J. Am. Chem. Soc.*, **88**, 2922 (1966).
2. L. F. Filiminova and A. P. Tomilov, *Zh. Vses. Khim. Obshchest.*, **15**, 352 (1970).
3. K. S. V. Santhanam and A. J. Bard, *J. Electroanal. Chem.*, **25**, App. 6-9 (1970).
4. K. S. V. Santhanam and A. J. Bard, *J. Am. Chem. Soc.*, **90**, 1118 (1968).
5. S. Wawzonek and J. H. Wagenknecht, Proc. 3rd International Polarography Congress, Southampton, 1965, MacMillan, London, 1966, p. 1017.
6. J. Grimshaw and J. S. Ramsey, *J. Chem. Soc. B*, **1968**, 63.
7. E. L. Colichman, *Anal. Chem.*, **26**, 1204 (1954).
8. K. D. Berlin, D. S. Rulison, and P. Arthur, *Anal. Chem.*, **41**, 1554 (1969).
9. P. Arthur and H. Lyons, *Anal. Chem.*, **24**, 1422 (1952).
10. L. Horner and J. Haufe, *J. Electroanal. Chem.*, **20**, 245 (1969).

27. SULFUR FUNCTIONS

Oxidation

Thioethers

Compound	Solvent System	Working Electrode	$E_{1/2}$	Reference Electrode	Reference
Thioanisole	CH_3CN, 0.1M (n-Propyl)$_4$N$\overline{C}$lO$_4$	Pt	+1.56	sce	1
o-Methylthioanisole	CH_3CN, 0.1M (n-Propyl)$_4$N$\overline{C}$lO$_4$	Pt	+1.35	sce	1
m-Methylthioanisole	CH_3CN, 0.1M (n-Propyl)$_4$N$\overline{C}$lO$_4$	Pt	+1.45	sce	1
p-Methylthioanisole	CH_3CN, 0.1M (n-Propyl)$_4$N$\overline{C}$lO$_4$	Pt	+1.22	sce	1
o-Bis(methylthio)benzene	CH_3CN, 0.1M (n-Propyl)$_4$N$\overline{C}$lO$_4$	Pt	+1.35	sce	1
m-Bis(methylthio)benzene	CH_3CN, 0.1M (n-Propyl)$_4$N$\overline{C}$lO$_4$	Pt	+1.45	sce	1
p-Bis(methylthio)benzene	CH_3CN, 0.1M (n-Propyl)$_4$N$\overline{C}$lO$_4$	Pt	+1.19	sce	1
1,3,5-Tris(methylthio)benzene	CH_3CN, 0.1M (n-Propyl)$_4$N$\overline{C}$lO$_4$	Pt	+1.43	sce	1
1,2,4,5-Tetrakis(methylthio)benzene	CH_3CN, 0.1M (n-Propyl)$_4$N$\overline{C}$lO$_4$	Pt	+1.08	sce	1
	CH_3CN(n-Propyl)	Pt	+1.32	sce	2

Compound	Solvent, Electrolyte	Electrode	E	Reference	
2-(Methylthio)naphthalene	CH_3CN, $(\underline{n}\text{-Propyl})_4\text{NClO}_4$	Pt	$+1.36_5$	sce	2
1,4-Bis(methylthio)naphthalene	CH_3CN, $(\underline{n}\text{-Propyl})_4\text{NClO}_4$	Pt	$+1.07$	sce	2
1,5-Bis(methylthio)naphthalene	CH_3CN, $(\underline{n}\text{-Propyl})_4\text{NClO}_4$	Pt	$+1.26_5$	sce	2
1,8-Bis(methylthio)naphthalene	CH_3CN, $(\underline{n}\text{-Propyl})_4\text{NClO}_4$	Pt	$+1.09$	sce	2
2,3-Bis(methylthio)naphthalene	CH_3CN, $(\underline{n}\text{-Propyl})_4\text{NClO}_4$	Pt	$+1.35_5$	sce	2
2,6-Bis(methylthio)naphthalene	CH_3CN, $(\underline{n}\text{-Propyl})_4\text{NClO}_4$	Pt	$+1.10$	sce	2
2,7-Bis(methylthio)naphthalene	CH_3CN, $(\underline{n}\text{-Propyl})_4\text{NClO}_4$	Pt	$+1.33$	sce	2
1,5-Dimethoxy-4,8-bis(methylthio)naphthalene	CH_3CN, $(\underline{n}\text{-Propyl})_4\text{NClO}_4$	Pt	$+0.70$	sce	2
9,10-Bis(methylthio)anthracene	CH_3CN, $(\underline{n}\text{-Propyl})_4\text{NClO}_4$	Pt	$+1.11$	sce	2
2,2'-Bis(methylthio)biphenyl	CH_3CN, $(\underline{n}\text{-Propyl})_4\text{NClO}_4$	Pt	$+1.39$	sce	2
3,3'-Bis(methylthio)biphenyl	CH_3CN, $(\underline{n}\text{-Propyl})_4\text{NClO}_4$	Pt	$+1.47_5$	sce	2
4,4'-Bis(methylthio)biphenyl	CH_3CN, $(\underline{n}\text{-Propyl})_4\text{NClO}_4$	Pt	$+1.25_5$	sce	2
1,6-Bis(methylthio)pyrene	CH_3CN, $(\underline{n}\text{-Propyl})_4\text{NClO}_4$	Pt	$+0.96$	sce	2

Sulfides

Dimethyl sulfide	CH_3CN, $0.1M$ $NaClO_4$	Pt	$+1.41$ E_p	Ag/0.01M Ag+	3
Dimethyl disulfide	CH_3CN, $0.1\underline{M}$ $NaClO_4$	Pt	$+0.91$, $+1.59$	Ag/0.01$\underline{M}$ Ag+	4

(continued)

979

Compound	Solvent System	Working Electrode	$E_{1/2}$	Reference Electrode	Reference
Ethylene Sulfide	CH_3CN, $0.1\underline{M}$ $NaClO_4$	Pt	+1.51 Ep	Ag/$0.01\underline{M}$ Ag+	3
Diethyl sulfide	CH_3CN, $0.1\underline{M}$ $NaClO_4$	Pt	+1.50 Ep	Ag/$0.01\underline{M}$ Ag+	3
Propylene sulfide	CH_3CN, $0.1\underline{M}$ $NaClO_4$	Pt	+1.69 Ep	Ag/$0.01\underline{M}$ Ag+	3
Diallyl sulfide	CH_3CN, $0.1\underline{M}$ $NaClO_4$	Pt	+1.74 Ep	Ag/$0.01\underline{M}$ Ag+	3
Di-i-propyl sulfide	CH_3CN, $0.1\underline{M}$ $NaClO_4$	Pt	+1.47 Ep	Ag/$0.01\underline{M}$ Ag+	3
Di-n-butyl sulfide	CH_3CN, $0.1\underline{M}$ $NaClO_4$	Pt	+1.45 Ep	Ag/$0.01\underline{M}$ Ag+	3
Di-s-butyl sulfide	CH_3CN, $0.1\underline{M}$ $NaClO_4$	Pt	+1.43 Ep	Ag/$0.01\underline{M}$ Ag+	3
Di-t-butyl sulfide	CH_3CN, $0.1\underline{M}$ $NaClO_4$	Pt	+1.06, +1.7 Ep	Ag/$0.01\underline{M}$ Ag+	3
Pentamethylene sulfide	CH_3CN, $0.1\underline{M}$ $NaClO_4$	Pt	+0.55, +1.42 Ep	Ag/$0.01\underline{M}$ Ag+	3
Dibenzyl sulfide	CH_3CN, $0.1\underline{M}$ $NaClO_4$	Pt	+1.48 Ep	Ag/$0.01\underline{M}$ Ag+	3

Mercaptans Etc

Compound	Solvent System	Working Electrode	$E_{1/2}$	Reference Electrode	Reference
n-Butyl mercaptan	CH_3CN, $0.1\underline{M}$ $NaClO_4$	Pt	+1.49 Ep	Ag/$0.01\underline{M}$ Ag+	3
s-Butyl mercaptan	CH_3CN, $0.1\underline{M}$ $NaClO_4$	Pt	+1.33, +1.69 Ep	Ag/$0.01\underline{M}$ Ag+	3
t-Butyl mercaptan	CH_3CN, $0.1\underline{M}$ $NaClO_4$	Pt	+1.59 Ep	Ag/$0.01\underline{M}$ Ag+	3
n-Propyl mercaptan	CH_3CN, $0.1\underline{M}$ $NaClO_4$	Pt	+1.14	Ag/$0.01\underline{M}$ Ag+	4
Naphthalene-1,8-di-sulfide	CH_3CN, $0.1\underline{M}$ (n-Propyl)$_4$N$\overline{C}$lO$_4$	Pt	+0.95	sce	5
1,8-Bis(methylthio)-naphthalene	CH_3CN, $0.1\underline{M}$ (n-Propyl)$_4$N$\overline{C}$lO$_4$	Pt	+1.09	sce	5
9,10-Dithiophenanthrene	CH_3CN, $0.1\underline{M}$ (n-Propyl)$_4$N$\overline{C}$lO$_4$	Pt	+1.47	sce	5
2,2'-Bis(methylthio)-biphenyl	CH_3CN, $0.1\underline{M}$ (n-Propyl)$_4$N$\overline{C}$lO$_4$	Pt	+1.39	sce	5

Compound	Medium	Electrode	Potential	Reference	Ref.
Thianthrene	CH_3CN, 0.1M (n-Propyl)$_4$NClO$_4$	Pt	+1.28	sce	5
Ethylenetrithiocarbonate	CH_3CN, 0.1M NaClO$_4$	Pt	+1.53 E_p	Ag/0.01M Ag+	3
Mercaptobenzothiazole	B-R buffer, 25% Ethanol, pH 2.86	Hg	-0.015	sce	6
	B-R buffer, 25% Ethanol, pH 5.57	Hg	-0.165	sce	6
	B-R buffer, 25% Ethanol, pH 10.0	Hg	-0.255	sce	6
Thiobenzamide	Borate buffer, 1M KCl, 40% Alcohol, pH 10.6	Hg	-0.33	sce	7
Thiobenzanilide	1N NaOH, 1M KCl, 40% Alcohol	Hg	-0.52	sce	7
	1N NaOH, 1M KCl, 40% Alcohol	Hg	-0.37	sce	7
Thiobenzomorpholide	1N NaOH, 1M KCl, 40% Alcohol	Hg	-1.53$_5$	sce	7
	CH_3CN, 0.1M (Et)$_4$NClO$_4$	Pt	+0.68, +1.12	sce	8

(continued)

Reduction

Thioethers

Compound	Solvent System	Working Electrode	$E_{1/2}$	Reference Electrode	Reference
1-(Methylthio)-naphthalene	CH_3CN, $(n\text{-Propyl})_4$-$NClO_4$	Hg	-2.25	sce	2
2-(Methylthio)-naphthalene	CH_3CN, $(n\text{-Propyl})_4$-$NClO_4$	Hg	-2.28	sce	2
1,4-Bis(methylthio)-naphthalene	CH_3CN, $(n\text{-Propyl})_4$-$NClO_4$	Hg	-2.10	sce	2
1,5-Bis(methylthio)-naphthalene	CH_3CN, $(n\text{-Propyl})_4$-$NClO_4$	Hg	-2.15	sce	2
1,8-Bis(methylthio)-naphthalene	CH_3CN, $(n\text{-Propyl})_4$-$NClO_4$	Hg	-2.22	sce	2
2,3-Bis(methylthio)-naphthalene	CH_3CN, $(n\text{-Propyl})_4$-$NClO_4$	Hg	-2.21	sce	2
2,6-Bis(methylthio)-naphthalene	CH_3CN, $(n\text{-Propyl})_4$-$NClO_4$	Hg	-2.24	sce	2
2,7-Bis(methylthio)-naphthalene	CH_3CN, $(n\text{-Propyl})_4$-$NClO_4$	Hg	-2.25	sce	2
1,5-Dimethoxy-4,8-bis-(methylthio)naphthalene	CH_3CN, $(n\text{-Propyl})_4$-$NClO_4$	Hg	-2.42	sce	2
9,10-Bis(methylthio)-anthracene	CH_3CN, $(n\text{-Propyl})_4$-$NClO_4$	Hg	-1.55	sce	2
2,2'-Bis(methylthio)-	CH_3CN, $(n\text{-Propyl})_4$-$NClO_4$	Hg	-2.58	sce	2

Compound	Electrolyte	Electrode	E	Reference	Ref.
3,3'-Bis(methylthio)-biphenyl	CH$_3$CN, (n-Propyl)$_4$-NClO$_4$	Hg	-2.35	sce	2
4,4'-Bis(methylthio)-biphenyl	CH$_3$CN, (n-Propyl)$_4$-NClO$_4$	Hg	-2.29	sce	2
1,6-Bis(methylthio)-pyrene	CH$_3$CN, (n-Propyl)$_4$-NClO$_4$	Hg	-1.83	sce	2
Thioketones					
Benzothione	CH$_3$CN, 0.05M (n-Bu)$_4$-NBr	Hg	-1.10	sce	9
Xanthione	CH$_3$CN, 0.05M (n-Bu)$_4$-NBr	Hg	-1.11	sce	9
Dimethoxybenzothione	CH$_3$CN, 0.05M (n-Bu)$_4$-NBr	Hg	-1.16	sce	9
Michler's thione	CH$_3$CN, 0.05M (n-Bu)$_4$-NBr	Hg	-1.42	sce	9
3,3'-Diaminobenzothione	CH$_3$CN, 0.05M (n-Bu)$_4$-NBr	Hg	-1.19	sce	9
Thiofluorenone	CH$_3$CN, 0.05M (n-Bu)$_4$-NBr	Hg	-0.92	sce	9
Camphorthione	CH$_3$CN, 0.05M (n-Bu)$_4$-NBr	Hg	-1.89	sce	9
Alkyl Tosylates					
Methyl tosylate	CH$_3$CN, 0.1M (n-Propyl)$_4$NClO$_4$	Pt	-2.61 E_p	Ag/0.1M Ag+	10
Ethyl tosylate	CH$_3$CN, 0.1M (4-n-Propyl)$_4$NClO$_4$	Pt	-2.52 E_p	Ag/0.1M Ag+	10

(continued)

983

Compound	Solvent System	Working Electrode	$E_{1/2}$	Reference Electrode	Reference
n-Butyl tosylate	CH_3CN, 0.1M (n-Propyl)$_4$NClO$_4$	Pt	-2.61 E_p	Ag/0.1M Ag+	10
Neopentyl tosylate	CH_3CN, 0.1M (n-Propyl)$_4$NClO$_4$	Pt	-2.52 E_p	Ag/0.1M Ag+	10
Cyclohexyl tosylate	CH_3CN, 0.1M (n-Propyl)$_4$NClO$_4$	Pt	-2.57 E_p	Ag/0.1M Ag+	10
Allyl tosylate	DMSO, 0.1M (Et)$_4$NClO$_4$	Hg	-2.13	sce	11
2-Carbethoxyallyl tosylate	DMSO, 0.1M (Et)$_4$NClO$_4$	Hg	-1.59	sce	11
tosylate	DMF, 0.1M (Et)$_4$NClO$_4$	Hg	-1.83	sce	11
Sulfides					
Pentamethylene sulfide	CH_3CN, NaClO$_4$	Pt	+0.12 E_p	sce	3
Tetramethylthiuram disulfide (Arasan)	Aq buffer, 50%-Dioxane, pH 4.8	Hg	-0.90	Hg, Hg$_2$SO$_4$	12
Diphenyl sulfide	DMF, (n-Bu)$_4$NI	Hg	-2.03	Hg pool	13
S$_8$	DMSO, (Et)$_4$NClO$_4$	Au	-0.62, -1.29 E_p	sce	15
Triphenylsulfonium bromide	Aq buffer, pH 11.5	Hg	-1.2	sce	14
Ethylenetrithiocarbonate	CH_3CN, 0.1M NaClO$_4$	Pt	+0.17 E_p	Ag/0.01M Ag+	3
Thiourea	0.05M H$_2$SO$_4$	Hg	-0.024	sce	16
Thiobenzamide	Phosphate-Citrate buffer, 1M KCl, 40% Alcohol, pH 3.55	Hg	-1.26	sce	7

Compound	Conditions	Electrode	E	Ref. electrode	Ref.
	Phosphate-Citrate buffer, 1M KCl, 40% Alcohol, pH 6.30	Hg	-1.32, -1.55	sce	7
Thiobenzanilide	Borate buffer, 1M KCl, 40% Alcohol, pH 10.6	Hg	-1.65	sce	7
	Phosphate-Citrate buffer, 1M KCl, -40% Alcohol, pH 3.55	Hg	-1.14	sce	7
	Phosphate-Citrate buffer, 1M KCl, 40% Alcohol, pH 6.30	Hg	-1.22, -1.51	sce	7
Thiobenzomorpholide	Phosphate-Citrate buffer, 1M KCl, 40% Alcohol, pH 3.60	Hg	-1.23	sce	7
	Phosphate-Citrate buffer, 1M KCl, 40% Alcohol, pH 6.30	Hg	-1.30, -1.52	sce	7

Benzensulphonamides

$X-C_6H_4SO_2NH_2$

Compound	Conditions	Electrode	E	Ref. electrode	Ref.
X=4-COOH	Borate buffer, pH 9.3	Hg	-1.827	sce	17
X=4-COOCH$_3$	Borate buffer, pH 9.3	Hg	-1.589	sce	17
X=3-CONH$_2$	Borate buffer, pH 9.3	Hg	-1.847	sce	17
X=4-CONH$_2$	Borate buffer, pH 9.3	Hg	-1.608	sce	17
X=3-CN	Borate buffer, pH 9.3	Hg	-1.839	sce	17
X=4-CN	Borate buffer, pH 9.3	Hg	-1.678	sce	17
X=3-SO$_2$NH$_2$	Borate buffer, pH 9.3	Hg	-1.860	sce	17
X=4-SO$_2$NH$_2$	Borate buffer, pH 9.3	Hg	-1.713	sce	17
X=3,5-(SO$_2$NH$_2$)$_2$	Borate buffer, pH 9.3	Hg	-1.702, -1.860	sce	17

(continued)

Compound	Solvent System	Working Electrode	$E_{1/2}$	Reference Electrode	Reference
X=3-SO$_2$CH$_3$	Borate buffer, pH 9.3	Hg	-1.812	sce	17
Sulphonic Acid Derivatives					
Naphthalene-1-sulphonic acid	0.5M KCl	Hg	-0.94	sce	18
Naphthalene-2-sulphonic acid	0.5M KCl	Hg	-0.94	sce	18
Naphthalene-1,5-disul- phonic acid	0.5M KCl	Hg	-0.89	sce	18
Naphthalene-2,7-disul- phonic acid	0.5M KCl	Hg	-0.88	sce	18
Naphthalene-1,3,6-tri- sulphonic acid	0.5M KCl	Hg	-0.92	sce	18
1-Naphthol-2-sulphonic acid	0.5M KCl	Hg	-1.08	sce	18
1-Naphthol-4-sulphonic acid	0.5M KCl	Hg	-0.92	sce	18
1-Naphthol-5-sulphonic acid	0.5M KCl	Hg	-0.08, -0.89	sce	18
1-Naphthol-3,6-disul- phonic acid	0.5M KCl	Hg	-0.90	sce	18
2-Naphthol-1-sulphonic acid	0.5M KCl	Hg	-0.93	sce	18
2-Naphthol-6-sulphonic acid	0.5M KCl	Hg	-0.87	sce	18
2-Naphthol-3,6-disul- phonic acid	0.5M KCl	Hg	-0.86	sce	18
	0.5M KCl	Hg	-0.88	sce	18

1-Naphthylamine-2-sulphonic acid	0.5M KCl	Hg	-0.96	sce	18
1-Naphthylamine-4-sulphonic acid	0.5M KCl	Hg	-0.90	sce	18
1-Naphthylamine-5-sulphonic acid	0.5M KCl	Hg	-0.90	sce	18
1-Naphthylamine-6-or-7-sulphonic acid	0.5M KCl	Hg	-0.93	sce	18
2-Naphthylamine-5,7-disulphonic acid	0.5M KCl	Hg	-0.85, -1.60	sce	18
2-Naphthylamine-6,8-disulphonic acid	0.5M KCl	Hg	-1.08	sce	18
Benzene sulphonic acid	0.5M KCl	Hg	-0.89	sce	18
Benzene-m-sulphonic acid	0.5M KCl	Hg	-0.89	sce	18
Toluene-p-sulphonic acid	0.5M KCl	Hg	-0.85	sce	18
p-Xylene-sulphonic acid	0.5M KCl	Hg	-0.88	sce	18
Toluidine-p-sulphon-amide[a]	0.5M KCl	Hg	-0.98	sce	18
p-Toluidine-2-sulphonic acid	0.5M KCl	Hg	-0.87, -1.62	sce	18
p-Toluidine-3-sulphonic acid	0.5M KCl	Hg	-0.93, -1.67	sce	18
m-Toluidine-4-sulphonic acid	0.5M KCl	Hg	-0.86, -1.61	sce	18
Coumarin[a]	0.5M KCl	Hg	-1.55	sce	18

Sulfones
___

Methylvinylsulfone	0.1M (Et)$_4$NI, 50% Ethanol	Hg	-1.700	sce	19

(continued)

987

Compound	Solvent System	Working Electrode	$E_{1/2}$	Reference Electrode	Reference
Ethylvinylsulfone	0.1M (Et)₄NI, 50% Ethanol	Hg	-1.735	sce	19
Propylvinylsulfone	0.1M (Et)₄NI, 50% Ethanol	Hg	-1.800	sce	19
Isopropylvinylsulfone	0.1M (Et)₄NI, 50% Ethanol	Hg	-1.825	sce	19
t-Isobutylvinylsulfone	0.1M (Et)₄NI, 50% Ethanol	Hg	-2.065	sce	19
n-Amylvinylsulfone	0.1M (Et)₄NI, 50% Ethanol	Hg	-1.835	sce	19
s-Isoamylvinylsulfone	0.1M (Et)₄NI, 50% Ethanol	Hg	-1.815	sce	19
t-Isoamylvinylsulfone	0.1M (Et)₄NI, 50% Ethanol	Hg	-2.025	sce	19
n-Octylvinylsulfone	0.1M (Et)₄NI, 50% Ethanol	Hg	-1.815	sce	19
s-Octylvinylsulfone	0.1M (Et)₄NI, 50% Ethanol	Hg	-2.050	sce	19
Decrylvinylsulfone	0.1M (Et)₄NI, 50% Ethanol	Hg	-1.950	sce	19

Aryl Methyl Sulfones

$X-C_6H_4SO_2CH_3$

Compound	Solvent System	Working Electrode	$E_{1/2}$	Reference Electrode	Reference
X=H	Borate buffer, 5%- Ethanol, pH 9.3	Hg	No wave	sce	20
X=3-COC₆H₅	Borate buffer, 45%- Ethanol, pH 9.3	Hg	-1.212	sce	20

X		Electrode		Reference	Temp.
X=4-COC$_6$H$_5$	Borate buffer, 45%- Ethanol, pH 9.3	Hg	-1.155	sce	20
X=3-COOH	Borate buffer, 5%- Ethanol, pH 9.3	Hg	No wave	sce	20
X=4-COOH	Borate buffer, 5%- Ethanol, pH 9.3	Hg	-1.783	sce	20
X=3-COOCH$_3$	Borate buffer, 5%- Ethanol, pH 9.3	Hg	-1.726	sce	20
X=4-COOCH$_3$	Borate buffer, 5%- Ethanol, pH 9.3	Hg	-1.523	sce	
X=4-COOC$_2$H$_5$	Borate buffer, 5%- Ethanol, pH 9.3	Hg	-1.520	sce	20
X=4-COOC$_3$H$_7$-$\underline{i}$	Borate buffer, 5%- Ethanol, pH 9.3	Hg	-1.531	sce	20
X=3-CONH$_2$	Borate buffer, 5%- Ethanol, pH 9.3	Hg	-1.792	sce	20
X=4-CONH$_2$	Borate buffer, 5%- Ethanol, pH 9.3	Hg	-1.566	sce	20
X=3-CN	Borate buffer, 5%- Ethanol, pH 9.3	Hg	-1.786	sce	20
X=4-CN	Borate buffer, 5%- Ethanol, pH 9.3	Hg	-1.586	sce	20
X=4-CH$_2$OH	Borate buffer, 5%- Ethanol, pH 9.3	Hg	No wave	sce	20
X=4-CH$_2$NH$_2$	Borate buffer, 5%- Ethanol, pH 9.3	Hg	No wave	sce	20
X=4-NH$_2$	Borate buffer, 5%- Ethanol, pH 9.3	Hg	No wave	sce	20
X=4-NHCOCH$_3$	Borate buffer, 5%- Ethanol, pH 9.3	Hg	No wave	sce	20

(continued)

989

Compound	Solvent System	Working Electrode	$E_{1/2}$	Reference Electrode	Reference
X=3-SO$_3$H	Borate buffer, 5%-Ethanol, pH 9.3	Hg	No wave	sce	20
X=4-SO$_3$H	Borate buffer, 5%-Ethanol, pH 9.3	Hg	No wave	sce	20
X=3-SO$_2$NH$_2$	Borate buffer, 5%-Ethanol, pH 9.3	Hg	-1.812	sce	20
X=4-SO$_2$NH$_2$	Borate buffer, 5%-Ethanol, pH 9.3	Hg	-1.632	sce	20
X=3-SO$_2$CH$_3$	Borate buffer, 5%-Ethanol, pH 9.3	Hg	-1.777	sce	20
X=4-SO$_2$CH$_3$	Borate buffer, 5%-Ethanol, pH 9.3	Hg	-1.598	sce	20
X=3-SO$_2$H	Borate buffer, 5%-Ethanol, pH 9.3	Hg	No wave	sce	20
X=4-SO$_2$H	Borate buffer, 5%-Ethanol, pH 9.3	Hg	No wave	sce	20
X-3,5-Cl$_2$	Borate buffer, 5%-Ethanol, pH 9.3	Hg	-1.814	sce	20

a Included for comparison.

References for Sulfur Functions

1. A. Zweig and J. E. Lehnsen, J. Am. Chem. Soc.,
    87, 2647 (1965).
2. A. Zweig, A. H. Maurer, and B. G. Roberts, J.
    Org. Chem., 32, 1322 (1967).
3. P. T. Cottrell and C. K. Mann, J. Electrochem.
    Soc., 116, 1499 (1969).
4. J. W. Loveland and G. R. Dimeler, Anal. Chem.,
    33, 1196 (1961).
5. A. Zweig and A. K. Hoffmann, J. Org. Chem., 30,
    3977 (1965).
6. G. Sartori and A. Liberti, J. Electrochem. Soc.,
    97, 20 (1950).
7. H. Lund, Collect. Czech. Chem. Commun., 25, 3313
    (1960).
8. N. D. Canfield, J. Q. Chambers, and D. L. Coffen,
    J. Electroanal. Chem., 24, App. 7-9 (1970).
9. R. M. Elofson, F. F. Gadallah, and L. A. Gadallah,
    Can. J. Chem., 47, 3979 (1969).
10. P. Yousefzadeh and C. K. Mann, J. Org. Chem., 33,
    2716 (1968).
11. J. P. Petrovich and M. M. Baizer, Electrochim.
    Acta, 12, 1249 (1967).
12. D. K. Gullstrom and H. P. Burchfield, Anal. Chem.,
    20, 1174 (1948).
13. S. Wawzonek and J. H. Wagenknecht, Proc. 3rd
    International Polarography Congress, Southampton,
    1965, MacMillan, London, 1966, p. 1017.
14. P. S. McKinney and S. Rosenthal, J. Electroanal.
    Chem., 16, 261 (1968).
15. M. V. Merritt and D. T. Sawyer, Inorg. Chem., 9,
    211 (1970).
16. C. J. Nyman and E. P. Parry, Anal. Chem., 30,
    1255 (1958).
17. O. Manousek, O. Exner, and P. Zuman, Collect.
    Czech. Chem. Commun., 33, 4000 (1968).
18. P. A. Brook and J. A. Crossley, Electrochim. Acta,
    11, 1189 (1966).
19. A. A. Pozdeeva, S. G. Mairanovskii, and L. K.
    Gladkova, Elektrokhim., 3, 1127 (1967).
20. O. Manousek, O. Exner, and P. Zuman, Collect.
    Czech. Chem. Commun., 32, 3988 (1968).

28. BIOLOGICALLY IMPORTANT COMPOUNDS

Oxidation

Purines and Pyrimidines

Compound	Solvent System	Working Electrode	$E_{1/2}$	Reference Electrode	Reference
Purine	Aq buffer	Graphite	No wave	sce	1
6-Aminopurine (Adenine)	OAc buffer, pH 5.7	Graphite	+1.01 $E_{p/2}$	sce	1
2-Amino-6-hydroxypurine (Guanine)	2$\underline{M}$ H_2SO_4	Graphite	+1.02 $E_{p/2}$	sce	1
6-Amino-2-hydroxypurine (Isoguanine)	2$\underline{M}$ H_2SO_4	Graphite	+1.05 $E_{p/2}$	sce	1
6-Hydroxypurine (Hypoxanthine)	0.25$\underline{M}$ Na_2SO_4 + H_2SO_4, pH 2.3	Graphite	+1.26 $E_{p/2}$	sce	1
2,6-Dihydroxypurine (Xanthine)	OAc buffer, pH 5.7	Graphite	+1.04 $E_{p/2}$	sce	1
	2$\underline{M}$ H_2SO_4	Graphite	+1.01 $E_{p/2}$	sce	1
2,6,8-Trihydroxypurine (Uric Acid)	OAc buffer, pH 5.7	Graphite	+0.71 $E_{p/2}$	sce	1
	2$\underline{M}$ H_2SO_4	Graphite	+0.62 $E_{p/2}$	sce	1
6-Thiopurine	OAc buffer, pH 3.7	Graphite	+0.45 $E_{p/2}$	sce	1
	OAc buffer, pH 5.7	Graphite	+0.33 $E_{p/2}$	sce	1
	1$\underline{M}$ HOAc, pH 2.3	Graphite	+0.35, +0.70, +1.57 E	sce	2

Compound	Buffer	Electrode	E	Reference	Ref.
Purine-6-sulfinic acid	McIlvaine buffer, pH 6.1	Graphite	$+0.25$, $+0.48$, E_p	sce	2
	Ammonia buffer, pH 9.9	Graphite	$+0.30$, $+0.62$, E_p	sce	2
	Ammonia buffer, pH 9.1	Graphite	$+0.70$ $E_{p/2}$	sce	2
2,5-Diaminopyrimidine	Universal buffer[a]	Pt	$+1.09$,[b] -0.077pH	Ag, AgCl	2A
4,5-Diaminopyrimidine	Universal buffer[a]	Pt	$+1.28$,[c] -0.08pH	Ag, AgCl	2A
	Universal buffer[a]	Pt	$+0.96$,[d] -0.03pH	Ag, AgCl	2A
4,5-Diamino-6-methyl-pyrimidine	OAc buffer, pH 2.94	Pt	$+1.03$	Ag, AgCl	2A
2,4,5-Triaminopyrimidine	OAc buffer, pH 2.94	Pt	$+0.69$	Ag, AgCl	2A
4,5-Diamino-6-hydroxy-pyrimidine	OAc buffer, pH 2.94	Pt	$+0.52$	Ag, AgCl	2A
4,5,6-Triaminopyrimidine	OAc buffer, pH 2.94	Pt	$+0.63$	Ag, AgCl	2A
2,4-Dihydroxy-5-methylaminopyrimidine	OAc buffer, pH 2.94	Pt	$+0.62$	Ag, AgCl	2A
4,5-Diamino-2,6-dihydroxypyrimidine	OAc buffer, pH 2.94	Pt	$+0.34$	Ag, AgCl	2A

Xanthines

Compound	Buffer	Electrode	E	Reference	Ref.
Xanthine (2,6-Dihydroxypurine)	Aq buffer, pH 0 to 12.5	Graphite	$+1.07$, -0.060pH[e]	sce	3
1-Methylxanthine	Aq buffers, pH 0 to 12.5	Graphite	$+1.05$, -0.049pH[e]	sce	3

(continued)

993

Compound	Solvent System	Working Electrode	$E_{1/2}$	Reference Electrode	Reference
3-Methylxanthine	Aq buffers, pH 0 to 11.9	Graphite	$+1.27$, -0.050pH[e]	sce	3
7-Methylxanthine	Aq buffers, pH 0 to 12.5	Graphite	$+1.22$, -0.042pH[e]	sce	3
1,3-Dimethylxanthine (Theophylline)	Aq buffers, pH 2.3 to 8.5	Graphite	$+1.45$, -0.056pH[e]	sce	3
1,7-Dimethylxanthine	Aq buffers, pH 0 to 12.5	Graphite	$+1.31$, -0.059pH[e]	sce	3
3,7-Dimethylxanthine (Theobromine)	Aq buffers, pH 2.3 to 5.5	Graphite	$+1.67$, -0.064pH[e]	sce	3
1,3,7-Trimethylxanthine (Caffeine)	Aq buffers, pH 2.3 to 5.5	Graphite	$+1.59$, -0.042pH[e]	sce	3

Phenothiazine Tranquilizers

Compound	Solvent System	Working Electrode	$E_{1/2}$	Reference Electrode	Reference
10-(3-Dimethylamino-propyl)phenothiazine	$0.1N$ H_2SO_4 $0.1\underline{N}$ H_2SO_4, Alcohol	Au Au	$+0.473$ $+0.545$	NCE NCE	4 4
10-[(1-Methyl-3-piperidy)methyl]pheno-thiazine	$0.1N$ H_2SO_4 $0.1\underline{N}$ H_2SO_4, Alcohol	Au Au	$+0.503$ $+0.600$	NCE NCE	4 4
10-(3-dimethylamino-propyl)-2-chlorophenothiazine	$0.1N$ H_2SO_4 $0.1\underline{N}$ H_2SO_4, Alcohol	Au Au	$+0.541$ $+0.636$	NCE NCE	4 4
10-[3-(4-methyl-1-piperazinyl)propyl]-2-chlorophenothiazine	$0.1N$ H_2SO_4 $0.1\underline{N}$ H_2SO_4, Alcohol	Au Au	$+0.547$ $+0.620$	NCE NCE	4 4

Compound	Medium	Electrode	E	Reference	
[3-(4-β-hydroxyethyl-piperazinyl)propyl]-2-chlorophenothiazine	$0.1N$ H_2SO_4	Au	+0.550	NCE	4
	$0.1\underline{N}$ H_2SO_4, Alcohol	Au	+0.619	NCE	4
10-[2-(1-pyrrolidinyl)-ethyl] phenothiazine	$0.1N$ H_2SO_4	Au	+0.567	NCE	4
	$0.1\underline{N}$ H_2SO_4, Alcohol	Au	+0.635	NCE	4
10-(3-dimethylamino-propyl)-2-acetylpheno-thiazine	$0.1N$ H_2SO_4	Au	+0.583	NCE	4
	$0.1\underline{N}$ H_2SO_4, Alcohol	Au	+0.684	NCE	4
10-(2-dimethylamino-propyl) phenothiazine	$0.1N$ H_2SO_4	Au	+0.619	NCE	4
	$0.1\underline{N}$ H_2SO_4, Alcohol	Au	+0.696	NCE	4
10-(2-diethylamino-propyl) phenothiazine	$0.1N$ H_2SO_4	Au	+0.619	NCE	4
	$0.1\underline{N}$ H_2SO_4, Alcohol	Au	+0.696	NCE	4
2-diethylaminoethyl-10-phenothiazine-carboxylate	$0.1N$ H_2SO_4	Au	+0.620	NCE	4
	$0.1\underline{N}$ H_2SO_4, Alcohol	Au	+0.724	NCE	4

a Acetate, phosphate, borate.
b pH 1 to 6.
c pH 3 to 6.
d pH 6 to 10.
e E_p.

(continued)

Reduction

Purines and Pyrimidine

Compound	Solvent System	Working Electrode	$E_{1/2}$	Reference Electrode	Reference
Purine	Py, 0.1$\underline{M}$ (Et)$_4$NClO$_4$	Hg	-1.66	Ag/1$\underline{N}$ Ag+, Py	5
	Aq buffer, pH 2-6	Hg	-0.697, -0.083pH, -0.902, -0.080pH	sce	1
6-Hydroxypurine (Hypoxanthine)	Py, 0.1$\underline{M}$ (Et)$_4$NClO4	Hg	-1.72, -2.15	Ag/1$\underline{N}$ Ag+, Py	5
2,6-Dihydroxypurine	Aq buffer, pH 5.7	Hg	-1.61	sce	1
	Py, 0.1$\underline{M}$ (Et)$_4$NClO4	Hg	-1.70, -2.11	Ag/1$\underline{N}$ Ag+, Py	5
6-Methoxypurine	Py, 0.1$\underline{M}$ (Et)$_4$NClO4	Hg	-1.74	Ag/1$\underline{N}$ Ag+, Py	5
6-Aminopurine (Adenine)	Py, 0.1$\underline{M}$ (Et)$_4$NClO4	Hg	-1.81	Ag/1$\underline{N}$ Ag+, Py	5
	Aq buffer, pH 5.7	Hg	-0.975, -0.090 pH	sce	1
6-Mercaptopurine	Py, 0.1$\underline{M}$ (Et)$_4$NClO4	Hg	-1.61, -2.02	Ag/1$\underline{N}$ Ag+, Py	5
6-Thiopurine	Aq buffers, pH 0-2.3	Hg	-0.79, -0.116 pH,	sce	2

	Conditions	Electrode	Potential	Reference	Ref.
	Aq buffers, pH 5.5-8	Hg	-1.29, -0.027 pH	sce	2
	Ammonia buffer, pH 9.1	Hg	-1.740	sce	2
Purine-6-sulfinic acid	Aq buffer, pH 2-9.1	Hg	-0.37, -0.094 pH	sce	2
	Aq buffer, pH 8-13	Hg	-0.79, -0.075 pH	sce	2
	Aq buffer, pH 3-9	Hg	-0.99, -0.080 pH	sce	2
Purine-6-sulfonic acid	Aq buffer, pH 1-7	Hg	-0.45, -0.078 pH	sce	2
	Aq buffer, pH 3.6-12.5	Hg	-0.68, -0.079 pH	sce	2
	Aq buffer, pH 1-9	Hg	-0.98, -0.064 pH	sce	2
Purine-6-sulfonamide	Ammonia buffer, pH 9	Hg	-1.04, -1.47, -1.58	sce	2
Bis(6-purinyl) disulfide	1M HOAc, pH 2.3	Hg	-0.05, -1.05	sce	2
Pyrimidine	CH$_3$CN, 0.1M (Et)$_4$-NClO$_4$	Hg	-2.628	Ag/0.01N Ag$^+$	6
	CH$_3$CN, 0.1M (Et)$_4$-NClO$_4$	Hg	-2.340	sce	6
	DMF, 0.1M (Et)$_4$NClO$_4$	Hg	-2.340	Ag, AgCl	6
	Aq buffers	Hg	-0.576, -0.105 pH	sce	6
	Aq buffers, pH 0.5-5	Hg	-0.576, -0.105 pH	sce	1
	Aq buffers, pH 3-5	Hg	-1.142, -0.011 pH	sce	1
	Aq buffers, pH 5-8	Hg	-0.680, -0.089 pH	sce	1

(continued)

Compound	Solvent System	Working Electrode	$E_{1/2}$	Reference Electrode	Reference
	Aq buffers, pH 7-8	Hg	-1.000, -0.005, pH	sce	1
	Aq buffers, pH 9-13	Hg	-0.805, -0.079, pH	sce	1
2-Aminopyrimidine	Aq buffers, pH 2-3	Hg	-0.685, -0.049, pH	sce	1
	Aq buffers, pH 4-7	Hg	-0.425, -0.121 pH, -1.360, -0.004 pH	sce	1
	Aq buffers, pH 7-9	Hg	-0.6800, -0.090, pH	sce	1
2-Amino-4-methyl-pyrimidine	Aq buffers, pH 2-4	Hg	-0.770, -0.063, pH	sce	1
	Aq buffers, pH 4-7	Hg	-0.550, -0.113, pH	sce	1
	Aq buffers, pH 5-7	Hg	-1.424, -0.008, pH	sce	1
	Aq buffers, pH 7-9	Hg	-0.745, -0.094, pH	sce	1
2-Hydroxypyrimidine	Aq buffers, pH 2-8	Hg	-0.530, -0.078, pH	sce	1
4-Amino-2-hydroxypyri-midine (Cytosine)	Aq buffers, pH 4-6	Hg	-1.125, -0.075, pH	sce	1
4-Amino-6-hydroxypyri-midine	Aq buffer, pH 1.2	Hg	-1.18	sce	1
4-Amino-2,6-dimethyl-pyrimidine	Aq buffer, pH 6.8	Hg	-1.62	sce	1
	Aq buffer, pH 2-8	Hg	-1.130, -0.073 pH	sce	1

Compound	Supporting electrolyte	Electrode	E	Reference	Note
4-Amino-2,5-dimethyl-pyrimidine	Aq buffer, pH 3-6	Hg	-1.11, -0.076 pH	sce	1
4-Hydroxypyrimidine	Aq buffer, pH 1.2	Hg	-1.16	sce	1
	Aq buffer, pH 6.8	Hg	-1.52	sce	1
4,5,6-Triaminopyrimidine	Aq buffer, pH 1.2	Hg	-1.17	sce	1
	Aq buffer, pH 6.8	Hg	-1.57	sce	1
1,4-o-Dimethylpyrimi-dine (Thymine)	Aq buffer, pH 1.2	Hg	-1.18	sce	1
	Aq buffer, pH 6.8	Hg	-1.62	sce	1

Adenine Derivatives

Compound	Supporting electrolyte	Electrode	E	Reference	Note
Adenine	Aq buffers, pH 1.0-6.5	Hg	-0.975, -0.084 pH	sce	7
Adenosine	Aq buffers, pH 2.0-4.5	Hg	-1.040, -0.070 pH	sce	7
	Aq buffers, pH 4.5-6.0	Hg	-1.180, -0.041 pH	sce	7
Adenine-1-N-oxide	0.05M HClO4	Hg	-0.84, -1.11	sce	7
Deoxyadenosine	Aq buffer, pH 2.5-4.6	Hg	-1.060, -0.069 pH	sce	7
Adenylic acid	Aq buffer, pH 4.6-6.5	Hg	-1.205, -0.037 pH	sce	7
	Aq buffer, pH 1.0-4.3	Hg	-1.015, -0.083 pH	sce	7
Deoxyadenylic acid	Aq buffer, pH 4.3-6.5	Hg	-1.115, -0.060 pH	sce	7
	Aq buffer, pH 2.0-6.5	Hg	-0.985, -0.080 pH	sce	7
Adenosine triphosphate	Aq buffer, pH 2.5-4.5	Hg	-1.035, -0.083 pH	sce	7

(continued)

999

Compound	Solvent System	Working Electrode	$E_{1/2}$	Reference Electrode	Reference
	Aq buffer, pH 4.5–5.5	Hg	-1.175, -0.052 pH	sce	7
Cytosine Derivatives					
Cytosine	Aq buffers, pH 2.5–7	Hg	-1.070, -0.084 pH	sce	7
Cytidine	Aq buffers, pH 2.5–7	Hg	-1.105, -0.072 pH	sce	7
Deoxycytidine	Aq buffers, pH 2.5–7	Hg	-1.154, -0.068 pH	sce	7
Cytidylic acid	Aq buffers, pH 6.5	Hg	-1.68	sce	7
Deoxycytidylic acid	Aq buffers, pH 3.5–7.3	Hg	-0.908, -0.110 pH	sce	7
5-Methylcytosine	Aq buffers, pH 7.3–8.7	Hg	-1.350, -0.050 pH	sce	7
	Aq buffers, pH 5.6–6.5	Hg	-1.65 to -1.73	sce	7
5-Methyldeoxycytidine	Aq buffers, pH 4–7.3	Hg	-0.775, -0.118 pH	sce	7
	Aq buffers, pH 7.3–10	Hg	-1.325, -0.042 pH	sce	7
5-Hydroxymethylcytosine	Aq buffers, pH 5.6–6.5	Hg	-1.61 to -1.70	sce	7
Pharmaceuticals					
Acetylsalicylic acid	CH₃CN, (n-Bu)₄NClO₄	Hg	-1.64	Ag/(n-Bu)₄NI	8
Atropine	CH₃CN, (n-Bu)₄NClO₄	Hg	-2.14	Ag/(n-Bu)₄NI	8

Compound	Solvent / Electrolyte	Electrode	Potential	Reference Electrode	Ref.
...serpine methyl nitrate	CH_3CN, $(n\text{-}Bu)_4NClO_4$	Hg	-2.04	$Ag/(n\text{-}Bu)_4NI$	8
Colchiceine	CH_3CN, $(n\text{-}Bu)_4NClO_4$	Hg	-0.96	$Ag/(n\text{-}Bu)_4NI$	8
Colchicine	CH_3CN, $(n\text{-}Bu)_4NClO_4$	Hg	-1.47	$Ag/(n\text{-}Bu)_4NI$	8
Deserpidine	CH_3CN, $(n\text{-}Bu)_4NClO_4$	Hg	-2.14	$Ag/(n\text{-}Bu)_4NI$	8
Diethylstilbestrol	CH_3CN, $(n\text{-}Bu)_4NClO_4$	Hg	No wave	$Ag/(n\text{-}Bu)_4NI$	8
Estradiol	CH_3CN, $(n\text{-}Bu)_4NClO_4$	Hg	-0.64	$Ag/(n\text{-}Bu)_4NI$	8
Estrone	CH_3CN, $(n\text{-}Bu)_4NClO_4$	Hg	No wave	$Ag/(n\text{-}Bu)_4NI$	8
Hydrochlorothiazide	CH_3CN, $(n\text{-}Bu)_4NClO_4$	Hg	-1.56	$Ag/(n\text{-}Bu)_4NI$	8
Hydrocortisone	CH_3CN, $(n\text{-}Bu)_4NClO_4$	Hg	-1.58	$Ag/(n\text{-}Bu)_4NI$	8
Methylchlorothiazide	CH_3CN, $(n\text{-}Bu)_4NClO_4$	Hg	-1.58	$Ag/(n\text{-}Bu)_4NI$	8
Nitroglycerine	CH_3CN, $(n\text{-}Bu)_4NClO_4$	Hg	-0.59	$Ag/(n\text{-}Bu)_4NI$	8
Phenobarbital	CH_3CN, $(n\text{-}Bu)_4NClO_4$	Hg	-2.07	$Ag/(n\text{-}Bu)_4NI$	8
Prednisolone	CH_3CN, $(n\text{-}Bu)_4NClO_4$	Hg	-1.64	$Ag/(n\text{-}Bu)_4NI$	8
Progesterone	0.1M HCl, 50% Ethanol	Hg	-1.11, -0.063 pH	NCE	9
Reserpine	CH_3CN, $(n\text{-}Bu)_4NClO_4$	Hg	-1.85	$Ag/(n\text{-}Bu)_4NI$	8
Salicylic Acid	CH_3CN, $(n\text{-}Bu)_4NClO_4$	Hg	-2.14	$Ag/(n\text{-}Bu)_4NI$	8
Sulfacetimide	CH_3CN, $(n\text{-}Bu)_4NClO_4$	Hg	-1.54	$Ag/(n\text{-}Bu)_4NI$	8
Sulfadiazine	CH_3CN, $(n\text{-}Bu)_4NClO_4$	Hg	-1.85	$Ag/(n\text{-}Bu)_4NI$	8
Sulfamethazine	CH_3CN, $(n\text{-}Bu)_4NClO_4$	Hg	-1.56, -1.95	$Ag/(n\text{-}Bu)_4NI$	8
	CH_3CN, $(n\text{-}Bu)_4NClO_4$	Hg	-1.45	$Ag/(n\text{-}Bu)_4NI$	8
	CH_3CN, $(n\text{-}Bu)_4NClO_4$	Hg	-2.02	$Ag/(n\text{-}Bu)_4NI$	8
Testosterone	0.1M HCl, 50% Ethanol	Hg	-1.13, -0.061 pH	NCE	9
Methyltestosterone	0.1M HCl, 50% Ethanol	Hg	-1.13, -0.059 pH	NCE	9
Testosterone propionate	0.1M HCl, 50% Ethanol	Hg	-1.13, -0.059 pH	NCE	9
17-Hydroxyprogesterone	0.1M HCl, 50% Ethanol	Hg	-1.14, -0.059 pH	NCE	9
11-Deoxycorticosterone	0.1M HCl, 50% Ethanol	Hg	-1.12, -0.059 pH	NCE	9

(continued)

Compound	Solvent System	Working Electrode	$E_{1/2}$	Reference Electrode	Reference
11-Deoxycorticosterone acetate	0.1M HCl, 50% Ethanol	Hg	-1.11, -0.059 pH	NCE	9
11-Deoxy-17-hydroxy-corticosterone	0.1M HCl, 50% Ethanol	Hg	-1.15, -0.059 pH	NCE	9
Tetracycline HCl	0.2M C-L buffer, pH 4.10 0.5M KCl	Hg	-1.04, -1.22, -1.33	sce	10
	0.1M OAc buffer, pH 4	Hg	-1.00, -1.18, -1.31 E_p^a	sce	10A
	0.2M C-L buffer, pH 7.60 0.5M KCl	Hg	-1.17, -1.25, -1.42	sce	10
	0.2M C-L buffer, pH 9.90 0.5M KCl	Hg	-1.33, -1.51	sce	10
	CH3CN, 0.1M (n-Bu)4-NClO4	Hg	-1.00, -1.33	sce	10
	DMF, 0.1M (n-Bu)4NClO4	Hg	-1.08, -1.36	sce	10
	DMSO, 0.1M (n-Bu)4-NClO4	Hg	-1.10, -1.365	sce	10
Tetracycline	DMF, 0.1M (n-Bu)4NClO4	Hg	-1.35, -1.60	sce	10
	DMSO, 0.1M (n-Bu)4-NClO4	Hg	-1.365	sce	10
Dedimethylaminotetra-cycline	Phosphate buffer, 10% Methanol, pH 4.1	Hg	-1.21, -1.39	sce	10
	DMF, 0.1M (n-Bu)4NClO4	Hg	-1.36	sce	10
	DMSO, 0.1M (n-Bu)4-NClO4	Hg	-1.39, -1.87	sce	10
Nialamid	Phosphate buffer, pH 5.9	Hg	-0.87, -1.01	sce	11
Conteben	Phosphate buffer, pH 5.9	Hg	-1.05	sce	11

Compound	Conditions		Potential (V)	Ref. electrode	Ref.
Iversal	pH 5.9 Phosphate buffer, pH 5.9	Hg	-0.17	sce	11
Adrenoxyl	Phosphate buffer, pH 5.9	Hg	-0.36	sce	11
Furacin	Phosphate buffer, pH 5.9	Hg	-0.24, -1.28	sce	11
Nardil	Phosphate buffer, pH 5.9	Hg	-1.33	sce	11
Chloramphenicol	B-R buffer, pH 2.2	Hg	-0.35, -1.05	sce	11
	0.1M OAc buffer, pH 4	Hg	-0.27, E_p[a]	sce	10A
	B-R buffer, pH 10.1	Hg	-0.75	sce	11
Streptomycin sulfate	0.1M NaOH, 0.001M EDTA	Hg	-1.52, -1.64 E_p[a]	sce	10A
Cortisone	OAc buffer, 50%- Methanol	Hg	-1.36	sce	12
Aureomycin	Phosphate buffer, pH 8.1	Hg	-1.16, -1.41	sce	12
Local Anesthetics					
Protocain	Phosphate buffer, pH 6.91	Hg	-1.65	--b	14
Procain	Phosphate buffer, pH 6.91	Hg	-1.68	--b	14
Intracain	Phosphate buffer, pH 6.91	Hg	-1.75	--b	14
Nupercain	Phosphate buffer, pH 6.34	Hg	-1.0, -1.5	--b	14

(continued)

Compound	Solvent System	Working Electrode	$E_{1/2}$	Reference Electrode	Reference
Amino Acids					
Asparagine	DMSO, (Et)4NClO4	Hg	-2.382	sce	16
Hydroxyproline	DMSO, (Et)4NClO4	Hg	-2.317	sce	16
Isoleucine	DMSO, (Et)4NClO4	Hg	-2.465	sce	16
Methionine	DMSO, (Et)4NClO4	Hg	-2.435	sce	16
Phenylalanine	DMSO, (Et)4NClO4	Hg	-2.398	sce	16
Proline	DMSO, (Et)4NClO4	Hg	-2.307	sce	16
Threonine	DMSO, (Et)4NClO4	Hg	-2.368	sce	16
Tryptophan	DMSO, (Et)4NClO4	Hg	-2.388	sce	16
Tyrosine	DMSO, (Et)4NClO4	Hg	-2.379	sce	16
Glutamine	DMSO, (Et)4NClO4	Hg	-2.090,c / -2.428	sce	16
Dehydroxyphenylalanine (DOPA)	DMSO, (Et)4NClO4	Hg	-2.375, / -2.625,c	sce	16
Glutamic acid	DMSO, (Et)4NClO4	Hg	-2.123,c / -2.604	sce	16
Aspartic acid	DMSO, (Et)4NClO4	Hg	-2.141,c / -2.608	sce	16
Cysteine HCl	DMSO, (Et)4NClO4	Hg	-0.168, / -1.321, / -1.788,c / -2.014 / -2.660	sce	16
Related Compounds					
n-Butylamine	DMSO, (Et)4NClO4	Hg	-2.23	sce	16
n-Propylamine	DMSO, (Et)4NClO4	Hg	-2.20	sce	16

Aminophenylsulfone	DMSO, (Et)$_4$NClO$_4$	Hg	-2.24	sce	16
Propionic acid	DMSO, (Et)$_4$NClO$_4$	Hg	-2.36	sce	16
Butyric acid	DMSO, (Et)$_4$NClO$_4$	Hg	-2.36	sce	16
Glutaric acid	DMSO, (Et)$_4$NClO$_4$	Hg	-2.39	sce	16
n-Propylamine	DMSO, (Et)$_4$NClO$_4$	Hg	-2.20	sce	16
Aminophenylsulfone	DMSO, (Et)$_4$NClO$_4$	Hg	-2.24	sce	16
Propionic acid	DMSO, (Et)$_4$NClO$_4$	Hg	-2.36	sce	16
Butyric acid	DMSO, (Et)$_4$NClO$_4$	Hg	-2.36	sce	16
Glutaric acid	DMSO, (Et)$_4$NClO$_4$	Hg	-2.39	sce	16
Succinic acid	DMSO, (Et)$_4$NClO$_4$	Hg	-2.29	sce	16
Methylglutaric acid	DMSO, (Et)$_4$NClO$_4$	Hg	-2.40	sce	16
Methylsuccinic acid	DMSO, (Et)$_4$NClO$_4$	Hg	-2.41	sce	16
Malic acid	DMSO, (Et)$_4$NClO$_4$	Hg	-2.25	sce	16

Vitamins

Vitamin B$_1$ (Thiamine)	0.1N KCl	Hg	-1.25	sce	17
	Phosphate buffer, pH 5.9	Hg	-1.9	sce	18
Vitamin B$_2$ (Riboflavin)	0.1N NaOH	Hg	No wave	sce	17
	B-R buffer, pH 1.81	Hg	-0.202	NCE	19
	B-R buffer, pH 6.80	Hg	-0.472	NCE	19
	B-R buffer, pH 11.98	Hg	-0.681	NCE	19
	Phosphate buffer, pH 7.38	Hg	-0.498	NCE	20
Lumichrom	DMSO, NaClO$_4$	Hg	-0.71, -1.02, -1.25	sce	21
	B-R buffer, pH 1.81	Hg	-0.30	NCE	22
	B-R buffer, pH 7.96	Hg	-0.64	NCE	22
	B-R buffer, pH 11.98	Hg	-0.85	NCE	22
Vitamine B$_6$ (Pyridoxin)	0.1N KCl	Hg	-1.8, -2.0	sce	17

(continued)

Compound	Solvent System	Working Electrode	$E_{1/2}$	Reference Electrode	Reference
Pantothenic acid	0.1M (Me)$_4$NBr OAc buffer, pH 4.7	Hg	-2.0	sce	17
Vitamin B12		Hg	-1.66	sce	23
	0.1M EDTA, pH 9.6	Hg	-1.00	sce	24
	0.1M EDTA, pH 9.6	Hg	$-1.246, -1.496, -1.616, E_p$	sce	24
	0.1M KCl	Hg	$-1.02, -1.50, -1.61 \; E_p$	sce	24
	0.1M KCN	Hg	-1.33	sce	18
	0.1M K$_2$SO$_4$	Hg	-1.11	sce	18
	1M K$_2$C$_2$O$_4$	Hg	-0.98	sce	24
	1M K$_2$C$_2$O$_4$	Hg	$-1.02, -1.50, -1.64 \; E_p$	sce	24
Vitamin B12r	0.1M EDTA, pH 9.6	Hg	-0.89	sce	24
	0.1M EDTA, pH 9.6	Hg	$-0.944, -1.450, -1.610 \; E_p$	sce	24
Vitamin B12S	0.1M EDTA, pH 9.6	Hg	-0.83	sce	24
	0.1M EDTA, pH 9.6	Hg	$-0.938, -1.485, -1.57 \; E_p$	sce	24
Vitamin B (Folic acid)	Borate buffer, KCl, pH 9	Hg	-0.91	NCE	25
Vitamin C	B-R buffer, pH 1.8	Hg	-0.41	NCE	26
	B-R buffer, pH 6.0	Hg	-0.68	NCE	26
	B-R buffer, pH 11.2	Hg	-0.93	NCE	26
	2% Metaphosphoric acid	Hg	-1.7	NCE	27
	Aq buffer, pH 1.50	Hg	$+0.21d$	NCE	28
	Aq buffer, pH 4.45	Hg	$+0.05d$	NCE	28

Compound	Conditions	Electrode	E	Ref. electrode	Ref.
Vitamin E (α-Tocopherol)	Aq buffer, pH 8.30	Hg	-0.11[d]	NCE	28
	0.1N anilinuim per-chlorate, 0.1N aniline buffer, 75%-Ethanol, pH 4.02				
Vitamin K₁	CH₃CN, LiClO₄	Pt	+0.76 E_p[d]	sce	30
Vitamin K₅	KCl, 2-Propanol	Hg	-0.54	sce	23
Nicotinic acid	B-R buffer, pH 6.3	Hg	-0.07	sce	23
	0.1N NaHCO₃, pH 8.4	Hg	-1.6	sce	31
Nicotinamide	Aq (Me)₄NBO₃, pH 9	Hg	-1.6	sce	17
	0.1N NaOH	Hg	-1.82	sce	32
Alkaloids					
Quinine	B-R buffer, pH 5.1	Hg	-1.31	NCE	33
	B-R buffer, pH 11.4	Hg	-1.64	NCE	33
Quinotoxol	B-R buffer, pH 7.7	Hg	-1.47	NCE	33
	B-R buffer, pH 10.7	Hg	-1.60	NCE	33
Quinotoxine	B-R buffer, pH 3.0	Hg	-0.50	NCE	33
	B-R buffer, pH 11.0	Hg	-1.10	NCE	33
Quininone	B-R buffer, pH 3.0	Hg	-0.36	NCE	33
	B-R buffer, pH 11.0	Hg	-1.06, -1.21	NCE	33
Quinidine	B-R buffer, pH 12.0	Hg	-1.59	NCE	34
Cinchonine	B-R buffer, pH 12.0	Hg	-1.65	NCE	34
Cinchonidine	B-R buffer, pH 12.0	Hg	-1.60	NCE	34
Vuzine	B-R Buffer, pH 12.0	Hg	-1.60	NCE	34
Optochine	B-R buffer, pH 12.0	Hg	-1.60	NCE	34
Eucupine	B-R buffer, pH 12.0	Hg	-1.60	NCE	34
Cinchotoxine	B-R buffer, pH 1.8	Hg	-0.43	NCE	34
	B-R buffer, pH 12.0	Hg	-1.00	NCE	34
Vuzinotoxine	B-R buffer, pH 1.8	Hg	-0.43	NCE	34
	B-R buffer, pH 12.0	Hg	-1.00	NCE	34

(continued)

Compound	Solvent System	Working Electrode	$E_{1/2}$	Reference Electrode	Reference
Narceine	OAc buffer, pH 5.0	Hg	-1.36	NCE	35
	Phosphate buffer, pH 7.0	Hg	-1.45	NCE	35
	Borate buffer, pH 8-10	Hg	-1.54	NCE	35
Hydrastinine	OAc buffer, pH 5.0	Hg	-1.14	NCE	35
	Phosphate buffer, pH 7.0	Hg	-1.15	NCE	35
	Borate buffer, pH 10.0	Hg	-0.89, -1.10	NCE	35
Cotarnine	OAc buffer, pH 5.0	Hg	-1.15	NCE	35
	Phosphate buffer, pH 7.0	Hg	-1.18	NCE	36
Berberine	OAc buffer, pH 5.0	Hg	-1.05, -1.21	NCE	36
	Phosphate buffer, pH 7.0	Hg	-1.03, -1.25, -1.50	NCE	36
Codeinone	B-R buffer, pH 7.5	Hg	-0.98, -1.62	NCE	37
Hydroxycodeinone	B-R buffer, pH 7.5	Hg	-0.94, -1.34	NCE	37
Acetoxycodeinone	B-R buffer, pH 7.5	Hg	-0.94, -1.34	NCE	37
Pseudocodeinone	B-R buffer, pH 7.5	Hg	-1.05, -1.64	NCE	37
Thebainone	B-R buffer, pH 7.5	Hg	-1.27	NCE	37
Metathebainone	B-R buffer, pH 7.5	Hg	-1.29	NCE	37
Hydroxythebainone	B-R buffer, pH 7.5	Hg	-1.34	NCE	37
Dihydromorphinone	B-R buffer, pH 7.5	Hg	-1.49	NCE	37
Dihydrocodeinone	B-R buffer, pH 7.5	Hg	-1.41	NCE	37
Dihydroxycodeinone	B-R buffer, pH 7.5	Hg	-1.42	NCE	37
Trigonelline	B-R buffer, pH 7.5	Hg	-1.43	NCE	38
Anabasine	0.1M Na$_2$HPO$_4$, 0.1M KCl	Hg	-1.74	Hg pool	39
Piperine	B-R buffer, pH 2.0	Hg	-1.33	NCE	40

Compound	Medium	Electrode	E	Ref. electrode	Page
	B-R buffer, pH 5.0	Hg	-1.56, -1.77	NCE	40
Lobeline	B-R buffer, pH 10.0	Hg	-1.75	NCE	40
	B-R buffer, pH 1.8	Hg	-1.12, -1.16	NCE	41
Lobelanine	B-R buffer, pH 8.0	Hg	-1.35, -1.44	NCE	41
	B-R buffer, pH 1.8	Hg	-1.17	NCE	41
Norbelanine	B-R buffer, pH 8.0	Hg	-1.36	NCE	41
	B-R buffer, pH 1.8	Hg	-1.12	NCE	41
	B-R buffer, pH 8.0	Hg	-1.35	NCE	41
Arecaidinbenzhydrazide	0.1M NH4Cl	Hg	-1.20	NCE	42
Arecaidinamide	B-R buffer, pH 1.8	Hg	-1.26	NCE	43
	B-R buffer, pH 6.6	Hg	-1.62	NCE	43
Erythrofleine	B-R buffer, pH 2.5	Hg	-1.3	NCE	44
	B-R buffer, pH 7.5	Hg	-1.5, -1.7	NCE	44
Pelletierine	0.1N HCl	Hg	-0.80, -1.10	NCE	45
	0.1N LiCl	Hg	-1.45	NCE	45
Jervine	McIlvaine buffer, pH 0.2	Hg	-1.31	NCE	46
	McIlvaine buffer, pH 5.5	Hg	-1.53	NCE	46
Veratramine	McIlvaine buffer, pH 2.2	Hg	-1.24	NCE	46
	McIlvaine buffer, pH 5.2	Hg	-1.47	NCE	46
Veratrosine	McIlvaine buffer, pH 2.2	Hg	-1.28	NCE	46
	McIlvaine buffer, pH 4.2	Hg	-1.40	NCE	46
Bilirubin	Aqueous buffer, pH 7.0	Hg	-1.29	sce	47
	Aqueous buffer, pH 11.0	Hg	-1.41	sce	47

(continued)

Compound	Solvent System	Working Electrode	$E_{1/2}$	Reference Electrode	Reference
Haematin (α-chlorhae-min)	Borate buffer, pH 7.9	Hg	-0.41	sce	47
	Borate buffer, pH 12.2	Hg	-0.66	sce	47
Haematoporphyrin	0.1M (Me)$_4$NOH	Hg	-1.16, -1.46	sce	47
Porphyrin C	0.1M (Me)$_4$NOH	Hg	-1.22, -1.52	sce	47
Saccharin	0.05N HCl	Hg	-0.96	sce	47
Parabanic acid	OAc buffer, pH 4.0	Hg	-0.75	sce	48
	Aq buffer, pH 0-7	Hg	-0.926, -0.058 pH	Hg, Hg$_2$SO$_4$	49
Methyl parabanic acid	Aq buffer, pH 0-6	Hg	-0.902, -0.061 pH	Hg, Hg$_2$SO$_4$	49
Dimethyl parabanic acid	Aq buffer, pH 0-6.6	Hg	-0.937, -0.043 pH	Hg, Hg$_2$SO$_4$	49

Polarography of single - stranded homopolynucleotides
Polarography of nucleic acids
See also Heterocycles (Section 30)

a Differential pulse polarography.
b Not specified.
c $E_{1/2}$ versus calomel containing saturated LiCl solution.
d Anodic wave.

References for Biologically Important Compounds

1. D. L. Smith and P. J. Elving, Anal. Chem., 34, 930 (1962).

2. G. Dryhurst, Anal. Chim. Acta, 47, 275 (1969).

2A. D. Cohen, M. Koenigsbuch, and M. Sprecher, Israel J. Chem., 6, 615 (1968).

3. B. H. Hansen and G. Dryhurst, J. Electroanal. Chem., 30, 417 (1971).

4. P. Kabasakalian and J. McGlotten, Anal. Chem., 31, 431 (1959).

5. K. Tsuji and P. J. Elving, Anal. Chem., 41, 286 (1969).

6. J. E. O'Reilly and P. J. Elving, J. Am. Chem. Soc., 93, 1871 (1971).

7. B. Janik and P. J. Elving, Chem. Rev., 68, 295 (1968).

8. A. L. Woodson and D. E. Smith, Anal. Chem., 42, 242 (1970).

9. P. Zuman, Substituent Effects of Organic Polarography, Plenum Press, New York, 1967, p. 315.

0. M. E. Caplis, Electroreduction of the Tetracycline Antibiotics, Ph.D. Thesis, Purdue University, 1970.

A. H. Siegerman, unpublished results.

1. L. Schlitt, M. Rink, and M. Von Stackelberg, J. Electroanal. Chem., 13, 10 (1967).

2. L. Meites, Polarographic Techniques, 2nd ed., Interscience, New York, 1965, p. 687.

3. Reference 12, p. 688.

4. A. R. McIntyre and R. E. King, Federation Proc., 1, 160 (1942).

5. J. Heyrovsky and J. Kuta, Principles of Polarography, Academic, New York, 1966, p. 552.

6. T. R. Koch and W. C. Purdy, Anal. Chim. Acta., 54, 271 (1971).

7. J. J. Lingane and O. L. Davis, J. Biol. Chem., 137, 567 (1941).

8. L. Meites, Polarographic Techniques, 2nd ed., Interscience, New York, 1965, p. 711.

9. R. Brdicka and E. Knobloch, Z. Elektrochem., 47, 721 (1941).

0. R. C. Kaye and K. J. Stonehill, J. Chem. Soc., 1952, 3244.

. S. V. Tatwawadi, K. S. V. Santhanam, and A. J. Bard, J. Electroanal. Chem., 17, 411 (1968).

. R. Brdicka, Chem. Listy, 36, 286 (1942).

. J. Heyrovsky and J. Kuta, Principles of Polarography, Academic, New York, 1966, p. 559.

24. S. L. Tackett and J. W. Ide, J. Electroanal. Chem., 30, 510 (1971).
25. J. B. Duncan and J. E. Christian, J. Am. Pharmac Assoc., 37, 507 (1948).
26. P. Zuman, Collect. Czech. Chem. Commun., 15, 1149 (1950).
27. M. M. Kirk, Anal. Chem., 13, 625 (1941).
28. Z. Vavrin. Collect. Czech. Chem. Commun., 14, 367 (1949).
29. L. I. Smith, I. M. Kolthoff, S. Wawzonek, and P. M. Ruoff, J. Am. Chem. Soc., 63, 1018 (1941).
30. V. D. Parker, J. Am. Chem. Soc., 91, 5380 (1969)
31. M. Shikata and I. Tachi, Mem. Coll. Agr. Kyoto Imp. Univ., No. 4, 35 (1927).
32. K. Wenig and M. Kopecky, Casopis Ceskoslov. Lekarnictva, 56, 49 (1943).
33. J. Bartek, M. Cernoch, and F. Santavy, Chem. Listy, 48, 1123, (1954).
34. F. Santavy, Spisy lekarske fak. Masaryk. Univ., 19, 29 (1945).
35. J. Hobza and F. Santavy, Casopis ceskeho lekarnictva, 62, 86 (1949).
36. H. F. W. Kirkpatrick, Quart. J. Pharm. and Pharm 18, 245, 338 (1945); 19, 8, 127, 526.
37. F. Santavy and M. Cernoch, Chem. Listy, 46, 81 (
38. F. Sorm and Z. Somorva, Chem. Listy, 42, 82 (194
39. R. H. Linnell, J. Am. Chem. Soc., 76, 1391 (1954
40. F. Santavy and O. Hanc, Casopis ceskeho lekarnic 57, 75 (1944).
41. F. Santavy, Casopis ceskeho lekarnictva, 57, 109 (1944).
42. V. Sapara, Chem. Listy, 43, 225 (1949).
43. V. Sapara, Chem. Listy, 45, 454 (1951).
44. M. Talas and F. Santave, Chem. Listy, 45, 457 (
45. O. F. Uffelic, Thesis, University of Groeningen,
46. E. J. Walaszek and A. Pircio, J. Am. Pharm. Assc Sci. Ed., 41, 270 (1952).
47. Reference 15, p. 563.
48. W. A. Struck and P. J. Elving, Anal. Chem., 36, 1374 (1964).
49. G. Dryhurst, B. H. Hansen, and E. B. Harkins, J. Electroanal. Chem., 27, 375 (1970).
50. E. Palecek, J. Electroanal. Chem., 22, 347 (1969
51. E. Palecek in J. N. Davidson and W. E. Cohn (Eds Progress in Nucleic Acid Research and Molecular Biology, Vol. 9, Academic, New York, 1969, p. 3

Oxidation

Metallocenes

Compound	Solvent System	Working Electrode	E1/2	Reference Electrode	Reference
Ferrocene	CH_3CN, 0.2M $LiClO_4$	Pt	+0.307 E1/4	sce	1
	CH_3CN, 0.2M $LiClO_4$	Pt	+0.315 E1/4	sce	2
	CH_3CN, 0.1M $LiClO_4$	Pt	+0.341 E1/4	sce	3
	CH_3CN, 0.1M $NaClO_4$	Pt	+0.310	sce	4
	CH_3CN, 0.2M $NaClO_4$	Pt	+0.063 E1/4	$Ag/AgClO_4$	5
	90% Ethanol, 0.1M $NaClO_4$, 0.01M $HClO_4$	Hg	+0.31	sce	6
Ethylferrocene	CH_3CN, 0.2M $LiClO_4$	Pt	+0.245 E1/4	sce	1
1,1'-Diethylferrocene	CH_3CN, 0.2M $LiClO_4$	Pt	+0.194 E1/4	sce	1
Vinylferrocene	CH_3CN, 0.2M $LiClO_4$	Pt	+0.325 E1/4	sce	1
Ferrocene carboxylic acid	CH_3CN, 0.2M $LiClO_4$	Pt	+0.550 E1/4	sce	1
Ferrocene phenyl ketone	CH_3CN, 0.2M $LiClO_4$	Pt	+0.571 E1/4	sce	1
Acetylferrocene	CH_3CN, 0.2M $LiClO_4$	Pt	+0.573 E1/4	sce	1
1,1'-Dicarbomethoxy-ferrocene	CH_3CN, 0.2M $LiClO_4$	Pt	+0.796 E1/4	sce	1
49 o-, m-, and p-Arylsubstituted ferrocenes	CH_3CN				2

(continued)

Compound	Solvent System	Working Electrode	$E_{1/2}$	Reference Electrode	Reference
Osmocene	CH_3CN, $0.2\underline{M}$ $LiClO_4$	Pt	$+0.633$, $+1.50$ $E_{1/4}$	sce	1
Ruthenocene	CH_3CN, $0.2\underline{M}$ $LiClO_4$	Pt	$+0.693$ $E_{1/4}$	sce	1
	90% Ethanol, $0.1\underline{M}$-$NaClO_4$, $0.01\underline{M}$ $H\underline{C}lO_4$	Hg	$+0.26$	sce	6
Phthalocyanines					
Tetrasulphonated cobalt phthalocyanine tetra-sodium salt	DMSO, $0.1\underline{M}$ $(Et)_4NClO_4$	Pt	$+0.455$, $+1.09$	sce	7
Tetrasulphonated nickel phthalocyanine tetra-sodium salt	DMSO, $0.1\underline{M}$ $(Et)_4NClO_4$	Pt	$+0.98$	sce	7
Tetrasulphonated copper phthalocyanine tetra-sodium salt	DMSO, $0.1\underline{M}$ $(Et)_4NClO_4$	Pt	$+0.872$	sce	7
Tetrasulphonated phthalocyanine tetra-sodium salta	DMSO, $0.1\underline{M}$ $(Et)_4NClO_4$	Pt	$+0.90$	sce	7
Porphins and Porphyrins					
$\alpha,\beta,\gamma,\delta,$-Tetraphenyl porphin	Butyronitrile, $0.1m$ $LiClO_4$	Pt	$+0.97$, $+1.12$	sce	8
Magnesium(II)-Tetra-phenylporphin	Butyronitrile, $0.1m$ $LiClO_4$	Pt	$+0.54$, $+0.86$	sce	8
Barium(II)-Tetraphenyl-	Butyronitrile, $0.1m$ $LiClO_4$	Pt	$+0.46$, $+0.75$	sce	8

Compound	Solvent	Electrode	Potentials	Reference	
Zinc(II)-Tetraphenyl-porphin	Butyronitrile, 0.1m LiClO4	Pt	+0.71, +1.03	sce	8
Cadmium(II)-Tetraphenyl-porphin	Butyronitrile, 0.1m LiClO4	Pt	+0.63, +0.93	sce	8
Lead(II)-Tetraphenyl-porphin	Butyronitrile, 0.1m LiClO4	Pt	+0.63, +0.96	sce	8
Copper(II)-Tetra-phenylporphin	Butyronitrile, 0.1m LiClO4	Pt	+0.90, +1.16	sce	8
Silver(II)-Tetraphenyl-porphin	Butyronitrile, 0.1m LiClO4	Pt	+0.54	sce	8
Nickel(II)-Tetrephenyl-porphin	Butyronitrile, 0.1m LiClO4	Pt	+0.95	sce	8
Cobalt(II)-Tetraphenyl-porphin	Butyronitrile, 0.1m LiClO4	Pt	+0.32, +1.06, +1.26	sce	8
Etioporphyrin I	Butyronitrile, 0.1m LiClO4	Pt	+0.77	sce	8
Magnesium(II)-Etiopor-phyrin I	Butyronitrile, 0.1m LiClO4	Pt	+0.40, +0.77	sce	8
Zinc(II)-Etioporphyrin I	Butyronitrile, 0.1m LiClO4	Pt	+0.51	sce	8
Lead(II)-Etioporphyrin I	Butyronitrile, 0.1m LiClO4	Pt	+0.50, +0.76	sce	8
Silver(II)-Etiopor-phyrin I	Butyronitrile, 0.1m LiClO4	Pt	+0.30, +0.58	sce	8
Platinum(II)-Etiopor-phyrin I	Butyronitrile, 0.1m LiClO4	Pt	+0.75, +1.37	sce	8
Palladium(II)-Etiopor-phyrin I	Butyronitrile, 0.1m LiClO4	Pt	+0.70, +1.40	sce	8
Nickel(II)-Etiopopor-phyrin I	Butyronitrile, 0.1m LiClO4	Pt	+0.70, +1.28	sce	8
Colbalt(II)-Etiopor-phyrin I	Butyronitrile, 0.1m LiClO4	Pt	+0.30, +0.87, +1.18	sce	8

(continued)

Compound	Solvent System	Working Electrode	$E_{1/2}$	Reference Electrode	Reference
Copper(II)-Etioporphyrin	Butyronitrile, 0.1m LiClO$_4$	Pt	+0.63	sce	8
N-Methyletioporphyrin I	Butyronitrile, 0.1m LiClO$_4$	Pt	+1.20	sce	8
Zinc(II)-Etioporphyrin-I	Butyronitrile, 0.1m LiClO$_4$	Pt	+1.09	sce	8
20 Metal Porphyrin esters	Butyronitrile, 0.1m LiClO$_4$				8
Organomercurials					
n-Propylmercuric chloride	CH$_3$CN, 0.1M NaClO$_4$	Pt	+1.28 $E_{p/2}$	Ag/0.01M Ag+	9
Benzylmercuric chloride	CH$_3$CN, 0.1M NaClO$_4$	Pt	+1.53 $E_{p/2}$	Ag/0.01M Ag+	9
Dibenzylmercury	CH$_3$CN, 0.1M NaClO$_4$	Pt	+1.58 $E_{p/2}$	Ag/0.01M Ag+	9
Phenylmercuric tetra-fluoroborate	CH$_3$CN, 0.1M NaClO$_4$	Pt	+1.95 $E_{p/2}$	Ag/0.01M Ag+	9
Diphenylmercury	CH$_3$CN, 0.1M NaClO$_4$	Pt	+1.84 $E_{p/2}$	Ag/0.01M Ag+	9

a Included for comparison.

Reduction

Metallocenes

Compound	Solvent System	Working Electrode	$E_{1/2}$	Reference Electrode	Reference
$[R-C_5H_4 \ Fe \ C_6H_6]^+ X^-$					
$X^-=BF_4, \ PF_6;$ R=H	50% Alcohol	Hg	-1.59	--a	10
$=C_2H_5$	50% Alcohol	Hg	-1.62	--a	10
$=\underline{n}-C_3H_7$	50% Alcohol	Hg	-1.58	--a	10
$=\underline{i}-C_3H_7$	50% Alcohol	Hg	-1.55	--a	10
$=CH_2C_6H_5$	50% Alcohol	Hg	-1.58	--a	10
$[C_5H_5 \ Fe \ C_6H_5R]^+ X^-$					
$X^-=BF_4, \ PF_6;$ R=H	50% Alcohol	Hg	-1.59	--a	10
$=CH_3$	50% Alcohol	Hg	-1.63	--a	10
$=C_2H_5$	50% Alcohol	Hg	-1.57	--a	10
$=\underline{n}-C_3H_7$	50% Alcohol	Hg	-1.54	--a	10
$=\underline{i}-C_3H_7$	50% Alcohol	Hg	-1.56	--a	10
$=\underline{C}(CH_3)_3$	50% Alcohol	Hg	-1.56	--a	10
$=C_6H_5$	50% Alcohol	Hg	-1.44	--a	10
Dicyclopentadienyl titanium(IV) difluoride	DMF, $0.1\underline{N}$ $(Et)_4NClO_4$	Hg	-1.21, -2.17	--a	11
Dicyclopentadienyl titanium(IV) dichloride	DMF, $0.1\underline{N}$ LiCl	Hg	-0.84	--a	11
	50% Methanol, $0.1\underline{N}$ KCl	Hg	-0.78	sce	11
	DMF, $0.1\underline{N}$ $(Et)_4NClO_4$	Hg	-0.63, -1.94	sce	11

(continued)

Compound	Solvent System	Working Electrode	$E_{1/2}$	Reference Electrode	Reference
	50% Methanol, 0.1N KCl	Hg	-0.50, -1.48	sce	11
	DMF, 0.1N LiCl	Hg	-0.63	sce	11
	90% Dioxane, 0.1N-(Et)4NClO4	Hg	-0.52, -1.70	sce	11
	CH3CN, 0.1N (Et)4NClO4	Hg	-0.67, -1.90	sce	11
	DMF, 0.1N (Et)4NClO4	Hg	-0.22, -1.95	sce	11
Dicyclopentadienyl titanium(IV) dibromide	DMF, 0.1N LiCl	Hg	-0.65	sce	11
	50% Methanol, 0.1N KCl	Hg	-0.50, -1.48	sce	11
	DMF, 0.1N (Et)4NClO4	Hg	-0.26, -1.99	sce	11
Dicyclopentadienyl titanium(IV) diiodide	DMF, 0.1N LiCl	Hg	-0.68	sce	11
	50% Methanol, 0.1N KCl	Hg	-0.52, -1.52	sce	11

Phthalocyanine

Compound	Solvent System	Working Electrode	$E_{1/2}$	Reference Electrode	Reference
Tetrasulphonated cobalt phthalocyanine sodium salt	DMSO, 0.1M (Et)4NClO4	Hg	-0.547, -1.346	sce	7
Tetrasulphonated nickel phthalocyanine sodium salt	DMSO, 0.1M (Et)4NClO4	Hg	-0.672, -1.165, -1.933	sce	7
Tetrasulphonated copper phthalocyanine sodium salt	DMSO, 0.1M (Et)4NClO4	Hg	-0.727, -1.111, -1.895	sce	7
Tetrasulphonated phthalocyanine tetrasodium saltb	DMSO, 0.1M (Et)4NClO4	Hg	-0.525, -0.970, -1.810	sce	7

Compound	Solvent/Electrolyte	Metal	E (V)	Ref. electrode	Ref.
α,β,γ,δ,-Tetraphenyl-porphin	DMF, 0.1M (n-Propyl)₄NClO₄	Hg	-1.08, -1.52, -2.38, -2.53	sce	12
Tetraphenylporphin	DMF, 0.1M (Et)₄NClO₄	Hg	-1.08, -1.45, -2.36, -2.48	sce	13
Tetraphenylchlorin	DMF, 0.1M (Et)₄NClO₄	Hg	-1.12, -1.52, -2.43	sce	13
Tetraphenylbacterio-chlorin	DMF, 0.1M (Et)₄NClO₄	Hg	-1.10, -1.55	sce	13
Zinc α,β,γ,δ-tetra-phenylporphin [N(n-Propyl)₄]₂	DMF, 0.1M (n-Propyl)₄NClO₄	Hg	-1.32, -1.73, -2.45, -2.67	sce	12
α,β,γ,δ,-tetraphenylporphin	DMF, 0.1M (n-Propyl)₄NClO₄	Hg	-1.45, -1.87, -2.26, -2.44	sce	12
Etioporphyrin I	DMF, 0.1M (n-Propyl)₄NClO₄	Hg	-1.37, -1.80, -2.67	sce	12
Zinc etioporphyrin I	DMF, 0.1M (n-Propyl)₄NClO₄	Hg	-1.62, -2.00, -2.77	sce	12
Copper etioporphyrin IV	DMF, 0.1M (n-Propyl)₄NClO₄	Hg	-1.48, -1.99, -2.70	sce	12
Zinc tetrabenzoporphin	DMF, 0.1M (n-Propyl)₄NClO₄	Hg	-1.47, -1.84, -2.49, -2.70	sce	12
Magnesium octaphenyl-tetraazaporphin	DMF, 0.1M (n-Propyl)₄NClO₄	Hg	-0.68, -1.11, -1.81, -2.18	sce	12
Deuteroporphyrin IX dimethyl ester	DMF, (Et)₄NClO₄	Hg	-1.29, -1.68, -2.53	sce	14
Mesoporphyrin IX di-methyl ester	DMF, (Et)₄NClO₄	Hg	-1.34, -1.73, -2.57	sce	14

Organomercurials

Compound	Solvent/Electrolyte	Metal	E (V)	Ref. electrode	Ref.
Phenylmercuric nitrate	B-R buffer, pH 1.5	Hg	-0.780	sce	15

(continued)

Compound	Solvent System	Working Electrode	$E_{1/2}$	Reference Electrode	Reference
	B-R buffer, pH 5.5	Hg	-0.984	sce	15
	B-R buffer, pH 10.1	Hg	-0.269, -1.159	sce	15
$\underline{i}$-C_3H_7HgCl	Dimethoxyethane, 0.1$\underline{N}$ $(\underline{n}$-Bu$)_4$NI	Hg	-3.27^c	Ag/0.001$\underline{N}$ Ag+	16
C_2H_5HgCl	Dimethoxyethane, 0.1$\underline{N}$ $(\underline{n}$-Bu$)_4$NI	Hg	-3.25^c	Ag/0.001$\underline{N}$ Ag+	16
CH_3HgCl	Dimethoxyethane, 0.1$\underline{N}$ $(\underline{n}$-Bu$)_4$NI	Hg	-3.10^c	Ag/0.001$\underline{N}$ Ag+	16
$C_6H_5CH_2CH_2HgCl$	Dimethoxyethane, 0.1$\underline{N}$ $(\underline{n}$-Bu$)_4$NI	Hg	-3.04^c	Ag/0.001$\underline{N}$ Ag+	16
$CH_2=CHHgCl$	Dimethoxyethane, 0.1$\underline{N}$ $(\underline{n}$-Bu$)_4$NI	Hg	-3.01^c	Ag/0.001$\underline{N}$ Ag+	16
C_6H_5HgCl	Dimethoxyethane, 0.1$\underline{N}$ $(\underline{n}$-Bu$)_4$NI	Hg	-2.94^c	Ag/0.001$\underline{N}$ Ag+	16
C_6Cl_5HgCl	Dimethoxyethane, 0.1$\underline{N}$ $(\underline{n}$-Bu$)_4$NI	Hg	-2.92^c	Ag/0.001$\underline{N}$ Ag+	16
CCl_3HgCl	Dimethoxyethane, 0.1$\underline{N}$ $(\underline{n}$-Bu$)_4$NI	Hg	-2.36^c	Ag/0.001$\underline{N}$ Ag+	16
$C_6H_5CH_2HgCl$	Dimethoxyethane, 0.1$\underline{N}$ $(\underline{n}$-Bu$)_4$NI	Hg	-2.1^c	Ag/0.001$\underline{N}$ Ag+	16
$OHCCH_2HgCl$	Dimethoxyethane, 0.1$\underline{N}$ $(\underline{n}$-Bu$)_4$NI	Hg	-2.08^c	Ag/0.001$\underline{N}$ Ag+	16
$CH_3OOCHgCl$	Dimethoxyethane, 0.1$\underline{N}$ $(\underline{n}$-Bu$)_4$NI	Hg	-2.08^c	Ag/0.001$\underline{N}$ Ag+	16
$ClCH=CHHgCl$	Dimethoxyethane, 0.1$\underline{N}$ $(\underline{n}$-Bu$)_4$NI	Hg	-1.92^c	Ag/0.001$\underline{N}$ Ag+	16
CH_3COCH_2HgBr	50% CH_3CN, 0.1$\underline{N}$ KNO_3	Hg	-0.122, -0.255	Ag/0.001$\underline{N}$ Ag+	16

Compound	Solvent, electrolyte	Electrode	Potentials (V)	Ref. electrode	Ref.
...COCH$_2$HgBr	50% CH$_3$CN, 0.1N KNO$_3$	Hg	-0.070, -0.265	Ag/0.001N Ag+	17
t-C$_4$H$_9$COCH$_2$HgBr	50% CH$_3$CN, 0.1N KNO$_3$	Hg	-0.123, -0.285	Ag/0.001N Ag+	17
C$_6$H$_5$COCH$_2$HgBr	50% CH$_3$CN, 0.1N KNO$_3$	Hg	-0.125, -0.240	Ag/0.001N Ag+	17
C$_6$H$_5$CH$_2$HgI	CH$_3$CN, 0.2N (n-Bu)$_4$NI	Hg	-0.31, -1	Ag/0.001N Ag+	1
C$_2$H$_5$HgCl	50% Ethanol, 0.1N KNO$_3$	Hg	-0.85, -1.37	Ag/0.001N Ag+	1
C$_2$H$_5$HgBr	50% Ethanol, 0.1N KNO$_3$	Hg	-0.875, -1.31	Ag/0.001N Ag+	1
C$_2$H$_5$HgI	50% Ethanol, 0.1N KNO$_3$	Hg	-0.905, -1.34	Ag/0.001N Ag+	1
C$_2$H$_5$HgSCN	50% Ethanol, 0.1N KNO$_3$	Hg	-0.46, -1.41	sce	19
XC$_6$H$_4$HgBr X=H	50% Methanol, 0.05M- B-R buffer, 0.01N KCl, pH 7.00	Hg	-0.125, -0.995	sce	17
=p-Me	50% Methanol, 0.05M- B-R buffer, 0.01M KCl, pH 7.00	Hg	-0.150, -0.965	sce	17
=p-Br	50% Methanol, 0.05M- B-R buffer, 0.01M KCl, pH 7.00	Hg	-0.155, -0.919	sce	17
=p-I	50% Methanol, 0.05M- B-R buffer, 0.01M KCl, pH 7.00	Hg	-0.130, -0.870	sce	17
=p-COOEt	50% Methanol, 0.05M- B-R buffer, 0.01M KCl, pH 7.00	Hg	-0.140, -0.865	sce	17

(continued)

Compound	Solvent System	Working Electrode	$E_{1/2}$	Reference Electrode	Reference
R_2Hg $R=CH_3$	DMF, $0.1\underline{N}$ (Hexyl)$_4$NClO$_4$	Hg	-2.880	sce	20
$=CCl_3$	Dimethoxyethane, $0.1M$ (n-Bu)$_4$NClO$_4$	Hg	-1.43	Ag/0.001$\underline{N}$-Ag$^+$	16
$=C_2H_5$	DMF, $0.1\underline{N}$ (Hexyl)$_4$NClO$_4$	Hg	-2.859	Ag/0.001$\underline{N}$-Ag$^+$	20
$=CH=CH_2$	Dimethoxyethane, $0.1\underline{N}$ (n-Bu)$_4$NClO$_4$	Hg	-3.34	Ag/0.001$\underline{N}$-Ag$^+$	16
	60% DMF, $0.1\underline{N}$ (Et)$_4$NClO$_4$	Hg	-2.301	sce	20
$=CH=CHCl$	60% DMF, $0.1\underline{N}$ (Et)$_4$NClO$_4$	Hg	-1.681	sce	20
$=CCl=CCl_2$	60% DMF, $0.1\underline{N}$ (Et)$_4$NClO$_4$	Hg	-0.538	sce	20
	50% Methanol, $0.1\underline{N}$-LiClO$_4$	Hg	-0.63	sce	21
$=n-C_3H_7$	DMF, $0.1\underline{N}$ (Hexyl)$_4$NClO$_4$	Hg	-2.976	sce	20
$=\underline{i}-C_3H_7$	DMF, $0.1\underline{N}$ (Hexyl)$_4$NClO$_4$	Hg	-2.861	sce	20
$=CH_2CH=CH_2$	Aq sol, pH 7.85	Hg	-1.02	sce	22
$=\underline{n}-C_4H_9$	DMF, $0.1\underline{N}$ (Hexyl)$_4$NClO$_4$	Hg	-2.970	sce	20
$=C\equiv CC_3H_7$	Dimethoxyethane, $0.1\underline{N}$ (n-Bu)$_4$NClO$_4$	Hg	-2.90	Ag/0.001$\underline{N}$-Ag$^+$	16
$=Cyclo-C_5H_5$	60% DMF, $0.1\underline{N}$ (Et)$_4$NClO$_4$	Hg	-0.862	sce	20
	DMF, $0.1\underline{N}$ (Hexyl)...	Hg	-2.486	sce	20

R group	Solvent, electrolyte	Electrode	Potential	Reference electrode	Ref.
	60% DMF, 0.1N (Et)4- NClO4	Hg	-2.199	sce	[20]
	DMF, 0.1N (Et)4NClO4	Hg	-2.21	Hg, Hg2I2/ 0.1N (Et)4- NI	[21]
=C6F5	Dimethoxyethane, 0.1N (n-Bu)4NClO4	Hg	-3.32	Ag/0.001N- Ag+	[16]
	Dimethoxyethane, 0.1N (n-Bu)4NClO4	Hg	-1.81	Ag/0.001N- Ag+	[20]
	60% DMF, 0.1N (Et)4- NClO4	Hg	-0.862	sce	[20]
=C6Cl5	60% DMF, 0.1N (Et)4- NClO4	Hg	-1.687	sce	[20]
=CH2C6H5	Dimethoxyethane, 0.1N (n-Bu)4NClO4	Hg	-2.63	Ag/0.001N- Ag+	[16]
	60% DMF, 0.1N (Et)4- NClO4	Hg	-1.719	sce	[20]
=C≡CC6H5	DMF, 0.1N (Hexyl)4- NClO4	Hg	-2.053	sce	[20]
	60% DMF, 0.1N (Et)4- NClO4	Hg	-1.073	sce	[20]
=CH2C6H4X X=o-F	Dimethoxyethane, 0.1N (n-Bu)4NClO4	Hg	-2.25	Ag/0.001N- Ag+	[16]
	CH3CN, 0.1N (Et)4NF	Hg	-1.26	Hg pool	[20]
=m-F	CH3CN, 0.1N (Et)4NF	Hg	-1.14	Hg pool	[20]
=p-F	CH3CN, 0.1N (Et)4NF	Hg	-1.39	Hg pool	[20]
R2Hg R=CH2C6H4X X=p-Cl	CH3CN, 0.1N (Et)4NF	Hg	-1.14	Hg pool	[21]
=m-Br	CH3CN, 0.1N (Et)4NF	Hg	-1.03	Hg pool	[21]
=p-Br	CH3CN, 0.1N (Et)4NF	Hg	-1.09	Hg pool	[21]

(continued)

1023

Compound	Solvent System	Working Electrode	$E_{1/2}$	Reference Electrode	Reference
$=C\equiv CC_6H_4X$					
X=p-Br, =H	CH$_3$CN, 0.1$\underline{N}$ (Et)$_4$NF	Hg	-1.39	Hg pool	21
=p-CH$_3$	CH$_3$CN, 0.1$\underline{N}$ (Et)$_4$NF	Hg	-1.54	Hg pool	21
X=p-Br	40% Dioxane, 0.1$\underline{N}$-(n-Bu)$_4$NF	Hg	-1.24	sce	21
=p-Cl	40% Dioxane, 0.1$\underline{N}$-(n-Bu)$_4$NF	Hg	-1.21	sce	21
=H	40% Dioxane, 0.1$\underline{N}$-(n-Bu)$_4$NF	Hg	-1.31	sce	21
=p-CH$_3$	40% Dioxane, 0.1$\underline{N}$-(n-Bu)$_4$NF	Hg	-1.39	sce	21
=p-OCH$_3$	40% Dioxane, 0.1$\underline{N}$-(n-Bu)$_4$NF	Hg	-1.31	sce	21
=CN	OAc buffer, pH>7	Hg	-0.18	sce	23
	60% DMF, 0.1$\underline{N}$ (Et)$_4$-NClO$_4$	Hg	-0.291	sce	20
=COOCH$_3$	50% Methanol, 0.1$\underline{N}$-LiClO$_4$	Hg	-1.31	sce	20
=$\underline{t}$-C$_4$H$_9$COCH$_2$	60% DMF, 0.1$\underline{N}$ (Et)$_4$-NClO$_4$	Hg	-0.44	sce	20
=CH$_3$OOCCH$_2$	DMF, 0.1$\underline{N}$ (Hexyl)$_4$-NClO$_4$	Hg	-1.471	sce	20
RHg$^+$ Cation					
R=CH$_3$	60% DMF, 0.1$\underline{N}$ (Et)$_4$-NClO$_4$	Hg	-0.288	sce	20
	DMF, 0.1$\underline{N}$ (Et)$_4$NClO$_4$	Hg	-0.612	sce	20
=C$_2$H$_5$	60% DMF, 0.1$\underline{N}$ (Et)$_4$-NClO$_4$	Hg	-0.217	sce	20
	DMF, 0.1$\underline{N}$ (Et)$_4$NClO$_4$	Hg	-0.594	sce	20

Compound	Medium	Electrode	E	Ref. electrode	Temp. (°C)
$=C_3H_7$ (continued)	$60\% \text{ DMF, } 0.1N \text{ } (Et)_4NClO_4$	Hg	-0.198	sce	20
$=i\text{-}C_3H_7$	$\text{DMF, } 0.1N \text{ } (Et)_4NClO_4$	Hg	-0.613	sce	20
	$60\% \text{ DMF, } 0.1N \text{ } (Et)_4NClO_4$	Hg	-0.183	sce	20
$=C_4H_9$	$\text{DMF, } 0.1N \text{ } (Et)_4NClO_4$	Hg	-0.592	sce	20
	$60\% \text{ DMF, } 0.1N \text{ } (Et)_4NClO_4$	Hg	-0.193	sce	20
$=C_5H_{11}$	$\text{DMF, } 0.1N \text{ } (Et)_4NClO_4$	Hg	-0.613	sce	20
	$60\% \text{ DMF, } 0.1N \text{ } (Et)_4NClO_4$	Hg	-0.174	sce	20
$=C_7H_{15}$	$\text{DMF, } 0.1N \text{ } (Et)_4NClO_4$	Hg	-0.535	sce	20
	$60\% \text{ DMF, } 0.1N \text{ } (Et)_4NClO_4$	Hg	-0.130	sce	20
$=C_8H_{17}$	$\text{DMF, } 0.1N \text{ } (Et)_4NClO_4$	Hg	-0.578	sce	20
	$60\% \text{ DMF, } 0.1N \text{ } (Et)_4NClO_4$	Hg	-0.118	sce	20

RHg^+ Cation

Compound	Medium	Electrode	E	Ref. electrode	Temp. (°C)
$R=cyclo\text{-}C_5H_{11}$	$\text{DMF, } 0.1N \text{ } (Et)_4NClO_4$	Hg	-0.596	sce	20
	$60\% \text{ DMF, } 0.1N \text{ } (Et)_4NClO_4$	Hg	-0.178	sce	20
$=cyclo\text{-}C_6H_{13}$	$\text{DMF, } 0.1N \text{ } (Et)_4NClO_4$	Hg	-0.573	sce	20
	$60\% \text{ DMF, } 0.1N \text{ } (Et)_4NClO_4$	Hg	-0.103	sce	20
$=C_6H_5$	$\text{DMF, } 0.1N \text{ } (Et)_4NClO_4$	Hg	-0.570	sce	20
	$60\% \text{ DMF, } 0.1N \text{ } (Et)_4NClO_4$	Hg	-0.180	sce	20
$=C_6H_5CH_2$	$\text{DMF, } 0.1N \text{ } (Et)_4NClO_4$	Hg	-0.544	sce	20
	$60\% \text{ DMF, } 0.1N \text{ } (Et)_4NClO_4$	Hg	$+0.023$	sce	20
	$\text{DMF, } 0.1N \text{ } (Et)_4NClO_4$	Hg	-0.370	sce	20

(continued)

Compound	Solvent System	Working Electrode	$E_{1/2}$	Reference Electrode	Reference
$=C_6H_5CH_2CH_2$	60% DMF, 0.1N (Et)$_4$NClO$_4$	Hg	-0.233	sce	20
$=C_6Cl_5$	DMF, 0.1N (Et)$_4$NClO$_4$	Hg	-0.585	sce	20
	60% DMF, 0.1N (Et)$_4$NClO$_4$	Hg	+0.027	sce	20
$=CH_2=CH$	60% DMF, 0.1N (Et)$_4$NClO$_4$	Hg	-0.090	sce	20
$=CH_2=CH-CH_2$	DMF, 0.1N (Et)$_4$NClO$_4$	Hg	-0.862	sce	20
	60% DMF, 0.1N (Et)$_4$NClO$_4$	Hg	-0.010	sce	20
$=CH_3OCO-$	DMF, 0.1N (Et)$_4$NClO$_4$	Hg	-0.343	sce	20
	60% DMF, 0.1N (Et)$_4$NClO$_4$	Hg	-0.100	sce	20
	DMF, 0.1N (Et)$_4$NClO$_4$	Hg	-0.426	sce	20
$XC_6H_4CH_2HgCl$ X=o-F	50% Methanol, 0.05M- B-R buffer, 0.1N KCl, pH 7.00	Hg	-0.175, -0.906	sce	17
=m-F	50% Methanol, 0.05M- B-R buffer, 0.1N KCl, pH 7.00	Hg	-0.145, -0.860	sce	17
=o-Cl	50% Methanol, 0.05M- B-R buffer, 0.1N KCl, pH 7.00	Hg	-0.155, -0.935	sce	17
=m-Cl	50% Methanol, 0.05M- B-R buffer, 0.1N KCl, pH 7.00	Hg	-0.165, -0.850	sce	17
=p-Cl	50% Methanol, 0.05M- B-R buffer, 0.1N KCl,	Hg	-0.215, -0.920	sce	17

Compound	Solvent / Electrolyte	Electrode	E	Ref. electrode	Ref.
=o-Br	50% Methanol, 0.05M-$\underline{K}$Cl, B-R buffer, 0.1N $\underline{K}$Cl, pH 7.00	Hg	-0.220, -0.195	sce	17
=m-Br	50% Methanol, 0.05M-, B-R buffer, 0.1N $\underline{K}$Cl, pH 7.00	Hg	-0.208, -0.825	sce	17
=o-Me	50% Methanol, 0.05M-, B-R buffer, 0.1N $\underline{K}$Cl, pH 7.00	Hg	-0.150, -1.045	sce	17
=p-Me	50% Methanol, 0.05M-, B-R buffer, 0.1N $\underline{K}$Cl, pH 7.00	Hg	-0.165, -1.100	sce	17
=H	50% Methanol, 0.05M-, B-R buffer, 0.1N $\underline{K}$Cl, pH 7.00	Hg	-0.115, -1.035	sce	17

Magnesium Compounds

Compound	Solvent / Electrolyte	Electrode	E	Ref. electrode	Ref.
Methylmagnesium	Dimethoxyethane, 0.1M (n-Bu)4NClO4	Hg	-2.49	Ag/0.001N-Ag+	24
Ethylmagnesium bromide	Dimethoxyethane, 0.1M (n-Bu)4NClO4	Hg	-2.44, -2.70	Ag/0.001N-Ag+	24
i-Propylmagnesium bromide	Dimethoxyethane, 0.1M (n-Bu)4NClO4	Hg	-2.44, -2.75	Ag/0.001N-Ag+	24
i-Butylmagnesium bromide	Dimethoxyethane, 0.1M (n-Bu)4NClO4	Hg	-2.46, -2.75	Ag/0.001N-Ag+	24
Diallylmagnesium	Dimethoxyethane, 0.1M (n-Bu)4NClO4	Hg	-2.65	Ag/0.001N-Ag+	24
Dicyclopentadienyl bromide	Dimethoxyethane, 0.1M (n-Bu)4NClO4	Hg	-2.50	Ag/0.001N-Ag+	24
Phenylmagnesium bromide	Dimethoxyethane, 0.1M (n-Bu)4NClO4	Hg	-2.46, -2.80	Ag/0.001N-Ag+	24

(continued)

1027

Compound	Solvent System	Working Electrode	$E_{1/2}$	Reference Electrode	Reference
Dibenzylmagnesium bromide	Dimethoxyethane, 0.1M (n-Bu)4NClO4	Hg	-2.74	Ag/0.001N Ag+	24
Magnesium bromide[b]	Dimethoxyethane, 0.1M (n-Bu)4NClO4	Hg	-2.47	Ag/0.001N Ag+	24
Magnesium perchlorate[b]	Dimethoxyethanem 0.1M (n-Bu)4NClO4	Hg	-2.30	Ag/0.001N Ag+	24
Organotin Compounds					
RSnCl3					
R=Methyl	Methanol, 0.5M LiClO4	Hg	-0.41, -0.56 Ep	Ag/0.001N Ag+	25
=Ethyl	Methanol, 0.5M LiClO4	Hg	-0.50 Ep	Ag/0.001N Ag+	25
=Propyl	Methanol, 0.5M LiClO4	Hg	-0.50 Ep	Ag/0.001N Ag+	25
=Butyl	Methanol, 0.5M LiClO4	Hg	-0.50 Ep	Ag/0.001N Ag+	25
=Octyl	Methanol, 0.5M LiClO4	Hg	-0.43 Ep	Ag/0.001N Ag+	25
=Phenyl	Methanol, 0.5M LiClO4	Hg	-0.27 Ep	sce	25
R2SnCl2					
R=Methyl	Methanol, 0.5M LiClO4	Hg	-0.72 Ep	sce	25
=Ethyl	Methanol, 0.5M LiClO4	Hg	-0.67 Ep	sce	25
=Propyl	Methanol, 0.5M LiClO4	Hg	-0.65 Ep	sce	25
=Butyl	Methanol, 0.5M LiClO4	Hg	-0.63 Ep	sce	25
=Octyl	Methanol, 0.5M LiClO4	Hg	-0.61 Ep	sce	25

Compound	Solvent/Electrolyte	Electrode	Potential	Ref. electrode	Ref.
R=Phenyl	Methanol, 0.5M LiClO$_4$	Hg	-0.52 Ep	sce	25
R$_3$SnCl					
R=Methyl	Methanol, 0.5M LiClO$_4$	Hg	-1.08	sce	25
=Ethyl	Methanol, 0.5M LiClO$_4$	Hg	-0.98	sce	25
=Propyl	Methanol, 0.5M LiClO$_4$	Hg	-0.94	sce	25
=Butyl	Methanol, 0.5M LiClO$_4$	Hg	-0.92	sce	25
=Phenyl	Methanol, 0.5M LiClO$_4$	Hg	-0.61	sce	25
Triethyl(methacroyloxy)-tin	1N NaOH	Hg	-1.50, -1.74	sce	26
Tributyl(methacroyloxy)-tin	1N NaOH	Hg	-1.41, -1.78	sce	26
Oxybis(triethyltin)	1N NaOH	Hg	-1.49, -1.76	sce	26
Oxybis(tributyltin)	1N NaOH	Hg	-1.40, -1.78	sce	26
Triethylhydroxytin	1N NaOH	Hg	-1.49, -1.76	sce	26
Organothallium Compounds					
Diphenylthallium cation	Phosphate buffer, pH 2.5	Hg	-0.62, -0.85, -1.24	sce	27
	Phosphate buffer, pH 5.1	Hg	-0.62, -0.92, -1.33	sce	27
	Phosphate buffer, pH 8.0	Hg	-0.60, -0.94, -1.43	sce	27

Mann (28) and Headridge (29) have compiled extensive tables of voltammetric data for organometallics

a Not specified.
b Included for comparsion.
c Second wave.

(continued)

References of Organometallics

1. T. Kuwana, D. E. Bublitz, and G. Hoh, J. Am.
    Chem. Soc., 82, 5811 (1960).
2. W. F. Little, C. N. Reilley, J. D. Johnson,
    K. N. Lynn, and A. P. Sanders, J. Am. Chem. Soc.,
    86, 1376 (1964).
3. G. L. K. Hoh, W. E. McEwen, and J. Kleinberg,
    J. Am. Chem. Soc., 83, 3949 (1961).
4. H. Henning and O. Gurtler, J. Organometal. Chem.,
    11, 307 (1968).
5. D. W. Hall and C. D. Russell, J. Am. Chem. Soc.,
    89, 2316 (1967).
6. J. A. Page and G. Wilkinson, J. Am. Chem. Soc.,
    74, 6149 (1952).
7. L. D. Rollmann and R. T. Iwamoto, J. Am. Chem.
    Soc., 90, 1455 (1968).
8. A. Stanienda and G. Biebl, Z. Phys. Chem. N. F.,
    52, 254 (1967).
9. M. Fleischmann, D. Pletcher, and G. Sundholm,
    J. Electroanal. Chem., 31, 51 (1971).
10. D. Astruc, R. Dabard, and E. Laviron, C. R. Acad.
    Sci. Paris, 269, 608 (1969).
11. S. P. Gubin and S. A. Smirnova, J. Organometal.
    Chem., 20, 229 (1969).
12. D. W. Clack and N. S. Hush, J. Am. Chem. Soc., 87,
    4238 (1965).
13. G. Peychal-Heiling and G. S. Wilson, Anal. Chem.,
    43, 550 (1971).
14. G. Peychal-Heiling and G. S. Wilson, Anal. Chem.,
    43, 545 (1971).
15. W. L. Wuggatzer and J. M. Cross, J. Am. Pharm.
    Assoc., 41, 80 (1952).
16. R. E. Dessy, W. Kitching, T. Psarras, R. Salinger
    A. Chen, and T. Chivers, J. Am. Chem. Soc., 88,
    460 (1966).
17. I. P. Beletskaya, K. P. Butin, and O. A. Rentov,
    Zh. Org. Khim., 3, 231 (1967).
18. S. Wawzonek, R. C. Duty, and J. H. Wagenknecht,
    J. Electrochem. Soc., 111, 74 (1964).
19. R. Barbieri and J. Bjerrum, Acta Chim. Scand.,
    19, 469 (1965).
20. O. A. Reutov and I. P. Beletskaya, Reaction
    Mechanisms of Organometallic Compounds, translate
    from the Russian by A. M. A. Mincer, North
    Holland, Amsterdam, Holland, 1968.
21. K. P. Butin, I. P. Beletskaya, and O. A. Reutov,
    Elektrokhim., 2, 635 (1966).

22. A. Koirrmann and M. Kleine-Peter, _Bull. Soc. Chim. Fr._, 894 (1957).
23. L. Newman, J. O. Carbal, and D. N. Hume, _J. Am. Chem. Soc._, 80, 1814 (1958).
24. T. Psarras and R. E. Dessy, _J. Am. Chem. Soc._, 88, 5132 (1966).
25. H. Mehner, H. Jehring, and H. Kriegsmann, _J. Organometal. Chem._, 15, 97 (1968).
26. D. A. Kochkin, T. L. Shkorbatova, L. D. Pegusova, and N. A. Voronkov, _Zh. Obshch. Khim._, 39, 1777 (1969).
27. J. S. Digregorio and M. D. Morris, _Anal. Chem._, 40, 1286 (1968).
28. C. K. Mann and K. K. Barnes, Electrochemical Reactions in Nonaqueous Systems, Marcel Dekker, New York, 1970.
29. J. B. Headridge, Electrochemical Techniques for Inorganic Chemists, Academic, New York, 1969.

30. HETEROCYCLES

Oxidation

Sulfur Heterocycles

Compound	Solvent System	Working Electrode	$E_{1/2}$	Reference Electrode	Reference
Thiazole	CH_3CN, $LiClO_4$	Pt	+1.980	Ag/0.01N Ag+	1
2-Phenylthiazole	CH_3CN, $LiClO_4$	Pt	+1.475	Ag/0.01N Ag+	1
4-Phenylthiazole	CH_3CN, $LiClO_4$	Pt	+1.380	Ag/0.01N Ag+	1
5-Phenylthiazole	CH_3CN, $LiClO_4$	Pt	+1.410	Ag/0.01N Ag+	1
2,4-Diphenylthiazole	CH_3CN, $LiClO_4$	Pt	+1.115	Ag/0.01N Ag+	1
2,5-Diphenylthiazole	CH_3CN, $LiClO_4$	Pt	+1.185	Ag/0.01N Ag+	1
4,5-Diphenylthiazole	CH_3CN, $LiClO_4$	Pt	+1.365	Ag/0.01N Ag+	1
2,4,5-Triphenylthiazole	CH_3CN, $LiClO_4$	Pt	+1.230	Ag/0.01N Ag+	1
Phenyl-2-biphenylyl-4-thiazole	CH_3CN, $LiClO_4$	Pt	+1.030	Ag/0.01N Ag+	1
Phenyl-4-biphenylyl-2-thiazole	CH_3CN, $LiClO_4$	Pt	+1.070	Ag/0.01N Ag+	1
Bis-(biphenylyl)-2,4-thiazole	CH_3CN, $LiClO_4$	Pt	Near 1	Ag/0.01N Ag+	1
Phenothiazone	CH_3CN, Et_4NClO_4	Pt	+0.56, +1.00	sce	2
	CH_3CN, $0.1M$ $LiClO_4$	Pt	+0.27, +0.77	Ag/0.01N Ag+	3,4
	$10N$ H_2SO_4, 30% Ethanol	Hg	-0.24	Hg, Hg_2SO_4	5
	CH_3CN, $0.1M$ $LiClO_4$	Pt	+0.270	Ag/0.01N Ag+	6
N-Methylphenothiazine	$0.1M$ $HClO_4$	Pt	+0.40, +0.97	Ag/0.01N Ag+	4

Compound	Electrolyte	Electrode	Potential	Reference	Ref.
Chlorpromazine hydro- chloride	1N H$_2$SO$_4$	Pt	+0.6	sce	7
Thiophene	9N H$_2$SO$_4$	Pt	+0.37, +0.95	sce	7,8
	CH$_3$CN, 0.1M NaClO$_4$	Pt	+1.84, Ep	Ag/0.1N Ag+	6
	CH$_3$CN, 0.1N NaClO$_4$	Pt	+1.60	Ag/0.1N Ag+	9
	HOAc, NaOAc	Pt	+1.91	sce	10
Tetrahydrothiophene	CH$_3$CN, 0.1M NaClO$_4$	Pt	+1.45, Ep	Ag/0.1N Ag+	6
Dibenzylthiophene	CH$_3$CN, 0.1M NaClO$_4$	Pt	+1.35, +1.64 Ep	Ag/0.1N Ag+	6
Phenoxathiin	CH$_3$CN, Perchlorate salt	Pt	+0.825, +1.32	Ag/0.01N Ag+	11
	CH$_3$CN, Perchlorate salt	Pt	+0.865, +1.19	Ag/0.01N Ag+	11
P-Dithiane	CH$_3$CN, 0.1M NaClO$_4$	Pt	+1.46, +1.91 Ep	Ag/0.01N Ag+	6
sym-Trithiane	CH$_3$CN, 0.1M NaClO$_4$	Pt	+1.30 Ep	Ag/0.01N Ag+	6
Nitrogen Heterocycles					
Pyridine	CH$_3$CN, NaClO$_4$	Pt	+1.82	Ag/0.1N Ag+	9
2-Vinylpyridine	CH$_3$CN, (Et)$_4$NClO$_4$	Pt	+2.72 to +2.80	Ag, AgCl	12
4-Vinylpyridine	CH$_3$CN, (Et)$_4$NClO$_4$	Pt	No wave	Ag, AgCl	12
Pyrrole	CH$_3$CN, NaClO$_4$	Pt	+0.76	Ag/0.1N Ag+	9
	CH$_3$CN, LiClO$_4$	Pt	+1.2	sce	13
N-Methylpyrrole	DMF, (n-Bu)$_4$NClO$_4$	Pt	+1.19	sce	14
N-Vinylpyrrolidone	CH$_3$CN, (Et)$_4$NClO$_4$	Pt	+1.67	Ag, AgCl	12
Quinoline	CH$_3$CN, (Et)$_4$NClO$_4$	Pt	+1.97 Ep/2	sce	15
Isoquinoline	CH$_3$CN, (Et)$_4$NClO$_4$	Pt	+1.84 Ep/2	sce	15
3-Methylisoquinoline	CH$_3$CN, (Et)$_4$NClO$_4$	Pt	+1.67 Ep/2	sce	15
Acridine	CH$_3$CN, (Et)$_4$NClO$_4$	Pt	+1.58 Ep/2	sce	15
Quinoxalaine	CH$_3$CN, (Et)$_4$NClO$_4$	Pt	+2.19 Ep/2	sce	15

(continued)

Compound	Solvent System	Working Electrode	$E_{1/2}$	Reference Electrode	Reference
5,6-Benzoquinoline	CH_3CN, $(Et)_4NClO_4$	Pt	+1.69 Ep/2	sce	15
7,8-Benzoquinoline	CH_3CN, $(Et)_4NClO_4$	Pt	+1.72 Ep/2	sce	15
Phenanthridine	CH_3CN, $(Et)_4NClO_4$	Pt	+1.80 Ep/2	sce	15
3,4-Benzacridine	CH_3CN, $(Et)_4NClO_4$	Pt	+1.73 Ep/2	sce	15
Phenazine	CH_3CN, $(Et)_4NClO_4$	Pt	+1.91 Ep/2	sce	15
3,4-Benzocinnoline	CH_3CN, $(Et)_4NClO_4$	Pt	+1.72 Ep/2	sce	15
5,6,7,8-Dibenzoquinoxaline	CH_3CN, $(Et)_4NClO_4$	Pt	+1.85 Ep/2	sce	15
1,2,3,4-Dibenzophenazine	CH_3CN, $(Et)_4NClO_4$	Pt	+1.82 Ep/2	sce	15
5,10-Dihydro-5,10-dimethylphenazine	CH_3CN, $0.1\underline{M}$ $(Et)_4NClO_4$	Pt	+0.11, +0.83 Ep/2	sce	16
5,10-Dihydro-5-methyl-10-phenylphenazine	CH_3CN, $0.1\underline{M}$ $(Et)_4NClO_4$	Pt	+0.13, +0.87 Ep/2	sce	16
5,10-Dihydro-5,10-diphenylphenazine	CH_3CN, $0.1\underline{M}$ $(Et)_4NClO_4$	Pt	+0.20, +0.94 Ep/2	sce	16
6-Amino-N-ethyl-1,2,3,4-Tetrahydroquinoline	Aq Diethylglycine buffer, pH 10	Pt	+0.33	sce	17

R=H	CH_3CN, $(Et)_4NClO_4$	Pt	+1.16 Ep/2	sce	18
=CH$_3$	CH_3CN, $(Et)_4NClO_4$	Pt	+1.10 Ep/2	sce	18
=C$_2$H$_5$	CH_3CN, $(Et)_4NClO_4$	Pt	+1.12 Ep/2	sce	18
=i-C$_3$H$_7$	CH_3CN, $(Et)_4NClO_4$	Pt	+1.14 Ep/2	sce	18
=C$_6$H$_5$	CH_3CN, $(Et)_4NClO_4$	Pt	+1.21 Ep/2	sce	18
=NO	CH_3CN, $(Et)_4NClO_4$	Pt	+1.76 Ep/2	sce	18
=COCH$_3$	CH_3CN, $(Et)_4NClO_4$	Pt	+1.64 Ep/2	sce	18
=COC$_6$H$_5$	CH_3CN, $(Et)_4NClO_4$	Pt	+1.60 Ep/2	sce	18

| | CH$_3$CN, (Et)$_4$NClO$_4$ | Pt | +1.08 E$_{p/2}$ | sce | 18 |

Structure: pyrrole ring with C_6H_5 substituents at 2,3,4,5 positions and N-substituted with a 4-R-phenyl group.

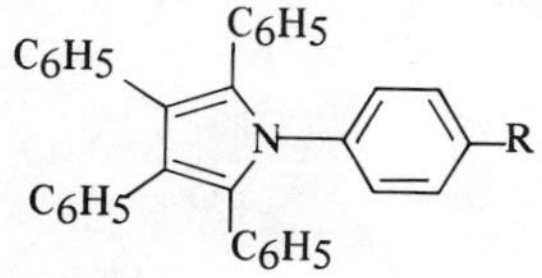					
R=H					
=CH$_3$	CH$_3$CN, LiClO$_4$	Pt	+0.755, +1.3	Ag/0.01$\underline{N}$ Ag+	19
=OCH$_3$	CH$_3$CN, LiClO$_4$	Pt	+0.745, +1.3	Ag/0.01$\underline{N}$ Ag+	19
=N(CH$_3$)$_2$	CH$_3$CN, LiClO$_4$	Pt	+0.735, +1.3	Ag/0.01$\underline{N}$ Ag+	19
	CH$_3$CN, LiClO$_4$	Pt	+0.520, +1.3	Ag/0.01$\underline{N}$ Ag+	19
			+0.93, +1.3		
5,6-Benzpyrido-(2',3'-1,2)-carbazole	CH$_3$CN, (Et)$_4$NClO$_4$	Pt	+1.06, +1.41	sce	2
5,6-Benzpyrido-(3',2',-1,2)-carbazole	CH$_3$CN, (Et)$_4$NClO$_4$	Pt	+1.04, +1.39	sce	2
2,5-Dimethyl-4-methyl-oxazole	Acetone, (n-Bu)$_4$NClO$_4$	Pt	+1.60	sce	20
See also Heterocyclic Amine N-Oxides (Section 25)					

Oxygen Heterocycles

Furan	HOAc, NaOAc	Pt	+1.70	sce	10
2,5-Dimethylfuran	HOAc, NaOAc	Pt	+1.20	sce	10
1,3,4,7-Tetraphenyl-isofuran	DMF, Perchlorate salt	Pt	+0.98	Ag/0.1$\underline{N}$ Ag+	21

(continued)

Reduction

Sulfur Heterocycles

Compound	Solvent System	Working Electrode	$E_{1/2}$	Reference Electrode	Reference
Dibenzothiophene	DMF, 0.15$\underline{M}$ ($\underline{n}$-Bu)$_4$NI	Hg	-2.432	Ag, AgCl	22
2,8-Dimethyldibenzo-thiophene	DMF, 0.15$\underline{M}$ ($\underline{n}$-Bu)$_4$NI	Hg	-2.452	Ag, AgCl	22
3,7-Dimethyldibenzo-thiophene	DMF, 0.15$\underline{M}$ ($\underline{n}$-Bu)$_4$NI	Hg	-2.563	Ag, AgCl	22
4,6-Dimethyldibenzo-thiophene	DMF, 0.15$\underline{M}$ ($\underline{n}$-Bu)$_4$NI	Hg	-2.501	Ag, AgCl	22
Rhodanine	Aq Solution, pH 7	Hg	-0.10	sce	23
	75% Alcohol, 0.1$\underline{M}$ (Et)$_4$NClO4	Hg	-1.81	sce	24
	75% Alcohol, 0.1$\underline{M}$ (Et)$_4$NClO4	Hg	-1.76	sce	24
	75% Alcohol, 0.1$\underline{M}$ (Et)$_4$NClO4	Hg	-1.82	sce	24
	75% Alcohol, 0.1$\underline{M}$ (Et)$_4$NClO4	Hg	-1.96	sce	24
	75% Alcohol, 0.1$\underline{M}$ (Et)$_4$NClO4	Hg	-2.00	sce	24
	75% Alcohol, 0.1$\underline{M}$ (Et)$_4$NClO4	Hg	-2.14	sce	24
	75% Alcohol, 0.1$\underline{M}$	Hg	-1.88	sce	24

Compound	Solvent, Electrolyte	Electrode	E	E	Reference	Ref.
(cyclic thioether ketone structure)	75% Alcohol, 0.1M (Et)4NClO4	Hg	-2.15		sce	24
Cyclopentanone[a]	90% Alcohol, 0.1M (Et)4NClO4	Hg	-2.46		NCE	24
Cyclohexanone[a]	90% Alcohol, 0.1M (Et)4NClO4	Hg	-2.40		NCE	24
Cycloheptanone[a]	90% Alcohol, 0.1M (Et)4NClO4	Hg	-2.48		NCE	24
Cyclooctanone[a]	90% Alcohol, 0.1M (Et)4NClO4	Hg	-2.43		NCE	24

Nitrogen Heterocycles

Compound	Solvent, Electrolyte	Electrode	E	E	Reference	Ref.
Pyridine	DMF, 0.1M (n-Bu)4NI	Hg	-2.15		Hg pool[c]	25
	DMF, 0.1M (n-Bu)4NI	Hg	-2.76		Ag, AgCl	26
	DMF, (Et)4NI	Hg	-2.10		Hg pool	27,28
	CH3CN, 0.1M (Et)4NClO4	Hg	-2.622		sce	29
2,2'-Bipyridyl	DMF, 0.1M (n-Bu)4NI	Hg	-2.19,	-2.76	Ag, AgCl	26
4,4'-Bipyridyl	DMF, 0.1M (n-Bu)4NI	Hg	-1.91,	-2.47	Ag, AgCl	26
2-Phenylpyridine	DMF, 0.1M (n-Bu)4NI	Hg	-2.30,	-2.78	Ag, AgCl	26
4-Phenylpyridine	DMF, 0.1M (n-Bu)4NI	Hg	-2.24,	-2.80	Ag, AgCl	26
2,2'-Azopyridine	DMF, 0.1M (n-Bu)4NI	Hg	-1.04,	-1.65	Ag, AgCl	26
3,3'-Azopyridine	DMF, 0.1M (n-Bu)4NI	Hg	-1.21,	-1.85	Ag, AgCl	26
1,2-Bis-(2-pyridyl)-ethylene	DMF, 0.1M (n-Bu)4NI	Hg	-1.92,	-2.32	Ag, AgCl	26
1,2-Bis-(4-pyridyl)-ethylene	DMF, 0.1M (n-Bu)4NI	Hg	-1.69,	-2.11	Ag, AgCl	26
Pyridazine (1,2-Diazine)	DMF, 0.1M (Et)4NClO4	Hg	-2.15		Ag, AgBr	30
	DMF, 0.1M (n-Bu)4NI	Hg	-2.20		Ag, AgCl	26
	DMF, 0.1M (Et)4NI	Hg	-1.61		Hg pool[c]	25

(continued)

Compound	Solvent System	Working Electrode	$E_{1/2}$	Reference Electrode	Reference
	CH_3CN, 0.1M $(Et)_4NClO_4$	Hg	-2.120	sce	29
Pyrimidine (1,3-Diazine)	DMF, 0.1M $(n\text{-}Bu)_4NI$	Hg	-2.35	Ag, AgCl	26
	DMF, 0.1M $(Et)_4NClO_4$	Hg	-2.32	Ag, AgBr	30
	DMF, 0.1M $(Et)_4NI$	Hg	-1.78	Hg pool[c]	25
	CH_3CN, 0.1M $(Et)_4NClO_4$	Hg	-2.080	sce	29
4-Phenylpyrimidine	DMF, 0.1M $(n\text{-}Bu)_4NI$	Hg	-2.00, -2.67	Ag, AgCl	26
Pyrazine (1,4-Diazine)	DMF, 0.1M $(n\text{-}Bu)_4NI$	Hg	-2.17	Ag, AgCl	26
	CH_3CN, 0.1M $(Et)_4NClO_4$	Hg	-2.080	sce	29
s-Triazine (1,3,5-Triazine)	DMF, 0.1M $(Et)_4NI$	Hg	-1.57	Hg pool[c]	25
	DMF, 0.1M $(Et)_4NClO_4$	Hg	-2.11	Ag, AgBr	30
	DMF, 0.1M $(n\text{-}Bu)_4NI$	Hg	-2.105	Ag, AgCl	26
as-Triazine	DMF, 0.1M $(Et)_4NI$	Hg	-1.47	Hg pool[c]	25
s-Tetrazine	DMF, 0.1M $(Et)_4NI$	Hg	-1.04	Hg pool[c]	25
Naphthalene[b]	DMF, 0.1M $(Et)_4NI$	Hg	-0.29	Hg pool[c]	25
Biphenyl[b]	DMF, 0.1M $(n\text{-}Bu)_4NI$	Hg	-1.98	Hg pool[c]	25
Quinoline	DMF, 0.1M $(Et)_4NI$	Hg	-2.03	Hg pool[c]	25
	DMF, 0.1M $(n\text{-}Bu)_4NI$	Hg	-1.59	Hg pool[c]	25
	DMF, $(n\text{-}Bu)_4NI$	Hg	-2.175	Ag, AgCl	26
	DMF, $(Et)_4NI$	Hg	-1.53	Hg pool	27,28
2,2'-Biquinoline	CH_3CN, 0.1M $(Et)_4NClO_4$	Hg	-2.105	sce	29
Isoquinoline	DMF, 0.1M $(n\text{-}Bu)_4NI$	Hg	-1.77, -2.20	Ag, AgCl	26
	DMF, 0.1M $(n\text{-}Bu)_4NI$	Hg	-2.22	Ag, AgCl	26
	DMF, 0.1M $(Et)_4NI$	Hg	-1.62	Hg pool[c]	25
	DMF, $(Et)_4NI$	Hg	-1.60	Hg pool	27,28
	CH_3CN, 0.1M $(Et)_4NClO_4$	Hg	-2.220	sce	29

Compound	Solvent, electrolyte		Potentials	Reference electrode	Ref.
Cinnoline (1,2-Diaza-naphthalene)	DMF, 0.1$\underline{M}$ (Et)$_4$NClO$_4$	Hg	-1.63, -2.18	Ag, AgBr	30
	CH$_3$CN, 0.1$\underline{M}$ (Et)$_4$NClO$_4$	Hg	-1.686, -2.134	sce	29
Phthalazine (2,3-Diaza-naphthalene)	DMF, 0.1$\underline{M}$ (n-Bu)$_4$NI	Hg	-1.68, -2.62	Ag, AgCl	26
	DMF, 0.1$\underline{M}$ (Et)$_4$NI	Hg	-0.982	Hg pool[c]	25
	DMF, 0.1$\underline{M}$ (Et)$_4$NClO$_4$	Hg	-1.97, -2.55	Ag, AgBr	30
	DMF, 0.1$\underline{M}$ (Et)$_4$NI	Hg	-1.41	Hg pool[c]	25
	DMF, 0.1$\underline{M}$ (n-Bu)$_4$NI	Hg	-2.02	Ag, AgCl	26
	CH$_3$CN, 0.1$\underline{M}$ (Et)$_4$NClO$_4$	Hg	-1.976, -2.315, -2.498	sce	29
Quinazoline (1,3-Diaza-naphthalene)	CH$_3$CN, 0.1$\underline{M}$ (Et)$_4$NClO$_4$	Hg	-1.799, -2.478	sce	29
	DMF, 0.1$\underline{M}$ (Et)$_4$NI	Hg	-1.22	Hg pool[c]	25
	DMF, 0.1$\underline{M}$ (Et)$_4$NClO$_4$	Hg	-1.74	Ag, AgBr	31
	DMF, 0.1$\underline{M}$ (Et)$_4$NClO$_4$	Hg	-1.63, -2.26	Ag, AgBr	30
Quinoxaline (1,4-Diaza-naphthalene)	CH$_3$CN, 0.1$\underline{M}$ (Et)$_4$NClO$_4$	Hg	-1.702, -2.163	sce	29
	DMF, 0.1$\underline{M}$ (n-Bu)$_4$NI	Hg	-1.80	Ag, AgCl	26
	DMF, 0.1$\underline{M}$ (Et)$_4$NI	Hg	-1.09	Hg pool[c]	25
	0.40$\underline{M}$ KHSO$_4$, pH 0.96	Hg	-2.54, -0.809	sce	32
	0.05$\underline{M}$ Phthalate buffer, pH 4.99	Hg	-0.554, -1.053	sce	32
	0.05$\underline{M}$ Borate buffer, pH 9.00	Hg	-0.863	sce	32
1,5-Naphthyridine(1,5-Diazanaphthalene)	DMF, 0.1$\underline{M}$ (Et)$_4$NClO$_4$	Hg	-1.81, -2.33	Ag, AgBr	31
Pyrido[2,3,$\underline{b}$]-pyrazine	DMF, 0.1$\underline{M}$ (n-Bu)$_4$NI	Hg	-1.86	Ag, AgCl	26
	DMF, 0.1$\underline{M}$ (Et)$_4$NI	Hg	-0.85	Hg pool[c]	25

(continued)

Compound	Solvent System	Working Electrode	$E_{1/2}$	Reference Electrode	Reference
Pteridine	DMF, 0.1M (Et)$_4$NI	Hg	-0.52	Hg pool c	25
Acridine	DMF, 0.1M (Et)$_4$NI	Hg	-1.04, -1.59	Hg pool c	28
	DMF, 0.1M (n-Bu)$_4$N1	Hg	-1.62, -2.38	Ag, AgCl	26
	CH$_3$CN, 0.1M (Et)$_4$NClO$_4$	Hg	-1.620, -1.994	sce	29
Phenanthridine	CH$_3$CN, 0.1M (Et)$_4$NClO$_4$	Hg	-2.118, -2.415	sce	29
5,6-Benzoquinoline (Benzo[f]-quinoline)	DMF, 0.1M (n-Bu)$_4$NI	Hg	-2.12, -2.64	Ag, AgCl	26
	DMF, (Et)$_4$NI	Hg	-1.52	Hg pool	28
	DMF, (Et)$_4$NI	Hg	-1.58	Hg pool	28
Benzo[f]-isoquinoline	DMF, 0.1M (n-Bu)$_4$NI	Hg	-2.20, -2.72	Ag, AgCl	26
	CH$_3$CN, 0.1M (Et)$_4$NClO$_4$	Hg	-2.140	sce	29
7,8-Benzoquinoline (Benzo[h]-quinoline)	95% DMF, (Et)$_4$NI	Hg	-1.41, -1.78	Hg pool	28
	CH$_3$CN, 0.1M (Et)$_4$NClO$_4$	Hg	-2.208	sce	29
Phenazine (9,10-Diazo-anthracene)	DMF, 0.1M (n-Bu)$_4$NI	Hg	-2.23, -2.72	Ag, AgCl	26
	DMF, 0.1M (Et)$_4$NClO$_4$	Hg	-1.15, -1.73	Ag, AgBr	30
	CH$_3$CN, 0.1M (Et)$_4$NClO$_4$	Hg	-1.227, -1.681	sce	29
	DMF, 0.1M (n-Bu)$_4$NI	Hg	-1.20, -2.01	Ag, AgCl	26
	Aq buffer, 10%- Ethanol, pH 3.95	Hg	-0.20	sce	33
	Aq buffer, 10%- Ethanol, pH 13.0	Hg	-0.76	sce	33
o-Phenanthroline	CH$_3$CN, 0.1M (Et)$_4$NClO$_4$	Hg	-2.053, -2.269	sce	29
	DMF, 0.1M (n-Bu)$_4$NI	Hg	-2.12, -2.70	Ag, AgCl	26

Compound	Solvent/electrolyte	Electrode	Potential (V)	Reference	Ref.
m-Phenanthroline	CH$_3$CN, 0.1M (Et)$_4$NClO$_4$	Hg	-2.092, -2.287	sce	29
p-Phenanthroline	CH$_3$CN, 0.1M (Et)$_4$NClO$_4$	Hg	-2.044, -2.29	sce	29
Benzo[c]-cinnoline	CH$_3$CN, 0.1M (Et)$_4$NClO$_4$	Hg	-1.554, -1.863, -2.396	sce	29
Benzo[f]-quinoxaline	DMF, 0.1M (n-Bu)$_4$NI	Hg	-1.55, -2.40	Ag, AgCl	26
	CH$_3$CN, 0.1M (Et)$_4$NClO$_4$	Hg	-1.744, -2.128, -2.673	sce	29
1-Azaphenanthrene	DMF, (Et)$_4$NI	Hg	-1.58	Hg pool	34
4-Azaphenanthrene	DMF, (Et)$_4$NI	Hg	-1.60	Hg pool	34
9-Azaphenanthrene	DMF, (Et)$_4$NI	Hg	-1.52	Hg pool	34
9-Aza-anthracene	DMF, (Et)$_4$NI	Hg	-1.04, -1.55	Hg pool	34
1-Azafluoranthene	DMF, (Et)$_4$NI	Hg	-0.94, -1.49	Hg pool	34
7-Azafluoranthene	DMF, (Et)$_4$NI	Hg	-0.94, -1.20	Hg pool	34
3,4-Benzacridine	DMF, 0.1M (n-Bu)$_4$NI	Hg	-1.73, -2.38	Ag, AgCl	26
5,6,7,8-Dibenzoquinoxaline	DMF, 0.1M (n-Bu)$_4$NI	Hg	-1.78	Ag, AgCl	26
Dibenzo-[a,c]-phenazine	DMF, 0.1M (n-Bu)$_4$NI	Hg	-1.35, -2.12	Ag, AgCl	26
60 Phenazine derivatives					35
2,2'-Bipyridyl	DMF, 0.1M (Et)$_4$NI	Hg	-1.59	Hg pool	25
4,4'-Bipyridyl	DMF, 0.1M (Et)$_4$NI	Hg	-1.31	Hg pool	25
Pyridinium iodide	DMF, 0.1M (Et)$_4$NI	Hg	-0.750	Hg pool	25
Pyrimidinium iodide	DMF, 0.1M (Et)$_4$NI	Hg	-0.465	Hg pool	25
Pyridazinium iodide	DMF, 0.1M (Et)$_4$NI	Hg	-0.250	Hg pool	25
Pyrazinium iodide	DMF, 0.1M (Et)$_4$NI	Hg	-0.192	Hg pool	25
1,1'-Dimethyl-4,4'-bipyridylium diiodide (Paraquat)	Borate buffer, pH 8.3	Hg	-0.69	sce	36
1,1'-Ethylene-2,2'-bipyridylium diiodide (Diquat)	Borate buffer, pH 8.3	Hg	-0.61	sce	36

(continued)

Compound	Solvent System	Working Electrode	$E_{1/2}$	Reference Electrode	Reference
1,1'-Trimethylene-2,2'-bipyridylium diiodide	Borate buffer, pH 8.3	Hg	-0.80	sce	36
1,1'-Tetramethylene-2,2'-bipyridylium diiodide	Borate buffer, pH 8.3	Hg	-0.88	sce	36
Morphamquat	Phosphate buffer, pH 6.9	Hg	-0.535, -0.91	sce	36
Ethylpyridinium bromide	Py, 0.1M LiClO4	Hg	-1.34	Ag/0.1N Ag+, Py	37
n-Butylpyridinium bromide	Py, 0.1M LiClO4	Hg	-1.38	Ag/0.1N Ag+, Py	37
2-Methoxyazocine	DMF, 0.1M (n-Bu)4NI	Hg	-1.84	sce	38
8-Methoxyazocine	DMF, 0.1M (n-Bu)4NI	Hg	-1.92	sce	38
3,8-Dimethyl-2-methoxy-azocine	DMF, 0.1M (n-Bu)4NI	Hg	-2.17	sce	38
4,6,8-Trimethyl-2-methoxyazocine	DMF, 0.1M (n-Bu)4NI	Hg	-2.22	sce	38
3,5,6,8-Tetramethyl-2-methoxyazocine	DMF, 0.1M (n-Bu)4NI	Hg	-2.46	sce	38
Indole Alkaloids					
Ajmalicine	Methanol, 0.2M LiCl	Pt	+0.852	sce	39
Alloyohimbane	Methanol, 0.2M LiCl	Pt	+0.805	sce	39
Corynanthine	Methanol, 0.2M LiCl	Pt	+0.874	sce	39
Deserpidine	Methanol, 0.2M LiCl	Pt	+0.849	sce	39
Deserpidine N-Oxide	Methanol, 0.2M LiCl	Pt	No wave	sce	39
Methyl Deserpidate	Methanol, 0.2M LiCl	Pt	+0.781	sce	39
Yohimbine	Methanol, 0.2M LiCl	Pt	+0.838	sce	39
3-epi-α-Yohimbine	Methanol, 0.2M LiCl	Pt	+0.859	sce	39
Corynantine Nitrate	Methanol, 0.2M LiCl	Pt	No wave	sce	39

Compound	Solvent / Electrolyte	Electrode	E	Ref. electrode	Ref.
Ibogaine	Methanol, 0.2M̲ LiCl	Pt	+0.547	sce	39
Ibogamine	Methanol, 0.2M̲ LiCl	Pt	+0.577	sce	39
Tabernanthine	Methanol, 0.2M̲ LiCl	Pt	+0.460	sce	39
Reserpine	Methanol, 0.2M̲ LiCl	Pt	+0.618	sce	39
Reserpine N-Oxide	50% Methanol, 0.2M̲ LiCl	Pt	+0.714	sce	39
Methyl Reserpate	Methanol, 0.2M̲ LiCl	Pt	+0.650	sce	39
Reserpic Acid	Methanol, 0.2M̲ LiCl	Pt	+0.853	sce	39
Iso-Reserpine	Methanol, 0.2M̲ LiCl	Hg	+0.631	sce	39
Rescinnamine	Methanol, 0.2M̲ LiCl	Hg	+0.611	sce	39
Reserpinine	Methanol, 0.2M̲ LiCl	Hg	+0.655	sce	39

Substituted 3-Phenylcoumarins

Substituent	Solvent / Electrolyte	Electrode	E	Ref. electrode	Ref.
Substitutent=7-OMe	68.1% Methanol, 0.1m̲ KCl	Hg	-1.641	Ag, AgCl	40
=7-t̲-Bu	68.1% Methanol, 0.1m̲ KCl	Hg	-1.599	Ag, AgCl	40
=6-Me	68.1% Methanol, 0.1m̲ KCl	Hg	-1.567	Ag, AgCl	40
=H	68.1% Methanol, 0.1m̲ KCl	Hg	-1.552	Ag, AgCl	40
=8-OMe	68.1% Methanol, 0.1m̲ KCl	Hg	-1.542	Ag, AgCl	40
=7-Cl	68.1% Methanol, 0.1m̲ KCl	Hg	-1.417	Ag, AgCl	40
=6-Br	68.1% Methanol, 0.1m̲ KCl	Hg	-1.443	Ag, AgCl	40
=6,8-Br$_2$	68.1% Methanol, 0.1m̲ KCl	Hg	-1.286	Ag, AgCl	40
=4'-OMe	68.1% Methanol, 0.1m̲ KCl	Hg	-1.585	Ag, AgCl	40

(continued)

C	Compound	Solvent System	Working Electrode	$E_{1/2}$	Reference Electrode	Reference
	=3',4'-(OMe)2	68.1% Methanol, 0.1m KCl	Hg	-1.563	Ag, AgCl	40
	=4-i-Propyl	68.1% Methanol, 0.1m KCl	Hg	-1.573	Ag, AgCl	40
	=3'-OMe	68.1% Methanol, 0.1m KCl	Hg	-1.540	Ag, AgCl	40
	=3'-Cl	68.1% Methanol, 0.1m KCl	Hg	-1.511	Ag, AgCl	40
	Chlorophyll	DMF, LiCl	Hg	-1.12, -1.56	sce	41
39	Coumarin, psoralene, and angelicin derivatives	5% (Et)4N1, 50% Methanol			sce	42

a Included for comparison.
b Arranged in increasing number of fused rings.
c -0.516 V versus sce in this electrolyte.

References for Heterocycles

1. J. Bonnier, P. Arnaud, and M. Maurey-Mey, Compt. Rend., 267, 10 (1968)
2. T. A. Gough and M. E. Peover, Proc. 3rd International Polarography Congress, Southampton, 1965, Macmillan, London, 1966, p. 1017.
3. J. P. Billon, Bull. Soc. Chim. Fr., 1923 (1961).
4. J. P. Billon, Ann. Chim., 7, 196 (1962).
5. F. H. Merkle and C. A. Discher, J. Pharm. Sci., 53, 620 (1964).
6. P. T. Cottrell and C. K. Mann, J. Electrochem. Soc., 116, 1499 (1969).
7. J. P. Billon, Ann. Chim., 7, 190 (1962).
8. W. Kemula and M. K. Kalinowski, Fresenius' Z. Anal. Chem., 224, 383 (1967).
9. J. W. Loveland and G. G. Dimeler, Anal. Chem., 33, 1196 (1961).
10. L. Eberson and K. Nyberg, J. Am. Chem. Soc., 88, 1686 (1966).
11. C. Barry, G. Cauquis, and M. Maurey, Bull. Soc. Chim. Fr., 2510 (1966).
12. J. W. Breitenbach and F. Sommer, Abstract CITCE Meeting, Prague, Czechoslovakia, 1970.
13. A. Stamenda, Z. Naturforsch., 22b, 1107 (1967).
14. N. L. Weinberg, Preprints of Papers, Durham Symposium on Electro-Organic Chemistry, Durham, North Carolina, 1968, p. 263.
15. R. N. Adams, Electrochemistry At Solid Electrodes, Marcel Dekker, Inc., New York, 1969, p. 320.
16. R. F. Nelson, D. W. Leedy, E. T. Seo, and R. N. Adams, Fresenius' Z. Anal. Chem., 224, 184 (1967).
17. D. B. Julian and W. R. Ruby, J. Am. Chem. Soc., 72, 4719 (1950).
18. J. F. Ambrose and R. F. Nelson, J. Electrochem. Soc., 115, 1159 (1968).
19. G. Cauquis and M. Genies, Bull. Soc. Chim. Fr., 3220 (1967).
20. N. L. Weinberg, unpublished results.
21. A. Zweig, G. Metzler, A. Maurer, and B. G. Roberts, J. Am. Chem. Soc., 88, 2864 (1966).
22. R. Gerdil and E. A. C. Lucken, J. Am. Chem. Soc., 88, 733 (1966).
23. K. Okamoto, Bull. Chem. Soc. Japan, 34, 920 (1961).
24. R. Herzschuk and R. Borsdorf, J. Electroanal. Chem., 23, 55 (1969).
25. K. B. Wiberg and T. P. Lewis, J. Am. Chem. Soc., 92, 7154 (1970).

1046 Appendix

26. B. J. Tabner and J. R. Yandle, J. Chem. Soc. A, 1968, 381.

27. G. Anthoine, G. Coppens, J. Nasielski, and E. Vander Donckt, Bull. Soc. Chim. Belges, 73, 65 (1964).

28. C. Parkanyi and R. Zahradnick, Bull. Soc. Chim. Belges, 73, 57 (1964).

29. S. Millefiori, J. Heterocycl. Chem., 7, 145 (1970).

30. D. Van Der Meer and D. Feil, Rec. Trav. Chim., 87, 746 (1968).

31. D. Van Der Meer, Rec. Trav. Chim., 88, 1361 (1969).

32. M. P. Strier and J. C. Cavagnol, J. Am. Chem. Soc., 79, 4331 (1957).

33. R. C. Kaye and H. I. Stonehill, J. Chem. Soc., 1952, 3244.

34. G. Anthoine, G. Coppens, J. Nasielski, and E. Vander Donckt, Bull. Soc. Chim. Belges, 73, 65 (1964).

35. Yu S. Rosum, S. B. Sebryanyi, E. F. Karaban, V. P. Chernetskii, and M. I. Drankina, J. General Chem. USSR, 34, 2622 (1964).

36. J. Volke, Collect. Czech. Chem. Commun., 33, 3044 (1968).

37. M. S. Spritzer, J. M. Costa, and P. J. Elving, Anal. Chem., 37, 211 (1965).

38. L. A. Paquette, J. F. Hansen, T. Kakihana, and L. B. Anderson, Tetrahedron Letters, 7, 533 (1970).

39. M. J. Allen and V. J. Powell, J. Electrochem. Soc., 105, 541 (1958).

40. J. F. Archer and J. Grimshaw, J. Chem. Soc. B, 1969, 266.

41. B. A. Kiselev, Yu. N. Kozlov, and V. B. Evstigne, Biofizika, 15, 594 (1970); Chem. Abstract, 73, 115771b (1970).

42. Yu. E. Orlov., A. P. Prokopenko, Zh. Obshch. Khi, 40, 1159 (1970).

31. DYES

Oxidation

Triphenylmethane Dyes

Compound	Solvent System	Working Electrode	$E_{1/2}$	Reference Electrode	Reference
Crystal violet[a]	1N H$_2$SO$_4$, 1N Na$_2$SO$_4$	Pt	+0.632, +0.886 E$_{p/2}$	sce	1
Leuco crystal violet	SO$_2$[b]	Pt	+0.87	Ag, AgBr	2
Malachite green[a]	SO$_2$[b]	Pt	+1.05	Ag, AgBr	2
	SO$_2$	Pt	+0.75	Ag, AgBr	2
p,p'-Methylenebis (N,N-dimethylaniline)	1N H$_2$SO$_4$, 1N Na$_2$SO$_4$	Pt	+0.690 E$_{p/2}$	sce	1
Ethyl violet[a]	1N H$_2$SO$_4$, 1N Na$_2$SO$_4$	Pt	+0.775 E$_{p/2}$	sce	1
	1N H$_2$SO$_4$, 1N Na$_2$SO$_4$	Pt	+0.796, +0.954 E$_{p/2}$	sce	1
	SO$_2$[b]	Pt	+0.87		
Brilliant green[a]	1N H$_2$SO$_4$, 1N Na$_2$SO$_4$	Pt	+0.798 E$_{p/2}$	Ag, AgBr	2
Brilliant green[c]	SO$_2$[b]	Pt	+0.83	sce	1
Alizarin S	Universal buffer	Pt	+0.82, +0.059pH[d]	Ag, AgBr	2
				sce	3

a Chloride salt.
b At -20°C.
c Bisulfate salt.
d pH 2.0 to 6.0.

(continued)

1047

Reduction

Triphenylmethane Derivatives

Compound	Solvent System	Working Electrode	$E_{1/2}$	Reference Electrode	Reference
Dibenzoquinomethane (Fuchsone)	Citrate-Phosphate buffer, 50% Methanol, pH 8.0	Hg	-0.52, -0.925	sce	4
	1N HCl	Hg	-0.645	sce	4
p,p'-Dihydroxyfunchsone (Aurine)	Citrate-Phosphate buffer, 50% Methanol, pH 8.0	Hg	-0.75, -1.18	sce	4
Triphenylmethanol	2M NaOAc, pH ~2	Hg	-0.40	sce	4
	21.5N H2SO4	Hg	-0.525, -0.825	Hg, HgSO4/- 98% H2SO4	5

Triphenylmethylperchlorates

$R_1 = R_2 = R_3 = Cl$	CH₃CN, (n-Bu)₄NClO₄	Hg	+0.38	sce	6
= F	CH₃CN, (n-Bu)₄NClO₄	Hg	+0.33	sce	6
= H	CH₃CN, (n-Bu)₄NClO₄	Hg	+0.27	sce	6

	Medium	Electrode	E	Ref. electrode	Ref.
=C$_6$H$_5$	CH$_3$CN, (n-Bu)$_4$NClO$_4$	Hg	+0.19	sce	6
=t-C$_4$H$_9$	CH$_3$CN, (n-Bu)$_4$NClO$_4$	Hg	+0.13	sce	6
=i-C$_3$H$_7$	CH$_3$CN, (n-Bu)$_4$NClO$_4$	Hg	+0.07	sce	6
R$_1$=OCH$_3$, R$_2$=R$_3$=H	CH$_3$CN, (n-Bu)$_4$NClO$_4$	Hg	+0.07	sce	6
R$_1$=R$_2$=R$_3$=CH$_3$	CH$_3$CN, (n-Bu)$_4$NClO$_4$	Hg	+0.05	sce	6
=cyclo-C$_3$H$_5$	CH$_3$CN, (n-Bu)$_4$NClO$_4$	Hg	-0.01	sce	6
=OCH$_3$	CH$_3$CN, (n-Bu)$_4$NClO$_4$	Hg	-0.20	sce	6
R$_1$=H, R$_2$=R$_3$=N(C$_2$H$_5$)$_2$	CH$_3$CN, (n-Bu)$_4$NClO$_4$	Hg	-0.64	sce	6
R$_1$=R$_2$=R$_3$=N(CH$_3$)$_2$	CH$_3$CN, (n-Bu)$_4$NClO$_4$	Hg	-0.79	sce	6
Azines					
Bindschedler Green	Buffer, pH 7.0	Hg	-0.02	sce	7
Toluylene blue	Buffer, pH 7.0	Hg	-0.13	sce	7
Diazines					
Phenosafranine	Buffer, pH 7.0	Hg	-0.48	sce	7
Induline B	Phosphate buffer, 1% Ethanol, pH 7.0	Hg	-0.38	sce	7
Neutral red	B-R buffer, pH 7.0	Hg	-0.57	sce	7
Neutral blue	Phosphate buffer, 1% Ethanol, pH 7.0	Hg	-0.57	sce	7
2-Hydroxyphenazine	B-R buffer, pH 4.0	Hg	-0.24	sce	7
Pyocyanine	Phosphate buffer, pH 7.16	Hg	-0.26	sce	7
Rosindulin G	McIlvaine buffer, pH 6.17	Hg	-0.46	sce	7
Gallocyanine	Buffer, pH 7.0	Hg	-0.22	sce	7
Capri blue	Phosphate buffer, 1% Ethanol, pH 7.0	Hg	-0.22	sce	7

(continued)

Compound	Solvent System	Working Electrode	$E_{1/2}$	Reference Electrode	Reference
Cresyl blue	Phosphate buffer, 1% Ethanol, pH 7.0	Hg	-0.21	sce	7
Sky blue	Buffer, pH 7.0	Hg	-0.13	sce	7
Thiazines					
Methylene blue	B-R buffer, pH 4.9	Hg	-0.15	sce	7
Methylene green	Phosphate buffer, 1% Ethanol, pH 7.0	Hg	-0.12	sce	7
Indophenols					
2,6-Dibromophenolindo-phenol	Phosphate buffer, pH 6.67	Hg	+0.00	sce	7
Indigo Dyes					
Indigo disulphonate	Buffer, pH 7	Hg	-0.37	sce	7
Indigo trisulphonate	Buffer, pH 7	Hg	-0.33	sce	7
Indigo tetrasulphonate	Buffer, pH 7	Hg	-0.30	sce	7
19 Anthraquinones	Aq buffers with 1% Ethanol				8
6 Azo dyes	B-R buffer				9
32 Azo dyes	OAc buffer, 90% Ethanol				10
27 Malachite green derivatives	Aq buffer, 2% Acetone				11

References for Dyes

1. Z. Galus and R. N. Adams, J. Am. Chem. Soc., 86, 1666 (1964).
2. D. A. Hall, M. Sakuma, and P. J. Elving, Electrochim. Acta, 11, 337 (1966).
3. L. S. Reishakrit, T. B. Argova, and L. V. Vesheva, Elektrokhim., 4, 653 (1969).
4. K. A. Harper, Electrochim. Acta, 15, 563 (1970).
5. P.H. Plesch and I. Sestahova, J. Chem. Soc. B, 1970, 87.
6. H. Volz and W. Lotsch, Tetrahedron Letters, 27, 2275 (1969).
7. J. Heyrovsky and J. Kůta, Principles of Polarography, Academic, New York, 1966, p. 549.
8. N. H. Furman and K. G. Stone, J. Am. Chem. Soc., 70, 3055 (1948).
9. M. Kozeny and V. Velich, Collect. Czech. Chem. Commun., 25, 1031 (1960).
10. L. Moelants and R. Janssen, Bull. Soc. Chem. Belg., 66, 209 (1957).
11. G. Bengtsson, Acta Chem. Scand., 24, 2868 (1970).

32. MISCELLANEOUS

Oxidation

Anti-Oxidants

Compound	Solvent System	Working Electrode	$E_{1/2}$	Reference Electrode	Reference
Age-Rite Stalite (Heptylated Diphenylamine)	95% Ethanol, LiCl	Graphite	+0.65	sce	3
Flexzone 3C (N-Isopropyl-N'-phenylphenylene-diamine)	95% Ethanol, LiCl	Graphite	+0.14, +0.43	sce	3
Dalpac (2,6-Di-t-butyl-cresol)	95% Ethanol, LiClO$_4$	Graphite	+0.82, +1.21	sce	3
Santowhite Powder (4,4'-Butylidene-bis-(3-methyl-6-t-butylphenol)	95% Ethanol, LiClO$_4$	Graphite	+0.80	sce	3
Age-Rite Powder (N-Phenyl--naphthylamine	95% Ethanol, LiCl	Graphite	+0.73	sce	3
Age-Rite Alba (Hydroquinone monobenzyl ether)	95% Ethanol, LiCl	Graphite	+0.66	sce	3
2246 (2,2'-Methylene-bis-(6-t-butyl-4-ethylphenol)	95% Ethanol, LiClO$_4$	Graphite	+0.80, +1.22	sce	3
UOP-88	95% Ethanol, LiCl	Graphite	+0.04, +0.39	sce	3

Compound	Solvent, electrolyte	Electrode	E	Ref. electrode	Ref.
Flexzone 6H (N'-Phenyl-N'-cyclohexylphenylene-diamine)	95% Ethanol, LiCl	Graphite	+0.41	sce	3
Santovar A(2,5-Di-t-amylhydroquinone)	95% Ethanol, LiCl	Graphite	+0.43	sce	3
Sodium tetraphenylborate	Aqueous buffers	Graphite	+0.216, +0.92, -0.057 pH	sce[a]	1
Diphenylborinic acid	CH_3CN, $LiClO_4$	Graphite	+0.32, +1.04	sce[a]	1
	DMF, $LiClO_4$	Graphite	+0.23, +0.8	sce[a]	1
	Aqueous buffers	Graphite	+0.54 $E_{p/2}$	sce[a]	1
Dimethyl sodiomalonate	50% Methanol, 50%-Benzene, $LiClO_4$ or $NaClO_4$	Graphite	+0.87	Ag, AgCl	2
Sodioacetylacetone	50% Methanol, 50%-Benzene, $LiClO_4$ or $NaClO_4$	Graphite	+0.83	Ag, AgCl	2
Methyl sodiocyano-acetate	50% Methanol, 50%-Benzene, $LiClO_4$ or $NaClO_4$	Graphite	+0.65	Ag, AgCl	2
Nitroethane (sodium salt)	50% Methanol, 50%-Benzene, $LiClO_4$ or $NaClO_4$	Graphite	+0.87	Ag, AgCl	2
2-Nitropropane (sodium salt)	50% Methanol, 50%-Benzene, $LiClO_4$ or $NaClO_4$	Graphite	+0.74	Ag, AgCl	2

[a] Made with saturated NaCl.

(continued)

Reduction

Quaternary Ammonium Ions

Compound	Solvent System	Working Electrode	$E_{1/2}$	Reference Electrode	Reference
NH_4^+	Aq Solution	Hg	-2.21	sce	4
$N(CH_3)_3H^+$	Aq Solution	Hg	-2.23	sce	4
$N(C_2H_4OH)_3C_2H_5^+$	Aq Solution	Hg	-2.38	sce	4
$N(C_3H_7)_4^+$	Aq Solution	Hg	-2.52	sce	4
$N(C_4H_9)_4^+$	Aq Solution	Hg	-2.57	sce	4
$N(CH_3)_3C_2H_4OCOCH_3^+$	Aq Solution	Hg	-2.64	sce	4
$N(C_2H_5)_4^+$	Aq Solution	Hg	-2.67	sce	4
$N(CH_3)_3C_2H_4OH^+$	Aq Solution	Hg	-2.72	sce	4
$N(CH_3)_4^+$	Aq Solution	Hg	-2.93	sce	4

Insecticides

Compound	Solvent System	Working Electrode	$E_{1/2}$	Reference Electrode	Reference
Nitrosated Carbaryl	Aq Solution, pH 13–14a	Hg	-0.45 Ep	Hg pool	5
Systox-thiol	2% (Et)$_4$NOH	Hg	-0.65 Ep	Ag, AgCl	6
Systox-thiol sulfoxide	2% (Et)$_4$NOH	Hg	-0.58 Ep	Ag, AgCl	6
Systox-thiol-sulfone	2% (Et)$_4$NOH	Hg	-0.64 Ep	Ag, AgCl	6
Systox-thione	2% (Et)$_4$NOH	Hg	-0.55 Ep	Ag, AgCl	6
Systox-thiono-sulfoxide	2% (Et)$_4$NOH	Hg	-0.70 Ep	Ag, AgCl	6
Systox-thiono-sulfone	2% (Et)$_4$NOH	Hg	No wave	Ag, AgCl	6
Thimet	2% (Et)$_4$NOH	Hg	-0.80 Ep	Ag, AgCl	6
Di-Syston	2% (Et)$_4$NOH	Hg	-0.62 Ep	Ag, AgCl	6

Parathion	OAc buffer, 0.2M NaCl, pH 5.0	Hg	-0.62	Ep	Ag	7
p-Nitrophenol[b]	OAc buffer, 0.2M NaCl, pH 5.0	Hg	-0.80	Ep	Ag	7
8 Insecticides						8

[a] 1 part NaNO$_2$, 1 part HOAc, 3 parts 50% KOH.
[b] Included for comparison.

(continued)

1056 Appendix

References for Miscellaneous

1. W. R. Turner and P. J. Elving, _Anal. Chem._, __37__, 207 (1965).
2. H. Schafer, _Chem. Ing. Techn._, __41__, 179 (1969).
3. G. A. Ward, _Talanta_, __10__, 261 (1963).
4. P. Van Rysselberghe, and J. M. McGee, _J. Am. Chem. Soc._, __67__, 1038 (1945).
5. R. J. Gajan, W. R. Benson, and J. M. Finocchiaro, _J. Ass. Offic. Agr. Chem._, __48__, 958 (1965).
6. R. J. Gajan, _J. Ass. Offic. Agr. Chem._, __45__, 401 (1962).
7. R. J. Gajan, _J. Ass. Offic. Agr. Chem._, __46__, 216 (1963).
8. R. J. Gajan, _J. Ass. Offic. Agr. Chem._, __48__, 1027 (1965).

INDEX